MODERN TECHNIQUES FOR CHARACTERIZING MAGNETIC MATERIALS

MODERN TECHNIQUES FOR CHARACTERIZING MAGNETIC MATERIALS

Edited by

YIMEI ZHU

KLUWER ACADEMIC PUBLISHERS
Boston / Dordrecht / London

Library of Congress Cataloging-in-Publication Data

A C.I.P. Catalogue record for this book is available
from the Library of Congress.

Modern techniques for characterizing magnetic materials / edited by Yimei Zhu.
     p. cm
  Includes bibliographical references and index.
  ISBN 1-4020-8007-7 (alk, paper) -- ISBN 0-387-23395-4 (e-book)
   1. Magnetic materials—Analysis. 2. Magnetic materials –Microscopy. 3. Electrons –
Scattering, I. Zhu, Yimei.

QC765.M63 2005
538'.4—dc22

2005042533

Printed in the United States of America.   (BS/DH)

9 8 7 6 5 4 3 2 1   SPIN 11052838

springeronline.com

Authors

Rolf Allenspach
IBM Research
Rüschlikon, Switzerland

Ernst Bauer
Arizona State University
Arizona, USA

Marco Beleggia
Brookhaven National Laboratory
New York, USA

Matthias Bode
University of Hamburg
Hamburg, Germany

Byoung-Chul Choi
University of Victoria
British Columbia, Canada

Dan Dahlberg
University of Minnesota
Minnesota, USA

Michael Fitzsimmons
Los Alamos National Laboratory
New Mexico, USA

Mark Freeman
University of Alberta
Edmonton, Alberta, Canada

Paul Fumagalli
Free University of Berlin
Berlin, Germany

Daniel Haskel
Argonne National Laboratory
Illinois, USA

Burkard Hillebrands
Technical University of Kaiserslautern
Kaiserslautern, Germany

Joachim Kohlbrecher
Paul Scherrer Institute
Villigen, Switzerland

Jeffery B. Kortright
Lawrence Berkeley National Laboratory
California, USA

Alexander Krichevsky
Naval Research Laboratory
Washington, DC, USA

Andre Kubetzka
University of Hamburg
Hamburg, Germany

Jonathan C. Lang
Argonne National Laboratory
Illinois, USA

Seung-Hun Lee
National Institute of Standards and Technology
Maryland, USA

Charles F. Majkrzak
National Institute of Standards and Technology
Maryland, USA

Oswald Pietzsch
University of Hamburg
Hamburg, Germany

Roger Proksch
Asylum Research
California, USA

George Srajer
Argonne National Laboratory
Illinois, USA

Elio Vescovo
Brookhaven National Laboratory
New York, USA

Werner Wagner
Paul Scherrer Institute
Villigen, Switzerland

Roland Wiesendanger
University of Hamburg
Hamburg, Germany

Igor Zaliznyak
Brookhaven National Laboratory
New York, USA

Yimei Zhu
Brookhaven National Laboratory
New York, USA

Brief Contents

Preface

Magnetism and magnetic phenomena surround us. Magnetic materials pervade our lives far beyond just compasses and common household magnets. Magnetic materials and magnetism are used inconspicuously in many complex gadgets, ranging from magnetic recording disks and cellular phones, to magnetic resonance imaging instruments and spintronic devices. Understanding magnetic phenomena and developing new functional magnetic materials pose a big challenge, but also great opportunities, to our scientists and engineers. New characterization techniques help us to understand the fascinating behavior of newly discovered magnetic materials, while new materials stimulate the further development of novel methods.

We understand the physical and chemical behaviors of materials through directly or indirectly measuring a material's structure and properties. The field of structural characterization is often too wide; identifying the most appropriate method to employ can be difficult. One objective of this book is to introduce the reader to various modern techniques in characterizing magnetic materials at different length scales, focusing on neutron, x-ray, electron, and laser-light scattering as well as proximal probes, their principles, applicability, limitations, and relationship to competing methods.

Neutrons, photons, and electrons are three major classes of modern probes for characterizing structures of materials. Since all materials absorb and emit electromagnetic radiation, the material's characteristics frequently manifest in the way it interacts with incident particles. Thus, the information gleaned not only depends on the wavelength of the radiation, but also on the nature of these interactions. Neutrons interact with atomic nuclei, x-ray photons with electron clouds, and electrons with electromagnetic potentials, i.e., both electrons and nuclei of the solid. For magnetic structural characterization, both electrons and nuclei in materials have a magnetic moment and, in principle, all the three sources can reveal magnetic information. In particular, because the neutron has spin ½, the orientation of its spin is easily manipulated and when combined with scattering geometry, it yields an opportunity to measure the spatial dependence of the vector magnetization. However, neutrons are not handily and copiously produced. In contrast, x-rays are easy to generate, but due to their weak spin interaction with matter, magnetic scattering can only be observed using extraordinarily intense synchrotron radiation.

The past two decades have witnessed significant advancement in instrumentation and technique development in characterization of magnetic materials, especially in proximal probes and laser-light scattering, along with the newly constructed next-generation neutron and synchrotron photon sources. In domain imaging, for example, an important branch of the field, we see an ever-increasing spatial resolution. There is a multiplicity of techniques, ranging from magneto-optical imaging to spin polarized scanning tunneling microscopy, that have different mechanisms of image formation and are suitable, hence, for different measurements. For instance, magneto-optic microscopy is based on the contrast produced by Kerr rotation of linearly polarized light reflected off domains with different magnetization. It has a typical spatial resolution of 500-1000nm (which can be much improved by near-field optical microscopy), but superior time resolution of up to 10-9 sec for dynamic observations. Magnetic force microscopy (MFM), on the other hand, measures the force that the stray magnetic field of the surface exerts on a tiny magnetic tip attached to a flexible cantilever. Unlike other methods, it is mainly sensitive to field gradients and its resolution is on the order of 40-100 nm. Scanning electron microscopy with polarization analysis (SEMPA) measures the spin polarization of secondary electrons emitted from a magnetized sample with resolution of about 30-200 nm. Spin polarized low energy electron microscopy (SPLEEM) is a surface-imaging technique using spin polarized electrons, and is sensitive to the interaction between the spins of the incident electron and the spins in the sample. Thus, surface magnetic behavior can be directly observed over a large area of view in real time with a resolution of 20nm. High-energy transmission electron microscopy techniques push resolution even further. A remarkable example is electron holography which is based on retrieving the phase shift of a coherent electron wave passing though the sample as encoded in a hologram, and provides a direct measure of the electrostatic and magnetic potentials of a local area of interest. Although a few nm resolution can be routinely achieved, separating magnetostatic potentials from electrostatic ones at the nanoscale can be challenging. To date, the most promising technique for attaining atomic resolution of local magnetic structure is spin-polarized scanning tunneling microscopy (SP-STM) that can reveal the magnetic lattice arrangement on an antiferromagnetic sample surface. This new breed of scanning tunneling microscopy measures the spin-polarized tunneling electrons between the tip and the sample.

Of course, the techniques for magnetic imaging exemplified here are not meant to be inclusive, as several others are available, notably the recently developed technique of synchrotron-based x-ray magnetic microscopy, i.e., the x-ray magnetic circular dichroism (XMCD). This method probes the transfer of the angular momentum of the x-ray photon to the photoelectron excited from a spin-orbit split core level with a spatial resolution of about 5 nm. The main advantage of the technique is its elemental specificity that derives from the process being tied to an absorption event at the core level, thus providing information on the spin orbital moments, site symmetry, and chemical state of the sample under study, which is not available from the desktop- or laboratory-instruments mentioned earlier. It is important to realize that these techniques are complementary; they all have their own advantages and drawbacks. Some are highly penetrating and non-destructive probes, while others require a high-quality surface or tedious sample preparation. If you are interested in these technologies and would like to know more about them, you will find this book, Modern Techniques for Characterizing Magnetic Materials, an invaluable tool for expanding your research capabilities.

This book is organized in the following way. The first three chapters deal with neutron scattering methods, including triple-axis spectrometry, small-angle scattering, and reflectometry. Chapter 4-6 focus on synchrotron-radiation based techniques, ranging from magnetic soft and hard x-ray scattering to photon-emission spectroscopy. Chapter 7-9 discuss electron scattering, with transmission electron microscopy focusing on Lorentz microscopy, electron holography and other phase-retrieval methods, and scanning electron microscopy with spin polarized analysis and spin polarized low energy electron microscopy. Chapter 10 and 11 describe proximal probes, covering spin-polarized scanning tunneling microscopy and magnetic force microscopy. The last three chapters deal with light scattering including the use of monochromatic laser light in magnetic imaging, such as scanning near-field optical microscopy, time-resolved scanning Kerr microscopy, and Brillouin light-scattering spectroscopy.

This book does not attempt to cover all aspects of magnetic structural characterization, but focuses on major modern techniques. Owing to the complexity of magnetic behavior, to tackle one single material problem often necessitates bringing to bear various characterization tools within our arsenal of techniques. Usually, this is not a trivial task. Different research communities employ different research techniques and, often, there is little communication between them. It is my hope that this book will bridge this gap, and make the combined use of various techniques in materials research a reality.

Although this book leans more toward the research laboratory than the classroom, it can serve as a methodological reference book for graduate students, university faculties, scientists and engineers who are interested in magnetic materials and their characterization. Expositions within individual chapters are largely self-contained without having been sequenced with any specific pedagogical thread in mind. Each chapter, therefore, has its own introduction, principle, instrumentation and applications, and references for further study. The level of presentation is intended to be intermediate between a cursory overview and detailed instruction. The extent of coverage is very much dictated by the character of the technique described. Many are based on quite complex concepts and instrumentation. Others are less so, and can be based on commercial products. Researchers working on non-magnetic materials may also find this book useful since many techniques and principles described in the book can be used for characterizing other materials.

Finally, I would like to express my appreciation to the many expert authors who have contributed to this book. On the production side, special thanks go to Lisa Jansson, the type-setting editor at Brookhaven, for her significant role in finalizing the book format, and to my colleague Marco Beleggia, who spent significant amounts of time in helping Lisa on the technical aspects of type-setting, such as the conversion of equations and symbols, and to my student June Lau for checking the Appendices. I am also grateful to the staff at Kluwer Academic Publisher, especially senior editor Greg Franklin, for their help and advice.

Yimei Zhu

Brookhaven National Laboratory
Long Island, New York

Table of Contents

Neutron Scattering

X-ray Scattering

Electron Scattering

Proximal Probe

Light Scattering

Neutron Scattering

CHAPTER 1

Magnetic neutron scattering

1.1. INTRODUCTION

Much of our understanding of the atomic-scale magnetic structure and the dynamical properties of solids and liquids was gained from neutron-scattering studies. Elastic and inelastic neutron spectroscopy provided physicists with an unprecedented, detailed access to spin structures, magnetic-excitation spectra, soft-modes and critical dynamics at magnetic phase transitions, which is unrivaled by other experimental techniques. Because the neutron has no electric charge, it is an ideal weakly interacting and highly penetrating probe of matter's inner structure and dynamics. Unlike techniques using photon electric fields or charged particles (*e.g.*, electrons, muons) that significantly modify the local electronic environment, neutron spectroscopy allows determination of a material's intrinsic, unperturbed physical properties. The method is not sensitive to extraneous charges, electric fields, and the imperfection of surface layers. Because the neutron is a highly penetrating and non-destructive probe, neutron spectroscopy can probe the microscopic properties of bulk materials (not just their surface layers) and study samples embedded in complex environments, such as cryostats, magnets, and pressure cells, which are essential for understanding the physical origins of magnetic phenomena.

Neutron scattering is arguably the most powerful and versatile experimental tool for studying the microscopic properties of the magnetic materials. The magnitude of the cross-section of the neutron magnetic scattering is similar to the cross-section of nuclear scattering by short-range nuclear forces, and is large enough to provide measurable scattering by the ordered magnetic structures and electron spin fluctuations. In the half-a-century or so that has passed since neutron beams with sufficient intensity for scattering applications became available with the advent of the nuclear reactors, they have became indispensable tools for studying a variety of important areas of modern science, ranging from large-scale structures and dynamics of polymers and biological systems, to electronic properties of today's technological materials. Neutron scattering developed into a vast field, encompassing many different experimental techniques aimed at exploring different aspects of matter's atomic structure and dynamics.

Modern magnetic neutron scattering includes several specialized techniques designed for specific studies and/or particular classes of materials. Among these are magnetic reflectometry aimed at investigating surfaces, interfaces, and multilayers, small-angle scattering for the large-scale structures, such as a vortex lattice in a superconductor, and neutron spin-echo spectroscopy for glasses and polymers. Each of these techniques and many others offer exciting opportunities for examining magnetism and warrant extensive reviews, but the aim of this chapter is not to survey how different neutron-scattering methods are used to examine magnetic properties of different materials. Here, we concentrate on reviewing the basics of the magnetic neutron scattering, and on the recent developments in applying one of the oldest methods, the triple axis spectroscopy, that still is among the most extensively used ones. The developments discussed here are new and have not been coherently reviewed. Chapter 2 of this book reviews magnetic small-angle scattering, and modern techniques of neutron magnetic reflectometry are discussed in Chapter 3.

In the first part of this chapter, we give an extensive, coherent introduction to magnetic neutron scattering. It includes an overview of the scattering problem with the derivation of the differential cross-section and its application to the neutron's magnetic interaction with an atom, the evaluation and properties of the magnetic form factors, and, finally, the general properties of the magnetic elastic and inelastic neutron scattering for the spin system of localized atomic electrons in the crystal. We describe magnetic neutron scattering at the "top level", concentrating on the highest-level formulae, but not giving particulars, which can be found in several books [1-5]. Further, rather than being exhaustive, we attempt to summarize those results that are general yet simple, and which, therefore, are most commonly used in everyday research.

The important issue of the magnetic form factors deserves special mention. A very complete theory was developed, accounting quite generally for the spin and the orbital magnetization density of atomic electrons, [3]. However, the general expressions in Ref. [3] are cumbersome so that they are rarely used in practice, and are replaced by the simple, but often highly inaccurate, "dipole approximation". Here, we derive simple formulae for the atomic spin magnetic form factors that accurately account for their angular anisotropy, a tremendous improvement over the dipole approximation. Although these expressions are not as completely general as those of Ref. [3], they accurately describe most situations encountered in magnetic neutron scattering. An example of where using the correct, anisotropic magnetic form factor is crucial for interpreting the experimental results is that of Cu^{2+} spins in topical cuprate materials. This issue gains more importance as magnetic neutron scattering conquers new heights in accessible energy transfers with the development of pulsed spallation neutron sources, such as ISIS in the UK and SNS in the United States. With energy transfers of 0.5 eV and above (see Fig. 1-1 for an example) the measured intensity is collected at very large wave vectors, where the magnetic form factor is small and often pronouncedly anisotropic.

In the second part, we describe the modern uses of the triple-axis spectrometer based on employing a large, multicrystal analyzer and/or the position-sensitive detector (PSD) to analyze the neutrons scattered by the sample. In many instances, the volume of the sample's phase space probed at each spectrometer setting can be increased by about an order-of-magnitude by using the PSD, thereby raising the rate of data collection. These advanced techniques, as known to the authors, were conceived and implemented on SPINS triple axis neutron spectrometer at the NIST Center for Neutron Research (NCNR) in Gaithersburg, MD, United States. Collin Broholm pioneered the PSD setup at the NCNR, with our active participation. It is a natural extension of SPINS capabilities based on employing a large multicrystal analyzer, originally designed for horizontal monochromatic (Rowland) focusing. Reportedly, a similar PSD setup was implemented on RITA spectrometer at the Risoe National Laboratory, Denmark. However, because the Risoe research reactor was permanently shutdown, the possibilities of RITA were not adequately explored. Subsequently, the spectrometer was moved to SINQ's continuous spallation neutron source at the Paul Sherrer Institute in Switzerland, where it now operates.

While an extensive literature addresses various aspects of neutron-scattering techniques, including several excellent books and monographs on magnetic neutron scattering [1-5], the advances outlined above are recent enough not to be described elsewhere. The general outline of this chapter is as follows. First, we review the fundamentals of neutron scattering: neutron interactions with matter, and magnetic scattering cross-section. We give a detailed exposition on magnetic form factors, deriving some simple and general formulae for the anisotropic form factors of the atomic orbitals that are not readily available elsewhere. Then, we summarize the properties of the two-point magnetization correlation functions in different classes of magnetic materials, paying special attention to pure spin scattering, where we derive the sum rules for the spin correlation function and review the single-mode approximation. Finally, we describe recent advances in triple axis spectroscopy, probably the most powerful technique for studying the dynamical properties of magnetic materials.

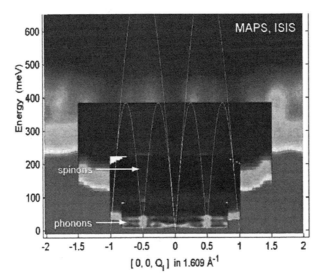

Figure 1-1: Color contour maps of the raw neutron-scattering intensity from a sample of the high-T_c-relative, chain cuprate $SrCuO_2$. The data was collected on MAPS time-of-flight neutron spectrometer at the ISIS pulsed spallation neutron source. Four measurements with the incident neutron energy $E_i \approx 100$, 250, 500 and 850 meV are shown stacked in the figure. They probe the energy transfers up to ≈ 80, 220, 400, and 650 meV, respectively. *Also see the color plate.*

1.2. Neutron interaction with matter and scattering cross-section

In this section, we review some important facts about the neutron, its properties, interaction with matter, and scattering cross-section.

The neutron is one of the basic constituents of matter. Together with its charged relative, the proton, it is a building block of the atomic nuclei (neutrons and protons are fermionic hadrons that, according to the "standard model", are the baryons, respectively composed of one "up" and two "down" quarks, and two "up" and one "down" quarks). Table 1-1 summarizes the basic properties of a neutron. Although the neutron is electrically neutral, it has a non-zero magnetic moment, similar in magnitude to that of a proton ($\mu_n \approx 0.685\mu_p$), but directed opposite to the angular momentum, so that the neutron's gyromagnetic ratio is negative.

Table 1-1: Basic properties of a neutron (mainly in Gauss CGS units). σ_n denotes the neutron's angular momentum, $\mu_N = e\hbar/(2m_p c) = 5.0508\bullet10^{-24}$ erg/Gs is the nuclear magneton.

Electric charge	Spin $S_n = \sigma_n/\hbar$	Mass m_n (g)	$m_n c^2/e$ (V)	Magnetic moment μ_n (erg/Gs)	Gyromag-netic ratio $\gamma_n, \mu_n = \gamma_n \sigma_n$ (s^{-1}/Gs)	g-factor g_n, $\mu_n = -g_n \mu_N S_n$	Life-time (s)	Decay reaction
0	1/2	$1.675\bullet10^{-24}$	$0.94\bullet10^9$	$9.662\bullet10^{-24}$	$-1.832\bullet10^4$	3.826	887	$n \to p\ e^-\ \underline{\nu}_e$

Outside the nucleus, a free neutron's lifetime is only about 15 minutes, after which it undergoes a β–decay into a proton, an electron, and an antineutrino. Nevertheless, this lifetime is long enough for neutron-scattering experiments. A neutron extracted through the beam-tube in a nuclear reactor typically has reached thermal equilibrium with the water that cools the reactor in a number of collisions on its way out (such neutrons usually are called thermal neutrons). Assuming the water has "standard" temperature of 293 K, the neutron's most probable velocity would be about 2200 m/s. It would spend only a fraction of a second while it travels in the spectrometer, is scattered by the sample, and arrives in the detector.

Generally, as widely accepted in the neutron-scattering literature, particle-physics notation is followed, and the energies both of a neutron and that of an excitation created in the scattering process are measured in millielectronvolts (meV). To ease comparison with the notations used in other techniques and in theoretical calculations, we list several different ways of representing the neutron's energy, $E_n = 1$ meV, in Table 1-2. The different energy notations shown in the Table can be used interchangeably, as a matter of convenience.

Table 1-2: Different notations used to represent the neutron's energy. e is the electron charge, h is the Plank's constant, c is the velocity of light, $\mu_B = e\hbar/(2m_e c) = 0.927\bullet10^{-20}$ erg/Gs is the Bohr's magneton, k_B is the Boltzman constant. Also shown are the corresponding neutron wave vector and deBroglie wavelength.

E_n (erg)	E_n/e (meV)	E_n/h (THz)	$E_n/(hc)$ (cm^{-1})	$E_n/(2\mu_B)$ (Gauss)	E_n/k_B (K)	λ_n (Å)	k_n (Å^{-1})
$1.602\bullet10^{-15}$	1	0.2418	8.0655	$8.638\bullet10^4$	11.604	9.0437	0.69476

Neutrons used in scattering experiments are non-relativistic. Therefore, the neutron's energy, E_n, is related to its velocity, v_n, wave vector, $k_n = (m_n v_n)/\hbar$, and the (de Broglie) wavelength, $\lambda_n = (2\pi)/k_n$, through

$$E_n = \frac{m_n v_n^2}{2} = \frac{\hbar^2 k_n^2}{2m_n} = \frac{h^2}{2m_n \lambda_n^2}.$$

In a typical experiment, neutrons with energies well in sub-eV range are used, although in some recent ones, the incident neutron energies were as high as 1 eV and more, Fig. 1-1. The neutron's wavelength and its wave vector are usually measured in Å (1 Å = 0.1 nm = 10^{-8} cm) and Å⁻¹, respectively. A useful relation connecting these quantities with the energy in meV follows from Table 2-2,

$$E_n = 2.0717\, k_n^2 = \frac{81.79}{\lambda_n^2}.$$

1.2.1. Basic scattering theory and differential cross-section

The general idea of a (direct geometry) scattering experiment is to place a sample in the beam of incident particles of mass m, with a well-defined wave vector $\boldsymbol{k}_i$ and known incident flux $\Phi_i(\boldsymbol{k}_i)$, and to measure the partial current, $\delta J_f(\boldsymbol{k}_f)$, scattered into a small ($\approx$ infinitesimal) volume of the phase space, $d^3\boldsymbol{k}_f = k_f^2 dk_f d\Omega_f = \left(mk_f/\hbar^2\right) dE_f d\Omega_f$, at a wave vector $\boldsymbol{k}_f$ (Fig.1-2).

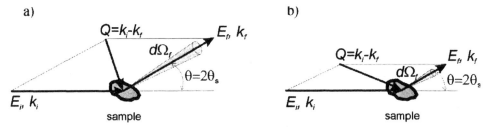

Figure 1-2: Typical geometry of a scattering experiment, (a) elastic, (b) inelastic.

The phase space density of the scattered current, normalized to the incident flux, defines the differential scattering cross-section with respect to the corresponding phase variables. The one most commonly measured and calculated is the double differential scattering cross-section,

$$\frac{d^2\sigma(\boldsymbol{Q},E)}{dEd\Omega} = \frac{1}{\Phi_i(\boldsymbol{k}_i)}\frac{\delta J_f(\boldsymbol{k}_f)}{dE_f d\Omega_f} = \frac{mk_f}{\hbar^2}\frac{1}{\Phi_i(\boldsymbol{k}_i)}\frac{\delta J_f(\boldsymbol{k}_f)}{d^3\boldsymbol{k}_f}. \tag{1.1}$$

Here, the laws of conservation determine the energy, E, and the wave vector, $\boldsymbol{Q}$, transferred to the sample,

$$E = \frac{(\hbar k_i)^2}{2m} - \frac{(\hbar k_f)^2}{2m}, \quad \boldsymbol{Q} = \boldsymbol{k}_i - \boldsymbol{k}_f. \tag{1.2}$$

Lippmann and Schwinger [5,6] most elegantly formulated the general solution of the scattering problem. Let η denote the complete set of variables that describe the state of the scatterer, and let the state of the scattered particle be described by its momentum, $\hbar\boldsymbol{k}$, and its spin quantum number, S^z. The state of the composite system, target sample (scatterer) + scattered particle that satisfies the boundary conditions of the scattering problem and has the energy $E_f^{(tot)} = E_i^{(tot)} = E_i(\eta_i) + (\hbar k_i)^2/(2m)$, is called the scattering state, $\left|\boldsymbol{k}_f, S_f^z, \eta_f\right\rangle$. It is obtained from the initial state, $\left|\boldsymbol{k}_i, S_i^z, \eta_i\right\rangle$, by applying the evolution operator $(1+\mathbf{GT})$,

$$\left|\boldsymbol{k}_f, S_f^z, \eta_f\right\rangle = (1+\mathbf{GT})\left|\boldsymbol{k}_i, S_i^z, \eta_i\right\rangle. \tag{1.3}$$

Here, $\mathbf{T}$ is the so-called *transition operator*, or *T-matrix*, and $\mathbf{G}$ is the retarded Green's function,

$$\mathbf{G}^{-1} = \left(E_i^{(tot)} - \mathbf{H}_0 + i \cdot 0 \right). \tag{1.4}$$

Here, $\mathbf{H}_0 = \mathbf{H}_s + (\hbar k_i)^2/(2m)$ is the part of the total Hamiltonian, $\mathbf{H} = \mathbf{H}_0 + \mathbf{V}$, which describes the sample and the scattered particle in the absence of their interaction, $\mathbf{V}$. The rate of transition, $\Gamma_{i \to f}$, from the initial to the final state $(i \to f)$, $\left| k_i, S_i^z, \eta_i \right\rangle \to \left| k_f, S_f^z, \eta_f \right\rangle$, is given by the appropriate matrix element of the transition operator,

$$\Gamma_{i \to f} = \frac{2\pi}{\hbar} \left| \left\langle k_f, S_f^z, \eta_f \left| \mathbf{T} \right| k_i, S_i^z, \eta_i \right\rangle \right|^2 \delta\left(E_i^{(tot)} - E_f^{(tot)} \right) = (2\pi)^3 \frac{\delta J_f(k_f)}{d^3 k}. \tag{1.5}$$

It determines the scattered current, $\delta J_f(k_f)$, and, therefore, the scattering cross-section. $E_f^{(tot)} = E_f(\eta_f) + (\hbar k_f)^2/(2m)$ is the energy of the system in the scattered state, so that the energy transfer to the sample is

$$E = E_f(\eta_f) - E_i(\eta_i) = \frac{(\hbar k_i)^2}{2m} - \frac{(\hbar k_f)^2}{2m}, \tag{1.6}$$

as required by the laws of energy conservation, Eq.(1.2). For the initial state of the incident particle in the form of a plane wave normalized to unity probability density, $\left\langle r | k_i \right\rangle = e^{ik_i \cdot r}$, the incident flux is $\Phi_i(k_i) = \hbar k_i / m$. Substituting this in Eqs.(1.1) and (1.5), the following general expression is straightforwardly obtained for the partial differential scattering cross-section corresponding to the transition $i \to f$,

$$\frac{d^2\sigma(\mathbf{Q}, E)}{dE d\Omega} = \frac{k_f}{k_i} \left| \left\langle S_f^z, \eta_f \left| \mathbf{b}(-\mathbf{Q}) \right| S_i^z, \eta_i \right\rangle \right|^2 \delta\left(E_i(\eta_i) - E_f(\eta_f) + E \right). \tag{1.7}$$

Here, the numerical pre-factor in front of $\mathbf{T}$ was conveniently absorbed into the definition of the *scattering length operator* $\mathbf{b}$,

$$\mathbf{b} = -\frac{m}{2\pi\hbar^2} \mathbf{T}. \tag{1.8}$$

By definition, $\mathbf{b}(-\mathbf{Q}) \equiv \mathbf{b}(-\mathbf{Q}, \mathbf{S}, \eta)$ in Eq.(1.7) is the Fourier transform of the matrix element of the scattering length with respect to the coordinate of the scattered particle,

$$\mathbf{b}(q) \equiv \int e^{-iqr'} \left\langle r' \left| \mathbf{b}(\mathbf{r}, \mathbf{S}, \eta) \right| r' \right\rangle d^3 r' = \left\langle k_f \left| \mathbf{b}(\mathbf{r}, \mathbf{S}, \eta) \right| k_i \right\rangle, \tag{1.9}$$

for the wave vector $q = k_f - k_i = -\mathbf{Q}$ that is transferred to that particle.

Finally, the $\mathbf{T}$-matrix operator satisfies the Lippmann-Schwinger equation, $\mathbf{T} = \mathbf{V} + \mathbf{TGV}$. Its iterative solution can be found in the form of the Born perturbation series (more generally, the von Neumann series) [5],

$$\mathbf{T} = \mathbf{V} + \mathbf{VGV} + \mathbf{VGVGV} + ... = \mathbf{V}\left(1 + \sum_n (\mathbf{GV})^n \right), \tag{1.10}$$

that completes the general solution of the scattering problem (provided the perturbation series converge). In many important cases, it appears sufficient to retain only the first-order term in this expansion, and use

$$\mathbf{T} = \mathbf{V}, \quad \mathbf{b} = -\frac{m}{2\pi\hbar^2}\mathbf{V},$$ (1.11)

that is known as the *Born approximation*. An expression for the transition rate in this approximation, obtained by substituting $\mathbf{T}=\mathbf{V}$ into Eq.(1.8) is one of the cornerstone results of Quantum Mechanics [7,8], and is universally used to describe scattering processes. Following Fermi, this expression often is called the "golden rule" [1,4].

1.2.2. Neutron interactions and scattering lengths

Two fundamental interactions govern the scattering of neutrons by an atomic system and define the neutron scattering cross-section measured in an experiment. The residual strong interaction, also known as the nuclear force, gives rise to scattering by the atomic nuclei (nuclear scattering). The electromagnetic interaction of the neutron's magnetic moment with the sample's internal magnetic fields gives rise to magnetic scattering. The sample's internal magnetic fields mainly originate from unpaired electrons in the atomic shells.

1.2.2.1. Nuclear scattering length

While magnetic interaction is relativistic and extremely weak, the nuclear force is not (as it is responsible for holding together protons and neutrons in the nucleus). However, it has extremely short range, 10^{-13} cm to 10^{-12} cm, comparable with the size of the nuclei, and much smaller than the typical neutron's wavelength. Consequently, away from the conditions of the resonance neutron capture, the probability of a neutron being scattered by an individual nucleus is very small, and can be treated in the scattering theory on par with the probability of magnetic scattering. In fact, it appears that nuclear scattering length, b_N, for the majority of natural elements is close in magnitude to the characteristic magnetic scattering length, $r_m = -(g_n/2)r_e = -5.391$ fm (1 fm $= 10^{-13}$ cm, $r_e = e^2/(m_e c^2)$ is the classical electron radius).

To describe the neutron's interaction with the atomic system in which the typical distances are about 1 $\mathring{A}$, the nuclear scattering length operator can be effectively treated as a delta-function in the coordinate representation,

$$\mathbf{b}_N = b_N \delta(r_n - R),$$ (1.12)

where r_n is a coordinate of a neutron and R is that of a nucleus. Alternatively, in the momentum representation it is just a number (for the nucleus fixed at the origin), $b_N(q) = b_N$, independent of the incident neutron's wave-vector and of the wave-vector transfer, q. This again indicates that the applicability of such treatment is limited to neutrons whose wavelength is large enough compared to the size of the nuclei. In the Born approximation, Eq.(1.12) for the scattering length would correspond to the neutron-nucleus interaction,

$$V_N(r_n, R) = -\frac{2\pi\hbar^2}{m_n} b_N \delta(r_n - R),$$ (1.13)

generally known as the Fermi pseudopotential [1,9]. In Eqs.(1.12) and (1.13), the scattering length refers to the fixed nucleus and is called the bound scattering length. Usually, it is treated as a phenomenological parameter that is determined experimentally [10]. In general, the bound scattering length is considered to be a complex quantity, $b_N = b' - ib''$, defining the total scattering cross-section, σ_s, and the absorption cross-section far from the nuclear resonance capture, σ_a, through

$$\sigma_s = 4\pi|b|^2, \quad \sigma_a = \frac{4\pi}{k_i}b''. \tag{1.14}$$

Ref. [11] tabulates the bound scattering lengths and cross sections of the different elements and their isotopes.

1.2.2.2. Magnetic scattering length

Because the magnetic interaction of a neutron with a single atom is very weak, the Born approximation, Eq.(1.11), very accurately describes the magnetic scattering length. The main contribution to magnetic scattering arises from the neutron's interaction with the total dipole magnetic moment of the atomic electrons; all other electromagnetic interactions are at least two orders-of-magnitude smaller and can be safely neglected [5]. The fundamental starting point for evaluating the neutron magnetic scattering length is the Hamiltonian of the electrons in the atom in the presence of the neutron's magnetic field [2,4]. The interaction Hamiltonian is

$$V_m(r_n, r_e) = \sum_e \left\{ \frac{2\mu_B}{\hbar} (A_n(r_e) \cdot p_e) + 2\mu_B (s_e \cdot H_n(r_e)) \right\} = \sum_e \{V_{le} + V_{se}\}, \tag{1.15}$$

where the sum extends over all electrons in the atom, indexed by e. r_n and r_e are the position of the neutron and that of the electron, respectively, p_e is the momentum, and $\hbar s_e$ is the spin angular momentum of the electron. $A_n(r_e)$ is the vector-potential, so that

$$H_n(r_e) = [\nabla_{r_e} \times A_n(r_e)] \tag{1.16}$$

is the magnetic field of the neutron at the position of the eth electron, r_e. The first term in Eq.(1.15), V_{le}, describes the interaction of the neutron magnetic field, $H_n(r)$, with the electric current produced by the electron's orbital motion. The second term, V_{se}, accounts for the neutron's magnetic interaction with the spin magnetic moment of the electrons.

The characteristic size of the inner structure of a neutron is extremely small, so that in describing the magnetic interaction with an electron in an atom it can be treated as a point dipole with the magnetic moment $\mu_n = \gamma_n \sigma_n$, γ_n is the neutron's gyromagnetic ratio, and $\sigma_n = \hbar s_n$ is its spin angular momentum (see Table 1-1). The corresponding expression for the neutron's magnetic field vector potential at the position of the electron is

$$A_n(r_e) = \left[\mu_n \times \frac{r_e - r_n}{|r_e - r_n|^3} \right] = \left[\nabla_{r_e} \times \frac{\mu_n}{|r_e - r_n|} \right] = \left[\nabla \times \frac{\mu_n}{r} \right], \tag{1.17}$$

$r = r_e - r_n$ is the spacing between the neutron and the electron [2,3-5,12].

On account of Eq.(1.17), the orbital part of the interaction Hamiltonian Eq.(1.15) can be recast in the following form,

$$\mathbf{V}_{le} = -\left(\boldsymbol{\mu}_n \cdot \left[\nabla \times \left(\frac{1}{r}\frac{e}{m_e c}\boldsymbol{p}_e\right)\right]\right) = 2\mu_B \frac{(\boldsymbol{\mu}_n \cdot \boldsymbol{l}_e)}{r^3}, \tag{1.18}$$

which also could be semirigorously derived from the Biot-Savart law [1,12]. Here $\boldsymbol{p}_e$ is the momentum of the electron, and $\hbar \boldsymbol{l}_e = [\boldsymbol{r} \times \boldsymbol{p}_e]$ is its orbital angular momentum in the neutron's rest frame. Eq.(1.18) is just the energy of the neutron's dipole magnetic moment, $\boldsymbol{\mu}_n$, in the magnetic field,

$$H_{le}(\boldsymbol{r}_n) = \left[\frac{(\boldsymbol{r}_e - \boldsymbol{r}_n)}{|\boldsymbol{r}_e - \boldsymbol{r}_n|^3} \times \frac{1}{c}\boldsymbol{I}_e\right] = \left[\nabla_{r_n}\left(\frac{1}{r}\right) \times \frac{1}{c}\boldsymbol{I}_e\right] = \left[\nabla \times \left(-\frac{1}{cr}\boldsymbol{I}_e\right)\right], \tag{1.19}$$

of the electron's orbital electric current $\boldsymbol{I}_e$, [12]. The latter is formally defined by $\boldsymbol{I}_e = -(e/m_e)\boldsymbol{p}_e$ [note, that $\nabla_{r_e} f(r) = \nabla f(r) = -\nabla_{r_n} f(r)$].

The second term in Eq.(1.15), describing the neutron's interaction with the spin magnetic moment of the electron, $\boldsymbol{\mu}_{se} = -2\mu_B \boldsymbol{s}_e$, can be rewritten symmetrically as the interaction of the two magnetic point dipoles at a distance $r = |\boldsymbol{r}_e - \boldsymbol{r}_n|$ from each other,

$$\mathbf{V}_{se}(r) = -\left(\boldsymbol{\mu}_{se} \cdot \left[\nabla \times \left[\nabla \times \frac{\boldsymbol{\mu}_n}{r}\right]\right]\right) = -\left(\boldsymbol{\mu}_n \cdot \left[\nabla \times \left[\nabla \times \frac{\boldsymbol{\mu}_{se}}{r}\right]\right]\right). \tag{1.20}$$

This expression contains essential singularity at $r = 0$ and needs to be treated carefully when evaluating the derivatives. By using $\nabla^2(1/r) = -4\pi\delta(r)$, Eq.(1.20) can readily be transformed to the form perhaps most commonly used for the interaction between two point dipoles [13-15],

$$\mathbf{V}_{se}(r) = -\left\{\frac{8\pi}{3}(\boldsymbol{\mu}_n \cdot \boldsymbol{\mu}_{se})\delta(r) - \frac{(\boldsymbol{\mu}_n \cdot \boldsymbol{\mu}_{se})}{r^3} + \frac{3(\boldsymbol{\mu}_n \cdot \boldsymbol{r})(\boldsymbol{\mu}_{se} \cdot \boldsymbol{r})}{r^5}\right\}. \tag{1.21}$$

The first, singular term here is called the Fermi contact interaction. The rest is the potential part that describes the interaction between the dipoles at large distances. Because the neutron's wave function overlaps with those of the electrons, it is essential to account for the contact term in the magnetic scattering length. Although less conventional, Eq.(1.20) is more convenient for evaluating the scattering cross-section. Not only does it correctly contain the singular part of the dipole-dipole interaction, but it also can be readily Fourier-transformed to obtain the spin contribution to the neutron's magnetic scattering length in the momentum representation

$$\mathbf{b}_{se}(\boldsymbol{q}) = \int e^{-i\boldsymbol{q}\boldsymbol{r}_n} \mathbf{b}_{se}(\boldsymbol{r}_n, \boldsymbol{r}_e)d^3\boldsymbol{r}_n = -\frac{m}{2\pi\hbar^2}\frac{4\pi}{q^2}(\boldsymbol{\mu}_n \cdot [\boldsymbol{q} \times [\boldsymbol{q} \times e^{-i\boldsymbol{q}\boldsymbol{r}_e}\boldsymbol{\mu}_{se}]]). \tag{1.22}$$

This expression is an important, fundamental result that governs the essential properties of the magnetic neutron scattering cross-section.

In many important cases, the contribution of the orbital currents to the magnetic scattering cross-section Eq.(1.7) is zero, or small, and can be neglected. This happens when the corresponding matrix elements of the orbital contribution Eq.(1.18) to the magnetic interaction are small, or vanish, as is the case, for example, for scattering by the s-electrons that are in the $l_e = 0$ state and, consequently, $\langle \eta_f | \mathbf{V}_{le} | \eta_i \rangle = 0$. For atoms of the transition elements in the crystal, the local crystal electric field typically quenches orbital angular momentum [14]. Hence, the orbital contribution to the magnetic scattering cross-section also is very small. On the other hand, accurately accounting for the orbital scattering is rather cumbersome, much more so than for spin-only scattering. This is because the matrix elements of the orbital part of the magnetic interaction, Eq.(1.18), depend significantly on the electron's wave functions and, in general, require specific calculations for each particular case of electronic configuration in the atom [17-22]. On these grounds, the orbital contribution is often discarded in the textbook treatments of the magnetic neutron scattering cross-section, [2,5,15].

Accounting for the orbital magnetic moment is important for the scattering by the $4f$- and $5f$-electrons in the rare earths. In this case, the crystal field is usually well screened by the filled outer atomic shells, and the total angular momentum, $J = L + S$, is a good quantum number. Fortunately, the useful general expressions for the magnetic neutron scattering length and for the corresponding cross-section can be derived without first evaluating the matrix elements of the orbital part of the neutron's magnetic interaction with the electrons. This task can be postponed till the end, where it becomes a part of the general problem of evaluating the atom's magnetic form factor.

One proceeds as follows. Under very general assumptions, the orbital contribution to the magnetic neutron-scattering length can be transformed to a form similar to the spin part, Eqs. (1.20) and (1.22). Consequently, they can be combined and treated together. The simplest way to do this is to assume that the main contribution to the matrix elements of the interaction of the neutron in the plane-wave state with the orbital electron current, Eq.(1.18), comes from the region $r_n >> r_e$. This approximation clearly holds if the neutron's wavelength is much greater than the characteristic size of the atomic wave functions, *i.e.,* for slow neutrons. Then, $1/|r_n-r_e|$ can be expanded in the power series and, to the leading order, the matrix element of the orbital magnetic field at the neutron's position becomes [12]

$$\langle \eta_f | H_{le}(r_n) | \eta_i \rangle = -\frac{2\mu_B}{\hbar} \langle \eta_f | \left[\nabla_{r_n} \times (1 - (r_e \cdot \nabla_{r_n})) \frac{1}{r_n} p_e \right] | \eta_i \rangle. \qquad (1.23)$$

The first term in the inner brackets here does not contribute to the result because, for an electron that remains localized on an atomic orbital, the average momentum is zero, $\langle \eta_f | p_e | \eta_i \rangle = 0$, [12,13]. The second term can be transformed by separating the full time derivative, whose matrix element for an electron in a stationary state is also zero, and using $m_e \dot{r}_e = p_e$, [13],

$$(r_e \cdot \nabla_{r_n}) p_e = \frac{1}{2} \left\{ m_e \frac{d}{dt} ((r_e \cdot \nabla_{r_n}) r_e) - r_e (p_e \cdot \nabla_{r_n}) + (r_e \cdot \nabla_{r_n}) p_e \right\}. \qquad (1.24)$$

It then follows that,

$$\langle \eta_f | (r_e \cdot \nabla_{r_n}) p_e | \eta_i \rangle = -\langle \eta_f | [\nabla_{r_n} \times [r_e \times p_e]] | \eta_i \rangle,$$

and, as a result, Eq.(1.23) becomes

$$\langle \eta_f | H_{le}(r_n) | \eta_i \rangle = -\mu_B \langle \eta_f | \left[\nabla_{r_n} \times \left[\nabla_{r_n} \times \frac{1}{r_n} l_e \right] \right] | \eta_i \rangle. \qquad (1.25)$$

This brings the matrix element of the orbital part of the magnetic interaction to the same form as that for the spin part, Eq.(1.20), but with r_n in place of r and with the orbital magnetic moment,

$$\mu_{le} = -\frac{\mu_B}{\hbar} [r_e \times p_e] = -\mu_B l_e, \qquad (1.26)$$

replacing the spin magnetic moment, $\mu_{se} = -2\mu_B s_e$.

Both contributions can be combined into a simple final expression for the matrix element of the atom's magnetic neutron scattering length,

$$\langle k_f, \eta_f | \mathbf{b}_m | k_i, \eta_i \rangle = -\frac{m}{2\pi\hbar^2} \frac{4\pi}{q^2} (\boldsymbol{\mu}_n \cdot [\boldsymbol{q} \times [\boldsymbol{q} \times \langle \eta_f | \boldsymbol{M}(\boldsymbol{q}) | \eta_i \rangle]]), \tag{1.27}$$

where $\boldsymbol{q} = \boldsymbol{k}_f - \boldsymbol{k}_i$ is the neutron's wave vector change, as in Eq.(1.9). The approximation adopted above in deriving the Eq.(1.25) gives only the lowest-order, $\boldsymbol{q}$-independent orbital contribution to the operator $\boldsymbol{M}(\boldsymbol{q})$. In this approximation $\boldsymbol{M}(\boldsymbol{q}) \approx \boldsymbol{M}(0) = -\mu_B \sum_e \{\boldsymbol{l}_e + 2\boldsymbol{s}_e\} = -\mu_B(\boldsymbol{L} + 2\boldsymbol{S})$ [16]. Trammel [17] developed a more accurate accounting for the orbital part of the magnetic interaction. His treatment is essentially similar to the above, but the terms of all orders are consistently retained in the series expansion. Consequently, $\boldsymbol{M}(\boldsymbol{q})$ in the right-hand side of Eq.(1.27) becomes

$$\boldsymbol{M}(\boldsymbol{q}) = \sum_e \left\{ -2\mu_B \boldsymbol{s}_e e^{-i\boldsymbol{q}\boldsymbol{r}_e} - \mu_B \frac{1}{2} (\boldsymbol{l}_e f(\boldsymbol{q} \cdot \boldsymbol{r}_e) + f(\boldsymbol{q} \cdot \boldsymbol{r}_e) \boldsymbol{l}_e) \right\}, \tag{1.28}$$

where

$$f(\boldsymbol{q} \cdot \boldsymbol{r}_e) = 2 \sum_{n=0}^{\infty} \frac{(i\boldsymbol{q} \cdot \boldsymbol{r}_e)^n}{n!(n+2)}. \tag{1.29}$$

Eq.(1.25) retains only the first, zero-order, $\sim O(q^0)$, term in this expression.

Clearly, the first term in Eq.(1.28) is simply the Fourier-transform of the density of the spin magnetic moment of the atomic electrons

$$\boldsymbol{M}_S(\boldsymbol{q}) = -2\mu_B \sum_e \int e^{-i\boldsymbol{q}\boldsymbol{r}_e} \boldsymbol{s}_e \delta(\boldsymbol{r}' - \boldsymbol{r}_e) d^3 \boldsymbol{r}' = \int e^{-i\boldsymbol{q}\boldsymbol{r}'} (-2\mu_B \boldsymbol{S}(\boldsymbol{r}')) d^3 \boldsymbol{r}'. \tag{1.30}$$

It also can be shown, [18,19], that the second (orbital) term in Eq.(1.28) is the Fourier-transform of the atom's orbital magnetization density

$$\boldsymbol{M}_L(\boldsymbol{q}) = \sum_e \int e^{-i\boldsymbol{q}\boldsymbol{r}'} \boldsymbol{\mu}_{el}(\boldsymbol{r}') d^3 \boldsymbol{r}'. \tag{1.31}$$

Here, the density of the orbital magnetization for an electron in the atom, $\boldsymbol{\mu}_{el}(\boldsymbol{r})$, is defined by the relation $\boldsymbol{j}_e(\boldsymbol{r}) = c[\nabla \times \boldsymbol{\mu}_{el}(\boldsymbol{r})]$, so that it determines the density of the orbital electric current

$$\boldsymbol{j}_e(\boldsymbol{r}) = -\frac{e}{2m_e} \{\boldsymbol{p}_e \delta(\boldsymbol{r} - \boldsymbol{r}_e) + \delta(\boldsymbol{r} - \boldsymbol{r}_e) \boldsymbol{p}_e\} = c[\nabla \times \boldsymbol{\mu}_{el}(\boldsymbol{r})], \tag{1.32}$$

and accounts for the magnetic field arising from the electron's orbital motion. Consequently, the contribution of the orbital electric currents to the magnetic interaction in Eq.(1.15) can be recast in the form of the double cross product, as in Eq.(1.27), using

$$e^{-i\boldsymbol{q}\boldsymbol{r}_e} \frac{1}{c} \boldsymbol{I}_e = \int e^{-i\boldsymbol{q}\boldsymbol{r}'} \frac{1}{c} \boldsymbol{j}_e(\boldsymbol{r}') d^3 \boldsymbol{r}' = \int e^{-i\boldsymbol{q}\boldsymbol{r}'} [\nabla_{\boldsymbol{r}'} \times \boldsymbol{\mu}_{el}(\boldsymbol{r}')] d^3 \boldsymbol{r}'. \tag{1.33}$$

Therefore, the matrix element of the neutron magnetic scattering length is expressed by the Eq.(1.27), where $\boldsymbol{M}(\boldsymbol{q})$ is the Fourier-transform of the total, spin and orbital, electronic magnetization density in the atom,

$$\boldsymbol{M}(\boldsymbol{q}) = \boldsymbol{M}_S(\boldsymbol{q}) + \boldsymbol{M}_L(\boldsymbol{q}) = \int e^{-i\boldsymbol{q}\boldsymbol{r}'} \sum_e \{-2\mu_B \boldsymbol{s}_e \delta(\boldsymbol{r}' - \boldsymbol{r}_e) + \boldsymbol{\mu}_{el}(\boldsymbol{r}')\} d^3 \boldsymbol{r}'. \tag{1.34}$$

1.2.3. Factorization of the magnetic scattering length and the magnetic form factors

By applying the Wigner-Eckart theorem, a matrix element of the atom's magnetization density operator Eq.(1.34) can be factorized into the product of the reduced matrix element that does not depend on the direction of the atom's angular momentum, and the Wigner $3j$-symbol, which entirely accounts for such dependence [8,13]. The first factor contains the q-dependence of the matrix element, while the second describes its symmetry with respect to rotations and relates them to the magnetic neutron scattering cross-section. Such factorization is extremely useful in understanding magnetic neutron scattering by macroscopic samples. It splits the task of calculating the scattering cross-section for a system of many atoms in two separate major parts that address different aspects of the problem. One is that of evaluating the neutron magnetic form factor, which describes the q-dependence of the scattering by a single atom and is determined by the reduced matrix element(s). The other one is that of properly adding the contributions from the correlated (and/or the uncorrelated) rotations of the magnetic moments of different atoms in the sample to obtain the total scattering cross-section.

Because $M(q)$ in Eq.(1.27) contains both spin and orbital contributions [cf Eq.(1.34)], its matrix elements must be expressed through those of the atom's *total* angular momentum, $J = L + S = \sum_e \{l_e + s_e\}$. Consequently, the Wigner-Eckart theorem applies directly to $\langle \eta_f | M(q) | \eta_i \rangle$ only if $|\eta_i\rangle$ and $|\eta_f\rangle$ are approximately the eigenstates of J and J^z, i.e., if J is an integral of motion for the scattering atom. In practice, this is the case if the spin-orbit interaction (LS-coupling) is much larger than any other interaction that depends on the atom's orbital and/or spin angular momentum, such as the interaction with the crystal field. We consider such a situation first.

From Eq.(1.28) we see that the matrix elements of the operators $M_S(q)$ and $M_L(q)$ between the eigenstates of the atom's total angular momentum, J, satisfy the "dipole" selection rules, [7,8]. Hence, for each of the two operators only the matrix elements between the states with $\Delta J = J(\eta_f) - J(\eta_i) = 0, \pm 1$ can differ from zero. Therefore, only such transitions are allowed in the magnetic neutron scattering. This also is evident from the conservation of the total, neutron's and atom's angular momentum, since ΔJ^z has to be offset by the change in the neutron's spin, which can only be $\Delta S_n^z = 0, \pm 1$.

While the Wigner-Eckart decomposition of the matrix element is quite tedious for a general tensor and for an arbitrary states $|\eta_i\rangle$ and $|\eta_f\rangle$, it is greatly simplified for a vector operator such as $M(q)$ that is a tensor of rank one [3,13]. As discussed above, the matrix elements of a vector satisfy the "dipole" selection rules, *i.e.*, they can only be non-zero between the states whose angular momentum quantum numbers differ by no more than 1 [13]. Therefore, no more than three different reduced matrix elements appear in the decomposition of $\langle \eta_f | M(q) | \eta_i \rangle$ in Eq.(1.27) and, consequently, in the magnetic neutron scattering cross-section. These reduced matrix elements completely account for the q-dependence of magnetic neutron scattering from a single atom. Normalized to 1 at $q = 0$, they define the atom's neutron magnetic form factors for the corresponding scattering channels, in complete analogy with the usual x-ray atomic form factors.

In most cases of practical importance for magnetic neutron scattering, both the initial and the final states of the atom, $|\eta_i\rangle$ and $|\eta_f\rangle$, belong to the *same* angular momentum multiplet, $|\eta_{i,f}\rangle = |\eta', J, J_{i,f}^z\rangle$. There are no transitions between atomic states with different angular momenta, *i.e.*, $J(\eta_f) - J(\eta_i) = 0$. Hence, the cross-section involves only a single reduced matrix element, that for the ground-state multiplet. Normalized appropriately, it defines what is usually called the neutron magnetic form factor of an atom. In this case, the expression for $\langle \eta_f | M(q) | \eta_i \rangle$ is simple. The $3j$-symbols are just the matrix elements of a vector operator J, and the statement of the Wigner-Eckart theorem is reduced to the well-known relation for the matrix elements of a vector [7,8],

$$\langle \eta_f | A | \eta_i \rangle = \langle \eta', J, J_f^z | J | \eta', J, J_i^z \rangle \frac{\langle \eta', J | (A \cdot J) | \eta', J \rangle}{J(J+1)}, \tag{1.35}$$

that also is expected from general symmetry arguments [13]. This expression is valid for any vector-operator A that has appropriate commutation relations with J, in particular, for $A = M(q)$. Applying it to $A = L + 2S = J + S$ and for the states and that also belong to the same spin and orbital multiplets, $S^2 = S(S+1)$ and $L^2 = L(L+1)$, the famous Lande result for the spectroscopic g-factor in the theory of the Zeeman effect is immediately obtained,

$$g = 1 + \frac{J(J+1) + S(S+1) - L(L+1)}{2J(J+1)}. \tag{1.36}$$

Upon applying Eq.(1.35) to $A = M(q)$, and introducing the atom's magnetic form factor, $F_J(q)$, we obtain

$$\langle \eta_f | M(q) | \eta_i \rangle = -g\mu_B F_J(q) \langle \eta_f | J | \eta_i \rangle. \tag{1.37}$$

$F_J(q)$ is the normalized average expectation value of the Fourier-transform of the atom's net magnetization density within the $J^2 = J(J+1)$ multiplet,

$$F_J(q) = \frac{\langle \eta', J | (M(q) \cdot J) | \eta', J \rangle}{-g\mu_B J(J+1)}. \tag{1.38}$$

These expressions can be directly applied to describe the neutron's magnetic scattering length in cases where the total angular momentum of the atom is a good quantum number, and the matrix elements of $M(q)$ for $\Delta J \neq 0$ are zero, or negligible. Typically, this is a good approximation for the rare earths where the spin-orbit interaction is strong, while the unpaired electrons in the unfilled $4f$ and $5f$ shells are well screened from the crystal field by the filled outer shells. Hence, the splitting between the different J-multiplets is much larger than the level splitting in the crystal field, and the mixing of states with different J and transitions between them can be ignored [4,13,14,17].

With an intermediate, or strong (with respect to the spin-orbit coupling) crystal field, J is not a good quantum number, and the simple relation Eq.(1.37) cannot be used. Here, $|\eta_i\rangle$ and $|\eta_f\rangle$ are not even approximately the eigenstates of J, but are the mixtures of states from the different J multiplets, so Eq.(1.35) does not apply. This situation is typical for the ions of the $3d$, $4d$, and $5d$ groups where the unpaired magnetic electrons occupy the outer valence d-shells. Hence, their interaction with the ligand crystal field is comparable in strength to, or stronger than, the spin-orbit LS-coupling [14]. Fortunately, in many important cases, a strong crystal field also fully lifts the orbital degeneracy, so that the orbital moment is quenched in the atom's ground state, and the orbital contribution to the magnetic scattering length is small and can be neglected [1-5,14,18].

In the general case there is no simple relation between the matrix element of the magnetic scattering length Eq.(1.27), and those of the atom's angular momentum, J. Nevertheless, a simple, independent factorization for the spin- and the orbital-contributions, similar to Eqs. (1.37), (1.38), is nearly always possible. In the general case, $|\eta_i\rangle$ and $|\eta_f\rangle$ are not (approximate) eigenstates of J and J^z, but are the superpositions of states described by the full set of spin and orbital quantum numbers, $\{S,S^z,L,L^z\}$, as in the Russel-Saunders LS-scheme,

$$|\eta_{i,f}\rangle = \sum_{S,S^z,L,L^z} C_{i,f}(S,S^z,L,L^z)|\eta',S,S^z,L,L^z\rangle. \tag{1.39}$$

Consequently, the matrix element of the atom's magnetization density in Eq.(1.27) is the sum of the contributions

$$\langle \eta',\{S,S^z,L,L^z\}_f \, |M_S(q)+M_L(q)|\eta',\{S,S^z,L,L^z\}\rangle. \tag{1.40}$$

Here, the spin part of the atom's magnetization density operator, $M_S(q)$, acts only on the spin variables, while the orbital part, $M_L(q)$, acts only on the coordinate part of the atomic wave function. Therefore, the only non-zero matrix elements of $M_S(q)$ are those between the states with the same orbital quantum numbers, $\Delta L = 0$, $\Delta L^z = 0$, that satisfy the selection rules on spin angular momentum, $\Delta S = 0, \pm 1$. On the other hand, the only non-zero matrix elements of $M_L(q)$ are between the states with the same spin quantum numbers, $\Delta S = 0$, $\Delta S^z = 0$, and satisfy $\Delta L = 0, \pm 1$. Hence, the Wigner-Eckart theorem can be applied separately to the spin and to the orbital contributions in Eq.(1.40). The matrix elements of $M_S(q)$ are expressed through those of the atom's spin operator, S, while the matrix elements of $M_L(q)$ through those of its orbital angular momentum, L.

In nearly all cases of interest both $|\eta_i\rangle$ and $|\eta_f\rangle$ are, to a good approximation, the eigenstates of S and L (although the S and L eigenvalues in the initial and the final states might differ). In particular, this holds for the atomic configurations that satisfy Hund's rule. In other words, there is little or no mixing of states from different spin or orbital multiplets in the initial and final states of the scattering atom, $|\eta_i\rangle$ and $|\eta_f\rangle$. Consequently, there is only one term in the sum over S and L in Eqs.(1.39), (1.40). Therefore, the Wigner-Eckart theorem can be applied directly to the matrix elements of the spin and the orbital parts of the atom's magnetization density in Eq.(1.27). The spin part is factorized into a reduced matrix element, $\langle\eta',S_f \,|M_S(q)|\eta',S_i\rangle$, and a $3j$-symbol that depends only on the atom's spin quantum numbers. Similarly, the orbital part is a product of a reduced matrix element, $\langle\eta',L_f \,|M_L(q)|\eta',L_i\rangle$, and a $3j$-symbol that depends only on the atom's angular momentum quantum numbers. Consequently, the magnetic neutron scattering length is expressed as a sum of the function of the atom's spin operator, S, and the function of its orbital angular momentum operator, L, each weighted with its own, q-dependent magnetic form factor. The latter quantifies the probability of the transition between the S and L multiplets to which the atom's initial and final states belong. In either case, the relations essentially repeat those between the matrix elements of $M(q)$ and J in the case of strong LS-coupling discussed above.

1.2.3.1. Magnetic form factors for Hund's ions: vector formalism

While the matrix elements between the states from the different atomic multiplets, *i.e.*, those with $\Delta S = \pm 1$, or $\Delta L = \pm 1$ can be factorized in Eq.(1.27) using the Wigner-Eckart theorem, both the calculation itself and the results are very unwieldy [3,17,20], and are of limited practical importance. Fortunately, such a calculation is rarely needed. Firstly, the neutron's energy typically is not sufficient to cause the transitions between different atomic multiplets. Secondly, both the initial and the final states of the scattering atom usually satisfy the Hund's rule and, therefore, belong to the same L and S multiplets. Among the notable exceptions to this rule are the ions with the electronic configuration $3d^6$, such as Co^{3+}, in a strong crystal field. There, the energies of states from the different spin multiplets are close, and the ground state may have a low spin, $S = 0$, or an intermediate spin, $S = 1$, in a clear violation of Hund's rule. In such a situation there might be significant mixing of the different spin states in $|\eta_i\rangle$ and $|\eta_f\rangle$. Transitions between these spin states can contribute to inelastic magnetic neutron scattering.

In the overwhelming majority of experimental situations, though, atoms in the sample remain in the states that belong to the same Hund's multiplet, $|\eta_{i,f}\rangle = |\eta'_{i,f}, L, S\rangle$, i.e., $\Delta L = 0$, $\Delta S = 0$. Restricting our attention to such cases we can write

$$\langle \eta_f | M(q) | \eta_i \rangle = -2\mu_B F_S(q) \langle \eta_f | S | \eta_i \rangle - \mu_B F_L(q) \langle \eta_f | L | \eta_i \rangle. \tag{1.41}$$

Here the spin, $F_S(q)$, and the orbital, $F_L(q)$, magnetic form factors of the atom are defined in the same way as $F_J(q)$, Eq.(1.37), but with $g=2$ and S replacing J in the first case, and with $g=1$ and L replacing J in the second. Substituting $M_S(q)$ and $M_L(q)$ from Eq.(1.28) gives the following explicit expressions,

$$F_S(q) = \frac{\langle \eta', L, S | \sum_e e^{-iqr_e} (s_e \cdot S) | \eta', L, S \rangle}{S(S+1)}, \tag{1.42}$$

and,

$$F_L(q) = \frac{\langle \eta', L, S | \sum_e (\{l_e f(q \cdot r_e) + f(q \cdot r_e) l_e\} \cdot L) | \eta', L, S \rangle}{2L(L+1)}. \tag{1.43}$$

The spin form factor Eq.(1.42) is the Fourier-transform of the atom's total spin density, projected on its total spin. It describes the q-distribution of the unpaired electron density in the atom. The orbital form factor Eq.(1.43) describes the q-distribution of the net orbital angular momentum of the atomic electrons. Both satisfy the important normalization condition

$$F_S(0) = F_L(0) = 1, \tag{1.44}$$

that is evident from the definitions Eq.(1.42), and (1.43).

When J is conserved $|\eta_i\rangle$ and $|\eta_f\rangle$, belong to the same J-multiplet, the matrix elements of S and L are proportional to those of J, $\langle \eta_f | S | \eta_i \rangle = (g-1)\langle \eta_f | J | \eta_i \rangle$, $\langle \eta_f | L | \eta_i \rangle = (2-g)\langle \eta_f | J | \eta_i \rangle$, where g is the Lande g-factor defined by Eq.(1.36). Upon substituting these into Eq.(1.41), the following relation is obtained for the magnetic form factors,

$$F_J(q) = \left(2 - \frac{2}{g}\right) F_S(q) + \left(\frac{2}{g} - 1\right) F_L(q). \tag{1.45}$$

In a great variety of important cases, the matrix elements of S and L within the lowest atomic multiplet to which the scattering states $|\eta_i\rangle$ and $|\eta_f\rangle$ belong, can be described by the effective spin operator $\tilde{S}$,

$$\langle \eta_f | 2S | \eta_i \rangle = g_S \langle \eta_f | \tilde{S} | \eta_i \rangle, \quad \langle \eta_f | L | \eta_i \rangle = (g - g_S) \langle \eta_f | \tilde{S} | \eta_i \rangle. \tag{1.46}$$

Here, g is the effective g-factor that describes the Zeeman splitting of the multiplet, and g_S is the effective spin-only g-factor. Consequently, Eq.(1.41) for the matrix element of the magnetization becomes

$$\langle \eta_f | \boldsymbol{M}(\boldsymbol{q}) | \eta_i \rangle = -g\mu_B F(\boldsymbol{q}) \langle \eta_f | \widetilde{\boldsymbol{S}} | \eta_i \rangle . \tag{1.47}$$

This is the same as Eq.(1.37), but with $\widetilde{S}$ in place of $\boldsymbol{J}$, and with $F_J(\boldsymbol{q})$ replaced by the generalized, effective magnetic form factor $F(\boldsymbol{q})$,

$$F(\boldsymbol{q}) = \frac{g_S}{g} F_S(\boldsymbol{q}) + \left(1 - \frac{g_S}{g}\right) F_L(\boldsymbol{q}) . \tag{1.48}$$

Eqs.(1.41)-(1.48), together with Eq.(1.27), rather accurately describe the neutron magnetic scattering length in nearly all cases of interest, and are the most widely used ones. They are exact descriptions in cases of a J-multiplet and a pure spin multiplet (e.g., with $L = 0$, where $g = g_S$ and the orbital contribution to the magnetic scattering is absent). They give the leading-order approximation in other cases. Eq.(1.45) follows from Eq.(1.48) upon substituting $\widetilde{S} = J$ and $g_S = 2(g - 1)$, as is appropriate for the J-multiplet. If the orbital moment is nearly quenched, as it is for the ions of the magnetic d-elements in strong crystal field, then $\widetilde{S} \approx S$, $g_S \approx 2$, and the orbital contribution to $F(\boldsymbol{q})$, proportional to $(g - g_S)$, is small, [18]. Usually, Eqs.(1.47), (1.48) are adopted, with S in place of $\widetilde{S}$, implying that it denotes an effective spin.

The simple equations derived above, starting with the expression for the matrix elements of a vector operator, Eq.(1.35), are usually referred to as *vector formalism*, [3,21]. It is nearly always sufficient for understanding the neutron magnetic scattering cross-section, which can be expressed in terms of the atom's magnetic form factors, Eqs.(1.42)-(1.45). In a few important cases, the multiplet mixing in the states of the scattering atom is essential, and the vector approach does not suffice. Then, the scattering cross-section cannot be simply factorized, and a tensor formalism based on Racah algebra for the tensor operators must be employed. Although the calculations are very tedious, explicit expressions can be obtained for the spin- and orbital-contributions to the magnetic scattering cross-section [3,20-22].

1.2.3.2. Evaluating the form factors and dipole approximation

Within Hund's spin-S atomic multiplet, the operator $(\boldsymbol{s}_e \cdot \boldsymbol{S})$ can have only two different values. They are equal to $(S +1)/2$ for the "spin-up" electrons, whose spin adds to the total spin of the atom, and $-(S +1)/2$ for the "spin-down" electrons, whose spin subtracts from the total. Substituting these values into the reduced matrix element in Eq.(1.42), the expression for the spin magnetic form factor becomes,

$$F_S(\boldsymbol{q}) = \frac{1}{2S} \sum_{e'=1}^{2S} \int e^{-iqr'} |\psi_{e'}(r')|^2 d^3 r' = \frac{1}{2S} \sum_{e'=1}^{2S} F_{S,e'}(\boldsymbol{q}) . \tag{1.49}$$

where e' numbers only the unpaired electrons. The contributions from the paired electrons cancel out, leaving $2S$ terms in the sum. The right-hand side of Eq.(1.49) is simply the average of the Fourier-transforms of the unpaired electron densities in the atom. Similarly, the orbital form factor is the Fourier-transformed average density of the uncompensated orbital currents.

The typical shapes of the spin magnetic form factor for $3d$ electrons are illustrated by the equal-level surfaces shown in the right-hand columns of all four panels in Fig.1-3, (a)-(d). These were obtained by the numeric fast Fourier transformation of the "hydrogenic" electron wave functions whose density distributions appear in the left columns of the corresponding panels in this figure. Although the radial wave functions in the real multi-electron magnetic ions may differ significantly from those of the hydrogen atom, the angular dependence of the electron density distribution is similar. It is determined by the eigenfunctions of the orbital angular momentum, the spherical harmonics $Y_m^l(\theta,\varphi)$.

The radial distribution of the $3d$ electron density in the Ni^{2+} ion was calculated using the Hartree-Fock approximation in Ref. [23]. Its maximum is at a distance that is about 1.5 times smaller than for the corresponding hydrogen-like orbital, the full width at half maximum (FWHM) is about 10% larger, and the decay at larger distances is noticeably slower. Therefore, the calculation for the hydrogen-like wave functions shown in the Fig. 1-3 is unsuitable to quantitatively analyze the magnitudes of spin magnetic form factors. Nevertheless, it correctly describes the form factors' anisotropy arising from the particular shape of an electronic orbital.

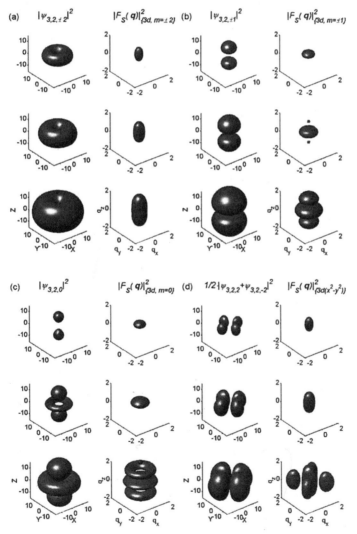

Figure 1-3: Electron densities (left columns) and the corresponding spin form factors (right columns) of Eq.(1.49) for the hydrogen-like $3d$ orbitals ($n=3$, $l=2$). (a) $m=\pm2$; (b) $m=\pm1$; (c) $m=0$, e_g ($3z^2-r^2$) orbital; (d) e_g (x^2-y^2) orbital. The three rows in (a)-(d) show the iso-surfaces at a level e^{-1}, e^{-2}, and e^{-4} of the corresponding maximum, from top to bottom, respectively.

Anisotropic magnetic form factor of a single $5d$ hole of the magnetic Ir^{4+} ion localized on the t_{2g} orbital in the cubic K_2IrCl_6 was studied by Lynn, Shirane and Blume in Ref. [24]. They found that the anisotropy of the magnetic form factor is very large, with an additional enhancement coming from the hybridization of the Ir $5d$-orbital with the Cl p-orbitals. For the e_g, $d(3z^2-r^2)$ orbital, which is elongated in the z-axis direction, the form factor is squashed along z and has a slower decay within the xy plane. On the other hand, the spin form factor for an electron in the e_g, $d(x^2-y^2)$ orbital is extended in the z-direction, and has a faster decay in the xy plane. Such is the situation for the Cu^{2+} ions in La_2CuO_4, $YBa_2Cu_3O_{6+y}$, and related cuprate materials, including the high-T_c superconductors, where a single unpaired magnetic electron occupies a $3d(x^2-y^2)$ orbital. Shamoto and colleagues [25] showed that properly accounting for the anisotropy of the Cu^{2+} magnetic form factor is essential for understanding the magnetic Bragg intensities measured in $YBa_2Cu_3O_{6+y}$ at large wave vectors q, and can also explain the peculiar q-dependence of the inelastic magnetic cross-section in this material. Accounting for the anisotropic Cu^{2+} form factor also was very important in analyzing neutron scattering by the high-energy spin waves in La_2CuO_4, [26]. The bandwidth of the magnetic excitations in this material exceeds 300 meV. Consequently, the measurements require very large wave vector transfers, for which the anisotropy of the Cu^{2+} form factor is very pronounced.

The single-electron density in Eq.(1.49), $|\psi_e(r)|^2$, is most generally determined from the multi-electron atomic wave function as

$$|\psi_e(r')|^2 = \langle \eta', L, S | \delta(r' - r_e) | \eta', L, S \rangle. \tag{1.50}$$

In calculating the magnetic form factors, the multi-electron atomic wave function is usually assumed to take the form of a Slater determinant made of the single-electron wave functions that then are used in Eq.(1.49). For unpaired electrons, these functions are usually considered to be the eigenfunctions of the electron's angular momentum, $\hbar l_e = [r \times p_e]$, with the same eigenvalue of $l_e^2 = l(l+1)$, and are tagged by the $n, l, m = l^z$ quantum numbers. Consequently, the radial and the angular parts of $\psi_e(r)$ are separated, as in the case of the Hartree-Fock one-electron wave functions,

$$\psi_{n,l,m}(r) = R_{n,l}(r) Y_l^m(\theta, \varphi), \tag{1.51}$$

where $R_{n,l}(r)$ is the appropriately normalized radial wave function, and $Y_m^l(\theta, \varphi)$ is the normalized spherical harmonics.

For such separation to be rigorous, it is necessary that the average potential acting on the electron in the atom on the level of the Hartree-Fock self-consistent mean field theory has spherical symmetry; this is a familiar *central field approximation*. It is a well-justified or 'mild' approximation for the unpaired electrons that belong to a single incomplete atomic shell, [8]. In fact, it is exact for an almost-filled shell with only a single electron, or a single hole, because the average net potential of the closed shell is, indeed, spherically symmetric. For the atomic configurations that obey the Hund's rule, the latter is also true for a half-filled shell, such as the $3d^5$ shells of the Mn^{2+} or Fe^{3+} ions.

Although Eq.(1.51) is sufficient in nearly all cases of interest, when it is not, a single-electron wave function can always be expanded in a series in spherical harmonics. In each term of such an expansion, the radial and the angular parts are again factorized, and the right-hand side of the equation simply becomes a sum of terms with different l and m. The same kind of an expansion is encountered in calculating the orbital contribution to the magnetic form factor. A single-electron orbital current density is defined in the same way as a single-electron density of Eq.(1.50), but with the current density operator, Eq.(1.32), instead of the delta-function. It is also expanded in a series in spherical harmonics (this is known as a *multipole expansion*, [20-22]), and then Fourier-transformed.

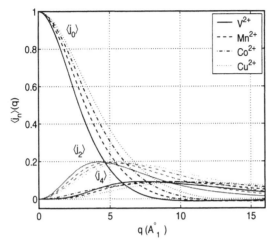

Figure 1-4: The wave vector dependence of the $j_0, j_2,$ and j_4 radial integrals for some typical magnetic $3d$ ions, calculated using the 3-Gaussian approximation of Ref. [11].

In the general case, the calculation of the magnetic form factors is ion-specific [3]. The general expressions can be obtained only for the leading contributions within the limit of small wave vector transfer q,

$$F_S(\mathbf{q}) = \langle j_0(q) \rangle, \quad F_L(\mathbf{q}) = \frac{1}{2}\{\langle j_0(q) \rangle + \langle j_2(q) \rangle\}, \tag{1.52}$$

known as the *dipole approximation*. However, although this approximation is very widely used in analyzing neutron scattering experiments, it is extremely crude. In particular, it does not account for the anisotropy of the magnetic form factors that is apparent in Fig. 1-3, and can be very important for ions with only one or two unpaired electrons. For a given orbital, $\psi_{n,m,l}(\mathbf{r})$, the dependence of the form factor on the length of q is contained in the so-called *radial integrals*,

$$\langle j_{2k}(q) \rangle_{n,l} = \int j_{2k}(qr)|R_{n,l}(r)|^2 r^2 dr, \tag{1.53}$$

where $R_{n,l}(r)$ is the radial part of the wave function, Eq.(1.51), and $j_{2k}(qr)$ are the spherical Bessel functions. While only $\langle j_0(q) \rangle$ and $\langle j_2(q) \rangle$ appear in the dipole approximation of Eq.(1.52), generally, the magnetic form factor of a shell with the orbital quantum number l is expressed as a sum of the radial integrals $\langle j_{2k}(q) \rangle$ with $k = 0,1,2,...,l$, and with the coefficients that depend on the direction of $\mathbf{q}$, *i.e.*, on the polar angles (θ_q, φ_q) (discussed in more detail in the next section).

The radial integrals for most known magnetic atoms and ions were calculated numerically from the appropriate Hartree-Fock or Fock-Dirac wave functions. Practically, they can be rather accurately approximated by the sum of the three Gaussians, multiplied by q^2 for $k > 0$, [11]. The coefficients and accuracy of such analytical approximations are tabulated in Ref. [11]. Figure 1-4 shows the first three radial integrals ($2k = 0,2,4$) for several typical magnetic $3d$ ions, calculated using these approximate expressions. An increase in the ion's size, from the smaller Cu^{2+} to the larger V^{2+}, is apparent from the correspondingly smaller extent of the radial integral $\langle j_0(q) \rangle$.

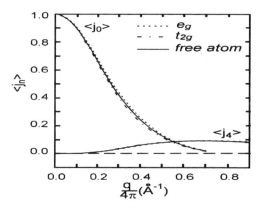

Figure 1-5: Splitting of the radial integrals in the spin magnetic form factor of Mn^{2+} ion in a cubic crystal field. The data is adapted from the early Hartree-Fock calculation by Freeman and Watson [27].

Finally, the single-electron radial wave function for an atom in the crystal environment may also depend on m because the crystal field splitting the l-multiplet causes expansion, or contraction, of the corresponding orbitals (in addition to the overall expansion/contraction of the outer electron wave functions) [27]. Typically, this is a small effect. It was first quantified in the early Hartree-Fock calculation of Ref. [27]. For the Mn^{2+} ion in an external cubic field of point charges, the authors obtained a small splitting of the magnetic spin form factors for e_g and t_{2g} orbitals, illustrated in Fig. 1-5. There is a small, barely distinguishable difference corresponding to the net expansion of the t_{2g} and the contraction of the e_g orbitals.

1.2.3.3. One-electron spin form factor beyond dipole approximation; anisotropic form factors for 3d electrons

The magnetic form factors are calculated [3,4,20-22] by expanding the exponent e^{-iqr_e} under the integral in the Fourier-transform of the atom's magnetization density in the series in spherical Bessel functions, $j_n(qr_e)$, [13]. The coefficients that contain the angular dependence in the expansion are the products of spherical harmonics, $Y_m^{l*}(\theta_q, \varphi_q) Y_m^l(\theta, \varphi)$. Therefore, the dependencies on the polar angles, (θ, φ) and (θ_q, φ_q), that respectively parameterize the directions of r and q in the corresponding spherical coordinates are separated in each term of the expansion. This is convenient because upon substituting this expansion and the wave functions of Eq.(1.51) into Eqs.(1.40)-(1.43), the angular part of the d^3r integration can be performed explicitly, using the orthogonality and the normalization of the spherical harmonics [8].

Consequently, a general analytical expression can be obtained for the one-electron spin magnetic form factor. Allowing for the transitions between the m and m' electronic states of the l-multiplet, we generalize the definition of the form factor as follows,

$$F_{S,l,m \to m'}(q) = \int e^{-iqr} \psi_{n,l,m}^*(r) \psi_{n,lm'}(r) d^3r . \qquad (1.54)$$

For the wave functions given by Eq.(1.51), the result is expressed in the form of a finite series with $\leq l+1$ terms, where each term is a product of the radial integral, $\langle j_{2k}(q)\rangle$, and a spherical harmonic, $Y_{2k}^{m'-m}(\theta_q,\varphi_q)$, accounting for the form factor's angular dependence,

$$F_{S,J,m\to m'}(q)=\sum_{k=\frac{|m-m'|}{2}}^{l}\sqrt{4\pi(4k+1)}\,\langle j_{2k}(q)\rangle_{n,J}(-1)^{m'-m}C_{l,m,J,m'}^{2k}Y_{2k}^{m'-m}(\theta_q,\varphi_q).\qquad(1.55)$$

The coefficients $C_{l,m,J,m'}^{2k}$ can be expressed as the product of two Clebsch-Gordan coefficients, $C_{l_1,m_1,J_2,m_2}^{l,m}$, which appear in the theory of addition of angular momentum [8],

$$C_{l,m,J,m'}^{r}=(-1)^{m'}\sqrt{\frac{2l+1}{2l'+1}}C_{l,0,J,0}^{l',0}C_{l,-m,J,m'}^{l',0}.\qquad(1.56)$$

Racah [28] obtained a closed analytical formula for the Clebsh-Gordan coefficients, which are related in a simple way to the $3j$-symbols. It is readily available in many textbooks [8,13], and the values of the Racah coefficients are tabulated in the literature. $C_{l,0,J,0}^{l',0}$ is the particular case for which there is an explicit expression [8,13].

An important particular case of Eq.(1.55) is that of $m=m'$. It defines the diagonal form factor, $F_{S,l,m}(q)$, that describes scattering by an electron whose wave function is an eigenstate of l and $l^z=m$. In this case,

$$F_{S,l,m}(q)=\sum_{k=0}^{l}(-1)^{k}(4k+1)\langle j_{2k}(q)\rangle_{n,J}C_{l,m,J,m}^{2k}P_{2k}(\cos\theta_q),\qquad(1.57)$$

where $P_{2k}(\cos\theta_q)$ are the Legendre polynomials [13]. This form factor does not depend on φ_q, and is axially symmetric. Further, the coefficient $C_{l,m,J,m}^{0}=1$ for all m. Therefore, the leading $k=0$ term in the series Eqs.(1.55), (1.57) is just $\langle j_0(q)\rangle$; this is exactly the result obtained in the dipole approximation. It also coincides with the angle-averaged form factor,

$$\langle F_{S,l,m\to m'}(q)\rangle=\frac{1}{4\pi}\iint F_{S,l,m\to m'}(q)\sin\theta_q\,d\theta_q\,d\varphi_q=\delta_{m,m'}\langle j_0(q)\rangle.\qquad(1.58)$$

This result is a straightforward consequence of the orthogonality of the spherical harmonics. Only the term with $m=m'$ and $k=0$ in Eq.(1.55) (i.e., only the $k=0$ term in Eq.(1.57) survives the spherical averaging.

For the d-electrons ($l=2$, $m=0,1,2$), there are only three terms in the series in Eq.(1.55) and (1.57). In this case, the following explicit expression for the anisotropic spin magnetic form factors can be obtained,

$$F_{S,2,m}(q)=\langle j_0(q)\rangle+$$
$$+\tilde{C}_{2,m}^{2}(1-3\cos^2\theta_q)\langle j_2(q)\rangle+\tilde{C}_{2,m}^{4}\left(1-10\cos^2\theta_q+\frac{35}{3}\cos^4\theta_q\right)\langle j_4(q)\rangle.\qquad(1.59)$$

Table 1-3 lists the coefficients $\tilde{C}_{2,m}^{2k}$ in this equation. The same formulas can also be derived from the expressions given in Ref. [29]. In this way, Eq.(1.59) for $m=2$ was obtained in Ref. [25].

Table 1-3: The coefficients in Eq.(1.59) for the form factors of the d-electrons.

	$m=0$	$m=1$	$m=2$
$\tilde{C}_{2,m}^{2}$	5/7	5/14	-5/7
$\tilde{C}_{2,m}^{4}$	27/28	-9/14	9/56

Figure 1-6 shows the $|F_{S,2,m}(q)|^2$ equal-level surfaces, at a level $e^{-2} \approx 0.14$, depicting the axially anisotropic spin form factors of Eq.(1.59) for a $3d$ electron in Cu^{2+}, and for $m=0$, 1, and 2. Their shapes are similar to the "hydrogenic" form factors depicted in Fig. 1-3. Here, however, the form factors are obtained from the realistic single-electron wave functions and, therefore, the axes now quantify the realistic wave vector transfer q in $Å^{-1}$, which is relevant for experiments.

$$Cu^{2+} \ (S=1/2)$$

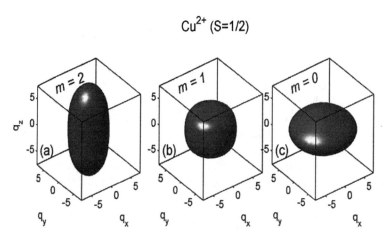

Figure 1-6: The isosurface of the anisotropic magnetic spin form factor (squared) for the Cu^{2+} ion with the single electron in (a) m=2 state, (b) m=1 state, and (c) m=0 state.

Figure 1-7 emphasizes the importance of the anisotropy of the spin magnetic form factor for a single unpaired d-electron for the topical Cu^{2+} ion. The difference between the form factor squared, $|F(q)|^2$, for two q orientations grows rapidly at non-zero q, as the form factor decreases. At a wave vector $q = 2.66 \ Å^{-1}$, which is typical for the thermal-neutron measurements, the anisotropy of the form factor already is about a factor two, and cannot be ignored.

Finally, for a $3d$-ion in a crystal the electronic wave functions are not necessarily the axially symmetric eigenfunctions of l^z, but are their linear combinations, often with lower symmetry, corresponding to the energy levels in a crystal field. In a cubic crystal field they correspond to the so-called e_g and t_{2g} orbitals [14,15]. One of the two e_g orbitals, $d(3z^2-r^2)$, corresponds to $l^z = m = 0$ eigenfunction, Fig.1-3 (c). Therefore, its form factor is axially symmetric, and is given by Eq.(1.59). For the other e_g orbital, $d(x^2-y^2)$, the wave function is proportional to $Y_2^2(\theta,\varphi)+Y_2^{-2}(\theta,\varphi)$, and the electron's density depends on both angles, θ and φ. It has a four-fold rotational symmetry around the z-axis, shown in Fig.1-3 (d).

The spin magnetic form factor for the electron in the $d(x^2-y^2)$ orbital can be straightforwardly derived from Eq.(1.55). Compared with the $l = 2$, $m = 2$ form factor of Eq.(1.59), it contains an additional, φ_q-dependent term, given by $(F_{S,2,-2\to2}(q)+F_{S,2,2\to-2}(q))/2$,

$$F_{S,d(x^2-y^2)}(q) = F_{S,2,2}(q)+\frac{15}{8}\langle j_4(q)\rangle \sin^4\theta_q \cos(4\varphi_q). \tag{1.60}$$

This formula describes the four-fold anisotropy in the xy-plane. With the appropriately redefined coordinate axes, it also describes the anisotropic spin magnetic form factors for the t_{2g}, xy, yz, and xz orbitals. This is because both the electron densities and form factors for these orbitals are related to those for the (x^2-y^2) orbital through the reflections and/or 90° rotations.

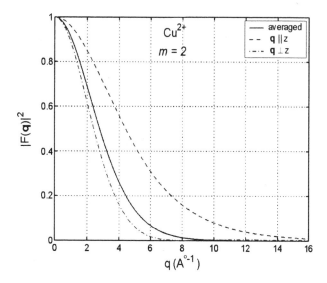

Figure 1-7: The wave vector dependence of the spin magnetic form factor squared for the $m=2$ 3d-electron in Cu^{2+} for two directions of the wave vector q. The solid line is $\langle j_0(q)\rangle^2$, corresponding to the angle-averaged form factor. Except for $q \to 0$, the anisotropy is large.

1.3. MAGNETIC SCATTERING BY A CRYSTAL

The Hamiltonian of the neutron's magnetic interaction with the crystal is the sum of the interactions Eq.(1.15) where r_e is replaced by $R_j + r_e$ over the lattice positions R_j where the magnetic atoms are located. Here e_j labels the electrons that belong to the jth atom; the adiabatic approximation for the electrons is adopted in assuming that they can instantly follow any change in R_j in a non-rigid crystal lattice. Hence, the neutron's total magnetic scattering length in the Born approximation is the sum of the atomic scattering lengths and, using $\mathbf{b}_m(r_n, r_e) = \mathbf{b}_m(r_n - r_e)$,

$$\mathbf{b}_m(\mathbf{q}) = \int e^{-i\mathbf{q}r_n} \sum_j \mathbf{b}_{m,j}(r_n, R_j + r_{e,j}) d^3 r_n = \sum_j e^{-i\mathbf{q}R_j} \mathbf{b}_{m,j}(\mathbf{q}). \qquad (1.61)$$

The matrix elements of the scattering length $\mathbf{b}_{m,j}(\mathbf{q})$ are defined by Eqs.(1.27) and (1.34); in the latter, the sum encompasses the unpaired electrons e_j that belong to the atom at R_j. If there are several magnetic atoms in the unit cell of the crystal, the sum in Eq.(1.61) can be separated into a sum over the sites of the Bravais lattice of the crystal, and another over the different magnetic atoms within the crystal's unit cell.

Alternatively, the sum in Eq.(1.61) can be thought of as running only over the sites of the Bravais lattice, and the scattering length $\mathbf{b}_{m,j}(\mathbf{q})$ as being the total magnetic scattering length for the crystal's unit cell, determined by the Fourier-transform of the total magnetization density of the unit cell, $M_j(\mathbf{q})$. The latter is defined by Eq.(1.34) that includes in the sum all unpaired electrons that belong to the unit cell. Where necessary, the density of the unit cell's magnetization can be expressed as a sum of the magnetizations of its individual atoms. Evidently, for the Bravais crystal, the magnetization of the unit cell is that of the single magnetic atom.

On account of Eqs.(1.61) and (1.27), the matrix element of the neutron magnetic scattering length for a crystal can be expressed as

$$\left\langle k_f, \eta_f \left| \mathbf{b}_m \right| k_i, \eta_i \right\rangle = \frac{2m}{\hbar^2} \left(\boldsymbol{\mu}_n - \frac{\boldsymbol{q}(\boldsymbol{q} \cdot \boldsymbol{\mu}_n)}{q^2} \right) \left\langle \eta_f \left| \sum_j e^{-i q R_j} M_j(\boldsymbol{q}) \right| \eta_i \right\rangle, \qquad (1.62)$$

where $\boldsymbol{q} = \boldsymbol{k}_f - \boldsymbol{k}_i$ is the wave vector transfer to the neutron. Hence, the magnetic scattering length is determined by the appropriate matrix element of the Fourier-transformed magnetization density in the crystal,

$$M_q = \sum_j e^{-i q R_j} M_j(\boldsymbol{q}). \qquad (1.63)$$

The polarization-dependent pre-factor in Eq.(1.62) selects only those components of magnetization that are perpendicular to the wave vector transfer $\boldsymbol{q}$, and parallel to the direction of the neutron spin, $\boldsymbol{\mu}_n = -g_n \mu_N \boldsymbol{S}_n$. If the neutron's spin polarization changes during scattering, then the appropriate matrix element of the pre-factor between the neutron's initial and final polarization states must be used; see Ref. [3] for details. Because the corresponding scattering length is zero, the neutron's magnetic scattering cross-section is not sensitive to magnetization directed along the wave vector transfer, $\boldsymbol{q}$. This fact is widely used in analyzing the polarizations of the magnetic densities and fluctuations measured by neutron scattering.

To obtain the total differential scattering cross-section for the scatterer in a stationary state $\left| \eta_i \right\rangle$, the cross-sections of Eq.(1.7) for all possible stationary final states, $\left| \eta_f \right\rangle$, must be added. If the scattering sample is not initially in a stationary state (e.g., if it is in thermal equilibrium at a finite temperature), an appropriate statistical averaging over $\left| \eta_i \right\rangle$ must be performed. Van Hove [29] proposed an elegant and general way to proceed with these tasks, based on employing the integral representation for the delta-function in Eq.(1.7). Then, the exponents that depend on the energies E_i and E_f of the scatterer's initial and final states can be absorbed into the time-dependent (Heisenberg) magnetization operator,

$$e^{-\frac{i}{\hbar}(E_i - E_f)t} \left\langle \eta_f \left| M_q \right| \eta_i \right\rangle = \left\langle \eta_f \left| M_q(t) \right| \eta_i \right\rangle = \left\langle \eta_f \left| \sum_j e^{-i q R_j(t)} M_j(\boldsymbol{q}, t) \right| \eta_i \right\rangle. \qquad (1.64)$$

This allows the summation/averaging over $\left| \eta_i \right\rangle$ and $\left| \eta_f \right\rangle$ described above.

While the cross-sections for scattering with particular polarizations of the incident and scattered neutrons can be derived straightforwardly from Eqs.(1.7) and (1.62), the corresponding expressions are lengthy and beyond our present scope [1,3]. Restricting our attention to scattering experiments with unpolarized neutron beams, we can average the expression for the squared magnetic scattering length over the neutron polarization. Then, the following general expression is obtained for the total differential cross-section of magnetic scattering of the unpolarized neutrons by a crystal,

$$\frac{d^2\sigma(\boldsymbol{q}, E)}{dE d\Omega} = \frac{k_f}{k_i} \left(\frac{2m}{\hbar^2} \mu_n \right)^2 \sum_{\alpha,\beta} \left(\delta_{\alpha\beta} - \frac{q_\alpha q_\beta}{q^2} \right) \int_{-\infty}^{\infty} e^{-\frac{i}{\hbar}Et} \left\langle M_q^\alpha M_{-q}^\beta(t) \right\rangle \frac{dt}{2\pi\hbar}. \qquad (1.65)$$

Here, $\boldsymbol{q} = \boldsymbol{k}_i - \boldsymbol{k}_f$ is the wave vector transfer to the sample (cf. Eq.(1.7); also, note the sign change compared to e.g., Eqs.(1.9), (1.62)). The cross-section is determined by the *scattering function* in the form of the temporal Fourier transform of the time-dependent pair-correlation function of the Fourier-transformed magnetization density in the crystal,

$$S_{(MM)}^{\alpha\beta}(\boldsymbol{q}, E) = \int_{-\infty}^{\infty} e^{-\frac{i}{\hbar}Et} \left\langle M_q^\alpha M_{-q}^\beta(t) \right\rangle \frac{dt}{2\pi\hbar} = \int_{-\infty}^{\infty} e^{-\frac{i}{\hbar}Et} \left\langle M_q^\alpha M_q^{\beta*}(t) \right\rangle \frac{dt}{2\pi\hbar}. \qquad (1.66)$$

It is also often called the *dynamical structure factor*, or the *dynamical correlation function*, although the latter term, strictly speaking, is only correct in the absence of static order (see the discussion in the next section).

For the magnetization carried by the atoms in the crystal the expression Eq.(1.66), as well as the cross-section Eq.(1.65), incorporate dependence on disorder and on the dynamics of the crystal lattice. It is contained in the exponent $e^{-i q R_j(t)}$ in the lattice Fourier transform of the unit cell magnetization density that appears inside the matrix element, *cf.* Eq.(1.64). Properly treating this dependence allows us to describe the effects of magneto-vibrational scattering and other interesting magneto-structural interferences in magnetic neutron scattering. However, in most cases, such effects can be neglected, and the rigid lattice approximation adopted. The leading correction to this approximation is obtained simply by multiplying the rigid-lattice cross-section by the Debye-Waller factor, $e^{-2W(q)}$, where $2W(q) = \langle (qu)^2 \rangle / 3$ quantifies the average squared displacement of an atom from its equilibrium position at a lattice site R_j.

With the definition Eq.(1.66), then Eq.(1.65) can be rewritten in its usual, more compact form,

$$\frac{d^2\sigma(\mathbf{q},E)}{dEd\Omega} = \frac{k_f}{k_i}(r_m)^2 \sum_{\alpha,\beta}\left(\delta_{\alpha\beta} - \frac{q_\alpha q_\beta}{q^2}\right)\frac{1}{(2\mu_B)^2}S^{\alpha\beta}_{(MM)}(\mathbf{q},E) \tag{1.67}$$

where

$$r_m = -2\mu_B \frac{2m}{\hbar^2}\mu_n = -\frac{1}{2}g_n r_e = -5.391 \cdot 10^{-13}\,\text{cm}, \tag{1.68}$$

is the characteristic magnetic scattering length. With Eq.(1.67) the analysis of neutron magnetic scattering cross-section in a crystal is reduced to considering the corresponding correlation function Eq.(1.66), the quantity that theorists typically calculate.

We note that although the probability of magnetic scattering of a neutron by a single atom is small, its coherent interaction with a large number of atoms in a perfect crystal can result in a scattering probability that is not small. Experimentally, this means that a significant part of the incident neutron beam is coherently reflected out as it penetrates the sample, extinguishing the beam inside the sample. This process is known as *primary extinction*; the beam's penetration depth is called the extinction length. Then, the first-order perturbation theory of the Born approximation is not applicable to describing the sample's scattering cross-section. In practice, the situation is usually remedied because large crystalline samples typically consist of many tiny perfect crystallites that are slightly misaligned, at random, to each other (within the "mosaic" angular spread, η). Because scattering from the different crystallites is not coherent, the Born approximation accurately describes the crystal's total scattering cross-section provided that the individual crystallites are small enough (*i.e.,* much smaller than the extinction length). While primary extinction can be described by the dynamical theory of diffraction [5], in practice, it is almost never a concern in magnetic neutron scattering. With few notable exceptions (such as the coherent Bragg scattering in a ferromagnet), the combined magnetic cross-section of the magnetic ions in the crystal's unit cell typically is too small to give an extinction length comparable to the size of the (magnetic) crystallites in the sample.

1.3.1. Elastic and quasi-elastic magnetic scattering

Elastic scattering is the characteristic defining feature of a solid. Apart from the trivial $q = 0$ forward-scattering channel, it is absent in a liquid or a gas [1]. Elastic magnetic scattering at a wave vector $q = Q$ exists provided the correlation function $\left\langle M_Q^\alpha M_{-Q}^\beta(t) \right\rangle$ has a non-zero time average

$$\overline{\left\langle M_Q^\alpha M_{-Q}^\beta(t) \right\rangle} = \left\langle M_Q^\alpha \overline{M_{-Q}^\beta(t)} \right\rangle = \left\langle M_Q^\alpha \right\rangle \left\langle \overline{M_{-Q}^\beta(t)} \right\rangle = \left\langle M_Q^\alpha \right\rangle \left\langle M_{-Q}^\beta \right\rangle \neq 0. \tag{1.69}$$

The bar over M here denotes time averaging. Decoupling the correlation function in Eq.(1.69) is usually justified by noting that the time average over the large time interval $\Delta t \to \infty$ can be carried out starting at any finite, although arbitrarily large, time t_0. If t_0 is larger than any characteristic relaxation time in the system, then $M_Q = M_{-Q}{}^* (t = 0)$ and $M_{-Q}(t > t_0)$ are uncorrelated and can be replaced in Eq.(1.69) by their averages [1-4].

Formally, decoupling in Eq.(1.69) is obviously permitted for the correlation function with the system in its non-degenerate ground state (or in any pure, non-degenerate quantum mechanical state). This is because the time-averaged operator $\overline{M_{-Q}^\beta(t)}$ is diagonal; the non-diagonal, oscillating matrix elements vanish upon time averaging. For a correlator $\left\langle M_Q^\alpha M_{-Q}^\beta(t) \right\rangle$ with the system in thermal equilibrium, straightforward quantum mechanical time-averaging leads to

$$\left\langle M_Q^\alpha \overline{M_{-Q}^\beta(t)} \right\rangle = \sum_{\lambda_n, \lambda_n'} e^{-\frac{E_n}{T}} \left\langle \lambda_n \left| M_Q^\alpha \right| \lambda_n' \right\rangle \left\langle \lambda_n' \left| \overline{M_{-Q}^\beta(t)} \right| \lambda_n \right\rangle. \tag{1.70}$$

The right-hand side of Eq.(1.70) becomes a product of thermal averages, as in Eq.(1.69), upon assuming ergodicity, *i.e.,* if the time-averaged operator is replaced with its expectation value in thermal equilibrium, $\overline{M_{-Q}(t)} = \left\langle M_{-Q} \right\rangle$. Note that ergodicity relies on the existence of the dissipative interactions which are not included in the quantum mechanical Hamiltonian of the pure isolated system.

It is clear that Eq.(1.69) implies the existence of static magnetic order in the system. It corresponds to a non-zero expectation value of the Fourier-component of the magnetization, $\left\langle M_Q^* \right\rangle = \left\langle M_{-Q} \right\rangle = \left\langle \overline{M_{-Q}(t)} \right\rangle \neq 0$, at a wave vector $\pm Q$. Consequently, the static and the fluctuating parts of the scattering function Eq.(1.66) can be separated,

$$S_{(MM)}^{\alpha\beta}(Q, E) = S_{(M)}^{\alpha\beta}(Q)\delta(E) + S_{(mm)}^{\alpha\beta}(Q, E), \tag{1.71}$$

where $S_{(M)}^{\alpha\beta}(Q) = \left\langle M_Q^\alpha \right\rangle \left\langle M_{-Q}^\beta \right\rangle$. The first term here describes the elastic scattering resulting from the static order of magnetic moments in the system. The second term describes the inelastic magnetic scattering arising from their motion.

Substituting $S_{(M)}^{\alpha\beta}(Q) = \left\langle M_Q^\alpha \right\rangle \left\langle M_{-Q}^\beta \right\rangle = \left\langle M_Q^\alpha \right\rangle \left\langle M_Q^{\beta*} \right\rangle$ for $S_{(MM)}^{\alpha\beta}(Q)$ in Eq.(1.67) we obtain the following general expression for the cross-section of magnetic elastic neutron scattering,

$$\frac{d^2\sigma_{el}(Q, E)}{dEd\Omega} = (r_m)^2 \left| \left[\frac{Q}{Q} \times \left\langle \frac{M_Q}{2\mu_B} \right\rangle \right] \right|^2 \delta(E) = \frac{(r_m)^2}{(2\mu_B)^2} \left| \left\langle M_Q^\perp \right\rangle \right|^2 \delta(E). \tag{1.72}$$

Here $M_Q^\perp$ is the projection of the Fourier-transformed magnetization density [Eqs.(1.63), (1.34)] on the plane perpendicular to the wave vector transfer, Q. Eq.(1.72) shows that elastic magnetic neutron scattering directly measures the magnitude of the system's average inhomogeneous static magnetization.

The elastic term in Eq.(1.71), being a delta-function in energy, usually dominates the scattering at zero energy transfer, *i.e.* $S^{\alpha\beta}_{\langle MM \rangle}(Q,0)$, in systems with static magnetic order. However, it is not the only source of scattering at $E = 0$. In many cases, there is also a quasi-elastic contribution, $S^{\alpha\beta}_{\langle mm \rangle}(Q,0)$. It may coexist with the elastic scattering, or may replace it, as happens in a critical region, with the system near the phase transition to the magnetically ordered state. When static magnetic order is absent, or weak, the quasi-elastic scattering is the dominant $E = 0$ contribution, and may even diverge as $E \to 0$. Such is the situation for a S = 1/2 Heisenberg antiferromagnetic chain, which is critical at $T = 0$ [33,34].

The commonest type of quasi-elastic scattering corresponds to the relaxational motion of magnetization described by a simple exponential decay in the time evolution of the correlation function [31],

$$\left\langle M^{\alpha}_Q M^{\beta}_{-Q}(t) \right\rangle = C^{\alpha\beta}_Q e^{-\frac{|t|}{\tau}} + \ldots . \tag{1.73}$$

Substituting this expression into Eq.(1.66) and subsequently integrating, straightforwardly gives a Lorentzian contribution to the scattering function

$$S^{\alpha\beta}_{\langle mm \rangle}(Q,E) = C^{\alpha\beta}_Q \frac{1}{\pi} \frac{\Gamma_E}{E^2 + \Gamma_E^2} + \ldots . \tag{1.74}$$

It is centered at $E = 0$; the energy width is determined by the characteristic decay time of the magnetization correlation function Eq.(1.73), $\Gamma_E = \hbar/\tau$.

This relaxation-type time-dependence Eq.(1.73) is typical when the temperature is in the critical region above the transition to a magnetically ordered phase. It naturally arises where the motion of the real-space magnetization density obeys the diffusion equation,

$$\frac{\partial}{\partial t} M(t) = D \nabla^2 M(t), \quad \frac{\partial}{\partial t} M_Q(t) = -DQ^2 M_Q(t). \tag{1.75}$$

The corresponding quasi-elastic scattering is often called diffuse scattering. Its presence at non-zero q is a characteristic feature of a classical liquid (there is only quasi-elastic scattering at $q = 0$ in a gas, while, in an incompressible quantum liquid, there is a spectral gap, and hence, no low-E scattering apart from an acoustic mode at $q = 0$).

On approaching the transition temperature, T_c, to the "magnetic solid" phase with static magnetic order, the diffusion constant tends toward zero and the magnetization relaxation time diverges,

$$D \sim \left(\frac{T}{T_c} - 1\right)^x \to 0, \quad \tau = \left(DQ^2\right)^{-1} \to \infty , \tag{1.76}$$

where the exponent $x > 0$. This is usually called a critical slowing down of magnetic fluctuations. Consequently, the quasi-elastic energy width, $\Gamma_E = \hbar/\tau$, vanishes, and the Lorentzian in Eq.(1.74) transforms into a delta-function. The theoretical value of the critical exponent x in Eq.(1.76) must be calculated from the microscopic model. For example, for a three-dimensional ferromagnet near the Curie point D obeys Eq.(1.76) with $x = 1$, while in the paramagnetic region far from T_c, $D \sim T_c / \sqrt{S(S+1)}$ [2].

1.3.2. Dynamical correlation function and dynamical magnetic susceptibility

The inelastic part of the scattering function Eq.(1.66), (1.71), is recognized as the *dynamical correlation function* of the Fourier-transformed magnetization density. It is defined as the correlation of its fluctuations around the equilibrium expectation value, $m_q(t) = M_q(t) - \langle M_q \rangle$,

$$S_{(mm)}^{\alpha\beta}(q,E) = \int_{-\infty}^{\infty} e^{-\frac{i}{\hbar}Et} \left\langle m_q^\alpha m_{-q}^\beta(t) \right\rangle \frac{dt}{2\pi\hbar} = S_{(MM)}^{\alpha\beta}(q,E) - \left\langle M_q^\alpha \right\rangle\left\langle M_{-q}^\beta \right\rangle \delta(E). \tag{1.77}$$

$m_q(t)$ is the appropriate dynamical variable for describing the system's response to the external magnetic field because its expectation value is zero when such a field is absent, $\langle m_q(t) \rangle = 0$. The thermal expectation value $\langle M_q(t) \rangle = \langle M_q \rangle$ is time-independent, as the system's density matrix in thermal equilibrium is diagonal and stationary.

The dynamical correlation function Eq.(1.77) has two important properties that are derived in response theory [3,4,31]. First, is the *detailed balance constraint* that relates the energy gain and the energy loss scattering,

$$S_{(mm)}^{\alpha\beta}(q,-E) = e^{-\frac{E}{T}} S_{(mm)}^{\alpha\beta}(q,E), \tag{1.78}$$

where T is the system's temperature. The second is the *fluctuation-dissipation theorem* that relates the scattering intensity with the imaginary part of the dynamical magnetic susceptibility,

$$\left(1 - e^{-\frac{E}{T}}\right) S_{(mm)}^{\alpha\beta}(q,E) = \frac{1}{\pi} \chi_{\alpha\beta}''(q,\omega). \tag{1.79}$$

The dynamical magnetic susceptibility, $\chi_{\alpha\beta}(q,\omega)$, describes the system's linear response to a small, inhomogeneous magnetic field with a wave vector q, oscillating in time with a frequency ω, $H(q,\omega)$,

$$\left\langle m_q^\alpha(\omega) \right\rangle = \int_{-\infty}^{\infty} e^{i\omega t} \left\langle m_q^\alpha(t) \right\rangle dt \equiv \chi_{\alpha\beta}(q,\omega) H_\beta(q,\omega). \tag{1.80}$$

Its imaginary part, $\chi_{\alpha\beta}''(q,\omega)$, determines the mean energy dissipation rate in the system under the action of such field [31,32].

The relations Eqs.(1.78) and (1.79) are extremely useful in analyzing neutron magnetic scattering. First, they establish a direct way of comparing the results with those of absorption measurements, such as electron spin resonance (ESR) that probe $\chi_{\alpha\beta}''(q,\omega)$ directly. Second, dynamical susceptibility often is the quantity that arises in theoretical calculations, for example, in the random phase approximation (RPA) [4]. Third, dynamical susceptibility often can be described by a very simple physical model, *e.g.*, a damped harmonic oscillator. Most importantly, dynamical susceptibility has several fundamental properties reflecting its analyticity and the causality principle, such as Kramers-Kronig relations, and the Onsager relation [4,31,32]. In fact, the condition of detailed balance Eq.(1.78) follows immediately from the fluctuation-dissipation theorem and the second of the causality relations below [31],

$$\chi_{\alpha\beta}'(q,-\omega) = \chi_{\alpha\beta}'(q,\omega), \quad \chi_{\alpha\beta}''(q,-\omega) = -\chi_{\alpha\beta}''(q,\omega). \tag{1.81}$$

Furthermore, it follows from Eqs.(1.78), (1.79), that at zero temperature the energy-gain scattering is absent, $S_{(mm)}^{\alpha\beta}(q,-|E|) = 0$, while for $E > 0$ the dynamical correlation function and the imaginary part of the dynamical susceptibility coincide up to a factor π, $\chi_{\alpha\beta}''(q,\omega) = \pi S_{(mm)}^{\alpha\beta}(q,E)$.

1.3.3. Magnetic Bragg scattering

Elastic Bragg scattering is a characteristic feature of a crystalline solid. It results from the breaking of translational invariance and the appearance of spatial periodicity associated with the crystal lattice. The latter is determined by the lattice's primitive translations, vectors $\boldsymbol{a}_\alpha$, $\alpha = 1, ..., D$, where D is the lattice dimension. Any equilibrium physical quantity in the crystal, $S(\boldsymbol{r})$, is lattice-periodic, $S(\boldsymbol{r} + n\boldsymbol{a}_\alpha) = S(\boldsymbol{r})$ (n is an integer), and can be represented as a Fourier series,

$$S(\boldsymbol{r}) = \sum_{\tau=1}^{N} e^{i\boldsymbol{Q}_\tau \cdot \boldsymbol{r}} S_{\boldsymbol{Q}_\tau}; \quad S_{\boldsymbol{Q}_\tau} = \frac{1}{V_0} \int_{V_0} e^{-i\boldsymbol{Q}_\tau \cdot \boldsymbol{r}} S(\boldsymbol{r}) d^D\boldsymbol{r} . \tag{1.82}$$

Here, the sum runs over the sites of the (dual) *reciprocal lattice*; the integral is over the volume V_0 of the unit cell of the direct lattice. For a three-dimensional, $D = 3$ lattice, $V_0 = (\boldsymbol{a}_1 \cdot [\boldsymbol{a}_2 \times \boldsymbol{a}_3])$. N is the total number of unit cells, which is the same for both lattices. The reciprocal lattice is determined by D primitive translations, $\boldsymbol{a}_\alpha{}^*$, such that $\boldsymbol{a}_\alpha \cdot \boldsymbol{a}_{\alpha'}{}^* = 2\pi\delta_{\alpha,\alpha'}$. Consequently, the *reciprocal lattice vectors*, $\boldsymbol{Q}_\tau$, which point to the sites of the reciprocal lattice, satisfy $(\boldsymbol{Q}_\tau \cdot \boldsymbol{a}_\alpha) = 2\pi n_\tau{}^\alpha$, where $n_\tau{}^\alpha$ is an integer.

Eq.(1.82) shows that any equilibrium physical property of the crystal, $S(\boldsymbol{r})$, can be defined by a discrete set of its Fourier components, $S_{\boldsymbol{Q}_\tau}$, on the sites of the crystal's reciprocal lattice, $\boldsymbol{Q}_\tau$. In particular, this is true for the spatial density in the crystal of the *equilibrium expectation value of the (microscopic) scattering length operator*, $\langle \mathbf{b}(\boldsymbol{r}) \rangle$, whose integral Fourier transform, $\langle \mathbf{b}(\boldsymbol{q}) \rangle$, determines the cross-section for elastic scattering by virtue of Eqs.(1.7), (1.9). Applying Eqs.(1.82) and (1.9) to $\langle \mathbf{b}(\boldsymbol{r}) \rangle$, we obtain the elastic Bragg scattering cross-section for the crystal

$$\frac{d^2\sigma_{el}(\boldsymbol{q}, E)}{dE d\Omega} = N \frac{(2\pi)^3}{V_0} \sum_{\tau=1}^{N} |F(\boldsymbol{Q}_\tau)|^2 \delta(\boldsymbol{q} - \boldsymbol{Q}_\tau) \delta(E) . \tag{1.83}$$

Here $F(\boldsymbol{Q}_\tau)$ is the Fourier integral of the scattering length within a unit cell,

$$F(\boldsymbol{Q}_\tau) = \int_{V_0} e^{-i\boldsymbol{Q}_\tau \cdot \boldsymbol{r}} \langle \mathbf{b}(\boldsymbol{r}) \rangle d^D\boldsymbol{r} , \tag{1.84}$$

usually called the *unit cell structure factor*. It accounts for the effects of interference on scattering by a single unit cell. Note the absence of the $1/V_0$ pre-factor, which is present in the second part of Eq.(1.82). It was absorbed in the pre-factor before the sum in Eq.(1.83), together with the multipliers arising from the different normalization of the Fourier *integral* for $\langle \mathbf{b}(\boldsymbol{r}) \rangle$, Eq.(1.9), and Fourier *series*, Eqs.(1.82). In obtaining Eq.(1.83) we used that for large N,

$$\left| \int_V e^{-i(\boldsymbol{q} - \boldsymbol{Q}_\tau)\boldsymbol{r}} d^D\boldsymbol{r} \right|^2 \cong V^2 \delta_{\boldsymbol{q}, \boldsymbol{Q}_\tau} \cong (2\pi)^3 V \delta(\boldsymbol{q} - \boldsymbol{Q}_\tau), \tag{1.85}$$

where $V = NV_0$ is the total volume of the crystal, [8].

In the simple and very frequent case where the unit cell scattering cross-section is associated with a number of point scatterers (*e.g.*, atoms, magnetic moments), $F(\boldsymbol{Q}_\tau)$ is reduced to the usual, commonly cited form [1-5,47],

$$\langle \mathbf{b}(\boldsymbol{r}) \rangle = \sum_\nu b_\nu \delta(\boldsymbol{r} - \boldsymbol{u}_\nu) \quad \Rightarrow \quad F(\boldsymbol{Q}_\tau) = \sum_\nu b_\nu e^{-i\boldsymbol{Q}_\tau \cdot \boldsymbol{u}_\nu} . \tag{1.86}$$

Here, ν numbers the scatterers, and $\boldsymbol{u}_\nu$ and b_ν are their position in the unit cell and the scattering length, respectively. In principle, we could have also considered directly the spatial density of the (microscopic) elastic scattering cross-section in the crystal. Hence, Eqs.(1.83) and (1.84) establish the important relationship between the Fourier-transformed scattering cross-section of a unit cell and its Fourier-transformed scattering length, $F(Q_\tau)$.

When the system's equilibrium magnetization density possesses spatial periodicity, we can write, following Eq.(1.82),

$$\langle M(r) \rangle = \frac{1}{V_m} \sum_{Q_m} M_{Q_m} e^{iQ_m \cdot r}; \quad M_{Q_m} = \int_{V_m} e^{-iQ_m \cdot r} \langle M(r) \rangle d^D r. \tag{1.87}$$

Here Q_m are the reciprocal lattice vectors for the *magnetic lattice*, whose unit cell of volume V_m contains one period of magnetic structure. $|\langle M_Q \rangle|^2$ can be straightforwardly calculated from the above series expansion for $\langle M(r) \rangle$, and using Eq.(1.85). Substituting the result into expression Eq.(1.72), the following cross-section is obtained for magnetic Bragg scattering,

$$\frac{d^2 \sigma_{el}(q,E)}{dEd\Omega} = N r_m^2 \frac{(2\pi)^3}{V_m} \sum_{Q_m} \frac{1}{(2\mu_B)^2} |M_{Q_m}^\perp|^2 \delta(Q - Q_m) \delta(E), \tag{1.88}$$

which is similar to Eq.(1.83). In fact, Eq.(1.88) also follows directly from Eq.(1.83) if the magnetic scattering length Eq.(1.27) is used in the expression for the unit cell form factor Eq.(1.84). For scattering by point-like magnetic moments μ_v localized at the positions u_v in the unit cell, the magnetic unit cell form factor is

$$F_m(Q) = -\frac{r_m}{\mu_B} \left(S_n - \frac{Q(Q \cdot S_n)}{Q^2} \right) \cdot \sum_v \mu_v e^{-iQ \cdot u_v} = -r_m \sum_v \frac{(S_n \cdot \mu_v^\perp)}{\mu_B} e^{-iQ \cdot u_v}, \tag{1.89}$$

where $\mu_v^\perp$ is the component of the magnetic moment perpendicular to Q, and S_n is the neutron's spin operator. In order to obtain the cross-section, matrix elements between the appropriate neutron spin states must be considered, and appropriate averaging has to be performed.

The expressions obtained above describe magnetic Bragg scattering by spatially periodic magnetic structures in terms of the corresponding magnetic lattice and do not rely on the existence of crystalline atomic order. In principle, they apply to such exotic cases as magnetically ordered structural glasses or structurally disordered alloys, where the crystal lattice is absent but there is a periodic magnetic structure. Although this description is general, allowing the use of magnetic symmetry groups to analyze magnetic structures, in many cases it is not the best choice. Indeed, describing Bragg scattering in terms of the magnetic lattice is natural where the magnetic unit cell is not significantly larger, or more complicated, than that of the underlying, paramagnetic crystal lattice. In fact, for materials with a complex crystal structure, the magnetic lattice may be even simpler and of higher symmetry than the crystal's lattice. Then, using the magnetic lattice may actually simplify indexing of the magnetic Bragg reflections.

On the other hand, with a simple crystal structure, the magnetic lattice's period may contain a number of paramagnetic crystal unit cells. This number can be very large, so that the magnetic unit cell is huge and contains many atoms that are equivalent in the paramagnetic phase. Then, switching to the magnetic reciprocal lattice with correspondingly smaller unit cell clearly is not a satisfactory way to describe the magnetic Bragg scattering. Indeed, the lower translational symmetry of the magnetically ordered phase arises from the spontaneous symmetry breaking associated with that order, and does not reflect the symmetry of the underlying magnetic Hamiltonian. Furthermore, switching to a large magnetic unit cell compatible with the periodicity of the underlying crystal lattice is impossible if magnetic order is incommensurate. Therefore, often it is desirable to describe magnetic Bragg scattering using the higher translational symmetry of the crystal lattice in the paramagnetic phase (which determines the symmetry of the magnetic Hamiltonian).

Magnetic ordering usually is associated with the appearance of a superlattice modulation of the magnetization density in a crystal with a single wave vector, Q. Such breaking of lattice's translational invariance is most generally described by adding the appropriate irreducible representation of the translation group to the magnetization density [31]. In other words, the spatial structure associated with the magnetic order parameter has a modulation with wave vector Q, superimposed on a modulation with the period of the underlying crystal lattice,

$$\langle M(r)\rangle = m_0(r) + m(r)e^{iQ\cdot r} + m^*(r)e^{-iQ\cdot r}, \tag{1.90}$$

where the "Bloch amplitudes" $m_0(r)$ and $m(r)$ are lattice-periodic vector-functions. $m_0(r)$ describes the density of the "homogeneous" magnetization, which is real and the same for all unit cells in the crystal. $m(r)$ is complex, in general, and describes the "staggered" part of the magnetization that is modulated from one unit cell to another. Since $m_0(r)$ and $m(r)$ are lattice-periodic, they can be expanded in the Fourier *series*,

$$m(r) = \frac{1}{V_0}\sum_{Q_\tau} m(Q_\tau)e^{iQ_\tau \cdot r}; \quad m(Q_\tau) = \int_{V_0} e^{-iQ_\tau \cdot r} m(r) d^D r . \tag{1.91}$$

These are similar to Eq.(1.87), but are based on the *paramagnetic* crystal lattice. Note, that the $m_0(r)$ term describes the uniformly magnetized and the ferromagnetic cases, where there is a non-zero homogeneous magnetization in the crystal. Hence, only wave vectors $Q \neq 0$ **mod** (Q_τ) are considered.

Substituting the above Fourier series for $m(r)$ and $m_0(r)$ in Eq.(1.90), we obtain the following, equivalent expression for $\langle M(r)\rangle$,

$$\langle M(r)\rangle = \frac{1}{V_0}\sum_{Q_\tau}\left\{ m_0(Q_\tau)e^{iQ_\tau \cdot r} + m(Q_\tau)e^{i(Q+Q_\tau)r} + m^*(Q_\tau)e^{-i(Q+Q_\tau)r} \right\}. \tag{1.92}$$

Its integral Fourier transform, $\langle M_q\rangle$, as well as $\left|\langle M_Q\rangle\right|^2$, are straightforwardly evaluated using Eq.(1.85). Substituting the result in Eq.(1.72), the following expression for the magnetic Bragg scattering cross-section for a single-Q magnetic structure in the crystal is obtained,

$$\frac{d^2\sigma_{el}(q,E)}{dEd\Omega} = N r_m^2 \frac{(2\pi)^3}{V_0}\sum_{\tau=1}^{N}\left\{\left|\frac{m_0^\perp(Q_\tau)}{2\mu_B}\right|^2 \delta(q - Q_\tau) + \right.$$
$$\left. \left|\frac{m^\perp(Q_\tau)}{2\mu_B}\right|^2 (\delta(q - Q - Q_\tau) + \delta(q + Q + Q_\tau))\right\}\delta(E) \tag{1.93}$$

Here, the sum extends over the sites of the crystal's paramagnetic reciprocal lattice. The intensities of the magnetic Bragg satellites are determined by the Fourier-series coefficients of the magnetization "Bloch amplitudes", $m_0(r)$ and $m(r)$, for the corresponding wave vectors, Q_τ [*i.e.* by the Fourier integrals over a single unit cell, Eq.(1.91)]. If the unit cell's magnetization is carried by point magnetic moments localized at positions u_v,

$$m_0(r) = \sum_{j,v} m_{0v}\delta(r - R_j - u_v), \quad m(r) = \sum_{j,v} m_v \delta(r - R_j - u_v), \tag{1.94}$$

the magnetic Bragg intensities are determined by the usual unit cell magnetic form factors, commonly cited in the literature [1-4],

$$\left|m_0^\perp(Q_\tau)\right|^2 = \left|\sum_\nu m_{0\nu}^\perp e^{-iQ_\tau \cdot u_\nu}\right|^2, \quad \left|m^\perp(Q_\tau)\right|^2 = \left|\sum_\nu m_\nu^\perp e^{-iQ_\tau \cdot u_\nu}\right|^2. \tag{1.95}$$

We note that Eqs.(1.90)-(1.93) apply in the most general case of the magnetic ordering with a single wave vector Q. They describe equally well a variety of particular situations, such as a flat spiral, a longitudinal spin-density wave, a helimagnet, an antiferromagnet [e.g., for $Q = (\pi, \pi, \pi)$], and a ferromagnet [for $m(r) \equiv 0$]. Finally, the fundamental reason for the practical importance of single-Q modulated structures is that only such states arise in a single second-order phase transition. A single Q corresponds to a unique magnetic order parameter associated with spontaneous breaking of the magnetic symmetry [31].

1.3.4. Scattering from short-range nanoscale correlations

In many important cases, the crystal's magnetic phase has only short-range order. Specifically, the crystal possesses static magnetic density (at least on the time scale probed by neutron scattering), but long-range, macroscopic coherence of the magnetic superstructure is lacking. In the absence of macroscopic spatial coherence, a non-trivial representation of the translation group (such as that with wave vector Q in Eq.(1.90)) does not appear in the crystal's equilibrium magnetization density. Consequently, elastic magnetic scattering does not have the form of Bragg peaks as in Eq.(1.93), but rather, is diffuse and broad.

In practice, such situations are usually described by replacing the delta-functions in the Bragg cross-section Eq.(1.93) with normalized Lorentzians,

$$\delta(q) \to L_\xi(q) \equiv \frac{1}{\pi} \frac{\xi}{1+(q\xi)^2}, \tag{1.96}$$

where ξ is the correlation length. In one dimension, this is expected on the basis of general physical arguments, such as the analogy with the response function of the harmonic oscillator with damping. However, for several reasons, such an *ad hoc* approach is not completely satisfactory. Firstly, the Lorentzians Eq.(1.96) are not periodic in the crystal's reciprocal lattice. Therefore, apart from the limit of vanishing width, they do not correspond to the crystal's real physical quantities. Nor do $m^\perp(Q_\tau)$, which have a physical meaning only when they weight delta-functions in Eq.(1.93). More fundamentally, for systems in more than one dimension it is not obvious whether a delta-function should be replaced with the product of Lorentzians, as follows

$$\delta(q) = \delta(q_x)\delta(q_y)\delta(q_z) \to L_{\xi_x}(q_x)L_{\xi_y}(q_y)L_{\xi_z}(q_z), \tag{1.97}$$

or with a single, "multi-dimensional" Lorentzian,

$$\delta(q) \to \frac{C}{1+\sum_\alpha (q_\alpha \xi_\alpha)^2} = \frac{C}{1+(q_x\xi_x)^2+(q_y\xi_y)^2+(q_z\xi_z)^2}, \tag{1.98}$$

where C is a constant, and $\xi_\alpha = \xi_x, \xi_y, \xi_z$ are the correlation lengths.

Eq.(1.98) can be derived in the Ornstein-Zernike-type theory that describes critical scattering at small wave vectors q in ferro- and antiferro-magnets above the ordering temperature [2,3].

While the expression is also commonly employed to describe scattering in the presence of short-range static magnetic order, this use has an important inconsistency. The scattering function of Eq.(1.98) cannot be normalized in more than one dimension: the corresponding integral on the right-hand side of Eq.(1.98) diverges. This problem does not occur if the 2D or 3D Lorentzian expression is truncated to within the region of validity of the Orstein-Zernike theory, $(q\xi) \ll 1$, and set to zero outside it. However, this restricts the applicability of Eq.(1.98) to the immediate vicinity of the peak position, and essentially negates its practicality. In this region, the multi-dimensional Lorentzian of Eq.(1.98) does not differ from the "factorized Lorentzian" of Eq.(1.97), whereupon multiplying the denominators of the 1D Lorentzians the terms of the fourth and higher orders in $(q_{\alpha}\xi_{\alpha})$ can be neglected in the Orstein-Zernike approximation.

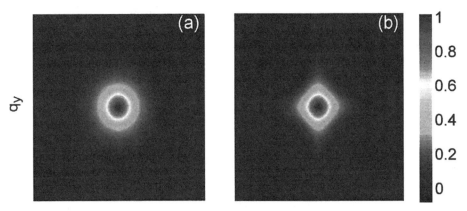

Figure 1-8: Two-dimensional contour maps of the intensity described by (a) a "2D Lorentzian", as in Eq.(1.98), and (b) a product of two 1D Lorentzians, as in Eq.(1.97). *Also see the color plate.*

Two-dimensional intensity distributions described by Eqs.(1.98) and (1.97) with $\xi_x = \xi_y$ are illustrated in panels (a) and (b) of the Fig. 1-8, respectively. The difference between them arises for wave vectors sufficiently far away from the peak position. The Ornstein-Zernike-like "multidimensional Lorentzian" retains full rotational symmetry even outside the long-wavelength region where it is rigorously expected. The symmetry of the "factorized Lorentzian" is lower. As we discuss below, it corresponds to the point symmetry of the order-destroying defects in the crystal lattice.

The scattering cross-section in the form of the product of 1D Lorentzians naturally arises in the problem of structural disorder introduced by antiphase domains in binary alloys, *e.g.,* in Cu_3Au [35]. It is free from the normalization problem, as both sides of Eq.(1.97) are rigorously normalized to 1 in any dimension. It is also straightforward to extend the derivation of Ref. [35] to the antiphase disclinations in the antiferromagnet, introduced by dislocations in the crystal lattice [36]. Furthermore, the factorized-Lorentzian cross-section describes a broad class of problems where disorder arises from independent linear phase slips that occur along different crystallographic directions. Below we show, under rather general assumptions, how the corresponding correlation function can be derived for a short-range correlated magnetic structure.

Consider the situation where the static magnetic structure exists locally (on the nanoscale), but macroscopically, the average translational symmetry of the crystal remains unbroken. We thus envision long-range coherence is destroyed by independent linear topological defects, such as disclinations, domain or grain boundaries, along the principal crystallographic directions.

Note, that the effective "disorder potential" describing these defects only has the rotational symmetry of the point group of the crystal lattice. It results in a glass-like, short-range correlated magnetic structure that can be anisotropic, with different correlation ranges along different crystallographic directions.

Where the magnetic correlations are between localized magnetic moments, it is more appropriate to express the magnetization in terms of "Wannier functions" describing the magnetization density of a crystal's unit cell, rather than in terms of "Bloch waves", as in Eq.(1.90). In the presence of the long-range magnetic order, magnetization is written as a lattice sum,

$$\langle M(r)\rangle = \sum_j \langle M_j(r - R_j)\rangle, \tag{1.99}$$

$$\langle M_j(r)\rangle = m_0(r) + m_Q(r)e^{iQ\cdot R_j} + m_Q^*(r)e^{-iQ\cdot R_j}.$$

Using M_q from Eq.(1.63), the magnetic elastic scattering cross-section of Eq.(1.72) can be recast in terms of the crystal-averaged correlation of the Fourier-transformed, equilibrium magnetization densities of the unit cell,

$$\frac{d^2\sigma_{el}(q,E)}{dEd\Omega} = N\left(\frac{r_m}{2\mu_B}\right)^2 \left(\sum_j^N e^{-iqR_j} \langle\langle M_0^\perp(-q)\rangle\langle M_j^\perp(q)\rangle\rangle\right)\delta(E). \tag{1.100}$$

Here, N is the number of unit cells in the crystal, and

$$\langle\langle M_0^\perp(-q)\rangle\langle M_j^\perp(q)\rangle\rangle = \frac{1}{N}\sum_{j'}^N \langle M_{j'}^\perp(-q)\rangle\langle M_{j+j'}^\perp(q)\rangle, \tag{1.101}$$

denotes the crystal-averaged product.

The existence of a system of magnetic disclinations in the crystal generates an additional random phase difference between the unit cell magnetizations in Eq.(1.101), which is not included in Eq.(1.99). This can be generally described by replacing

$$\langle\langle M_0^\perp(-q)\rangle\langle M_j^\perp(q)\rangle\rangle \rightarrow \langle e^{-i\varphi_j}\langle M_0^\perp(-q)\rangle\langle M_j^\perp(q)\rangle\rangle, \tag{1.102}$$

while still using Eq.(1.99) for $\langle M_j(q)\rangle$. The averaging of the phase multiplier can be decoupled in view of its randomness. Then, assuming that system self-averages, the average over the sample's volume can be replaced with the Gaussian statistical average, *i.e.*, using the Bloch identity,

$$\langle e^{-i\varphi_j}\rangle = e^{-\frac{1}{2}\langle\varphi_j^2\rangle}. \tag{1.103}$$

Here $\langle\varphi_j^2\rangle$ is the mean-square phase deviation between two unit cells at a distance R_j on the lattice, introduced by the disorder-generating defects.

Finally, we assume that this random phase-difference is described by independent random walks along the principal crystallographic directions (labeled by α). Consequently, we write

$$\langle\varphi_j^2\rangle = \sum_\alpha D_\alpha|n_j^\alpha|, \quad \langle e^{-i\varphi_j}\rangle = \prod_\alpha e^{\frac{|n_j^\alpha|}{\xi_\alpha}}, \tag{1.104}$$

where n_j^α label positions of the lattice sites in lattice units, $R_j = \sum_\alpha n_j^\alpha a_\alpha$, and D_α are the phase "diffusion coefficients". This is certainly an appropriate approach for modeling the independent linear disclinations considered here, which are a natural generalization of the "antiphase domains" picture described in Ref. [35]. The random phase disorder Eq.(1.102) causes an exponential decay of magnetic correlations. The correlation function Eqs.(1.101), (1.102) is factorized into a product of decaying harmonic modulations along the principal lattice directions. The correlation lengths in dimensionless lattice units (lu) are $\xi_\alpha = 2D_\alpha^{-1}$.

Substituting Eqs.(1.101)-(1.104) and (1.99) into (1.100) and summing over the crystal lattice, the following expression is obtained for the elastic scattering cross-section in the presence of such short-range magnetic order,

$$\frac{d^2\sigma_{el}(\boldsymbol{q},E)}{dEd\Omega} = N r_m^2 \left\{ \left| \frac{m_0^\perp(\boldsymbol{q})}{2\mu_B} \right|^2 \prod_\alpha \frac{\sinh \xi_\alpha^{-1}}{\cosh \xi_\alpha^{-1} - \cos(\boldsymbol{q}\cdot\boldsymbol{a}_\alpha)} \right.$$
$$\left. + \sum_{+,-} \left| \frac{m_Q^\perp(\pm\boldsymbol{q})}{2\mu_B} \right|^2 \prod_\alpha \frac{\sinh \xi_\alpha^{-1}}{\cosh \xi_\alpha^{-1} - \cos((\boldsymbol{q}\mp\boldsymbol{Q})\cdot\boldsymbol{a}_\alpha)} \right\} \delta(E) \tag{1.105}$$

The elastic scattering cross-section is described by the product of the independent *lattice-Lorentzian* functions,

$$\widetilde{L}_{\xi_\alpha}(q_\alpha) \equiv \frac{\sinh \xi_\alpha^{-1}}{\cosh \xi_\alpha^{-1} - \cos(\boldsymbol{q}\cdot\boldsymbol{a}_\alpha)}, \tag{1.106}$$

along the principal crystallographic directions. Function $\widetilde{L}_{\xi_\alpha}(q_\alpha)$ has peaks for $q_\alpha = Q_\tau^\alpha$, whose half width at half maximum is $\sim(a_\alpha\xi_\alpha)^{-1}$. For large ξ_α it is approximated by a set of Lorentzians Eq.(1.96) centered at $q_\alpha = Q_\tau^\alpha$, and transforms into a set of delta-functions for $\xi_\tau \to \infty$. For $\xi_\alpha \gg 1$ lu the scattering cross-section Eq.(1.105) adopts the form implied by Eq.(1.97). Note that the lattice-Lorentzians Eq.(1.106) are co-periodic with the reciprocal lattice. Hence, Eq.(1.105) lacks the summation over the reciprocal lattice sites $\boldsymbol{Q}_\tau$.

1.3.5. Spin scattering and spin correlation function

If the magnetization density in the crystal is carried by the localized electrons of the magnetic ions, the magnetic scattering cross-section can be expressed in terms of the lattice correlation between the atomic spin and orbital variables using the results of Section 1.2.3. In the commonest case, the magnetization of an atom is described by an effective spin, Eqs.(1.46)-(1.48). Substituting Eq.(1.47) into Eq.(1.65) we obtain the following expression for the magnetic scattering cross-section,

$$\frac{d^2\sigma(\boldsymbol{q},E)}{dEd\Omega} = \frac{k_f}{k_i} r_m^2 \sum_{\alpha,\beta} \left(\delta_{\alpha\beta} - \frac{q_\alpha q_\beta}{q^2} \right) \sum_{j,j'} \left(\frac{g_{\alpha,j}}{2} F_j^*(\boldsymbol{q}) \frac{g_{\beta,j'}}{2} F_{j'}(\boldsymbol{q}) \right)$$
$$\times \int_{-\infty}^{\infty} e^{-\frac{i}{\hbar}Et} \left\langle e^{-i\boldsymbol{q}(\boldsymbol{R}_j-\boldsymbol{R}_{j'})} S_j^\alpha S_{j'}^\beta(t) \right\rangle \frac{dt}{2\pi\hbar} \tag{1.107}$$

Here j either labels the lattice sites where, in this case g_j, $F_j(\boldsymbol{q})$ and S_j^α denote the corresponding effective values for the unit cell, or labels the magnetic atoms to which g_j, $F_j(\boldsymbol{q})$ and S_j^α refer.

For a system of identical magnetic atoms Eq.(1.107) becomes

$$\frac{d^2\sigma(\boldsymbol{q},E)}{dEd\Omega} = N \frac{k_f}{k_i} r_m^2 |F(\boldsymbol{q})|^2 \sum_{\alpha,\beta} \frac{g_\alpha}{2} \frac{g_\beta}{2} \left(\delta_{\alpha\beta} - \frac{q_\alpha q_\beta}{q^2} \right) S^{\alpha\beta}(\boldsymbol{q},E). \tag{1.108}$$

$S^{\alpha\beta}(q,E)$ denotes the Fourier transform of the two-point spin correlation function and is usually called the *dynamical spin structure factor*,

$$S^{\alpha\beta}(q,E)= \int_{-\infty}^{\infty} e^{-\frac{i}{\hbar}Et} \frac{1}{N}\sum_{j,j'} \left\langle e^{-iq(R_j-R_{j'})} S_j^{\alpha} S_{j'}^{\beta}(t)\right\rangle \frac{dt}{2\pi\hbar}.$$ (1.109)

It is also known as the dynamical spin correlation function, or the Van Hove scattering function [4]. As emphasized at the beginning of this section, Eqs.(1.107) and (1.108) refer to the ideal rigid lattice, and do not account for the lattice's thermal vibrations or the structural disorder. These effects can be roughly accounted for by multiplying the magnetic cross-section of Eqs.(1.107) and (1.108) with the Debye-Waller factor, $e^{-2W(q)}$.

$S^{\alpha\beta}(q,E)$ is a fundamental characteristic of the dynamical properties of the spin system. It satisfies the detailed balance constraint Eq.(1.78) and is related to the *dynamical spin susceptibility* by the fluctuation-dissipation theorem, given by Eq.(1.79). In many cases, a single $S^{\alpha\beta}(q,E)$ can also describe the spin dynamics of a system with non-equivalent magnetic atoms, where the magnetic form-factor $F_j(q)$ and the g-factor g_j are site-dependent. Indeed, for the Bravais lattice populated by atoms with different $F_j(q)$ and g_j, we can usually write

$$g_j F_j(q)= \langle gF(q)\rangle \left\{ 1+\sum_{Q_f} \left(f_{Q_f} e^{iQ_f \cdot R_j} + f_{Q_f}^* e^{-iQ_f \cdot R_j}\right)\right\},$$ (1.110)

where the wave vector(s) Q_f define the superlattice(s) of equivalent magnetic sites. Hence, the magnetic scattering cross-section is determined by the linear combination of the dynamic structure factors for wave vector q and wave vectors $q \pm Q_f$. It is defined by the same Eq.(1.108), in which the scattering function is replaced by

$$S^{\alpha\beta}(q,E) \to S^{\alpha\beta}(q,E)+ \sum_{Q_f} \left| f_{Q_f}\right|^2 \left\{ S^{\alpha\beta}(q+Q_f,E)+ S^{\alpha\beta}(q-Q_f,E)\right\}.$$ (1.111)

This situation, in particular, is pertinent to materials with orbital ordering, such as the pseudocubic perovskites, LaMnO$_3$ and KCuF$_3$ [37]. A similar approach can be applied to complex spin systems consisting of several non-equivalent simpler subsystems, such as spin chains [38].

1.3.6. Sum rules for the dynamic spin structure factor

The definition Eq.(1.109) of the dynamic spin structure factor generates the exact relations for its frequency moments, $\int_{-\infty}^{\infty} (\hbar\omega)^n S^{\alpha\beta}(q,\hbar\omega) d(\hbar\omega)$. For zero and first moments, $n = 0, 1$, these relations are known as *sum rules* [39]. The $n = 0$ moment simply defines the *static structure factor* $S^{\alpha\beta}(q)$, which is given by the space Fourier transform of the *equal-time pair correlation* function,

$$S^{\alpha\beta}(q) \equiv \int_{-\infty}^{\infty} S^{\alpha\beta}(q,\hbar\omega) d(\hbar\omega)= \frac{1}{N}\sum_{j,j'} \left\langle e^{-iq\cdot R_{jj'}} S_j^{\alpha} S_{j'}^{\beta}\right\rangle = \frac{1}{N}\left\langle S_q^{\alpha} S_{-q}^{\beta}\right\rangle.$$ (1.112)

Here $S_q^{\alpha}=\sum_j e^{-iq\cdot R_j}S_j^{\alpha}$ is the lattice Fourier transform of the lattice spin operators. Integrating the above expression over the Brillouin zone and taking the trace over the spin indices α, β, yields an important sum rule

$$\sum_{\alpha} \int S^{\alpha\alpha}(q) \frac{V_0 d^3q}{(2\pi)^3} = \frac{1}{N}\sum_{j,\alpha}\left\langle \left(S_j^{\alpha}\right)^2\right\rangle.$$ (1.113)

For a system of identical spins, S, the right-hand side of Eq.(1.113) is simply the square of the spin operator, $\sum_a \left\langle \left(S_j^a \right)^2 \right\rangle = \left\langle S_j^2 \right\rangle = S(S+1)$. Consequently, the sum rule for the integral spectral weight of $S^{\alpha\beta}(q,\hbar\omega)$ reads

$$\sum_\alpha \int \int_{-\infty}^{\infty} S^{\alpha\alpha}(q,\hbar\omega) \frac{V_0 d^3 q}{(2\pi)^3} d(\hbar\omega) = \sum_\alpha \int S^{\alpha\alpha}(q) \frac{V_0 d^3 q}{(2\pi)^3} = S(S+1). \qquad (1.114)$$

A general expression for the n-th moment of the dynamic structure factor Eq.(1.109) is obtained from the standard expression for the Fourier transform of the n-th time derivative of the two-point correlation function,

$$\frac{1}{N} \sum_{j,j'} \int_{-\infty}^{\infty} e^{-i\omega t} \frac{d^n}{dt^n} \left\langle e^{-iq(R_j - R_{j'})} S_j^\alpha S_{j'}^\beta(t) \right\rangle \frac{dt}{2\pi\hbar} = (i\omega)^n S^{\alpha\beta}(q,\hbar\omega). \qquad (1.115)$$

Interchanging the sides and integrating both sides in $d(\hbar\omega)$ we obtain

$$\int_{-\infty}^{\infty} (\hbar\omega)^n S^{\alpha\beta}(q,\hbar\omega) d(\hbar\omega) = \left(\frac{\hbar}{i} \right)^n \frac{1}{N} \sum_{j,j'} \left\langle e^{-iq(R_j - R_{j'})} S_j^\alpha \frac{d^n}{dt^n} S_{j'}^\beta(t) \bigg|_{t=0} \right\rangle, \qquad (1.116)$$

because the integral in $\hbar\omega$ on the right-hand side gives the delta-function in time, and therefore, both $\hbar\omega$ and t can be integrated out.

Using equation of motion for Heisenberg operators and introducing the system's Hamiltonian, **H**, the time derivatives in Eq.(1.116) can be replaced by the appropriate commutators. Consequently, in the simplest case, $n = 1$, the following expression for the first moment sum rule is obtained [39],

$$\int_{-\infty}^{\infty} (\hbar\omega) S^{\alpha\beta}(q,\hbar\omega) d(\hbar\omega) = \frac{1}{N} \sum_{j,j'} \left\langle e^{-iq R_{jj'}} S_j^\alpha \left[\mathbf{H}, S_{j'}^\beta \right] \right\rangle = \frac{1}{N} \left\langle S_q^\alpha \left[\mathbf{H}, S_{-q}^\beta \right] \right\rangle. \qquad (1.117)$$

Eqs.(1.116), (1.117) are quite general, but, practically, are not very useful. Indeed, for a nonlinear (e.g., quadratic) Hamiltonian even the first moment is expressed through an expectation value of a three-operator product.

A useful simplification of the diagonal, $\alpha = \beta$ part of Eq.(1.117) was suggested in Ref. [39]. Firstly, the Hamiltonian in the statistical average on the right-hand side can be moved from the right to the left of the three-operator product. Then, it can be recast in the following, equivalent form

$$\int_{-\infty}^{\infty} (\hbar\omega) S^{\alpha\beta}(q,\hbar\omega) d(\hbar\omega) = \frac{1}{N} \left\langle S_q^\alpha \left[\mathbf{H}, S_{-q}^\beta \right] \right\rangle = \frac{1}{N} \left\langle \left[S_q^\alpha, \mathbf{H} \right] S_{-q}^\beta \right\rangle. \qquad (1.118)$$

Secondly, for a centro-symmetric lattice, invariant with respect to the inversion, $R_j \to -R_j$, the Fourier transform of the spin operators is even in q, $S_q^\alpha = S_{-q}^\alpha$. Hence, we can symmetrize the right-hand side of Eq.(1.118) and obtain a usual, simpler expression for the first-moment sum rule with the double commutator on the right [39],

$$\int_{-\infty}^{\infty} (\hbar\omega) S^{\alpha\alpha}(q,\hbar\omega) d(\hbar\omega) = \frac{1}{2N} \left\langle \left[S_q^\alpha, \left[\mathbf{H}, S_{-q}^\alpha \right] \right] \right\rangle = \frac{1}{2N} \left\langle \left[\left[S_q^\alpha, \mathbf{H} \right], S_{-q}^\alpha \right] \right\rangle. \qquad (1.119)$$

We note that there is no implied summation over the repeating index, α, that, if needed, would be written out explicitly. The commutators are often easily evaluated, giving extremely useful expressions for the first moment of the dynamical structure factor. In the single-mode approximation, these expressions define the q-dependent static spin structure factor of the system.

1.3.6.1. Static structure factor and spectrum averaged energy

It is often useful to split the dynamic structure factor Eq.(1.109) into a product of static structure factor Eq.(1.112), and a normalized spectral function $f_q^{\alpha\beta}(\omega)$, which, in general, is polarization- and wave-vector-dependent,

$$S^{\alpha\beta}(q,\hbar\omega) = S^{\alpha\beta}(q)f_q^{\alpha\beta}(\omega) \quad , \quad \int_{-\infty}^{\infty} f_q^{\alpha\beta}(\omega)d(\hbar\omega) = 1. \tag{1.120}$$

The normalization condition on $f_q^{\alpha\beta}(\omega)$ follows from the total spectral weight sum rule Eq.(1.114). Perhaps the commonest choice of the spectral function is a Lorentzian of Eq.(1.96), $L_{\Gamma_q}(\hbar\omega-\hbar\omega_q)$ with a q-dependent width Γ_q, and centered at some energy $\hbar\omega_q$, which describes the dispersion of an excitation. To describe the asymmetric line shapes characteristic of the excitation continua it is often convenient to use a "half-Lorentzian" spectral function, *i.e.*, a Lorentzian, truncated on one side of the peak [40,53].

The energy, averaged over the fluctuation spectrum, is defined by

$$\langle\hbar\omega\rangle_q^{\alpha\beta} = \frac{\int_{-\infty}^{\infty} \hbar\omega S^{\alpha\beta}(q,\omega)d(\hbar\omega)}{\int_{-\infty}^{\infty} S^{\alpha\beta}(q,\omega)d(\hbar\omega)} = \int_{-\infty}^{\infty} \hbar\omega f_q^{\alpha\beta}(\omega)d(\hbar\omega), \tag{1.121}$$

where the last equality follows from the normalization condition Eq.(1.120). Then, according to the average value theorem, $\langle\hbar\omega\rangle_q^{\alpha\beta}$ is some energy from within the support of the spectral function (in fact, from the range where it is defined and, in addition, takes the non-zero values).

In view of the definition Eq.(1.121), the sum rule Eq.(1.119) leads to a simple and very useful expression for the static structure factor,

$$S^{\alpha\alpha}(q) = -\frac{1}{2N} \frac{\langle[S_q^\alpha,[S_{-q}^\alpha,\mathbf{H}]]\rangle}{\langle\hbar\omega\rangle_q^{\alpha\alpha}}. \tag{1.122}$$

Consequently, the dynamical structure factor can be expressed as

$$S^{\alpha\alpha}(q,\hbar\omega) = -\frac{1}{2N}\langle[S_q^\alpha,[S_{-q}^\alpha,\mathbf{H}]]\rangle \frac{f_q^{\alpha\alpha}(\omega)}{\langle\hbar\omega\rangle_q^{\alpha\alpha}}. \tag{1.123}$$

This is valid for an arbitrary, not necessarily normalized, spectral function $f_q^{\alpha\alpha}(\omega)$, because the ratio of the spectral function and the spectrum averaged energy, $\langle\hbar\omega\rangle_q^{\alpha\alpha}$, is independent of the normalization.

The *single mode approximation* (SMA) simply consists in identifying the spectrum-averaged frequency with an eigenfrequency of a single excitation at the corresponding wave vector q, $\langle\hbar\omega\rangle_q^{\alpha\alpha} \cong \hbar\omega(q)$. It is justified when a spectral function takes the form of a sharp peak whose width is small compared to its position, which is approximately given by $\langle\hbar\omega\rangle_q^{\alpha\alpha}$. Then, Eq.(1.123) directly relates the q-dependent intensity of the peak with its dispersion, and with the expectation value of the product of the Fourier-transformed spin operators whose q-dependence is known. For a spectral function with a single, infinitely narrow delta-function peak, SMA is exact.

1.3.6.2. First moment sum rule for Heisenberg spin Hamiltonian with anisotropy and magnetic field

It is useful to apply the sum rules Eqs.(1.118)-(1.123) to a typical spin system with a Hamiltonian $\mathbf{H}$ consisting of a Heisenberg exchange coupling, $\mathbf{H}_E$, a quadratic single-ion anisotropy energy, $\mathbf{H}_A$, and a Zeeman energy in the magnetic field, $\mathbf{H}_Z$,

$$\mathbf{H} = \sum_{j,j'} J_{jj'} \mathbf{S}_j \cdot \mathbf{S}_{j'} + \sum_{j,\beta} D_\beta \left(S_j^\beta\right)^2 - \mu_B \sum_{j,\beta} g_\beta H_\beta S_j^\beta \equiv \mathbf{H}_E + \mathbf{H}_A + \mathbf{H}_Z. \tag{1.124}$$

Here $J_{jj'}$ is the exchange coupling between the sites j and j', D_α and g_α ($\alpha = x, y, z$) are the axial anisotropy constants and the g-factors for the three directions. To calculate the commutators in Eqs.(1.118)-(1.123) we rewrite the above Hamiltonian in terms of the Fourier transformed spin operators S_q^α,

$$\mathbf{H} = \frac{1}{N} \sum_q J_q \mathbf{S}_q \cdot \mathbf{S}_{-q} + \frac{1}{N} \sum_{q,\beta} D_\beta S_q^\beta S_{-q}^\beta - \mu_B \sum_\beta g_\beta H_\beta S_{q=0}^\beta, \tag{1.125}$$

J_q being the lattice Fourier transform of exchange interaction.

We also rewrite the spin commutation relations in Fourier components,

$$\left[S_q^\alpha, S_{q'}^\beta\right] = i e_{\alpha\beta\gamma} S_{q+q'}^\gamma \quad \Leftrightarrow \quad \left[S_j^\alpha, S_{j'}^\beta\right] = i e_{\alpha\beta\gamma} \delta_{jj'} S_j^\gamma. \tag{1.126}$$

The different contributions to the double commutator in Eq.(1.119) are easily evaluated, resulting in the following expression for the first-moment sum rule for the Hamiltonian Eq.(1.124) in q-representation,

$$\int_{-\infty}^{\infty} (\hbar\omega) S^{\alpha\alpha} (q, \hbar\omega) \, d(\hbar\omega) = \frac{1}{N} \sum_\beta (1 - \delta_{\alpha\beta}) \frac{1}{2} \mu_B g_\beta H_\beta \left\langle S_{q=0}^\beta \right\rangle - $$
$$- \frac{1}{N^2} \sum_{q',\beta} (1 - \delta_{\alpha\beta}) \left[D_\alpha + 2D_\beta - D + \left(J_{q'} - \frac{J_{q+q'} + J_{q-q'}}{2} \right) \right] \left\langle S_{q'}^\beta S_{-q'}^\beta \right\rangle, \tag{1.127}$$

where $D = \sum_\alpha D_\alpha = D_x + D_y + D_x$. We see, that the first moment of the dynamical structure factor is determined by the expectation value of the *equal-time* two-point spin correlation and, for the non-zero magnetic field H, by the uniform static spin polarization.

It is useful to rewrite Eq.(1.127) in terms of the original, lattice spin operators S_j^α. Performing the inverse Fourier transformation we obtain,

$$\int_{-\infty}^{\infty} (\hbar\omega) S^{\alpha\alpha} (q, \hbar\omega) \, d(\hbar\omega) = -\frac{1}{N} \sum_{j,j',\beta} (1 - \delta_{\alpha\beta}) J_{jj'} (1 - \cos(q \cdot R_{jj'})) \left\langle S_j^\beta S_{j'}^\beta \right\rangle$$
$$+ \frac{1}{N} \sum_{j,\beta} (1 - \delta_{\alpha\beta}) \left[D_\beta \left(S(S+1) - \left\langle \left(S_j^\alpha\right)^2 \right\rangle - 2 \left\langle \left(S_j^\beta\right)^2 \right\rangle \right) + \frac{1}{2} \mu_B g_\beta H_\beta \left\langle S_j^\beta \right\rangle \right]. \tag{1.128}$$

This expression relates the first moment of the dynamical structure factor with the equilibrium expectation values of the contributions to the exchange bond energies, $J_{jj'} \left\langle S_j^\beta S_{j'}^\beta \right\rangle$, the single-site anisotropy energy, $D_\beta \left\langle \left(S_j^\beta\right)^2 \right\rangle$, and the Zeeman energy, $\mu_B g_\beta H_\beta \left\langle S_j^\beta \right\rangle$.

For the Heisenberg Hamiltonian with only nearest-neighbor exchange the coupling $J_{jj'}$ is the following sum,

$$J_{jj'} = \sum_n J_n \delta(R_{jj'} - a_n). \tag{1.129}$$

Here a_n are the vectors connecting the site j to its nearest neighbors, which are numbered by n, and J_n are the corresponding coupling constants. Then, the first-moment sum rule adopts a very simple form [39],

$$\int_{-\infty}^{\infty} (\hbar\omega) S^{\alpha\alpha}(q,\hbar\omega)\, d(\hbar\omega) = -\sum_{n,\beta} J_n [1 - \cos(q \cdot a_n)](1 - \delta_{\alpha\beta}) \left\langle\!\left\langle S_{R_j}^{\beta} S_{R_j + a_n}^{\beta} \right\rangle\!\right\rangle, \tag{1.130}$$

where $\left\langle\!\left\langle S_{R_j}^{\beta} S_{R_j + a_n}^{\beta} \right\rangle\!\right\rangle$ denotes the site-independent lattice average. The situation is further simplified when there are only one or two different coupling constants, as often found in one- or two-dimensional systems [40].

1.4. MEASURING ELASTIC AND QUASI-ELASTIC MAGNETIC SCATTERING IN EXPERIMENT

In a growing number of important cases, static magnetic order in materials of fundamental interest to modern condensed matter physics and its technological applications is not long-range, but has short-range, nanoscale correlations, as were described in the previous section. Such is the well-known case of the magnetic "stripe" order in the doped cuprates [41], nickelates [42,43], and in many other doped oxides [44,45]. It also occurs in the frustrated spin systems, where high ground-state degeneracy is responsible for the short-range correlated, often anisotropic, spin-ordered state [46]. A half-doped perovskite oxide $La_{1.5}Sr_{0.5}CoO_4$ represents a beautiful generic example of such an anisotropic (2D) glassy magnetic state, which was extensively characterized by magnetic neutron scattering [45].

1.4.1. Short-range magnetic order in $La_{1.5}Sr_{0.5}CoO_4$

Magnetic properties of the doped, strongly correlated transition metal oxides, such as super-conducting cuprates, magneto-resistive manganites, and others, are at the focus of the modern condensed matter research. The interplay of charge, spin and/or orbital degrees of freedom in these systems generates non-trivial ground states and many fascinating physical properties, such as a magnetic-field-driven metal-insulator transition. Frequently, the magnetic ordering observed in the charge-ordered phases of these materials has short-range, nanoscale correlations of the type discussed in section 1.3.4. Here we describe the neutron scattering study of the magnetic ordering in the half-doped cobaltate, $La_{1.5}Sr_{0.5}CoO_4$. It is one of a series of the half-doped layered perovskite oxides, also including $La_{0.5}Sr_{1.5}MnO_4$ and $La_{1.5}Sr_{0.5}NiO_4$, which recently were studied to gain insight into the physics of the colossal-magneto-resistive manganites [44] (the strongest response to the magnetic field there occurs at half-doping).

The crystal structure of $La_{1.5}Sr_{0.5}CoO_4$ is very simple. Like the other half-doped layered perovskites discussed, it remains at all temperatures in the tetragonal phase with space group *I4/mmm* and has the low-T lattice parameters $a = 3.83$Å, and $c = 12.5$Å. This is the famous "HTT" (high-temperature tetragonal) structure of the high-T_c parent cuprate, La_2CuO_4. At half-doping, there is a natural tendency towards a checkerboard charge/valence order in each CoO_2 square-plaquet layer in this structure. While in the manganite and the nickelate such charge order (CO) is intimately coupled with the magnetic degrees of freedom, in the cobaltate it is a robust structural feature independent of them [45]. Even though the CO in $La_{1.5}Sr_{0.5}CoO_4$ is very short-range (the in-plane correlation length is $\xi_{ab} \approx 23\text{Å}$, the inter-plane correlation is only between the nearest planes, $\xi_c \approx 8\text{Å}$), it occurs at a strikingly high temperature, $T_{co} \approx 825$ K, and shows no anomaly at a temperature of the magnetic spin order, $T_{so} \approx 30$ K.

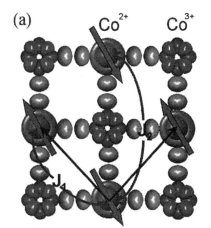 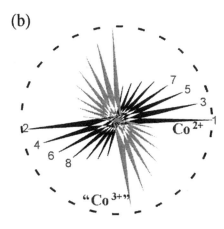

Figure 1-9: (a) Checkerboard Co^{2+}/Co^{3+} valence order in $La_{1.5}Sr_{0.5}CoO_4$ and schematics of the resulting magnetic subsystem; only Co^{2+} effectively carries spin at low temperatures. (b) Schematic depiction of the short-range-correlated, "damped spiral" magnetic structure for $La_{1.5}Sr_{0.5}CoO_4$. Black arrows illustrate the average spins at the consecutive, $n=1,2,3,...$, sites of the Co^{2+} sub-lattice in the diagonal direction, parallel to the spin structure propagation vector $Q = (\zeta,\zeta,1)$. Gray arrows show the average spins that would be expected on the respective Co^{3+} sites in such a single-Q structure.

The magnetic properties of $La_{1.5}Sr_{0.5}CoO_4$ are essentially determined by the checkerboard order of the Co^{2+}/Co^{3+} valence. The electronic configuration of the Co^{2+} ion is $3d^7$ and, although there is a significant crystal-field splitting of the e_g and t_{2g} levels, the total spin of its electrons adds up to S=3/2, in accordance with the Hund's rule. The situation differs for the $3d^6$ Co^{3+} ion, where there is a close competition between the Hund's energy and the crystal field. As a result, Co^{3+} ions in $La_{1.5}Sr_{0.5}CoO_4$ are in the S=1, "intermediate spin" (IS) state. At low temperatures, they are quenched to a $S^z = 0$ singlet state by a strong single-ion spin anisotropy arising from the crystal field, and do not participate in magnetic ordering [45]. Nevertheless, Co^{3+} sites are essential in defining the magnetic properties of $La_{1.5}Sr_{0.5}CoO_4$. They bridge the magnetic Co^{2+} ions, creating a peculiar pattern of exchange couplings on the Co^{2+} square sub-lattice, illustrated in Fig. 1-9 (a). A simple counting of the exchange pathways suggests that, in addition to the nearest-neighbor coupling along the side, J_1, there is a frustrating diagonal coupling J_2, such that $J_1/J_2 \approx 2$.

This peculiar, frustrated coupling geometry is the most probable cause for the short-range and incommensurate nature of the magnetic ordering occurring below $T_{so} \approx 30$ K. Figure 1-10 shows the patterns of the intensity of neutron elastic magnetic scattering measured well below the magnetic ordering temperature, at $T \approx 10$ K, for the wave vector transfers in the ab plane and perpendicular to that plane. The data were collected on BT2 thermal neutron triple axis spectrometer at NIST Center for Neutron Research [45]. A fixed energy for the incident and the scattered neutrons, $E_i = E_f = 14.7$ meV, was selected by using (002) Bragg reflection from the pyrolytic graphite (PG). Unwanted neutrons from the higher-order, (004) and (006) reflections, were filtered out of the beam using several inches of a PG with rather broad mosaics. The sample was mounted in a closed-cycle refrigerator with the (*h*,*h*,*l*) reciprocal lattice plane horizontal. The horizontal angular divergences of the incident and the scattered neutron beams were 60′ - 20′ - 20′ - 100′ for the 3 crystals in the setup (PG-sample-PG). Beam collimations around the sample were defined by Soller collimators (multi-channel transmission devices with neutron-absorbing cadmium coating on the channel's horizontal walls); the corresponding apertures controlled the other two.

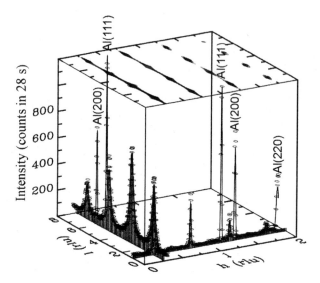

Figure 1-10: Elastic magnetic neutron scattering in $La_{1.5}Sr_{0.5}CoO_4$ for wave vector transfers in the (*hhl*) plane. Points are the intensities measured on BT2 spectrometer at NIST with $m \approx 0.5$ g sample at $T \approx 10$ K; the error bars are smaller than the symbol size and are not visible. Solid lines with gray shading are the fits to the "lattice Lorentzian" cross-section discussed in text. The grey-scale map on the top represents the corresponding calculated magnetic intensity.

The magnetic peaks are observed at $Q = (\zeta,\zeta,1) + \tau$, where $\zeta \approx 0.258$, *l* is an odd-integer, and τ is a crystal reciprocal lattice vector. The peaks have finite width indicating a finite correlation length, and are much broader in the direction perpendicular to the *ab* plane than in the plane. This fact reflects a quasi-two-dimensional nature of the magnetic order, where the correlations in the *ab* planes are significantly better developed than between them. The magnetic intensity arises on cooling below $T_{so} \approx 30$ K; this is discussed in more detail in the next section. Clearly, there is non-zero magnetic intensity for all values of *l* in the $Q = (0.258,0.258,l)$ scan. It is much larger than the background arising from neutrons scattered off the air, the cryostat, and other elements of the spectrometer setup, seen in the *h*-scan. This magnetic intensity is inherent to the short-range order, and is accurately described by the "lattice Lorentzian" cross-section of Eq.(1.105).

Solid curves with gray shading in Fig. 1-10 show the best fit of all data to the cross-section Eq.(1.105), where the localized spin model with the magnetic form factor appropriate for the cobalt ions was used [45]. The strong, sharp peaks (not shaded) arising from the scattering from the polycrystalline aluminum of the sample holder and the cryostat were also included in the fit. The in-plane and the inter-plane correlation lengths were refined to be $\xi_{ab} \approx 79$Å and $\xi_c \approx 10$Å, respectively. The "lattice-Lorentzian" cross-section Eq.(1.105) provides a perfect description of the short-range magnetic order in $La_{1.5}Sr_{0.5}CoO_4$, which is illustrated in Fig. 1-9 (b).

1.4.2. Temperature dependence of quasi-elastic magnetic fluctuations

The practical limit on the smallest intrinsic width Γ_E of the quasi-elastic peak (and, correspondingly, on the longest relaxation time of magnetic correlations) that neutron scattering can measure is imposed by the neutron spectrometer's energy resolution. Several specialized techniques were developed for such studies, based on the time-of-flight- and neutron backscattering- methods. The elastic energy resolution of the corresponding instruments can be as small as, or less than, $\Delta E \approx 1\ \mu eV$, that measures relaxation times up to about $\tau \sim 1$ ns. Neutron spin-echo spectroscopy extends this limit up to 0.1 ms [47]. However, formidable energy resolution is usually achieved by restricting the numbers of incident and scattered neutrons that the spectrometer uses, and, consequently, proportionally lowers sensitivity. Therefore, these high-resolution techniques are mainly used for measuring spatially incoherent quasi-elastic scattering, so that the intensity obtained at different scattering angles can be combined.

Significant improvements can be made in the accuracy with which Γ_E and τ can be determined without a loss of sensitivity if the phase space structure of the instrument's resolution function is known in great detail. Then, a highly precise determination of the parameters of the scattering cross-section, e.g., of Eq.(1.74), can be achieved by fitting the experimental data with an intensity obtained from this cross-section upon appropriately accounting for the resolution effects. Consequently, the smallest quasi-elastic energy width that can be measured experimentally is determined not by the instrument's elastic energy resolution ΔE, but by the accuracy with which spectrometer resolution function is known. We caution, though, that high precision in determining the parameters of a model cross-section does not imply the same precision in distinguishing between the model cross-sections. Indeed, upon convolution with a typically Gaussian resolution function, the information on the intrinsic fine structure of the cross-section on the scale $< \Delta E$ is lost. With the finite error bars on the data, it may not be possible to distinguish whether the broadening of the quasi-elastic peak results from the intrinsic Lorentzian width Γ_E, or from its splitting into the inelastic features. The only way to obtain this information would be to tighten the energy resolution.

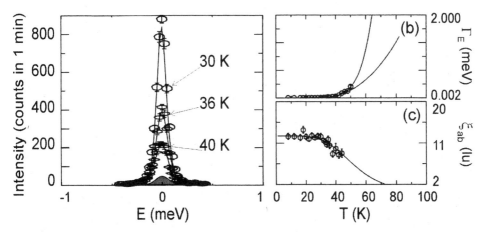

Figure 1-11: Quasi-elastic neutron magnetic scattering in $La_{1.5}Sr_{0.5}CoO_4$ for temperatures at and above the spin-freezing transition to the spin-ordered phase, $T_{so} \approx 30$ K [45]. The wave vector is $Q = (0.258, 0.258, 3)$, measured in reciprocal units of the I4/mmm crystal lattice. (a) E-scans through the peak at three temperatures; the non-magnetic, incoherent elastic background (not subtracted) accounts for a small peak which is shown in gray. (b) the temperature dependence of the intrinsic Lorentzian energy width (FWHM), and (c) the in-plane spin correlation length. Curves in (b), (c) are guides for the eye.

An outstanding case where the instrument's resolution function can be precisely accounted for is that of a triple axis neutron spectrometer [47]. This instrument, whose invention was honored by the Nobel Prize, is a workhorse of neutron spectroscopy, and is often used to study elastic and quasi-elastic magnetic scattering. A very accurate description of its resolution function was developed over the years [47-51]. It adopts a Gaussian approximation for describing the transmission of the instrument's individual components. Since the resulting errors are uncorrelated, the accuracy of the spectrometer's Gaussian resolution function follows from the central limit theorem, and relies on a large number, ~10, of these components. Here we give an example where an accurate account of the spectrometer's resolution allows measuring the quasi-elastic energy width as small as $\Gamma_E \leq 0.1 \, \Delta E$.

Figure 1-11 illustrates the evolution of quasi-elastic magnetic neutron scattering on approaching the transition to the phase with static magnetic order in the strongly correlated, doped layered perovskite oxide $La_{1.5}Sr_{0.5}CoO_4$ [43]. The energy scans through the peak of the incommensurate magnetic scattering at wave vector $Q = (0.258, 0.258, 3)$, in reciprocal lattice units of the *I4/mmm* crystal lattice, are shown in panel (a). The measurements were made with a SPINS triple axis neutron spectrometer at NIST Center for Neutron Research. With the energy of the scattered neutrons fixed at $E_f = 3.7$ meV, the incoherent elastic energy resolution of the experiment was $\Delta E = 0.16(1)$ meV. This determines the full width at half maximum (FWHM) of the non-magnetic Gaussian peak arising from the incoherent elastic scattering that contaminates the intensities in Fig. 1-11 (a). This peak (shown in gray) was measured at 100 K and added to the calculated resolution-corrected magnetic scattering intensity to obtain the solid lines shown in this figure.

At temperatures above $T_{so} \approx 30$ K, the magnetic scattering is very well described by a Lorentzian centered at $E = 0$, whose width increases with rising temperature. The correction for resolution is extremely important for accurately analyzing the experimental data, as the Lorentzian intrinsic energy width Γ_E, Fig. 1-11 (b), is comparable with, or smaller than, the energy resolution, ΔE. For temperatures 30 K, 36 K, and 40 K, shown in Fig. 1-11 (a), Γ_E is 0.007 meV, 0.019 meV, and 0.035 meV, respectively. In fact, to accurately refine such small intrinsic energy widths it is important to follow the change in the peak q-width, which reflects the concomitant decrease of the magnetic correlation length with increasing T, and to account for it in the procedure for resolution correction. It is also important that the resolution function is very accurately known, as it is for a triple axis spectrometer [51]. The solid curves in Fig. 1-11, (a) were obtained from the global fit of the energy scans given in the figure, and the q-scans measured in the same experiment, to the resolution-corrected scattering function in the form of a product of Eq.(1.74) and a lattice-Lorentzians describing the short-range spatial correlations. We discussed the latter in the previous section. Consequently, the in-plane spin correlation length, ξ_{ab}, was also refined, and is shown in Fig. 1-11 (c). At temperatures of 30 K, 36 K, and 40 K [*cf.* Fig. 1-11 (a)], ξ_{ab} is 12.4, 10.5, and 9.8 (diagonal) lattice units (lu), respectively. Unlike relaxation time, this length changes only a little, and saturates below ≈ 30 K at 14(1) lu.

At $T_{so} \approx 30$ K, the intrinsic energy width refined from the fit is smaller than the uncertainty of the energy resolution. In fact, by following the T-dependence of Γ_E in Fig. 1-11 (b), this temperature was identified as the critical temperature of the transition to the state with the static spin order [45]. In this reference, Γ_E was determined following the same procedure as here, but from measurements on the thermal neutron spectrometer that has about an order-of-magnitude broader energy resolution, ΔE. In the region where they overlap, both the present measurement and that of Ref. [45] yield similar values for the refined quasi-elastic FWHM, Γ_E, even though it is significantly smaller than the energy resolution of the thermal neutron measurement in this region. This finding strongly supports our confidence in the resolution correction procedure, which can be used to refine the intrinsic quasi-elastic energy widths as small as one tenth of the FWHM of the instrument's incoherent elastic energy resolution.

I.5. MODERN TECHNIQUES IN THE TRIPLE AXIS NEUTRON SPECTROSCOPY

In a conventional triple axis spectrometer, consisting of a single crystal-analyzer and a single detector, the scattering cross-section can be measured only at a single combination of energy and momentum transfer at a time. This limits the instrument's rate of data collection. The combination of a Position-Sensitive Detector (PSD) and/or a multiple-crystal analyzer affords several possibilities for increasing this rate by allowing us to simultaneously probe scattering events at different energy and momentum transfers, or to integrate the energy/momentum transfer over the range of interest.

I.5.1. Inelastic neutron scattering setups with horizontally focusing analyzer

In what is now a standard technique for increasing the throughput of a triple axis neutron spectrometer, the analyzer crystal is segmented, and the individual segments (analyzer blades) are aligned to geometrically focus the neutron beam emanating from the sample on the detector, Fig.1-12. The idea is to use a larger reflecting area of the crystal-analyzer for the measurement.

By appropriately choosing the rotation angle of the analyzer as a whole (the angles for the individual segments are then determined from the Bragg condition), a good energy resolution can be maintained, while the wave vector, Q-resolution can be significantly relaxed along a particular direction in Q-space [monochromatic focusing, Fig.1-12 (b)]. Then, the volume of the sample reciprocal space probed by the analyzer-detector system is stretched along that direction, and the rate of data collection is increased correspondingly. This approach is very efficient for samples where the dependence on the wave vector transfer along some direction(s) is weak, or absent. Such situation is typical of low-dimensional systems and systems with short-range correlations, such as spin glasses, and frustrated magnets. In practice, about a four-fold increase in the measured intensity is routinely achieved in such cases.

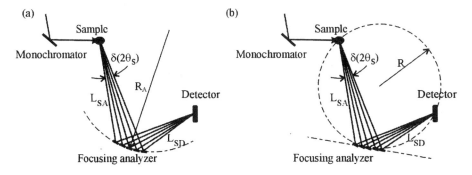

Figure 1-12: Schematics of the two horizontally focusing multi-crystal analyzer setups. (a) The polychromatic (in general), real-space geometric focusing by the cylindrically curved multi-crystal device. (b) The monochromatic "Rowland" focusing by the analyzer with independently aligned individual crystals. The dashed lines show the Rowland circle and a tangent to it at the analyzer position.

A typical design of the multi-crystal analyzer has several individually rotating, thin, 10- to 20 mm-wide aluminum alloy blades mounted along a single line on a rotating platform, and with co-aligned analyzer crystals attached to them. The blades are then aligned with respect to the common rotation axis of the platform as required by the experimental setup, *e.g.*, parallel to the circumference of a cylinder of a given radius, Fig.1-12 (a), thus approximating a cylindrically curved mirror.

Analyzers with a simple cylindrical focusing are currently the most common. In such a "GMI"-type device, a single motor that controls the curvature drives the rotations of all analyzer blades. The curvature radius, R_A, is defined by the distances from the analyzer's axis to the sample, L_{SA}, and to the detector, L_{AD}, and the analyzer's Bragg angle θ_A (*i.e.*, the selected neutron energy), according to

$$\frac{2}{R_A \sin \theta_A} = \frac{1}{L_{SA}} + \frac{1}{L_{AD}}.$$
(1.131)

This simply is a condition for a geometrical focusing of the paraxial incident beam emanating from the point sample by a cylindrical mirror with the symmetry axis at an angle θ_A to that beam (*i.e.*, with the device's symmetry point aligned for the selected Bragg reflection). It is important to realize, that, except for a symmetric setup where $L_{SA} = L_{AD}$, the reflection angle changes along the surface of a cylindrically curved analyzer. Hence, in general, such a device provides a polychromatic geometrical focusing, where the intensity gain is linked with the relaxed energy resolution.

In the general case where the distances differ (usually, $L_{SA} > L_{AD}$), the reflection angle along the analyzer does not change provided that the reflecting surface of the device follows the circumference of the "Rowland" circle, connecting the sample, the analyzer, and the detector axes. This geometry, known as the monochromatic, or "Rowland" focusing, is illustrated in the Fig.1-12 (b). Only in a symmetric setup, where the distances from the analyzer to the detector and to the sample are equal, $L_{SA} = L_{AD}$, is monochromatic focusing achieved with a cylindrically curved device shown in Fig.1-12 (a). In practice, Rowland focusing is approximated by aligning the platform of the multi-crystal segmented analyzer at a tangent to the Rowland circle. Then, the individual blades are aligned for the Bragg reflection at a given θ_A [47]. This setup, Fig. 1-12 (b), is successfully implemented on SPINS cold neutron triple axis spectrometer at the NIST Center for Neutron Research.

1.5.2. Inelastic neutron scattering using the high count rate setups with the PSD

Employing a Position-Sensitive Detector provides additional flexibility in inelastic triple axis neutron measurements. It permits a different, wave-vector-resolved mode of using the large reflecting area of the analyzer crystal. The main idea is that each individual pixel on the PSD acts as a separate, small single detector that views a particular, small segment of a large multi-crystal analyzer. According to the Bragg's law, only scattered neutrons with a particular energy and wave vector are reflected by this small analyzer segment, and then detected in the corresponding PSD pixel. In optimized conditions, the size of the PSD matches that of the analyzer. This is clear, in particular, from Fig.1-13 that illustrates two such high-efficiency PSD setups successfully implemented on SPINS at NCNR.

To measure typically weak inelastic scattering intensity, it is crucial that the real-space view of each PSD pixel is restricted to the monochromatic neutron image of the sample alone (obtained

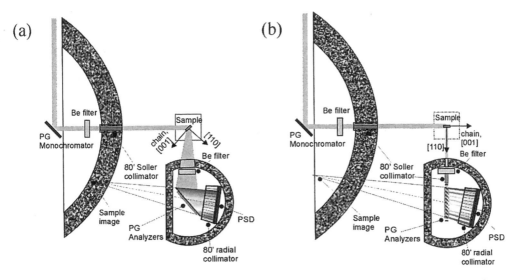

Figure 1-13: Two high-throughput setups with a PSD and a large, segmented PG crystal analyzer implemented on SPINS. The imaginary neutron source illuminating the detector in the real space is a polychromatic image of the sample. The reciprocal lattice directions refer to the $CsNiCl_3$ measurements described below. (a) The wave-vector-dispersive setup with flat analyzer. (b) The energy-dispersive setup with consecutive analyzers. *Also see the color plate.*

by reflection from the analyzer of neutrons with a particular energy). The simplest way to do this is by placing the radial collimator (RC) in front of the PSD. It restricts the real-space scattering volume seen by the PSD pixels to the near neighborhood of the collimator's focal point (it is the same for all pixels), reducing it dramatically. The RC shields the PSD from the polychromatic images of the sample's environment and other incoherently scattering instrument components illuminated by the neutron beam, and from the incoherent scattering by the analyzer's support structures, which otherwise generate a prohibitively high background. Hence, the RC is the essential element of the inelastic setups shown in Fig.1-13.

With a RC placed in front of the PSD, the images obtained by the appropriate Bragg reflections from all analyzer segments must coincide, thus forming a single, polychromatic image of the sample. Moreover, for it to be seen by the detector pixels un-obscured, this image must be at the focal point of the radial collimator. Figure 1-13 shows two possible ways to achieve this. With the flat analyzer, Fig.1-13 (a), the neutron Bragg reflection at any point across the analyzer's surface produces a mirror image of the sample in the real space (for neutrons of appropriate energy). In the alternative setup of Fig.1-13 (b), the image is created at the crossing point of the roughly plane-parallel paraxial beams, produced by the reflections from the different analyzer blades.

The energy measured at each PSD pixel and the relative sensitivity of the pixels across the PSD are determined in a calibration experiment, using the incoherent elastic scattering from an axially symmetric standard sample, such as a hollow cylinder made of Vanadium or a hydrogen-rich plastic. The incident energy is scanned within the range that includes the PSD acceptance window, and the energy/sensitivity curve is obtained by fitting the intensity profiles observed in the different PSD pixels (*cf.* Fig.1-14 and below). The advantage of the PSD mode is that the instrument's energy and wave-vector resolution, as well as the signal-to-noise ratio remain as good as for a conventional triple-axis spectrometer with the same collimations. The rate of data collection, however, is increased by about an order-of-magnitude.

1.5.2.1. Setup with a flat crystal analyzer

The instrument operation in the typical multiplexing-detection mode with a PSD and a large flat crystal analyzer (the wave-vector-dispersive setup) is illustrated in the Fig. 1-14. On SPINS, the multiple-crystal pyrolytic graphite (PG) analyzer consists of 11 blades, each 2.1 cm wide and 15 cm high. The PSD that we used is a two-dimensional, position-sensitive ^{3}He proportional counter manufactured by ORDELA. The active counting area is 26.4 cm $\times$ 22.6 cm (width (W) $\times$ height (H)) with 256 $\times$ 256 pixels, whereas the actual spatial resolution is 0.5 cm $\times$ 0.5 cm ($W \times H$). The detection efficiency for 5 meV neutrons is about 80%. An 80$'$ radial collimator was placed between the analyzer and the PSD. When the analyzer is set to be flat, each blade reflects different neutron energy to a different location on the PSD. Figure 1-15 (a) is a photograph of this setup on SPINS.

The geometry of the analyzer's reflection is illustrated in Fig.1-14 (a). The neutron trajectories are colored in accordance with the neutron energy selected by the Bragg reflection at the corresponding point on the analyzer. Blue marks the slower neutrons, reflected at larger angles, and yellow represents the fastest, corresponding to the smallest θ_A; a somewhat similar color scheme was used in Fig.1-13. For a given energy of the incident neutrons, E_i, and a sample scattering angle (with respect to the analyzer's platform axis), $2\theta_S$, the volume in the ($\hbar\omega, Q$)-space of energy and wave vector transfers probed by the PSD pixels follows a trajectory illustrated in Fig.1-14 (b). This produces a coupled scan in the sample's phase space, where both the $\hbar\omega$ and the wave vector transfer, q, change as a function of pixel position on the detector.

It is important, however, that only the component of the wave vector parallel to the analyzer's surface, $q_{\|A} \equiv Q$, varies in a scan across the PSD for a given k_i and $2\theta_S$. The component perpendicular to the analyzer surface, $q_{\perp A}$, does not change because the projection of the scattered wave vector, k_f, on the analyzer reciprocal lattice vector, τ_A, is fixed by the Bragg condition, $(k_f \cdot \tau_A) = \tau_A^2/2$; hence, $q_{\perp A} \approx (k_f \cdot \tau_A)/\tau_A - \tau_A/2$. This allows a "quasi" constant-Q scans, where the projection of the wave vector transfer is constant along a particular direction in the sample's reciprocal space. Such scans are very useful, as they are identical with the usual constant-Q scans for samples where the dispersion and/or the other wave vector dependence(s) in one direction are unimportant, or absent, if that direction is aligned parallel to the analyzer' surface (*cf.* Section 1.5.3.1).

The results of the calibration measurements for the wave-vector-dispersive setup with the flat analyzer and using the standard vanadium sample are presented in Fig.1-14 (c) and Fig.1-15 (b,c). The former shows the intensity observed in five different PSD pixels as a function of the incident neutron energy when the analyzer's center is set to $E_f^{(0)} = 3.15$ meV. The circles of different colors indicate different scattered neutron energies, following the same convention as in (a). The energy range covered by the whole flat analyzer in this configuration is 2.6 meV $< E_f <$ 3.7 meV, and the energy resolution FWHM varies from 0.1 meV to 0.15 meV. The correspondence between the pixel position on the PSD and the scattered neutron energy assigned to it, as well as the relative sensitivity of the pixels, are obtained by fitting these intensity profiles to the Gaussians. Typical calibration curves obtained in this way for $E_f^{(0)} = 4.2$ meV (used in the measurements described in Section 1.5.3.1) are shown in Fig.1-15 (b,c).

By changing the incident neutron energy and/or the sample scattering angle, we can survey the scattering cross-sections in a large portion of the ($\hbar\omega, Q$)-space. In fact, even when exploring the ($\hbar\omega, Q$) window that can be covered at a single instrument setting, it is still useful to split the measurement into a scan with varying k_i. Then, the intensity at each energy and wave vector transfer is measured at different k_f and by different PSD pixels. Combining these measurements appropriately evens out any systematic fluctuations arising from the variation of the pixel sensitivity across the PSD.

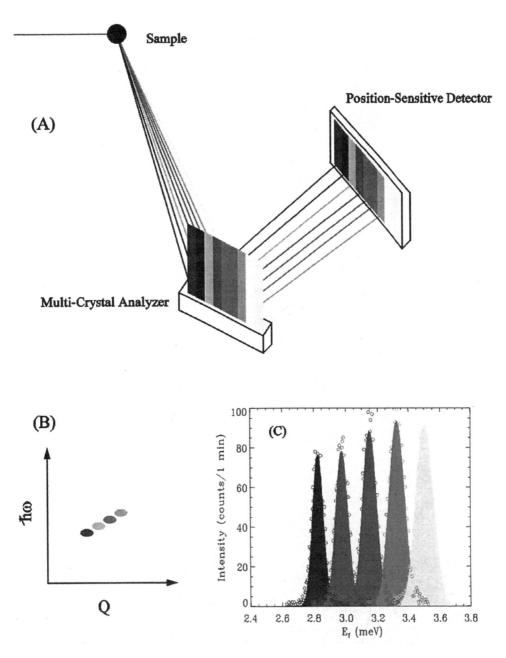

Figure 1-14: Schematics of the multiplexing detection mode with a PSD and a flat analyzer. (a) the reflection geometry, (b) the region of the wave vector and energy probed by different pixels across the PSD. (c) Results of the standard vanadium calibration measurement. The energy at the center of the analyzer was 3.15 meV. Peaks show the measured intensity profiles for five different PSD pixels detecting, respectively, five different neutron energies. *Also see the color plate.*

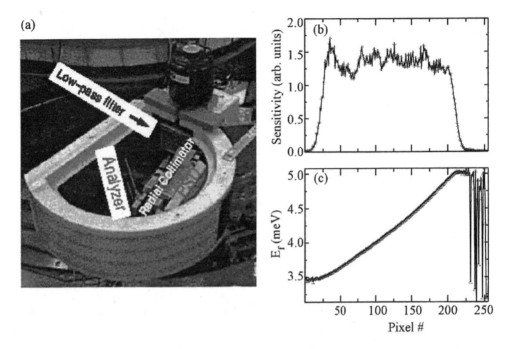

Figure 1-15: (a) Photograph of the actual wave-vector-dispersive setup with the PSD and a large flat PG analyzer on SPINS. Panels (b) and (c) show the results of the calibration measurement for the setup with the PSD central energy fixed at $E_f^{(0)} = 4.2$ meV: the relative scattering intensity (b) and the scattered energy (c) measured by the different PSD pixels.

1.5.2.2. Setup with consecutive analyzers

In an alternative, final-energy-dispersive setup illustrated in Fig.1-12 (b), the analyzer platform is aligned parallel to the line connecting the axes of the sample and the analyzer, in a transmission-, or Laue-like geometry. The neutron beam then sequentially passes through the analyzer blades, with each consecutive blade reflecting a portion of the scattered neutron spectrum onto the PSD. This setup requires relatively fine tuning. Indeed, for best performance, all analyzer blades must be illuminated by neutrons and must reflect the appropriate monochromatic beams on the detector. Therefore, the angular offsets between the consecutive blades have to be optimized so that the overlaps of their energy-dependent reflectivity do not significantly shield one blade by another. This essentially requires that the offsets exceed the FWHM of the mosaic spread of the analyzer's crystals. On the other hand, the offsets should not be too large so that the variation of the sensitivity across the PSD is not excessively strong and, in particular, to avoid "blind spots" on the detector, which are not illuminated by the analyzer. Finally, the condition of the sample's geometrical imaging in the focus of the radial collimator also has to be satisfied, at least approximately.

Figure 1-16 (a) is a photograph of a setup with the consecutive analyzers on SPINS, which was employed in one of the measurements described in Section 1.5.3.1. The PSD central energy was $E_f^{(0)} \approx 4.57$ meV. The angles $\delta\theta_A \approx 35' - 40'$ between the consecutive PG crystals with mosaic of $\approx 30'$ (and the corresponding difference in the reflected neutron energy) were chosen to satisfy the fine-tuning conditions described above. Nine of the 11 analyzer blades provided full coverage of the PSD, reflecting the neutron energies in the 4.03 meV $< E_f <$ 5.13 meV range. The corresponding PSD calibration curves obtained using the Vanadium standard are shown in Fig.1-16 (b,c). The variation of the pixel sensitivity across the PSD is much stronger than for the flat analyzer, with nine peaks corresponding to the reflections from the consecutive blades clearly visible.

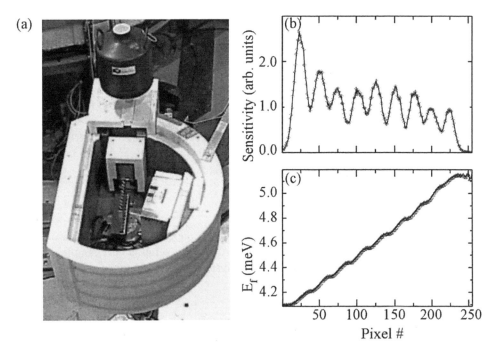

Figure 1-16: (a) Photograph of the E_f-dispersive setup with consecutive analyzers on SPINS. (b) and (c) show the pixel sensitivity and the detected neutron energy as a function of the position on the PSD.

A very important feature of this setup is the possibility of restricting the analyzer's angular acceptance by inserting a Soller collimator after the sample, thus defining a single direction for the scattered neutrons wave vector. A constant-Q scan can be effectively performed for systems with at least one non-dispersive direction by co-aligning this direction with k_f, as shown in Fig.1-12 (b). Using the Soller collimator also allows very significant reductions in the amount of the sample's incoherent scattering seen by the PSD, as well as shaping/tightening of the instrument's wave vector resolution in the direction perpendicular to k_f.

1.5.2.3. Energy-integrating configuration in the two-axis mode

Figure 1-17 illustrates a high-count-rate setup without an analyzer and with a polycrystalline low-pass neutron filter followed by the PSD in the so-called two-axis mode. The filter only transmits the scattered neutrons with energies less than a cut-off energy E_c, and within the angular range determined by the beam-defining aperture. On SPINS, polycrystalline Be or BeO filters with the energy cutoffs $E_c = 5.1$ meV, or $E_c = 3.7$ meV, respectively, are typically used. The horizontal angular acceptance of this configuration on SPINS is currently about 11°. It is determined by the detector's angular size at the sample's position, *i.e.* the ratio w/L, where w is the width of the PSD and L is the distance between the sample and the PSD.

Sample   Low-pass filter   PSD

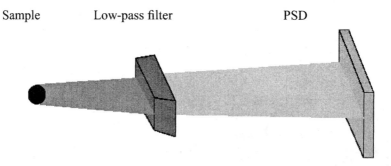

Figure 1-17: Energy-integrating setup without an analyzer, and with the position-sensitive detector in. the "two-axis" mode.

Without an analyzer, a position-sensitive detector placed behind the filter detects the scattering intensity integrated over the energy transfers in the range $E_i - E_c < \hbar\omega < E_i$,

$$I(\boldsymbol{Q}) = \int_{E_i - E_c}^{E_i} \frac{k_f}{k_i} S(\boldsymbol{Q}, \omega) d(\hbar\omega) \qquad (1.132)$$

The energy integration and the wide angular acceptance increase the rate of data collection in this mode by more than an order-of-magnitude. Albeit this setup is very simple, it is very efficient for studying low dimensional systems and systems with short-range correlations.

1.5.3. Experimental examples

1.5.3.1. Spin excitation continuum in the Haldane spin chain

Quantum liquids are among the most interesting and important fundamental condensed matter systems. Despite having been studied for about a century since Dewar and Kamerling-Onnes first liquified hydrogen and helium, the quantum-liquid state continues to amaze physicists with new discoveries. Recently, several remarkable new examples of the quantum liquid state were found in the systems of quantum spins in magnetic crystals. Perhaps, the most important was the prediction and the experimental discovery of the spin-liquid, Haldane-gap state in an integer-spin, one-dimensional Heisenberg antiferromagnet (HAFM spin chain), see Ref. [55] for an early overview.

In a striking neglect of the naive "common sense" expectation, the ground state (GS) of a Haldane spin chain does not connect the Neel-ordered "spin solid" GS of the semiclassical, S>>1 HAFM, with the almost-ordered "marginal liquid" state of the S=1/2 chain [33,34]. Instead, it is a quantum liquid with a finite correlation length, and a gap in the spin excitation spectrum. The spectral weight of spin fluctuations is concentrated in a long-lived massive triplet mode near the Brillouin zone's (BZ) boundary, $q = \pi$, reminiscent of a roton in ^{4}He. Any remainder of the spectacular continuum observed at $q \approx \pi$ in the S=1/2 1D HAFM (*cf.* Fig.1-1) is extremely faint. On the other hand, continuum two-magnon states are predicted to be the lowest-energy excitations at $q \approx 0$, and to dominate the spin fluctuation spectrum close to the Brilloun zone's center. For a non-interacting magnons, the two-magnon continuum starts above a q-dependent energy threshold, $\varepsilon_{2m}(q) = \mathbf{min}\{2\varepsilon(\pi + q/2), \Delta_H + \varepsilon(\pi+q)\}$, where $\varepsilon(q)$ is the dispersion of the single-magnon excitation, and $\Delta_H = \varepsilon(\pi)$ is the Haldane gap in the excitation spectrum at $q = \pi$.

Experimentally observing the spin excitation continuum in a Haldane chain is an extremely challenging task. Primarily, this is due to the rapid decrease at small q of the static structure factor, $S(q)$, which determines the energy-integrated intensity of scattering by spin fluctuations. In the single-mode approximation (SMA), $S(q) \sim (1 - \mathbf{cos}q)/\varepsilon(q)$, cf. Eq.(1.130), and vanishes $\sim q^2$ as $q \to 0$. However, using the high-count-rate setups described in the previous sections opens the possibility of performing such a measurement. Efficiently using the SPINS large-area segmented PG analyzer and a matching large PSD is a key to success.

We studied a large, $\sim$6 g sample of a quasi-1D HAFM $CsNiCl_3$, composed of two single crystals that were co-aligned with an effective mosaic less than 1°. Our sample had its longer dimension parallel to the chains (hexagonal c-axis), and was mounted on an Al plate in the standard "ILL orange" 70 mm cryostat, with (h,h,l) zone in the scattering plane and in a transmission, "Laue" geometry, as illustrated in Fig.1-18 (a,b) [this is rather important in view of a rather large absorption of the natural Cl]. Magnetic scattering was measured at T=1.5 K, while the non-magnetic background (BG) was collected in the identical scans at T=150 K.

$CsNiCl_3$ is probably the least anisotropic and best-studied Haldane model compound. It has a hexagonal crystal structure (space group $P63/mmc$), with the chains of chlorine-linked Ni^{2+} ions parallel to the c-axis, and with two equivalent ions per c-spacing, so that the wave vector transfer $Q = (h,k,l)$, in reciprocal lattice units (rlu), corresponds to $q = \pi l$ in the 1D BZ of a chain. In $CsNiCl_3$ a supercritical interchain exchange coupling, $J' \approx 0.03J$, leads to a long-range order below $T_N \approx 4.8$ K. However, as temperature rises above T_N, a gap opens in the spin excitation spectrum, and it quickly recovers the properties of an isolated S=1 HAFM chain [55]. Moreover, even at $T = 0$ the interchain coupling modifies only the low-energy part of the excitation spectrum and, therefore, even at $T < T_N$ the dynamic spin response of $CsNiCl_3$ in the better part of the BZ, specifically around the top of the 1D dispersion, is identical with that of an individual chain [40,53]. Quantitatively, for energies $E > 2$ meV the dependence of the spin scattering function on the wave vector transfer perpendicular to the chains (*i.e.* in the ab-plane), $q_\perp$, is small (less than 10%). Thus, to a good approximation it can be ignored [40].

Figure 1-18 shows the spin excitation spectrum of $CsNiCl_3$ measured on SPINS cold neutron triple axis spectrometer using two high-count-rate instrument configurations described in Sections 1.5.2.1 and 1.5.2.2. The scattering geometry used in the two experiments is illustrated in the upper panels, (a) and (b) of the figure. The bottom panels (c) and (d) show the contour maps of the normalized magnetic scattering intensity, $I(q,E)/\int I(q,E)dE$, with the non-magnetic, (q,E)-dependent linear background subtracted. The intensity maps are constructed from the raw data (the normalization integrals are performed via point-by-point summation), and are slightly distorted by the instrument's resolution. Refs. 40 and 53 have more detailed description of the experiments, and the (arbitrary) intensity scales used in the contour maps.

It is clear in both panels (c) and (d) of Fig.1-18 that the spin excitation spectrum has a finite width in energy at $l < 0.5$. In the single-mode part of the spectrum, at $l \geq 0.6$, the measured line-shape is completely determined by the interplay of the dispersion and the instrument's resolution. The resolution is quite different in two setups. Although the accepted phase volume is smaller in setup (b), the "focusing" effect [the longer axis of the FWHM ellipse is parallel to the dispersion in Fig.1-18 (c)] produces sharper peaks in setup (a). In principle, an opposite, "de-focusing" effect is of concern for $l \leq 0.5$ measurements in the latter, as it could cause quite significant broadening, even in a single-mode spectrum. However, carefully accounting for the resolution shows that the non-zero *intrinsic* width at $l \leq 0.5$ accounts for $\geq 2/3$ of the spectral width measured in setup (a). "De-focusing" is absent in setup (b), where the FWHM ellipse is approximately round, Fig.1-18 (d). Another important distinction between the two setups is that $q_{\parallel} \approx const$ geometry imposes different choices of $q_{\perp}$.

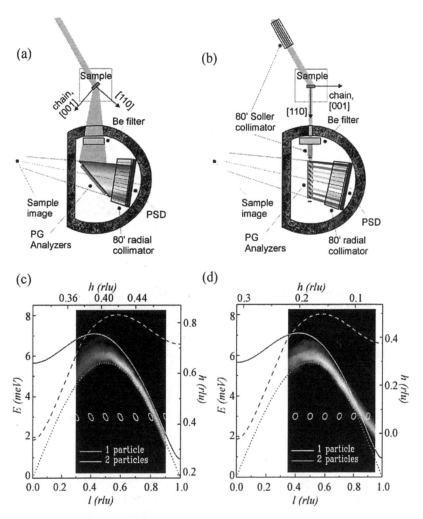

Figure 1-18: (a) and (b) on the top illustrate the experimental setups used for the measurements shown in (c) and (d) on the bottom, respectively (the data is shown below the corresponding setup). Contour plots of the spectral density of magnetic scattering in (c) and (d) were reconstructed using linear interpolation from the constant-$q_{\parallel}$ scans, measured with the chain perpendicular to the analyzer surface in setup (a), and perpendicular to k_f in setup (b). Scale on the right shows variation with energy of the wave vector transfer perpendicular to the chain at $l=0.5$, scale on the top - its variation with l at $E = 3$ meV. Ellipses are the calculated half-maximum contours of the instrument resolution function at $E = 3$ meV. The solid curve is the single-magnon dispersion discussed in text, the dashed line shows the lowest energy of two non-interacting magnons with a total $q_{\parallel} = \pi l$, and the dotted line is $\varepsilon(q_{\parallel}) = 2.49J \sin(q_{\parallel})$. *Also see the color plate.*

The spectacular agreement of the excitation spectrum at $q_\parallel \leq 0.5\pi$ measured in two setups shows explicitly that the observed crossover from a single mode to a continuum with a significant intrinsic width is not an instrumental/resolution effect. In fact, very similar behavior is found in the "classical" quantum liquid - superfluid ^{4}He. There, the "maxon" excitation turns into a broad, continuum-like feature under pressure when the "roton" spectral gap is suppressed below the half the maxon energy.

1.5.3.2. Spin fluctuations in a geometrically frustrated magnet

Systems encompassing a large diversity of states are common in biology, chemistry and physics, such as glasses, liquids, and proteins. An essential concept in understanding those systems is frustration, *i.e.*, due to competing interactions, the degrees of freedom cannot be optimized simultaneously. Magnetic systems offer extreme examples in the form of spin lattices where all interactions between spins cannot be simultaneously satisfied. Such geometrical frustration can lead to macroscopic degeneracy and produce qualitatively new states of matter.

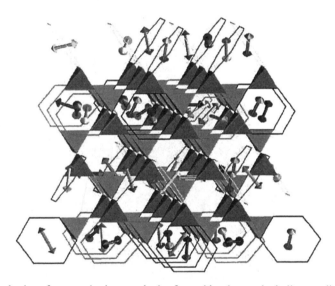

Figure 1-19: The lattice of corner-sharing tetrahedra formed by the octahedrally coordinated B sites in a spinel structure with chemical formula AB_2O_4. A periodic assignment of all spins in the pyrochlore lattice is made to four different types of non-overlapping hexagons, colored in blue, green, red, and gold. Every spin belongs to just one hexagon, and each such hexagon carries a six-spin director. The resulting tetragonal structure of these hexagons has a unit cell of $2a \times 2a \times 3c$ and can be described by a stacking of two different types of three-layer slabs along the c-axis. The hexagon coverage on consecutive slabs is uncorrelated, so that a macroscopic number of random slab-sequences are generated. *Also see the color plate.*

We explored some of these possibilities by examining magnetic fluctuations in $ZnCr_2O_4$. The B-site of this spinel lattice occupied by spin-3/2 Cr^{3+} leads to a magnet with dominant nearest-neighbor coupling on the lattice of corner-sharing tetrahedra, as shown in Fig.1-19 [56]. Because the spin interaction energy is minimized when the four spins on each tetrahedron add up to zero, the interactions do not call for a long-range order, but simply define a restricted phase space for fluctuations. Just as composite fermions can emerge from degenerate Landau levels in a two-dimensional electron gas, the near-degenerate manifold of states in a frustrated magnet is fertile ground for emergent behavior.

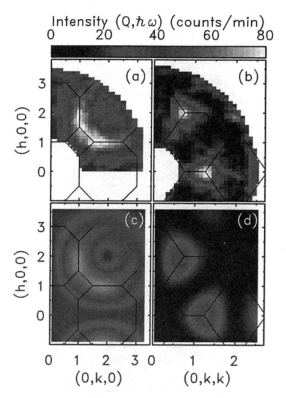

Figure 1-20: (a), (b) Color images of inelastic neutron scattering intensities from single crystals of ZnCr$_2$O$_4$ in the (hk0) and (hkk) symmetry planes obtained at T = 15K for $\hbar\omega$ = 1 meV. The data are a measure of the dynamic form factor for self-organized nanoscale spin clusters in this material. (c), (d) Color images of the form factor squared, calculated for antiferromagnetic hexagon spin loops averaged over the four hexagon orientations in the spinel lattice. The excellent agreement between model and data identifies the spin clusters as hexagonal spin loops. *Also see the color plate.*

Neutron scattering is the most effective tool to study possible composite spin degrees of freedom by directly probing the form factor of such entities. Figure 1-20 (a) and (b) demonstrate the wave-vector dependence of the low-energy inelastic neutron scattering cross-section in the spin liquid phase of ZnCr$_2$O$_4$ [56]. The data exhibit broad maxima at the Brillouin zone's boundaries, signaling the emergence of confined nanoscale spin clusters. Rather than Fourier-inverting the data, we consider potential spin clusters and test the corresponding prediction for the form factor against the data. Individual tetrahedra would be prime candidates, as they constitute the basic motif of the pyrochlore lattice. However, a tetrahedron is too small to account for the observed features.

The next smallest symmetric structural unit is the hexagonal loop formed by a cluster of six tetrahedra, Fig.1-21. Two spins from each tetrahedron occupy the vertices of a hexagon while the other two spins from each tetrahedron belong to different hexagons. Therefore, all spins on the spinel lattice can be simultaneously assigned to hexagons, thus producing N/6 weakly interacting degrees of freedom, Fig.1-19. An outstanding fit is achieved for the antiferromagnetic hexagonal spin loops, as evidenced by Fig.1-20 (c) and (d). Thus, rather than scattering from individual spins, neutrons scatter from antiferromagnetic hexagonal spin clusters. In effect, ZnCr$_2$O$_4$ at low temperatures is not a system of strongly interacting spins, but a protectorate of weakly interacting spin-loop directors. Since the six hexagon spins are anti-parallel, the staggered magnetization vector for a single hexagon, which shall be called the spin loop director, is decoupled from the 12 outer spins. Hence, its reorientation embodies the long-sought local zero-energy mode for the pyrochlore lattice [56].

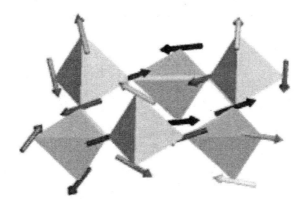

Figure 1-21: Spin cluster surrounding a hexagon (shown in gold) in the pyrochlore lattice of Fig. 1-19.

1.5.4. Neutron polarization analysis with PSD

Polarized neutron diffraction is a well-established method for investigating the spin configuration of a system. It distinguishes between the magnetic- and the nuclear-structural contributions because of the different selection rules for non-spin-flip (NSF) and spin-flip (SF) scattering processes. Figure 1-22 (a) shows a conventional geometry for polarized neutron diffraction. A polarized monochromatic beam with neutrons in a particular spin eigenstate (-) is obtained using the forward transmission polarizer. Subsequently, a flipper can rotate the polarization adiabatically to the other spin state (+). The spin state of the beam scattered from the sample is then analyzed using this combination. Each of the four possible channels, (off,off), (on,on), (off,on), and (on,off), is measured by appropriately turning on and off the front and the rear flippers, before and after the sample. With perfect efficiencies of the forward and rear polarizers and flippers, the measured intensities would be in a one-to-one correspondence to the spin-dependent cross sections, σ, I(off,off) $\rightarrow$ $\sigma(--)$; I(on,on) $\rightarrow$ $\sigma(++)$; I(off,on) $\rightarrow$ $\sigma(-+)$; I(on,off) $\rightarrow$ $\sigma(+-)$. In this technique, each channel is measured sequentially, which significantly reduces the rate at which data can be collected at a given scattering angle.

In this section, we describe a new technique for polarized neutron diffraction that was recently developed [57], utilizing a two-dimensional position-sensitive detector. It increases the rate of data-collection by a factor of two. The idea is simple: since a transmission polarizer passes neutrons in one spin state (-) straight through and without deviations, but deflects neutrons in the other spin state (+) by a few degrees, then by placing a two-dimensional position sensitive detector (PSD) in a two-axis configuration as shown in Fig.1-22 (b), we can simultaneously measure neutrons scattered by the sample into both spin states.

The separation of the deflected beam and the transmitted straight beam on the PSD, Δ, is determined by the distance from the polarizer to the PSD, D, and the critical reflection angle of the polarizer's supermirror, θ, that depends on the neutron's wavelength, λ: $\Delta \sim \theta D$. Obviously,

Δ should exceed the width of the beam. In our case, the polarizer was optimized for a λ of 4 Å and longer. To measure all four spin-dependent scattering processes, this technique requires only a single flipper, located in front of the sample (upstream). If the flipper is placed in the incident beam as shown in Fig.1-22 (b)-I, then, with the flipper off, the straight beam measures the (off,off) channel, and the deflected beam the (off,on) channel. In this case, the PSD measures both (off,off) and (off,on) channels simultaneously, thereby eliminating the need for a rear flipper used in the conventional setup. When the front flipper is on, Fig.1-22 (b)-II, the straight beam corresponds to the (on,off) channel, and the deflected beam to the (on,on) channel.

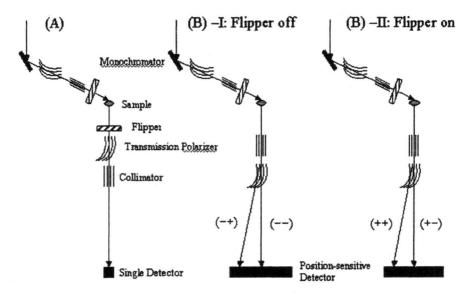

Figure 1-22: (a) Conventional experimental setup with transmission polarizers for polarized neutron diffraction, and, (b) setup for polarized neutron diffraction with a PSD.

The curved stack, Fe/Si supermirror transmission polarizers that we used in the measurements discussed below, are described in Refs. [57,58]. The flippers are made of Al wire coils and are also described in Ref. [57]. The incident neutron energy was fixed at 5 meV and a cooled polycrystalline Be filter eliminated the short-wavelength neutrons from the higher-order reflections in the PG monochromator. Two 20′ collimators were placed before and after the sample. In this set-up, $D = 79$ cm and $\theta = 2.4°$, yielding a deflected beam well separated from the straight-transmitted, by about $\Delta = 3.3$ cm (corresponding to 32 horizontal pixels on the PSD).

1.5.4.1. Nuclear and magnetic Bragg scattering in $La_{5/3}Sr_{1/3}NiO_4$

$La_{5/3}Sr_{1/3}NiO_4$ exhibits both charge and spin ordering in real space, which is related to the magnetic incommensurate peaks observed in the high-T_c superconducting cuprates. However, polarization analysis of the related incommensurate peaks in the cuprates is difficult because of their relatively low intensities. The incommensurate peaks are predominantly static and intense in the isostructral system $La_{2-x}Sr_xNiO_4$, making this nickelate a very good model system to study them. In our measurements, the $La_{5/3}Sr_{1/3}NiO_4$ crystal was mounted with the ($hk0$) reciprocal lattice plane (in which the Ni^{2+} spins lie) in the scattering plane, and a vertical magnetic guide field was applied perpendicular to it. In this configuration, the magnetic and nuclear scattering is purely SF and NSF, respectively. Detailed analysis of these measurements and supplementary data for ($h0l$) scattering plane were reported in Ref. [43]. Only data for one specific nuclear reflection and one magnetic Bragg reflection are presented here.

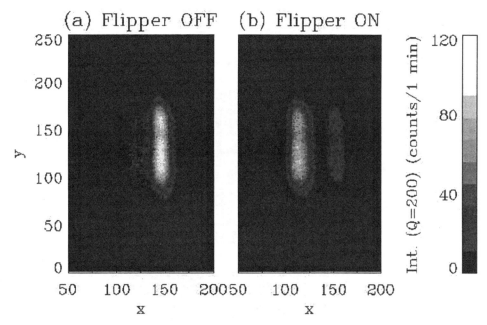

Figure 1-23: PSD images of the scattering from (200) nuclear Bragg reflection of $La_{5/3}Sr_{1/3}NiO_4$; (a) with the front flipper off, and (b) with the front flipper on.

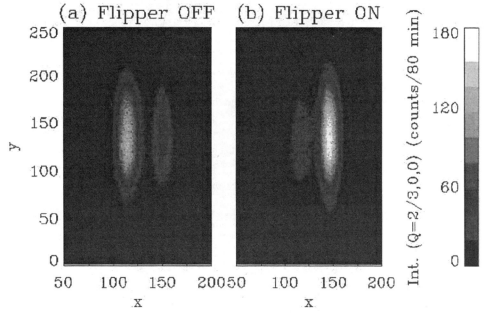

Figure 1-24: PSD images of scattering from the magnetic superlattice reflection (2/3,0,0) in $La_{5/3}Sr_{1/3}NiO_4$; (a) with the front flipper off, and (b) with the front flipper on. *Also see the color plate.*

Figure 1-23 shows the color PSD images of the scattering intensity for a nuclear Bragg reflection in $La_{5/3}Sr_{1/3}NiO_4$. When the flipper is off, Fig. 1-23 (a), the bright spot should correspond to the (—) NSF channel, the straight-transmitted beam. For this purely nuclear peak, we expect no scattering in the SF channel. The faint signal on the left is a contamination due to the polarizer's imperfect efficiency. Using the top five points around the peak for each of the two beams, we find that the polarizing efficiency is 0.790(1). When the front flipper is on, as shown in Fig. 1-23 (b), the positions of the bright and the faint spots on the PSD are switched. The bright peak is now the (on,on) channel. The (on,on) peak is slightly weaker than the (off,off) peak, indicating that the reflectivity of the polarizer is less than one, or 0.798(1). For systems such as $La_{5/3}Sr_{1/3}NiO_4$, in which scattering cross-sections in both NSF channels are identical to each other, as are the SF cross sections, the polarizer's reflectivity cancels out, and the (off,off) and (on,on) signals can be combined. Ignoring the deviation of the flipper's efficiency from unity, the (off,on) and (on,off) intensities can also be combined, which is a good approximation in the present example.

Figure 1-24 shows how this technique works for a magnetic reflection. In our configuration, the magnetic scattering is expected to contribute only in the SF channel. The color contour maps of the resulting scattered intensities are shown. As expected, with the front flipper off, the PSD image shows a bright spot at the position of the deflected beam, the (off,on) channel, Fig. 1-24 (a). With the flipper on, the straight-transmitted beam, corresponding to the (on,off) channel, is brightest, Fig. 1-24 (b). As expected, the ratio of the (on,off) and (off,on) peak intensities matches the reflectivity of the rear (behind the sample) polarizer.

ACKNOWLEDGEMENTS

The authors would like to express their sincere gratitude to many colleagues and collaborators they had over the years, many of whom have contributed to establishing author's experience in magnetic neutron scattering and their experimental expertise. First of all, our thanks go to Collin Broholm and Chuck Majkrzak, who spearheaded, with our modest involvement, development of the advanced triple axis neutron scattering techniques that we have used and described in this chapter. We also thank R. W. Erwin, J. W. Lynn, S. Park, L.-P. Regnault, J. M. Tranquada, and A. I. Zheludev for many interesting and useful discussions and for occasional help with the measurements. We are also indebted to A. Abanov, J. Bhaseen, F. Essler, and R. Konik for reading the parts of this manuscript, and for the discussions and useful comments. Finally, it is a pleasure to acknowledge the hospitality and support of the NIST Center for Neutron Research where the experiments described here were performed. This work was supported by the US Department of Energy under the Contract DE-AC02-98CH10886. The work on SPINS at NCNR was supported by the NSF through DMR-9986442.

REFERENCES

[1]   G. L. Squires, Introduction to the theory of thermal neutron scattering. Cambridge: Cambridge University Press, 1978; New York: Dover Publications (1996).

[2]   Yu. A. Izyumov, and R. P. Ozerov, Magnetic Neutron Diffraction. New York: Plenum Press (1970).

[3]   S. W. Lovesey, Theory of Neutron Scattering from Condensed Matter. Oxford: Clarendon Press (1984).

[4] Jensen, Mackintosh, R. Alan, Rare Earth Magnetism. Oxford: Clarendon Press (1991).

[5] Varley F. Sears, Neutron Optics. Oxford: Oxford University Press (1989).

[6] B. A. Lippmann, and J. Schwinger, Phys. Rev. **79** (1950) 469.

[7] P. A. M. Dirac, The Principles of Quantum Mechanics. Oxford: Clarendon Press (1930, 1958).

[8] Hans A. Bethe, Roman, Jackiw, Intermediate Quantum Mechanics. Massachusets, USA: Addison-Wesley (1997).

[9] E. Fermi, Ric. Sci. **7** (1936) 13.

[10] E. Fermi, and L. Marshall, Phys. Rev. **71** (1947), 666; Phys. Rev. **72** (1947), 408.

[11] P. J. Brown, "Magnetic scattering of neutrons". In International Tables for Crystallography, Volume C: Mathematical, Physical and Chemical Tables, A. J. C. Wilson, ed. Dordrecht: Kluwer Academic Publishers (1995).

[12] L. D. Landau, and E. M. Lifshitz, The Classical Theory of Fields. Oxford: Pergamon Press (1987).

[13] L. D. Landau, and E. M. Lifshitz, Quantum Mechanics. Oxford: Pergamon Press (1977).

[14] A. Abragam, B. Bleaney, Electron Paramagnetic Resonance of Transition Ions. New York: Dover Publications (1986).

[15] K. Yosida, Theory of Magnetism. Berlin: Springer-Verlag (1998).

[16] J. S. Schwinger, Phys. Rev. **51** (1937) 544.

[17] G. T. Trammell, Phys. Rev. **92** (1953) 1387.

[18] M. Blume, Phys. Rev. **124** (1961) 96.

[19] O. Steinsvoll, G. Shirane, R. Nathans, M. Blume, H. A. Alperin, S. J. Pickart, Phys. Rev. **161** (1967) 499.

[20] C. Stassis, and H. W. Deckman, Phys. Rev. B **12** (1975) 1885.

[21] S. W. Lovesey, D. E. Rimmer, Rep. Prog. Phys. **32** (1969) 333.

[22] S. W. Lovesey, J. Phys. C: Solid State Phys. **11** (1978) 3971.

[23] A. J. Freeman, R. E. Watson, Phys. Rev. **120** (1960) 1125; *ibid.* 1134.

[24] J. W. Lynn, G. Shirane, and M. Blume, Phys. Rev. Lett. **37** (1976) 154.

[25] S. Shamoto, M. Sato, J. M. Tranquada, B. J. Sternlieb, and G. Shirane, Phys. Rev. B **18** (1993) 13817.

[26] R. Coldea, S. M. Hayden, G. Aeppli, T. G. Perring, C. D. Frost, T. E. Mason, S.-W. Cheong, and Z. Fisk, Phys. Rev. Lett. **86** (2001) 5337.

[27] A. J. Freeman, R. E. Watson, Phys. Rev. **118** (1960) 1168.

[28] G. Racah, Phys. Rev. **62** (1942) 438.

[29] A. J. Freeman, Phys. Rev. **113** (1959) 169.

[30] L. Van Hove, Phys. Rev. **95** (1954) 1374.

[31] L. D. Landau, and E. M. Lifshitz, Statistical Physics. Oxford: Pergamon Press (1958).

[32] White, R. M., Quantum Theory of Magnetism. Berlin: Springer-Verlag (1983).

[33] H. A. Bethe, Z. Phys. **71** (1931) 265.

[34] A. H. Bougourzi, M. Couture, M. Kacir, Phys. Rev. B **54** (1996) R12669.

[35] B. E. Warren, X-Ray Diffraction. New York: Dover Publications (1990).

[36] G. E. Volovik, and I. E. Dzyaloshinskii, Sov. Phys. JETP **48** (1978) 555 [Zh. Eksp. Teor. Fiz. 75 (1978) 1102].

[37] K. I. Kugel, and D. I. Khomskii, Sov. Phys. Usp. **25** (1982) 231.

[38] A. Zheludev, M. Kenzelmann, S. Raymond, T. Masuda, K. Uchinokura, and S.-H. Lee, Phys. Rev. B **65** (2002) 014402.

[39] P. C. Hohenberg, and W. F. Brinkman, Phys. Rev. B **10** (1974) 128.

[40] I. A. Zaliznyak, S.-H. Lee, S. V. Petrov, Phys. Rev. Lett. **87** (2001) 017202.

[41]   J. M. Tranquada, N. Ichikawa, S. Uchida, Phys. Rev. B **59** (1999) 14712.

[42]   S.-H. Lee, Tranquada, K. Yamada, D. J. Buttrey, Q. Li, and S.-W. Cheong,
       Phys. Rev. Lett. **88** (2002) 126401.

[43]   S.-H. Lee, S.-W. Cheong, Y. Yamada, and C. F. Majkrzak, Phys. Rev. B **63** (2001)
       060405(R).

[44]   B. J. Sternlieb, J. P. Hill, U. C. Wildgruber, G. M. Luke, B. Nachumi, Y. Moritomo,
       and Y. Tokura, Phys. Rev. Lett. **76** (1996) 2169.

[45]   I. A. Zaliznyak, J. P. Hill, J. M. Tranquada, R. Erwin, Y. Moritomo, Phys. Rev. Lett. **85**
       (2000) 4353.

[46]   I. A. Zaliznyak, C. L. Broholm, M. Kibune, M. Nohara, H. Takagi,
       Phys. Rev. Lett. **83** (1999) 5370.

[47]   Currently, description and operational characteristics of the neutron spectrometers
       mentioned in the text are available on the web page of the NIST Center for Neutron
       Research, Gaithersburg, MD, at http://www.ncnr.nist.gov.

[48]   G. Shirane, S. M. Shapiro, J. M. Tranquada, Neutron Scattering with a Triple-axis
       Spectrometer. Cambridge: Cambridge University Press (2002).

[49]   M. J. Cooper, and R. Nathans, Acta Cryst. **23** (1967) 357.

[50]   N. J. Chesser and J. D. Axe, Acta Cryst. **A29** (1973) 160.

[51]   M. Popovici, Acta Cryst. **A31** (1975) 507.

[52]   C. Broholm, Nucl. Instr. and Meth. in Physics Res. A **369** (1996) 169.

[53]   I. A. Zaliznyak, J. Appl. Phys. **91** (2002) 812210MMM.

[54]   S.-H. Lee, C. Broholm, T. H. Kim, W. Ratcliff, and S.-W. Cheong,
       Phys. Rev. Lett. **84** (2000) 3718.

[55]   I. A. Zaliznyak, L.-P. Regnault, and D. Petitgrand, Phys. Rev. B **50** (1994) 15824.

[56]   S.-H. Lee, C. Broholm, W. Ratcliff, G. Gasparovic, Q. Huang, T. H. Kim,
       and S.-W. Cheong, Nature **418** (2002) 856.

[57]   S.-H. Lee, and C. F. Majkrzak, Physica B **267-268** (1999) 341.

[58]   C. F. Majkrzak, Physica B **213-214** (1995) 904.

CHAPTER 2

Small-angle neutron scattering

2.1. INTRODUCTION

Small Angle Neutron Scattering (SANS) is a technique that allows characterizing structures or objects on the nanometer scale, typically in the range between 1 nm and 150 nm. The information one can extract from SANS is primarily the average size, size distribution and spatial correlation of nanoscale structures, as well as shape and internal structure of particles (e.g. core-shell structure). Further, the scattering intensity on an absolute scale contains the product of scattering contrast of the investigated structures in the surrounding medium, and number or volume density. If one of both quantities is known, the other one can be derived in addition to the information mentioned before. All in all, SANS is a valuable technique, widely used in many fields, to characterize particles (in solution or in bulk), clusters, (macro-)molecules, voids and precipitates in the nanometer size range. Further, in-situ measurements allow following the temporal development and dynamics of such structures, on a time scale ranging from microseconds (stroboscopic) to hours.

Besides the nuclear interaction, due to their magnetic moment neutrons undergo a magnetic interaction with matter, approximately equally strong (in terms of order of magnitude) as the nuclear interaction. This property distinguishes neutrons markedly from x-rays where magnetic interaction is very weak and difficult to access experimentally. With this dual interaction of neutrons with matter they offer the opportunity to study both, compositional and magnetic structures and correlations. Thus, a strong area of application of neutrons traditionally was and still is the area of magnetism in solid state physics and condensed matter research. SANS in particular, probing structures on the nanometer scale, finds applications in micromagnetism, to magnetic clusters embedded in a solid nonmagnetic matrix, magnetic clusters suspended in fluids (e.g. ferrofluids), magnetism in nanostructured materials, vortex lattices in superconductors and many others. Further, by using a polarized neutron beam, very specific information on the magnetic structure or alignment of nanoparticles can be obtained, as well as on their response to an external magnetic field. In general, neutrons are the only probe which give direct access to magnetic moments and magnetic interactions and alignment down to the atomic scale, and the probing does not by itself impose a magnetic perturbance, as e.g. by an external magnetic field.

2.1.1. Neutron beams

Neutron beams for materials science and condensed matter research are produced either by nuclear reactors, with the most prominent representative being the ILL, Grenoble (F), or by neutron spallation sources, like ISIS, Abington (UK), the latter operating in pulsed mode. The neutron energy spectra provided at these sources, generated by specially tailored moderation systems, are primarily in the range of thermal energies with a Maxwellian distribution around 320 K (thermal neutrons) or 30 to 40 K (cold neutrons). The corresponding wavelengths of the neutron beams peak around 0.13 nm (thermal) or 0.3 to 0.4 nm (cold), both with a considerable tail towards higher wavelengths, in the case of cold neutrons reaching far beyond 1 nm. Thus, the wavelength range of neutron beams overlaps well with that of x-ray beams, being the basis of the similarities and complementarities of both types of probing beams for structural analysis. The difference lies in the type of response or interaction with materials: the x-rays are scattered by the electron clouds, resulting in a linear dependence of the scattering strength (scattering amplitude) on the atomic number. In contrast, neutrons are scattered at the atomic nuclei, and the nuclear scattering lengths vary more or less unsystematically for the different elements, and also distinguish different isotopes of one element. Further, as outlined before, neutrons undergo magnetic interaction with matter, which makes them a valuable probe for magnetic structures.

2.1.2. Atomic scattering amplitudes

The scattering amplitudes for nuclear scattering are widely tabulated, for the elements in natural isotopic abundance as well as for the individual isotopes [1]. The atomic scattering amplitudes for magnetic scattering will depend upon the atomic magnetic moment which includes implicitly the magnetic form factor descriptive of the spatial origin of the atomic magnetic moment. Values for the most prominent magnetic elements, like Fe, Co, Ni, Gd, can be found in [2]. For more complex systems, like alloys containing magnetic elements, the individual atomic magnetic moments may depend on the local magnetic surrounding and would have to be determined experimentally for each case. Examples for the Fe-Cr-, Ni-Fe- and Co-Cr-series can be found in [3].

2.1.3. The classical SANS instrument

The classical concept of a SANS instrument at a continuous neutron source was first realized in the early 1970's at the Jülich Research Centre, Germany, the ILL Grenoble, France [4] and the HFIR reactor in Oak Ridge, USA [5]. Modern instruments of this type, like the D22 instrument at the ILL, the SANS at HMI, Berlin, Germany, or the SINQ-SANS at PSI, Switzerland [6], still follow the same principle concept, although using state-of-the-art components and advanced technical concepts. For measurements of magnetic structures polarization and spin-flipping of the incident beam is a viable option at HMI [7], GKSS, Geesthacht (D) [8], NIST, Gaithersburg (USA) LLB, Saclay (F) [9] and PSI, Villigen (CH).

The basic layout of a classical SANS instrument is illustrated in Fig. 2-1. The preferential position of the instrument is at the end of a neutron guide supplying a spectrum of cold neutrons. The neutron energy, or wavelength, respectively, is selected by a mechanical velocity selector with a resolution of typically 10% FWHM. Double pin-hole collimation tailors the beam for the necessary angular resolution, and a two-dimensional position sensitive detector registers the neutrons which are scattered to small angles around the incoming beam. The favored instrument is typically 40 m in length, 20 m for the collimation and 20 m for the secondary flight path with a flexible distance between sample position and detector. The detector sizes nowadays reach 96x96 cm^2 with about 16000 pixels of 7.5x7.5 mm^2 resolution. Electromagnets, cryomagnets, furnaces and cryostats, alone or in combination, belong generally to the standard equipment for sample environments.

At a pulsed source, like ISIS (UK) or IPNS (USA) [10], the concept of a SANS instrument is different, making use of a time-of-flight selection of the 'white' incoming beam. Besides that, the operational concept is very similar to instruments at continuous sources.

In the experiment the scattered intensity is registered as a function of the radial distance from the beam center, i.e. as a function of the scattering angle 2θ, or, more general, as function of the scattering vector $\mathbf{Q}$ or of its modulus Q. The latter is related to 2θ via $Q = (4\pi/\lambda) \sin \theta$, with λ, the neutron wavelength. By appropriate calibration one obtains the intensity in absolute units of the differential scattering cross section $(d\sigma/d\Omega)$ (Q). When the scattering is isotropic around the central beam, it may be averaged azimuthally for each value of Q (so-called "radial average"). If the scattering is non-isotropic, as often observed in the case of magnetic scattering, one has to consider the scattering in different azimuthal directions by sectional averaging.

For characterizing magnetic structures it is mostly necessary, or at least helpful, to analyze the response of the scattering to an externally applied magnetic field. In the examples treated in the present chapter such an external field, when applied, is assumed to be homogeneous, directing horizontally and perpendicular to the incident neutron beam. Other configurations are possible and can be adapted if appropriate, for instance, a beam-parallel field for investigating vortex lattices in superconductors. Such examples are not considered here, although the same theoretical principles as outlined in section 2.2 apply.

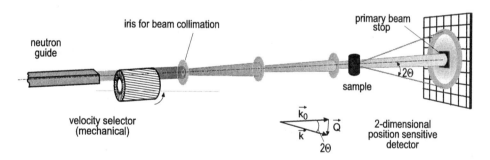

Figure 2-1: Schematic layout of a classical SANS facility at a continuous neutron source.

2.2. NEUTRON SCATTERING

This section gives a general theoretical introduction to neutron scattering focusing to magnetic scattering and small angle scattering, with the objective to introduce the established notations and present and explain the most important basic formulae needed to evaluate a scattering experiment characterizing magnetic structures. For further reading and more details we refer to [11-20].

In a scattering experiment, the primary goal is a detailed analysis of the measured scattering pattern in relation to the properties of the incoming neutron beam. Monochromator and collimator define the energy (wavelength) and divergence of the incoming neutrons. Those interact with a sample and thereby undergo a momentum transfer $\hbar\mathbf{Q}$, with the scattering vector $\mathbf{Q}=\mathbf{k}_0-\mathbf{k}$, and $\mathbf{k}_0$, $\mathbf{k}$ being the wave vectors of the incoming and scattered neutrons, respectively. This process in principle can be elastic or inelastic. For small angle scattering inelastic scattering events are of minor importance and will be neglected in the present treatise.

2.2.1. SCATTERING POTENTIAL

The scattering of a neutron, interacting through a scalar potential $V(\mathbf{r})$ with material, can be described by the asymptotic solution of the Schrödinger equation, see e.g. [11,12], which results in the wave function

$$\varphi_{\mathbf{k}_0}(\mathbf{r})\xrightarrow[r\to\infty]{}\frac{1}{(2\pi)^{3/2}}\left[e^{i\mathbf{k}_0\mathbf{r}}+f(\theta,\varphi)\frac{e^{ik_0r}}{r}\right] \tag{2.1}$$

where $f(\theta,\varphi)$ in first Born approximation is given by

$$f(\theta,\varphi)=-\frac{m_N}{2\pi\hbar^2}\int d^3r\, e^{i\mathbf{Q}\mathbf{r}}V(\mathbf{r}) \tag{2.2}$$

For low-energy and short-ranging (i.e. nuclear) interactions the potential $V(\mathbf{r})$ can well be approximated by the Fermi pseudo-potential [13]

$$V(\mathbf{r})=\frac{2\pi\hbar^2}{m_N}b_j\delta^3(\mathbf{r}-\mathbf{r}_j) \tag{2.3}$$

thus that $f(\theta,\varphi)$ reduces to

$$f(\theta,\varphi)=-b_j \tag{2.4}$$

b_j being the so-called atomic scattering length of atom j.

In this approximation, the differential scattering cross section per atom can be written as [14]

$$\frac{d\sigma}{d\Omega}(\mathbf{Q})=\frac{1}{N}|f(\theta,\varphi)|^2=\frac{1}{N}\left|\sum_{j=1}^{N}b_j\exp(i\mathbf{Q}\mathbf{r}_j)\right|^2 \tag{2.5}$$

where N is the number of atoms exposed to the beam.

2.2.2. MAGNETIC SCATTERING

In the case of magnetic moments in the sample, the neutron undergoes a magnetic interaction in addition to the nuclear interaction. The corresponding interaction potential is given by [15]

$$V(\mathbf{r})=-\boldsymbol{\mu}_N\cdot\mathbf{B}(\mathbf{r})\ \text{ with }\ \boldsymbol{\mu}_N=\gamma\frac{e\hbar}{2m_N}\boldsymbol{\sigma} \tag{2.6}$$

where $\boldsymbol{\mu}_N$ is the magnetic dipole moment of the neutron, $\boldsymbol{\sigma}$ the Pauli spin operator, $\gamma=-1.913$ the gyromagnetic ratio and $\mathbf{B}(\mathbf{r})$ the magnetic field induced by an atom at the position of the neutron. The latter has two components, one induced by the magnetic dipole moment $\boldsymbol{\mu}_S$ of the electrons denoted $\mathbf{B}_S(\mathbf{r})$, and one by their orbital moment $\boldsymbol{\mu}_L$, denoted $\mathbf{B}_L(\mathbf{r})$. The (weak) magnetic interaction $V(\mathbf{r})=\boldsymbol{\mu}_N(\mathbf{B}_S(\mathbf{r})+\mathbf{B}_L(\mathbf{r}))$ can as well be treated in first Born approximation, resulting in the magnetic scattering amplitude, in analogy to the nuclear scattering amplitude, given by the Fourier transform of the magnetic interaction potential

$$b_M = -\frac{m_N}{2\pi\hbar^2} \int d^3r\, e^{i\mathbf{Qr}}\, \boldsymbol{\mu}_N \cdot (\mathbf{B}_S(\mathbf{r}) + \mathbf{B}_L(\mathbf{r})). \tag{2.7}$$

An additional static magnetic field $\mathbf{H}(\mathbf{r})$ at the point of local magnetization $\mathbf{M}(\mathbf{r})$ (originating from $\mathbf{B}_S(\mathbf{r}) + \mathbf{B}_L(\mathbf{r})$) induces a total local magnetic induction of

$$\mathbf{B}(\mathbf{r}) = \mu_0 (\mathbf{H}(\mathbf{r}) + \mathbf{M}(\mathbf{r})) \tag{2.8}$$

and the Fourier transform of $\mathbf{B}(\mathbf{r})$ yields [cf. 16-18]

$$\mathbf{B}(\mathbf{Q}) = \mu_0\, \frac{\mathbf{Q} \times [\mathbf{M}(\mathbf{Q}) \times \mathbf{Q}]}{Q^2} = \mu_0 \mathbf{M}_\perp(\mathbf{Q}) = \mu_0 \mathbf{M}(\mathbf{Q}) \sin(\angle(\mathbf{Q}, \mathbf{M})) \tag{2.9}$$

where $\mathbf{M}(\mathbf{Q}) = \int d^3r \exp(i\mathbf{Q} \cdot \mathbf{r})\mathbf{M}(\mathbf{r})$, with $\mathbf{M}(\mathbf{r})$ given in units of Am.

$\mathbf{M}_\perp(\mathbf{Q}) = \mathbf{Q} \times [\mathbf{M}(\mathbf{Q}) \times \mathbf{Q}] / Q^2 = |\mathbf{M}(\mathbf{Q})| \sin(\angle(\mathbf{Q}, \mathbf{M}))$ is the magnetization component perpendicular to the scattering vector $\mathbf{Q}$. The magnetic scattering length then is [16].

$$b_M = \frac{\gamma e \mu_0}{2\pi\hbar} \boldsymbol{\sigma} \cdot \mathbf{M}_\perp(\mathbf{Q}) = D_M \mu_0 \boldsymbol{\sigma} \cdot \mathbf{M}_\perp(\mathbf{Q}). \tag{2.10}$$

For the differential scattering cross section one finally obtains

$$\frac{d\sigma_M}{d\Omega}(\mathbf{Q}) = \frac{D_M^2}{N} |\mu_0 \mathbf{M}_\perp(\mathbf{Q})|^2 \tag{2.11}$$

2.2.3. POLARIZED NEUTRON SCATTERING

In the presence of a preferred direction, for example induced by an external magnetic field, the magnetic scattering depends on the spin state σ of the neutrons. Let the z-axis be the preferred direction, and let (+) and (−) denote the neutron spin polarizations parallel and antiparallel to the z-axis, then the scattering is described by four scattering processes: two processes where the incident states (+) and (−) remain unchanged (++ and − −), the so-called 'non-spin-flip' processes, and two processes where the spin is flipped (+− and −+), the 'spin-flip' processes. Keeping in mind that the nuclear scattering does not flip the neutron spin, the four related scattering lengths are [19]

$$b_{\pm\pm} = b_N \mp D_M \mu_0 M_{\perp z} \tag{2.12a}$$

$$b_{\pm\mp} = -D_M \mu_0 (M_{\perp y} \pm i M_{\perp x}) \tag{2.12b}$$

It is evident that non-spin-flip scattering only contains magnetic contributions from effective magnetization components along the z-axis. If the scattering vector $\mathbf{Q}$ is parallel to the z-axis, $M_{\perp z}$ is zero. On the other hand, if spin-flip scattering is present, it is exclusively due to effective magnetic components deviating from the z-axis, the axis of magnetic polarization.

For an unpolarized neutron beam (which may be taken composed of 50% (+) and 50% (−) polarization) the square of the modulus of the scattering length is

$$|b|^2 = \frac{1}{2}\left(b_{++}^2 + b_{--}^2 + |b_{+-}|^2 + |b_{-+}|^2\right) = b_N^2 + D_M^2 |\mu_0 \mathbf{M}_\perp|^2 \tag{2.13}$$

The differential cross section of the unpolarized neutron beam can therefore be described by the sum of the nuclear and the magnetic cross section, without any cross terms.

2.2.4. Small angle neutron scattering

2.2.4.1. Scattering by individual magnetic particles

Small angle scattering does not resolve individual atoms, but structures of sizes in the nanometer range. Therefore the discrete atomic scattering lengths b_j can be replaced by a scattering length density $\rho(\mathbf{r})$ of the sample. The differential scattering cross-section (c.f. eq. (5)) is than given by the Fourier transform of $\rho(\mathbf{r})$:

$$\frac{d\sigma}{d\Omega}(\mathbf{Q}) = \frac{1}{V} \left| \int_V d^3 r \rho(\mathbf{r}) e^{i\mathbf{Q}\cdot\mathbf{r}} \right|^2 \tag{2.14}$$

with V being the sample volume. Assuming that this volume contains N particles embedded in a surrounding matrix of constant scattering cross section ρ_{matrix}, we define the scattering length distribution inside each particle by $\rho_{P_j}(\mathbf{r}) = \Delta\eta_j(\mathbf{r}) + \rho_{\text{matrix}}$. Than, with $\mathbf{R}_j$ being the vector pointing to the centre of the particle j, the related scattering cross section can be written as

$$\frac{d\sigma}{d\Omega}(\mathbf{Q}) = \frac{1}{V} \left| \sum_{j=1}^{N} F_j(\mathbf{Q}) e^{i\mathbf{Q}\mathbf{R}_j} \right|^2 \text{ with } F_j(\mathbf{Q}) = \int_{V_j(\mathbf{R}_j)} d^3 r \Delta\eta_j(\mathbf{r}) e^{i\mathbf{Q}(\mathbf{r}-\mathbf{R}_j)} \tag{2.15}$$

For particles of constant scattering length density, the scattering amplitude for nuclear scattering can be expressed as

$$F_N(\mathbf{Q}) = \Delta b_N V_P f(\mathbf{Q}). \tag{2.16}$$

where the constant $\Delta\eta_j$ was replaced by Δb_N, the contrast for nuclear scattering between the particles and the surrounding matrix. V_P is the particle volume. $f(\mathbf{Q})$ denotes the so-called particle form factor and can be calculated analytically for many simple particle shapes, as tabulated in [20]. For spherical particles of radius R, it is the well-known expression, dating back to Lord Rayleigh [21]

$$f(QR) = 3 \frac{\sin(QR) - QR\cos(QR)}{(QR)^3}. \tag{2.17}$$

We now assume the same particles being of homogenous magnetisation $\mathbf{M}_P$, embedded in a homogenously magnetized surrounding $\mathbf{M}_M$. The magnetic scattering of these particles then depends on the magnetic contrast vector $\Delta\mathbf{M} = \mathbf{M}_P - \mathbf{M}_M$ relative to the scattering vector $\mathbf{Q}$:

$$\mathbf{M}_\perp = \frac{\mathbf{Q} \times (\Delta\mathbf{M} \times \mathbf{Q})}{Q^2} \tag{2.18}$$

To calculate the magnetic scattering amplitude one has to consider the spin state before and after the scattering process according to Eqs.(2.12a,b)

$$F_{\pm\pm}(\mathbf{Q}) = \Delta b_{\pm\pm} V_P f(\mathbf{Q}) , \qquad F_{\pm\mp}(\mathbf{Q}) = \Delta b_{\pm\mp} V_P f(\mathbf{Q}) \tag{2.19}$$

2.2.4.2. Scattering by groups of particles

The scattering from an accumulation of many particles is obtained by summing up the scattering amplitudes of all particles weighted by a phase shift at each particle position. The general expression for the scattering cross section then is given by

$$\frac{d\sigma_j}{d\Omega}(Q) = \left\langle F_j^2(Q) \right\rangle + \left\langle F_j(Q) \right\rangle^2 (S(Q)-1) \tag{2.20}$$

(j standing for $\pm\pm$, $\pm\mp$, or N and M in the case of unpolarized neutrons, respectively). The right-hand-side of eq.(2.20) consist of two terms: The first one depends only on the particle structure and corresponds to the independent scattering of N particles, while the second one considers their spatial distribution and reflects the interparticle interference described by S(**Q**). The $\langle\,\rangle$ indicates an average over all possible configurations and sizes of the particles.

As postulated by eq.(2.13), in the case of an unpolarized incident neutron beam the scattering of both contributions, nuclear and magnetic, is linearly superposed

$$\frac{d\sigma_{unp}}{d\Omega}(\mathbf{Q}) = \frac{d\sigma_N}{d\Omega}(\mathbf{Q}) + \frac{d\sigma_M}{d\Omega}(\mathbf{Q}) \tag{2.21}$$

In both contributions two averages are involved: the average of the squared scattering function and the square of the average scattering function. For monodisperse and radially symmetric particles, the averages for nuclear scattering $\left\langle F_N^2(Q) \right\rangle \equiv \left\langle F_N(Q) \right\rangle^2$ are identical, so that

$$\frac{d\sigma_N}{d\Omega}(Q) = F_N^2(Q)S(Q) \tag{2.22}$$

To evaluate the averages for the magnetic scattering cross-section we have to consider the angular orientations of $\Delta\mathbf{M}$, parameterized by the angular alignment probability $p(\varphi_M, \Theta_M)$ of $\Delta\mathbf{M}$. Following up eq.(2.11), the averages over all $p(\varphi_M, \Theta_M)$ are than given by the general expressions

$$\left\langle F_M^2(Q) \right\rangle = V_P^2 f^2(Q,R) \int p(\varphi_M, \Theta_M) \left[D_M \mu_0 \mathbf{M}_\perp(\mathbf{Q}) \right]^2 d\varphi_M d\Theta_M \tag{2.23a}$$

$$\left\langle F_M(Q) \right\rangle^2 = \left[\int p(\varphi_M, \Theta_M) V_P f(Q,R) D_M \mu_0 \mathbf{M}_\perp(\mathbf{Q}) d\varphi_M d\Theta_M \right]^2 \tag{2.23b}$$

For two extremes, i.e. a demagnetised sample ($\Delta\mathbf{M}$ at random orientation) and a sample in magnetic saturation ($\Delta\mathbf{M}$ all parallel), the averages can readily be performed:

In the case of random orientation of $\Delta\mathbf{M}$ the square of the average formfactor $\left\langle F_M(\mathbf{Q}) \right\rangle^2$ is zero and

$$\frac{d\sigma_M}{d\Omega}(\mathbf{Q}) = \left\langle F_M^2(\mathbf{Q}) \right\rangle = \frac{2}{3}(D_M\mu_0\Delta M)^2 V_P^2 f^2(\mathbf{Q}) \tag{2.24}$$

which is independent of interparticle interference effects.

In the case of magnetic saturation, the averages $\left\langle F_M^2(\mathbf{Q}) \right\rangle \equiv \left\langle F_M(\mathbf{Q}) \right\rangle^2$ are identical, as for nuclear scattering, and we obtain

$$\frac{d\sigma_M}{d\Omega}(\mathbf{Q}) = (D_M\mu_0\Delta M)^2 V_P^2 f^2(\mathbf{Q}) \sin^2(\Psi) \tag{2.25}$$

where $\Psi = \angle(\mathbf{Q}, \Delta\mathbf{M})$, the angle between the direction of the magnetic contrast $\Delta\mathbf{M}$, and the scattering vector $\mathbf{Q}$, in practice the azimuthal angle on the two-dimensional SANS detector.

Averaging with regard to the azimuthal angle results in

$$\frac{d\sigma_M}{d\Omega}(\mathbf{Q}) = \frac{1}{2}(D_M \mu_0 \Delta M)^2 V_P^2 f^2(\mathbf{Q}) \tag{2.26}$$

which differs from eq.(2.24) only by the prefactor 1/2 instead of 2/3.

If the particles are not monodisperse in size, and/or not equal in shape, each size/shape class has to be considered individually. When interparticle interference can be neglected, the scattering contributions of the different classes are incoherently superposed. In the particular case of a finite size distribution of equally shaped particles, neglecting interparticle interference, the scattering cross section can be calculated as the integral over all individual contributions from the size interval between R and $R + dR$. For the nuclear scattering, this results in

$$\frac{d\sigma_N}{d\Omega}(\mathbf{Q}) = \Delta b_N^2 \int f^2(\mathbf{Q}) V_P^2(R) N(R) dR \tag{2.27}$$

with $N(R)\ V_p(R)\ dR$ being the incremental volume fraction with $V_p(R)$, the particle volume in the related size interval.

For the magnetic scattering of particles of finite size distribution and in magnetic saturation, the scattering cross section is given in analogy to eq.(2.27) by

$$\frac{d\sigma_M}{d\Omega}(\mathbf{Q}) = (D_M \mu_0 \Delta M)^2 \sin^2 \Psi \int f^2(\mathbf{Q}) V_P^2(R) N(R) dR \tag{2.28}$$

2.2.4.3. SANS from superparamagnetic particles

A further example where the average over all orientations of the magnetic contrast can be treated analytically is the case of superparamagnetic particles. Since the same model, in adapted versions, holds to explain many of the scattering patterns in the successional examples we will discuss this case here in some more detail.

The superparamagnetic state is characterized as an ensemble of non-interacting magnetic particles, the magnetic orientation of each particle being governed by the balance between the thermal energy $E_{th}=kT$ and the potential energy of a magnetic particle imposed by an external magnetic field. The potential energy of a homogenously magnetized, single domain particle of magnetic moment $\mathbf{\mu}_p=\mathbf{M}_p V_p$ in a magnetic field $\mathbf{H}$ is $E_{pot}= -\mu_0 \mathbf{H}\cdot\mathbf{M}_p V_p$. Since the relaxation time for reaching the thermodynamic equilibrium is short compared to the SANS measuring time, the scattering cross-section of an ensemble of superparamagnetic particles can be calculated as thermodynamic equilibrium state following the Langevin statistics, as outlined in the following.

To calculate the averages according to Eqs.(2.23a,b), the orientation probability of the magnetic contrast $\Delta\mathbf{M}$ and its magnitude needs to be known. In the case of a classical superparamagnet, where magnetic single domain particles are embedded in a non-magnetic matrix, the magnetic contrast is identical to the magnetisation of the particle $\Delta\mathbf{M}=\mathbf{M}_p$, and the orientation distribution of the magnetic moments are following the Boltzmann statistics. When a magnetic field $\mathbf{H}$ is applied, it forces the magnetic moments $\mathbf{\mu}_p=\mathbf{M}_p V_p$ of the particle (see Fig. 2-2a) to rotate towards the direction of $\mathbf{H}$. This tendency is opposed to disturbances by the thermal excitation. The probability p for a given orientation is then given by the Boltzmann factor

$$p = p_0 \exp(\mu_0 \mathbf{H} \cdot \mathbf{M}_P V_P / kT). \tag{2.29}$$

This model of a classical superparamagnet can be extended by allowing for a magnetic matrix surrounding the magnetic particle. For our treatment we assume a homogeneously magnetized matrix, i.e. a matrix in which the magnetisation is constant in magnitude and always parallel to the applied magnetic field $\mathbf{H}$, as illustrated in Fig. 2-2b. The magnetic property of the matrix enters the scattering problem at two points. Firstly, the scattering contrast is now defined as the vector difference between the particle and the matrix $\Delta \mathbf{M} = \mathbf{M}_P - \mathbf{M}_M$, and secondly, the magnetic matrix influences the potential energy of the particle and therefore the magnetic orientation distribution. The influence on the orientation distribution depends strongly on the magnetic coupling between matrix and particle. If we assume that there is only magneto-static coupling, and no exchange coupling between particle and matrix, the magnetization of the matrix simply amplifies the external magnetic field such that the Boltzmann factor has to be modified accordingly, and the orientation probability is given as

$$p = p_0 \exp(\mu_0 (\mathbf{H} + \mathbf{M}_M) \cdot \mathbf{M}_P V_P / kT). \tag{2.30}$$

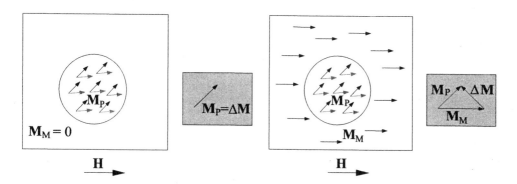

Figure 2-2: Schematic illustration of a superparamagnetic particle in an external external field $\mathbf{H}$, embedded in a nonmagnetic matrix (a) or in a homogeneous magnetic matrix with the magnetization direction aligned parallel to $\mathbf{H}$ (b). The corresponding magnetic scattering contrast vectors are illustrated in the shaded boxes.

Describing the orientation distribution of the magnetization of the particle $\mathbf{M}_P$ by the Boltzmann factor as in eq.(2.30), the averages of the formfactors $\langle F_i^2(Q) \rangle$ and $\langle F_j(Q) \rangle^2$ can be calculated. For this purpose it is convenient to define the scattering vector $\mathbf{Q}$ and the magnetization of the particle $\mathbf{M}_P$ in polar coordinates as illustrated in Fig. 2-3. Here, the $\mathbf{e}_x$-direction is assumed to be the direction of the incident neutron beam. The applied magnetic field $\mathbf{H}$ (and hence the magnetization of the matrix $\mathbf{M}_M$) are assumed to be parallel to $\mathbf{e}_z$. The 2-dimensional neutron detector is placed in the $\mathbf{e}_y \mathbf{e}_z$-plane. Then the relevant vectors are defined as

$$\mathbf{Q} = Q \begin{pmatrix} \cos \delta \\ \sin \delta \sin \Psi \\ \sin \delta \cos \Psi \end{pmatrix}, \quad \mathbf{M}_P = M_P \begin{pmatrix} \sin \Theta \cos \Phi \\ \cos \Theta \cos \Phi \\ \cos \Theta \end{pmatrix}, \quad \mathbf{M}_M = \begin{pmatrix} 0 \\ 0 \\ M_M \end{pmatrix}, \quad \mathbf{H} = \begin{pmatrix} 0 \\ 0 \\ H \end{pmatrix},$$

The averages of the formfactors for the different spin states (cf. eq.(2.19)) can be written as

$$\left\langle F_{\genfrac{}{}{0pt}{}{\pm\pm}{\pm\mp}}^{2}(\mathbf{Q})\right\rangle = \int_{0}^{2\pi}\int_{0}^{\pi}4\pi\sin\Theta\, p_{0}e^{\mu_{0}(H+M_{M})M_{P}V_{P}\cos\Theta/kT}\left(\Delta b_{\genfrac{}{}{0pt}{}{\pm\pm}{\pm\mp}}V_{P}f(\mathbf{Q})\right)^{2}d\Theta\, d\Phi \qquad (2.31a)$$

$$\left\langle F_{\genfrac{}{}{0pt}{}{\pm\pm}{\pm\mp}}(\mathbf{Q})\right\rangle^{2} = \left(\int_{0}^{2\pi}\int_{0}^{\pi}4\pi\sin\Theta\, p_{0}e^{\mu_{0}(H+M_{M})M_{P}V_{P}\cos\Theta/kT}\Delta b_{\genfrac{}{}{0pt}{}{\pm\pm}{\pm\mp}}V_{P}f(\mathbf{Q})d\Theta\, d\Phi\right)^{2} \qquad (2.31b)$$

The Boltzmann factor p is normalized to unity by choosing the normalization factor $p_{0}=\left[(4\pi)^{2}\sinh(\alpha)/\alpha\right]^{-1}$ with the α–parameter representing the ratio between magnetic (potential) and thermal energy:

$$\alpha = \mu_{0}(H+M_{M})M_{P}V_{P}/kT \qquad (2.32)$$

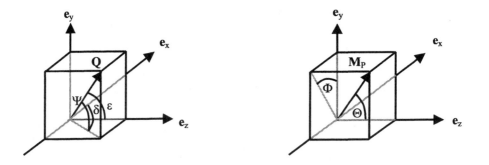

Figure 2-3: Definition of the scattering vector $\mathbf{Q}$ and the particle magnetization $\mathbf{M}_{p}$ in polar coordinates.

The averages can be calculated analytically: combining Eqs.(2.12,2.18-2.20,2.31) and introducing the classical Langevin-function $L(\alpha)=\coth\alpha-1/\alpha$, after some calculations one obtains the scattering cross-section of a superparamagnetic particle system in a homogeneous magnetic matrix described by the expressions

$$\frac{d\sigma_{\pm\pm}}{d\Omega}(\mathbf{Q}) = \left(F_{N}(\mathbf{Q})\mp\widetilde{F}_{M}(\mathbf{Q})\left[L(\alpha)-\gamma\right]^{2}\sin^{2}\varepsilon\right)^{2}S(\mathbf{Q})$$
$$+\widetilde{F}_{M}^{2}(\mathbf{Q})\left(\frac{L(\alpha)}{\alpha}\sin^{2}\varepsilon - \left(L^{2}(\alpha)-1+3\frac{L(\alpha)}{\alpha}\right)\sin^{4}\varepsilon\right) \qquad (2.33a)$$

$$\frac{d\sigma_{\pm\mp}}{d\Omega}(\mathbf{Q}) = \left(\sin^{2}\varepsilon-\sin^{4}\varepsilon\right)\left[L(\alpha)-\gamma\right]^{2}\widetilde{F}_{M}^{2}(\mathbf{Q})\,S(\mathbf{Q})$$
$$+\widetilde{F}_{M}^{2}(\mathbf{Q})\left(\left(\sin^{2}\varepsilon-\sin^{4}\varepsilon\right)\left(L^{2}(\alpha)-1+3\frac{L(\alpha)}{\alpha}\right)+\left(2-\sin^{2}\varepsilon\right)\frac{L(\alpha)}{\alpha}\right) \qquad (2.33b)$$

$$\frac{d\sigma_{unp}}{d\Omega}(\mathbf{Q}) = \frac{1}{2}\left(\frac{d\sigma_{++}}{d\Omega}(\mathbf{Q})+\frac{d\sigma_{--}}{d\Omega}(\mathbf{Q})+\frac{d\sigma_{+-}}{d\Omega}(\mathbf{Q})+\frac{d\sigma_{-+}}{d\Omega}(\mathbf{Q})\right)$$

$$= \left(\widetilde{F}_{M}^{2}(\mathbf{Q})\left[L(\alpha)-\gamma\right]^{2}\sin^{2}\varepsilon+F_{N}^{2}(\mathbf{Q})\right)S(\mathbf{Q}) \qquad (2.33c)$$

$$+\widetilde{F}_{M}^{2}(\mathbf{Q})\left(2\frac{L(\alpha)}{\alpha}-\left(L^{2}(\alpha)-1+3\frac{L(\alpha)}{\alpha}\right)\sin^{2}\varepsilon\right)$$

$\tilde{F}_M(\mathbf{Q}) = \mu_0 D_M M_P V_P f(\mathbf{Q})$ and $F_N(\mathbf{Q}) = \Delta b_N V_P f(\mathbf{Q})$ are the magnetic and nuclear scattering amplitudes, respectively. γ is defined as $\gamma = M_M / M_P$, and ε is the angle between $\mathbf{Q}$ and $\mathbf{e}_z$, which in practice is the same as Ψ ($\cos\varepsilon = \sin\delta\cos\Psi \approx \cos\Psi$ for $\delta \approx \pi/2$).

If all formfactors $f(\mathbf{Q})$ only depend on the modulus of the scattering vector, Q, the scattering cross-sections can be written in the form

$$\frac{d\sigma}{d\Omega}(Q) = A(Q) + B(Q)\sin^2\Psi + C(Q)\sin^4\Psi \tag{2.34}$$

whereby for unpolarized neutrons the last term vanishes, i.e. $C(Q)=0$. Eqs.(2.33) describe the transition of the magnetic scattering contribution from an anisotropic (ψ-dependent) to an istotropic (ψ-independent) scattering behavior when increasing the disorder of the magnetic moments. From Eqs.(2.33) one obtains immediately the limiting cases of saturation and complete disorder:

For large values of α, i.e. high magnetic fields and/or low temperatures such that all magnetic moments are uniformly aligned, we obtain $\lim_{\alpha\to\infty} L(\alpha)/\alpha = 0$ and $\lim_{\alpha\to\infty} L(\alpha) = 1$, and the isotropic magnetic scattering term $\tilde{F}_M^2(Q)2L(\alpha)/\alpha$ vanishes. Hence, the magnetic scattering gets fully anisotropic, only the nuclear scattering remains isotropic. For unpolarized neutrons the two remaining terms in eq.(2.34), $A(Q)$ and $B(Q)$, then are given by the expressions

$$\lim_{\alpha\to\infty} A_{unp}(Q) = |F_N(Q)|^2 S(Q)$$

$$\lim_{\alpha\to\infty} B_{unp}(Q) = D_M^2 \mu_0^2 \Delta M^2 V_P^2 f^2(Q) S(Q) \tag{2.35}$$

For small values of α, i.e. high temperatures and/or low magnetic fields, we get the other limiting case where $\lim_{\alpha\to 0} L(\alpha)/\alpha = 1/3$ and $\lim_{\alpha\to 0} L(\alpha) = 0$ and hence, the magnetic scattering contributes by 2/3 of its magnitude to the isotropic scattering:

$$\lim_{\alpha\to 0} A_{unp}(Q) = |F_N(Q)|^2 S(Q) + \frac{2}{3}\tilde{F}_M^2(Q). \tag{2.36}$$

For unpolarized neutrons and for random orientation of the magnetic moments of the particles the unisotropic term converges towards

$$\lim_{\alpha\to 0} B_{unp}(Q) = \gamma^2 \tilde{F}_M^2(Q) S(Q) = D_M^2 \mu_0^2 M_M^2 V_P^2 f^2(Q) S(Q) \tag{2.37}$$

This contribution only vanishes for random magnetic orientation of particles in a nonmagnetic (or paramagnetic) matrix ($M_M = 0 \Rightarrow \gamma = 0$), because only in that case the average magnetic contrast is zero, i.e. $<\Delta M> = 0$.

The scattering behavior of a system of superparamagnetic particles in many aspects is very typical for other systems of micromagnetism, as treated by the examples in Section 2.3, and can be a valuable basis for the interpretation of the observed scattering. Therefore, in preparation of Section 2.3 Fig.2-4 shows simulated examples of magnetic scattering patterns from superparamagnetic particles on a two-dimensional SANS detector. The simulations were made on the basis of eq.(2.33c), i.e. for the case of an unpolarized neutron beam. For the simulations nuclear scattering has been neglected. The latter, if present, would be isotropic and linearly superposed. Figure 2-4 distinguishes examples for magnetic particles in a nonmagnetic matrix, cf. Fig.2-2a ($\gamma=0$), and in a magnetic matrix, cf. Fig.2-2b, for the latter assuming in magnitude the same magnetization as for the particles ($\gamma=1$). Varying the Langevin parameter α from 0 to ∞ simulates the transition from fully random magnetic orientation of the particles to a strong field-parallel alignment.

For $\gamma=0$ (nonmagnetic matrix) the scattering patterns show the classical transition from being isotropic for $\alpha=0$ to passing a vertical (field-perpendicular) elliptical distortion for intermediate α (note that the hypothetical magnetic field is directing horizontally) and finally, for large α, showing the $\sin^2\psi$ behavior expected from eq.(2.25). On the other hand, in a magnetic matrix ($\gamma=1$) the patterns are quite different: Here, the resulting net magnetic contrast $\Delta M = M_p - M_M$, initially governed by the field-parallel matrix, develops a residual field-perpendicular component (cf. shaded inset in Fig. 2-2b). Consequently, the scattering patterns show a vertical $\sin^2\psi$ behavior for small α, passing nearly isotropic patterns for intermediate α, and finally converting into a horizontal (field-parallel) elliptical elongation. When approaching saturation ($\alpha \to \infty$) the magnetic contrast reduces to zero in this case and the magnetic scattering vanishes.

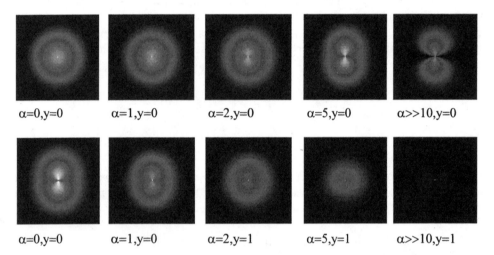

| $\alpha=0, y=0$ | $\alpha=1, y=0$ | $\alpha=2, y=0$ | $\alpha=5, y=0$ | $\alpha\gg10, y=0$ |

| $\alpha=0, y=0$ | $\alpha=1, y=0$ | $\alpha=2, y=1$ | $\alpha=5, y=1$ | $\alpha\gg10, y=1$ |

Figure 2-4: Simulated 2-dimensional magnetic SANS patterns of superparamagnetic particles in a non-magnetic matrix ($\gamma=0$) and in a magnetic matrix of the same magnetic moment in magnitude as that of the particles ($\gamma=1$), in both cases for an unpolarized incident neutron beam. Varying the Langevin parameter α from 0 to ∞ simulates the transition from fully random magnetic orientation of the particles to a strong field-parallel alignment (the hypothetic field is directing horizontally). *Also see the color plate.*

2.3. EXAMPLES

2.3.1. Ferrofluids

'Ferrofluids' or 'Magnetic Liquids' are two common expressions for suspensions in which nanometer sized particles are dispersed in a carrier liquid. Figure 2-5 gives a schematic illustration. The particle material can be ferri- or ferromagnetic, and is often coated with a stabilizing dispersing agent (surfactant). The particle size is smaller than the size of magnetic domains of the corresponding bulk material. This means that every particle is single-domain ferri- or ferromagnetic. The surfactant molecules consist of a head and a polar tail. The head can coalesce with the magnetic particle, and the tail protrudes into the carrier liquid and can dissolve in it. This polar tails protruding into the liquid lead to repulsion between the colloids. Different substances like organic acids or polymers usually serve as surfactants. The carrier liquid is selected to meet the requirements for a particular application: It can be hydrocarbon, ester, water or others. As a result of the small size of the particles they are very mobile by Brownian motion. The homopolar outer surfactant layer avoids an agglomeration of particles in the liquid. The surfactant must be matched to the carrier liquid and must overcome the attractive van-der-Waals and magnetic forces between the particles. Thereby a sedimentation or a de-mixing, induced by gravity or applied external fields, is prevented.

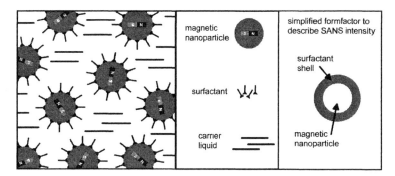

Figure 2-5: Schematic illustration of magnetic nanoparticles dispersed in a carrier liquid. The particles are coated with surfactant molecules. The polar tails of the surfactants protruding into the carrier liquid lead to a repulsion between the colloidal particles and avoid agglomeration. At the right, a simplified scheme to describe the SANS formfactor by means of a core-shell model is illustrated.

The (at a first glance) simple micromagnetic structure of ferrofluids makes them an ideal model system to illustrate the basics of magnetic Small Angle Neutron Scattering. In a dilute state ferrofluids behave like ideal superparamagnets. The scattering behavior of such systems, introduced in Paragraph 2.4.3, is straightforward. As a special feature of the magnetic particles coated by a layer of surfactant molecules, the formfactor for the nuclear scattering contribution needs to be described by a core-shell model [20]: The scattering amplitude for a spherical core-shell system of core radius R, shell thickness ΔR, and related scattering contrasts to the surrounding carrier liquid, $\Delta\eta_{core}$ and $\Delta\eta_{shell}$, respectively, is given by

$$F_N(Q,R,\Delta\eta_{core},\Delta R,\Delta\eta_{shell}) = \frac{4}{3}\pi R^3 \Delta\eta_{core}\ f(Q,R)$$
$$+ \frac{4}{3}\pi(R+\Delta R)^3(\Delta\eta_{shell} - \Delta\eta_{core})\ f(Q,R+\Delta R) \tag{2.38a}$$

Since only the core of the ferrofluid-particle is magnetic, the magnetic formfactor $\widetilde{F}_M(\mathbf{Q})$, referring to the definitions given for Eqs.(2.33a-c), is given by

$$\widetilde{F}_M(\mathbf{Q}) = \mu_0 D_M M_P \frac{4}{3}\pi R^3 f(QR) \tag{2.38b}$$

where $f(QR)$ is the formfactor of a sphere introduced by eq.(2.17). For ferrofluids the carrier liquid is nonmagnetic and hence $M_M = 0 \Rightarrow \gamma = 0$. In the case the applied magnetic field is large enough to bring the sample in magnetic saturation ($\alpha \to \infty$), Eqs.(2.33a-c) simplify to

$$\frac{d\sigma_{\pm\pm}}{d\Omega}(\mathbf{Q}) = \left(F_N(\mathbf{Q}) \mp \widetilde{F}_M(\mathbf{Q})\sin^2\varepsilon\right)^2 S(\mathbf{Q}) \tag{2.39a}$$

$$\frac{d\sigma_{\pm\mp}}{d\Omega}(\mathbf{Q}) = \left(\sin^2\varepsilon - \sin^4\varepsilon\right)\tilde{F}_M^2(\mathbf{Q})S(\mathbf{Q}) \tag{2.39b}$$

$$\frac{d\sigma_{unp}}{d\Omega}(\mathbf{Q}) = \frac{1}{2}\left(\frac{d\sigma_+}{d\Omega}(\mathbf{Q}) + \frac{d\sigma_-}{d\Omega}(\mathbf{Q})\right) = \left(\tilde{F}_M^2(\mathbf{Q})\sin^2\varepsilon + F_N^2(\mathbf{Q})\right)S(\mathbf{Q}) \tag{2.39c}$$

Since polarization analysis at SANS instruments is difficult and mostly not available, the favorably measured differential scattering cross-section is that of polarized incident neutrons without polarization analysis of the scattered neutrons. This obeys the equation

$$\frac{d\sigma_\pm}{d\Omega}(\mathbf{Q}) = \frac{d\sigma_{\pm\pm}}{d\Omega}(\mathbf{Q}) + \frac{d\sigma_{\pm\mp}}{d\Omega}(\mathbf{Q})$$

$$= \left(F_N^2(\mathbf{Q}) \mp 2\tilde{F}_M(\mathbf{Q})\sin^2\varepsilon + \tilde{F}_M^2(\mathbf{Q})\sin^2\varepsilon\right)S(\mathbf{Q}) \tag{2.39d}$$

Figure 2-6 illustrates the scattering length density profile for a cobalt ferrofluid suspended in a H_2O/D_2O solvent [22,23]. For the scattering intensity perpendicular to $\Delta\mathbf{M}$ (in this case perpendicular to $\mathbf{H}$) one has to consider that, depending on the polarization state of the incident neutrons, the magnetic scattering length density of the core has either to be subtracted from the nuclear scattering length density (neutrons polarized parallel to $\mathbf{H}$), or added to it (neutrons polarized anti-parallel to $\mathbf{H}$). By switching the polarization of the neutron spin state only the scattering length density of the core is changed. The experimental two-dimensional scattering patterns of such a system, measured with polarized incident neutrons in spin-up (+) and spin-down (-) state [23], are shown in Fig. 2-7. The experiments were done in magnetic saturation such that eq.(2.39d) holds. Aside of the experimental data, Fig. 2-7 also shows simulated patterns applying eq.(2.39d), based on the scattering contrast profiles shown in Fig. 2-6.

This example illustrates that, using polarized neutrons, an easy contrast variation of the inner core is possible. This generates two independent scattering data sets and thus improves the quality of the analysis and allows to extract specific details for the shell structure of the Co-colloids which otherwise are difficult or impossible to be obtained. By that Wiedenmann et al. [24] could show that the shell is impenetrable for the solvent molecules. Although this result in principle could also be obtained by using unpolarized neutrons, the fitting parameters then are highly correlated, and in practice the fitting results are not unambiguous [25].

Apart from the possibility of performing contrast variation experiments as in the above given example, the use of polarized neutrons allows experimentally the separation of scattering contributions from particles which have both, nuclear and magnetic contrast, from those which have only nuclear contrast. Only scattering contributions from those particles which show both, nuclear and magnetic scattering, depend on the polarization state of the neutrons. The contribution of those which show only nuclear scattering do not. The scattering cross section of the latter we denominate $d\sigma_2/d\Omega$. Then the spin-dependent scattering cross section of such a combined system is given by

$$\frac{d\sigma_\pm}{d\Omega}(\mathbf{Q}) = \left(F_N^2(\mathbf{Q}) \mp 2\tilde{F}_M(\mathbf{Q})F_N(\mathbf{Q})\sin^2\varepsilon + \tilde{F}_M^2(\mathbf{Q})\sin^2\varepsilon\right)S(\mathbf{Q}) + \frac{d\sigma_2}{d\Omega}(\mathbf{Q}) \tag{2.40}$$

The difference of the scattering intensity for the two polarization states is independent of $d\sigma_2/d\Omega$ and only depends on the magnetic-nuclear cross term $\tilde{F}_M(\mathbf{Q})F_N(\mathbf{Q})$:

$$\frac{d\sigma_-}{d\Omega}(\mathbf{Q}) - \frac{d\sigma_+}{d\Omega}(\mathbf{Q}) = 4\tilde{F}_M(\mathbf{Q})F_N(\mathbf{Q})\sin^2\varepsilon\, S(\mathbf{Q}) \tag{2.41}$$

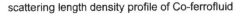

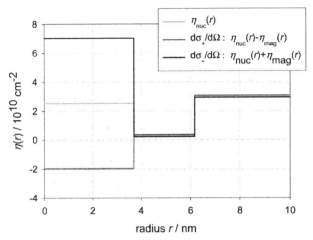

Figure 2-6: Scattering length density profile of a cobalt ferrofluid: The ferromagnetic Co-particles (core radius 3.7 nm) are coated with $C_{21}H_{39}$-N-O_3 surfactants (the tails extending to 6.2 nm) and dissolved in a H_2O/D_2O mixture. *Also see the color plate.*

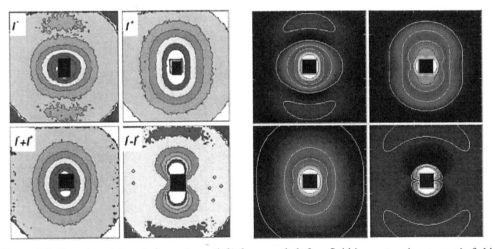

Figure 2-7: Experimental scattering patterns (left) from a cobalt ferrofluid in a saturating magnetic field (directing horizontally), coated with $C_{21}H_{39}$-N-O_3 surfactants and dissolved in a H_2O/D_2O mixture (measured at the V4 instrument of HMI, Berlin, courtesy of A. Wiedenmann). Simulated patterns (right) based on eq.(2.39d) with the scattering contrast profiles of Fig. 2-6. *Also see the color plate.*

For the case of a concentrated suspension, the Co ferrofluid contains not only magnetic core-shell particles, as in the above given example, but in addition two other species: non-magnetic free surfactant molecules and aggregates of magnetic Co-particles without a surfactant shell. Hoell et al. [26] could show that in this case the use of polarized neutrons was capable to separate all three scattering contributions. The scattering curves in Fig. 2-8a shows the magnetic-nuclear cross term I-I^+ (eq.(2.41)) which, as outlined above, necessarily must be attributed to the scattering of magnetic particles, i.e. Co-core-shell particles, and agglomerations of Co particles without a surfactant shell. The reconstruction of I^+ and I^- from this model (Fig. 2-8b) shows that there is an additional nonmagnetic scattering contribution in the experimental data at large Q-values, which consequently must attributed to a non-magnetic species. A quantitative analysis of this nonmagnetic scattering component yields information on size and shape giving evidence that it originates from free surfactants molecules. The related size distributions of all three species are plotted in Fig. 2-9, showing that they are well separated and distinguished.

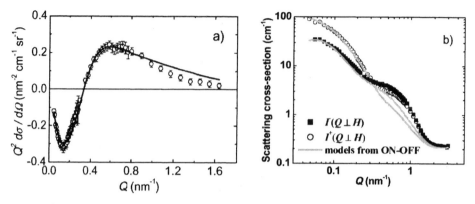

Figure 2-8: (a) Magnetic-nuclear cross term $I(Q) - I^+(Q)$ of a concentrated Co ferrofluid. The fitted curve shown in (*a*) requires two magnetic structures to be present, i.e. core-shell particles and aggregates of uncoated Co-particles. Using this fit model the reconstruction of I and I^+ yields the grey lines in (*b*). The squares and circles represent the measured scattering cross section data. Discrepancies between model and measurements obvious at larger Q-values, indicate the existence of additional nonmagnetic structures in the ferrofluid.

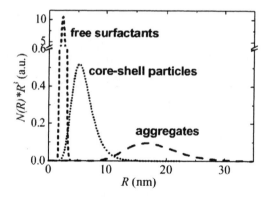

Figure 2-9: Volume-weighted size distributions of the three identified structural components in a concentrated Co-ferrofluid: free surfactants, core-shell particles with a magnetic Co-core, aggregates of magnetic Co-particles without a surfactant shells. For the core-shell composites the core radius is plotted.

2.3.2. Nanostructured ferromagnetic materials

Nanostructured or nanocrystalline materials are bulk-polycrystals, consisting of quite conventional substances like metals, alloys or ceramics, with the specific feature that the grain size is of the order of a few to some tens of nanometers. The overall properties of that type of material, such as electrical, mechanical and magnetic, are to a great extent determined by the presence and the nature of the large amount of grain boundaries or interfaces. As a second aspect, the grain size frequently is in the range of or even beyond the critical length scale for physical properties, like e.g. the domain wall width in magnetism, free path length in electronic motion etc., giving these materials the character of mesoscopic systems. As a consequence, the properties of these materials can be significantly altered compared to their coarse grained counterparts, making them interesting for potential novel applications in technology as well as in the field of basic science. For the latter, the high volume fraction of interfaces and the mesoscopic length scale allows to study specific phenomena where the influence of the one or the other dominates the properties. Numerous publications and conference proceedings document the attention and relevance meanwhile attributed to nanocrystalline materials [27-29].

Besides others, this particular morphology has multiple influences on the magnetic properties. Whereas isolated magnetic clusters above the onset of superparamagnetism are magnetically hard [30], nanostructured magnets consisting of consolidated grains often have superior soft magnetic properties [31]. Further, it was found that the coercivity in nanostructured materials can drastically change and even revert their trend by a variation of the grain size. This was observed for instance in nanostructured Fe where the coercivity shows a distinct maximum at grain sizes of about 30 nm, falling off sharply towards smaller grain sizes [32]. This behaviour is documented in Fig. 2-10. These results are based on magnetization measurements, which, as a gross-averaging technique, do not allow direct insight into the magnetization behaviour on the nanometer scale. It was evident, however, that the interplay of structure and magnetic correlations on the nanoscale play a key roll for the explanation of this behavior. Therefore the same system was studied extensively by SANS [33-36], with or without an external field applied in order to distinguish the spontaneous magnetic correlations and their response to an external field.

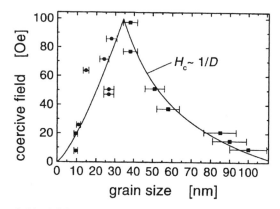

Figure 2-10: Coercive field of different samples of nanocrystalline Fe as a function of grain size, measured at a temperature of 300 K. The solid curve is primarily a guide to the eye visualizing the general trend of the data, but also represent a 1/D behaviour for grain sizes larger than 30 nm.

The samples used for this study were prepared by the inert-gas condensation technique (IGC), introduced by Gleiter [37] already in the mid-1980's: Hereby, the original material (99.99% pure Fe or Co) is evaporated in a high-vacuum system filled with typically 100 Pa pure He-gas. In the inert gas atmosphere the vapour condenses to nanometer sized clusters which are collected on a cold finger. The collected powder is finally consolidated by pressing at a pressure of 1 to 2 GPa at room temperature or at elevated temperatures up to 300°C, resulting in a disk-shaped, metallic shining sample of 100-300 µm thickness. This technique guaranties a close-to-random crystallographic orientation of the individual grains without pronounced texture or accumulation of small-angle grain boundaries. For grain sizes below 100 nm, the grains are single domain ferromagnets. One special feature of ICG samples is the relatively high portion of free volume or nanometer-sized porosity: After consolidation the density of the samples hardly exceeds 90% of the theoretical bulk density. In practice this insufficient densification cannot be avoided even not by hot-pressing without inducing considerable grain growth. When interpreting our scattering data we have to consider the possible influence of these microstructural features.

Figure 2-11 shows examples of two-dimensional intensity patterns from a nanocrystalline Fe sample, measured with unpolarized neutrons and thus linearly superposing the nuclear and the magnetic scattering (cf. eq.2.21). The measurements stem from the SANS instrument V4 at the Hahn-Meitner-Institute, Berlin. In zero field, both contributions add up isotropic. Hence, besides the isotropic grain orientation, in zero-field we have in average arbitrarily directing individual magnetic moments of the grains. In the (saturating) field of 10 kOe the scattering pattern shows a vertical (field-perpendicular) extension combined with a 'necking' in field direction. This can well be attributed to a $\sin^2\Psi$-dependence of the magnetic scattering (cf. eq.2.25), as expected for saturated magnetization of the individual grains. At the lower field of 1.5 kOe, the scattering pattern is found elongated in field direction, which at the first glance is unexpected in the simplified view of a $\sin^2\Psi$- dependence. However, as we will see further below, this observation can well be related to specific magnetic alignments of the individual grains.

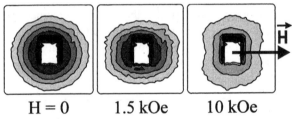

$$H = 0 \qquad 1.5 \text{ kOe} \qquad 10 \text{ kOe}$$

Figure 2-11: SANS iso-intensity contours on a two-dimensional detector, for a nanocrystalline Fe-sample (average grain size $\approx$20 nm) measured at zero field and in external magnetic fields as indicated, applied in horizontal direction (Q-range 0.02-0.16 nm^{-1}).

A further example from another Fe sample is shown in Fig. 2-12 in more detailed gradation of the magnetic field, increasing and decreasing the field following the course of one branch of a hysteresis loop. Qualitatively the same behavior is observed as that of the sample shown in Fig. 2-11, i.e. distortion parallel to the field of the intensity contours at intermediate fields, and the indication of a vertical $\sin^2\Psi$- dependence in strong fields. Same observations were made in other Fe samples and also in samples of different material (e.g. Co). It seems that the reaction of the micromagnetic correlations in nanostructured ferromagnetic materials to an external field is kind of universal, and the field-induced magnetic alignment is reversible when the field is reduced or switched off.

Before going for an explanation for the response of the two-dimensional SANS pattern to an external field, we focus to an answer for the observed grain-size dependence of the coercivity (Fig. 2-10) by a quantitative evaluation of the radial scattering cross section. Those are obtained by averaging the two-dimensional scattering patterns with regard to the azimuthal angle around the centre, regardless of the anisotropy of the scattering in non-zero fields. An example of the results for a nanostructured Fe-sample is shown in Fig. 2-13a, along with another equivalent example for a nanocrystalline Co sample (Fig. 2-13b).

For the range of larger Q-values ($Q > 0.2$ nm^{-1}) the scattering curves measured with and without magnetic field lie close together and are parallel. Here they obey dynamical scaling [38] and hence originate from the same structural objects, i.e. grains and/or pores. Note that (nonmagnetic) pores in a magnetic environment do also show magnetic scattering. In the high field, the intensity is lower due to the weighting factor of ½ for the magnetic scattering compared to 2/3 for random magnetization, c.f. Eqs.(2.24 and 2.26).

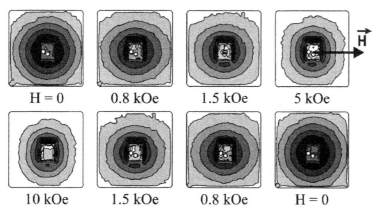

Figure 2-12: SANS iso-intensity contours on a two-dimensional detector, for a second nanostructured Fe sample (average grain size ≈12 nm) measured at zero-field and in external fields of different (increasing and decreasing) strengths applied in horizontal direction (Q-range 0.01-0.1 nm⁻¹).

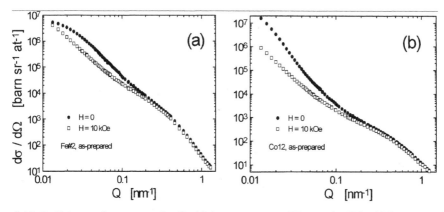

Figure 2-13: Radial scattering cross section for (a) the nanostructured Fe sample of Fig. 12 (average grain size ≈12 nm) and (b) a nanostructured Co sample (average grain size ≈8 nm), measured at H = 0 and H = 10 kOe.

In contrast, at smaller Q ($Q < 0.1$ nm⁻¹) a significant splitting between the scattering measured in zero field and in the strong field is observed (note the logarithmic scale!). This splitting obviously is due to an extra magnetic scattering in zero field (and at low fields). Hence, the external field has not just redirected initially randomly oriented magnetic moments of the grains, because in this case we had found parallel curves in Fig. 2-13 all across, i.e. curves obeying scaling over the entire covered Q-range. The Q-range of the extra magnetic scattering relates it to magnetic correlations in real space of spatial dimensions L_{mag} of around 100 to 200 nm (derived via the generic relation $L_{mag} = 2\pi/Q$). Hence, this scattering gives direct evidence for extended magnetic correlations in zero field (and low fields), exceeding the grain size, thus confining many grains to a correlated magnetic alignment. This is exactly the situation described by the so-called 'Random Anisotropy Model', RAM [39,40].

Figure 2-14 illustrates schematically the situation: The single-domain grains of random crystallographic orientation have random orientation of their magnetic anisotropies **K** as well, tied to specific crystalline directions. In the case of exchange coupling across the grain boundaries within a limited volume of dimension L_{mag}, the magnetic moments **M** of the individual grains align correlated within this volume, thus partly overcoming the anisotropy energy and inducing a reduced effective anisotropy K_{eff}. Then the response to an external field is governed by this reduced effective anisotropy, which can drastically reduce the coercivity and makes the material to appear magnetically soft. This is the observation of Fig. 2-10 for grain sizes below about 30 nm.

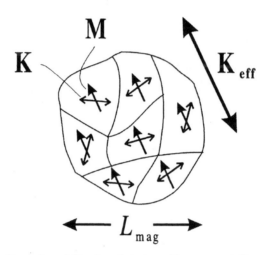

Figure 2-14: Schematic illustration of 'Random Anisotropy' in nanocrystalline materials: The single-domain grains have random orientation of their magnetic anisotropies (**K**), tied to the random crystallographic orientation. In the case of exchange coupling across the grain boundaries within a limited volume of dimension L_{mag}, the magnetic moments (**M**) align parallel, thus inducing a reduced effective anisotropy K_{eff}.

Exploiting the SANS data in a further step, a quantitative evaluation of the extra magnetic scattering was accomplished based on eq.(2.28), averaged with regard to the azimuthal angle ψ. The finite distribution in spatial dimension of the conglomerates of correlated grains was parameterized by a log-normal function. For details on the evaluation procedure we refer to [41]. Another approach for analysing similar data via the correlation of the spin misalignment have recently be published by Michels et al [42-44]. The fit of the theoretical scattering cross section to the data results in the characteristic length scale for the magnetic correlations, L_m (which is equivalent to the above defined L_{mag}). L_m is plotted in Fig. 2-15 for samples of different grain sizes. Shortest correlations are found for grain sizes around 30 nm. Comparing to Fig. 2-10, this is exactly the position of the highest coercivity. For larger grain sizes, L_m is found increasing approximately proportional to D, and the coercivity follows the proportionality $H_c \sim 1/D$. Such a behaviour was observed in 'classical' polycrystalline materials, e.g. Fe [45] and NiFe and CoFe alloys [46], and is referred to domain-wall pinning at grain boundaries [47].

The transition around 30 nm occurs in a range of grain sizes which correlates with the bulk domain-wall width. For coarse-grained Fe, this was calculated to 46 nm [32]. As outlined above, the strong magnetic softening occurring below this grain size is well explained by an effect of random anisotropy. This is underlined in Fig. 2-15 by the observation of a steep increase of the magnetic correlation length L_m towards smaller grain sizes for grain sizes below 30 nm. Even quantitatively, the L_m – data in this range are well described by the $1/D^3$ behaviour predicted by the RAM [32], although the limited number of data points would also allow for other functions to fit this region. No magnetic correlations of length scale L_m below a critical length L_{crit} of about

46 nm are found. In general, such direct evidence for random anisotropy can only be obtained from SANS as a tool of nanometer scale resolution of the magnetic structures.

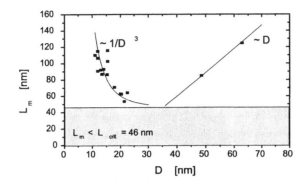

Figure 2-15: Average spatial magnetic correlation length L_m versus average grain size D, for nanostructured Fe. The solid line in the low grain size regime is primarily a guide to the eye, but also represents a fit proportional to $1 / D^3$ predicted by the RAM. No magnetic correlations with $L_m < L_{crit}$ are found.

Coming back to the two-dimensional SANS patterns (Figs. 2-11 and 2-12), the intensity contours at the intermediate fields around 1 to 2 kOe were found elongated *parallel* to the external field, before they merge into a field-perpendicular $\sin^2\Psi$- dependence. The intermediate field-parallel distortion reflects a net magnetic component of the correlated grains directing *perpendicular* to the field.

In order to achieve a plausible interpretation of this observation we attempt to model the nanostructured ferromagnetic samples and compute the expected two-dimensional scattering patterns for a given probability distribution of magnetic moments of single-domain grains. For this modeling we can closely refer to the modeling of superparamagnetic particles outlined in Paragraph 2.4.3: in analogy, we here consider the case of ferromagnetic domains confining several grains, embedded in a ferromagnetic matrix (the latter being represented by the surrounding grains or domains as a whole). In addition we consider the presence of nonmagnetic pores embedded in the same matrix. This situation is illustrated in Fig. 2-16.

The nuclear scattering can be computed straightforward, following eq.(2.27), although there is some uncertainty on the contrast between grains and surrounding grain boundaries. For the magnetic scattering, the probabilities of all possible orientations of ΔM needs to be known.

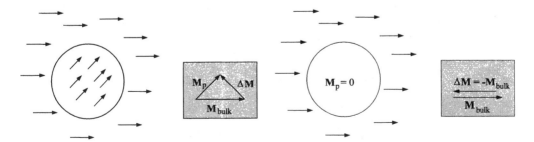

Figure 2-16: Schematic illustration of a ferromagnetic grain or domain (left) or a non-magnetic pore (right) embedded in a matrix of homogeneous magnetization.

As for the superparamagnetic particles we assume that here the orientation distribution of the magnetic domains is described by the Langevin statistics. The probability p for a given orientation then is given by the Boltzmann factor eq.(2.29). Now, for our specific case we further assume that the bulk magnetization is always parallel to an applied magnetic field for all field strengths, including small fields. This assumption seems justified in view of the low coercivity and high permeability of the material compared to the magnetic fields applied in our experiment. Again, the scattering contrast is the vector difference of the magnetization of the domain or pore and the matrix. The ferromagnetic matrix influences the orientation distribution of the magnetic moments. This influence strongly depends on the type of magnetic coupling between the nanocrystaline grains. Formally, the coupling can be expressed by an effective magnetization $\mathbf{M}_{eff}$ which amplifies the external magnetic field $\mathbf{H}$ such that the Boltzmann factor (eq. (2.29)) has to be modified to $p = p_0 \exp\left(\mu_0 \left(\mathbf{H} + \mathbf{M}_{eff}\right) \cdot \mathbf{M}_p V_P / kT\right)$.

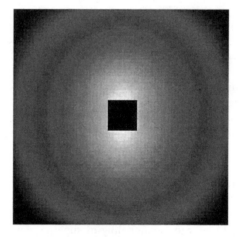

Figure 2-17: Computed scattering pattern (nuclear and magnetic scattering superposed) of a system of pores embedded in a homogeneously magnetized bulk material (magnetization directing horizontally). Nuclear and magnetic scattering contrast were assumed to be that of bulk Fe against vacuum.

With this expression for the orientation distribution one can calculate the scattering cross section $\frac{d\sigma}{d\Omega}(\varrho)$ following eq.(2.33c).

Figure 2-17 shows the computed SANS pattern of a modeled system of pores, the pores being surrounded by a homogeneously magnetized matrix with the magnetic moments aligned parallel to a (horizontally directing) external field. Referring to the example of nanostructured Fe, the contrast for nuclear scattering was taken to 9.45 fm ($9.45 \cdot 10^{-13}$ cm) and for magnetic scattering to 5.99 fm.

Figure 2-18 shows the computed scattering pattern of a magnetic domain (in the absence of pores) with an orientation distribution following the Langevin statistics, embedded in a homogeneously magnetized matrix parallel to the applied field (i.e. horizontally), for a stepwise increased Langevin parameter α. Further assumptions were:

- Equal modulus of magnetic moment per atom of the domain and the surrounding matrix.

- No nuclear contribution from the domains.

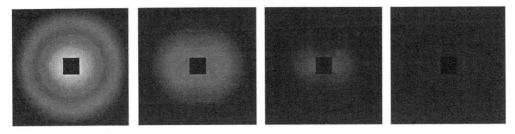

Figure 2-18: Computed scattering patterns of magnetic domains of orientation distribution following the Langevin statistic, for an increasing Langevin parameter α, embedded in a homogeneously magnetized matrix (magnetization directing horizontally). No pore scattering considered. *Also see the color plate.*

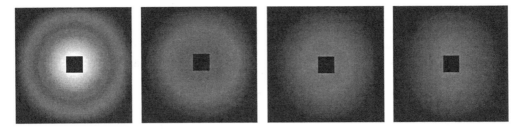

Figure 2-19: Computed scattering patterns of magnetic domains of orientation distribution following the Langevin statistic, for an increasing Langevin parameter α (as for the previous Figure), superposed by the nuclear and magnetic scattering of 5% volume fraction of pores, both embedded in a homogeneously magnetized matrix. Nuclear and magnetic scattering contrast of the pores were assumed to be that of bulk Fe against vacuum. *Also see the color plate.*

For Fig. 2-19, additionally the presence of 5% volume fraction of pores is considered.

Comparing the simulated patterns (Figs.2-17 – 2-19) with the measured ones (Figs.2-11 and 2-12), it is obvious that pores cannot solely be responsible for the observed scattering, since their scattering in a magnetic matrix shows only field-perpendicular, not field-parallel distortions, in contradiction to the observations made at intermediate fields. On the other hand, the domain scattering (Fig.2-18) does show field-parallel distortions, instead, for strong fields approaching saturation the magnetic scattering disappears completely, and hence would not generate a field-perpendicular $\sin^2\Psi$- dependence as observed. The conclusion is that most likely the superposition of both, pore and domain scattering (Fig.2-19), creates the observed sequences of scattering patterns.

Very similar studies as that reported here were conducted by Przenioslo et al. on nanocrystalline Ni and Co [48] and on nanocrystalline and amorphous chromium [49]. For these studies the materials were prepared by a special mode of pulsed electro deposition, which also is known to produce nanostructured grain morphologies. There are, however, some significant differences: in electrodeposited materials, the grain orientation is not necessarily at random but rather afflicted with a pronounced texture, the grain size distribution can be rather broad, and the samples have usually higher density than the cluster-consolidated ones, approaching the bulk density of the coarse-grained material. Therefore it can be expected that the grain boundary structure is different, which should also have influences on the magnetic properties.

Despite these differences in morphology, the authors make very similar observations as reported above for the IGC-material: The magnetic SANS contributions measured at zero field indicate that both, nanocrystalline Ni and Co of grain sizes between 10 and 60 nm, have magnetic domains with sizes exceeding 150 nm, i.e. the domains must confine several crystallite grains. As well, the field-induced distortions of the scattering pattern for Co are found very similar as in IGC-nanocrystalline materials: At zero field, the scattering is approximately isotropic. At small and intermediate fields (from ≈ 0.1 to 3 kOe) the patterns show an elliptical elongation parallel to the external field, which in higher fields (approaching 10 kOe) converts to a field-perpendicular elongation. The interpretation given by the authors is consistent with the one outlined in detail above.

2.3.3. Magnetic Nanostructure in Fe-Si-B-based alloys

The samples treated in subsection 2.3.2 where produced from nanometer-sized crystalline clusters by the IGC-technique. The route of synthesis followed here starts from metallic glasses which undergo a decomposition and nano-crystallisation upon annealing at temperatures near their glass transition temperature T_g. The example refers to nanocrystalline materials of the FeSiBNbCu system, well known for their soft magnetic properties, i.e. high permeability, low coercivity and low magnetostriction [31,50]. They are prepared from the melt spun amorphous alloy by thermal treatment: Annealing at 823 K for 1 h produces nano-sized α-Fe(Si) crystallites of about 5 nm in radius embedded in an amorphous matrix. The excellent soft magnetic properties of these materials have attracted much attention for structural and magnetic studies. It is known that the combined addition of copper and niobium to the amorphous Fe-Si-B alloy is crucial for the formation of the nanostructured morphology: Whereas the addition of Cu enhances the nucleation of α-Fe(Si) grains, the slow-diffusing Nb hinders a rapid growth.

The low coercivity of these alloys in the literature is frequently attributed to effects of random anisotropy RAM [39,40]. As outlined in Subsection 2.3.2, one of the basic assumptions for applying the RAM is the presence of a magnetic exchange coupling across the interface, here across the amorphous-crystalline interface. This was never directly proven, which leaves some scepticism on the interpretation. In order to clarify the real underlying physical reason for the soft magnetic properties of these materials, again the SANS technique was applied to characterise the magnetic microstructure and correlations in FeSiBNbCu on the nanometer scale.

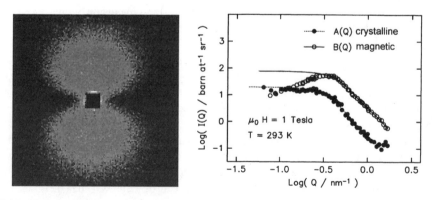

Figure 2-20: 2D intensity pattern (left) of nanacrystalline FeSiBNbCu measured at 300 K with unpolarized neutrons in a horizontal magnetic field of 1 Tesla which fully saturates the sample. Right: Nuclear and magnetic scattering contribution extracted from the pattern by applying eq.(2.34). *Also see the color plate.*

A scattering pattern from this material measured in an external magnetic field of 1 Tesla, which fully saturates the sample, is shown in Fig.2-20 (left). The pattern was measured at room temperature with unpolarized neutrons. For unpolarized neutrons and for a sample in magnetic saturation, the isotropic and anisotropic scattering contribution can separately be attributed to nuclear and magnetic scattering, respectively, and can be separated by fitting the intensity pattern with eq.(2.34) in the shortened version, i.e. $A(Q)+B(Q)\sin^2\Psi$. The results for $A(Q)$ and $B(Q)$ are plotted in Fig.2-20 (right). In contrast to the nuclear scattering, the magnetic scattering shows a maximum at about $Q = 0.4$ nm^{-1}. We will see further below that this can well be explained by different nuclear and magnetic formfactors of the nanocrystalline ferromagnetic precipitates.

We first focus to field dependent SANS measurements performed at different temperature in the regimes around the crystallisation temperature (T_x=750 K) and Curie temperatures of the nanocrystalline precipitates (T_c^{cr}~900 K) and the amorphous matrix phase (T_c^{am}~590 K in the as-prepared state), the latter being lower than the crystallisation temperature T_x. Fig.2-21a shows a schematic illustration of the temperature dependent magnetization of precipitates and matrix, and the magnetic contrast ΔM resulting from the superposition of both. At room temperature the material shows ferromagnetic behaviour in both phases, and the resulting contrast ΔM is relatively low. With increasing temperature the saturation magnetisation of the two phases reduces and ΔM increases until the amorphous phase becomes paramagnetic at $T = T^{am}$. At this point ΔM is highest. At still higher temperatures ($T_c^{am} < T < T_c^{cr}$) we observe ferromagnetic single-domain crystallites showing superparamagnetic properties. In this regime the contrast ΔM decreases towards T_c^{cr} and is zero above.

The measured integral magnetic scattering intensity is plotted in Fig.2-21b. The data show a maximum at a temperature of ~650 K, which, based on the considerations behind Fig.2-21a, is attributed to the Curie temperature of the amorphous phase. This value is about 60 K higher than the Curie temperature of the as-prepared, untreated amorphous ribbons, which is at 590 K as cited above. The deviation is attributed to a change in chemical composition of the matrix during the annealing.

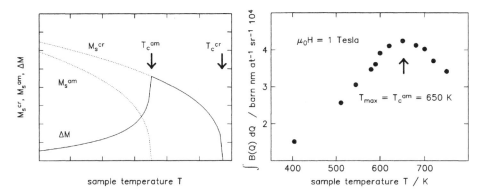

Figure 2-21: (a) Schematic illustration of the temperature dependent magnetization of the crystalline precipitates M_s^{cr} and the amorphous matrix M_s^{am}, and the magnetic contrast ΔM resulting from the superposition of both. The highest magnetic contrast is obtained at the Curie temperature of the magnetic matrix. (b) Integral anisotropic scattering intensity of FeSiBNbCu, as measured in a saturating magnetic field of 1 Tesla. The maximum at 650 K is attributed to the Curie temperature of the amorphous matrix phase.

For a general quantitative analysis of the temperature and field dependent scattering of this alloy, eq.(2.34) (in shortened version) applies as above, however, now with different definitions for the coefficients $A(Q)$ and $B(Q)$ in order to account for the specific conditions of the two-phase alloy when it is not in magnetic saturation: We consider ferromagnetic single-domain precipitates of magnetisation M_s^{cr} in a ferromagnetic amorphous matrix of magnetization M_s^{am}. Again, there are two scattering contributions of the precipitates: the nuclear scattering $F_N^2(Q)$ and the magnetic scattering $\tilde{F}_M^2(Q)$, the latter measuring fluctuations in both, amplitude and orientation, of the local magnetisation. The two coefficients of eq.(2.34) then become:

$$A(Q) = F_N^2(Q) + \tilde{F}_M^2(Q)\, 2\frac{L(\alpha)}{\alpha} \qquad (2.42a)$$

$$B(Q) = \tilde{F}_M^2(Q)\left\{[L(\alpha) - \gamma]^2 + \hat{L}(\alpha)\right\} \qquad (2.42b)$$

with the nominations as defined in Paragraph 2.4.3, adapted to the here discussed alloy: $\gamma = M_s^{am}/M_s^{cr}$, $\alpha = \mu_0(H + M_{eff})M_s^{cr} V_P / kT$ and $\tilde{F}_M(Q) = M_s^{cr} V_P f(Q)$, where M_{eff} is the effective magnetization of the combined system of matrix and precipitates. Further, the abbreviation $\hat{L}(\alpha) = 1 - L^2(\alpha) - 3L(\alpha)$ has been introduced. The terms with in Eqs.(2.42) describe the transition from anisotropic magnetic scattering for large values of α (high fields and low temperatures) to isotropic scattering when the values of α become small. For large α the alignment is perfect and the isotropic magnetic contribution $\tilde{F}_M^2(Q)2L(\alpha)/\alpha$ in $A(Q)$ vanishes; for small α the anisotropic magnetic term $B(Q)$ vanishes and the isotropic term approaches $F_N^2(Q) + 2/3\tilde{F}_M^2(Q)$.

The results on the integrated intensities $\int A(Q)dQ$ and $\int B(Q)dQ$ extracted from the SANS measurements are shown in Fig. 2-22. The solid lines are the results of fits of Eqs.(2.42) to the data.

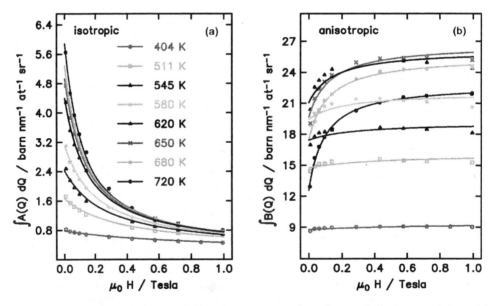

Figure 2-22: Experimental data on field and temperature dependence of the integrated intensities $\int A(Q)dQ$ (a) and $\int B(Q)dQ$ (b), fitted by Eqs.(2.42) (solid curves).

The temperature dependent fit parameters $\mu_0 M_{eff}(T)$ and $\gamma(T)$ are shown in Fig. 2-23. For the product $\mu_0 M_{eff}(T)$ the fit always yields values below 1 Tesla, although in the case of ferromagnetic exchange coupling between matrix and precipitates one would expect values of about 40-50 Tesla [51]. The much lower magnitude of $\mu_0 M_{eff}(T)$ derived from SANS implies that the ferromagnetic single domain α-Fe(Si) nanocrystals are not coupling to the ferromagnetic amorphous matrix by exchange interaction but by the much weaker magnetic dipole-dipole interaction. Hence, in this temperature range the random anisotropy model [39,40], although frequently propagated, does not apply to explain the soft magnetic properties of such materials at least at elevated temperatures.

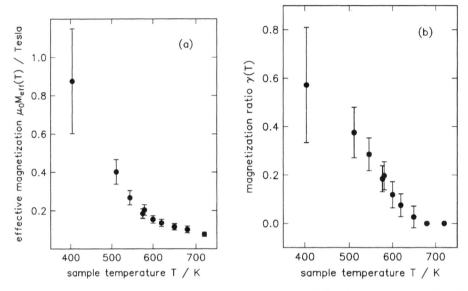

Figure 2-23: Effective magnetisation $\mu_0 M_{eff}$ and magnetisation ratio $\gamma(T) = M_s^{am}(T)/M_s^{cr}(T)$ resulting from a fit of Eqs.(2.42) to the experimental data. The error bars correspond to the 68.3% confidence interval.

The breakdown of exchange interaction between the two ferromagnetic phases suggests the presence of a nonmagnetic or paramagnetic interface between the nanocrystals and the matrix. In fact, in situ SAXS measurements gave evidence for a concentration pile-up of Nb around the α-Fe(Si) crystals during the growth of the nanocrystals, the latter being controlled by the Nb-diffusion. The Nb-concentration has a strong influence on the Curie temperature of the amorphous phase: For instance, it was found to induce a 40 K decrease for only 1 at.% increase in Nb concentration [52]. In Fig. 2-24 the proposed Nb concentration profile (c_{Nb}) is plotted schematically as a function of the distance x from the centre of the Fe_3Si precipitates, together with the local change in Curie temperature, calculated via $T_c(x) = 650K - \{40K(c_{Nb}(x) - c_M)\}$. .

To verify the existence of a layer of reduced magnetization around the precipitates, again the use of polarized neutrons was of great importance [22,53]. The proposed profile in Nb-concentration surrounding the precipitates will have an influence on both, the nuclear and magnetic scattering length densities at the interface. As already outlined in Subsection 2.3.1, the use of polarized neutrons in SANS allows to differentiate experimentally whether the nuclear and magnetic scattering stem from the same or from different structural features. This differentiation is achieved by analysing the difference intensities $I^- - I^+$, which depend only on the cross-term $4F_N^2(Q)\widetilde{F}_M^2(Q)$ [22,23]. From an experiment with polarized neutrons [53] the scattering cross sections I_{mag}, I_{nuc}, $I^+ (\mathbf{Q} \perp \mathbf{H})$ and $I^- (\mathbf{Q} \perp \mathbf{H})$ from an FeSiBNbCu alloy are obtained, as plotted in Fig. 2-25a. Both polarization states show a pronounced maximum at $Q \sim 0.4$ nm^{-1}. As the sample was prepared with a very low volume fraction of precipitated (~2%), interparticle interference effects which would lead to a similarly looking maximum in the scattering curve can be excluded, and the maximum in the scattering curve must be related to the formfactor itself. To calculate the formfactor analytically, the contrast profile in the diffusion zone around the precipitates was described by an exponential function [53], i.e.

$$\eta(r) = \begin{cases} \eta_c & \text{for } r \leq R \\ \eta_a - (\eta_a - \eta_i) \exp([R-r]/l) & \text{for } r < R \end{cases}$$

illustrated in Fig. 2-25b-c. Heinemann et al [53] could show by this model that the magnetic and nuclear SANS signal have the same origin and indeed yield a contrast profile as shown in Fig.2-25c.

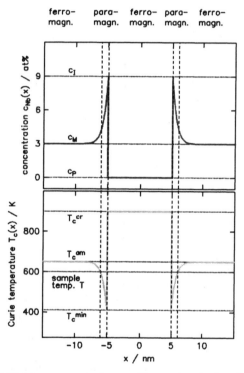

Figure 2-24: Proposed Niobium concentration profile crossing the interface between the matrix and a nanocrystalline precipitate (top) and its influence on the local Curie temperature $T_c(x)$. For temperatures $T_c^{min} < T < T_c^{am}$ both, matrix and precipitates, are ferromagnetic but separated by a thin paramagnetic layer.

In conclusion, it has been found that the low diffusivity of Nb results in a thin niobium enriched layer at the interface, which, combined with a reduction in Curie temperature, becomes paramagnetic and prevents exchange coupling between ferromagnetic matrix and ferromagnetic precipitates, at least at elevated temperatures. Thus, the precipitates are essentially uncoupled from the matrix and their magnetic orientation can easily follow an externally applied magnetic field. This is the physical reason for the soft magnetic properties rather than an effect of random anisotropy. The example again shows that, when using polarized neutrons, the simultaneous fit for both polarization states I^- and I^+ entails uncorrelated fit parameters, in the presented example a necessary prerequisite for a sound interpretation of the Nb-controlled mechanism of nanocrystallization in these amorphous alloys.

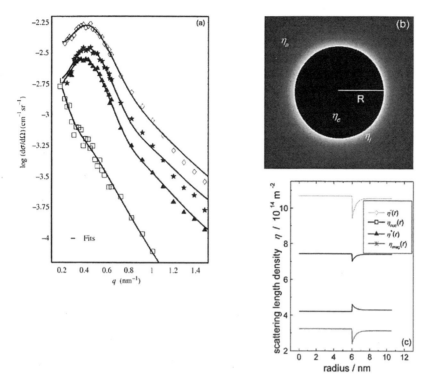

Figure 2-25. (a) SANS data from FeSiBNbCu, measured with polarized neutrons: nuclear scattering I_{nuc} (□), magnetic scattering I_{mag} (★), and scattering for parallel and antiparallel polarization, I^+(**Q**⊥**H**)(▲) and I^-(**Q**⊥**H**) (◊). (b) contrast illustration $\eta(r)$ of a spherical $Fe_{80}Si_{20}$ precipitate of radius R, and a diffusion zone around it, and (c) the related contrasts profiles for nuclear and magnetic scattering, $\eta_{nuc}(r)$ and $\eta_{mag}(r)$, and for the two polarized cases $\eta^+(r)$ and $\eta^-(r)$. *Also see the color plate.*

2.3.4. Micromagnetism in a two-phase CuNiFe alloy

The objective of this study was the investigation of the micromagnetic properties of an alloy in which nanometer sized ferromagnetic clusters are embedded in a nonmagnetic (or paramagnetic) crystalline bulk matrix. The medium chosen which provides such kind of structure is a two-phase alloy of the CuNiFe system. This system has a broad miscibility gap; upon annealing at temperatures within the miscibility gap the alloy decomposes, segregating precipitates of a ferromagnetic Fe/Ni-rich phase (α'-phase) embedded in a paramagnetic Cu-rich matrix (α-phase) [54]. As in the examples presented before, the study can be taken as a demonstration of the potential of magnetic SANS for characterizing complex magnetic microstructures by the use of theoretical and computational tools modelling the two-dimensional SANS patterns.

2.3.4.1. The material

Because preparation and treatment of the samples has an important influence on the thereby established microstructure, some details are given in the following: The alloy (nominal composition Cu-24at.%Ni-8at%Fe) was prepared by levitation melting in an induction furnace. The ingot was 50% cold-worked by pressing, solution annealed (1300 K for 2 h) and water quenched. The pressed ingot of about 3 mm in thickness was then cold rolled to foils of 1 mm in thickness in 5 consecutive rolling passes, and cut into pieces of 13 x 15 mm². As we will see further below, the rolling treatment is crucial for the specific precipitate arrangement.

After the rolling all samples were solution annealed at 1073 K for 2 h, followed by a water quench. For the samples of relevance for the here reported study, the final annealing, which generates the precipitation and coarsening of the Fe/Ni-rich α'-phase, was done in argon atmosphere at a temperature of 823 K, for periods between 16 and 1000 h. These annealing treatments yield precipitates of sizes ranging from about 5 to 40 nm in diameter, occupying a volume fraction of 22%. The related compositional data, determined from Field Ion Microscopy (FIM) atom probe analysis, are given in Table 1.

Table 1: Composition data of the equilibrium phases establishing in the CuNiFe alloy at 823 K, determined by selected area FIM atom probe analysis, together with the nominal composition of the homogeneous alloy.

	Cu (at.%)	Ni (at.%)	Fe (at.%)	C	others
nominal homogeneous	68.0	24.0	8.0		
ferromagnetic phase	6.7±0.8	58.5±2.4	34.8±1.8	<0.2	<0.2
paramagnetic matrix	85.5±2.9	13.6±0.3	0.9±0.3	0.5±0.3	<0.2

2.3.4.2. Microstructural Characteristics

The two-phase morphology of the CuNiFe alloy system was extensively studied in the past by means of Transmission Electron Microscopy (TEM), analytical Field Ion Microscopy and scattering techniques, mostly SANS and SAXS.

FIM-image recording during continuous field evaporation (the so called 'persistence technique' [55]) enables a three-dimensional reconstruction of the precipitate microstructure. This procedure revealed that in the respective alloy the α'-phase forms isolated precipitates, in the early stages approximately spherical in shape, with the tendency of having approximately equal distances to their neighbours [56]. The precipitates are coherent with the fcc Cu-rich matrix, i.e. within one grain (which is in the micrometer-size range) all precipitates have the same crystallographic structure and orientation. Indications for linear pile-ups in short rows aligning typically 4 to 5 precipitates were observed, preferentially with orientation along the <100> directions.

An early TEM study [57], characterizing the morphology on a somewhat coarser scale than FIM, suggests the existence of periodic concentration modulations along the <100> directions (so-called tweed structure). An extensive SAXS study on CuNiFe single crystals by Lyon and Simon [58,59] determined most detailed features of the anisotropic microstructure. Their scattering data, two-dimensional intensity maps, showed a fourfold symmetric peak structure. These were quantitatively explained in terms of a non-interacting pile-up model of alternating Cu- and Ni-Fe-rich platelets along <100> directions. 'Non-interacting' in this context means that the individual pile-ups are independent from each other.

In our alloy, although being polycrystalline, the fourfold symmetric peak structure reported by Lyon and Simon [58,59] is comparably observed as well: Fig. 2-26 shows a two-dimensional SANS pattern of the alloy measured at the Curie point, 550°C, where the magnetic scattering has vanished, leaving only nuclear scattering which represents the compositional microstructure. The SANS measurements were carried out at the SANS facility of SINQ at PSI [6].

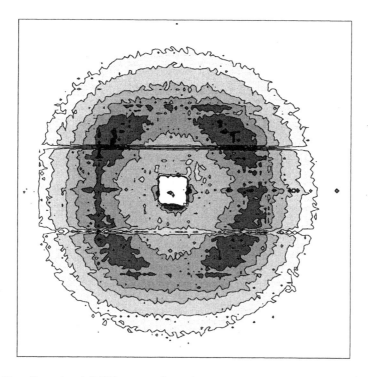

Figure 2-26: Two-dimensional SANS pattern from the Cu-24at.%Ni-8at.%Fe alloy after annealing at 823 K for 16 h (average precipitate size 9.6 nm). The pattern, recorded with the sample heated to 550°C which is at the Curie point, shows the pure nuclear scattering. The fourfold peak structure is caused by a preferential pile-up of the precipitates along the <100> directions in combination with a pronounced {100}<110> rolling texture (here the rolling direction is horizontal).

The preferential direction of these interference peaks towards the diagonal is related to the texture of the samples. Rolling the material is known to induce pronounced texture. We must assume that in our samples the shear component dominates: In fcc metals the shear component has a {100} plane parallel to the rolling plane and a <110> direction along the rolling direction [60]. Then the two other {100} planes necessarily are standing perpendicular to the rolling plane, and in 45° direction relative to the rolling direction (in Fig. 2-26, the rolling direction is horizontal). This combined with the preferential regular pile-up of the precipitates along the <100> directions, as observed by FIM and TEM, distinguishes the 45° direction to pin down a fourfold symmetric peak pattern, the same as observed by Lyon and Simon [58,59] in single crystals.

2.3.4.3. Magnetic SANS from CuNiFe

The data of Fig. 2-26 were taken at the Curie point where the magnetic scattering had vanished. Measuring at room temperature, the magnetic precipitate scattering is present and superposed to the nuclear scattering. From the comparison of nuclear and magnetic scattering contrast of the precipitates in the surrounding nonmagnetic matrix, i.e. $\Delta b^2_N = 0.0316$ barn/(sr at.), calculated from [1] for the compositional data given in Table 1, and $\Delta b^2_M = 0.127$ barn/(sr at.), derived from the FeNi-series data given in Ref. [3], we expect that the magnetic scattering is about three times more intense than the nuclear scattering and therefore dominating.

With an external field applied, here in horizontal direction perpendicular to the neutron beam, the response of the magnetic scattering allows to extract information on the field–induced rearrangement of the magnetic moments. This is shown by means of two examples in Figs. 2-27 and 2-28. Significant changes are observed in the intensity contours: At zero field, the scattering is fairly isotropic. The fourfold peak structure of the nuclear scattering can hardly be distinguished. With increasing field accretive distortions occur approaching the fourfold peak structure observed before. However, one should keep in mind that now it is the magnetic scattering which dominates the intensity, also at the peak positions.

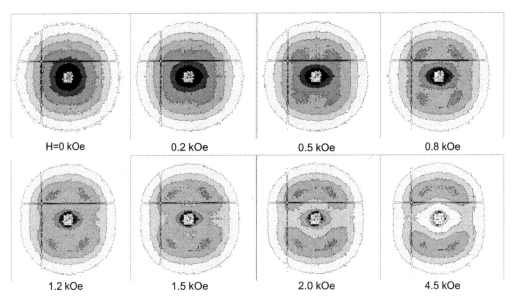

Figure 2-27: Sequence of two-dimensional SANS patterns, recorded at room temperature, from Cu-24at.%Ni-8at.%Fe after annealing 823K/16h. For the measurements an incrementally increased external magnetic field (as indicated) was applied in horizontal direction to the sample. Q-range: 0.1 nm^{-1} < Q < 0.8 nm^{-1}. Average precipitate diameter $2<R>_p$ =9.6 nm (note that the two crossing lines in the Figs. are artefacts originating from imperfections of the detector readout system).

Contrary to the nuclear scattering, the magnetic scattering depends on the vectorial alignment of the magnetic moments. Given the above described texture, and given that the <100> directions correlate with the axis of easy magnetisation, the magnetisation vectors of the precipitates may be trapped in <100> direction when an external field not too high in strength is applied. This explains the magnetic peak structure following the nuclear peaks when the rolling direction of the samples is parallel to the field direction as for Figs. 2-27 and 2-28.

Figure 2-29 shows the development of the radial scattering curves, averaged over 18°-wide sectors in the direction of the peaks (45° direction), with increasing external field, by means of the example of Fig. 2-28. The figure illustrates quantitatively the evolution of the interference peak out of the initially unstructured scattering at low Q.

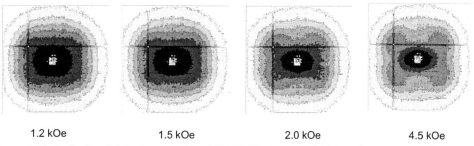

1.2 kOe   1.5 kOe   2.0 kOe   4.5 kOe

Figure 2-28: As for Fig.2-27, after annealing 823K/500h. Q-range: 0.025 nm^{-1} < Q < 0.2 nm^{-1}. 2<R>$_p$ =28 nm.

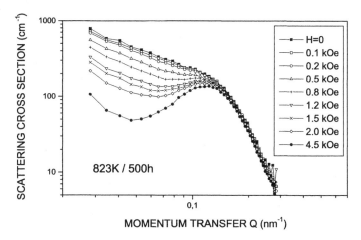

Figure 2-29: Variation of the scattering as a function of the external field of the Cu-24at.%Ni-8at.%Fe sample after annealing 823K/500h, for the sections in 45°-direction relative to the external field.

2.3.4.4. Modelling the 2D SANS pattern

In order to allow a more detailed interpretation of the two-dimensional SANS patterns and their response to the external field, it is necessary in the present case to consider the interparticle interference contribution in magnetic scattering and its dependence on the orientation distribution of the magnetic moments. For that we assume the case of ferromagnetic, single domain precipitates in a paramagnetic matrix, as treated in Paragraph 2.4.2. There, two extreme cases were already outlined explicitly: a demagnetised sample (**M** at random), described by eq.(2.24), and a sample in magnetic saturation (**M** always parallel **H**), described by eq.(2.25), both in combination with eq.(2.20). We here pick up these two cases and denominate them case i. and case ii., respectively.

We further consider two more cases which are specifically tailored to our alloy, referring to the <100> directions of easy magnetisation pointing towards 45° relative to the rolling direction (=field direction). Case iii.: the magnetisation vectors of the precipitates are trapped in one of the easy directions, but only in those which possess one component parallel to the **H**, i.e. at angles of ±45°. Case iv.: the magnetization vectors are trapped in the easy directions, but distributed randomly to all possible easy directions.

For case iii. the two averages of the formfactor in eq. (2.20) result in

$$\left\langle F_M^2(\mathbf{Q})\right\rangle \equiv 1/2\Delta b_M^2 V_P^2 f^2(Q) \text{ and } \left\langle F_M^2(\mathbf{Q})\right\rangle^2 \equiv 1/4\Delta b_M^2 V_P^2 f^2(Q)$$

such that the scattering cross section is given by

$$\frac{d\sigma_M}{d\Omega}(Q) = \frac{1}{4}\Delta b_M^2 V_P^2 f^2(Q)(S(Q)+1)$$

In case iv. the average magnetisation of the sample is zero and we get for the two averages of the formfactor $\left\langle F_M^2(\mathbf{Q})\right\rangle \equiv 1/2\Delta b_M^2 V_P^2 f^2(Q)$ and $\left\langle F_M(\mathbf{Q})\right\rangle^2 \equiv 0$, and hence for the scattering cross section

$$\frac{d\sigma_M}{d\Omega}(Q) = \frac{1}{2}\Delta b_M^2 V_P^2 f^2(Q)$$

which, as in case i, is independent of interparticle interference contributions and also independent of ψ.

Table 2 summarizes the results for cases i to iv. Now the variation of the magnetic scattering in the external field shown in Fig. 2-29 by means of the scattering in the direction of the most pronounced interference effect (45°) can readily be interpreted: In zero field (and small fields) the interference modulation of the magnetic scattering is absent, i.e. no intensity peak is developed (note that the nuclear scattering is by more than a factor of 3 lower than the magnetic scattering). Missing interference modulation is consistent only with the cases i. and iv., i.e. the magnetization directions of the precipitates are either at random, or randomly distributed to the easy axes. With increasing field the upcoming interference effect increasingly suppresses the scattering at low Q, thus developing the intensity peak, a clear indication for a correlated (or forced) magnetic alignment of the precipitates, as in cases ii. and iii. Such behaviour, expressed by the equations collected in Table 2, allows a very general conclusion, namely that the presence or absence of the interference modulation in the magnetic scattering of a dense array of magnetic particles is a clear indication for the presence or absence of a correlated alignment.

Figures 2-30 – 2-32 show computational results of 2D SANS patterns for the above presented cases, along with a scheme of squares representing the <100> aligned precipitates in 45°-directions. The spin directions lying in the easy axes are shown projected to that scheme. The origin of the fourfold peak structure in the nuclear SANS pattern (Fig. 2-26), i.e a preferential pile-up of the precipitates along the <100> directions in combination with a pronounced {100}<110> rolling texture, was the basis for the structural assumptions made for the modelling. A 45° alignment or pile-up of the precipitates, as illustrated in the simulation scheme (cf. Figs. 2-30 to 2-32), was indeed necessary to successfully simulate the observed patterns.

Figure 2-30 presents two possible configurations of demagnetization, with the magnetic moments of the precipitates trapped in the easy axes (case iv.): in one configuration the easy axes are occupied at random (Fig. 2-30a), in the other one they are occupied confined to domains of uniform alignment (Fig. 2-30b). The corresponding computed scattering patterns are quite different: whereas for the random case the pattern is fairly isotropic, for the domain case the fourfold peak pattern arises.

Figure 2-31 represents case iii. in two different configurations, one with the magnetization vectors independently in the two possible easy magnetization directions with a field-parallel component, the other one grouped in domains of common magnetization direction in the easy axes with field-parallel component. Again, for both configurations the fourfold peak pattern shows up, very similar to Fig. 2-30b. The difference for the two configurations of Fig. 2-31 is found most pronounced in the centre of the scattering patterns: For the magnetization vectors being distributed independently to the two possible easy axes the computed scattering pattern shows a pair of 'wings' in the centre, stretched parallel to the net magnetization direction. For the confinement in domains, no such wings show up and the scattering in the centre is essentially flat.

Table 2: Illustration of the different magnetization cases considered, along with the two related averages of the magnetic scattering function (the hypothetic magnetic field is directing horizontally to the right).

	Case i. random orientation	$\left\langle F_M^2(\mathbf{Q}) \right\rangle \equiv 2/3 \Delta b_M^2 V_P^2 f^2(Q)$ $\left\langle F_M(\mathbf{Q}) \right\rangle^2 = 0$
	Case ii. Magnetic saturation	$\left\langle F_M^2(\mathbf{Q}) \right\rangle = \Delta b_M^2 V_P^2 f^2(Q) \sin^2 \Psi$ $\left\langle F_M(\mathbf{Q}) \right\rangle^2 = \Delta b_M^2 V_P^2 f^2(Q) \sin^2 \Psi$
	Case iii. Trapped in easy directions $\pm 45°$ to the field	$\left\langle F_M^2(\mathbf{Q}) \right\rangle \equiv 1/2 \Delta b_M^2 V_P^2 f^2(Q)$ $\left\langle F_M(\mathbf{Q}) \right\rangle^2 \equiv 1/4 \Delta b_M^2 V_P^2 f^2(Q)$
	Case iv. Trapped in easy directions at random, i.e. aligned $\pm 45°$ and $180°\pm 45°$ to the field	$\left\langle F_M^2(\mathbf{Q}) \right\rangle \equiv 1/2 \Delta b_M^2 V_P^2 f^2(Q)$ $\left\langle F_M(\mathbf{Q}) \right\rangle^2 = 0$

Figure 2-32 finally presents case ii., i.e. magnetic saturation with all magnetization vectors forced in field direction. The simulation still yields a fourfold peak structure, although forced narrower towards the 90° direction. A $\sin^2 \Psi$-contribution is indicated in the centre of the simulated pattern by means of a pair of 'wings', now stretched perpendicular to the magnetization direction.

Viewing the 2D simulated SANS patterns, only the random case of Fig. 2-30, not the domain case, yields an isotropic pattern. Isotropic or fairly isotropic patterns were observed from all our samples in zero field or in low fields (cf. Figs. 2-27 and 2-28), and reversibly after switching off a high field. This observation clearly favours the configuration of Fig. 2-30a (case iv.) as the one establishing in zero field, regardless of the magnetic history of the material (the latter refers to relaxation times shorter than the turnaround time of the measurements, which was typically some minutes). For completeness, it should be mentioned that case i. (random magnetization) also yields an isotropic scattering pattern. The further discussion, however, favouring the easy axes in the 45° directions, makes case i. unlikely to occur in that alloy.

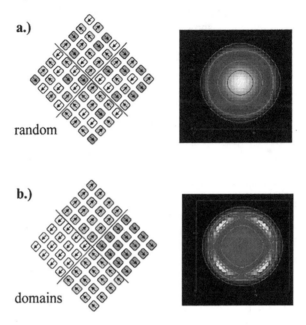

Figure 2-30: Computational results of 2D SANS pattern for possible spin configurations in the Cu-24at.%Ni-8at.%Fe alloy, here fore the case of a demagnetized sample in two different configurations, the upper one representing case iv. The modelled spin configurations are illustrated by a projected scheme (left) that represents the <100> aligned precipitates in 45°-directions. *Also see the color plate.*

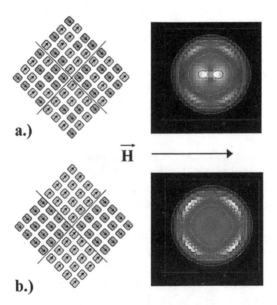

Figure 2-31: As for Fig. 2-30, for case iii. in two different configurations. *Also see the color plate.*

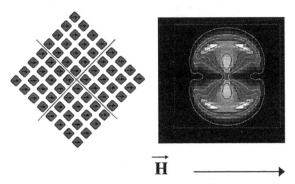

$\overrightarrow{\mathbf{H}} \longrightarrow$

Figure 2-32: As for Fig. 2-30, for case ii. (magnetic saturation). *Also see the color plate.*

Comparing the computed pattern for magnetic saturation (Fig. 2-32) with the experimental scattering patterns in Figs. 2-27 and 2-28, those which come closest to this result are the ones of Fig. 2-27 (823K/16h) for H ≥ 2.0 kOe, and the one of Fig. 2-28 (823K/500h) for H = 4.5 kOe: Those exhibit the fourfold peak structure merging into the 90° direction. However, the vertically stretched wings in the centre are not found in either pattern. Rather, for smaller fields (up to 2 kOe) in all three samples shown in Figs. 2-27 and 2-28 a horizontally elongated scattering is observed. Viewing our simulations, such type of pattern is found for case iii., Fig. 2-31a, with the precipitates being independently aligned in one of the two easy axes with field-parallel component. The alignment in domains (Fig. 2-31b) would cause the central scattering to vanish, according to the simulations, by keeping the fourfold peak structure unchanged. Actually, this type of pattern has not been observed in any of the samples measured in the reported study.

In conclusion, the comparison of the experimental data with the computed 2D scattering patterns suggests the following magnetic configurations of the precipitates: In zero field the magnetization vectors of the (single domain) precipitates are in the easy (<100>) axes, independent from each other and randomly distributed. The same configuration restores after switching off a strong field (at least after some minutes of relaxation). When an external field is applied, for intermediate fields (typically 0.5 to 2.0 kOe) the magnetic moments are switched into those easy axes which have a field-parallel component (situation of Fig. 2-31a). For stronger fields (typically 4.5 kOe and above), the magnetization directions are progressively forced into the field-direction, approaching saturation (Fig. 2-32), although full saturation may not have been reached, rather a state between that of Fig. 2-31a and Fig. 2-32.

The behaviour in general is qualitatively the same for the range of precipitate sizes investigated ($2<R>_p$ ranging from 9.6 to 38 nm), only the field dependence is slightly shifted: For instance, the configuration of Fig. 2-28 ($2<R>_p$ =28 nm) at 2.0 kOe corresponds approximately to that of Fig. 2-27 ($2<R>_p$ =9.6 nm) at about 1 kOe. Over the size range investigated the external field needed to reach a comparable spin configuration was found approximately proportional to the average precipitate size.

2.4. SUMMARY AND CONCLUSIONS

Neutron scattering is one of a series of modern, powerful techniques with the ability to characterize magnetic structures and correlations at a spatial resolution down to the atomic scale. Focussing to magnetic SANS, the key applications are probing magnetic structures on the nanometer scale, like magnetic clusters embedded in different media, solid or liquid, magnetic or nonmagnetic. The topics of interest span from basic scientific questions, understanding the micromagnetic physical origin for macromagnetic observations, to technical applications and developments like high-density magnetic recording media.

Compared to the nuclear scattering of neutrons, which in complement allows characterizing compositional structures, magnetic scattering is a vectorial interaction, i.e. it depends on the direction and magnitude of the (local) magnetization vector M of the structures to be analysed relative to the scattering vector Q. Thanks to this vectorial interaction the local alignment of magnetic structures can by analysed. This benefit, however, goes along with a more complicated theoretical background for describing magnetic scattering and for analysing scattering patterns.

The general theoretical introduction in section 2.2 aims to lay the basis for a quantitative description of magnetic SANS data, including polarized neutron scattering where the neutron spin has a defined directional relation to the magnetization vector and to any preferred direction in the sample. In order to illustrate the application of the basic equations to a real case in a general sense, a system of superparamagnetic particles was chosen. Such system can be considered a universal working model for a magnetic orientation distribution in a thermally activated environment, following the Langevin statistics. The results in many aspects are very typical for other systems of micromagnetism.

Following this general example, four more cases are reviewed and discussed to illustrate the potential of magnetic SANS applications to systems of specific properties:

(i)   Ferrofluids showing orientational distributions following the Langevin statistics, here with the special feature that the magnetic particles in their carrier liquid are stabilized by surfactant layers, which requires a core-shell model to compute the scattering formfactor. The example further illustrates the usefulness of polarized neutrons to distinguish magnetic and non-magnetic species.

(ii)   Nanostructured ferromagnetic materials synthesized by cluster consolidation. These materials show effects of random anisotropy. The direct evidence for that is obtained from magnetic SANS, allowing a quantitative analysis of the magnetic correlation of the ferromagnetic grains being confined to small domains, thus overcoming the magnetocrystalline anisotropy. A comparison of the two-dimensional SANS patterns with computed patterns based on a model of correlated grains in a homogeneous ferromagnetic matrix allows a well-founded understanding of the reaction of the micromagnetism on an externally applied field.

(iii)   Ferromagnetic crystalline precipitates embedded in a ferromagnetic amorphous matrix in the alloy system FeSiBNbCu, a candidate material with superior soft magnetic properties. Here magnetic SANS gives evidence for a paramagnetic Nb-rich shell establishing around the ferromagnetic precipitates, thus prohibiting fast coarsening and at the same time inhibiting exchange coupling between the precipitates and the matrix. This finding constitutes the physical origin for the soft-magnetic properties, by that giving a guideline for dedicated developments of such materials for technological applications. In this example, again, the use of polarized neutrons allows to distinguish microstructural and micromagnetic details which otherwise are not accessible.

(iv)   A two-phase alloy forming ferromagnetic precipitates embedded in a paramagnetic matrix. This gives an example for a rather complicated interplay of precipitate microstructure and magnetic orientation distribution, and the response of those to an external field. The influence of interference effects markedly stamps the scattering patterns, nuclear and magnetic. Here the usefulness of computational reconstruction of the two-dimensioal SANS patterns based on a microstructural model is essential. It most evidently illustrates the importance of the application of computational tools for a sound standing interpretation of scattering data in complicated systems.

REFERENCES

[1]   V. F. Sears, Neutron News **3** (1992) 26, L. Koester, H. Rauch, M. Herkens, K. Schröder, Report of the Research Centre Jülich (FZJ), Germany, ISSN 0366-0885, IAEA-Contract 2517/RB (1981), G.E.Bacon, *Neutron Diffraction*, Clarendon Press, Oxford (1975).

[2]   CRC Handbook of Chemistry and Physics, ed. R. C. Weast, CRC Press, Florida, E-107.

[3]   C.G. Shull and M.K. Wilkinson, *Phys. Rev.* **97** (1955) 304.

[4]   K. Ibel, *J. Appl. Cryst.* **9** (1976) 296.

[5]   H. R. Child, S. Spooner, *J. Appl. Cryst.* **13** (1980) 259.

[6]   J. Kohlbrecher and W. Wagner, *J. Appl. Cryst* **33** (2000) 804.

[7]   T. Keller, T. Krist, A. Danzig, U. Keiderling, F. Mezei, A. Wiedenmann, *Nuclear Instruments and Methods in Physics Research A* **451** (2000) 474.

[8]   J. Zhao, W. Meerwinck, T. Niinikoski, A. Rijllart, M. Schmitt, R. Willumeit and H. Stuhrmann, *Nuclear Instruments and Methods in Physics Research A* **356** (1995) 133.

[9]   H. Glättli, M. Eisenkrämer, M. Pinot and C. Fermon, Physica B **213-214** (1995) 887.

[10]   P. Thiyagarajan, J. E. Epperson, R. K. Crawford, J. M. Carpenter, T. E. Klippert and D. G. Wozniak, *J. Appl. Cryst.* **30** (1997) 280.

[11]   O. Hittmair, *Lehrbuch der Quantentheorie*, Karl Thiemig publishing (1972).

[12]   J. Byrne, *Neutrons*, Institute of Physics Publishing, Bristol (1994), Chap. 2.

[13]   E. Fermi, *Ric. Sci.* **7** (1936) 13.

[14]   G. Kostorz and S.W. Lovesey, in *Treatise on Materials Science and Technology* (Vol. **15**: Neutron Scattering), edited by G. Kostorz, Academic Press, New York, (1979), p. 22.

[15]   J. Rossat-Mignot, *Neutron Scattering*, Vol **23**, Part C (1997) Academic Press.

[16]   G.L. Squires, *Introduction to the Theory of Thermal Neutron Scattering*, Cambridge University Press (1978).

[17]   I.I. Gurevich and L.V. Tarasov, *Low-Energy Neutron Physics*, North Holland Publishing Company, Amsterdam (1968).

[18]   W. Schmatz, in *Treatise on Materials Science and Technology* (Vol. 2), edited by H. Herman Academic Press, New York (1973), 128.

[19]   R.M. Moon, T. Riste, W.C. Koehler, *Phys. Rev.* **181** (1969) 920.

[20]   J.S. Pedersen, *Advances in Colloid and Interface Science* **70** (1997) 171.

[21]   O. M. Lord Rayleigh, *Proc. R. Soc. London*, A **84** (1911) 25.

[22]   A. Wiedenmann, *J. Appl. Cryst.* **33** (2000) 428.

[23]   A. Wiedenmann, *Physica B* **297** (2001) 226.

[24]   A. Wiedenmann, A. Hoell, M. Kammel, *J. Magn. Magn. Mater.* **252** (2002) 83.

[25]   A. Heinemann, A. Wiedenmann, *J. Appl. Cryst.* **36** (2003) 845.

[26]   A. Hoell, M. Kammel, A. Heinemann, A. Wiedemann, *J. Appl. Cryst.* **36** (2003) 558.

[27]   J. R. Weertman, *Mechanical Behaviour of Nanocrystalline Metals in Nanostructured Materials: Processing, Properties, and Potential Applications.* (William Andrew Publishing, Norwich, 2002).

[28]   Papers in special issue dedicated to Prof. H. Gleiter of *Zeitschrift für Metallkunde*, 10 (2003).

[29]   H. Van Swygenhoven, *Science* **296,** 66 (2002).

[30]   W. Wernsdorfer, E.B. Orozco, K. Hasselbach, A. Benoit, B. Barbara, N. Demoncy, A. Loiseau, H. Pascard and D. Mailly, *Phys. Rev. Lett.* **78** (1997) 1791.

[31]   Y. Yoshizava, S. Oguma and K. Yamauchi, *J. Appl. Phys.* **64** (1988) 6044.

[32]   J.F. Löffler, J.P. Meier, B. Doudin, J.-P. Ansermet, W. Wagner, *Phys. Rev. B* **57** (1998) 2915.

[33]   J. F. Löffler, W. Wagner, H. van Swygenhoven, A. Wiedenmann, J. *Nanostructured Materials* **9** (1997) 331.

[34]   J.F. Löffler, H.B. Braun, and W. Wagner, *Phys. Rev. Lett.* **85** (2000) 1990.

[35]   J. F. Löffler, H. B. Braun, W. Wagner, G. Kostorz, A. Wiedenmann, *Mater. Sci. Eng. A* **14195,** (2000) 1.

[36]   W. Wagner, J. Kohlbrecher, J. F. Löffler, *Materials Week*, www.materialsweek.org/proceedings, mw2000_433.pdf (2001).

[37]   H. Gleiter, *Progress in Material Science* **33** (1989) 223.

[38]   P. Fratzl, J. L. Lebowitz, J. Marro, and M. H. Kalos, *Acta Metall* **31** (1983) 1849.

[39]   B. Hofmann, T. Reininger, H. Kronmüller, *Phys. Stat. Sol.* (a) **134** (1992) 247.

[40]   A. Hernando, M. Vazquez, T. Kulik, C. Prados, *Phys. Rev. B* **51** (1995) 3581.

[41]   W. Wagner, A. Wiedenmann, W. Petry, A. Geibel and H. Gleiter, *J. Mater. Res.* **6** (1991) 2305.

[42]   J.A. Michels et al, *Phys. Status Solidi A* **189,** (2002) 509.

[43]   J. Weissmüller, D. Michels, A. Michels, A. Wiedenmann, C.E. Krill, H.M. Sauer and R. Birringer *Scripta Materialia* **44,** (2001) 2357.

[44]   A. Michels, R. N. Viswanath, J. G. Barker, R. Birringer, and J. Weissmüller, *Phys. Rev. Lett.* **91,** No 267204 (2003).

[45]   J. Degauque, B. Astié, J. L. Porteseil and R. Vergne, *J. Magn. Magn. Mater.* **26** (1982) 261.

[46]   F. Pfeifer and C. Radeloff, *J. Magn. Magn. Mater.* **19** (1980) 190.

[47]   A. Mager, *Ann. Phys.* (Leipzig) **11** (1952) 15.

[48]   R. Przenioslo, R. Winter, H. Natter, M. Schmelzer, R. Hempelmann, W. Wagner, *Phys. Rev. B* **63** (2001) 054408.

[49]   R. Przenioslo, J. Wagner, H. Natter, R. Hempelmann, W. Wagner, *Journal of Alloys and Compounds* **328** (2001) 259.

[50]   G. Herzer, *IEEE Trans. Mag.* **25** (1989) 3327.

[51]   J. Kohlbrecher, A. Wiedenmann, H. Wollenberger, *Z. Phys. B* **104** (1997) 1

[52]   A. R. Yavari, G. Fish, S. K. Das, and L. A. Davis, Mat. Sci. Eng. **A182/A183,** 1415 (1994), A. R. Yavari and O. Drbohlav, *Mat. Trans. JIM* **36** (1995) 896.

[53]   A. Heinemann, H. Hermann, A. Wiedenmann, N. Mattern and K. Wetzig, *J. Appl. Cryst.* **33** (2000) 1386.

[54]   Y.-Y Chuang, R. Schmid, and Y. Austin-Chang, *Acta metall.* **33** (1985) 1369.

[55]   R. Wagner in *Crystals*, Vol. 6, Springer Verlag, Berlin (1982).

[56]   W. Wagner, R. Lang, H. Wollenberger, and W. Petry, in *Phase Transformation ‹87*, Ed. G.W.Lorimer, The Institute of Metals, 566 (1988); W. Wagner, *Mat. Sci. Forum* **27/ 28,** 413 (1988).

[57]   R. P. Wahi and J. Stajer, in *Decomposition of Alloys: the early stages*, eds. P. Hasen et al., Pergamon Press (1984) 165.

[58] O. Lyon and J. P. Simons, *J. Phys.: Condens. Matter* **4** (1992) 6073.

[59] O. Lyon and J. P. Simons, *J. Phys.: Condens. Matter* **6** (1994) 1627.

[60] L. Delannay, O. V. Mishin, D. Juul Jensen and P. van Houtte, *Acta Mater.* **49** (2001) 2441.

CHAPTER 3

Application of polarized neutron reflectometry to studies of artificially structured magnetic materials

3.1. INTRODUCTION

Reflectometry involves measurement of the intensity of a beam of electromagnetic radiation or particle waves reflected by a planar surface or by planar interfaces. The technique is intrinsically sensitive to the difference of the refractive index (or contrast) across interfaces. For the case of specular reflection, i.e., the case when the angle of reflection, α_r, equals the angle of incidence, α_i, see Figure 3-1(a), the intensity of the reflected radiation is related to the depth dependence of the index of refraction averaged over the lateral dimensions of the surface or interface. In this simplest example of reflectometry, the sharpness of an interface can be quantitatively measured, the distance between two or more planar interfaces can be obtained, and the strength of the scattering potential, i.e., the index of refraction, between the interfaces can be measured relative to that of the medium through which the radiation travels to reach the sample surface (in many cases the surrounding medium is air or vacuum—for neutron scattering there is little distinction). In more complex situations, variations of the refractive index within the plane of the interface may give rise to diffuse scattering or off-specular reflectivity, i.e., radiation reflected away from the specular condition, see Figure 3-1(b and c). From measurements of off-specular reflectivity, correlations between lateral variations of the scattering potential along an interface can be deduced. Off-specular scattering introduces a component of wavevector transfer in the plane of the sample mostly parallel to the incident neutron beam, Figure 3-1(b)] [1], or perpendicular to it, Figure 3-1(c) [2,3].

So far, the capabilities of reflectometry have been described without regard to the kind of radiation used. Many detailed discussions of X-ray [4-10] and unpolarized neutron reflectometry [10,11] from non-magnetic materials can be found in the literature. Treatments of X-ray reflectometry invariably use concepts of optics and Maxwell's equations. Treatments of neutron reflectometry can be optical in nature, but often treat the neutron beam as a particle wave and use quantum mechanics to calculate reflection and transmission probabilities across interfaces bounding potential wells. In the present chapter, we focus on reflectometry of magnetic thin films and artificially structured magnetic materials using polarized neutron beams.

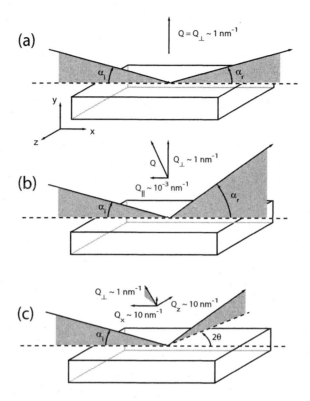

Figure 3-1: Scattering geometry for (a) specular reflectometry, where $\alpha_r = \alpha_i$, (b) off-specular reflectometry where $\alpha_r \neq \alpha_i$, and (c) glancing incidence diffraction where $2\theta \neq 0$. The components of wavevector transfer, $Q = k_f - k_i$ are shown for each scattering geometry.

Polarized neutron reflectometry is a tool to investigate the magnetization profile near the surfaces of crystals, thin films and multilayers. Surface (or interface) sensitivity derives from working in glancing incidence geometry near the angle for total external reflection. Polarized neutron reflectometry is highly sensitive, having measured the absolute magnetization of a monolayer of iron ($\sim 10^{-4}$ emu) with 10% precision [12], and magnetization density as small as 30 emu/cm^3 (e.g., as found in $Ga_{0.97}Mn_{0.03}As$) with comparable precision. Detection of small moments (from samples with surfaces measuring a ~ 4 cm^2 in area) is combined with excellent depth resolution—a fraction of a nanometer even for films as thick as several hundred nanometers. Reflectometry has enjoyed dramatic growth during the last decade and has been applied to important problems such as, the influence of frozen or pinned magnetization on the origin of exchange bias [13], the influence of exchange coupling on magnetic domain structures [14,15], and the identification of spatially inhomogeneous magnetism in nanostructured systems [16-18].

Several descriptions of polarized neutron reflectometry are available in the literature [19-25]. Recently, reviews of polarized neutron reflectometry, one that includes illustrative examples [26], and a second very detailed account of the scattering of polarized neutron beams, with copious mathematical derivations of formulae, have been published [27,28]. In this chapter, we present a tutorial on polarized neutron reflectometry, a description of a polarized neutron reflectometer at a pulsed neutron source, and examples of applications of the technique.

3.2. NEUTRON SCATTERING IN REFLECTION (BRAGG) GEOMETRY

3.2.1. Reflectometry with unpolarized neutron beams

In Figure 3-2, we show the general situation for a neutron beam with wavelength λ represented by a plane wave in air (Medium 0) with incident wavevector $\mathbf{k}_i$ ($|\mathbf{k}_i| = k_0 = 2\pi/\lambda$) and reflected wavevector $\mathbf{k}_f$ (reflected by the sample, Medium 1). A portion of the plane wave is transmitted across the reflecting interface with wavevector $\mathbf{k}_t$. Depending upon the distributions of chemical or magnetic inhomogenities in the plane of the sample, neutron radiation can be scattered in directions such that $2\theta \neq 0$ and/or $\alpha_r \neq \alpha_i$ [see Figure 3-1]. The case of elastic and specular ($2\theta = 0$ and $\alpha_r = \alpha_i$) reflection is the simplest to treat. Neutron scattering is called elastic when the energy $E = \hbar^2 k_0^2 / 2m_n$ of the neutron is conserved. Thus, the magnitudes of $\mathbf{k}_i$ and $\mathbf{k}_f$ are equal, i.e., $|\mathbf{k}_i| = |\mathbf{k}_f|$. The magnitude of $\mathbf{k}_t$ in Medium 1, $|\mathbf{k}_t| = k_1$, may be (and usually is) different than that of Medium 0.

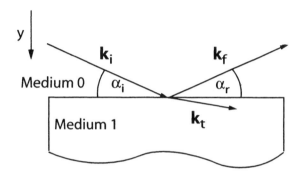

Figure 3-2: Schematic diagram showing the incident, reflected and transmitted wavevectors. The sample in this case is Medium 1.

The quantity measured in a neutron reflectometry experiment is the intensity of the neutron beam reflected from the surface. The probability of reflection or the reflectivity is given by the reflected intensity divided by the incident intensity. To calculate the reflectivity of an interface, we apply the time-independent Schrödinger equation [29] to obtain a solution for the wave function, Ψ, representing the neutron wave inside and outside of the reflecting sample. Dropping the parts of the wave function with wavevector components parallel to the interface (we consider a potential that varies in only one dimension which cannot change the neutron's wavevector parallel to the interface), the wave functions in Mediums 0 and 1 are given by:

$$\Psi_0(y) = e^{+ik_0 y} + r e^{-ik_0 y} \tag{3.1}$$
$$\Psi_1(y) = t e^{+ik_1 y}$$

Unless otherwise noted, k_i is the component of the wavevector k_i that is perpendicular to the sample surface.

The neutron reflectivity, R, of the interface is related to the reflection amplitude, r, by $R = rr^*$. Ψ is obtained by solving Schrödinger's equation:

$$\left[\frac{\hbar^2}{2m_n} \frac{\partial^2}{\partial y^2} + V(y) \right] \Psi(y) = E\Psi(y) \tag{3.2}$$

where $V(y)$ is the depth dependent scattering potential. For a planar sample, the neutron (nuclear) scattering potential is represented by the expression:

$$V_n = \frac{2\pi\hbar^2}{m_n}\rho(y) \tag{3.3}$$

where $\rho(y)$ is the neutron scattering length density in units of $Å^{-2}$. Owing to the decay in the strength of the reflected neutron beam with wavevector transfer (discussed later), neutron reflectometry usually involves measurements that are restricted to fairly small wavevector transfer, $Q_\perp < 0.3$ Å. Over this range of $Q_\perp$, the scattering medium can be considered to consist of a continuous scattering length density of N (scattering centers or formula units per unit volume) each with coherent neutron scattering length b. For systems composed of a mixture of elements or formula units,

$$\rho = \sum_i^J N_i b_i \tag{3.4}$$

where J is the number of distinct isotopes, and N_i and b_i are the number density and scattering length for the i-th species. Values of N, b and ρ are given for a number of common materials in Table 1 [30].

Invoking the condition of elastic scattering, Equation 3.2 can be rewritten as:

$$\left[\frac{\partial^2}{\partial y^2} + k_0^2 - 4\pi\rho(y)\right]\Psi(y) = 0 \tag{3.5}$$

In the language of ordinary light optics, the $\perp$-component of the wavevector in Medium 1, k_1, is related to the $\perp$-component of the wavevector in Medium 0, k_0, through the index of refraction, n, by

$$k_1 = nk_0 = \sqrt{1 - \frac{4\pi\rho}{k_0^2}}k_0 \tag{3.6}$$

During an experiment, the intensity of the reflected radiation is measured for selected values of k_0, which are chosen either by changing the angle of incidence of the beam to the sample surface, α_1, and/or by changing the wavelength, λ, of the neutron beam. For sufficiently small values of k_0, the index of refraction will be purely imaginary, so the neutron wave in Medium 1 is evanescent [31]. Therefore, the wave is reflected by the sample with unit probability. The wavevector transfer $Q_\perp$ at which n obtains a real component is the called the critical edge, Q_c. For $Q_\perp < Q_c$, the reflected intensity is unity, and provides a means to normalize the reflectivity to an absolute scale (in contrast to small angle neutron scattering). Since the reflectivity of the sample is unity below Q_c, the scattering in this region is strong, so a dynamical treatment of the scattering is required. By dynamical, we mean the wave function inside Medium 1 is not the same as that illuminating the sample. Because the Born approximation [29] is a perturbation theory, it is valid for weak scattering, e.g., small-angle neutron scattering in transmission geometry, so this approximation is not adequate for calculating reflection of neutrons or X-rays at glancing angles from planar or nearly planar interfaces.

Table 1: List of common elements and their neutron nuclear and magnetic scattering length densities.

Material	Number density, $N [10^{-2}\text{Å}^{-3}]$	Nuclear scattering length, $b [10^{-5}\text{Å}]$	Magnetic moment, $\mu [\mu_B]$	Nuclear scattering length density, $\rho_n [10^{-6}\text{Å}^{-2}]$	Magnetic scattering length density, $\rho_m [10^{-6}\text{Å}^{-2}]$
Ag	5.86	5.92		3.47	
Al	6.02	3.45		2.08	
Al_2O_3	2.13	24.4		5.21	
Au	5.90	7.90		4.66	
Co	9.09	2.49	1.715	2.26	4.12
Fe	8.47	9.45	2.219	8.00	4.97
FeF_2	2.75	20.76		5.71	
Fe_2O_3 hematite	2.00	36.32		7.26	
Fe_3O_4 magnetite	1.35	51.57	4.1	6.97	1.46
GaAs	2.21	13.87		3.07	
$LaAlO_3$	1.84	29.11		5.34	
$LaFeO_3$	1.65	35.11		5.78	
$LaMnO_3$	1.71	21.93		3.75	
MgF_2	3.07	16.68		5.12	
MgO	5.35	11.18		5.98	
MnF_2	2.58	7.58		1.96	
Nb	5.44	7.05		3.84	
Ni	9.13	10.3	0.604	9.40	1.46
^{58}Ni	9.13	14.4	0.604	13.14	1.46
^{62}Ni	9.13	-8.7	0.604	-7.94	1.46
$Ni_{81}Fe_{19}$	8.93	10.14	1.04	9.06	2.46
NiO	5.49	16.11		8.84	
Pd	6.79	5.91		4.01	
Pt	6.60	9.60		6.34	
Pu	4.88	5.8±2.3		2.8±1.1	
Si	4.99	4.15		2.07	
SiO_2	2.66	15.76		4.19	
$SrTiO_3$	1.68	21.00		3.54	
U	4.82	8.417		4.06	
V	6.18	-0.38Å		-0.23	

3.3. THEORETICAL EXAMPLE I: Reflection from a perfect interface surrounded by media of infinite extent

The goal of a reflection experiment is to determine the distribution of material within the sample from its reflectivity as a function of $Q_\perp$. To accomplish this goal, we need to determine the probabilities that the wave function is reflected and transmitted by the sample. Conservation of neutron intensity, i.e., $|\Psi|^2=1$, and conservation of momentum require that $\Psi(y)$ and its derivative, $\partial\Psi/\partial y$, be continuous across the interface. Thus,

$$\begin{pmatrix} \Psi_0(0) \\ \left.\dfrac{\partial \Psi_0}{\partial y}\right|_{y=0} \end{pmatrix} = \begin{pmatrix} \Psi_1(0) \\ \left.\dfrac{\partial \Psi_1}{\partial y}\right|_{y=0} \end{pmatrix} \Rightarrow \begin{pmatrix} 1+r \\ ik_0(1-r) \end{pmatrix} = \begin{pmatrix} t \\ ik_1 t \end{pmatrix} \tag{3.7}$$

Solving eq.(3.7) for r, the reflection amplitude of a single interface between two media of infinite extent, gives:

$$r = \left(\frac{k_0 - k_1}{k_0 + k_1} \right) \tag{3.8}$$

from which the reflectivity of a single interface is obtained:

$$R = rr^* = \left(\frac{k_0 - k_1}{k_0 + k_1} \right)\left(\frac{k_0 - k_1}{k_0 + k_1} \right)^* = \left(\frac{1-n}{1+n} \right)\left(\frac{1-n}{1+n} \right)^* \tag{3.9}$$

As an example to illustrate application of eq.(3.9), we consider the case of an unpolarized neutron beam reflecting from a perfectly smooth silicon substrate (surrounded by air). The neutron scattering length density for Si is $\rho_{Si}=2.07 \times 10^{-6}$ Å^{-2} (obtained from Table 1), and the depth dependence of the scattering length density profile for the sample is shown in Figure 3-3(a). The reflectivity versus $Q_\perp$ [Figure 3-3(b)] is calculated using eq.(3.9). The position of the critical edge, Q_c, is determined by the condition $n=0$, i.e., $Q_c = 4\sqrt{\pi \rho_{Si}}$.

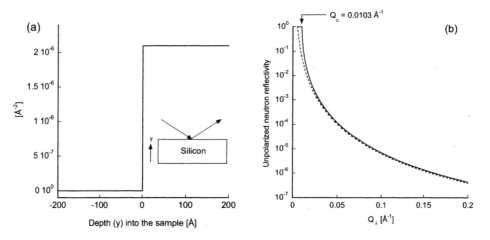

Figure 3-3: (a) Unpolarized neutron scattering length density profile of a perfect interface between air and a silicon substrate (inset). (b) The calculated reflectivity for the interface (a) is shown by the solid curve. The dashed curve represents a reflectivity curve calculated using the Born approximation (see text) and varies as $Q_\perp^{-4}$ normalized to 0.9 times the solid curve at $Q_\perp = 0.2$Å^{-1} (see text).

The dynamical calculation of the silicon substrate reflectivity [solid curve, Fig. 3-3(b)] in the region of $Q_\perp \sim 0.1$ Å^{-1} is similar to that obtained using the Born approximation (i.e., the kinematical case, dashed curve) from which the reflectivity is equated to the Fourier transform of the scattering length density profile:

$$R_{BA} \propto \frac{1}{Q_\perp^2} \left| \int_{-\infty}^{\infty} e^{iQ_\perp y} \rho(y) dy \right|^2 \tag{3.10}$$

However, for smaller values of $Q_\perp$ the two reflectivity curves diverge.

In the large $Q_\perp$ regime, the decay of the curve scales as $Q_\perp^{-4}$. This decay, called the Fresnel decay [6], is a property of reflection from a planar surface, and thus contains little information leading to a better understanding of the spatial representation of matter beneath the surface. However, the Fresnel decay rapidly diminishes the reflected neutron beam intensity until it can become swamped by sources of background, including incoherent scattering from the substrate.

3.4. THEORETICAL EXAMPLE 2: Reflection from perfectly flat stratified media

For the case of reflection from a single perfect interface, there is little additional information that can be obtained beyond that provided by the position of the critical edge (surface roughness can also be measured—a topic discussed later). More interesting and realistic cases involve reflection from stratified media. In these cases, the scattering length density is not constant with depth, and indeed abrupt changes of the scattering length density, such as those produced by buried interfaces, modulate the reflectivity.

Now consider the representation of a stratified sample in Fig. 3-4— one depicting reflection of a neutron beam from a perfect interface formed by the boundary between air and the surface of a thin film with thickness Δ that is in contact with a smooth Si substrate of infinite thickness.

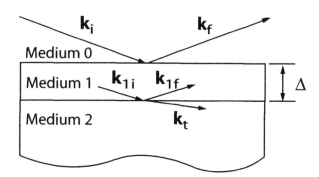

Figure 3-4: Schematic diagram showing the wavevectors in a stratified medium. The thickness of the thin film is Δ.

The wave functions in the different media are:

$$\Psi_0(y) = e^{+ik_0 y} + re^{-ik_0 y}$$
$$\Psi_1(y) = Ge^{+ik_1 y} + He^{-ik_1 y} \qquad (3.11)$$
$$\Psi_2(y) = te^{+ik_{21} y}$$

Again Schrödinger's equation is solved yielding a matrix equation from which the reflection and transmission amplitudes, r and t, can be obtained:

$$\overline{M_1(\Delta)}\begin{pmatrix} 1+r \\ ik_0(1-r) \end{pmatrix} = \begin{pmatrix} t \\ ik_2 t \end{pmatrix} e^{ik_2 \Delta} \qquad (3.12)$$

and

$$\overline{M_1(\Delta)} = \begin{pmatrix} \cos(k_1\Delta) & \dfrac{1}{k_1}\sin(k_1\Delta) \\ -k_1\sin(k_1\Delta) & \cos(k_1\Delta) \end{pmatrix}. \qquad (3.13)$$

Eq.(3.12) represents two simultaneous equations that can be solved to obtain r. (Note, eq.(3.7) is recovered for the case of a single (air/substrate) interface for the case of $\Delta=0$.)

$$r = \frac{r_{01} + r_{12}e^{ik_1 2\Delta}}{1 + r_{01}r_{12}e^{ik_1 2\Delta}}$$

$$r_{mn} = \frac{k_m - k_n}{k_m + k_n} \tag{3.14}$$

$$k_j = n_j k_0 = \sqrt{1 - \frac{4\pi\rho_j}{k_0^2}}k_0$$

To calculate a reflectivity curve, a value of $Q_\perp$ is chosen from which $k_0 (= Q_\perp/2$, a real quantity) is obtained. Next, the $\perp$-components of the wavevector in Medium 1 and Medium 2 are calculated, using eq.(3.14), from which the reflection amplitudes for a pair of interfaces are obtained. The reflection amplitudes for an ensemble of interfaces (in this case two, see eq.(3.14), r, is related to the reflection amplitudes of each interface, r_{01} and r_{12} (here, the amplitude of the wave reflected by the interface between Medium m and Medium n is called r_{mn}), in the ensemble after combination with a phase factor, $\exp(ik_1 2\Delta)$, as appropriate (the wave reflected by the interface between Medium 1 and Medium 2 is out of phase by the path length 2Δ with respect to the wave reflected by the interface between Medium 0 and Medium 1). This procedure was performed to obtain the reflectivity curve (Figure 3-5) for a sample consisting of a 20 nm thick perfectly flat layer of material with the nuclear scattering length density of Fe on a perfect Si substrate.

The most notable feature of the solid curve, Figure 3-5(b) is the oscillation of the reflectivity. The period of the oscillation in the kinematical limit (far from the critical edge where dynamic effects are most pronounced) is approximately equal to $2\pi/\Delta$. The amplitude of the oscillation is related to the contrast or difference between the scattering length densities of the iron film and silicon substrate. A second notable feature is the position of the critical edge, which for the 20 nm Fe/Si sample still occurs at a position coinciding with that of the silicon substrate and not at the position for an iron substrate [compare the dotted and dashed curves in Figure 3-5(b)]. Unlike the case for X-ray reflectometry, in which only a couple of nanometers of material is sufficient to be opaque, and thus create a well-defined critical edge, the critical edge for neutron reflectivity is often determined by the sample substrate, and not the thin film, owing to the fact that a neutron beam is a highly penetrating probe.

One strength of reflectometry is its ability to measure layer thickness with very high precision and accuracy (for a discussion of the distinction see Ref. [32]). An illustrative example is to compare the calculated reflectivity curves for iron films of 20.0 nm and 20.6 nm thickness—corresponding to a 3% change in film thickness (Fig. 3-6). The shift between the reflectivity curves at large wavevector transfer is easily distinguished, because the resolvable wavevector transfer is smaller than the shift. For small scattering angles, the resolution of a reflectometer, $\delta Q/Q$ is approximately given by:

$$\left(\frac{\delta Q}{Q}\right)^2 = \left(\frac{\delta\theta}{\theta}\right)^2 + \left(\frac{\delta\lambda}{\lambda}\right)^2 \tag{3.15}$$

The first term is determined by a combination of factors including sample size and the dimensions of slits that collimate the incoming neutron beam. For glancing angles of incidence (typically less than 5°), $\delta\theta/\theta$ is of order 2% (root-mean-square). The second term is determined by how well the wavelength of the incident neutron beam is measured. For situations in which a graphite monochromator selects the wavelength (as used for example at a nuclear reactor), $\delta\lambda/\lambda$ is typically 1 to 2% (rms). For situations in which the time-of-flight technique measures neutron wavelength (as used for example at a short pulsed neutron source) $\delta\lambda/\lambda$ is typically 0.2% (rms). So, with little effort, the resolution of a reflectometer in $\delta Q/Q$ can be made less than 3% (rms). Consequently, the change of fringe phase, which is about 3% for the case illustrated in Fig. 3-6, can be readily measured.

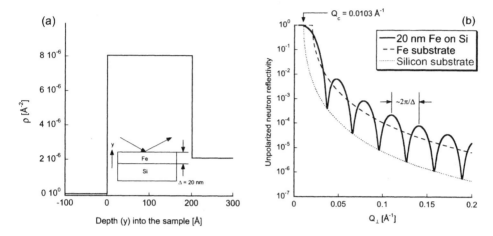

Figure 3-5: (a) The nuclear scattering length density profile of a perfect thin iron film on silicon (inset). (b) The reflectivity of the sample is shown as the solid curve. The reflectivities of a silicon substrate (dotted curve) and a substrate with the nuclear scattering length density of iron (dashed curve) are shown for comparison.

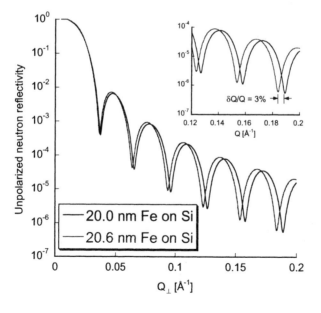

Figure 3-6: Comparison of reflectivity curves from Fe films with different thicknesses. The difference of 3% in thickness can be easily resolved. *Also see the color plate.*

In contrast to measuring sub-nanometer changes in film thickness, detection of a single sub-nanometer thick film is considerably more challenging. The Fresnel decay of the reflectivity restricts the degree to which perturbations in the scattering length density profile over thin layers can be measured. Let Q_{max} be the maximum value of $Q_\perp$ that can be measured before the reflectivity, R_{min}, is approximately equal to the instrumental background. Thin films having thickness $\Delta > 2\pi/Q_{max}$ can perturb the reflectivity (at Q_{max}) by superimposing oscillations on the Fresnel decay. In principle, by measuring the period and amplitude of the oscillations, information about the thickness of the thin film and its composition can be inferred. On the other hand, for films with thickness $\Delta < 2\pi/Q_{max}$ the perturbation to the reflectivity might well be missed on account that the first pair of fringe maximum and minimum occur at wavevector transfer so large that the intensity of the reflected beam is below R_{min} (in other words, the oscillations of the reflectivity curve might be swamped by instrumental background).

Neutron reflectivity has been measured to values of $R_{min}=10^{-8}$ under ideal conditions. In these conditions, Q_{max} might be on order of 0.3 Å^{-1}, so detection of films as thin as 2 nm might be possible. However, most experiments are not conducted under ideal circumstances. For example, experiments usually involve sample environment equipment, e.g., cryostats etc., or samples that are either not perfectly smooth or are themselves sources of incoherent scattering. In these situations neutron reflectivity measurements to less than 10^{-7} are often not achievable.

3.5. THEORETICAL EXAMPLE 3: Reflection from "real-world" stratified media

The first two examples of perfect interfaces illustrate the importance of the critical edge (providing a means to place the reflectivity on an absolute scale), fringe period (related to layer thickness) and fringe amplitude (related to change of, or contrast between, scattering length density across an interface). Since real systems can be less than perfect, we consider the case of rough or diffused interfaces. This case serves to show how reflectometry can be a useful tool to study systems that are imperfect (indeed reflectometry provides a useful measure of imperfection).

Consider the case where the diffusion of Fe and Si across the Fe/Si interface in the previous example obeys Fick's second law [33]. We further assume the characteristic diffusion length, σ, of Fe into the Si matrix is the same as Si into the Fe matrix (though this assumption is unlikely to be correct). In this case, the concentrations of Fe and Si with depth (in units of atoms/Å^3) are given by:

$$N_{Fe}(y) = N_{Fe}\left[1 - erf\left(\frac{y-\Delta}{\sqrt{2}\sigma}\right)\right]$$
$$N_{Si}(y) = N_{Si}\left[1 + erf\left(\frac{y-\Delta}{\sqrt{2}\sigma}\right)\right]$$

(3.16)

After substitution of eq.(3.16) into eq.(3.4), and using the appropriate values of the neutron scattering lengths and densities for Fe and Si (see Table 1), the neutron scattering length density profile is obtained:

$$\rho(y) = b_{Fe}N_{Fe} + \frac{b_{Si}N_{Si} - b_{Fe}N_{Fe}}{2}\left[1 + erf\left(\frac{y-\Delta}{\sqrt{2}\sigma}\right)\right]$$

(3.17)

The variation of the neutron scattering length density across the interface is represented by an error function connecting the scattering length densities of pure Fe and pure Si. We note the derivative of the error function with argument $(y-\Delta)/\sqrt{2}\sigma$ is proportional to a Gaussian function with root-mean-square width of σ [34]. The scattering length density profile for a 20 nm thick Fe layer bounded by a diffuse air/Fe surface (i.e., a rough surface) and diffuse Fe/Si interface with characteristic widths of σ=5 Å is shown in Fig. 3-7(a). The thickness of the film is the distance between the centers of the two Gaussian functions.

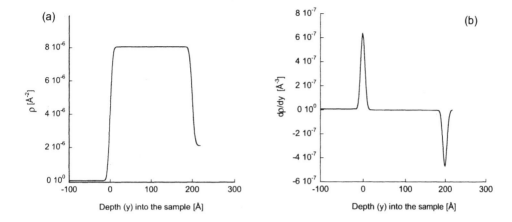

Figure 3-7: (a) Representation of the Fe/Si sample with rough and/or diffuse interfaces. (b) The derivative of the scattering length density profiles consisting of a pair of Gaussian profiles from which (a) is obtained upon integration.

While the scattering length density profile in Fig. 3-7(a) can be obtained using eq.(3.17), in fact the profile shown in the figure was obtained by integrating the derivative of the scattering length density profile with respect to depth (y-coordinate) Fig. 3-7(b). The peaks in Fig. 3-7(b) are Gaussian peaks whose positions, widths and integrals correspond to the positions, diffusion or roughness widths, and contrast across the interfaces, respectively. For example, the integral of the peak at y=0 Å in Fig. 3-7(b) is $\rho_{Fe}-\rho_{air}=8\times10^{-6}$ Å^{-2}. One motivation for constructing the derivative of the scattering length density profile (and then integrating it) is to allow the possibility for interfaces to be close. By close, we mean the thickness of one or both layers on either side of an interface is thinner than the rms width, σ, attributed to the interface. While arguments can be made whether such a situation is physically meaningful, mathematically the situation corresponds to one where tails of adjacent Gaussian peaks overlap, and certainly such a profile can be integrated. When the tails of two Gaussian peaks overlap (significantly), the profile obtained from integrating the derivative profile will not yield an error function variation between the two interfaces, but may nevertheless produce a calculated reflectivity curve that closely resembles a measured reflectivity. It should be emphasized that only in situations where $\sigma_{mn}\ll\Delta_{m}$ and $\sigma_{mn}\ll\Delta_{n}$, should the value of σ_{mn} be interpreted as an interface width and Δ as a layer thickness. Otherwise, the parameters of a density profile— ones that yield a well-fitting reflectivity curve, have little meaning, though the density profile might accurately represent the scattering potential of the system.

The process for calculating the reflectivity of the "roughened" sample first involves approximating the continuous profile in Fig. 3-7 by a discrete sequence of thin slabs of width δ with step-like changes in scattering length density. The choice of δ, i.e., the thickness over which ρ is constant, is made such that $\delta\ll2\pi/Q_{max}$ —a relation assuring the Sampling Theorem of Fourier analysis [35] is satisfied. An example of such an approximation for δ = 2 Å is shown in Figure 3-8.

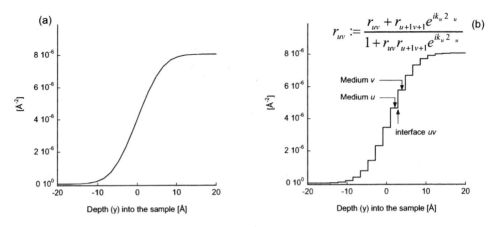

Figure 3-8: (a) Variation of the scattering length density profile of the air/Fe interface for $\sigma = 5$ Å is shown. (b) Approximation of the continuous function in (a) using discrete steps.

There are two common approaches to calculate the (dynamical) reflectivity using the approximate scattering length density profile shown in Fig. 3-8(b). The first approach, which is suitable for calculating the scattering length density profiles from scalar potentials (eq.(3.3) is an example of a scalar potential) is to use eq.(3.14) to calculate the reflection amplitude of the interface between the bottom-most thin slab and the infinitely thick substrate. Let the reflection amplitude of this interface be r_{mn} (bounded by Medium $m=n-1$ and Medium n—the substrate). Then, the reflection amplitude of the next higher interface— the $m-1m$-th interface, is computed using r_{mn} as the reflection amplitude of the phase-shifted term in eq.(3.14). This equation is applied recursively (as indicated in Fig. 3-8(b) for the uv-th interface) until the top interface (the air/sample interface) is reached. Calculation of the reflectivity by recursively applying eq.(3.14) (for a particular $Q_\perp$) is required in order to properly account for dynamical scattering of the neutron beam by the sample surface at glancing angles. In other words, were the Born approximation a good representation of the scattering, then a recursive calculation to obtain the reflectivity curve would not be necessary. The recursive calculation is often referred to as the Parratt formalism [4].

The second approach to calculate the reflectivity curve is to generalize the matrix relation, eq.(3.12), for an arbitrary number of thin slabs, and then to solve the simultaneous equations to obtain the reflection amplitude of the ensemble (i.e., the entire sample). The second approach is one that can be used to calculate the reflection amplitude of a sample that might be represented by a scalar or vector potential (an example of a vector potential is one that includes the vector magnetization of a sample). The matrix relation is generalized to the case of an any number of thin slabs as follows (for a detailed derivation see Ref. [28]):

$$\prod_{j=n-1}^{0}\overline{M_j(\delta_j)}\binom{1+r}{ik_0(1-r)}=\binom{t}{ik_n t}\prod_{j=n-1}^{0}e^{ik_n\delta_j}=\binom{t}{ik_n t}e^{ik_n\Delta}$$

$$\overline{M_j(\delta_j)}=\begin{pmatrix}\cos(k_j\delta_j) & \dfrac{1}{k_j}\sin(k_j\delta_j)\\ -k_j\sin(k_j\delta_j) & \cos(k_j\delta_j)\end{pmatrix}$$

$$\Delta=\sum_{j=1}^{n-1}\delta_j$$

(3.18)

The subscript "j" in eq.(3.18) represents the j-th medium or slab. So, for example, k_j is the magnitude of the $\perp$-component of the wavevector in the j-th medium, eq.(3.14), and δ_j is the thickness of the medium over which the scattering length density is considered constant [2 Å for the case of Figure 3-8(b)].

The reflectivity calculated for a 20 nm thick Fe film with roughened interfaces, whose scattering length density profile is shown in Figure 3-8(b), is the solid curve in Figure 3-9. The case for the ideal Fe film, whose scattering length density profile is shown in Figure 3-5(b), is the dashed curve in the figure.

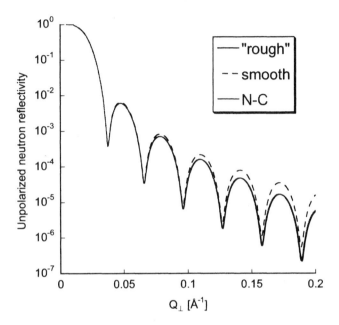

Figure 3-9: The influence of rough or diffuse interfaces with $Q_\perp$ on reflectivity. The "rough" and "N-C" (computed using the Nevot and Croce relation) curves are essentially identical. *Also see the color plate.*

The specular reflectivity curve of a sample with rough or diffuse interfaces is attenuated more so than that of a smooth sample. The attenuation increases with $Q_\perp$. In fact, for the case of a single interface, Nevot and Croce [36] have analytically shown that the reflection amplitude of a single rough interface, r_r, having short length-scale roughness ($Q_\perp\sigma\ll1$) is related to that of the ideal interface, r_i, by the relation:

$$r_r = r_i \exp\left(Q_\perp Q'_\perp \sigma^2 / 2\right) \tag{3.19}$$

where $Q'_\perp (= k_{lf} - k_{li})$ is the wavevector transfer in the sample. As the kinematical limit is approached (i.e., $Q'_\perp \rightarrow Q_\perp$), the attenuation factor is identical to a "static" Debye-Waller factor [37] (application of eq.(3.19) to the "smooth" curve in Figure 3-9 yields the red "N-C" curve). An important consequence of this observation is that interface roughness (or diffusion) will further limit the accessible region of wavevector transfer, and consequently the sensitivity of reflectometry to changes of the scattering length density profile over thin layers. The attenuation of the reflectivity with roughness is a strong function of σ and $Q_\perp$; thus, more information can be extracted from samples with smooth interfaces than those with rough interfaces (although the physics of rough interfaces is often interesting!). For many experiments, useful information can be obtained from samples with (rms) interface roughness on the order of 10 Å, whereas, for samples with interface roughness of 20+ Å, success of the experiment may be hopelessly compromised.

A second important consequence of rough interfaces is the redistribution of intensity from the specular reflectivity into diffuse scattering. Diffuse scattering is most easily recognized as elevated levels of intensity in off-specular directions, but diffuse scattering from rough interfaces is often peaked in the specular direction (much like how thermal diffuse scattering can be concentrated at Bragg reflections [37]). Thus, the intensity of the neutron (or X-ray) beam reflected into the specular direction contains the specular reflectivity (which provides information about the depth dependence of the scattering potential averaged over the sample's lateral dimensions), and diffuse scattering (which provides information about the correlation of roughness across the lateral dimensions of the sample, and is often modulated with $Q_\perp$ in the same manner the specular reflectivity is modulated). Estimates of the intensity of the diffuse scattering can be interpolated from measurements of the off-specular scattering on either side of the specular direction; obtained, for example, by rocking the sample or detector in a manner such that $\alpha_i \neq \alpha_r$. Most reflectivity fitting packages implement one-dimensional specular theory; therefore, these packages are not intended to correctly treat scattering from systems that produce non-negligible diffuse scattering.

The previous three theoretical examples have illustrated useful concepts and interpretations of reflectivity curves. The measurements and their interpretations are summarized in Table 2.

Table 2: List of measurements and the information yielded by the measurements.

Measurement feature	**Information obtained from a sample of cm^2 or so size**
Position of critical edge, Q_c	Nuclear (chemical) composition of the neutron-optically thick part of the sample, often the substrate.
Intensity for $Q < Q_c$	Unit reflectivity provides a means of normalization to an absolute scale.
Periodicity of the fringes	Provides measurement of layer thickness(es). Thickness measurement with uncertainty of 3% is routinely achieved. Thickness measurement to less than 1 nm can be achieved.
Amplitude of the fringes	Nuclear (chemical) contrast across an interface.
Attenuation of the reflectivity	Roughness of an interface(s) or diffusion across an interface(s). Attenuation of the reflectivity usually establishes a lower limit (typically of order 1-2 nm) for reflectometry to detect thin layers.

3.5.1. Reflectometry with polarized neutron beams

In the previous section, neutron reflectometry was discussed in terms of the reflection of neutron beams from scattering potentials that are purely nuclear in origin. Since the neutron possesses an intrinsic magnetic moment and spin, the scattering potential may be spin dependent. There are two reasons that the interaction between a neutron and matter may depend on the neutron's spin. In some scattering processes (e.g., incoherent scattering of neutrons by hydrogen), the nuclear spin of an atom can interact with the spin of a neutron. On other occasions, the nuclei in a material from which the neutron scatters, may possess net spin and be polarized. Examples include spin polarized ^{3}He nuclei [38], or spin polarized Ga or As nuclei in the presence of a magnetic material [39]. The spin dependence of the potential for these examples involves two neutron scattering lengths, b^+ and b^-, where the sign of the term indicates whether the spin of the nuclei is parallel or anti-parallel to the laboratory magnetic field of reference (see Fig. 3-10), which will later be identified with the polarization axis of the neutron beam.

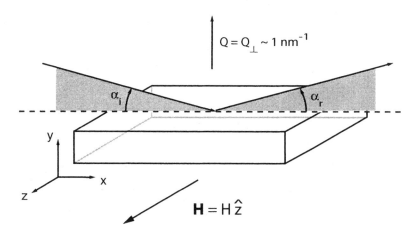

Figure 3-10: Diagram of a reflection experiment in which the sample is immersed in a magnetic field.

A commonly encountered second case involves the interaction of the neutron spin with atomic magnetism or other sources of magnetic induction. The modification to the scattering potential (including the nuclear potential) is given by:

$$V = V_n + V_m = V_n \mp \underline{\mu} \cdot \boldsymbol{B} \qquad (3.20)$$

Here, $\boldsymbol{B}$ is the (spatially dependent) magnetic induction vector, and $\underline{\mu}$ is the magnetic moment of the neutron, where $\underline{\mu} = \mu_n \underline{\sigma}$, $\mu_n = -1.913$ μ_N (the negative sign indicates that the neutron moment and its spin are anti-parallel), and $\underline{\sigma}$ is a linear combination of the (2 x 2) Pauli matrices [29] directed along each of the three orthogonal spatial axes with the magnetic field direction taken to lie along the $\hat{z}$-axis (Fig. 3-10). The "$\mp$"-sign in eq.(3.20) is taken to be negative (positive) if the neutron spin is parallel (anti-parallel) to the laboratory field of reference ($\boldsymbol{H}$ in Fig. 3-10). Since μ_n is negative, the quantity $-\mu_n B$ is positive, thus, adding to a normally positive nuclear scattering length—one for a repulsive potential (Mn, however, is an example of an atom with a negative nuclear scattering length—an attractive potential). Fundamentally, the neutron spin interacts with magnetic induction, $\boldsymbol{B}$, so a materials-property that gives rise to $\boldsymbol{B}$, e.g., orbital and/or spin moments of atoms, or accumulation of spin in electronic devices, in principle can affect the neutron scattering process. The fact that the neutron spin interacts with magnetic induction and not magnetic field [40-43] is fortunate, since were this not the case neutron scattering might not be a useful a tool in the study of magnetism.

Expressing the scattering potential V in matrix notation, we obtain [28]:

$$\overline{V(y)} = \frac{2\pi\hbar^2}{m_n}\begin{pmatrix} \rho_n & 0 \\ 0 & \rho_n \end{pmatrix} \mp \mu_n \begin{pmatrix} B_z & B_x - iB_y \\ B_x + iB_y & -B_z \end{pmatrix} \tag{3.21}$$

The elements of the matrices are understood to depend on position, i.e., the nuclear scattering length density term, $\rho_n = \rho_n(y)$, etc. (dependence on x and z is also possible to observe with off-specular reflectometry). It is important to recognize that while most often the nuclear scattering potential outside of the sample is zero, $\rho_n = 0$, this is not necessarily the case for the magnetic induction. For example, in a polarized neutron reflectometry experiment, some magnetic field (as little as a couple Oe may be needed) is nearly always applied to the sample, in order to maintain the polarization of the neutron beam. Since neutron reflection occurs across interfaces with different scattering length densities (nuclear or magnetic), the *field* applied to the sample and the *field* inside the sample being the same do not yield contrast across the interface. Setting $B = \mu_0 H + M$, where M is the intensity of magnetization, and for fields applied along $\hat{z}$, eq.(3.21) can be rewritten as:

$$\overline{\delta V(y)} = \frac{2\pi\hbar^2}{m_n}\begin{pmatrix} \rho_n & 0 \\ 0 & \rho_n \end{pmatrix} \mp \mu_n \begin{pmatrix} M_z & M_x - iM_y \\ M_x + iM_y & -M_z \end{pmatrix} \tag{3.22}$$

Eq.(3.22) is a relation for the potential difference, $\overline{\delta V(y)}$, between the sample and the surrounding medium (here, assumed to be air, but for cases in which the sample is not surrounded by air, the nuclear scattering length density of the surrounding medium must also be removed from ρ_n). The neutron magnetic scattering length density can be defined in terms similar to those used to define the neutron nuclear scattering length density, eq.(3.4).

$$\rho_m = \sum_i^J N_i p_i = C\sum_i^J N_i \mu_i = C'm = -\frac{m_n}{2\pi\hbar^2}\mu_n\sum_i^J M_i \tag{3.23}$$

The units of the magnetic scattering length, p, are Å. For the magnetic moment per formula unit, μ, expressed in units of μ_B, $C = 2.645 \times 10^{-5}$ Åμ_B^{-1}. If, rather, the volume magnetization density, m, is known in units of Tesla, then $C' = 2.9109 \times 10^{-5}/4\pi$ Å^{-2}T^{-1}; otherwise, for m in units of emu/cm^3, $C' = 2.853 \times 10^{-9}$ Å^{-2}cm^3/emu. Substituting eq.(3.23) into eq.(3.22) yields:

$$\overline{\delta V(y)} = \frac{2\pi\hbar^2}{m_n}\begin{pmatrix} \rho_n + \rho_{mz} & \rho_{mx} - i\rho_{my} \\ \rho_{mx} + i\rho_{my} & \rho_n - \rho_{mz} \end{pmatrix} \tag{3.24}$$

Finally, we associate the so-called non-spin-flip, ρ_{++} and ρ_{--}, and spin-flip scattering potentials, ρ_{+-} and ρ_{-+}, with the matrix elements in eq.(3.24).

$$\begin{aligned} \rho_{++} &= \rho_n + \rho_{mz} \\ \rho_{--} &= \rho_n - \rho_{mz} \\ \rho_{+-} &= \rho_{mx} - i\rho_{my} \\ \rho_{-+} &= \rho_{mx} + i\rho_{my} \\ \overline{\delta V(y)} &= \frac{2\pi\hbar^2}{m_n}\begin{pmatrix} \rho_{++} & \rho_{+-} \\ \rho_{-+} & \rho_{--} \end{pmatrix} \end{aligned} \tag{3.25}$$

The "+" ("-") sign is for the neutron spin parallel (anti-parallel) to the applied field, so the positive magnetic scattering potential adds to the normally positive (repulsive) nuclear scattering potential. So, for example, ρ_{++} is the element of the scattering potential attributed to the scattering of an incident neutron with spin-up that does not change the orientation of the neutron spin with respect to the magnetic field. Likewise, ρ_{+-} is the element of the scattering potential attributed to the scattering of an incident neutron that changes its spin from up to down, and so on.

We now desire a solution to Schrödinger's equation—one that takes into account the spin dependence of the scattering potential, eq.(3.25), and the spin dependence of the neutron wave function:

$$\Psi(y) = U_+ \begin{pmatrix} 1 \\ 0 \end{pmatrix} \Psi_+(y) + U_- \begin{pmatrix} 0 \\ 1 \end{pmatrix} \Psi_-(y)$$

$$\Psi_+(y) = e^{ik_+y} \tag{3.26}$$

$$\Psi_-(y) = e^{ik_-y}$$

The value of $k_\pm$ is for the $\perp$-component (or y-component in Fig. 3-10) of the wavevector for the different neutron spin states. The spin dependence of $k_\pm$ arises from the energy dependence of the neutron spin in the magnetic field. In the field, the refractive index becomes spin-dependent (i.e., birefringent).

$$k_\pm = n_\pm k_0 = \sqrt{1 - \frac{4\pi(\rho_n \pm |\rho_m|)}{k_0^2}} k_0 \tag{3.27}$$

The spin dependence of the incident neutron wave function contained in U_+ and U_- is determined by the polarization of the incident neutron beam.

3.6. THEORETICAL EXAMPLE 4: Reflection of a polarized neutron beam from a magnetic film

In this example, we consider the reflection of a polarized neutron beam from a magnetic thin film in which the direction of magnetic induction is uniform. This example illustrates how the Parratt formalism developed earlier for unpolarized neutron reflection can be straightforwardly applied to a (saturated) magnetic thin film. Since the direction of magnetic induction is assumed to be parallel to the applied field, (though the magnitude of the induction need not be uniform), the off-diagonal entries in the matrix of eq.(3.25) are zero. We now imagine performing an experiment involving two measurements of the sample reflectivity; first with spin-up neutrons (so $U_+=1$ and $U_-=0$), and then later with spin-down neutrons (so $U_+=0$ and $U_-=1$). A device called a spin-flipper (discussed later) flips the neutron spins from one state to the other. Eq.(3.18) is easily generalized to account for the spin dependence of the neutron scattering potential [28].

$$\prod_{j=n-1}^{0} \overline{M_j^\pm(\delta_j)} \begin{pmatrix} 1+r_\pm \\ ik_0(1-r_\pm) \end{pmatrix} = \begin{pmatrix} t_\pm \\ ik_{n\pm}t_\pm \end{pmatrix} \prod_{j=n-1}^{0} e^{ik_{n\pm}\delta_j} = \begin{pmatrix} t_\pm \\ ik_{n\pm}t_\pm \end{pmatrix} e^{ik_{n\pm}\Delta}$$

$$\overline{M_j^\pm(\delta_j)} = \begin{pmatrix} \cos(k_{j\pm}\delta_j) & \dfrac{1}{k_{\pm j}}\sin(k_{j\pm}\delta_j) \\ -k_{j\pm}\sin(k_{j\pm}\delta_j) & \cos(k_{j\pm}\delta_j) \end{pmatrix} \tag{3.28}$$

In the previous example of a thin Fe layer on Si, we had considered the 20 nm thick layer to be a non-magnetic material with the nuclear scattering length density of Fe. Now, we consider the Fe to be fully saturated with magnetization parallel to the field as shown in Fig. 3-10. The magnetic moment of an Fe atom is $\mu_{Fe} = 2.219\ \mu_B$, so the neutron magnetic scattering length density is

$\rho_m = N_{Fe}C\mu_{Fe} = 4.97x10^{-6}$ Å^{-2} (Table 1). The scattering length density profiles for spin-up and spin-down neutrons are shown in Fig. 3-11(a), as is the profile of the nuclear scattering length density, (Fig. 3-7(a)), for the sake of comparison. Depending upon whether the polarization of the neutron beam is parallel or anti-parallel to H, ρ_m either adds or subtracts from ρ_n. The reflectivities for spin-up neutrons, R^{++}, for which the blue curve in Fig. 3-11(a) is appropriate, and spin-down neutrons, R^{--}, for which the red curve in Fig. 3-11(a) is appropriate, are shown in Fig. 3-11(b). The dotted curve in Figure 3-11(b) is the reflectivity of a non-magnetic film with the nuclear scattering length density of Fe (Fig. 3-7), and would not be measured from a magnetized film of Fe with polarized neutron beams (having a polarization axis in the sample plane and perpendicular to Q). In this example, the splitting between the R^{++} and R^{--} is a measure of the depth profile of the sample magnetization projected onto the applied field direction.

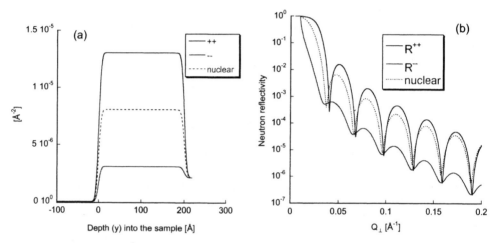

Figure 3-11: (a) Bifurcation of the magnetic scattering length densities profiles depending upon whether the neutron spin is parallel (++, blue curve) or anti-parallel (--, red curve) to the direction of the applied magnetic field. The dotted curve is the nuclear scattering length density profile. (b) The R^{++} and R^{--} reflectivity of the sample is shown as the blue and red curves, respectively. The dashed curve is the reflectivity curve for the case of a film with the nuclear scattering length density of Fe (and not magnetic). *Also see the color plate.*

3.6.1. Influence of imperfect polarization on the reflectivity

In the preceding discussions, reflectivity curves were calculated for neutron beams that were assumed to contain only spin-up neutrons or spin-down neutrons. In other words the neutron beams were ideally polarized. In practice, the polarization of a neutron beam,

$$P = \frac{I_+ - I_-}{I_+ + I_-} \tag{3.29}$$

where I_+ and I_- represent the numbers or fractions of spin-up and spin-down neutrons, respectively, is not 100%. Typically, polarizations of order 90+ % are available for reflectometry experiments.

In order to produce a polarized neutron beam, polarization devices (discussed later) are inserted into the beam line before and sometimes after the sample. A polarization device acts to suppress one spin state by either absorbing the undesired spin state (such a device is called a polarization filter), or by spatially separating the two spin states through reflection from magnetized materials. Nearly all polarized neutron beams contain some fraction of undesired spins. Assume the desired spin state is the spin-up state. The contamination of the polarized neutron beam is attributed to spin-down neutrons. The polarization of the neutron beam approaches 100%, when the ratio, called the flipping ratio $F=I_+/I_-$, of desired neutron spins to undesired neutrons spins becomes large, in fact:

$$P = \frac{F-1}{F+1} \qquad (3.30)$$

Since the transmission of a neutron beam through polarizing supermirrors is typically reduced by about 30% due to absorption of the beam by the Si substrates and Co in the coatings, experimentalists are best served by neutron beams with just enough polarization to obtain the data needed to solve a problem. Somewhat counter-intuitively, it may sometimes be more advantageous to study highly magnetic materials with higher neutron polarizations than used for materials that are only slightly magnetic. To understand this point, we assume that rather than using the perfectly polarized neutron beam in Theoretical Example 3.4, we use one having a flipping ratio of 10 (i.e., 1 in 11 neutrons has the wrong spin state, $P=82\%$). The as-measured spin-up reflectivity will be composed of $0.9R^{++}$, Figure 3-11(b) and $0.1R^{--}$, Fig. 3-11(b), which hardly changes the result (compare the solid and dashed blue curves in Fig. 3-12). However, since the spin-up reflectivity is so much larger than the spin-down reflectivity (in this example), the as-measured spin-down reflectivity will consist of $0.1R^{++}$ (a large source of contamination) and $0.9R^{--}$ (compare the solid and dashed red curves in Fig. 3-12). Failure to account for imperfection of the polarized neutron beam would lead one to mistakenly conclude that the Fe film was less magnetic than it actually is. Provided the polarization of the neutron instrument is known, the true reflectivity curves can be obtained from reflectivity measurements using neutron beams with less than 100% polarization [44].

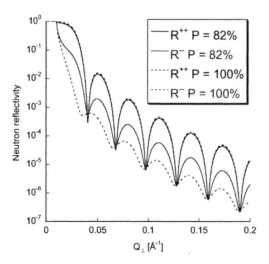

Figure 3-12: Reflectivity curves calculated for an ideally polarized neutron beam (dashed curves) are compared to those calculated for a neutron beam with 82% polarization (solid curves). The large spin-up reflectivity (blue curve) is hardly affected by a poorly polarized neutron beam. On the other hand, the contamination in the poorly polarized neutron beam greatly perturbs the much weaker spin-down reflectivity (red curve), because the contamination when measuring spin-down is spin-up and the spin-up reflectivity is much larger than the spin-down reflectivity. *Also see the color plate.*

In contrast, for the case of a material that is only weakly magnetic, e.g., a magnetic semiconductor with magnetization ~30 emu/cm^3, R^{++} and R^{--} will be little different, so the contamination posed by having one tenth of the wrong spin state in the as-measured reflectivity might be negligible. In this situation, a relatively poorly polarized neutron beam might be preferred over a highly polarized neutron beam, especially if the intensity of the poorly polarized neutron beam is larger than that of the highly polarized beam.

3.6.2. "Vector" magnetometry with polarized neutron beams

In the previous discussions, the neutron spin and magnetic induction have been treated as if they were always parallel (or anti-parallel) to the neutron spin direction. However, this constraint does not always exist. For example, a material with strong uniaxial anisotropy could be oriented with M at an angle of ϕ to H (Figure 3-13). Classically, when a neutron enters a region in which its spin is not parallel to the induction, its spin begins to precess. Depending upon the time the neutron spends in this region and the strength of the induction, it's spin may flip 180°— the intentional rotation of a neutron spin by 180° is the basis for operation of a so called Mezei spin-flipper [45]. Likewise, the magnetization of a material can rotate the spin of a neutron such that a beam with one polarization scatters from the sample with diminished polarization, i.e., some of the spin-up neutrons may be flipped to spin-down—so-called spin-flip scattering. In this situation, the scattering potential, $\overline{\delta V(z)}$, is not simply birefringent: in other words the off-diagonal elements in eq.(3.25) are non-zero.

For the geometry of the neutron reflectometry experiment shown in Fig. 3-13, a further simplification to the off-diagonal elements of eq.(3.22) can be made. One of Maxwell's equations (specifically $\nabla \cdot B = 0$ [46]) requires the out-of-plane component of B across the interface to be continuous, so the component of B parallel to Q or $\hat{y}$ will not yield a change in contrast across the interface. The important consequence of the dipolar interaction between neutron and magnetic moments is that magnetic scattering of the neutron is only produced by the component of the magnetization perpendicular to wavevector transfer. In the case of a (specular) neutron reflection experiment (Fig. 3-13), this requirement means that spin-dependence of the neutron reflectivity arises from the components of the sample magnetization projected onto the reflection (or sample) plane.

3.7. THEORETICAL EXAMPLE 5: Reflection from a medium with arbitrary direction of magnetization in the plane of the sample

We now calculate the scattering from the Fe film for the case when the Fe magnetization is rotated through an angle ϕ about the surface normal from the applied field direction (see Fig. 3-13). In order to account for the possibility that the sample changes the spin state of a neutron, a generalization of eq.(3.28) to include spin-flip scattering, is required [19,20,28].

$$\prod_{j=n-1}^{0} A_j \begin{pmatrix} I_+ + r_+ \\ I_- + r_- \\ ik_0(I_+ - r_+) \\ ik_0(I_- - r_-) \end{pmatrix} = \begin{pmatrix} t_+ \\ t_- \\ ik_{n+}t_+ \\ ik_{n-}t_- \end{pmatrix} \tag{3.31}$$

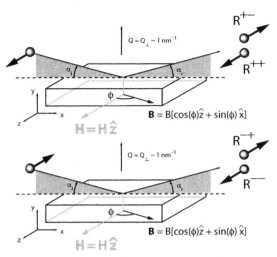

Figure 3-13: Schematic diagram showing (top) a spin-up polarized neutron beam reflecting from a sample with magnetic induction at an angle of ϕ from the applied field. The reflected beam has two components—the (R^{++}) non-spin-flip and (R^{+-}) reflectivities. (lower) The case is shown when the polarization of the incident neutron beam is spin-down.

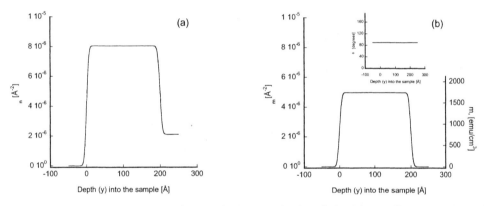

Figure 3-14: Plot of the nuclear (a) and magnetic (b) scattering length densities profiles. Inset: The angle about the surface normal of the magnetization from the applied field is $90°$.

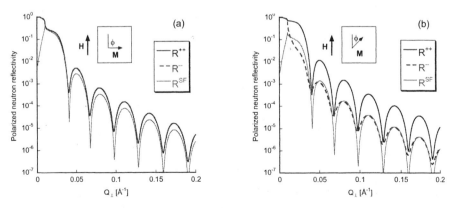

Figure 3-15: Polarized neutron reflectivity curves for Fe/Si (inset) with Fe magnetization rotated (a) $\phi = 90°$ and (b) $45°$ from the applied field and polarization axis of the neutron beam. *Also see the color plate.*

where the elements of $\overline{\overline{A_j}}$ are:

$$A_{11} = 2\eta[\gamma_3 \cosh(S_+\delta_j) - \gamma_1 \cosh(S_-\delta_j)]$$

$$A_{21} = 2\eta[\gamma_1\gamma_3 \cosh(S_+\delta_j) - \gamma_1\gamma_3 \cosh(S_-\delta_j)]$$

$$A_{31} = 2\eta[S_+\gamma_3 \sinh(S_+\delta_j) - S_-\gamma_1 \sinh(S_-\delta_j)]$$

$$A_{41} = 2\eta[S_+\gamma_1\gamma_3 \sinh(S_+\delta_j) - S_-\gamma_1\gamma_3 \sinh(S_-\delta_j)]$$

$$A_{12} = 2\eta[-\cosh(S_+\delta_j) + \cosh(S_-\delta_j)]$$

$$A_{22} = 2\eta[-\gamma_1 \cosh(S_+\delta_j) + \gamma_3 \cosh(S_-\delta_j)]$$

$$A_{32} = 2\eta[-S_+ \sinh(S_+\delta_j) + S_- \sinh(S_-\delta_j)]$$

$$A_{42} = 2\eta[-S_+\gamma_1 \sinh(S_+\delta_j) + S_-\gamma_3 \sinh(S_-\delta_j)]$$

$$A_{13} = 2\eta\left[\frac{\gamma_3}{S_+}\sinh(S_+\delta_j) - \frac{\gamma_1}{S_-}\sinh(S_-\delta_j)\right]$$

$$A_{23} = 2\eta\left[\frac{\gamma_1\gamma_3}{S_+}\sinh(S_+\delta_j) - \frac{\gamma_1\gamma_3}{S_-}\sinh(S_-\delta_j)\right]$$

$$A_{33} = A_{11}$$

$$A_{43} = A_{21}$$

$$A_{14} = 2\eta\left[-\frac{1}{S_+}\sinh(S_+\delta_j) + \frac{1}{S_-}\sinh(S_-\delta_j)\right]$$

$$A_{24} = 2\eta\left[-\frac{\gamma_1}{S_+}\sinh(S_+\delta_j) + \frac{\gamma_3}{S_-}\sinh(S_-\delta_j)\right]$$

$$A_{34} = A_{12}$$

$$A_{44} = A_{22}$$

(3.32)

and

$$S_\pm = in_\pm k_0 = ik_{n\pm}$$

$$\gamma_1 = \frac{|\rho_m| - \rho_{mz} + \rho_{mx} + i\rho_{my}}{|\rho_m| + \rho_{mz} + \rho_{mx} - i\rho_{my}}$$

$$\gamma_3 = \frac{|\rho_m| + \rho_{mz} - \rho_{mx} - i\rho_{my}}{|\rho_m| - \rho_{mz} - \rho_{mx} + i\rho_{my}}$$

$$2\eta = \frac{1}{\gamma_3 - \gamma_1}$$

To calculate the four neutron spin reflectivities, R^{++}, R^{+-}, R^{-+} and R^{--}, the nuclear (ρ_n) and magnetic (ρ_m, a vector) scattering length density profiles for the sample are computed. Examples of these profiles are shown in Figure 3-14, where $\rho_m = C'm_{Fe}(\hat{z}\cos\phi + \hat{x}\sin\phi)$.

Next, we assume the sample is illuminated by a spin-up polarized neutron beam, so $I_+ = 1$ and $I_- = 0$, and use eq.(3.31) to compute R^{++} and R^{+-} ($\equiv |r_+|^2$ and $|r_-|^2$; the probabilities that a neutron with spin-up is reflected with spin-up or spin-down, respectively). Then, the calculation is repeated for a spin-down polarized neutron beam ($I_+ = 0$ and $I_- = 1$) to obtain R^{-+} and R^{--} ($\equiv |r_+|^2$ and $|r_-|^2$; the probabilities that a neutron with spin-down is reflected with spin-up or spin-down, respectively). The result is plotted in Figure 3-15, where $R^{SF} = (R^{+-} + R^{-+})/2$, for the cases (a) $\phi = 90°$ and (b) $\phi = 45°$. For the case $\phi = 90°$, the net magnetization of the sample along the applied field is zero, so there is no splitting between the two non-spin-flip cross-sections (and a strong signal in the spin-flip cross-section). On the other hand, for $\phi = 45°$, the net magnetization of the sample along the applied field is non-zero, so splitting between R^{++} and R^- is observed along with a lower magnitude for R^{SF}.

3.8. A QUALITATIVE (AND INTUITIVE) UNDERSTANDING OF "VECTOR" MAGNETOMETRY

An intuitive understanding of spin-dependent reflection is most easily obtained by considering the kinematical equations that describe reflection, which so far has been treated using the dynamical (exact) formalism. By kinematical, we mean that effects such as the evanescence of the wave function below the critical edge, which greatly perturb the wave function inside the sample, are neglected. These effects are neglected when the transmitted wave function in eq.(3.1) is replaced with the incident wave function. Within the Born approximation, the spin-dependent specular reflection amplitudes for the scattering geometry shown in Fig. 3-13 are [28]:

$$r_{BA}^{\pm\pm}(Q_\perp) \propto \int_0^\Delta [\rho_n(y) \pm \rho_m(y)\cos\phi(y)]e^{iQ_\perp y}dy$$

$$r_{BA}^{\pm\mp}(Q_\perp) \propto \int_0^\Delta \rho_m(y)\sin\phi(y)e^{iQ_\perp y}dy \tag{3.33}$$

The reflectivities for the non-spin-flip processes are a sum of the squares of the nuclear and magnetic structure factors given in eq.(3.33) plus a term resulting from the interference between nuclear and magnetic scattering. The interference term is observed with polarized neutron beams. The spin-flip reflectivity is purely magnetic in origin. Note for the special case where $\phi = 90°$, as can be realized for samples with uniaxial anisotropy, the non-spin-flip reflectivities are purely nuclear (or chemical) in origin. In this special case, the magnetic and chemical profiles of the sample can be isolated from one another. By measuring both the non-spin-flip and spin-flip reflectivities as a function of $Q_\perp$, eq.(3.33) suggests that the variation of the magnetization vector, in amplitude and direction in the sample plane, can be obtained as a function of depth. This capability is an important reason why polarized neutron reflectometry complements conventional vector magnetometry, which is a technique that measures the net (or average) magnetization vector of a sample.

A second important example of the power of polarized neutron reflectometry is for detecting and isolating the magnetism of weakly magnetic materials from that of strongly magnetic materials through analysis of the Fourier components of the reflectivity. Situations in which this capability may be valuable include detecting coerced or proximal magnetism in materials that are normally non-magnetic in the bulk, e.g., Pd that becomes magnetic in proximity to Fe [47]. Polarized neutron reflectometry is also valuable in studies of weakly ferromagnetic thin films, e.g., (Ga, Mn)As [48], grown on substrates that contribute a strong diamagnetic or paramagnetic background to the signal measured in a conventional magnetometer.

For studies of films whose magnetization does not change with depth, but instead the magnetization changes along the film plane, as realized for example in films composed of magnetic domains, the sizes of the magnetic domains in relation to the coherence of the neutron beam (which is typically microns in size) determine whether off-specular or diffuse scattering of the neutron beam, in addition to specular scattering, is observed. Diffuse scattering can be observed when the lateral variation of the magnetization is small in comparison to the coherence of the neutron beam. On the other hand, if the domains are much bigger than the coherence of the beam, then information about the magnetism of the sample will be observed in the specular reflectivity.

Consider specular reflection of a neutron beam from a film of uniform thickness, Δ, containing a single domain with uniform magnetization and having a lateral size that is large in comparison to the coherent region of the neutron beam. In this example, the reflectivity of the domain is straightforwardly calculated using eq.(3.33).

$$R_{BA}^{\pm\pm}(Q_\perp) \propto \left(\rho_n^2 + \rho_m^2 \cos^2\phi \pm 2\rho_n\rho_m \cos\phi\right)\left(1 - \cos Q_\perp\Delta\right) \tag{3.34}$$
$$R_{BA}^{SF}(Q_\perp) = R_{BA}^{\pm\mp} \propto \rho_m^2 \sin^2\phi\left(1 - \cos Q_\perp\Delta\right)$$

Using the relation between ρ_m and $\boldsymbol{m}$ provided by eq.(3.23) we resolve $\boldsymbol{m}$ into components parallel and perpendicular to the applied field such that $m_\parallel \propto \rho_m \cos\phi$ and $m_\perp \propto \rho_m \sin\phi$, respectively. Then, using eq.(3.34) we obtain a physical meaning for the difference (or splitting) between the non-spin-flip reflectivities, Δ_{NSF}, and R^{SF}.

$$\Delta_{NSF}(Q_\perp) = R_{BA}^{++}(Q_\perp) - R_{BA}^{--}(Q_\perp) \propto m_\parallel\left(1 - \cos Q_\perp\Delta\right) \tag{3.35}$$
$$R_{BA}^{SF}(Q_\perp) \propto m_\perp^2\left(1 - \cos Q_\perp\Delta\right)$$

That is, the splitting between the non-spin-flip reflectivities is proportional to the projection of the domain magnetization onto the applied field, and the spin-flip reflectivity is proportional to the *square* of the domain magnetization perpendicular to the applied field.

Owing to the fact that neutron scattering is a statistical probe of a sample's potentially non-uniform distribution of magnetization, rather than a scanning probe of the magnetization at the atomic scale (which could be non-representative), there is an important complication to the interpretation of the neutron scattering results. The complication stems from, as discussed earlier, whether the non-uniformity of magnetization varies on a length scale that is small or large compared to the coherent region [49] of the neutron beam. If the fluctuations of magnetization are small compared to the coherent region of the neutron beam, then the reflectivity is obtained from the reflection amplitude of an ensemble of domains. Depending upon the details of the fluctuations, the scattering may consist of specular and off-specular (or diffuse scattering) components. On the other hand, if the fluctuations occur on a length scale larger than the coherent region of the neutron beam, then the reflectivity is the sum of the reflectivity of each component, and the reflectivity is specular.

It is the second case, one composed of domains that are large in comparison to the coherent region of the neutron beam that is easiest to treat. In this case, $\Delta_{NSF} \propto \langle m_\parallel \rangle$ and $R_{BA}^{SF} \propto \langle m_\perp^2 \rangle$, where $\langle\ \rangle$ denotes the average value of the ensemble of domains. The first term, Δ_{NSF}, provides a measure of the Fourier components of the net sample magnetization projected onto the applied field and is similar to the net magnetization of the sample as measured by a magnetometer. The second term contains qualitatively different information than that which can be measured by a vector magnetometer. Specifically, R^{SF} is a measure of the *mean square deviation* of the magnetization away from the applied field. For the examples of magnetic domain distributions shown in Fig.3-16, the net sample magnetization in any direction is zero. In this situation, a vector magnetometer would measure the zero-vector, yet, provided the domains are large in comparison to the coherence of the neutron beam, the mean square deviation of the magnetization away from the applied field is a (non-zero) quantity obtained from polarized neutron scattering as R^{SF}[50]. Note, f_1, $f_\perp = f_2 + f_4$, and f_3, [Figure 3-16(left)] and ϕ, [Figure 3-16(right)] can be chosen such that $\langle m_\perp^2 \rangle$ is the same for both models, so polarized neutron reflectometry cannot distinguish between these two particular domain distributions; nevertheless, the technique does provide information about related magnetic properties, e.g., anisotropy [51].

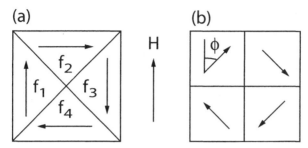

Figure 3-16: Examples of magnetic domains with magnetization directed as shown by the arrows. (a) In this closure domain model f_i represents the area fraction of the i-th domain, and the magnetization of the material reverses by changing the value of f_i. (b) In this model, the area fractions of the domains are equal and the magnetization of the material reverses as the angle between the magnetization and the applied field direction changes from 0 to 180°.

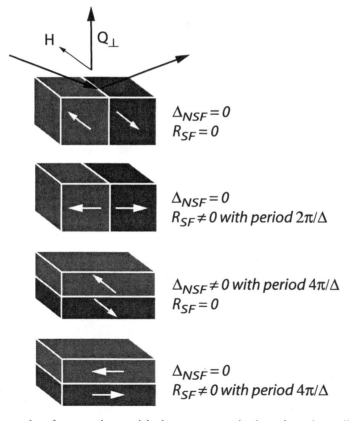

Figure 3-17: Four examples of a magnetic material whose net magnetizations along the applied field, H, (or any direction) are all zero (the volume fractions of red and blue domains are equal, and the domain sizes are assumed to be large in comparison to the coherent region of the neutron beam). The neutron scattering signature in the specular reflectivities from each model is unique.

Extreme cases of domain distributions—ones that yield no net magnetization along the applied field (as realized when the magnitude of the applied field is equal to the coercive field) are shown in Fig. 3-17, along with the associated (unique) features of the *specular* reflectivity for the particular domain structure. In the first case, the non-spin-flip reflectivities would be superimposed with amplitudes that contain nuclear and magnetic contributions [the reflectivity would not be the same as the purely nuclear case shown in Fig. 3-5(b)]. The period of the non-spin-flip reflectivities would be $2\pi/\Delta$, and the spin-flip reflectivity would be zero. In the second case, the non-spin-flip reflectivities would be purely nuclear in origin, and the spin-flip reflectivity would be non-zero with a period equal to the $2\pi/\Delta$. In the third case, the two non-spin-flip reflectivities would be different and have a period of $2\pi/(\Delta/2)$. The spin-flip reflectivity would be zero for this case. For the last case, the non-spin-flip reflectivities would be purely nuclear in origin (as in the second case) and the spin-flip reflectivity would be non-zero with a period of $2\pi/(\Delta/2)$.

3.9. DESCRIPTION OF A POLARIZED NEUTRON REFLECTOMETER

Three essential requirements for any polarized neutron reflectometer are:

(1) a priori knowledge of the polarization of the neutron beam illuminating the sample;

(2) the capability to measure the intensity and polarization of the neutron beam reflected by a sample;

(3) and the ability to make these measurements as a function of wavevector transfer parallel and perpendicular to the sample surface.

The first feature requires a device to polarize the neutron beam (a polarizer) and to flip the neutron beam polarization (a spin-flipper). The second feature requires a neutron detector and a device(s) to flip and measure the neutron beam polarization after reflection from the sample. Finally, wavevector transfer is obtained from measurements of the neutron wavelength and the angle through which the neutron has been scattered. Angles are measured using slits to define the path of the neutron beam that is allowed to strike a neutron detector, or by using a position sensitive neutron detector.

3.9.1. Preparation of the cold neutron beam for a reflectometer at a pulsed neutron source

We briefly describe a reflectometer/diffractometer (Fig. 3-18) designed for studies of magnetic materials at a source of pulsed neutrons. Sources of pulsed neutrons (e.g. LANSCE at Los Alamos National Laboratory) provide neutron pulses that are typically very short, on the order of 100-300 µs, and periodic—with periods ranging between $\tau \sim 10$ - 100 ms.

For neutron scattering measurements in the small-Q or large d-spacing regimes (neutron scattering measurements of magnetic materials are often in these regimes), neutrons with very low energies (long wavelengths) are desirable because the sine of the critical angle, $\sin \theta_c = \lambda Q_c / 4\pi$ is proportional to neutron wavelength. Since the relative influence of systematic or alignment errors decreases with increasing angle, $Q_\perp$ can be more accurately measured with long wavelength neutrons than with short wavelength neutrons. Also, by measuring relevant values of $Q_\perp$ at large angles, we can take advantage of the intrinsically large and divergent neutron beam (in comparison to X-ray beams which can be very small and highly collimated). Even though the wavelength of a cold neutron beam might be an order of magnitude larger than X-ray beams, the specular reflectivity is still measured through angles on order of 1° (as in X-ray reflectometry) because Q_c probed with neutrons is typically an order of magnitude smaller than that for X-rays. Cold neutron beams are obtained by viewing the neutron source through a material like l-H_2 that absorbs neutron energy through collisions with hydrogen atoms. The so-called moderator changes the energy of the neutron beam from MeV to meV energies. The spectrum of a cold neutron beam is shown in Fig. 3-19.

In order to preserve the intensity of a neutron beam, the beam may travel through a glass pipe, called a neutron guide, whose inside surfaces are coated with a highly reflecting material, e.g., ^{58}Ni, to neutrons. Neutrons striking the sides of the guide at sufficiently small wavevector transfer, i.e., $Q_\perp < Q_c$, (= 0.026 Å^{-1} for ^{58}Ni) are reflected and stay confined within the guide. The angular divergence of neutron trajectories emanating from the end of the guide is $2\alpha_c \cong 2\lambda/Q_c 4\pi$; thus, neutrons with trajectories within $\pm\alpha_c$ of the centerline of the neutron guide can, in principle, interact with a sample. Note, the divergence of the neutron beam from a guide increases linearly with wavelength.

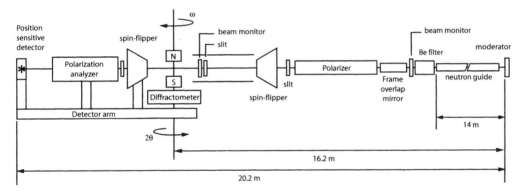

Figure 3-18: Schematic diagram of a polarized neutron reflectometer/diffractometer at a pulsed neutron source (LANSCE).

While the moderator greatly reduces the energy of the neutron beam, there still remain some highly energetic neutrons that may interact with components of the instrument and consequently lose energy through these interactions. Thermalization of energetic neutrons is an important source of instrumental background. In order to suppress this source of background, a filter is required, i.e., a device that is nearly opaque to high energy neutrons and transparent to low energy neutrons. One such device, called a Be filter, consists of a cryogenically cooled block of polycrystalline Be through which the neutron beam passes. Since, the lowest order Bragg reflection of Be corresponds to a d-spacing of approximately 4 Å, the portion of the neutron beam with $\lambda < 4$ Å will be scattered by the Be block thereby reducing the high energy content of the transmitted neutron beam. The spectrum shown in Fig. 3-19 was measured after the neutron beam passed through a Be filter (Fig. 3-18).

A second source of instrumental background arises from very slow (i.e., very long wavelength) neutrons that require more than one period of the neutron source to reach the detector. As mentioned earlier, neutron wavelength is measured at pulsed neutron sources by recording the time-of-flight of a neutron to travel a known distance. The velocity of a neutron is the slope of its trajectory in Fig. 3-20. Ambiguity in the time-of-flight exists, since the neutron detector can not distinguished between a neutron whose time-of-flight is t (e.g., t corresponding to a neutron with $\lambda = 9$ Å, Fig. 3-20, solid line) compared to one with $t + \tau$ (e.g., t corresponding to a neutron with $\lambda = 18$ Å, Fig. 3-20, dashed line). At first glance, the contamination of the neutron beam by very long wavelength neutrons might seem unimportant because there are so few of these neutrons. However, their probability of reflection from a sample is very great (because $Q_\perp$ is so small when λ is so large). In fact, it is a happy coincidence that the decay of the spectrum with wavelength is approximately counteracted by the approximate λ^4 increase (Fresnel) of the reflectivity curve, so the measured intensity of the sample reflectivity is reasonably comparable (within an order of magnitude or so) for all wavelengths.

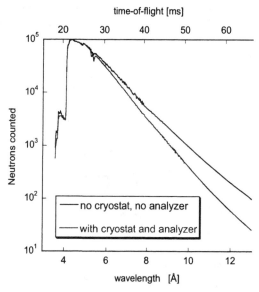

Figure 3-19: Neutron spectrum measured from a coupled l-H_2 moderator (blue curve) without a cryostat and without polarization analysis and (red curve) with a cryostat and polarization analyzer inserted. The spectrum is plotted for neutron events counted in wavelength increments of constant size. The Be filter is mostly transparent above ~4 Å. The decay of the spectrum with wavelength is reminiscent of that from a black-body radiator. *Also see the color plate.*

Fortunately, the possibility of an elevated instrumental background due to very long wavelength (i.e., the frame overlap) neutrons is easily addressed by placing a mirror in the neutron beam at a large angle such that neutrons with very long wavelengths (for example greater than 14 Å) are reflected out of the neutron beam, while the remaining neutrons are transmitted through the mirror. In the case of the instrument discussed here, so-called $3\theta_c$ (unpolarizing) supermirrors, consisting of a multilayer metallic coating whose critical edge is three times larger than that of Ni are deposited onto relatively transparent Si substrates. Extension of the critical edge is accomplished by arranging a variation in the thickness of the layers in the mirror such that a series of Bragg reflections are produced starting at Q_c (for a thick Fe film) out to large wavevector transfer [52,53]. After traveling through the neutron guide, Be filter and frame overlap mirrors, an intense beam of unpolarized neutrons with large cross-section (for the instrument whose schematic is shown in Fig. 3-18, the cross-section is 6 cm by 6 cm) and having a large wavelength band from 4 to 13 Å, is ready for polarization.

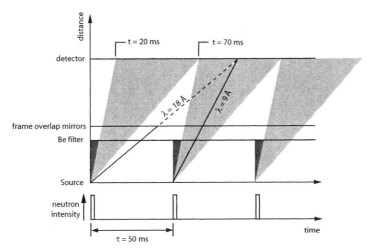

Figure 3-20: Distance vs. time graph of trajectories corresponding to neutrons with different velocities (or wavelengths). Since a detector can only distinguish times-of-flight within one source period, neutrons with $\lambda = 9$ Å and 18 Å are assigned the same times-of-flight; therefore, their wavelengths cannot be distinguished. Fortunately, mirrors, called frame overlap mirrors, can be placed in the neutron beam line to reflect the $\lambda = 18$ Å neutrons out of the beam while allowing the $\lambda = 9$ Å neutrons to pass through (to the detector). *Also see the color plate.*

3.9.2. Polarization of cold neutron beams

Production of cold neutron beams with polarization in excess of 90% is primarily accomplished using polarizing supermirrors. Much like the supermirror discussed previously, a polarizing supermirror consists of hundreds of layers of alternating non-magnetic and magnetic materials [52,53]. The key to a good polarizing mirror is to maximize the magnitude of the spin-up potential, while matching the spin-down potential to that of the substrate, typically Si. The spin-dependent reflectivity for such a polarizing supermirror is shown in Fig. 3-21(a). Spin-up neutrons are reflected from the mirror, while spin-down neutrons are transmitted through it. In the case of the instrument illustrated in Fig. 3-18, the transmitted spin-down neutron beam is used for experiments, and the spin-up beam is discarded (or absorbed). The polarization of the transmitted neutron beam is shown in Fig. 3-21(b) [54,55]. An important reason for using the transmitted polarized neutron beam is that the same beam line can be used for experiments that require either polarized or unpolarized neutrons (e.g., an unpolarized neutron beam is obtained by simply translating the polarizer out of the beam line).

The range in wavelength, $\Delta\lambda$, and angular divergence, $\Delta\theta$, for which the neutron beam is well-polarized is determined by the locations of the critical edge of the spin-down neutrons (below this edge the transmitted neutron beam has little intensity) and the critical edge for spin-up neutrons (above this edge both spin-down and spin-up neutrons are transmitted through the polarizer). The range in $\Delta Q_\perp$ over which the neutron beam is well-polarized is 0.055 Å^{-1} (Fig. 3-21)—a typical range for a supermirror. For an instrument using monochromatic radiation $\Delta\lambda/\lambda_0 = \delta\lambda/\lambda_0 \sim 2\%$ (rms), so the contribution to $\Delta Q/Q$ from the range in neutron wavelength is correspondingly small. Consequently, monochromatic neutron beams with relatively large divergence, $\Delta\theta$, can be easily polarized. For example, if $\lambda_0 = 5$ Å, a neutron beam with divergence of order $\Delta\theta = 1.25°$ can be polarized. However, for instruments which use the time-of-flight technique and therefore use a large wavelength band, e.g., $\Delta\lambda \sim 10$ Å, a challenging situation is encountered in that conventional techniques to polarize a monochromatic neutron beam can not efficiently polarize a neutron beam with a large wavelength band *and* achieve the large divergence of a monochromatic polarized neutron beam (within one period of the neutron source).

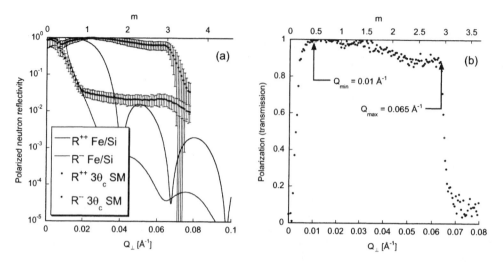

Figure 3-21: (a) Reflectivity from one of the polarizing supermirrors used in the Mezei polarization cavity. The error bars represent the 1-σ standard deviations of reflectivities measured from 192 of the 196 mirrored-surfaces comprising the polarizing supermirror wedge in the polarization cavity (see text). The calculated reflectivity curves from the Fe/Si example in Fig. 3-11 are shown for comparison. (b) The polarization obtained from the spin dependence of the neutron beam transmitted through the polarizing supermirror whose reflectivity is shown in (a). *Also see the color plate.*

For example, say $3\theta_c$ polarizing supermirrors are placed at an angle of $\theta = 0.9°$ relative to the neutron beam, so as to polarize $\lambda = 5$ Å neutrons (see Fig. 3-22). Provided the neutron beam is collimated so that neutrons strike the mirror with angles ranging from 0.25° to 1.5° the neutron beam will be polarized (for $\lambda = 5$ Å). Since collimation of the neutron beam is usually achieved with mechanical slits, neutrons with $\lambda = 15$ Å will also strike the mirror with angles ranging from 0.25° to 1.5°. For these longer wavelength neutrons, $Q_\perp$ ranges from 0.004 to 0.022 Å^{-1}; thus, when using the beam transmitted through a polarizer, the neutron beam will remain polarized (neutrons with $Q_\perp < Q_{min} = 0.01$ Å^{-1}, will be reflected out of the neutron beam regardless of their spin direction), but only a small fraction of the long wavelength neutrons (those satisfying 0.01 Å$^{-1} < Q_\perp < 0.022$ Å^{-1}) will be transmitted through the polarizer (about 20% of long wavelength neutrons are transmitted in comparison to short wavelength neutrons). In this configuration, neutrons with $\lambda > \lambda_{min}$ are inefficiently polarized since $\Delta\theta$ and θ could have been larger. What is needed is a method to achieve the optimum values of $\Delta\theta$ and θ *within one period of the neutron source* for $\lambda > \lambda_{min}$.

One solution that achieves these requirements is the so-called polarization cavity [illustrated in Figure 3-22(b)] [56-58]. Transmission polarizing supermirrors are placed inside a neutron guide to form the shape of a wedge [Fig. 3-22(b)]. The angle of the wedge, ε, is chosen such that $\varepsilon = 2(\theta_c^{mirror} - \theta_c^{guide})$ where θ_c is the critical angle for the shortest wavelength, λ_{min}, in the neutron beam whose polarization is desired. In this scheme, the divergence of the neutron beam illuminating the polarization cavity is that of the neutron guide, and as noted earlier, the divergence increases linearly with wavelength. So, the polarization cavity has the property that any spin-up neutron with $\lambda > \lambda_{min}$ will be reflected out of the cavity by the supermirrors and escape the neutron guide (to be absorbed by boron in the glass or in the borated-polyethylene surrounding the guide), because after reflection from the supermirrors, the angle of incidence between the neutron and the guide exceeds the critical angle of the guide. On the other hand, spin-down neutrons will travel straight through the wedge mirrors; thus, for $\lambda > \lambda_{min}$, the neutron beam is polarized spin-down with divergence that increases linearly with λ. Pictures of the interior and exterior of a polarization cavity are shown in Figure 3-23. The cross-section of the polarized neutron beam produced by the cavity is 130 mm tall by 25 mm wide. The length of the cavity is about 1.2 m.

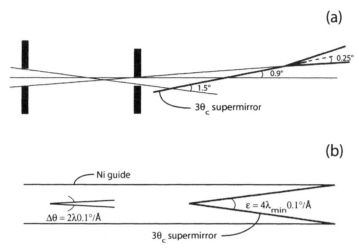

Figure 3-22: (a) A conventional arrangement for polarizing a neutron beam. Mechanical slits limit the range of trajectories that impinge on a $3\theta_c$ polarizing supermirror. The transmitted neutron beam is polarized spin-down. The min and max angles chosen to assure that the transmitted neutron beam with $\lambda \doteq 5$ Å is polarized spin-down. Some spin-down neutrons with $\lambda = 15$ Å (the blue colored trajectory) will nevertheless reflect from the supermirror and be lost from the transmitted polarized beam, thus, reducing the efficiency of this polarization device for multi-wavelength applications. The spatial and angular intensity distributions across the polarized neutron beam will exhibit undesirable wavelength dependence. (b) A polarization cavity consists of a wedge-shaped arrangement of polarizing supermirrors placed inside a Ni-coated neutron guide (at its end). The divergence of the neutron beam impinging upon the supermirrors is limited by the neutron guide and has a favorable linear dependence upon λ. The angle subtended by the supermirrors, ε, is chosen such that for $\lambda > \lambda_{min}$ only spin-down neutrons are transmitted through the cavity. *Also see the color plate.*

The mirrors in the polarization cavity are magnetized in a 315 Oe field produced by a solenoid in which the cavity resides, Fig. 3-23(b). The polarization axis of the neutron beam at the exit of the cavity is initially directed along the neutron beam line (the x-axis). At the exit of the solenoid, the magnetic field is about 200 Oe (along the x-axis) and decays with distance (Fig. 3-24). Permanent magnets after the solenoid generate a magnetic field transverse to the neutron beam (Fig. 3-24, inset). The variation of the angle, ϕ, of the magnetic field with respect to the neutron beam line with distance from the exit of the guide is shown in the inset of Fig. 3-24. The rate of change in ϕ with time, $d\phi/dt$, is shown in Fig. 3-25 (open symbols) for the case of a neutron with wavelength of $\lambda = 4$ Å. Provided $|d\phi/dt| < |\varpi_L|/4$ where $\omega_L = \gamma|B|$ (called the Larmor precession frequency) and $\gamma = -1.833 \times 10^4$ rad/Gs, the neutron beam polarization will follow the change in the direction of the magnetic field with minimal (<3%) depolarization —the neutron spin changes direction approximately adiabatically [59-61]. Specifically, the neutron spin precesses on the surface of a cone with frequency equal to ω_L. The axis of the cone changes direction at the rate of $d\phi/dt$. Since the condition for adiabatic rotation of spin is fulfilled for neutrons with wavelength λ_{min} (see Fig. 3-25) (and therefore for $\lambda > \lambda_{min}$), the polarization of the neutron beam rotates from the beam axis (i.e., the direction of neutron flight) to one perpendicular to the beam axis with negligible (<3%) loss in polarization [62].

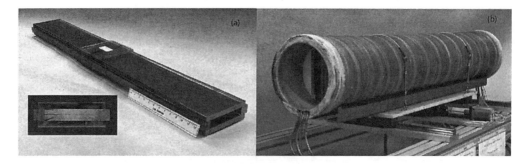

Figure 3-23: (a) Polarization cavity with a length of neutron guide containing a wedge-shaped arrangement of polarizing transmission supermirrors. (b) The polarization cavity is visible through the end of a solenoid used to magnetize the mirrors. The polarization cavity assembly is supported on a translation stage so that it can be reproducibly translated in and out of the neutron beam as needed. *Also see the color plate.*

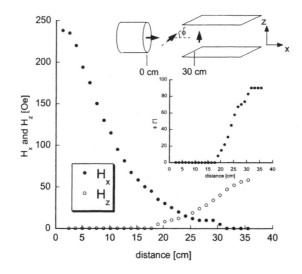

Figure 3-24: The variation of the x (along the beam axis) and z (in the vertical direction) components of the magnetic guide field in the region (picture inset) between the exit of the cavity and the beginning of the guide plates is shown. (Graph inset) The angle between the magnetic field and the beam axis is ϕ.

3.9.3. Spin-flippers

In order to separate nuclear and magnetic scattering, at least two measurements are needed—one with spin-up neutrons and the other with spin-down neutrons. Thus, a method of flipping the neutron beam polarization is required. A device that flips the neutron spin is called a spin-flipper. An example of such a flipper—a radio-frequency gradient field spin-flipper [63,64], is shown in Fig. 3-26 and Fig. 3-27. The flipper consists of a pair of permanent magnets that are tilted from front to back. The tilt introduces a gradient in the strength of the vertical field (a field that is parallel to the laboratory reference field) along the neutron beam line (x-axis) with a value of about 95 G (large in comparison to stray fields) in the center of the flipper. A magnetic shield consisting of μ-metal sandwiched between layers of steel surrounds the flipper [Fig. 3-27(b)]. The inner steel layer helps return the magnetic field lines created by the permanent magnets inside the flipper; thus, assuring the field, $B_0(x)\hat{z}$, changes linearly with position along the beam inside the flipper (Fig. 3-28). The outer steel shield shunts some of the stray field that may be produced by high field magnets (e.g., superconducting magnets) used to magnetize samples. The μ-metal decouples the fields inside the flipper from those outside it. The Larmor precession frequency is shown in Fig. 3-29 as a function of position and time-of-flight (for a neutron with λ = 4 Å).

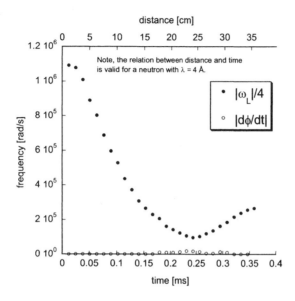

Figure 3-25: Rate of change in the angle between the magnetic field and the neutron beam line as a neutron with λ = 4 Å travels from the exit of the polarization cavity to the beginning of the magnetic guide. Provided |dφ/dt| (open symbols) < |ω$_L$|/4 (solid symbols), the neutron spin will follow the magnetic field.

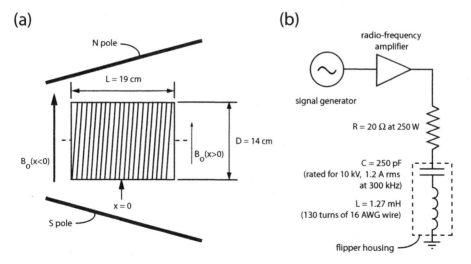

Figure 3-26: Schematic (a) mechanical and (b) electrical diagrams of the radio-frequency gradient field spin-flipper. The neutron beam line is shown by the dashed line in (a).

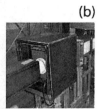

Figure 3-27: Photograph of a radio-frequency gradient field spin-flipper (a) without and (b) with magnetic shields installed. *Also see the color plate.*

By itself, a gradient *in the magnitude* of the field through the flipper will not change the polarization of the neutron beam; rather, a slow (adiabatic) change in direction is needed. The second part of the flipper (the part responsible for the change in direction of the field) is the radio-frequency coil that consists of copper wire wrapped around a ceramic (Al_2O_3) tube. Since the ceramic tube is an insulator, eddy currents (a source of loss) induced by the radio-frequency field are minimized. Connected to the coil are high power capacitors that form a resonance circuit [see Fig. 3-26(b)]. When radio-frequency power (~70 W) is applied to the flipper, the resultant magnetic induction inside the flipper consists of a component that rotates with frequency ω_0 in the x-y plane time and a static component, $B_0(x)\hat{z}$, (from the permanent magnets) transverse to the beam line (i.e., parallel to the z-axis). In the rotating frame of reference (one that rotates with the radio-frequency field about the z-axis), the effective field is a combination of two fields whose orientations are fixed in the rotating frame: one is the peak amplitude of the radio-frequency field ($B_1 \sim 13$ G) (Fig. 3-28), and the other is the spatially changing component in the vertical direction with a constant offset removed [60]. The effective induction, $\boldsymbol{B}_{eff}$, in the rotating frame of reference is:

$$\boldsymbol{B}_{eff}(x) = \left(B_0(x) - \frac{\varpi_0}{|\gamma|} \right)\hat{z} + B_1\hat{x} \qquad (3.36)$$

In the rotating frame of reference, $\boldsymbol{B}_{eff}$ changes direction smoothly with position (see Fig. 3-28, inset)—first directed up, then down from the front to the back of the flipper. The rate of change in angle of $\boldsymbol{B}_{eff}$, $d\phi/dt$, and $1/4\omega_L$ (for $\lambda = 4$ Å) with position along the beam line are shown in Fig. 3-29. Since $|d\phi/dt| < |\varpi_L|/4$, the polarization of the neutron beam rotates 180° adiabatically, provided the flipper is energized. By choosing $\varpi_0 = |\gamma B_0(x = 0)|$, which corresponds to about $2\pi \cdot 297$ kHz, the midpoint in the 180° rotation of the polarization occurs in the center of the flipper.

The measured flipping efficiency of the radio-frequency gradient field spin-flipper is shown in Fig. 3-30. This flipper performs nearly ideally approaching flipping efficiencies of 100% over a broad range of wavelengths ($\lambda > \lambda_{min}$). Besides efficiently flipping neutron spin, the flipper also has two other attractive properties: first, tuning, e.g., optimization of currents etc., is never required even if the stray field environment changes, and secondly, no material is introduced into the beam line (i.e., the flipper is hollow) that might otherwise increase the background of the instrument or absorb neutrons.

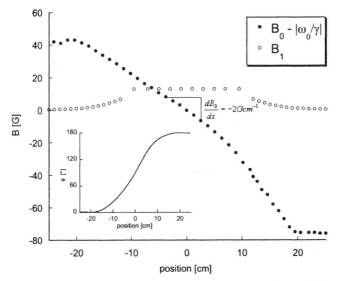

Figure 3-28: Plot of the variation in the static field component, B_0 (solid symbols), and the amplitude (when the flipper is "on") of the radio-frequency field component, B_1 (open symbols), with position along the centerline of the spin-flipper. Inset: The angle, ϕ, between the effective field, B_{eff}, and the z-axis, corresponding to the direction of the magnetic field applied to the sample (and the laboratory frame of reference). For the case when the spin-flipper is off, $B_1 = 0$, so $\phi = 0$ for all positions through the flipper.

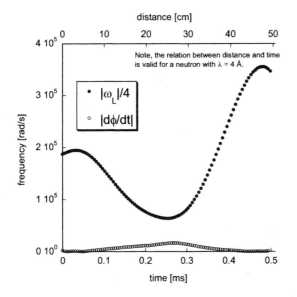

Figure 3-29: Rate of change in the angle between the magnetic field and the z-axis as a neutron with $\lambda = 4$ Å travels through the flipper. Provided $|d\phi/dt|$ (open symbols) $< |\omega_L|/4$ (solid symbols), the neutron spin will follow the effective magnetic field with minimal depolarization.

During an experiment, the radio-frequency coil is energized (de-energized) to obtain spin-up (down) neutrons. For example, the coil might be cycled on and off every couple minutes. For experiments requiring measurements of the four spin-dependent neutron cross-sections a second flipper (Figure 3-18) and another polarizing device (Figure 3-18, usually this device is a polarizing supermirror or stack of polarizing supermirrors) are required. For these experiments, the two flippers are cycled on and off in the four possible combinations. Figure 3-31 shows the detector arm on which the second spin-flipper, polarization analyzer and detector rests.

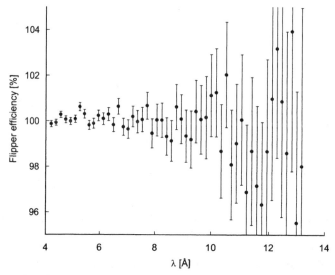

Figure 3-30: Variation of the efficiency of the flipper to flip a polarized neutron beam while maintaining good polarization is shown as a function of wavelength. The average flipper efficiency is 99.8%.

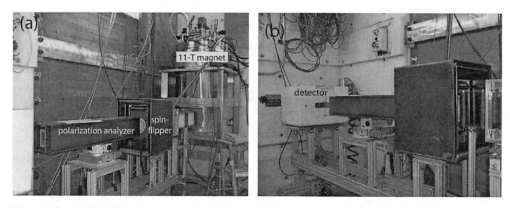

Figure 3-31: (a) View from the detector-end of the reflectometer. An 11-T superconducting magnet visible in the background. (b) View from the sample position towards the detector. *Also see the color plate.*

3.10. APPLICATIONS OF POLARIZED NEUTRON REFLECTOMETRY

In this section we describe two applications of polarized neutron reflectometry to the study of a simple magnetic system. In the first example, X-ray and polarized neutron reflectivity data are analyzed separately, and then jointly, to obtain the magnetic structure of a sample of FeCo on GaAs [65]. This example illustrates how the reflectivity obtained from a model of the chemical composition of a sample is fitted to X-ray data using the computer routine CO_REFINE. From this analysis, parameters for surface and interface roughness and film thickness are obtained. These parameters serve as initial guesses in the refinement of a second model to the neutron data that also uses CO_REFINE. Finally, the X-ray and neutron models are compared, and a new model—one that includes a reacted layer between the magnetic film and substrate, is fitted to the X-ray and neutron data at the same time using CO_REFINE to achieve a consistent explanation for all the data.

In the second example, we illustrate application of vector magnetometry using neutron scattering to isolate the magnetic scattering from the nuclear scattering of the same sample. In this example, a nuclear model is fitted to the non-spin-flip scattering and a magnetic model to the spin-flip scattering. Again, the magnetic and nuclear layer thicknesses of the film differ and are reconciled by the addition of a non-magnetic reacted layer between the magnetic layer and substrate.

3.10.1. Magnetic vs. chemical structures identified through X-ray and polarized neutron reflectometry

In this example, a detailed understanding of the magnetic structure of the interface in one prototypical spin injection heterostructure comprised of an alloy of FeCo on one side (the spin source) and GaAs(100) on the other side (the spin sink) is obtained using a combination of X-ray and polarized neutron reflectometry [65]. The example demonstrates use of the computer routine CO_REFINE to obtain a magnetic and nuclear model whose X-ray and polarized neutron (non-spin-flip) reflectivities fit the data. In fitting the model to the data, we find the magnetic thickness of the FeCo layer to be 6 Å less than its chemical thickness.

The sample consists of an $Fe_{48}Co_{52}$ film epitaxially grown onto semi-insulating GaAs(100) (2x4)/c(2x8)β2 –As rich surface [66] by molecular beam epitaxy under ultra high vacuum and characterized by *in-situ* electron diffraction and *ex-situ* X-ray diffraction, Rutherford backscattering spectrometry, transmission electron microscopy and magnetometry. The 20-nm thick FeCo layer was grown at 95°C, and a 3 nm thick Al capping layer was subsequently deposited to prevent oxidation during *ex situ* characterization. The detailed sample preparation and structural characterization results are described elsewhere [67]. The magnetization hysteresis loops of FeCo alloy grown on GaAs(100) (2x4)/c(2x8)β2 with magnetic fields applied along two perpendicular directions in the sample plane are shown in Figure 3-32. The magnetization measured along the [011] direction (solid curve) indicates this axis is an easy axis, while the sheared hysteresis loop measured along the [011] direction (dashed curve) suggests this direction is considerably harder.

The specular X-ray reflectivity of the sample after removal of diffuse scattering (Figure 33) was measured at room temperature with a conventional rotating anode X-ray generator, producing CuK_α radiation, and a position sensitive detector as described elsewhere [68]. One important distinction between how X-ray and neutron reflectivity data are collected involves the portion of the sample illuminated by the respective beam. Generally (with the exception of the very small $Q_\perp$ region), the X-ray beam illuminates only a portion of the sample, whereas, in the case of neutron reflectometry, the sample is most often completely bathed in the neutron beam. When comparing the X-ray and neutron reflectivity data from the same sample it is important to be cognizant of the possibility that the two techniques are perhaps measuring different quantities, since the sample may not be uniform over its entire surface. In order to minimize the influence of non-uniformity (if any) of the sample on the reflectivity data, the X-ray reflectivity of the sample was measured over several parts of the sample in order to make a more accurate comparison (by addition of the curves) with the neutron measurements.

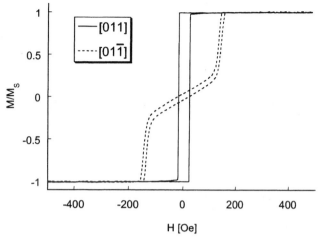

Figure 3-32: Magnetization of the FeCo/GaAs sample along the easy, [011], and hard, $[01\bar{1}]$, axes. The sample magnetization has a large uniaxial anisotropy. Figure adapted from Ref. [65].

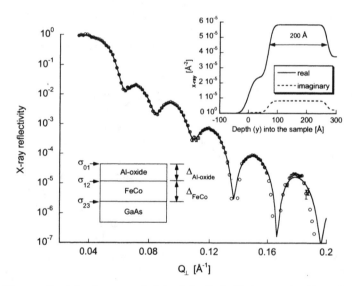

Figure 3-33: X-ray reflectivity (open symbols) of the FeCo/GaAs sample. The calculated reflectivity (solid curve) is from the model shown in the inset.

The X-ray reflectivity (solid curve, Fig. 3-33) of a model structure, Fig. 3-33, inset, was calculated using the optical formalism of Parratt [4]. The initial guesses for the parameters of the model from which the X-ray scattering length density profile was calculated, are given in Table 3. These values were chosen based on knowledge of the deposition process, and using literature values for the X-ray scattering length density, $\rho_x = NZr_e$, where Z is the atomic number (per formula unit) and r_e is the Bohr radius of 2.82×10^{-5} Å. In principle, ρ_x is a complex number with the imaginary component related to the mass absorption length of the X-ray beam in the material. The imaginary component of ρ_x was initially assigned a value of zero; however, its value will be optimized (the imaginary component of ρ_x for FeCo will be significant for CuK$_\alpha$ radiation).

The CO_REFINE routine optimized the parameters within the constraints of the lower and upper bounds listed in the Table 3 subject to the condition that χ^2—a measure of error between the fitted and observed curves was minimized [69]. The optimized parameters of the model are shown in the second to last column of Table 3. The last column of Table 3 lists the perturbations to the optimized parameters such that the reduced χ^2, $\chi_v^2 = \chi^2/v$ (where v is the number of data points minus adjustable parameters), is changed by one, i.e., the last column represents the perturbation to the optimized parameter that significantly worsens the fit by 1-σ [70]. The error bar is a measure of the sensitivity of the model function to yield a curve that represents the data. The error bar is not necessarily a representation of the uncertainty of the fitted parameter. As a mechanism to understand the accuracy and precision of the optimized parameters, it is instructive to compare them with the initial guesses. In the case of the Al-oxide layer thickness, the initial guess of 30 Å was chosen based on the quantity of Al deposited in the growth chamber. Naturally, upon exposure to air, the Al will oxidize and thus its layer thickness can be expected to swell. In this case the swelling of a 30 Å thick film is roughly a factor of two. The X-ray scattering length density attributed to the Al-oxide layer is greater than that for Al $[\rho_x(Al) = 2.2 \times 10^{-5}$ Å$^{-2}]$, but less than that for Al_2O_3 $[\rho_x(Al_2O_3) = 3 \times 10^{-5}$ Å$^{-2}]$. The intermediate value of 2.39×10^{-5} Å^{-2} might indicate the oxide layer is composed of elemental Al and Al_2O_3. Further support for this conclusion is found in the large value of surface roughness/diffusion (in comparison to the buried interfaces).

Table 3: List of initial guess values for the model, v_0, lower, v_-, and upper, v_+, bounds which constrain the optimization, the optimal values that yield a minimum in χ^2, and the perturbation, δv, of the value that produces an increase in χ^2 corresponding to a 1-σ error bar. The number of data points is 134.

Medium	Parameter	v_0	v_-	v_+	v_{opt}	δv
Vacuum	$Re(\rho_x)[10^{-5}$Å$^{-2}]$	0			0	
	$Im(\rho_x)[10^{-5}$Å$^{-2}]$	0			0	
	$\sigma_{01}[$Å$^2]$	10	5	20	16.4	0.8
	$\Delta[$Å$]$	0			0	
Al-oxide	$Re(\rho_x)[10^{-5}$Å$^{-2}]$	3	1	4	2.39	0.01
	$Im(\rho_x)[10^{-5}$Å$^{-2}]$	0	0	1	0.059	0.004
	$\sigma_{12}[$Å$^2]$	10	5	20	11.75	0.03
	$\Delta[$Å$]$	30	20	80	67.77	0.08
FeCo	$Re(\rho_x)[10^{-5}$Å$^{-2}]$	6.6	4	7	5.835	0.005
	$Im(\rho_x)[10^{-5}$Å$^{-2}]$	0	0	2	0.83	0.01
	$\sigma_{23}[$Å$^2]$	10	5	20	9.21	0.05
	$\Delta[$Å$]$	200	190	210	200.41	0.08
GaAs	$Re(\rho_x)[10^{-5}$Å$^{-2}]$	3.99	2	5	3.74	0.01
	$Im(\rho_x)[10^{-5}$Å$^{-2}]$	0	0	1	0.23	0.04
	$\chi^2(v=121)$	704614			382	

The FeCo layer thickness is certainly in good agreement with the thickness sought by the sample growers. The magnitude of $|\rho_x|$ is 90% of the initial guess for FeCo obtained from the lever-rule addition of ρ_x for bulk Fe and Co. The value of ρ_x obtained for GaAs, which has a small imaginary component, is very close to that obtained from the literature. The difference, $\Delta\rho_x = 0.25 \times 10^{-5}$[Å$^{-2}]$, can be related to misalignment of the sample (or measurement of α_j), by differentiating the relation $Q_c = 4\sqrt{\pi\rho} = (4\pi/\lambda)\sin\alpha_i$ with respect to α_i. A misalignment of the sample by $\Delta\alpha_i = 0.03°$ could account for the difference between the measured and literature values of ρ_x.

Next, we wish to combine the X-ray reflectivity study, from which information about the chemical or nuclear structure of the sample is learned, with polarized neutron reflectivity data taken from the sample in a large, 1 kOe, (saturating) magnetic field. The intent of the neutron study is to identify the Fourier components of the sample magnetization, in order to distinguish the magnetization of the FeCo/GaAs interface from the FeCo bulk.

The specular reflectivity (after removal of diffuse scattering) is shown in Fig. 3-34. The *SF* reflectivity was also measured, but no *SF* reflectivity was observed. The lack of *SF* (specular) reflectivity is consistent with the sample being saturated, i.e., the entire sample magnetization was parallel to the 1-kOe field.

The model shown in the inset of Figure 3-34 was fitted to the neutron data. The initial guesses for this model were obtained from the X-ray analysis with the exception that values of ρ_n and ρ_m were calculated based on literature values of the neutron scattering lengths and magnetization of FeCo (see Table 1). The routine CO_REFINE was once again used to obtain the parameters of the model (Table 4) such that the calculated reflectivities fitted the neutron data. The reflectivity curves obtained from the model are shown as the solid curves in Fig. 3-34. Interestingly, the fitted chemical layer thickness of FeCo obtained from the analysis of the X-ray data is larger than that obtained from the analysis of the (neutron) *NSF* reflectivities. One explanation for the difference is that the X-ray fitting is one involving only the chemical structure of the sample, while the neutron fitting weights the magnetic and nuclear contributions roughly equally. The implication is the magnetic thickness of the FeCo layer is less than its chemical thickness.

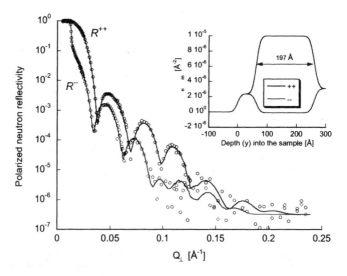

Figure 3-34: Polarized neutron reflectivity data (open symbols) for the FeCo/GaAs sample are shown along with reflectivity curves (solid curves) that best fit the data obtained from a model structure (inset). The initial guesses for the fitting procedure were those obtained from the fit to the X-ray data. This fit is to the neutron data only. *Also see the color plate.*

In order to test this implication, a new model was developed—one that included an extra interface layer between the FeCo film and the GaAs substrate. The FeCo layer thickness was constrained to be the value obtained from neutron scattering, i.e., $\Delta_{FeCo} = 197$ Å, and the thickness of the "reacted" layer was constrained to be the difference between the values of Δ_{FeCo} determined separately by X-ray and neutron fitting, or 3 Å. The new model was then fitted or co-refined to the X-ray and polarized neutron reflectivity data at the same time. Values intermediate between the optimized values in Table 3 and Table 4 were chosen as initial guesses for the new model. The initial guesses of ρ_x, ρ_n and ρ_m for the reacted layer were chosen to be ρ_x for the FeCo layer, and ρ_n and ρ_m (= 0) for GaAs, respectively (see Table 5). The X-ray and neutron reflectivities obtained from the model are shown in Figure 3-35. The magnetization refined for the reacted layer is not significantly different than zero; therefore, we conclude that the additional level of complexity achieved by adding a thin non-magnetic layer between the FeCo and GaAs, yields one model that explains the X-ray and neutron data in a self-consistent fashion.

Table 4: List of guess and optimized parameters for a fit of the FeCo/GaAs model structure to the polarized neutron reflectivity consisting of 348 measurements.

Medium	Parameter	v_0	v_-	v_+	v_{opt}	δv
Vacuum	ρ_n [10^{-6}Å^{-2}]	0			0	
	ρ_m [10^{-6}Å^{-2}]	0			0	
	σ_{01} [Å^2]	16.4	5	20	10.6	0.3
	Δ [Å]	0			0	
Al-oxide	ρ_n [10^{-6}Å^{-2}]	5.21	2	6	2.36	0.02
	ρ_m [10^{-6}Å^{-2}]	0			0	
	σ_{12} [Å^2]	11.75	5	20	10.8	0.1
	Δ [Å]	67.77	20	80	62.7	0.3
FeCo	ρ_n [10^{-6}Å^{-2}]	5.13	4	6	4.94	0.01
	ρ_m [10^{-6}Å^{-2}]	4.97	4	6	5.04	0.01
	σ_{23} [Å^2]	9.21	5	20	12.0	0.1
	Δ [Å]	200.41	190	210	197.5	0.1
GaAs	ρ_n [10^{-6}Å^{-2}]	3.07	2	4	3.07	0.02
	ρ_m [10^{-6}Å^{-2}]	0			0	
	$\chi^2(v = 337)$	26868			1102	

Table 5: Refinement of one model to X-ray and polarized neutron reflectivity data (consisting of 482 measurements) at the same time.

Medium	Parameter	v_0	v_-	v_+	v_{opt}	δv
Vacuum	$Re(\rho_x)[10^{-5}\text{Å}^{-2}]$	0			0	
	$Im(\rho_x)[10^{-5}\text{Å}^{-2}]$	0			0	
	$\rho_n[10^{-6}\text{Å}^{-2}]$	0			0	
	$\rho_m[10^{-6}\text{Å}^{-2}]$	0			0	
	$\sigma_{01}[\text{Å}^2]$	13.6	5	20	15.6	0.08
	$\Delta[\text{Å}]$	0			0	
Al-oxide	$Re(\rho_x)[10^{-5}\text{Å}^{-2}]$	2.39	2	3	2.17	0.01
	$Im(\rho_x)[10^{-5}\text{Å}^{-2}]$	0.06	0	1	0.08	0.01
	$\rho_n[10^{-6}\text{Å}^{-2}]$	2.36	2	6	2.68	0.02
	$\rho_m[10^{-6}\text{Å}^{-2}]$	0			0	
	$\sigma_{12}[\text{Å}^2]$	11.2	5	20	11.82	0.02
	$\Delta[\text{Å}]$	65.2	60	70	65.46	0.02
FeCo	$Re(\rho_x)[10^{-5}\text{Å}^{-2}]$	5.83	5	6	5.75	0.03
	$Im(\rho_x)[10^{-5}\text{Å}^{-2}]$	0.83	0	1	0.72	0.08
	$\rho_n[10^{-6}\text{Å}^{-2}]$	4.94	4	6	5.01	0.01
	$\rho_m[10^{-6}\text{Å}^{-2}]$	5.03	4	6	5.00	0.01
	$\sigma_{23}[\text{Å}^2]$	10.5	5	20	10.8	0.1
	$\Delta[\text{Å}]$	197.5			197.5	
Reacted	$Re(\rho_x)[10^{-5}\text{Å}^{-2}]$	5.83	3	6	5.13	0.02
	$Im(\rho_x)[10^{-5}\text{Å}^{-2}]$	0.83	0	1	0.31	0.02
	$\rho_n[10^{-6}\text{Å}^{-2}]$	3.08	3	7	6.1	0.3
	$\rho_m[10^{-6}\text{Å}^{-2}]$	0	0	6	0.00	0.02
	$\sigma_{34}[\text{Å}^2]$	10.5	5	20	7.63	0.04
	$\Delta[\text{Å}]$	2.9			2.9	
GaAs	$Re(\rho_x)[10^{-5}\text{Å}^{-2}]$	3.74	3	4	3.84	0.01
	$Im(\rho_x)[10^{-5}\text{Å}^{-2}]$	0.23	0.2	0.3	0.23	0.02
	$\rho_n[10^{-6}\text{Å}^{-2}]$	3.08	2	4	3.10	0.02
	$\rho_m[10^{-6}\text{Å}^{-2}]$	0			0	
	$\chi^2(\nu=459)$	4182			1617	

3.10.2. Magnetic and chemical structures obtained from vector magnetometry using neutron scattering

For magnetic systems with large remanent magnetization (or strong anisotropy), polarized neutron reflectometry with polarization analysis is a powerful tool for isolating the nuclear or chemical structure of a material from its magnetic structure. Indeed, artificially structured materials are examples of systems that often exhibit unusual magnetic anisotropies. Here, we show how the uniaxial anisotropy of a material and neutron scattering can be applied to rigorously separate the nuclear and magnetic structures of the sample discussed in the previous example.

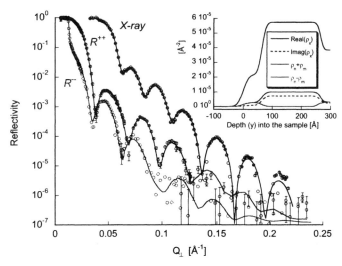

Figure 3-35: X-ray and neutron reflectivity curves of one model (inset) fitted to the X-ray and neutron data at the same time. *Also see the color plate.*

First, the magnetic history of the sample was prepared by applying a field along the [011] (easy) axis of the GaAs substrate (Step 1, Figure 3-36). The field was chosen to be large enough (1 kOe) in order to saturate the sample magnetization (Figure 3-32). Next, the field is reduced monotonically to a small value on the order of a few Oe (Step 2, Figure 3-36). The magnetization of the sample in this field remained nearly the same as its saturated value (see Figure 3-32). The sample was then rotated about its surface normal by $90°$ (Step 3, Figure 3-36), and the non-spin-flip and spin-flip reflectivities for the sample in a field of 9 Oe pointed along the $[01\bar{1}]$ axis were measured (Figure 3-37).

In order to investigate whether the nuclear and magnetic layer thicknesses of the FeCo layer were different, a second model was fitted to the data (symbols) shown in Figure 3-37. This model represents the nuclear and magnetic structures of the sample (inset, Figure 3-37) separately—the magnetic and nuclear layer thicknesses and interface roughnesses were optimized independently. Finally, to account for the vector property of the sample magnetization, an angle, ϕ, between the sample magnetization and the applied field (Figure 3-36) was also refined. The best fitting reflectivities are shown as the solid curves in Figure 3-37. The input guesses and the optimized parameters for the nuclear and magnetic models are listed in Table 6.

An important result of the data analysis is that the portion of the sample magnetization that rotated with the sample is about 6 Å thinner than the thickness attributed to the nuclear scattering from FeCo. In other words, the magnetic thickness of FeCo is 197 Å thick, while the chemical thickness is 203 Å. This result is consistent with the conclusion of the previous analysis of the X-ray and polarized neutron data which in order to account for a discrepancy between chemical thickness as determined by X-ray reflectometry and magnetic thickness as determined by polarized neutron reflectometry (of the sample taken in saturation), a thin non-magnetic reacted layer was needed between FeCo and GaAs.

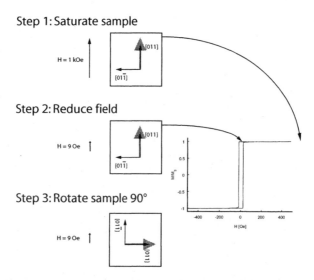

Figure 3-36: Procedure for preparing the magnetic history of the FeCo/GaAs sample prior to measurement with polarized neutrons. The magnetization of the sample is shown by the red arrow. The polarization of the neutron beam is parallel (or anti-parallel) to the applied field. The sample magnetization is first saturated, then the magnetic field is reduced to a small value—just large enough to maintain the polarization of the neutron beam, and finally the sample is rotated 90° about its surface normal, placing its magnetization perpendicular to the polarization axis of the neutron beam.

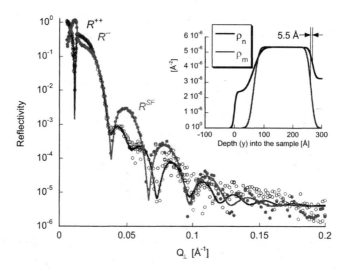

Figure 3-37: The polarized neutron reflectivities (symbols) of the FeCo/GaAs sample with its magnetization pointed along the neutron beam line (i.e., perpendicular to the applied field and in the plane of the sample) are shown. Note the R^{++} and R^- reflectivities are nearly superimposed. The fitted reflectivities (solid curves) were obtained from the nuclear and magnetic scattering length density profiles (inset). The magnetization vector has a magnitude given by the magnetic component in the figure inset and a direction rotated about the surface normal from the applied field by an angle of 89.7°. *Also see the color plate.*

Table 6: List of guesses (v_0) and optimized parameters (v_{opt}) for nuclear magnetic models for the scattering length density profiles. Unless otherwise noted, values in the magnetic model were constrained to be the same as those in the nuclear model. The parameters were optimized between lower and upper limits, v_- and v_+, respectively. If no limits are given the parameter was not optimized. The reflectivities contained 405 measurements.

Medium	Parameter	v_0	v_-	v_+	v_{opt}
Vacuum (nuclear)	$\rho_n [10^{-6}\text{Å}^{-2}]$	0			0
	$\sigma_{01} [\text{Å}^2]$	15.6	5	20	5.06
	$\Delta [\text{Å}]$	0			0
Al-oxide (nuclear)	$\rho_n [10^{-6}\text{Å}^{-2}]$	2.36	2	4	2.37
	$\sigma_{12} [\text{Å}^2]$	11.82	4	20	19.9
	$\Delta [\text{Å}]$	65.46	60	70	64.6
FeCo (nuclear)	$\rho_n [10^{-6}\text{Å}^{-2}]$	5.01	4	6	5.32
	$\sigma_{23} [\text{Å}^2]$	10.8	4	20	9.1
	$\Delta [\text{Å}]$	200.9	190	210	202.9
GaAs (nuclear)	$\rho_n [10^{-6}\text{Å}^{-2}]$	3.10	2	4	3.26
Al-oxide (magnetic)	$\rho_m [10^{-6}\text{Å}^{-2}]$				
	$\sigma_{12} [\text{Å}^2]$	11.82	5	20	12.65
	$\Delta [\text{Å}]$				
FeCo (magnetic)	$\rho_m [10^{-6}\text{Å}^{-2}]$	5.0	4	6	5.36
	ϕ	$-\pi/2$	-1.6	-1.53	-1.565
	$\sigma_{23} [\text{Å}^2]$	10.8	5	20	11.2
	$\Delta [\text{Å}]$	197	190	210	197.4
GaAs (magnetic)	$\rho_m [10^{-6}\text{Å}^{-2}]$				
	$\chi^2 (v = 391)$	3428			1511

3.11. SUMMARY AND CONCLUSIONS

The intent of this chapter is to provide a practical tutorial on polarized neutron reflectometry—one that provides reasonable limits to what can be learned from neutron reflectometry, a working knowledge of a polarized neutron reflectometer, a detailed understanding of how neutron scattering data are acquired and information obtained from the data, and an example problem solved in detail. Many other examples of solved problems exist in the literature, see for example Ref. [23] and references therein.

It is worth stressing the role polarized neutron reflectometry can play in solving problems involving artificially structured materials and nanomagnetism. Many sample fabrication techniques exist, *e.g.,* thin film growth, lithography, templating and self-assembly, to modulate the atomic, electronic and chemical structures of materials. Physical properties can be modulated via confinement in one, two or all dimensions to create multilayers, wires or dots that exhibit novel magnetic behavior. Confinement of physical structures can influence the magnetic properties of materials in ways that cannot be predicted from the averaging of constituent component properties, *e.g.,* giant magnetoresistance in Fe/Cr superlattices. These new nanocomposites are inhomogeneous materials with unique magnetic properties. To understand the magnetism of such artificially structured materials requires an understanding of the interplay between structure and magnetism at the nanometer length scale.

Bulk probes, such as magnetometry, are ill-suited to provide information about the spatial variation of magnetization in non-homogeneous artificially structured materials. Fortunately, the spatial length-scales of magnetism are precisely those that can be probed with neutron scattering. Particular strengths of polarized neutron reflectometry include its ability to measure the magnetic vector response of buried materials to extremes of magnetic and electric fields, temperature and (photon) irradiation. For example, polarized neutron reflectometry is a technique that can measure the depth dependent magnetization in thin films. Since polarized neutron reflectometry is inherently interface specific, the magnetization of the interfacial region (even ones that are deeply buried) can be measured with a great degree of accuracy in the presence of a strongly magnetic substrate. While not discussed in depth here, measurements of off-specular diffuse scattering in reflection geometry provide information about the lateral distribution of (inhomogeneous) magnetism across a sample surface or interface. For example, the lateral dimensions of magnetic domains can be determined from off-specular diffuse scattering and when these measurements are made as a function of $Q_\perp$, correlation lengths of lateral magnetic domains at one depth into the sample with those at another depth can be obtained [71]. Other examples include characterizing the flow of magnetic induction around patterned holes (antidots) or correlation of magnetism between discrete but close by neighbors [72].

While many truly nanopatterned systems cannot be conveniently made in cm^2 size areas (necessary for study with neutron scattering), polarized neutron reflectometry may still provide important information to understand magnetism in model systems that replicate certain structural features, e.g., interfaces, in these systems. For example, interest in ferromagnetic semiconductors is motivated by the prospect of spin-injection devices that *automatically imply the existence of buried interfaces* in the structures of interest. Characterization and understanding of interface quality is therefore a key issue for such devices to succeed. Here, polarized neutron reflectometry is expected to play an important role. For example, the flow of spin current, while most probably too small to be directly measured with neutron scattering, is profoundly affected by the magnetic properties of interfaces, which can be examined quite naturally with polarized neutron reflectometry.

ACKNOWLEDGMENTS

The FeCo/GaAs sample and its magnetization data were kindly provided to us by X.Y. Dong, B.D. Schultz and C.J. Palmstrøm (University of Minnesota). We thank S. Park (LANL) for the analyses of the X-ray and neutron data of the FeCo/GaAs sample. The polarizing supermirrors were made using facilities at the Paul Scherrer Institut provided to one of us (M.F.) by D. Clemens (HMI) and P. Böni (TU-München) for which we are grateful. We thank F. Mezei (HMI) for his advice, technical help and encouragement in the development of the polarization cavity and along with T. Keller (TU-München) and L. Mokrani (HMI) in the implementation of the radio-frequency gradient field spin flipper. Development of the polarized neutron reflectometer at LANSCE could not have been accomplished without the support of R. Pynn (LANL). This manuscript benefited greatly from vigorous discussions with N. Berk (NIST), G. Felcher (ANL), R. Pynn, I.K. Schuller (UCSD) and S. Sinha (UCSD). The neutron scattering facility of the Manuel Lujan Jr. Neutron Scattering Center is gratefully appreciated. The facility is supported by the U.S. Department of Energy, BES-DMS under Contract No. W-7405-Eng-36.

REFERENCES

[1]   V. Lauter-Pasyuk, H.J. Lauter, B.P. Toperverg, L. Romashev, V. Ustinov,
      Phys. Rev. Lett. **89** (2002) 167203.
[2]   J.F. Ankner, C.F. Majkrzak, D.A. Neumann, A. Matheny and C.P. Flynn,
      Physica B **173** (1991) 89.
[3]   D.R. Lee, G. Srajer, M.R. Fitzsimmons, V. Metlushko, S.K. Sinha,
      Appl. Phys. Lett. **82** (2003) 82.
[4]   L.G. Parratt, *Phys. Rev.* **95** (1954) 359.
[5]   T.P. Russell, *Annual Review of Materials Science* **21** (1991) 249.
[6]   J. Lekner, *Theory of Reflection of Electromagnetic and Particle Waves*,
      (Martinius Nijhoff Publishers, Dordrecht 1987).
[7]   H. Dosch, *Phys. Rev. B* **35** (1987) 2137.
[8]   G.H. Vineyard, *Phys. Rev. B* **26** (1982) 4146.
[9]   S. Dietrich and H. Wagner, *Z. Phys. B* **56** (1984) 207.
[10] M. Tolan and W. Press, *Z. Kristallogr.* **213** (1998) 319.
[11] G.S. Smith and C.F. Majkrzak, in the *International Tables of Crystallography*,
      ed. E. Prince, vol. B, 3rd edition (Kluwer Academic Publishers, Dordrecht 2004).
[12] Y. Y. Huang, C. Liu, and G. P. Felcher, *Phys. Rev. B* **47** (1993) 183.
[13] A. Hoffmann, J.W. Seo, M.R. Fitzsimmons, H. Siegwart, J. Fompeyrine, J.P. Locquet,
      J.A. Dura, C.F. Majkrzak, *Phys. Rev. B* **66** (2002) 220406.
[14] M.R. Fitzsimmons, P. Yashar, C. Leighton, I.K. Schuller, J. Nogues, C.F. Majkrzak,
      J.A. Dura, *Phys. Rev. Lett.* **84** (2000) 3986.
[15] M. Gierlings, M.J. Prandolini, H. Fritzsche, M. Gruyters, D.Riegel, *Appl. Phys. A* **74**
      (2002) S1523.
[16] K. Temst, M.J. Van Bael, H. Fritzsche, *J. Magn. Magn. Mater.* **226** (2001) 1840.
[17] H. Fritzsche, M.J. Van Bael, K. Temst, *Langmuir* **19** (2003) 7789.
[18] V. Leiner, K. Westerholt, A.M. Blixt, H. Zabel and B. Hjörvarsson, *Phys. Rev. Lett.* **91**
      (2003) 037202.
[19] G.P. Felcher, R.O. Hilleke, R.K. Crawford, J. Haumann, R. Kleb, and G. Ostrowski,
      Rev. Sci. Instrum. **58** (1987) 609.
[20] C.F. Majkrzak, *Physica* (Amsterdam) **221B** (1996) 342.
[21] H. Dosch, *Physica B* **192** (1993) 163.
[22] J.A.C. Bland in *Ultrathin Magnetic Structures*, J.A.C. Bland and B. Heinrich (eds.),
      (Springer Verlag, Berlin, 1994).
[23] M.R. Fitzsimmons, S.D. Bader, J.A. Borchers, G.P. Felcher, J.K. Furdyna, A. Hoffmann,
      J.B. Kortright, Ivan K. Schuller, T.C. Schulthess, S.K. Sinha, M.F. Toney, D. Weller,
      S. Wolf, *J. Magn. Magn. Mater.* **271** (2004) 103.
[24] S. Dietrich and H. Wagner, *Z. Phys. B* **59** (1985) 35.
[25] http://www.lpm.u-nancy.fr/webperso/mangin.p/hercules/courshercules.html
[26] J.F. Ankner and G.P. Felcher, *J. Magn. Magn. Mater.* **200** (1999) 741.
[27] W.G. Williams, *Polarized Neutrons*, (Clarendon Press, Oxford 1988).
[28] C.F. Majkrzak, K.V. O'Donovan and N.F. Berk, unpublished.
[29] E. Merzbacher, *Quantum Mechanics*, (John Wiley & Sons, Inc., New York 1970).
[30] Additional scattering lengths can be obtained at
      http://www.ncnr.nist.gov/resources/n-lengths/.
[31] M. Born and E. Wolf, *Principles of Optics, Electromagnetic Theory of Propagation,
      Interference and Diffraction of Light*, (Pergamon Press, Oxford, 1980) p. 563.
[32] P.R. Bevington and D. K. Robinson, *Data Reduction and Error Analysis for the
      Physical Sciences*, 3rd edition, (McGraw-Hill Book Company, New York 2003) p. 2.

[33] P.G. Shewmon, *Diffusion in Solids*, (McGraw-Hill Book Company, New York 1963).

[34] P.R. Bevington and D. K. Robinson, *Data Reduction and Error Analysis for the Physical Sciences*, 3rd edition, (McGraw-Hill Book Company, New York 2003) p. 252.

[35] R.N. Bracewell, *The Fourier Transform and its Applications*, (McGraw-Hill Book Company, New York 1978).

[36] L. Nevot and P. Croce, *Rev. Phys. Appl.* **15** (1980) 761.

[37] B.E. Warren, *X-ray Diffraction*, (Dover Publications, Inc., New York 1990).

[38] F. Radu, A. Vorobiev, J. Major, H. Humblot, K. Westerholt, H. Zabel, *Physica B* **335** (2003) 63.

[39] R.K. Kawakami, Y. Kato, M. Hanson, I. Malajovich, J.M. Stephens, E. Johnston-Halperin, G. Salis, A.C. Gossard, D.D. Awschalom, *Science* **294** (2001) 131.

[40] D.J. Hughes and M.T. Burgy, *Phys. Rev.* **76** (1949) 1413.

[41] J. Schwinger, *Phys. Rev.* **51** (1937) 544.

[42] F. Bloch, *Phys. Rev.* **50** (1936) 259.

[43] D.J. Hughes and M.T. Burgy, *Phys. Rev.* **81** (1951) 498.

[44] M.R. Fitzsimmons, M. Lütt, H. Kinder and W. Prusseit, *Nucl. Inst. and Methods A*, **411** (1998) 401.

[45] F. Mezei, *Z. Physik*, **255** (1972) 146.

[46] J.D. Jackson, *Classical Electrodynamics*, (John Wiley & Sons, Inc. New York 1975).

[47] E. E. Fullerton, D.M. Kelly, J. Guimpel, I.K. Schuller, and Y. Bruynseraede, *Phys. Rev. Lett.* **68** (1992) 859.

[48] H. Ohno, *Science* **281** (1998) 951.

[49] M. Born and E. Wolf, *Principles of optics: electromagnetic theory of propagation, interference and diffraction of light*, (Pergamon Press, New York 1980).

[50] W.T. Lee, S.G.E. te Velthuis, G.P. Felcher, F. Klose, T. Gredig, E.D. Dahlberg, *Phys. Rev. B* **65** (2002) 22417.

[51] S.G.E. teVelthuis, A. Berger, G.P. Felcher, B.K. Hill, E.D. Dahlberg, *J. Appl. Phys.* **87** (2000) 5046.

[52] F. Mezei, *Comm. on Physics* **1** (1976) 81.

[53] F. Mezei and P.A. Dagleish, *Comm. on Physics* **2** (1977) 41.

[54] D. Clemens, P. Böni, H.P. Friedli, R. Göttel, C. Fermon, H. Grimmer, H. van Swygenhoven, J. Archer, F. Klose, Th. Krist, F. Mezei, P. Thomas, *Physica B* **213-214** (1995) 942.

[55] P. Høghøj, I. Anderson, R. Siebrecht, W. Graf and K. Ben-Saidane, *Physica B* **267** (1999) 355.

[56] P. Dhez and C. Weisbuch, eds., *Physics, Fabrication and Applications of Multilayered Structures*, NATO ASI Series B, **182** (Plenum Press, New York 1988).

[57] C. Majkrzak, ed., Proc. Conf. On Thin-film Neutron Optical Devices, San Diego, USA, August 1988 SPIE Proc., **983** (Bellingham, WA 1989).

[58] F. Mezei in: Use and Development of Low and Medium Flux Research Reactors, eds., O.K. Harling, L. Clark and P. von der Hardt, Supplement to Atomenergie-Kerntechnik, **44** (Karl Thiemig, München 1984) p. 735.

[59] J.B. Hayter in *Neutron Diffraction*, ed. H. Dachs, (Springer Verlag, Berlin 1978).

[60] C.P. Slicther, *Principles of Magnetic Resonance*, (Springer Verlag, Berlin 1980).

[61] G. Badurek, *Nucl. Instr. and Methods* **189** (1981) 543.

[62] O. Scharpf in *Neutron Spin Echo*, ed. F. Mezei (Springer-Verlag, Berlin 1980) pp. 27-52.

[63] R. Herdin, A. Steyerl, A.R. Taylor, J.M. Pendlebury and R. Golub, *Nucl. Instr. and Methods* **148** (1978) 353.

[64] T. Keller, T. Krist, A. Danzig, U. Keiderling, F. Mezei, A. Wiedenmann, *Nucl. Instr. and Methods A* **451** (2000) 474.

[65]   S.Park, M.R. Fitzsimmons, X.Y. Dong, B.D. Schults and C.J. Palmstrøm, submitted to
        Phys. Rev. B.

[66]   L.C. Chen, J.W. Dong, B.D. Schultz, C.J. Palmstrøm, J. Berezovsky, A. Isakovic,
        P.A. Crowell and N. Tabat, *J. Vac. Sci. Technol. B* **18** (2000) 2057.

[67]   B.D. Schultz, H.H. Farrel, M.M.R. Evans, K. Lüdge, and C.J. Palmstrøm,
        J. Vac. Sci. Technol. B **20** (2002) 1600.

[68]   M. Lütt, M.R. Fitzsimmons, D. Li, *J. Chem. Phys. B* **102** (1998) 400.

[69]   P.R. Bevington and D. K. Robinson, *Data Reduction and Error Analysis for the
        Physical Sciences*, 3rd edition, (McGraw-Hill Book Company, New York 2003) p. 67.

[70]   W.H. Press, B.P. Flannery, S.A. Teukolsky, and W.T. Vetterling, *Numerical Recipes,
        the art of scientific computing*, (Cambridge University Press 1986) p. 536.

[71]   K. Temst, M.J. Van Bael, H. Fritzsche, *Appl. Phys. Lett.* **79** (2001) 991.

[72]   D.R. Lee, G. Srajer, M.R. Fitzsimmons, V. Metlushko, S.K. Sinha, *Appl. Phys. Lett.* **82**
        (2003) 82.

X-ray Scattering

CHAPTER 4

Resonant soft x-ray techniques to resolve nanoscale magnetism

4.1. INTRODUCTION

The soft x-ray spectral region contains important core levels of $3d$ transition and rare earth elements and so has emerged as a powerful spectral region in which to study magnetism in a variety of materials [1]. While some spectroscopic techniques are relatively mature in this spectral range, others are still under development. This chapter reviews a variety of evolving soft x-ray techniques as applied to the study of magnetism and magnetic materials. Emphasis is given to entirely photon based techniques that, compared to techniques detecting photoelectrons, can probe relatively deeply into samples and are compatible with strong and varying applied fields. Emphasis is also placed on techniques that can resolve magnetic structure either in depth or laterally in samples, rather than just providing spatially averaged properties.

The soft x-ray spectral range extends roughly from 100 eV to 2500 eV, and is often defined as that region where the path length of x-rays is insufficient to propagate in air at atmospheric pressure. This strong soft x-ray absorption has made this spectral range one of the last to be exploited to study magnetic materials, since specialized sources, optical elements, and instrumentation are necessary for such measurements [2]. Paradoxically perhaps, this strong absorption indicates large interaction cross-sections, especially at certain core levels, where magneto-optical effects can be larger than in any other spectral range. Synchrotron radiation sources generally provide the polarized soft x-rays needed for these studies, and are now common enough to provide reasonable access. Essentially all optical and scattering techniques common in the near-visible and x-ray spectral ranges have been extended into, or sometimes rediscovered in, the soft x-ray range. The coupling of these large, resonant magneto-optical effects with these various techniques is discussed here.

The following sections review fundamental characteristics of resonant magneto-optical spectra at x-ray core levels before introducing different approaches to apply these effects in different ways. Since modern magnetic materials are typically chemically and magnetically heterogeneous, often down to nanometer length scales, it is natural to categorize techniques

The submitted manuscript has been authored by a contractor of the U.S. Government under contract No. DEAC03-76SF00098. Accordingly, the U.S. Government retains a nonexclusive, royalty-free license to publish or reproduce the published form of this contribution, or allow others to do so, for U.S. Government purposes.

by their ability to resolve such structure both in depth and laterally. For example, direct measurements of transmitted (forward scattered) beams average both laterally and in-depth throughout the illuminated area. Specular reflection techniques average over lateral structure, but can provide depth resolution in different ways. Diffuse scattering and diffraction, in transmission or reflection geometry, can resolve lateral structure and can also have variable depth sensitivity. Partially coherent scattering provides an ensemble average over structure, while coherent scattering retains details of local structural information. Finally, zone-plate microscopy provides direct images of local chemical and magnetic structure. Recent advances in each of these areas are discussed and compared below.

4.2. CORE RESONANT MAGNETO-OPTICAL PROPERTIES

Optical and magneto-optical (MO) properties relevant to soft x-ray measurements are briefly reviewed here. First, geometrical conventions used in this chapter are defined in Fig.4-1 showing a generalized scattering event. Incoming and scattered wavevectors $\mathbf{k}_0$ and $\mathbf{k}_f$ define a scattering plane containing the scattering vector $q \equiv \mathbf{k}_f - \mathbf{k}_0$. The magnitude $|\mathbf{q}| = (4\pi \sin\theta)/\lambda$, where 2θ is the total scattering angle and λ is the x-ray wavelength, is the spatial frequency probed in a scattering measurement. Structural information can be obtained only along $\mathbf{q}$ and is averaged perpendicular to $\mathbf{q}$. In the soft x-ray range $\lambda \approx 0.5$–10 nm and we can probe spatial frequencies corresponding to real-space distances as small as $\sim 2\pi/q = \lambda/2$ or $\sim 0.2 - 2$ nm. This ability to resolve nanometer-scale structure is one important feature of soft x-ray magneto-optical measurements. The wavevectors have associated polarization unit vectors $\mathbf{e}_0$ and $\mathbf{e}_f$ that are important in determining how charge and magnetic effects manifest in measurements. We define the Cartesian coordinate system with $\mathbf{z} \| \mathbf{q}$, so the $\mathbf{x} - \mathbf{z}$ plane defines the scattering plane in Fig.4-1.

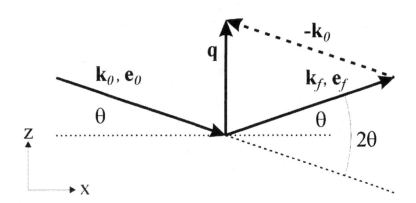

Figure 4-1: Vectors relevant to the scattering process. The sample can be of arbitrary form.

The generalized scattering event in Fig.4-1 can be measured in transmission, specular reflection, or off-specular reflection geometry. The terms Faraday and Kerr are used here to designate transmission and reflection geometries, respectively, partly as a reminder of the distinct geometries used by these pioneers of MO effects. The sample magnetization $\mathbf{M}(x, y, z)$ is considered to potentially exhibit spatial variation in both direction and magnitude with position the sample. From MO effects in the near visible spectral range we adopt the terminology longitudinal, transverse, and polar to describe scattering measurements in which $\mathbf{M}$ is predominantly along $\mathbf{x}$, $\mathbf{y}$ or $\mathbf{z}$, respectively.

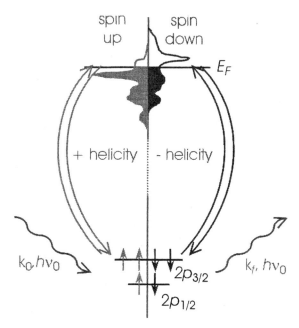

Figure 4-2: Quasi-elastic resonant scattering at with x-ray energies tuned to sharp, spin-orbit split core levels exhibit large magneto-optical effects through coupling to spin-polarized, empty intermediate states.

We can equally well describe the polarization (e_0 and e_f) of incident and scattered fields via orthogonal linear or circular components. Orthogonal linear components are referenced with respect to the scattering plane; e_σ has the electric field component always normal to the scattering plane ($\parallel y$), while for e_π it is in the scattering plane ($\parallel \pm\sin\theta\, x + \cos\theta\, z$). Orthogonal circular components are referenced with respect to the directions k_0 and k_f. For right and left circular polarization e_0 is given by $e_+=(e_\sigma+ie_\pi)/\sqrt{2}$ and $e_-=(e_\sigma+ie_\pi)/\sqrt{2}$, respectively. Below we adopt linear or circular bases to describe e_0 and e_f according to which provides the simplest description of the phenomena under consideration.

4.2.1. The resonant atomic scattering factors – theoretical description

The complex index of refraction $n(\lambda)$ and dielectric tensor $\varepsilon(\lambda)$ are valid descriptions of optical and MO properties of matter in the x-ray spectral range. However, the atomic scattering factor $f(\lambda,q)$ provides a more fundamental, microscopic description, since it resolves the optical properties into the contributions of individual atoms and ions, and indeed their individual electronic states that constitute the sample. This elemental, electronic state specificity is another powerful aspect of core-resonant x-ray scattering measurements. Explicit polarization effects associated with distinct resonant charge and magnetic effects are simply expressed in $f(\lambda,q)$, as seen below.

The resonant scattering process at an atomic core level is schematically depicted in Fig.4-2. While resonant second-order matrix elements allow for many complex processes such as resonant inelastic scattering and fluorescence [3] we are concerned here with quasi-elastic scattering in which scattered and incident photons have the same energy. Detectors that operate in the soft x-ray range typically cannot discriminate between elastic and inelastic scattering, and grating spectrometers are needed to clearly resolve inelastic events. The scattering events discussed below are dominated by elastic scattering.

The atomic scattering factor is generally expressed as

$$f(\lambda, q) = f^0(q) + f'(\lambda, q) + if''(\lambda, q) \equiv f_1(\lambda, q) + if_2(\lambda, q)$$

containing a non-resonant term (f^0) and real and imaginary resonant terms (f' and f'') that are related through a Kramers-Kronig dispersion relation [4]. Refractive (real) terms are collected in f_1 and absorptive terms in f_2 [5]. $f(\lambda, q)$ is essentially the Fourier transform of the atomic electron density as sensed by radiation with wavelength λ. The q dependence results from the shape of the atomic electron density, although for the relatively large λ in the soft x-ray range the atoms scatter as points and this dependence is typically ignored. The resonant terms represent the sum of all allowed transitions within the scattering atom, which are spread throughout the spectrum for a given element, and vary systematically with atomic number [5]. Sensitivity to the electronic structure of, and immediately surrounding, the scattering atom results from the resonant absorption, and often yields valuable information about the distribution of electrons as well as their spin and orbital moments, as discussed below.

Quantum mechanical calculations identify many distinct resonant contributions to scattering at a given atomic core level [6], and the general expression for $f(\lambda, q)$ above effectively groups these together in f' and f''. We consider only electric dipole transitions, as higher order terms that may be significant in the hard x-ray spectral range are expected to be negligible in the soft x-ray range. Retaining only non-resonant pure charge and resonant charge and magnetic terms, all of which are much larger than non-resonant magnetic terms [6,7] yields 3 leading resonant terms with distinct dependencies on $\mathbf{e}_0$ and $\mathbf{e}_f$. Together with the non-resonant charge term the scattering factor becomes

$$f(\lambda) = (\mathbf{e}_f^* \cdot \mathbf{e}_0)\left\{\frac{3}{8\pi}\lambda[F_{11} + F_{1-1}] - r_e Z\right\}$$

$$+ \frac{3}{8\pi}\lambda\{i(\mathbf{e}_f^* \times \mathbf{e}_0) \cdot \mathbf{m}[F_{1-1} - F_{11}] + (\mathbf{e}_f^* \cdot \mathbf{m})(\mathbf{e}_0 \cdot \mathbf{m})[2F_{10} - F_{11} - F_{1-1}]\} \qquad (4.1)$$

Here r_e is the electron radius, Z the atomic number, and $\mathbf{m}$ is a unit vector along the magnetization of the ion. The different F_{LM} terms are dipole matrix elements from initial to final states resolved into different spherical harmonics, and thus represent the spectral dependence of transitions between electronic states of specific symmetry, specific linear combinations of which are associated with distinct polarization dependencies. These polarization- and symmetry-specific contributions reveal how anisotropies in bonding as well as spin polarization can lead to anisotropic optical and magneto-optical properties. Indeed, in single crystals with reduced symmetry, measurements of the anisotropy of resonant scattering factors can provide information about the anisotropy in state-specific anti-bonding orbitals [7]. While the expression above was developed explicitly for localized atomic or ionic final states, these basic terms are also found in theoretical descriptions of atomic scattering factors in itinerant metallic systems described by band structure.

Most of the materials considered here are not single crystals, and we choose to simplify the scattering factor expression as

$$f(\lambda) = p_c(\mathbf{e}_0, \mathbf{e}_f) f_c(\lambda) + p_{m1}(\mathbf{e}_0, \mathbf{e}_f, \mathbf{m}) f_{m1}(\lambda) + p_{m2}(\mathbf{e}_0, \mathbf{e}_f, \mathbf{m}) f_{m2}(\lambda). \qquad (4.2)$$

Here f_c represents the resonant and non-resonant charge scattering, f_{m1} the resonant magnetic scattering 1st order in $\mathbf{m}$, and f_{m2} the resonant magnetic scattering 2nd order in $\mathbf{m}$. Corresponding polarization prefactors p_c, p_{m1}, and p_{m2} contain the distinct polarization dependence of these terms resulting from the interaction of $\mathbf{e}_0$ with the vector spherical harmonics describing the transitions in (1). $p_c = \mathbf{e}_f^* \cdot \mathbf{e}_0$ is the well-known Thompson polarization dependence for a free electron. First order $p_{m1} = -i(\mathbf{e}_f^* \cdot \mathbf{e}_0)$ depends on 2θ, $\mathbf{e}_0$ and $\mathbf{e}_f$, and $\mathbf{m}$, and is non-zero for circular polarization when $\mathbf{e}_0$ and $\mathbf{e}_f$ have the same helicity, or for linear polarization for $\sigma \to \pi$, $\pi \to \sigma$, or $\pi \to \pi$ scattering. Second order $p_{m2} = (\mathbf{e}_f^* \cdot \mathbf{m})(\mathbf{e}_0 \cdot \mathbf{m})$ depends on the projections of $\mathbf{m}$ with $\mathbf{e}_0$ and $\mathbf{e}_f$, and varies generally with orthogonal linear polarization. The terms optical, charge, and chemical scattering refer to f_c, while magneto-optical and magnetic scattering refer to f_{m1} and f_{m2}. However one must always consider how both charge and magnetic terms contribute to measured signals. First order MO effects include magnetic circular dichroism (MCD) and magnetic circular birefringence that yields rotation of linearly polarized light in the Faraday and Kerr geometries. These typically are the dominant MO effects in ferromagnetic (FM) materials, and are absent in compensated antiferromagnetic (AF) materials. Second order terms yield magnetic linear dichroism (MLD) that is present in compensated antiferromagnets and can be significant in ferrimagnets and high-anisotropy ferromagnets. Since f_{m2} is generally small compared to f_{m1} for metallic ferromagnets, it is common to ignore f_{m2}, as is done for the most part here.

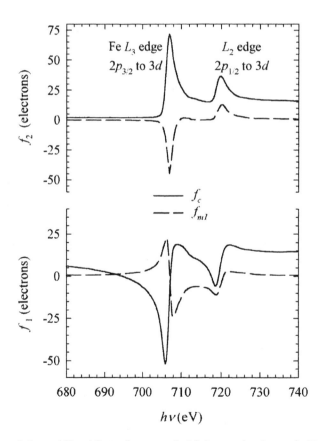

Figure 4-3: Measured charge (f_c) and first order magnetic (f_{m1}) scattering factors for Fe across its $L_{2,3}$ edges. The imaginary (f_2) and real (f_1) parts are related through Kramers-Kronig dispersion relations [10].

The linear basis ($\mathbf{e}_\sigma$, $\mathbf{e}_\pi$) for $\mathbf{e}_0$ and $\mathbf{e}_f$ is a common choice in describing magnetic scattering [7-9]. However the circular basis ($\mathbf{e}_+$, $\mathbf{e}_-$) can bring added simplicity, since, according to (2) there is no polarization mixing in the scattering process, while there is mixing when using the linear basis. Adopting the circular basis and ignoring f_{m2}, the scattering factor simplifies to $f_\pm = p_c f_c \pm p_{ml} f_{ml}$, where $\pm$ refers to opposite helicity. For small 2θ and longitudinal **m**, we have $f_\pm \cong f_c \pm f_{ml}$. The asymmetry $f_+ - f_- \cong 2 f_{ml}$ gives just the first order magnetic part and the average $(f_+ + f_-)/2 = f_c$ gives just the charge part of the scattering factor [10]. While these simple expressions hold for these scattering *amplitudes*, we see below that the asymmetry and average of scattered *intensities* measured with opposite helicity are not so simply related to magnetism and charge because of interference of their amplitudes in the scattering process.

4.2.2. The Fe scattering factors across the $L_{2,3}$ edges

Particularly large resonant effects in f_c are expected to occur at absorption edges that couple sharp core levels to partially filled empty levels at and above the Fermi level via dipole-allowed transitions. When the core levels are spin-orbit split and the empty states are spin-polarized, large resonant magnetic effects in is expected. For the $3d$ transition elements, relevant L_2 and L_3 edges couple initial $2p_{1/2}$ and $2p_{3/2}$ levels to the spin-polarized $3d$ levels in the $500 - 1000$ eV range. For rare-earth elements, relevant edges with strong resonances include M_4 and M_5 edges ($3d_{3/2}, 3d_{5/2} \rightarrow 4f$) in the $850 - 1600$ eV range, N_4 and N_5 edges ($4d_{3/2}, 4d_{5/2} \rightarrow 4f$) in the $100-200$ eV range. Weaker rare-earth resonances occur at the N_2 and N_3 edges ($4p_{1/2}, 4p_{3/2} \rightarrow 5d$) in the $200 - 400$ eV range, and M_2 and M_3 edges ($3p_{1/2}, 3p_{3/2} \rightarrow 5d$) in the $1200 - 2200$ eV range. All of these edges fall in the soft x-ray spectral range, making it especially attractive for resonant magnetic studies. Also interesting are the rare earth L_2 and L_3 edges in the $5,700 - 10,300$ eV hard x-ray spectral range (see Chapter 5).

Experimentally determined values of f_c and f_{ml} for elemental Fe across its $L_{2,3}$ edges [10] in Fig.4-3 are dominated by strong white line absorption at the sharp core levels, and corresponding strong resonances in the refractive contributions. The results for Fe are representative of the other $3d$ transition metals (Cr – Ni) of interest in magnetic materials, whose white line strengths decrease as the $3d$ states fill with increasing Z. Similar results for selected rare earth elements are available in the literature [11]. As will be seen, it is these large resonant changes in both f_c and f_{ml} that enable many new measurements in this spectral range. For elemental Fe, f_{m2} shows very weak resonant enhancement compared to f_{ml} [10].

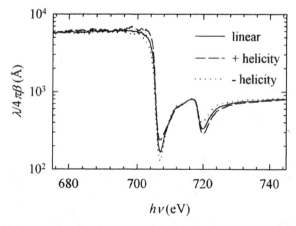

Figure 4-4: The $1/e$ penetration depth at normal incidence for Fe across its $L_{2,3}$ edges shows strong helicity dependence. Magnetization is assumed to be normal to the surface [10].

The resonant spectral behavior of f_c and f_{m1} can be important for many reasons, ranging from experiment planning to interpretation of various scattering results including absolute determination of spin and orbital moments via sum rules. Because their resonant values can depend strongly on local chemical environment and hence can vary significantly with sample, their careful determination for specific samples studied can be important. Here we use these values for Fe to draw some general conclusions for soft x-ray studies of magnetic materials by considering the behavior of the skin depth and critical angle for total external reflection vs. energy near the $2p$ core levels.

The relation between the complex index of refraction n and the atomic scattering factors is useful in some of these considerations. For a homogeneous, multicomponent phase the index is given by $1 - n(\lambda) = \sum_i N_i r_e \lambda^2 f_i(\lambda,0) / 2\pi$, where different species i have different $f_i(\lambda,0)$ and number density N_i, and r_e is the electron radius. For elemental Fe, considering only the first order magnetic term leads to three limiting cases for n [10]. Zero f_{m1} contribution (only charge scattering) yields n_0 and corresponds to propagation normal to **M**. Maximum f_{m1} contribution of opposite sign yields n_+ and n_-, corresponding to circular polarization propagating parallel and antiparallel to saturated **M**. Each can be written $n = 1 - \delta - i\beta$ where β is the absorption index and δ the refractive contribution. The penetration depth for the electric field intensity at normal incidence, $\lambda / 4\pi\beta$, is strongly dependent on $h\nu$ and polarization across the $2p$ spectrum, as seen in Fig.4-4. Below the L_3 edge radiation penetrates hundreds of nm into an Fe sample, independent of polarization. Strong L_3 absorption significantly reduces penetration in a polarization dependent way to 13 and 24 nm for the opposite helicity circular components. Away from normal incidence, skin depth scales as $\sin\theta$ until near the critical angle for total reflection where refractive effects come into play. Thus at $\theta = 15°$, commonly used to study samples with in-plane **M**, the penetration depth is only 3 and 6 nm for the circular components. Below it is seen that this energy dependent penetration depth can be used to obtain depth-resolved information.

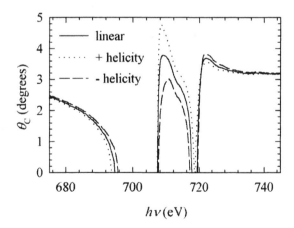

Figure 4-5: The critical angle for total external reflection evaluated for different polarizations assuming magnetization is saturated in the longitudinal direction in the surface plane [10].

Large optical and MO effects occur near the critical angle for total external reflection, $\theta_c = \sqrt{2\delta}$, that is plotted for +/- helicity (charge plus/minus magnetic) and linear (charge only) scattering in Fig.4-5. The dispersive resonances associated with the L_3 and L_2 lines are strong enough that θ_c vanishes when $f' < -Z$ electrons and $\text{Re}[n] > 1$, in which case incident wave fields refract *into* the sample rather than toward the sample surface. In the transition from total external to total internal reflection, the optical properties pass through the zero refraction condition when δ passes through 0. Not only are the resonant charge refractive effects quite large, but the magnetic counterparts are likewise large, as seen by the distinct differences in for the different polarizations. Both the reflected intensity and the phase change on reflection (which varies by π from $\theta = 0$ to $\theta = \theta_c$) are strong functions of polarization (or $\mathbf{M}$ direction), and so can produce striking MO effects in experiments operating near this angular range.

Especially when working in reflection geometry, changes in θ, $h\nu$, $\mathbf{e}_0$, and $\mathbf{M}$ can have a striking influence on the shape of q-resolved scattered signals because of changes in the effective sample volume resulting from these changes in penetration depth and θ_c, and from changes in magnetic scattering amplitudes. These optical and magneto-optical effects can overshadow effects due, for example, to spatial structure variation of interest in a specific measurement [10]. Thus it is important to utilize realistic resonant scattering factors in modeling experimental results, so that these effects can be clearly distinguished from possible structural signals of interest.

4.3. XMCD AND RELATED SPECTROSCOPIES

In this section spectroscopy is narrowly defined to represent the energy dependence of either the imaginary (f_2) or the real (f_1) part of f_c, f_{m1}, and f_{m2}. Specifically, we limit consideration first to measurements of $f_{2,c}$ and $f_{2,m1}$, and $f_{1,m1}$. Here we are primarily concerned with measuring the spatially averaged values, rather than their possible variation with position. In subsequent sections we consider spectroscopic aspects of techniques to explicitly resolve the spatial variation of magnetic structure. In the case of scattering techniques, measured energy spectra generally depend on both the real and imaginary parts of the scattering factors, often in complex ways.

Near edge x-ray absorption fine structure (NEXAFS) or x-ray absorption (XAS) spectra $(f_{2,c})$ are of interest because they provide a direct measure of the density of unoccupied states that can be used to study systematic changes in materials and can be compared with theoretical calculations of electronic structure [12]. For example, the strength of the Fe L_3 and L_2 white lines in Fig.4-3 is a measure of the unoccupied $3d$ and $3s$ states accessible from the $2p$ level via dipole selection rules. The high symmetry of Fe atoms in the bulk bcc structure yields relatively featureless L_3 and L_2 lines, although reduced crystal field symmetry generally introduces multiplet splitting in the $3d$ states that can show up as pronounced, characteristic NEXAFS features [13]. NEXAFS spectra thus provide a useful experimental indicator of general trends in local atomic or ionic configurations. Theoretical interpretation of such spectra must proceed with care, as the absorption process itself complicates the measurement of ground state (unexcited) electronic structure, since this structure relaxes to screen the core hole produced in the absorption event, and scattering effects of the outgoing photoelectrons can influence spectral shapes [12,7].

4.3.1. XMCD sum rules and applications

Magnetic dichroism spectra are measured in accordance with the polarization dependence of p_{m1} and p_{m2}. While x-ray MLD (XMLD) [14] was observed before x-ray MCD (XMCD) [15] in the x-ray spectral region, XMCD is sensitive to ferromagnetic moments and thus is more generally utilized than XMLD. Using circular polarization with reversed helicity or magnetization, XMCD is defined as $\text{Im}[f_+ - f_-] = 2p_{m1} f_{2,m1}$, where p_{m1} emphasizes that measurements sense the projection of $\mathbf{m}$ on $\mathbf{k}_0$. The XMCD spectrum $(f_{2,m1})$ for Fe is in the upper panel of Fig.4-3.

Beyond element-specific detection of ferromagnetic moments and their qualitative variation among samples, sum rules can yield quantitative measures for elemental spin [16] and orbital moments [17]. The sum rules use the areas under the L_3 and L_2 XMCD peaks, A_3 and A_2, respectively, to determine the effective spin moment $m_{spin}^{eff} \propto A_3 + A_2$ and the orbital moment $m_{orb} \propto A_3 - 2A_2$ [18]. Determination of the absolute value of these elemental moments requires a value for the number of holes in the $3d$ shell n_{3d}^h that is generally not known. Experimentally n_{3d}^h is proportional to the sum of the areas of the L_3 and L_2 white lines in f_c. The ratio m_{orb} / m_{spin}^{eff} is independent of n_{3d}^h. m_{spin}^{eff} is an effective spin moment because, according to the sum rules, the spectra leave unresolved a contribution from the magnetic dipole moment. The sum rules have been used to obtain values for spin and orbital moments consistent with those expected for Fe, Co, and Ni [18,19].

The ability to measure m_{orb} is a powerful aspect of XMCD spectroscopy, especially since orbital moments are sensitively related to magnetocrystalline anisotropy in the bulk and at interfaces in thin films. This sensitivity has advanced our understanding of the reorientation transition from in-plane to perpendicular magneto-crystalline anisotropy (PMA) in ultra thin films by revealing enhanced orbital moments associated with interfaces [20,21], as predicted originally by Néel [22]. This reorientation transition is generally thought to occur as the interfacial contribution to the total anisotropy, generally favoring perpendicular anisotropy, overcomes the bulk contribution favoring in-plane anisotropy, with decreasing film thickness. XMCD studies applying the sum rules have observed large enhancements in orbital moments in ultrathin ferromagnetic films grown on or sandwiched between normal metals [23-27]. Several studies suggest that this enhanced interfacial m_{orb} has a distinct perpendicular anisotropy [20,21] in films thin enough to exhibit PMA, while another points out the difficulty of resolving anisotropy of orbital from spin moments because of spin-obit coupling [28]. One study observed an enhanced interfacial m_{orb} with in-plane orientation in a thicker film with overall in-plane anisotropy [27].

In addition to the enhanced interfacial orbital moments, XMCD has revealed other important phenomena relating to magnetism in ultrathin films and at buried interfaces. One example is induced moments in nominally non-magnetic, ultrathin spacer layers such as Pt, Pd, and Cu when layered with ferromagnetic layers such as Co [29,24,30]. Another example is a significant increase (or change, more generally) in the size of the L_3 and L_2 white lines of $3d$ ferromagnetic layers at interfaces with non-metals [27,27], indicating an increase in consistent with charge transfer out of the $3d$ states. Yet another example is the observation of magnetic dead layers at interfaces. While interfacial Co in Co/Pt multilayers shows enhanced orbital and spin moments, Ni interfaces with Pt show regions of reduced or no magnetism [31,32].

All of these examples point to the conclusion that significant hybridization generally occurs at buried interfaces, with resulting redistribution of electrons in ways that radically alter electronic and magnetic properties potentially on both sides of the interface. It is also clear that XMCD and NEXAFS provide valuable capabilities to resolve these and other changes in electronic structure at buried interfaces. Of course, bulk properties of thin films are also studied beneficially with these spectroscopies. Magnetic semiconductors are one such class of materials, where induced moments of Ga and As host species are observed in $(Ga_{1-x}Mn_x)As$ [33], and Co moments are observed in dilute Co anatase $(Ti_{1-x}Co_x)O_2$ films [34].

4.3.2. Sensitivity of different absorption techniques

The above examples raise questions regarding the depth-sensitivity of soft x-ray XAS and XMCD measurements. How, for example, can sensitivity to buried interfaces be obtained? For the most part, these studies have determined that specific properties are intrinsic to interfaces by studying ultrathin films as their thickness decreases; interfacial effects are reasonably inferred to dominate in the thin limit. But the measured signals have only finite sensing depths, and, since subtle changes in spectral shapes and intensities are sometimes used to draw conclusions, it is important to consider the systematics of depth sensitivity. Three different modes are commonly used to measure XAS and XMCD in the soft x-ray range; total electron yield, fluorescence, and transmission.

By far the most common means to measure absorption in the soft x-ray range is by total electron yield, which is dominated by low energy secondary electron emission from the sample. In practice it is often easier to measure not the emitted electrons directly, but their complement given by the sample drain current flowing into the sample. Low energy secondary electrons have short escape depths $l_e \cong 2-4$ nm that limit the available information depth. The x-ray penetration length $l_x = (4\pi\beta/\lambda)\sin\theta$ for grazing incidence angle θ is generally large compared to l_e, but can be comparable to or less than l_e at the strong white lines in absorption spectra (above). Thus, even though l_e is small and essentially constant with λ, the strong energy dependence of $l_x(\lambda)$ imparts an energy dependence on the excitation rate of secondary electrons with depth than can significantly influence intensity of absorption features. Correction factors for these well-known saturation artifacts must be applied to obtain realistic values for spin and orbital moments from sum rules [19]. This in turn requires knowledge of $l_x(\lambda)$, as well as its polarization dependence. Electron yield detection can be compatible with varying applied magnetic fields, provided care is taken to ensure that all emitted electrons contribute to the signal.

Even though its cross-section is weak in the soft x-ray region, fluorescence detection mode is also used to obtain NEXAFS and XMCD spectra, and is clearly compatible with strong and varying applied fields. The fluorescence signal escape depth l_f is much greater than l_e, although similar considerations and corrections for the interplay between l_f and $l_x(\lambda)$ are applicable. A more fundamental concern is whether the L fluorescence signal is a true measure of the overall absorption, since the fluorescence cross-section is not necessarily equal for different multiplets [35,36]. A practical concern relates to the very low cross-section (< 1 %) for fluorescence in general in the soft x-ray range [37], that does not preclude the possibility that resonant elastically scattered photons may be detected along with fluorescence. While careful detector positioning based on p_c can minimize elastic charge scattering, p_{ml} shows that this is not effective for resonant magnetic scattering. Thus a grating spectrometer is the best means to ensure that only fluorescence photons are detected.

Standard transmission absorption measurements are possible provided samples can be synthesized with appropriate thickness [10]. Silicon nitride membranes provide semi-transparent substrates on which thin films can be deposited, and freestanding films can also be used. Fortuitously, the optimal thickness for transmission measurements (one or two absorption lengths) is in the thickness range of many magnetic thin film systems of current fundamental and technological interest. As in other spectral regions, saturation artifacts in transmission absorption measurements become important especially at strong absorption lines when samples are too thick [10]. Transmission measurements sense the entire sample thickness, and are compatible with applied fields.

4.3.3. Polarizing optical elements

Complete characterization of x-ray MO effects, just as in the near-visible spectral range, requires the ability to measure not just the intensity but also the phase (or polarization) of incident or scattered beams. At a minimum, rotating linear polarizers are needed to measure a beam's degree of linear polarization (P_L). Ideally quarter wave retarders would also be available to distinguish possible unpolarized (P_U) from circularly polarized radiation and thereby unambiguously determine the degree of circular polarization (P_C). In practice, synchrotron radiation is inherently polarized so that $P_U = 0$, and P_C can be determined from $P_L^2 + P_C^2 = 1$.

Optics for the measurement and manipulation of polarization in the soft x-ray range are typically based on the polarization dependence of charge scattering [38]. According to p_c, the Brewster angle (minimum in reflectivity for $\mathbf{e}_0 = \mathbf{e}_\pi$) in the x-ray range is $\theta_B \cong 45°$ where. Across the $100 - 2000$ eV range the extinction ratio, or the reflectivity ratio of σ to π polarization, at θ_B is high ($10^2 - 10^7$), providing good polarization rejection for a linear polarizer. Reflectivity for the σ component is generally quite low ($10^{-2} - 10^{-7}$) for highly polished, semi-infinite mirrors, and to boost efficiency multilayer structures are used to create an interference peak at θ_B. In the $100 - 2000$ eV range the period of such multilayers decreases from 8.8 nm to only 0.44 nm. Multilayer linear polarizers with periods down to ~ 0.6 nm have been fabricated and tested [39]. In the $500 - 1000$ eV range including the L edges of the $3d$ transition metals, these polarizers typically have component reflectivity $\sim 10^{-2} - 10^{-3}$ at the interference peak that is of order 1° (or $\Delta\lambda / \lambda \approx 0.01$) wide. To increase the polarizer bandwidth, translation along a period gradient [40], and operation at a range of angles near [41,42] are used. Fig.4-6 shows a schematic of a tunable linear polarizer based on graded multilayer reflectors [46]. The multilayer film is deposited on the reflecting surface with its period gradient normal to the scattering plane. Translating the multilayer along its gradient then tunes its interference peak at θ_B to occur at the desired x-ray energy. Alternatively, one can operate at fixed multilayer period or incidence angle and correct any energy dependent measurements for the reflectivity spectrum of the multilayer. Such linear polarizers are used both to measure the polarization of $\mathbf{e}_0$ and to monitor polarization changes in $\mathbf{e}_f$ resulting from magneto-optical effects.

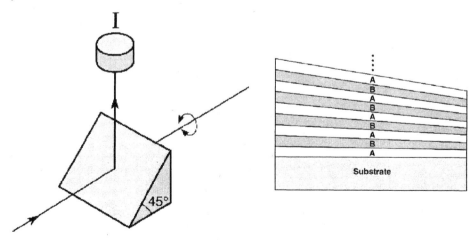

Figure 4-6: A Brewster angle polarizer in the soft x-ray ranges operates by reflection through 90° into an intensity detector (left). The entire assembly rotates around the incident beam. Efficiency is improved by operating at the interference maximum of a multilayer interference film. A lateral gradient in multilayer period (right) allows the interference peak to be tuned in energy by translating the gradient normal to the scattering plane [46,40].

Phase retarding optics, such as quarter wave plates, can also be made for use in the soft x-ray range. One approach uses multilayer interference structures either in transmission or reflection, operating near the interference peak where optical standing waves yield different phase changes on scattering for σ and π components [43,44]. Resonant MO effects themselves have also been used to create elliptically polarizing filters, whose utility is limited to very near the resonant core levels [45].

4.3.4. X-ray Faraday effect measurements

Transmission (Faraday) measurements of rotation of linearly polarized incident radiation provide a direct measure of magnetic circular birefringence, the refractive counterpart of MCD. Multilayer polarizers must be used to measure this rotation directly, and can also measure any ellipticity induced on transmission through the sample [46,10]. Faraday rotation spectra yield $f_{1,ml}$ directly, and $f_{2,ml}$ indirectly via the dispersion relation. While transmission measurements of XMCD can yield erroneous results especially for thick samples due to the presence of higher harmonics in the beam, multilayer polarizers suppress unwanted radiation outside of their narrow bandpass. The Faraday rotation signal can be large several eV below the L_3 edge, where absorption is minimum. These factors enable Faraday rotation spectra to sense element specific magnetization behavior in thicker samples than might be possible using the direct absorption channel.

Because the sample volume is well defined in Faraday rotation measurements, the magnetic rotary power or specific rotation (Verdet constant) thus measured can be compared across spectral ranges for a given material. Specific rotation measured at the L_3 white line in Fe is up to 6 x 10^5 deg/mm [47,10]. This is more than an order of magnitude larger than observed for Fe in any other spectral region, and attests to the value of x-ray MO studies using sharp core resonances.

4.3.5. Theoretical spectral calculations

Theoretical calculations of XMCD and Faraday effect signals have been made by different groups, and generally show good qualitative agreement with experimental measurements. Theories generally calculate the spin-polarized density of states using either an atomic or band formalism. Absorption spectra are obtained by summing transitions allowed by dipole selection rules ($\Delta l = \pm 1$ and $\Delta j = 0, \pm 1$ for l, j quantum numbers) weighted by the projected density of empty states and by the square of the radial matrix elements.

Enhanced (and suppressed) interfacial moments at interfaces have been observed in calculations [24,48]. Effects of induced moments at nominally non-magnetic sites due to magnetic neighbors have been calculated [49], as have field-induced magneto-optical effects in paramagnetic solids [50]. The importance of spin polarization and spin splitting in $2p$ core states has been established through comparisons of measurements with theory [51,52]. XAS and XMCD spectra of Fe_3O_4 and related structures have been calculated and found to be in good agreement with experimental measurements [53].

4.4. X-RAY MAGNETO-OPTICAL KERR EFFECT (XMOKE) – SPECULAR REFLECTION

The magneto-optical Kerr effect (MOKE) refers to specular reflection geometry. Lateral structure within the illuminated area is averaged in measurements of the specular beam. Core resonant x-ray MOKE (XMOKE) adds elemental specificity, shorter wavelengths for q-resolved studies of in-depth interference effects, and tunable penetration depth. These features imply that XMOKE effects should be better suited to study depth-variations in magnetic behavior in layered thin film systems than their near-visible MOKE counterparts. The term XMOKE is utilized here to refer to all resonant magnetic effects measured in the specularly reflected beam, acknowledging that many researchers have used different terminology.

4.4.1. Theoretical considerations

All theoretical considerations made for MOKE measurements in the near visible spectral range [54,55], such as intensity and phase effects in different settings (longitudinal, transverse, polar) generally extend into the x-ray range. Thus it is straightforward to extend formalisms for near-visible MOKE into the x-ray range, requiring only that optical and MO properties embodied by f_c, f_{m1}, and f_{m2} be properly translated into refractive indices and dielectric tensor elements used in these formalisms [10]. Again we limit consideration to f_c and f_{m1} contributions to these MO effects.

To describe XMOKE effects it is important to adopt formalisms that explicitly allow for depth variations in chemical and magnetic structure, *i.e.*, for layered structures [56,57]. In these descriptions, matrices are formulated to describe the complex amplitudes of reflected and transmitted fields at each interface, as well as matrices describing how complex field amplitudes change on propagation through each layer. Suitable matrix manipulation thus allows the evaluation of various MOKE effects for arbitrary $\mathbf{k}_0$, $\mathbf{e}_0$.

Such MOKE formalisms typically utilize the linear basis ($\mathbf{e}_\sigma$, $\mathbf{e}_\pi$), and yield a matrix

$$\begin{pmatrix} r_{\sigma\sigma} & r_{\sigma\pi} \\ r_{\pi\sigma} & r_{\pi\pi} \end{pmatrix}$$

describing the reflectance properties of the sample and containing the magnetic properties of the model structure. Operating on $\mathbf{e}_0$ then yields the amplitude of the reflected field. With $\mathbf{e}_0 = \mathbf{e}_\sigma$, *e.g.*, the scattered field is given by

$$\begin{pmatrix} r_{\sigma\sigma} & r_{\sigma\pi} \\ r_{\pi\sigma} & r_{\pi\pi} \end{pmatrix} \begin{pmatrix} 1 \\ 0 \end{pmatrix} = \begin{pmatrix} r_{\sigma\sigma} \\ r_{\pi\sigma} \end{pmatrix},$$

where $r_{\sigma\sigma}$ gives the non-rotated (charge) amplitude and $r_{\pi\sigma}$ the rotated (magnetic) amplitude. The reflected Kerr intensity is $r_{\sigma\sigma}^2 + r_{\pi\sigma}^2$ and the polarization of the reflected field is described by $r_{\pi\sigma}/r_{\sigma\sigma} = \varnothing_\sigma' + i\varnothing_\sigma''$ where $\varnothing_\sigma'$ is the Kerr rotation and $\varnothing_\sigma''$ the induced ellipticity of the linearly polarized incident beam.

Arbitrarily complex layered systems of magnetic and non-magnetic materials can be treated with such recursive formalisms by defining a sufficient number of layers to account, for example, for depth varying magnetic properties across a single FM layer. Parameters of such models can be varied to fit measured data to provide direct insight into depth resolved behavior.

Accurate values of resonant f_c and f_{ml} are essential in this modeling process, as deduced magnetic structures are meaningful only in reference to their assumed or measured values.

4.4.2. Exchange-spring heterostructures

Early studies of XMOKE effects include demonstrations of the various expected polarization and intensity effects [58-66,11]. Because of the added complexity of measuring polarization dependent effects, such as Kerr rotation and ellipticity, many early studies consider only intensity effects. These studies emphasize the large size of the Kerr effects across relevant core levels, and have primarily been concerned with first order MO effects, although second order effects have also been observed in reflection geometry [67]. The rapid variation of these large MO effects with λ and θ can be at first puzzling, or useful when understood in the context of model calculations using realistic values for resonant optical properties.

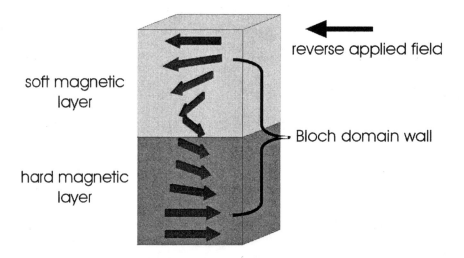

Figure 4-7: Exchange-spring heterostructures formed by exchange-coupled low (soft) and high (hard) anisotropy ferromagnetic films have useful properties. In a reverse applied field the soft layer switches more easily than the hard layer.

Here we review early studies of exchange-spring heterostructures that provide new information about their depth-resolved magnetic properties. Fig. 4-7 illustrates a magnetic bi-layer system in which a high anisotropy (hard) FM layer is exchange-coupled to a lower anisotropy (soft) FM layer. These systems are the basis of some of today's magnetic storage devices [68], and are of interest as model systems to study FM-FM exchange coupling and thereby composite magnets with potentially larger stored energy than single-phase magnets [69,70]. The illustration shows the expected response to a small reverse H; initially the top of soft layer reverses while the bottom is pinned by exchange coupling to the hard layer, producing a magnetization spiral (or Bloch domain wall) in depth in the film. At small H this partial switching is completely reversible (no hysteresis), hence the term exchange-spring. As H increases, eventually the hard layer switches completely. Theoretical models and macroscopic magnetization measurements observe a two-step reversal in such films [71], with low H_{rev} corresponding to the reversible initial switching of the soft layer, and with high H_{irr} corresponding to the switching of the hard layer. H_{irr} is reduced from the coercivity of an isolated hard layer, since the interfacial exchange softens the hard layer just as it hardens the soft layer.

While these general trends were clear prior to XMOKE studies, the details of how depth-dependent reversal proceeds was impossible to verify using techniques sensitive only to macroscopic behavior. Various aspects of resonant soft x-ray techniques make them useful in elucidating depth-dependent reversal in the different layers of such exchange-coupled systems.

Several measurements have focused on exchange-spring systems of general structure MgO(110)/buffer(t_A nm)/Sm-Co(t_B nm)/Fe(t_C nm)/cap(t_D nm), where the buffer and cap layers are either Cr or Fe and Cr or Ag, respectively, and thickness t_A, t_B, t_C, and t_D differ slightly [61,72]. The quasi-epitaxial nature of these samples yields an in-plane, uniaxial anisotropy. XMOKE measurements were made along the easy axis in longitudinal geometry (Fig.4-8) at small θ near both the Fe and Co $L_{2,3}$ resonances to probe the soft and hard layer **M** structure through minor and major hysteresis loops, always starting from saturation. Measured quantities include the Kerr intensity following reflection from sample as well as the raw Kerr rotation signal following reflection from a tunable linear polarizer.

The soft Fe layer is probed by tuning near the Fe $L_{2,3}$ resonances. Figure 4-9 shows XMOKE hysteresis loops from a sample with a 5 nm Cr cap layer and a 20 nm Fe layer measured at $\theta = 3.6°$ with energy tuned 2.3 eV below the peak in the L_3 absorption line, where penetration depths are relatively large [61]. Data collected using both linear and near-circular polarizations ($P_C = 0.9$) were used.

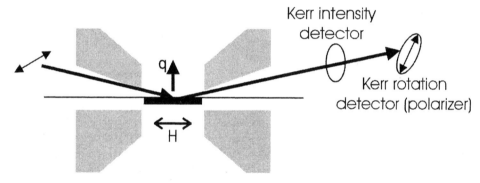

Figure 4-8: Experimental geometry for longitudinal XMOKE. The Kerr intensity is measured together with the Kerr rotation. If the intensity varies on magnetization reversal, the Kerr rotation signal must be normalized by this variation to obtain the true Kerr rotation.

We consider first the XMOKE results obtained with $\mathbf{e}_0 = \mathbf{e}_\sigma$ in the left panels of Fig.4-9. The raw Kerr rotation loop (top panel), measured with polarizer's scattering plane at 45° with respected to sample's scattering plane about k_f, shows a large intensity variation and asymmetric shape that, interpreted in isolation, could lead to mistaken conclusions. This is because the Kerr intensity signal (middle panel), measured before the polarizer, itself shows large changes through the loop. The Kerr intensity loop is symmetric with respect to **M(H)** reversal, showing a sharp jump at low H, followed by gradual changes and another jump at high H. These features are clearly associated with the onset of reversal by the top of the Fe layer at H_{rev}, a laterally coherent twist structure at intermediate H, and the high field switching of the hard layer at H_{irr}. The raw Kerr rotation signal, normalized by the Kerr intensity signal and converted to Kerr rotation angle, is in the bottom panel. This normalized Kerr rotation is now symmetric in H, and like the Kerr intensity shows the onset of soft-layer reversal, the evolution of the twist structure, and the switching of the hard layer with increasing H. The size of the Kerr angle is much larger than that typically observed in near-visible spectral regions.

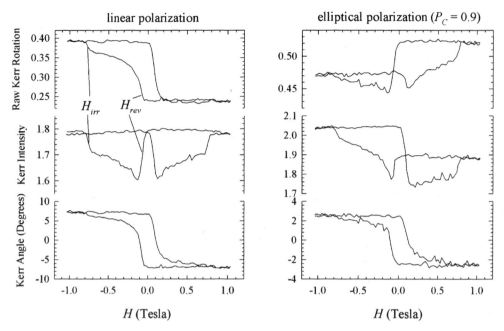

Figure 4-9: XMOKE data showing reversal of the soft Fe layer in a Sm-Co/Fe exchange-spring sample obtained by tuning near the Fe L_3 edge. Data at left are obtained with linear ($\mathbf{e}_\sigma$) incident polarization and at right with near circular polarization. Two-stage reversal of Fe is evident [61].

The observation of a Kerr intensity signal is significant for different reasons. It is well known that transverse MOKE effects yield intensity changes for $\mathbf{e}_0 = \mathbf{e}_\pi$ polarization with changes in net transverse moment. While the measurement had $\mathbf{e}_0 = \mathbf{e}_\sigma$, strong Kerr rotation yields an induced π component as radiation penetrates into the sample that in turn yields the observed Kerr intensity signal. The net transverse moment yielding the Kerr intensity signal implies a largely coherent magnetization spiral of specific chirality, consistent with the quasi-epitaxial nature of the Sm-Co/Fe system. Similar XMOKE studies of polycrystalline FePt/NiFe exchange-spring couples do not observe a Kerr intensity signal under similar conditions, meaning that incoherent spin spirals equally populate both chiralities during reversal [73]. Thus, buy measuring both Kerr intensity and Kerr rotation, we learn simultaneously about longitudinal and net transverse magnetization behavior.

It is instructive to compare the shapes of XMOKE signals measured using linear incident polarization with those measured using near-circular polarization. The right-hand panels in Fig.4-9 show the same signals as at left measured under the same conditions ($h\nu,\theta$), except using elliptical ($P_C = 0.9$) polarization. The Kerr intensity loop is now asymmetric in H, as expected since circular polarization has different reflectivity for oppositely oriented longitudinal M because the charge-magnetic interference term (below) is odd in M. The raw Kerr rotation loop is asymmetric in a different way than the Kerr intensity, but the normalized Kerr angle is again symmetric with an identical shape as that measured using $\mathbf{e}_0 = \mathbf{e}_\sigma$, although with reduced magnitude. The observation of any Kerr rotation might at first be surprising for nearly pure circular incident polarization. However, again the magnetization itself induces an increase in P_L as radiation propagates in the sample.

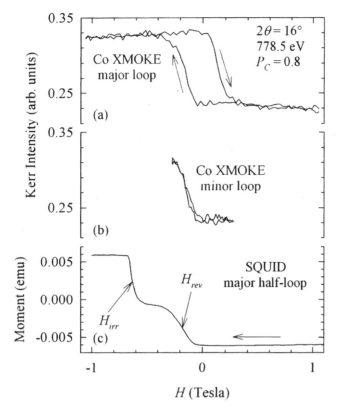

Figure 4-10: XMOKE loops (a & b) showing the reversal in a buried Sm-Co layer in a Sm-Co/Fe exchange-spring couple obtained with near circular polarization at the Co L_3 line. The SQUID major half-loop (c) shows the macroscopic reversal of the same sample. All loops start at positive saturation [72].

The sensitivity to longitudinal and net transverse **M** is clearly different for linear and circular polarization. The higher apparent symmetry of loops measured using linear polarization results because it is equivalent to a superposition of + and – helicity circular components. This allows for easier qualitative interpretation of linear polarization loops in terms of longitudinal and transverse M changes, provided that a polarizer is available to measure Kerr rotation.

These qualitative interpretations of the behavior of longitudinal and net transverse moments associated with the spiral structure within the Fe layer can be more quantitatively established by modeling. By describing the spiral in the Fe layer using multiple thinner layers, the recursive XMOKE formalism above reproduces the shapes of the Kerr rotation and intensity loops [61]. Such modeling easily rules out the (unlikely) possibility that the spiral is pinned at the top, rather than the bottom surface of the soft layer, thereby providing depth resolution throughout the 20 nm Fe layer through measurements made at a *single*, low θ and fixed hv. This relatively good penetration results because hv is tuned just below the L_3 line, where absorption is minimized and Re[n]>1 so that radiation refracts into the Fe layer rather than toward it's surface. Tuning to the peak of the Fe L_3 line maximizes absorption and changes the refractive conditions to substantially reduce penetration into the Fe layer. Loops measured under these conditions can be sensitive to just the upper region of the soft Fe layer, showing only the low H reversal as the spiral is formed, with no signal resulting from the pinned region near the interface [61].

Direct sensitivity to the buried hard layer can also be obtained by tuning to the Co $L_{2,3}$ edges to monitor the Co reversal in the Sm-Co layer. This is done for a sample having structure MgO(110)/Fe(20 nm)/Sm-Co(80 nm)/Fe(20 nm)/Ag(20 nm), operating at $\theta = 8.0°$, $P_C = 0.8$, and with tuned to the peak of the L_3 line where Co MCD is largest [72]. Figure 4-10 shows a major and minor Kerr intensity loop together with half of a major loop measured with a SQUID magnetometer from the same sample. The Co XMOKE loops are interesting for two reasons. First, these loops show a large, *reversible* Co signal at H_{rev} when the reversible spiral is formed in the soft Fe layer. Second, the major loop shows no signal at $H_{irr} \sim 0.6$ *Tesla* when the hard layer switches irreversibly. This later observation is understood to result from the limited (~ 2 nm) penetration in the thick Sm-Co layer with $h\nu$ tuned to the Co L_3 peak. Together the two XMOKE loops imply that the reversible twist structure penetrates significantly into the hard layer as soon as it is formed in the soft layer, as was also observed for another exchange-spring system [73]. Such information is not readily available from standard magnetometry.

4.4.3. Opportunities

The above examples obtain depth-resolved information either from modeling the shape of hysteresis loops at fixed θ, from element-specificity, and from varying the skin depth by tuning near a sharp resonance. Other approaches to gain depth resolution are available. One is to grow a compatible chemical marker layer at different depths within a magnetic multilayer, and tune to the core level of the marker layer [73]. Another is to measure systematic XMOKE effects vs. q and fit these variations with realistic models of magnetic variations in depth [61,65]. The resonant enhancements in chemical and charge scattering generally tend to enhance interference effects in layered films, enabling these q-resolved signals to be fit with relatively simple models involving magnetic variations with depth. An extreme example of this approach is to generate optical standing waves in a multilayer structure and tune the position of the standing waves to resolve magnetic structure within a fraction of the standing wave period [27,74].

XMOKE techniques are generally applicable to any layered system. Further systematic studies of exchange-coupled FM-FM systems can be expected. These techniques can sense uncompensated AF moments in exchange bias systems [75], and provide depth dependent reversal information in such systems, as shown by several early studies. Induced magnetism across non-magnetic interfaces, interface-induced changes in magnetic layers, and other proximity effects are examples of problems that will be beneficially studied with XMOKE in future studies.

4.5. DIFFUSE SCATTERING AND DIFFRACTION

While XMOKE provides depth resolution through different approaches, it averages over lateral structure (both chemical and magnetic) that is common in real systems. Topological interface roughness, grain boundaries, chemical segregation, and magnetic domains are distinct forms of heterogeneity, all of which give rise to scattering away from the specular beam in reflection geometry and away from the forward scattered beam in transmission geometry. The q resolved scattered intensity directly probes the spatial frequency spectrum of an ensemble of heterogeneities present in the sample. With q_{max} the highest spatial frequency measured, spatial resolution is given by $2\pi / q_{max}$, or $\lambda/2$ at backscattering. Tuning to core resonances dramatically enhances both chemical and magnetic contributions to scattered intensities, allowing the distinction of charge and magnetic structural that often co-exist without perfect correlation in real systems.

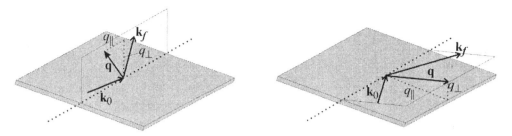

Figure 4-11: Reflection geometries of diffuse scattering utilize the in-plane component of the scattering vector to couple to lateral heterogeneity. At left the scattering plane (outlined) is normal to the surface, while at right it is strongly inclined toward the surface.

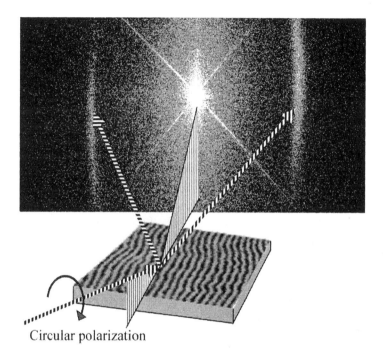

Circular polarization

Figure 4-12: A CCD detector positioned in the specularly reflected beam can measure diffuse scattering with a range of $q_\parallel$ and $q_\perp$ corresponding to a range of scattering planes. The sample consists of oriented magnetic stripe domains. Positive and negative diffraction orders from the domains are on either side of the bright specular beam on the CCD detector. From Ref. 9 (used with permission).

The first soft x-ray magnetic diffuse scattering studies used the off-specular reflection geometry to investigate magnetic and chemical surface roughness [76,77]. In specular reflection geometry, **q** is strictly perpendicular ($q_\perp$) to the surface, and off-specular geometries yield an in-plane **q** component ($q_\parallel$) that can be varied to study in-plane heterogeneity. For a fixed scattering angle 2θ or total **q**, the partition of **q** into $q_\perp$ and $q_\parallel$ depends on how the scattering plane is oriented with respect to the sample surface; two limiting cases are shown in Fig.4-11. With the scattering plane normal to the surface it is difficult to get to large $q_\parallel$ values for small 2θ, while with the scattering plane at a small angle from the surface **q** is predominantly along $q_\parallel$. For a given sample the two geometries have different sensitivity to magnetic structure through their different $\mathbf{e}_0$ and $\mathbf{e}_f$. If measuring diffuse scattering using an apertured detector, scans of detector or sample angles trace out an intensity trajectory through reciprocal space. Alternatively a CCD detector can be positioned in the scattered beam as in Fig.4-12 to measure a range of and for a given incident beam.

Such studies generally have used circular polarization with $e_0 = e_{0+/-}$ in conjunction with reversed **M** to measure intensities I_+ and I_-, where + and − refer either to reversed helicity for fixed **H**, or *vise versa*. Both of these quantities typically decrease monotonically with $q_{||}$. The difference $I_+ - I_-$ was interpreted initially as giving the magnetic intensity, and $(I_+ + I_-) / 2$ as the charge (or chemical) scattering. Variations of the difference and average with were then interpreted to represent the power spectra of magnetic and charge roughness. These initial studies found that $I_+ - I_-$ falls more quickly with $q_{||}$ than does $(I_+ + I_-) / 2$, and concluded that magnetic roughness has longer in-plane correlation lengths than chemical roughness. Studies have since pointed out that $I_+ - I_-$ actually corresponds to the charge-magnetic interference term [78]. In these measurements, $q_\perp$ can vary along with $q_{||}$. Such variations in $q_\perp$ introduce variations in the specular reflectivity that modulates the scattered intensity from heterogeneities either at or below the surface [79,10]. These variations can be particularly strong as $q_\perp$ approaches zero and either θ_i or θ_o approach θ_c. Since θ_c and penetration depths vary significantly with helicity and **M** (Figs.4-4 & 4-5), these optical and MO effects strongly influence the variation of $I_+ - I_-$ vs. $q_{||}$. Thus care should be taken in interpreting its shape relative to that of $(I_+ + I_-)$.

4.5.1. Theoretical considerations

Questions such as these concerning the relative charge and magnetic contributions to scattered intensities $I(\mathbf{q},\lambda)$ can be analyzed with relatively simple formalisms based on the scattering factor (2). While f_{m2} scattering has been observed from antiferromagnets [80], $f_{m1} \gg f_{m2}$ for ferromagnets and we limit consideration here to scattering from f_c and f_{m1} whose spatial variations generally differ. Here we adopt the circular basis and again use $f_\pm = p_c f_c \pm p_{m1} f_{m1}$.

A sample's scattering amplitude is generally written for opposite helicity circular polarization as

$$a_\pm = \sum_i f_{i\pm} \exp[i\mathbf{q} \cdot \mathbf{r}_i],$$

where the sum is over all atoms in the sample volume that may or may not be magnetic. For simplicity consider that all atoms are magnetic and of the same species. The intensity scattered in the Born approximation is then given by

$$
\begin{aligned}
I_\pm(q) &= \sum_i \sum_j f_{i\pm}^* f_{j\pm} \exp[i\mathbf{q} \cdot \mathbf{r}_{ij}] \\
&= f_c^2 s_{c-c} + f_m^2 s_{m-m} \pm 2(f_{2c} f_{1m} - f_{1c} f_{2m}) s_{c-m}
\end{aligned}
\tag{4.3}
$$

where partial structure factors $s_{c-c}(\mathbf{q})$, $s_{m-m}(\mathbf{q})$, and $s_{c-m}(\mathbf{q})$ describe the spatial distribution of charge-charge, magnetic-magnetic, and charge-magnetic correlations, respectively. Even for multi-component, mixed magnetic-nonmagnetic samples, the intensity can be grouped according to terms having p_c^2, $p_c p_{m1}$, and p_{m1}^2 polarization dependence, so the same 3 partial structure factors can be identified, although their weighting factors may be different. Thus there are always intensity contributions representing charge-charge, magnetic-magnetic, and charge-magnetic correlations. The orientations of the atomic moments $\mathbf{m}_i$ determine the strength of the magnetic contribution through p_{m1}. Since s_{c-m} and s_{m-m} are odd and even in helicity, respectively, it follows that $I_+ - I_- = 4(f_{2c} f_{1m} - f_{1c} f_{2m}) s_{c-m}$ contains only the cross-term, as noted in [78]. It also follows that $(I_+ + I_-)/2 = f_c^2 s_{c-c} + f_m^2 s_{m-m}$ contains the pure charge and pure magnetic contributions to the intensity [81]. Furthermore, if linear incident polarization were used, its scattered intensity $I_{lin} \equiv (I_+ + I_-)/2$ [81,88].

These expressions reveal the importance of distinguishing between the charge-magnetic and the pure magnetic intensity contributions, both of which are magnetic in origin but behave very differently as applied fields change the magnetization distribution. At resonance, $|f_c|$ is generally greater than $|f_{m1}|$. Thus, if measuring scattering using circular polarization, one can expect large changes in intensity with $\mathbf{M(H)}$ reversal resulting from the $p_c p_{m1}$ term implicit in s_{c-m} since p_{m1} is odd in M. If linear polarization is used, much smaller changes in intensity are expected from the p_{m1}^2 term in s_{m-m}, and these changes are expected to be symmetric in $\mathbf{M(H)}$. These features imply that the q dependence of the asymmetry $I_+ - I_-$ can provide information about magnetic structure, but in a form that is strongly field-modulated through cross-correlation with chemical structure information. Furthermore, I_{lin} vs. q provides information about pure magnetic structure, but only in the presence of a background of pure charge scattering. Thus, the difference in intensities $I_{lin} \equiv (I_+ + I_-)/2$ obtained at two different fields isolates changes in pure magnetic scattering, independent of any charge scattering whatsoever [81].

4.5.2. Perpendicular stripe domains in thin films

Magnetic domains and chemical grains, often present in thin films and bulk magnetic materials, are distinctly different forms of heterogeneity from magnetic and chemical topological surface roughness investigated in the earliest soft x-ray resonant magnetic scattering studies mentioned above. Such heterogeneities, in the form of up and down stripe domains in films with perpendicular anisotropy, illustrated in Fig.4-13, form an important class of samples in which resonant magnetic scattering has been extensively studied [82-89]. Film systems with relatively weak perpendicular anisotropy tend to form surface closure domains, as in Fig.4-13a, that become difficult to study as the stripe domain widths decrease.

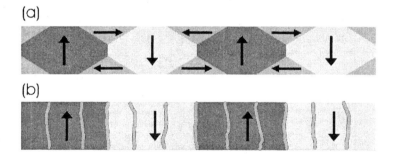

Figure 4-13: Schematic cross sections of magnetic stripe domains in thin films having perpendicular magnetic anisotropy, with arrows and shading indicating local magnetization directions. Relatively weak perpendicular anisotropy leads to surface closure domains as in (a). Possible chemical heterogeneity in the form of columnar grain boundaries or chemical segregation are indicated by irregular vertical lines in (b).

Off-specular reflection geometry studies of magnetic domains in FePt films with weak perpendicular anisotropy using detection geometries in Fig.4-11 (right) and Fig.4-12 reveal that the periodic domain structure forms a magnetic grating that scatters into a series of diffraction orders extending to + and − in $q_\parallel$ at fixed $q_\perp$ [82,88]. Using circular polarization, the +/- asymmetry of these magnetic diffraction peaks is observed to switch with helicity reversal [83], indicating that in-plane $\mathbf{M}$ components are located at the perpendicular grain boundaries as in

Fig.4-13a. The sum and difference of these magnetic intensities obtained with opposite helicity are observed to be quadric and linear in M[9], consistent with the simple model developed above. The incidence angle dependence of the asymmetry provides a measure of the depth of the surface closure domains and experimentally verifies the chiral nature of closure and stripe domains as in Fig.4-13a. These studies demonstrate how resonant magnetic scattering can resolve details of 3-dimensional magnetization structures in samples that are sufficiently ordered.

Co/Pt multilayer films have stronger perpendicular anisotropy (less pronounced surface closure domains), and provide a model system both to further test fundamental predictions of the simple scattering theory developed above and to study a wide range of magnetic phenomena that can be designed into these synthetic materials. Considering p_{m1} reveals that transmission geometry scattering measurements with $\mathbf{q}$ in the film plane, as in Fig.4-14, maximizes magnetic relative to charge scattering for studies of the spatial characteristics of magnetic domains in these systems. Such measurements are analogous to traditional small-angle x-ray scattering (SAS), but with added magnetic sensitivity.

Transmission SAS studies of Co/Pt multilayers reveal distinct magnetic and charge peaks resulting from magnetic domain and polycrystalline chemical grain structure, respectively [84]. Figure 4-15 shows $I_{lin}(q)$ measured with incident linear polarization tuned to the Co L_3 peak to yield resonant enhancements in both f_c and f_{m1}. At saturating field a peak at $q \cong 0.3$ nm^{-1} is observed. Near remanence a second peak appears at lower $q \cong 0.042$ nm^{-1}. Disordered stripe domains are known to proliferate in the reversal of such films from magnetic force microscopy (MFM) and x-ray microscopy (XRM) images. Figure 4-16 shows an XRM image of domains in a similar film measured using an imaging zone-plate microscope (below) at the Co L_3 edge. In addition to the scattered intensity vs. q, Fig.4-15 shows the scaled power spectral density (PSD) functions obtained from the Fourier transform of the XRM domain image and from an atomic force microscope (AFM) surface height distribution. Both the field dependence of low-q peak in I_{lin} and its coincidence in position with the domain PSD indicate that it originates from the magnetic domains. The spatial wavelength $2\pi/q_{peak} = 150$ nm corresponds to the average up-down domain pair length scale. The AFM image shows height variations characteristic of polycrystalline grains, and the high-q scattering peak results from interference of scattering from adjacent polycrystalline grains.

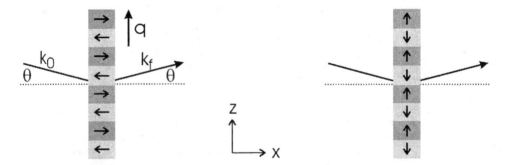

Figure 4-14: Transmission geometry to measure scattering from magnetic domains in thin films positions q in the film plane. Arrows indicate local magnetization directions. For small magnetic sensitivity is optimized for perpendicular magnetization as in (a). Reduced, but nonzero, sensitivity exists to in-plane magnetization structure as in (b).

A hysteresis loop of the diffuse scattering at the low-q peak is shown together with a visible MOKE hysteresis loop for this sample in Fig.4-17. The scattering is at a very small background level when the sample is saturated, rises abruptly at the nucleation of reverse domains, peaks near the coercive field, and falls to the background level at saturation. This comparison shows clearly that magnetic scattering is sensitive to a very different part of the hysteresis process than the MOKE loop, or any other hysteresis loop measuring nominally the *average* magnetization of the sample. Specifically, magnetic scattering measures *deviations from the average* magnetization with great sensitivity both to the field and spatial frequency dependence of these deviations. Scattering is thus an excellent tool to study magnetization structure present during complex reversal processes.

While the predominant magnetic and charge origin of the low- and high-q peaks, respectively, are without doubt, it is instructive to compare the energy dependence of the scattering at each peak with theoretical predictions (above) and measured values of f_c and f_{ml}. This is done in Fig.4-18, where the top and middle panels show energy scans (symbols) at the magnetic and charge peaks, respectively, collected both at saturation and at 0.1 T where the magnetic domain scattering is most intense. The bottom panel shows values of f_c and f_{ml} across the Co $L_{2,3}$ resonances, where the measured f_2 parts yield the f_1 parts via the dispersion relation.

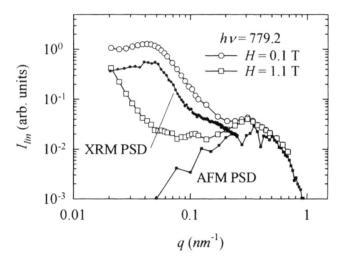

Figure 4-15: Scattered intensity from a Co/Pt multilayer (symbols) shows a low-q and high-q peak. Only the low-q peak is field dependent. Power spectral density functions obtained from an x-ray microscope domain image and an atomic force microscope height profile are scaled vertically and plotted on the same q scale [84].

Since this scattering was measured using linear incident polarization, the theoretical model developed above predicts that the scattering contains only magnetic-magnetic and charge-charge contributions, with no charge-magnetic scattering. Thus, the low-q magnetic domain peak near remanence should be well modeled by the spectrum of $4f^*_{ml}f_{ml}$ for Co, as illustrated in the inset of Fig.4-18a that shows the pure magnetic scattering contrast between oppositely oriented domains. This scattering contrast reverses with helicity. The resonant absorption by the sample must be included in the model, which means multiplying the calculated scattering spectrum by the sample's transmission spectrum. Using measured values for f_{ml} yields good agreement with the measured peaks in the scattering spectrum at remanence, provided a small, non-resonant charge scattering background is added to fit the measured background away from the Co L lines. At saturation, this weak charge background is the only contribution to calculated scattering at the low-q peak.

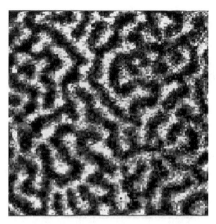

Figure 4-16: Imaging x-ray microscope domain image (4.5 μm field) from a Co/Pt multilayer similar to that whose scattering is in Fig.4-15.

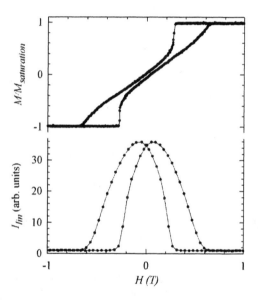

Figure 4-17: A SQUID easy axis hysteresis loop showing the average magnetization of a Co/Pt multilayer (top). Corresponding resonant scattering loop showing the strength of magnetization fluctuations (domains) at a fixed spatial frequency (bottom) [84].

To describe the charge scattering spectra from many possible models, it is reasonable to assume from the AFM results that topological interface roughness and/or density variations at polycrystalline grain boundaries are responsible for this scattering, and so that some linear combination of Co and Pt f_c should describe the scattering. Iterative modeling yields $a_c = f_{c,Co} + 3f_{c,Pt}$ for an amplitude that, when squared, produces good agreement with the measured spectrum at saturation in Fig.4-18b. (Note the characteristic bipolar shape of this pure charge scattering resonance, while the pure magnetic scattering resonance in Fig.4-18a has unipolar enhancements at each Co L line.) To model the scattering at the high-q peak near remanence, it was found that adding some pure magnetic intensity ($f^*_{m1} f_{m1}$) to this pure charge intensity produces good agreement. Such field-dependent pure magnetic intensity at the high-q peak is reasonably expected to be associated with the chemical density variation of Co. It was

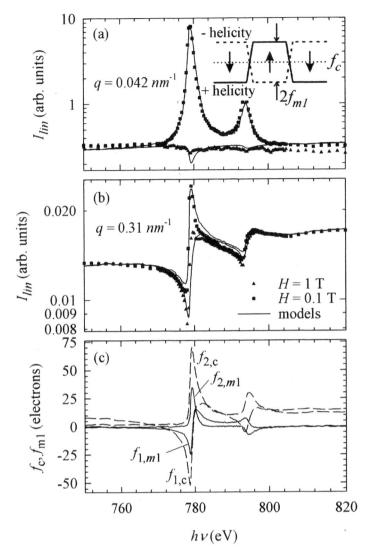

Figure 4-18: Energy spectra of scattering (symbols) at saturation and near remanence measured at the low-q domain peak and the high-q charge peak are in (a) and (b), respectively. The Co resonant charge and magnetic scattering factors in (c) are used to model the measured spectra above (lines). The inset in (a) shows the pure magnetic scattering contrast for oppositely oriented domains [84].

impossible to improve the agreement between model and data at the high-q peak with an added charge-magnetic interference term. Thus, these modeling results at both the low- and high-q peaks are consistent with the theoretical prediction that only pure magnetic and pure charge intensities contribute to I_{lin}.

Co/Pt multilayers and similar films with perpendicular anisotropy provide model systems to study **M** structure in films whose energetics are altered by introducing perpendicular exchange-bias [85], antiferromagnetic coupling [86], microstructural disorder [87], lithographic patterning [88], and domain order [89]. Resonant scattering has been used to follow the effects of these interactions on domain structure and reversal behavior. Such modifications may be important in future generations of perpendicular magnetic recording media and other devices.

4.5.3. Magnetic and chemical correlation lengths in recording media

While the transmission geometry is optimized to study magnetic structure in films with perpendicular anisotropy because of its overwhelming sensitivity to the $\mathbf{m}_x$ component, reduced magnetic sensitivity exists even when $\mathbf{M}$ is confined to the plane of the film as in Fig.4-14b. With $\mathbf{e}_0=\mathbf{e}_\sigma$, for example, scattering from the $\mathbf{m}_z$ component scales as $\sin\theta$ in $\mathbf{e}_\sigma \rightarrow \mathbf{e}_\pi$ scattering. With $\mathbf{e}_0=\mathbf{e}_\pi$, similar sensitivity exists to the $\mathbf{m}_z$ component in $\mathbf{e}_\pi \rightarrow \mathbf{e}_\sigma$ scattering, and to the $\mathbf{m}_y$ component in $\mathbf{e}_\pi \rightarrow \mathbf{e}_\pi$ scattering.

Scattering studies of recording media films with in-plane anisotropy, have demonstrated this sensitivity to in-plane magnetization and provided the first direct measurements of magnetic correlation lengths in such films [90-92]. Recording media films are often termed granular alloy films because they are designed to chemically phase separate into nanometer scale grains whose centers are magnetic and whose grain boundaries are nominally non-magnetic. The chemical heterogeneity associated with these films is readily observed in transmission electron microscopy (TEM) [93], and small-spot electron energy loss spectroscopy (EELS) reveals that Cr segregates to the grain boundaries and Co segregates to the grain-centers. This heterogeneous microstructure functions by reducing exchange interactions between the magnetic grain centers, thereby allowing sharper bit transitions to be written and, consequently, higher recording density. The grain sizes are small enough, however, that the length scale over which magnetism is correlated between grains had been difficult to measure directly.

In traditional recording media alloys having in-plane anisotropy the magnetic Co and non-magnetic Cr both have accessible L edges for resonant scattering in the soft x-ray spectral range. Thus, in addition to field-dependent measurements to resolve charge (chemical) from magnetic heterogeneity as demonstrated above, measurements at these different edges can be used for the same purpose [90]. This is demonstrated in Fig.4-19, that shows q scans for three distinct, in-plane granular alloy media films measured with x-ray energy tuned to the Co and Cr L_3 resonances at 778 and 574 eV, respectively. These data were obtained using bending magnet radiation having a range of incident polarization states, for which theory implies that charge-charge, charge-magnetic, and magnetic-magnetic terms all contribute to measured intensities. Even so, at least a partial, qualitative distinction between these terms is obtained from the combined resonant scattering results at these two edges. Specifically, since Cr is known to exhibit negligible f_{m1} (from XMCD measurements), only chemical (charge-charge) correlations are expected when tuned to the Cr edge. (The implicit assumption that the Co term is small near the Cr resonance is valid.) However, near the Co L_3 resonance Co exhibits large enhancements in both f_c and f_{m1}, so that both charge and magnetic amplitudes contribute to scattering at this edge.

The three panels of Fig.4-19 correspond to a series of three recording media film compositions in the historical development of granular alloy media over the last decade or more. All films were grown on an identical underlayer structure. One or more peaks are observed in the resonant q scans for each sample. At the Cr edge a single peak predominates for each sample at $q \cong 0.7$ nm^{-1}. This peak results from interference between adjacent scattering centers separated by $2\pi/q \cong 10$ nm, consistent with intergrain separations observed in TEM images. Since it is known that Cr segregates to the grain boundaries, this peak clearly originates predominantly from chemical heterogeneity. At the Co edge the same interference peak is observed at $q \cong 0.7$ nm^{-1}, and additional scattering is observed at significantly lower q. The high q scattering implies that this peak results from chemical segregation of Co and Cr. While at the Cr edge the scattering at the grain boundaries is enhanced, at the Co edge the scattering from the grain

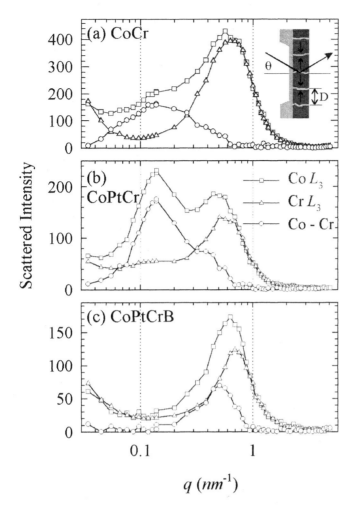

$q\ (nm^{-1})$

Figure 4-19: Scattered intensity vs. q for a series of Co-Cr based granular alloy recording media. Measurements are made at the Co and Cr L_3 edges, and are scaled to match at high q. The difference is also plotted. Data collected at the Cr edge result predominantly from pure charge correlations. Data collected at the Co edge contain both charge and magnetic contributions. The difference tends to isolate the magnetic contributions, revealing that magnetic correlation lengths are generally much longer than chemical correlation lengths in these materials, and that the addition of boron significantly reduces magnetic correlation lengths [90].

centers is enhanced. Thus these two complementary scattering structures produce essentially identical peaks, as they should since the Fourier transform of structure in a positive and negative black and white image is the same.

The additional scattering at lower q observed only at the Co edge must have a separate origin from the chemical segregation. In the CoCr and CoPtCr alloys a low q peak is observed at $q \cong 0.15 \text{ nm}^{-1}$, while in the CoPtCrB alloy the additional scattering essentially broadens the grain size peak to lower q. The data in Fig.4-18 are scaled so that Cr and Co edge data match on the high q side of the chemical grain peak. By subtracting the Cr from the Co edge scans theory implies that the remainder represents predominantly magnetic-magnetic and magnetic-charge correlations, both of which are expected to be strong for Co. These difference curves show a

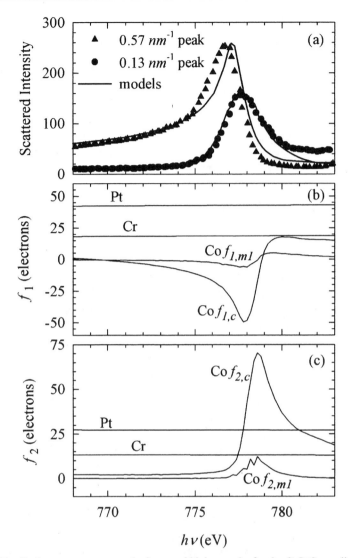

Figure 4-20: Co L_3 energy spectra at the low- and high-q peaks for the CoPtCr media sample are symbols in (a). The real and imaginary parts of scattering factors for Co, Cr, and Pt in (b) and (c), respectively, are used to produce the model spectra (lines) in (a) [91].

pronounced peak that is well separated from the chemical grain size peak that is reasonably inferred to represent magnetic correlations, *i.e.*, interference between regions that have different average orientations of magnetization, based on scattering theory. The observation of distinct peaks in these difference curves implies that there is, on average, a characteristic magnetic correlation length describing the size of these magnetic regions.

To experimentally confirm the magnetic and chemical origin of the distinct peaks observed in Fig.4-19, one could measure changes with applied fields. However, these q scans were collected at remanence before the first application of an external field in order to study magnetic correlations in the as-grown state. While subsequent field dependent measurements do confirm the magnetic origin of the low-q peaks, initially this was determined by modeling energy scans across the Co L_3 edge collected at the two peaks of the CoPtCr sample, as shown in Fig.4-20.

Like in the case of Co/Pt multilayers (above), the energy spectrum at the low-q peak of this granular alloy film can be well modeled by the shape of $f^*_{ml}f_{ml}$ plus a small, non-resonant background. No models involving added resonant chemical scattering fit the data as well as this pure resonant magnetic scattering model, thus establishing the low-q peak as pure magnetic in origin. The shape of this magnetic peak is determined solely by energy dependence of f_{ml}. It does not depend on the in-plane orientation of the magnetization of adjacent regions.

Modeling the high-q peak not only confirms its predominant origin from charge-charge correlations, but also provides a measure of the chemical segregation responsible for this scattering. This modeling assumes a chemically segregated grain structure with Co-rich grain centers, and Co-deficient grain boundaries, consistent with prior knowledge from electron microscopy. Designating these two phases A and B, then the scattering amplitude of each phase will be a_A and a_B, where each represents a linear combination of the elemental atomic scattering factors weighted by their density in each phase. Fig.4-20b and 4-20c give the relevant scattering factors; the resonant Co values are measured and the non-resonant Cr and Pt values are taken from tabulated data [5]. The scattering amplitude contrast $a_{charge} = a_A - a_B$ then represents the scattering power of the chemically heterogeneous ensemble. The energy dependence of the scatter intensity $a^*_{charge}a_{charge}$ is clearly very sensitive to the chemical makeup of the two phases. Iterative evaluation of this spectra adjusting the composition of the two phases led to a model in which the Co-rich grain centers have Co:Pt: Cr ratio of 20:2:1, and the Co-deficient grain boundaries contain no Pt and have Co:Cr ratio of 1:1. The resulting spectral model is the dashed line in Fig.4-20a that reasonably accounts for the shape of the high-q peak. While it is not claimed that this pure charge scattering model is unique or yields the best possible fit to these data [91], these compositions are in reasonable agreement with expectations of composition based on focused EELS studies.

Thus there is good reason to believe that the low- and high-q features in Fig.4-19 do represent predominantly magnetic-magnetic and charge-charge correlations, respectively. The high-q chemical grain peaks are essentially constant in position with alloy composition, and represent the average intergrain distance or chemical correlation length scale. The magnetic peak, especially as evidenced in the difference curves, indicates that the magnetic correlation length remains substantially longer than the chemical grain size for both the CoCr and CoPtCr alloys. This in turn suggests that the magnetization of several adjacent grains tends to be correlated, presumably by undesirable intergrain exchange interactions. It is then deviations in magnetization orientation between regions containing several correlated grains that gives rise to the magnetic scattering observed at the low-q peak. With the addition of boron, however, the magnetic correlation length decreases significantly so that the difference peak closely approaches the chemical grain size peak.

Thus resonant magnetic scattering reveals directly that boron addition is highly effective in reducing intergrain magnetic exchange interactions. This correlates with improved signal to noise and recording density of the more complex alloy system. In addition to resolving magnetic correlation lengths, chemical correlation lengths are observed simultaneously, and careful modeling can provide meaningful measures of composition differences between chemically segregated regions at lengths scales of order 10 nm and below.

4.5.4. Coherent magnetic scattering

An x-ray beam's coherence is characterized by its longitudinal or temporal coherence length (along the propagation direction) and its transverse coherence length (normal to the propagation direction). A monochromatic plane wave constitutes a fully coherent beam. The longitudinal coherence length $\lambda^2/\Delta\lambda$ is typically set by the resolving power of a grating monochromator in the soft x-ray region to have a value $\sim 1000 - 10000\ \lambda$, and tends to be large compared to absorption lengths in general. The transverse coherence scales as λ^2 over the source brightness, which for undulators at 3rd generation synchrotron sources can be appreciable [2]. While coherent x-ray scattering was first studied in the hard x-ray range [94], the λ^2 scaling makes the soft x-ray range especially attractive for coherent scattering studies [95].

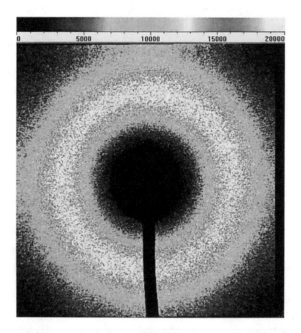

Figure 4-21: CCD image of the diffuse scattering ring produced by disordered stripe domains in a Co/Pt multilayer. An upstream spatial filter was used to enhance the speckle that can be seen within the smooth scattering envelop. A beam stop support is seen at the bottom [87]. *Also see color plate.*

The scattering discussed in the previous section is sometimes termed incoherent. However, it actually uses the coherence of the beam in two ways. The temporal coherence aids in tuning to specific resonant energies. The interference of scattering from inhomogeneities to produce peaks in reciprocal space can only occur for real space structure within the beam's coherence volume. Thus incoherent scattering is a misnomer, and partially coherent scattering is a more apt description. Even so, this partially coherent scattering ignores one important feature resulting from the coherence of the beam – the speckled intensity distribution of the scattering.

Coherent scattering refers to the use of incident radiation having a high degree of transverse coherence, *i.e.*, nearly plane wave fronts. The phases of waves scattered across the plane wave front interfere in the far field (at the detector) to produce a speckled intensity pattern within the envelope function of the partially coherent intensity peaks discussed above. The intensity variation or contrast of the speckle is in principle 100% for fully coherent illumination. The degree of coherence is often augmented by positioning a pinhole (spatial filter) between the source and sample. The speckle encodes the complete local details of the 2-dimensional spatial distribution of scattering centers, and thus contains far more information that just the position and width of the interference peaks as discussed above. To measure the speckled intensity distribution, a detector with spatial resolution finer than the speckle size is needed. Often a charge-coupled device (CCD) detector is used for this purpose.

Magnetic speckle has been observed in several x-ray studies [80,96,88,87] although is in its infancy at the time of this writing. Figure 4-21 shows an example of a magnetic speckle pattern obtained from Co/Pt multilayers identical to those discussed above. The diffuse ring of intensity in this CCD image is the same magnetic peak in Fig.4-15, resulting from labyrinth stripe domains like those in Fig.4-16. Examination of the scattering reveals speckles that extend over one or two CCD pixels, so the oversampling of the intensity peak is just sufficient. Just as the mechanisms of coherent magnetic scattering are direct extensions of visible coherent techniques, so are the general types of information obtainable from these techniques. The information contained in magnetic speckle can be used in several different ways.

In principle the speckle distribution can be inverted to obtain a real-space image of the scattering object. In practice, phasing the speckle pattern can be problematic in this form of lenseless imaging, although oversampling the speckle pattern and the use of known boundary conditions to the scattering object can overcome the phase problem [97]. Attempts to perform this inversion with magnetic speckle have been carried out using real [98] and simulated data [98,99]. Initial results are promising, although expected concerns about the uniqueness of resulting images are not currently resolved. Rapid advances in speckle inversion via holographic or heterodyning approaches, that interfere the speckle pattern with a known reference wave, can be anticipated [100,101]. The lenseless imaging approach is being investigated as an imaging mode for fourth generation synchrotron x-ray sources operating in single-shot mode.

The second approach is the x-ray analog of dynamic light scattering, time (or photon) correlation spectroscopy, or intensity fluctuation spectroscopy initially developed using lasers. Here, temporal intensity fluctuations in the solid angle of a single speckle provide information about sample dynamics at that spatial frequency. X-ray time correlation spectroscopy was first applied in the hard x-ray [102,103], where time resolution down to 50 nanoseconds has been obtained [104]. In the soft x-ray it has been used in dynamic, non-resonant, charge scattering down to microsecond time resolution in liquid crystals [95]. Studies of magnetization dynamics with microsecond or possibly better time resolution should be feasible.

A third approach to utilizing coherent magnetic scattering is similar to time correlation spectroscopy in that it relies on changes in magnetic speckle, but now with respect to field-induced, quasi-static changes in domain structure rather than rapid spontaneous fluctuations [87]. Very different from dynamic fluctuation spectroscopies, this speckle metrology approach uses a large portion of the coherent scattering pattern containing many speckles. Correlation coefficients are defined such that the auto-correlation of a static speckle pattern yields 1, while the cross-correlation of two uncorrelated speckle patterns yields 0. So defined, the cross-correlation coefficient of two speckle patterns obtained at different points around a hysteresis loop provides a direct statistical measure of the ensemble-averaged domain correlations between these two measurements. Speckle metrology is thus an ideal tool to measure microscopic magnetic domain memory in films. Compared to speckle inversion, this approach sidesteps the phase uniqueness question to obtain statistical microscopic correlations. Compared to soft x-ray microscopy (below) and other magnetic microscopies, this approach forgoes specific local real space information about individual domains in favor of a statistical value over all of real space sampled.

Domains in Co/Pt multilayer films were the first objects studied by magnetic speckle metrology. Specifically, two multilayers of similar nominal structure, $[Co(0.4 \text{ nm})/(Pt \ 0.7 \text{ nm})]_{50}$, but sputter deposited at 3 and 12 mTorr argon pressure were investigated [87]. The visible MOKE hysteresis loops for these two samples in Fig.4-22 show that the two films have significantly different reversal behavior. While domains mediate reversal in each film, their remanence and coercivity depend sensitively on the relative amount of microstructural disorder that in turn depends on the sputter pressure. Films deposited at lower pressure tend to be smoother than those deposited at higher pressure; the extreme roughness limit corresponds to columnar growth of polycrystalline grains with voided grain boundaries, while the smooth limit corresponds to a polycrystalline microstructure with dense grain boundaries and a relatively smooth surface topography [105]. Speckle metrology reveals that the extent of microscopic magnetic memory is dramatically different depending on the growth-induced disorder in these films. Return-point memory of the domain configurations is quantified by the cross-correlation coefficient between speckle patterns measured at the same field point on a hysteresis loop, but following one or many minor or major loop excursions away from the initial point.

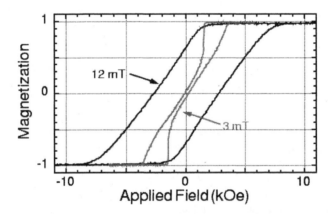

Figure 4-22: Visible MOKE hysteresis loops for Co/Pt multilayers sputter deposited at two different argon pressures reveal very different reversal mechanisms [87].

Consider first the return-point memory when the end points are at the coercive point, where M = 0 and many domains are present. It was found that the smooth sample exhibited finite memory for minor loops that stopped short of saturation [87]. However, once saturation was reached in minor or major loops, all return-point memory of domain structure was lost. This is consistent with essentially random nucleation and growth of domains from saturation in the smooth sample. The rougher, 12 mTorr sample, on the other hand, exhibited non-zero return-point memory for both minor loops to saturation and major loops. Figure 4-23 shows the RPM for major loops for this sample, where the x-axis corresponds to the starting and ending field at which the cross-correlations were evaluated. The sharp onset of memory at negative fields corresponds to the initial nucleation of domains in a loop starting from negative saturation. Following the nucleation peak, the cross-correlation remains high as domain growth and wall motion proceeds, before falling to zero as saturation is approached. Thus, while domain nucleation exhibits strong memory, as domains proliferate their distribution becomes more random. This characteristic shape of the memory vs. endpoint H is retained for cross-correlations between tens of major loops, indicating that the increased structural disorder contains numerous pinning sites that influence both nucleation and growth of domains in a very robust way.

Its relatively direct and quantitative statistical measure of microscopic magnetic memory of a large structural ensemble suggests that the speckle metrology technique will become a useful tool in studying a variety of magnetic systems where details of the reversal mechanism are of interest. Magnetic storage media are one example, where the interplay between structural and magnetic disorder are expected to have a strong bearing on signal-to-noise and related device performance issues. In addition to this new form of speckle metrology, it is expected that lenseless imaging and time correlation spectroscopy using coherent magnetic scattering will grow as the limits of their capabilities are more fully established.

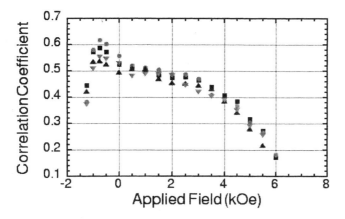

Figure 4-23: Cross correlation coefficients obtained at the same applied field point after one or more complete hysteresis loop cycles reveal the trend in microscopic return point domain memory for the 12 mTorr sample [87].

4.6. DIRECT MAGNETIZATION IMAGING

Several different approaches for direct (real space) magnetic imaging have been developed utilizing resonant XMCD and XMLD to gain sensitivity to ferromagnetic and antiferromagnetic structure. These are broadly categorized as electron imaging microscopes and zone-plate microscopes that utilize x-rays entirely to form images.

4.6.1. Photo-electron emission microscopes

One approach images low energy (secondary) electrons emitted from a magnetic surface, much as in the SPLEEM technique discussed in Chapter 9. Rather than using spin-polarization detection, however, magnetic contrast is gained through the XMCD or XMLD effects on the secondary emission, requiring incident tunable soft x-rays of suitable polarization [106, 107, 108,109]. Such photo-emission electron microscopes (PEEMs) have routine spatial resolution of ~20-50 nm set by chromatic aberrations of imaging secondary electrons. Elaborate schemes to improve this resolution down to ~ 1-2 nm are under investigation. The extreme sensitivity to surface magnetic properties of secondary emission enables the study of smooth surfaces and layers buried under only 2-3 nm of material. Short, pulsed fields applied between synchrotron x-ray pulses are feasible, and have been used to study magnetization dynamics in confined thin films [110,111]. Static applied fields orthogonal to the nominal electron trajectory are incompatible with low energy electron imaging. Other limitations include the inability to study layers and interfaces several nanometers or more below the surface and samples that are insulating or have surface topography that yields unwanted contrast.

These properties make PEEMs especially powerful to study exchange bias systems, where XMCD and XMLD contrast allows imaging of ferromagnetic and antiferromagnetic domains, respectively, in the same sample [112,113]. In addition to spatial resolution, quantitative spectroscopy can provide detailed information about the anisotropy axes in different layers [114] and interfacial spins [115].

4.6.2. Imaging and scanning zone-plate microscopes

Two distinct types of soft x-ray zone-plate microscopes, full field imaging and scanning, have been utilized to image magnetic structure. Their fundamental components are illustrated in Fig.4-24. High-resolution Fresnel zone plate lenses are at the heart these microscopes. These ultrafine diffracting structures require state-of-the-art electron beam lithography to precisely position concentric circular rings over large areas forming the physical aperture of the lens [2,116]. The diffraction-limited spatial resolution offered by a simple zone-plate lens approximately equals $1.22\delta r_N$ where δr_N is the width of the outermost zone. For current lithography technology the diffraction limited spatial resolution is 18-30 nm.

Imaging zone-plate microscope

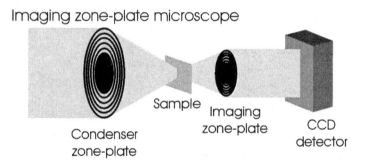

Condenser
zone-plate

Sample Imaging
zone-plate

CCD
detector

Scanning zone-plate microscope

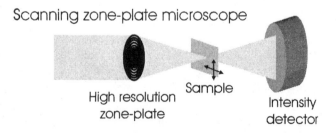

High resolution
zone-plate

Sample

Intensity
detector

Figure 4-24: Two different microscope configurations using zone-plate lenses to image magnetization.

Full field imaging microscopes require a condenser optic to illuminate the sample, followed by a high resolution zone plate to project an image onto a 2-dimensional detector [117,118].

Coarse resolution zone plate lenses are often used as the condenser, although focusing mirror condensers may also serve this purpose.

Scanning transmission zone-plate microscopes utilize a single high-resolution zone-plate lens to focus a high-brightness incident beam to a diffraction-limited spot. The sample is raster-scanned through the focal spot, and the transmitted intensity is monitored. An order-sorting aperture slightly smaller than the opaque central region of the lens is positioned to aperture most of the radiation not diffracted in first order (focused) by the zone plate. Compared to a limited field of view of an imaging microscope, the field of a scanning microscope is limited only by the scanning stages.

Both types of zone-plate microscopes obtain magnetic contrast via the XMCD effect in transmission absorption. Transmission geometry requires freestanding samples or samples on semitransparent substrates. For samples with magnetization in the substrate plane, the sample must be tilted away from normal incidence to obtain magnetic contrast ($\propto p_{m\parallel}$), while samples with perpendicular anisotropy are easily studied at normal incidence.

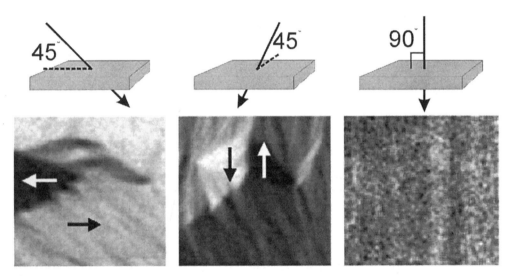

Figure 4-25: Scanning transmission x-ray microscope images of the same 60×60 nm region of a demagnetized, 33 nm thick Fe film taken at three viewing angles as indicated. Each image is actually the division of two images obtained with opposite helicity elliptical polarization tuned to the Fe L_3 line to enhance magnetic contrast. These projections of M variation along the beam direction reveal a blade-shaped domain growing from the left into a larger 180° domain. Arrows indicate M orientation along projected directions [126].

4.6.3. Zone-plate imaging of domain structure

The first SXR images of magnetic structure studied stripe domains in Gd/Fe multilayers having perpendicular anisotropy [119,120], much like those in Fig.4-16. Such studies have expanded to include patterned films with perpendicular anisotropy [118,121,122], field-dependent domain behavior [123,89], and extended [124] and patterned films having in-plane anisotropy.

The first magnetic images from a scanning transmission x-ray microscope (STXM) [125] were of domain and domain-boundary structures in demagnetized films having in-plane anisotropy [38,126]. Since tilting is required to obtain XMCD contrast for in-plane M, it is simple to extend tilting to multiple viewing angles in order to reconstruct the vector nature of magnetization resolved laterally across a film. This is demonstrated in Fig.4-25, where three images show black and white XMCD contrast images obtained at viewing angles as indicated from a demagnetized, 33 nm thick Fe film [126]. Seen in these images is a needle- or blade-shaped domain growing into a larger region of nominally reversed magnetization. Since the contrast in these images results from the projection of **M** along the viewing direction, the data is readily transformed to obtain the direction of **M**, as seen in Fig.4-26. The main image shows in-plane **M** components that are color-coded to represent directions as indicated. The inset shows perpendicular **M** components, with separate color scale indicating the angle out of plane (90 indicates in-plane M). A rich variety of structure is observed at and near the grain boundary, including hybrid Néel and Bloch domain walls, pronounced perpendicular magnetization components at the core of vortex structures, and vortex pairs extending well away from the main domain wall. While these structures were known to exist before this study, the improved spatial resolution coupled with quantitative vector imaging capabilities are expected to yield new information on a range of magnetic features as they become more routinely used.

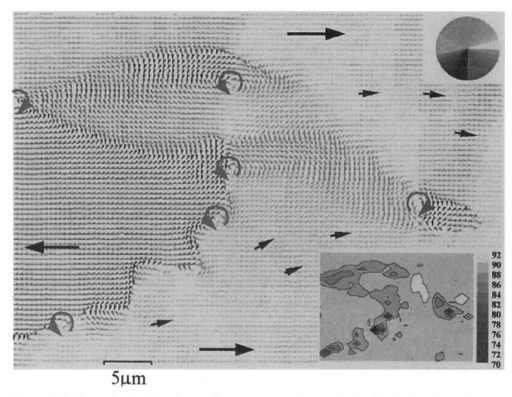

5µm

Figure 4-26: Composite 50×35 *nm* image of vector magnetization at the tip of blade domain growing into a 180° domain. The main image shows in-plane magnetization, with arrows and color wheel indicating in plane directions. The inset shows out-of-plane magnetization components, with 90° corresponding to in-plane [126]. *Also see color plate.*

To date, zone-plate microscopes have been developed primarily for polymer and wet chemical and biological studies, not for magnetic studies. Implementing limited sample tilting and applied magnetic field capabilities has been feasible in these existing microscopes. Future zone-plate microscopes specifically designed for magnetic studies would enable stronger applied fields, sample temperature control, more flexible transmission geometries, and reflection geometry studies. Dynamic studies using pump-probe techniques are compatible with both microscopes [127]. The high flux density in the focused STXM spot should enable interesting time correlation experiments analogous to dynamic light scattering using coherent scattering, only now in a small, well-defined sample volume.

4.6.4. Complementarity of microscopy and scattering

The results presented above raise questions concerning the relative utility of scattering vis-à-vis microscopy to study magnetic heterogeneity since, after all, microscopy was used to image (Fig.4-16) the same domains whose low q scattering was measured in both partially coherent (Fig.4-15) and more fully coherent (Fig.4-21) fashion. Of course, microscopy and scattering have complementary sensitivity to magnetic heterogeneity. Images provide direct, detailed information about local magnetic structure, especially isolated or individual features, that is unavailable from scattering. Thus real space images are tremendously valuable because of their specific information content and because of the ease with which it is grasped. The correlation functions from scattering, on the other hand, typically average over a larger ensemble than do images to provide (arguably) a statistically more robust measure of certain types of structure. Scattering is inherently a dark field technique, and its very origin makes it extremely sensitive to magnetic heterogeneity. Zone plate x-ray microscopy is a bright-field technique, and so has somewhat different, albeit large, sensitivity to magnetic heterogeneity.

In addition to these well-known attributes, specific differences are emerging as relevant in their comparative capabilities. Spatial resolution in zone-plate microscopy scales as the inverse of the zone-plate numerical aperture that is difficult to increase beyond ~ 0.1. Scattering can be measured out to backscattering with an effective numerical aperture ($\propto q$) of 1, and so can resolve features ten times smaller than microscopy. This larger spatial frequency range over which signals are inherently Fourier filtered, coupled with a larger dynamic range, gives scattering significantly greater sensitivity to mixed heterogeneity than microscopy. For example, simultaneous sensitivity to magnetic and chemical structure below the zone-plate resolution limit has been obtained in scattering measurements of granular recording media [91, 128,129]. The more open geometry of scattering experiments may make them more amenable to utilization in a variety of sample environments.

Of course, the combination of scattering and microscopy generally provide more information that either technique in isolation.

4.7. SUMMARY

In summary, relevant core resonances for $3d$ transition and $4f$ rare earth elements provide elemental, electronic state, spin, and orbital moment sensitivity to spectroscopies in the soft x-ray spectral region. These are driving the extension of the entire variety of spectroscopic, scattering, and microscopy techniques familiar from other spectral regions into the soft x-ray region, that in turn brings nanometer scale sensitivity to magnetically and chemically heterogeneous systems. While these developments are still in progress, numerous studies have already demonstrated that the attributes of these soft x-ray techniques can impact our understanding of a variety of important systems. Their general applicability to interfaces, thin films and near surface properties of bulk samples suggests that their application will expand.

Dynamic and time-resolved techniques, also familiar from other spectral ranges, are beginning to be combined with these unique capabilities offered by the soft x-ray spectral range. It is reasonable to expect that these techniques will all continue to evolve to provide increasing understanding of magnetic systems and phenomena of fundamental and technological interest.

ACKNOWLEDGEMENTS

The author would like to acknowledge collaborators E. E. Fullerton, O. Hellwig, S.-K. Kim, S. Jiang, S. D. Bader, L. Sorensen, S. D. Kevan, G. P. Denbeaux, T. Warwick and K. M. Chesnel for their contributions to work described in this chapter, and for many stimulating discussions. The work at LBNL was supported by the Director, Office of Energy Research, Office of Basic Energy Sciences, U. S. Department of Energy under Contract No. DE-AC03-76SF00098.

REFERENCES

[1]   J. B. Kortright, D. D. Awschalom, J. Stöhr, S. D. Bader, Y. U. Idzerda, S. S. P. Parkin, I. K. Schuller, and H.-C. Seigmann, J. Magn. Magn. Mater. **207** (1999) 7.

[2]   D. T. Attwood, Soft x-rays and extreme ultraviolet radiation: principles and applications, Cambridge University Press, Cambridge (2000).

[3]   J. J. Sakurai, Advanced quantum mechanics, Addison-Wesley, Reading, Mass. (1967).

[4]   R. W. James, The optical principles of the diffraction of x-rays, Ox Bow Press, Woodbridge, Connitecut, 1982.

[5]   B. L. Henke, E. M. Gullikson, and J. C. Davis, At. Data and Nucl. Data Tables, **54** (1993), 181, and X-Ray Data Booklet, http://xdp.lbl.gov.

[6]   J. P. Hannon, G. T. Trammel, M. Blume, and D. Gibbs, Phys. Rev. Lett. **61**, 1245 (1988), and errata, *ibid*, **62** (1989) 2644.

[7]   S. W. Lovsey and S. P. Collins, X-Ray Scattering and Absorption in Magnetic Materials, Oxford University Press, Oxford, 1996.

[8]   D. Gibbs, D. R. Harshman, E. D. Isaacs, D. B. McWhan, D. Mills, and C. Vettier, Phys. Rev. Lett. **61** (1989) 1241.

[9]   K. Chesnel, M. Belakhovsky, A. Marty, G. Beutier, G. van der Laan, and S. P. Collins, Physica B **345** (2004) 148.

[10]   J. B. Kortright and S.-K. Kim, Phys. Rev. B **62** (2000) 12216.

[11]   J. E. Prieto, F. Heigl, O. Krupin, G. Kaindl, and K. Starke, Phys. Rev. B **68** (2003) 134453 .

[12]   J. Stöhr, NEXAFS Spectroscopy, Springer-Verlag, Berlin, (1992).

[13]   F. M. F. de Groot, J. C. Fuggle, B. T. Thole, and G. A. Sawatzky, Phys. Rev. B **42** (1990) 5459 .

[14]   B. T. Thole, G. van der Laan, and G. A. Sawatzky, Phys. Rev. Lett. **55** (1985) 2086.

[15]   G. Schütz, W. Wagner, W. Wilhelm, P. Kienle, R. Zeller, R. Frahm, and G. Materlik, Phys. Rev. Lett. **58** (1987) 737.

[16]   B. T. Thole, P. Carra, F. Sette, and G. van der Laan, Phys. Rev. Lett. **68** (1992) 1943.

[17]   P. Carra, B. T. Thole, M. Altarelli, and X. Wang, Phys. Rev. Lett. **70** (1993) 694.

[18]   C. T. Chen, Y. U. Idzerda, H.-J. Lin, N. V. Smith, G. Meigs, E. Chaban, G. H. Ho, E. Pellegrin, and F. Sette, Phys. Rev. Lett. **75** (1995) 152.

[19]   R. Nakajima, J. Stöhr, Y. U. Idzerda, Phys. Rev. B. **59** (1999) 6421.

[20]   D. Weller, J. Stöhr, R. Nakajima, A. Carl, M. G. Samant, C. Chappert, R. Mégy, P. Beauvillain, P. Veillet, and G. A. Held, Phys. Rev. Lett. **75** (1995) 3752.

[21] J. Stöhr and H. König, Phys. Rev. Lett. **75** (1995) 3748.

[22] L. Néel, J. Phys. Radium **15** (1954) 376.

[23] Y. Wu, J. Stöhr, B. D. Hermsmeier, M. G. Samant, and D. Weller, Phys. Rev. Lett. **69** (1992) 2307.

[24] M. Samant, J. Stöhr, S. S. P. Parkin, G. A. Held, B. D. Hermsmeier, F. Herman, M. van Schilfgaarde, L.-C. Duda, D. C. Mancini, N. Wassdahl, and R. Nakajima, Phys. Rev. Lett. **72** (1994) 1112.

[25] M. Tischer, O. Hjortstam, D. Arvanitis, J. Hunter Dunn, F. May, K. Baberschke, J. Trygg, J. M. Wills, B. Johansson, and O. Eriksson, Phys. Rev. Lett. **75** (1995) 1602.

[26] N. Nakajima, T. Koide, T. Shidara, H. Miyauchi, H. Fukutani, A. Fujimori, K. Iio, T. Katayama, M. Nývlt, and Y. Suzuki, Phys. Rev. Lett. **81** (1998) 5229.

[27] S.-K. Kim and J. B. Kortright, Phys. Rev. Lett. **86** (2001) 1347.

[28] G. van der Laan, Phys. Rev. Lett. **82** (1999) 640.

[29] R. Wienke, G. Schütz, H. Ebert, J. Appl. Phys. **69** (1991) 6147.

[30] S. Pizzini, A. Fontaine, C. Giorgetti, E. Dartyge, J.-F. Bobo, M. Piecuch, B. Baudelet, Phys. Rev. Lett. **74** (1995) 1470.

[31] F. Wilhelm, P. Poulopoulos, G. Ceballos, H. Wende, K. Baberschke, P. Srivastava, D. Benea, H. Ebert, M. Angelakeris, N. K. Flevaris, D. Niarchos, A. Rogalev, and N. B. Brookes, Phys. Rev. Lett. **85** (2000) 413.

[32] S.-K. Kim, J.-R. Jeong, J. B. Kortright, and S.-C. Shin, Phys. Rev. B **64** 052406 (2001).

[33] D. J. Keavney, D. Wu, J. W. Freeland, E. Johnston-Halperin, D. D. Awschalom, and J. Shi, Phys. Rev. Lett. **91** (2003) 187203.

[34] J.-Y. Kim, J.-H. Park, B.-G. Park, H.-J. Noh, S.-J. Oh, J. S. Yang, D.-H. Kim, S. D. Bu, T.-W. Noh, H.-J. Lin, H.-H. Hsieh, and C. T. Chen, Phys. Rev. Lett. **90** (2003) 017401.

[35] F. M. F. de Groot, M. A. Arrio, Ph. Sainctavit, Ch. Cartier, and C. T. Chen, Solid State Com. **92** (1994) 991.

[36] M. van Veenendaal, J. B. Goedkoop, and B. T. Thole, Phys. Rev. Lett. **77** (1996) 1508.

[37] M. O. Krause, J. Phys. Chem. Ref. Data **8** (1979) 307.

[38] J. B. Kortright, M. Rice, S.-K. Kim, C. C. Walton, and T. Warwick, J. Magn. Magn. Mater. **191** (1999) 79.

[39] J. B. Kortright, J. Magn. Magn. Mater. **156** (1996) 271.

[40] J. B. Kortright, M. Rice, and K. D. Frank, Rev. Sci. Instrum. **66** (1995) 1567.

[41] M. Yanagihara, T. Maehara, H. Nomura, M. Yamamoto, T. Namioka, and H. Kimura, Rev. Sci. Instrum. **63** (1992) 1516.

[42] F. Schäfers, H. C. Mertins, A. Gaupp, W. Gudat, M. Mertin, I. Packe, F. Schmolla, S. Di Fonzo, G. Soullie, W. Jark, R. Walker, X. Le,Cann, R. Nyholm, M. Eriksson, Applied Optics **38** (1999) 4074.

[43] J. B. Kortright and J. H. Underwood, Nucl. Instrum. Meth. A **291** (1990) 272.

[44] J. B. Kortright, H. Kimura, V. Nikitin, K. Mayama, M. Yanagihara, and M. Yamamoto, Appl. Phys. Lett. **60** (1992) 2963.

[45] J. B. Kortright, S.-K. Kim, T. Warwick, and N. V. Smith, Appl. Phys. Lett. **71** (1997) 1446 .

[46] J. B. Kortright, M. Rice, and R. Carr, Phys. Rev. B **51** (1995) 10240.

[47] H. C. Mertins, F. Schäfers, X. Le Cann, A. Gaupp, and W. Gudat, Phys. Rev. B. **61** (2000) R874.

[48] R. Wu and A. J. Freeman, J. Appl. Phys. **79** (1996) 6500.

[49] A. Scherz, H. Wende, K. Babershke, J. Minár, D. Benea, and H. Ebert, Phys. Rev B, **66** (2002) 184401.

[50] H. Ebert and S. Man'kovsky, Phys. Rev. Lett. **90** (2003) 077404.

[51] H.-C. Mertins, P. M. Oppeneer, J. Kuneš, A. Gaupp, D. Abramsohn, and F. Schäfers, Phys. Rev. Lett. **87**1 (2001) 04740.

[52] J. Kuneš, P. M. Oppeneer, H.-C. Mertins, F. Schäfers, A. Gaupp, W. Gudat, and P. Novak, Phys. Rev. B **64** (2001) 174417.

[53] V. N. Antonov, B. N. Harmon and A. N. Yaresko, Phys. Rev. B **67** (2003) 024417.

[54] M. J. Frieser, IEEE Trans. Mag. **4** (1968) 152.

[55] H. Ebert, Rep. Prog. Phys. **59** (1996) 1665.

[56] J. Zak, E. R. Moog, C. Liu, and S. D. Bader, Phys. Rev. B **43** (1991) 6423.

[57] Z. Q. Qiu and S. D. Bader, Rev. Sci. Instrum. **71** (2000) 1243.

[58] C. Kao, J. B. Hastings, E. D. Johnson, D. P. Siddons, G. C. Smith, and G. A. Prinz, Phys. Rev. Lett. **65** (1990) 373.

[59] J. M. Tonnerre, L. Sève, D. Raoux, G. Soullié, B. Rodmacq, and P. Wolfers, Phys. Rev. Lett. **75** (1995) 740.

[60] J. B. Kortright, M. Rice, S.-K. Kim, C. C. Walton, and T. Warwick, J. Magn. Magn. Mater. **191** (1999) 79.

[61] J. B. Kortright, S.-K. Kim, E. E. Fullerton, J. S. Jiang, and S. D. Bader, Nucl. Instrum. Meth. A **467** (2001) 1396.

[62] P. M. Oppeneer, H.-C. Mertins, O. Zaharko, J. Phys. Cond. Mat. **15** (2003) 7803.

[63] H.-C. Mertins, S. Valencia, D. Abramsohn, A. Gaupp, W. Gudat, and P. M. Oppeneer, Phys. Rev. B **69** (2004) 064407.

[64] K. Starke, F. Heigl, A. Vollmer, M. Weiss, G. Reichardt, and G. Kaindl, Phys. Rev. Lett. **86** (2001) 3415.

[65] M. Sacchi, A. Mirone, C. F. Hague, P. Castrucci, R. Gunnella, and M. De Crescenzi, Phys. Rev. B **64** (2001) 012403.

[66] Y. U. Idzerda, V. Chakarian, and J. W. Freeland, Phys. Rev. Lett. **82** (1999) 1562.

[67] P. M. Oppeneer, H.-C. Mertins, D. Abramsohn, A. Gaupp, J. Kuneš, and C. M. Schneider, Phys. Rev. B **67** (2003) 052401.

[68] S. Gider, B.-U. Runge, A. C. Marley, and S. S. P. Parkin, Science **281** (1998) 797.

[69] E. F. Kneller and R. Hawig, IEEE Trans. Magn. **27** (1991) 3588.

[70] R. Skomski and J. M. D. Coey, Phys. Rev. B **48** (1993) 15812.

[71] E. E. Fullerton, J. S. Jiang, and S. D. Bader, J. Magn. Magn. Mater. **200** (1999) 392.

[72] J. B. Kortright, J. S. Jiang, S. D. Bader, O. Hellwig, D. T. Marguiles, and E. E. Fullerton, Nucl. Instrum. Meth. B **199** (2003) 301.

[73] O. Hellwig, J. B. Kortright, K. Takano, and E. E. Fullerton, Phys. Rev. B **62** (2000) 11694 .

[74] K. Amemiya, S. Kitagawa, D. Matsumura, T. Yokoyama, and T. Ohta, J. Phys.: Condens. Matter **15** (2003) S561.

[75] K.-S. Lee, S.-K. Kim, and J. B. Kortright, Appl. Phys. Lett. **83** (2003) 3764.

[76] J. F. MacKay, C. Teichert, D. E. Savage, and M. G. Lagally, Phys. Rev. Lett. **77** (1996) 3925 .

[77] J. W. Freeland, V. Chakarian, K. Bussmann, and Y. U. Idzerda, J. Appl. Phys. **83** (1998) 6290.

[78] R. M. Osgood III, S. K. Sinha, J. W. Freeland, Y. U. Idzerda, and S. D. Bader, J. Magn. Magn. Mater. **198-199** (1999) 698.

[79] S. K. Sinha, E. B. Sirota, S. Garoff, and H. B. Stanley, Phys. Rev. B **38** (1988) 2297.

[80] A. Rahmim, S. Tixier, T. Tiedje, S. Eisebitt, M. Lörgen, R. Scherer, W. Eberhardt, J. Lüning, and A. Scholl, Phys. Rev. B **65** (2002) 235421.

[81] J. B. Kortright, O. Hellwig, K. Chesnel, S. Sun, and E. E. Fullerton, Phys. Rev. B, in press.

[82] H. A. Dürr, E. Dudzik, S. S. Dhesi, J. B. Goedkoop, G. van der Laan, M. Belakhovsky, C. Mocuta, A. Marty, and Y. Samson, Science **284** (1999) 2166.

[83] E. Dudzik, S. S. Dhesi, S. P. Collins, H. A. Dürr, G. van der Laan, K. Chesnel, M. Belakhovsky, A. Marty, Y. Samson, and J. B. Goodkoep, J. Appl. Phys. **87** (2000) 5469.

[84] J. B. Kortright, S.-K. Kim, G. P. Denbeaux, G. Zeltzer, K. Takano, and E. E. Fullerton, Phys. Rev. B **64** (2001) 092401.

[85] O. Hellwig, S. Maat, J. B. Kortright, and E. E. Fullerton, Phys. Rev. B **65** (2002) 144418 .

[86] O. Hellwig, T. L. Kirk, J. B. Kortright, A. Berger, and E. E. Fullerton, Nature Materials, **2** (2003) 112.

[87] M. S. Pierce, R. B. Moore, L. B. Sorensen, S. D. Kevan, E. E. Fullerton, O. Hellwig, and J. B. Kortright, Phys. Rev. Lett. **90** (2003) 175502.

[88] K. Chesnel, M. Belakhovsky, F. Livet, S. P. Collins, G. van der Laan, S. S. Dhesi, J. P. Attané, and A. Marty, Phys. Rev. B **66** (2002) 172404.

[89] O. Hellwig, G. P. Denbeaux, J. B. Kortright, and E. E. Fullerton, Physica B **336** (2003) 136.

[90] O. Hellwig, D. T. Margulies, B. Lengsfield, E. E. Fullerton, and J. B. Kortright, Appl. Phys. Lett. **80** (2002) 1234.

[91] J. B. Kortright, O. Hellwig, D. T. Marguiles, and E. E. Fullerton, J. Magn. and Magn. Mater. **240** (2002) 325.

[92] E. E. Fullerton, O. Hellwig, K. Takano, and J. B. Kortright, Nucl. Instrum. Meth. B **200** (2003) 202.

[93] M. Doerner, X. Bain, M. Madison, K. Tang, Q. Peng, A. Polcyn, T. Arnoldussen, M. F. Toney, M. Mirzamaani, K. Takano, E. E. Fullerton, M. Schabes, K. Rubin, M. Pinarbase, S. Yuan, M. Parker, and D. Weller, IEEE Trans. Magn. **34** (2001) 1564.

[94] M. Sutton, S. G. J. Mochrie, T. Greytak, S. E. Nagler, L. E. Berman, G. A. Held, and G. B. Stephenson, Nature **352** (1991) 608.

[95] A. C. Price, L. B. Sorensen, S. D. Kevan, J. Toner, A. Poniewierski, and R. Holyst, Phys. Rev. Lett. **82** (1999) 755.

[96] F. Yakhou, A. Létoublon, F. Livet, M. de Boissieu, and F. Bley, J. Magn. Magn. Mater. **233** (2001) 119.

[97] J. Miao, P. Charalambous, J. Kirz, and D. Sayre, Nature **400** (1999) 342.

[98] A. Rahmim, M.Sc. Thesis, University of British Columbia (1999).

[99] T. O. Menteş, C. Sánchez-Hanke, and C. C. Kao, J. Synchrotron Radiation **9** (2002) 90.

[100] S. Eisebitt, M. Lörgen, W. Eberhardt, J. Lüning, S. Andrews, and J. Stöhr, Appl. Phys. Lett. 84, (2004) 3373.

[101] S. Eisebitt (private communication).

[102] G. Brauer, G. B. Stephenson, M. Sutton, R. Brüning, E. Dufrense, S. G. J. Mochrie, G. Grübel, J. Als-Nielsen, and D. L. Abernathy, Phys. Rev. Lett. **74** (1995) 2010.

[103] S. Dierker, R. Pindak, R. M. Flemming, I. K. Robinson, and L. Berman, Phys. Rev. Lett, **75** (1995) 449.

[104] I. Sikharulidze, I. P. Dolbnya, A. Fera, A. Madsen, B. I. Ostrovskii, and W. H. de Jeu, Phys. Rev. Lett. **88** (2002) 115503.

[105] E. E. Fullerton, J. Pearson, C. H. Sowers, S. D. Bader, X. Z. Wu, and S. K. Sinha, Phys. Rev. B **48** (1993) 17432.

[106] J. Stöhr, Y. Wu, B. D. Hermsmeier, M. G. Samant, G. R. Harp, D. Dunham, and B. P. Tonner, Science **259** (1993) 658.

[107] C. M. Schneider and G. Schonhense, Reports on Progress in Physics **65** (2002) R1785.

[108] S. Anders, H. Padmore, R. M. Duarte. T. Renner, T. Stammler, A. Scholl, M. R. Scheinfein, J. Stöhr, L. Sève, and B. Sinkovic, Rev. Sci. Instrum. **70** (1999) 3973.

[109] W. Kuch, L. I. Chelaru, F. Offi, M. Kotsugi, and J. Kirschner, J. Vac. Sci. Technol. B **20** (2002) 2543.

[110] J. Vogel, W. Kuch, M. Bonfim, J. Camarero, Y. Pennec, F. Offi, K. Fukumoto, J. Kirschner, A. Fontaine, and S. Pizzini, Appl. Phys. Lett. **82** (2003) 2299.

[111] S.-B. Choe, Y. Acermann, A. Scholl, A. Bauer, A. Doran, J. Stöhr, and H. A. Padmore, Science **304** (2004) 420.

[112] A. Scholl, J. Stöhr, J. Lüning, J. W. Seo, J. Fompeyrine, H. Siegwart, J. P. Locquet, F. Nolting, S. Anders, E. E. Fullerton, M. R. Scheinfein, and H. A. Padmore, Sience **287** (2000) 1014.

[113] W. Kuch, L. I. Chelaru, F. Offi, J. Wang, M. Kotsugi, and J. Kirschner, Phys. Rev. Lett. **92** (2004) 017201.

[114] H. Ohldag, A. Scholl, F. Nolting, S. Anders, U. Hillebrecht, and J. Stöhr, Phys. Rev. Lett. **86** (2001) 2878.

[115] H. Ohldag, T. J. Regan, J. Stöhr, A. Scholl, F. Nolting, J. Lüning, C. Stamm, S. Anders, and R. L. White, Phys. Rev. Lett. **87** (2001) 7201.

[116] R. Tatchyn, P. L. Csonka, and I. Lindau, J. Opt. Soc. Am. B **1** (1984) 806.

[117] G. Schmahl, D. Rudolph, P. Guttmann, G. Schneider, J. Thieme, and B. Niemann, Rev. Sci. Instr. **66** (1995) 1282.

[118] G. Denbeaux, P. Fischer, G. Kusinski, M. Le Gros, A. Pearson, and D. Attwood, IEEE Trans. Magn. **37** (2001) 2764.

[119] P. Fischer, G. Schütz, G. Schmahl, P. Guttmann, and D. Raasch, J. de Physique IV **7-C2** (1997) 467.

[120] P. Fischer, T. Eimüller, G. Schütz, P. Guttmann, G. Schmahl, K. Pruegl, G. Bayreuther, J. Phys. D: Appl. Phys. **31** (1998) 649.

[121] T. Eimüller, P. Fischer, G. Schütz, M. Scholz, G. Bayreuther, P. Guttmann, G. Schmahl, M. Köhler, J. Appl Phys. **89** (2001) 7162.

[122] G. J. Kusinski, K. M. Krishnan, G. Denbeaux, G. Thomas, G. Denbeaux, B. D. Terris, and D. Weller, Appl. Phys. Lett. **79** (2001) 2211.

[123] P. Fischer, G. Denbeaux, T. Eimuller, D. Goll, and G. Schütz, IEEE Trans. Magn., **38** (2002) 2427.

[124] P. Fischer, T. Eimüller, G. Schütz, M. Köhler, G. Bayreuther, G. Denbeaux, and D. Attwood, J. Appl. Phys. **89** (2001) 7159.

[125] T. Warwick, H. Ade, S.Cerasari, J. Denlinger, K. Franck, A. Garcia, S. Hayakawa, A. Hitchcock, J, Kikuma, S. Klingler, J. Kortright, G. Morisson. M. Moronne, E. Rightor, E. Rotenberg, S. Seal, H.J. Shin. W. F. Steele, and B. P. Tonner, Journal of Synchrotron Radiation **5** (1998) 1090.

[126] S.-K. Kim, J. B. Kortright, and S.-C. Shin, Appl. Phys. Lett. **78** (2001) 2724.

[127] P. Fischer (private communication).

[128] O. Hellwig, D. T. Marguiles, B. Lengsfield, E. E. Fullerton, and J. B. Kortright, Appl. Phys. Lett. **80** (2002) 1234.

[129] E. E. Fullerton, O. Hellwig, Y. Ikeda, B. Lengsfield, K. Takano, and J. B. Kortright, IEEE Trans. Magn. **38** (2002) 1693.

CHAPTER 5

Hard x-ray resonant techniques for studies of nanomagnetism

5.1. INTRODUCTION

The application of synchrotron radiation to the study of magnetic materials has grown rapidly in recent years, owing in part to the availability of high-brightness synchrotron sources around the world. Several characteristics of synchrotron radiation make the study of magnetic materials very attractive. First of all, the high brightness of the beam typically results in a flux of 10^{13} photons/sec in less than a 1 mm^2 area, which enables the study of very small or highly diluted samples. The naturally high scattering wave vector resolution due to high degree of x-ray collimation and monochromaticity allows for very precise determination of magnetic modulations. Furthermore, the well-defined polarization characteristics of synchrotron radiation (linear in the plane of the particle orbit), together with its relatively simple manipulation and analysis by crystal optics, can be used to study a variety of magnetization states. Lastly, by tuning the energy of the incident beam near absorption edges (or resonances) of constituent elements, one can study the magnetic contributions of individual components in heterogeneous structures.

X-rays interact with matter through scattering from both the electron's charge and its magnetic moment. The scattering from the charge is the dominant term and is the basis for most condensed matter studies using x-rays. Although small, the scattering from the magnetic moment is sufficient to extract valuable information on magnetic structures in single crystals [1-3]. Enhanced sensitivity to magnetic moments can be achieved by tuning the x-ray energy to selected resonances. These resonant enhancements have resulted in widespread applications of x-rays in the study of magnetism, both in the absorption (x-ray magnetic circular dichroism) [4-6] and scattering (x-ray resonant magnetic scattering) [7-8] channels. These include studies of interfacial magnetic roughness in multilayers [9-12], and morphology of magnetic domains in buried interfaces [13-14], just to mention a few examples. Furthermore, by performing polarization analysis of the scattered radiation [15] or applying sum rules to dichroic spectra of spin-orbit split absorption edges [16] one can distinguish between spin and orbital contributions to the magnetic moment in an element-specific way. This is in fact a unique attribute of magnetic

All or portions of the above manuscript have been created by the University of Chicago as Operator or Argonne National Laboratory ("Argonne") under Contract No. W-31-109-ENG-38 with the U.S. Department of Energy. The manuscript may include authors from other organizations as set forth therein. The U.S. Government retains for itself and others acting on its behalf, a paid-up, nonexclusive, irrevocable worldwide license in said article to reproduce, prepare derivative works, distribute copies to the public, and perform publicly and display publicly, by or on behalf of the Government.

scattering and spectroscopy techniques, and it is the primary reason why these techniques are powerful tools in magnetism studies. Recent reviews of general applications of x-rays to study magnetism are given by Gibbs and collaborators [17] and McWhan [18].

Most synchrotron studies of nanomagnetism have been performed using soft x-rays (loosely defined as less than 3 keV) since resonant dipolar transitions in this energy regime access electronic states carrying large magnetic moments in most materials (e.g., $3d$ states in transition metals and $4f$ states in rare-earth compounds). Hence, the magnetic signals are larger and easier to observe. Harder x-ray energies (>3 keV) access electronic states with smaller, yet significant, magnetic moments (e.g., $4p$ states in transition metals and $5d$ states in rare-earth compounds). While experimentally more challenging, hard x-ray studies of magnetism offer unique advantages. The higher penetrating power of these x-rays enables the study of buried structures and interfaces, which can be important in characterizing a wide variety of systems used in modern technologies, such as permanent magnetic materials and artificial thin film heterostructures. The penetrating power of hard x-rays yields a true bulk-measurement probe without the need for high-vacuum conditions, while soft x-ray measurements are surface sensitive and must be performed in UHV conditions. Furthermore, the short x-ray wavelengths permit diffraction studies to probe the magnetic order in both crystals and artificial, periodic nanostructures, such as multilayers and patterned dot/hole arrays.

The rich polarization dependence of magnetic scattering is commonly used to extract the magnetic ordering of a material. Antiferromagnetic (AFM) structures are commonly studied with linearly polarized radiation. In the absorption channel, the linear dichroism effect [19-20] results in absorption contrast for parallel and perpendicular alignments of x-ray's linear polarization and sample's magnetization in the presence of magneto-crystalline anisotropy, which can be used, e.g., to image AFM domains in exchange-biased systems [21]. In the diffraction channel, AFM ordering results in Bragg diffraction at the magnetic ordering's wave vector, since the x-rays' magnetic field couples to the ordered electron spins. As discussed below, while this coupling is relatively weak, synchrotron radiation brightness, together with resonant enhancement of the magnetic scattering cross section, results in easy detection of x-ray magnetic scattering from AFM-ordered systems at third-generation synchrotron sources. Circularly polarized (CP) radiation can also be useful in studies of AFM materials. An example where CP x-rays were used for real-space imaging of chiral domains by helicity-dependent Bragg scattering from the spiral AFM state of a holmium crystal is included in Section 5.2 below [22].

Ferri- or ferromagnetic (FM) structures are commonly studied with CP radiation. In the absorption channel, magnetic circular dichroism results in absorption contrast for parallel and antiparallel alignment of x-ray's helicity and sample's magnetization [4]. By measuring this absorption contrast through spin-orbit split core levels (e.g., L_2 and L_3 edges), element-specific magnetic moments in the final state of the absorption process (both spin and orbital components) can be extracted through the application of sum rules [23-24]. This contrast, in combination with focused x-ray beams, can be used to image FM domains in nanostructures. Examples of such imaging studies in buried spring-magnet structures are shown in Section 5.4. Diffraction contrast for opposite helicities of CP radiation can also be used to study FM structures. While this differential measurement removes pure chemical scattering, without polarization analysis it results in charge-magnetic interference scattering [25]. This interference scattering contains information on both magnitude and direction of magnetic moments. Examples wherein this interference signal is used to obtain magnetization depth profiles across buried interfaces in artificial Gd/Fe nanostructures, and site-specific magnetism in crystals, are presented in Sections 5.3 and 5.4 below [26-27].

Our chapter is organized along the commonly used synchrotron techniques. In Section 5.1, the cross section for x-ray scattering is outlined and its dependence on sample's magnetization state emphasized. In Section 5.2, studies of site-specific magnetism and spiral antiferromagnetic domains with diffraction techniques are presented. In Section 5.3, the use of reflectivity techniques to probe buried interfacial magnetism is demonstrated. Finally, in Section 5.4, spectroscopy was used to image magnetic domains, study inhomogeneous magnetization profiles and perform vector magnetometry in patterned arrays. Although there are many more examples that illustrate the utility of hard x-ray characterization techniques in magnetism studies, our goal is to familiarize our readers with some of the tools that might benefit their own research.

5.1.1. X-ray scattering cross section

The general expression for the scattering of x-rays from atomic electrons in a periodic medium is given by Blume and Gibbs [25],

$$\frac{d\sigma}{d\Omega} = r_o^2 \left| \sum_n e^{i\vec{Q}\cdot\vec{r}_n} f_n(\vec{k},\vec{k}',\hbar\omega) \right|^2 , \tag{5.1}$$

where $\vec{r}_n$ is the position of the nth atom in the crystal, $\hbar\omega$ is the incident photon energy and Q=k-k' is the scattering vector. The scattering amplitude per atom, f, consists of several contributions,

$$f(\vec{k},\vec{k}',\omega) = f^{ch\,arge}(\vec{Q}) + f'(\vec{k},\vec{k}',\omega) + if''(\vec{k},\vec{k}',\omega) + f^{spin}(\vec{k},\vec{k}',\omega). \tag{5.2}$$

Here f^{charge} is the usual Thomson scattering, f' and f'' are the energy-dependent anomalous contributions, and f^{spin} is the scattering from the spin of the electrons. For a typical hard x-ray energy of 10 keV, the pure spin magnetic scattering amplitude is down by 0.02 relative to Thomson scattering, i.e., four orders of magnitude reduction in the scattering cross section. The small ratio of magnetic to nonmagnetic electrons reduces the magnetic scattering even further, typically by 5-6 orders of magnitude depending on the net magnetic moment. Although small, this signal can be easily observed in systems where the magnetic modulation is different than the charge, such as in AFM structures. Further the strength of the magnetic scattering can be strongly enhanced near a resonance, as described below.

Depending on the energy of the incident x-ray radiation, two regimes can be distinguished: the nonresonant regime, where the incident energy is far away from the excitation energy of absorption edges of constituent atomic species, and the resonant regime, where the incident energy is close to an absorption edge. Although all examples given in this chapter belong to the latter, a formalism describing the nonresonant regime will be introduced first, followed by its extension into the resonant limit.

5.1.1.1. Nonresonant cross section

Far from a resonance, the magnetic dependence of the anomalous contributions can be separated, reducing the expression for the scattering amplitude into two terms, one containing the interaction with the charge of the atom and another containing its interaction with the magnetic moment. These two terms derived from second-order perturbation theory [25], [28-29] are given below,

$$f = f^{charge} + f^{magnetic} = \rho(Q)\hat{\varepsilon}' \cdot \hat{\varepsilon} + ir_o\left(\frac{\hbar\omega}{m_e c^2}\right)\left[\frac{1}{2}\vec{L}(Q) \cdot \vec{A} + \vec{S}(Q) \cdot \vec{B}\right]. \tag{5.3}$$

Here $\rho(Q)$, $S(Q)$, and $L(Q)$ are the Fourier transforms of the electrons charge, and the spin and orbital magnetic moments, respectively. A and B are matrices (see reference [25]) that contain the polarization dependence of the magnetic scattering, which differs from that of the charge scattering. In principle, this polarization dependence can be exploited to obtain quantitative information on the size of the magnetic moments responsible for the scattering. In practice, however, this information is typically limited to AFM structures, where the magnetic scattering is separated from the charge scattering in reciprocal space. Therefore, most studies of FM materials are performed using resonant scattering and absorption as described below.

5.1.1.2. Resonant cross section

When the energy of the incident photon is near an absorption edge, additional resonant terms contribute to the x-ray scattering [7-8], [30]. These resonances occur at energies sufficient to promote deep-core electrons into states at and above the Fermi level (Fig.5-1). This results in a large increase in the x-ray scattering cross section for a material due to an energy difference denominator in this second-order process, resulting in increased absorption and an enhancement of the magnetic scattering. In other words, the enhancement in the magnetic x-ray scattering results from virtual transitions to excited, intermediate, resonant states near the Fermi level. In magnetic materials, these states are spin polarized leading to an increase in the sensitivity of the scattering to the magnetism of the scattering atom. This additional scattering results in the anomalous term in eq.(5.1). The resonant scattering amplitude can be calculated using the following expression [30],

$$f_{EL} = 4\pi\lambda \sum_{M=-L}^{L} \left[\hat{\varepsilon}_f^* \cdot Y_{LM} Y_{LM}^* \cdot \varepsilon_i\right] F_{LM}. \tag{5.4}$$

Here ε is the beam polarization, Y_{LM}'s are vector spherical harmonics, and F_{LM} are the matrix elements involved in the transition. In general, different multipole order transitions can contribute to the resonant enhancement. In most cases, however, the electric dipole transitions dominate, yielding the following simple expression for the scattering amplitude [30-31],

$$f^{res} = F^0\left(\hat{\varepsilon}_f \cdot \hat{\varepsilon}_i\right) - iF^1\left(\hat{\varepsilon}_f \times \hat{\varepsilon}_i\right) \bullet \hat{m}_n + F^2\left(\hat{\varepsilon}_f \cdot \hat{m}_n\right)\left(\hat{\varepsilon}_i \cdot \hat{m}_n\right). \tag{5.5}$$

Here the F^n are complex quantities containing the amplitude of the scattering given by the matrix elements of the transitions involved, ε are the initial and final polarization vectors and $\hat{m}$ is the direction of the magnetic moment of the atom. The first term above is the charge anomalous scattering amplitude, whose polarization dependence is the same as that for Thomson scattering.

The second term is linear in the magnetization and therefore can be isolated by reversing the magnetization of the sample and measuring the difference in the absorption or scattering. Similarly this term leads to differences in the scattering and absorption for circularly polarized x-rays, and its imaginary component is responsible for circular magnetic x-ray dichroism (XMCD) in absorption measurements. The last term is quadratic in the magnetization and is typically much smaller than the F^1 term. This term is responsible for linear dichroism effects.

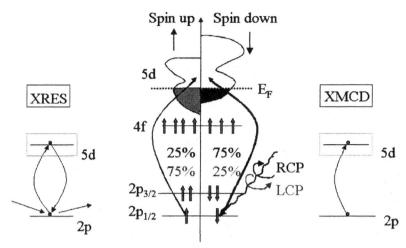

Figure 5-1: Illustration of the resonant scattering (left) and absorption (right) of x-rays. Absorption involves the photoexcitation of an electron to an energy above the Fermi level, while resonant scattering involves a virtual transition through a similar intermediate state. For circularly polarized incident x-rays, the transition probability to spin-up (-down) states is different, leading to magnetic sensitivity. *Also see the color plate.*

5.2. DIFFRACTION TECHNIQUES

The intrinsic periodicity d of atomic arrangements in crystals, interfaces in layered structures, and lithographically patterned arrays of periodic nanostructures (dots, antidots, rings) results in diffracted intensity at Bragg conditions associated with that periodicity ($2d\sin\theta=\lambda$). While the atomic-scale periodicity of crystals ($d \approx 1\text{Å}$) results in Bragg reflections at relatively large angles, the much larger periodicity of artificial multilayers or lithographically patterned arrays of magnetic elements ($d \approx 5$ nm-1 μm) results in small-angle diffraction. Nonetheless, since x-rays scatter from atomic electrons in crystals and from interfaces in layered structures, this diffracted intensity contains information about the electronic atomic charge density in crystals or electronic density contrast at the interfaces of artificial layered structures. As pointed out in the introduction, while the x-ray's magnetic field does not couple strongly to the magnetic moment of electrons, large resonant enhancements of magnetic-sensitive scattering of x-rays can be achieved near the absorption threshold for resonant transitions between a deep-core electron level and electronic orbitals near the Fermi level, which carry magnetic moments. This magnetic resonant scattering is much stronger, 10^{-1}-10^{-3} of the charge scattering (for hard x-rays), and contains information on the resonating element's magnetic moment. In the following examples, we exploit the periodicity of the magnetic structure under study and the enhanced sensitivity of resonant magnetic scattering to first zoom into the atomic origins of magnetocrystalline anisotropy in single crystals of $Nd_2Fe_{14}B$ permanent magnet, and, second, to image chiral antiferromagnetic domains in holmium single crystals.

5.2.1. Site-specific magnetism in ferro- (ferri-) magnetic crystals

The ability of x-ray spectroscopic techniques to separate the magnetic contributions from different elements in heterogeneous systems; i.e., element specificity, has proven remarkably useful in disentangling the complex magnetic behavior encountered in systems of current fundamental and technological interest.

The complexity of magnetic materials, however, goes beyond the presence of multiple elements: this includes materials where elements of the same specie reside in inequivalent crystal sites (such as magnetite Fe_3O_4, with octahedral and tetrahedral Fe sites), and nanocomposite materials, where elements of the same specie occur in more than one nanocrystalline phase (such as $Nd_2Fe_{14}B/\alpha$-Fe exchanged coupled nanocomposites). Current developments are aimed at extending the ability of x-ray-based techniques past element specificity towards site- and phase-specific magnetism. In particular, we show how basic crystallography can be combined with resonant scattering of CP x-rays to extract element- and site-specific magnetism in crystals.

Table 5-1: Calculated site-specific Nd structure factor for selected Bragg reflections. Resonance charge contributions to the structure factor at the Nd L_2 resonance are from tabulated values. Structural parameters are from Herbst [32].

Wyckoff Site	(110)	(220)	(440)
4f	2.2+0.5i	140+35i	86+29i
4g	69+15i	5.2+1.3i	92+31i

In magnetic materials, the resonant (anomalous) scattering of CP x-rays is modified from that in nonmagnetic materials. This is because the virtual photoelectron that is excited from the core state to the intermediate, resonant, state is partially spin polarized and therefore becomes sensitive to the spin imbalance in the density of states at the intermediate state near the Fermi level. The inherent element specificity of this resonant scattering can then be combined with structure factor effects in crystals to enhance/suppress scattering from selected lattice sites. Here we exploit the symmetry properties of a crystal of $Nd_2Fe_{14}B$ to study the magnetization reversal of the two inequivalent Nd sites in this structure (4f and 4g sites in Wyckoff notation). The permanent magnet of choice for many applications is $Nd_2Fe_{14}B$. Its magnetic hardness, i.e., its resistance to demagnetizing fields, has its origins at the atomic level and is due to the large orbital moment at Nd sites (predominantly from 4f atomic shells) interacting with the crystal field of the lattice. Since inequivalent Nd sites reside in quite distinct atomic environments, they experienced different crystal fields [32] and therefore are expected to display different local magnetocrystalline anisotropy. There is currently no technique that can directly measure the magnetic response of these distinct sites separately.

Site selectivity is achieved by exploiting the symmetry properties of the crystal; $Nd_2Fe_{14}B$ has a $P4_2/mnm$ tetragonal space group with four formula units per unit cell. The 56 Fe atoms are distributed among six inequivalent sites while the 8 Nd atoms occupy two other inequivalent sites (4f, 4g). As shown in Table 5-1, by selecting scattering vectors along the high-symmetry [110] direction, structure factor contributions from either one or the other Nd sites nearly vanish. Diffraction from (110) planes probes Nd at 4g sites since scattering from the four 4f sites interferes destructively. The opposite is true for a (220) diffraction condition, while nearly equal contributions of the two Nd sites are measured at a (440) Bragg reflection.

The $Nd_2Fe_{14}B$ single crystal was placed in the 6 kOe applied field of an electromagnet, and measurements were carried out at room temperature. Diffraction experiments were performed in reflection geometry with the [HHL] zone aligned in the vertical scattering plane. The magnetic field was applied along the [001] easy-axis direction, which was parallel to the sample surface. The surface normal was [110]. Resonant diffraction was measured through the Nd L_2 edge by

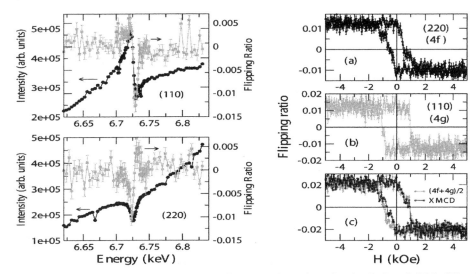

Figure 5-2: (Left) Resonant diffraction through the Nd L_2 absorption edge for (110) and (220) diffraction conditions. Black: charge (chemical) scattering; yellow: magnetic sensitive scattering. At each Bragg condition, only one inequivalent Nd site is visible; the scattering contributions from the other add destructively. (Right) Top two panels show element- and site-specific Nd hysteresis loops obtained on each reflection. Bottom panel shows the average of two reflections and the loop obtained from the XMCD signal. *Also see the color plate.*

switching the helicity of the incident CP x-rays at every energy point while maintaining a given diffraction condition (fix Q) at all energies. The XMCD measurements were simultaneously performed by measuring the difference in Nd L_β fluorescence for opposite helicities using Ge solid-state detectors. Element- and site-specific hysteresis loops were performed at Bragg peaks as indicated in Fig.5-2 by recording changes in scattering intensity as a function of applied field for opposite helicities of CP x-rays.

Figure 5-2 shows resonant diffraction data taken under (110) and (220) Bragg conditions. These reflections alternately probe *4g* and *4f* sites, respectively. The resonant charge scattering is obtained from (I^++I^-), while the charge-magnetic interference scattering [27] is obtained from the flipping ratio $(I^+-I^-)/(I^++I^-)$. This interference scattering is proportional to the magnitude of the magnetic moment and also contains information about the moment's direction relative to the x-ray polarization vectors. Here (I^+,I^-) are scattered intensities for opposite x-ray helicities.

Modeling of the charge-magnetic interference scattering should allow separation of chemical and magnetic anomalous scattering factors for each of the Nd sites. This will yield unique quantitative information on the size of the magnetic moment and on the chemical valence state at each site. Spectroscopic information, such as differences in the spin polarization of unoccupied Nd *5d* states at both sites, is included in the imaginary parts of the magnetic anomalous scattering factors. Current work is focused on developing the required algorithms needed to retrieve this information. While deriving magnetic information from the energy-dependent scattering requires accurate modeling of the interference of charge and magnetic scattering, field-dependent measurements relax this constraint and allow for studies of the magnetic response at each site to be recorded by working under either one diffraction condition for selected energies that maximize the magnetic contrast at each site. This is also shown in Fig.5-2, where site-specific hysteresis loops are shown. They show that the magnetic moment at Nd *4g* sites requires a significantly larger reversed applied field and reverses more sharply than those at Nd *4f* sites.

This clearly shows the Nd $4g$ sites are predominantly responsible for the magnetocrystalline anisotropy (MCA) providing unprecedented detail into the atomic origins of MCA in $Nd_2Fe_{14}B$. The same figure also compares the weighted average of the magnetic signals at each site with the XMCD measurement. Since the absorption-based XMCD signal intrinsically averages over the two sites, the good agreement provides self-consistency. In addition, we measured the same reversal curves in both diffraction and absorption channel for (440) Bragg reflections as expected due to the equal scattering contributions of Nd sites at these Q vector (Table 5-1). The different magnetic responses of Nd sites are likely due to the different crystal fields arising from the unequal crystalline environments. For the two reflections shown here, the polarization dependence of the charge-magnetic interference signal yields a near-zero signal when the magnetization is along [110] direction. This implies that at a reversed applied field of 500 Oe the magnetizations of Nd $4g$ and $4f$ sites are nearly orthogonal to one another (the $4g$ site is still magnetized along the [001] direction at this field). The ferromagnetic Nd-Nd coupling in this material is indirect and weak, through exchange interactions with the surrounding Fe ions and the spin polarization of conduction electrons. The magnetization reversal process, which is dominated by competing MCA at the two Nd sites and the Nd-Fe exchange, is nontrivial, as seen in Fig.5-2. The MCA strongly affects the reversal process, which includes largely static noncollinear configurations of Nd moments. It is likely that Fe moments mediating the Nd-Nd coupling participate in this unconventional reversal mechanism as well.

In summary, this example demonstrates the potential of combining the spectroscopic signatures inherent in resonance x-ray scattering with structure factor effects in crystals to obtain element- and site-specific magnetism in crystals. This method allows obtaining information on the atomic origins of MCA and can be applied to both crystals and epitaxially grown films. Extension of this method to phase-specific magnetism in nanocomposite magnetic structures with multiple crystalline phases is currently under way.

5.2.2. Imaging spiral magnetic domains

Many techniques have been developed to image magnetic domains [33], but most measure either ferromagnetic or linear antiferromagnetic structures. A wide variety of materials, however, exhibit more exotic magnetic ordering, particularly materials that contain rare-earth elements. These exotic magnetic structures can exhibit their own unique domain structures within a material. Hard x-rays can be used to image such domains through the use of microfocusing optics in conjunction with resonant magnetic x-ray scattering techniques.

Spiral antiferromagnets form one particular type of such structures, where the moments align in ferromagnetic planes within an atomic layer but rotate by a characteristic angle between successive layers along the magnetic propagation direction. The sense of this rotation can be either right or left handed leading to the formation of chirality domains within the sample. Holmium metal offers one example of such a magnetic structure. Holmium orders in the spiral structure below $T_N=133K$, with the propagation direction along the c-axis of the hexagonal unit cell. Below $T=19K$, the moments cant away from the basal plane forming a conical structure (Fig.5-3). This magnetic superstructure results in the appearance of satellite peaks on either side of the charge Bragg diffraction peaks at $(0,0,L\pm\tau)$. At these magnetic peaks, circularly polarized x-rays become sensitive to the handedness of such a helix (i.e., either right or left handed) [25], [34]. Therefore, contrast between magnetic domains of opposing handedness can be obtained by reversing the incident beam helicity and measuring the difference in the Bragg scattering intensity.

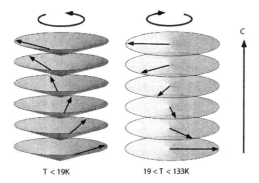

Figure 5-3: The low-temperature conical (left) and high-temperature basal plane spiral (right) magnetic structure of Ho.

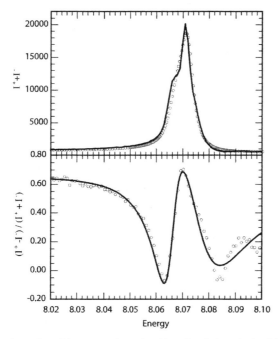

Figure 5-4: Scattering intensity of the magnetic peak with a circularly polarized incident beam. Top: Total counts. Bottom: Normalized difference. Lines: Theory.

A measure of this sensitivity to spiral helicity is shown in Fig.5-4, which shows the energy dependence of the intensity and helicity contrast at the $(0,0,4+\tau)$ peak near the Ho L_3 absorption edge resonance. Near this resonance, the intensity of the magnetic scattering is strongly enhanced (top of Fig.5-4) and the sensitivity of helicity reversal to spiral handedness varies dramatically. The maximum contrast of about 75% occurs about 1 eV below the peak of the scattering intensity, with contrast strongly suppressed on either side of the resonance. Away from the resonance, the contrast was similar to that at the peak, but, while it quickly approached this value below the edge, the contrast suppression persisted for over a hundred eV above the edge.

A 640μm×500μm image of the chiral domain structure of Ho obtained at the $(0,0,\tau)$ magnetic peak is shown in Fig.5-5. This image was obtained using slits to define a 25×25 μm² immediately before the sample. The Ho crystal was placed inside closed-cycle He refrigerator, mounted to a Huber psi-circle goniometer. The sample was oriented on a magnetic Bragg peak and then the sample was scanned through the beam, reversing the helicity at each point to obtain an image of the helicity domains. The domain features exhibit a characteristic length scale on the order of 100 µm and are uncorrelated with the crystal lattice (**a** axis was oriented up in the figure). Warming the sample past T_N and recooling nucleated a completely different domain pattern, indicating that crystalline defects play very little role in the nucleation of spiral domains.

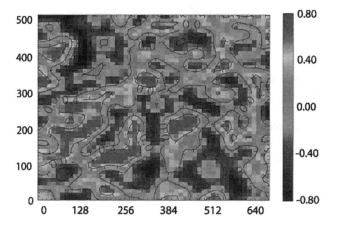

Figure 5-5: 640×500 µm^2 spiral domain pattern in Ho metal at 10K. *Also see the color plate.*

5.3. REFLECTIVITY TECHNIQUES

In this section, the study of inhomogeneous magnetic profiles in Fe/Gd multilayers with polarized hard x-rays is presented. Experimental results are supported by theoretical work on modeling interfacial magnetism. The question of magnetic roughness and its correlation with chemical roughness will be discussed, since it is one of the important questions in the design of new-generation electronics devices that utilize the spin of electrons [35].

5.3.1. Studies of interfacial magnetism with circularly polarized x-rays

Understanding chemical and magnetic properties of buried interfaces in layered systems is of great scientific and technological interest. For example, chemical interfacial roughness affects spin-polarized transport and related giant magnetoresistance effects in spin valves [36]. Interfacial magnetic disorder accompanying chemical disorder introduces uncompensated spins at ferromagnetic-antiferromagnetic exchange-biased interfaces affecting magnetization reversal processes [37]. Our goal in this example is to show how one can *quantify* fundamental properties of buried magnetic interfaces, including strength and extent of interlayer exchange coupling and chemical and magnetic roughness.

X-ray specular reflectivity has been widely used to extract charge density profiles in layered structures. Since, in the x-ray regime, a material's index of refraction is always slightly less than one, total external reflection below a critical angle is the norm contrary to the total internal reflection commonly found at interfaces between media in the optical regime. The reflected intensity is almost unity below the critical angle for total external reflection but decreases sharply with increased scattering angle as Q^{-4} (Q is the scattering vector: $Q=4\pi/\lambda \sin\theta$). In the hard x-ray regime, this results in significant reflected intensity only at small scattering angles $\theta \leq 10°$. Since the associated scattering wave vectors are much smaller than the inverse of typical interatomic distances, the atomic structure can be neglected and scattering occurs at the interfaces between media with different indices of refraction (charge density). The specular reflectivity is related to the Fourier transform of the charge-density profile along the scattering vector, i.e., the normal to the sample surface. X-ray resonance magnetic reflectivity (XRMR), which is a special case of x-ray resonance exchange scattering (XRES), measures the difference in specular reflectivity between left- and right-CP x-rays. Much like x-ray reflectivity yields charge-density profiles in layered structures, XRMR is related to the Fourier transform of the magnetization density profile along the normal direction. Modeling of both signals allows for chemical and magnetic density profiles to be retrieved, including the position of chemical and magnetic interfaces and their chemical and magnetic roughness. In contrast, the absorption-based XMCD measures the difference in absorption coefficient between opposite helicities of x-rays and averages over the magnetization depth profile provided the x-ray penetration depth at the measurement incident angle is larger than the sample thickness, which is usually the case for magnetic thin films and incident angles larger than a few degrees.

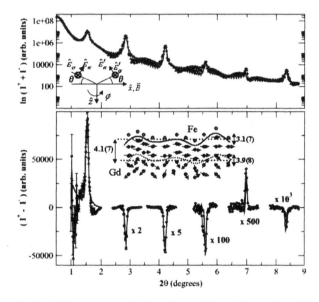

Figure 5-6: Charge (top) and charge-magnetic interference (bottom) specular reflectivity (points) and fits (lines) for E=7929 eV and 300K across six multilayer Bragg peaks. Top inset: scattering geometry and applied field direction; bottom inset: derived interfacial magnetic structure. The Gd/Fe interface has both charge and magnetic roughness (same within uncertainties); ferro-paramagnetic interface has only magnetic roughness (all units Å). This experiment probes the Gd magnetization only. *Also see the color plate.*

We illustrate the application of this techniques on a $[\text{Fe}(15\text{Å})\text{Gd}(50\text{Å})]_{15}$ multilayer sample that was sputtered in vacuum onto a Si substrate using Nb buffer (100Å) and cap (30Å) layers. Since Gd and Fe have similar bulk magnetizations, 2020 and 1750 emu/cm³, respectively, the much thicker Gd dominates the magnetization at low temperatures, and its magnetization aligns with an applied field. A strong antiferromagnetic interlayer exchange coupling forces the Fe into an antiparallel alignment. SQUID magnetometry shows that Gd dominates the magnetization up to at least 350K, i.e, well above its bulk Curie temperature of 293K. The markedly different Curie temperatures of Gd and Fe (1024K) allow us to distinguish Gd "bulk" and interfacial regions by tuning the sample temperature. More generally, it allows the investigation of proximity effects between low- and high-T_c ferromagnets at the atomic scale.

Figure 5-6 shows specular reflectivity curves measured near the Gd L_2 edge at T=300K, as obtained by adding (top) and subtracting (bottom) scattered intensities for opposite helicities of the incoming CP radiation. The sum yields the charge reflectivity curve related to interference between x-rays scattered from variations in the charge density depth profile; the difference is due to interference between x-rays scattered from both charge and magnetic density variations, as explained below.

The difference signal was modeled within the first Born approximation (BA), which assumes weak scattering and is valid away from the regime of total external reflection. Combining eq.(5.4) from Section 5.1 with the nonresonant charge scattering term from eq.(5.1), the coherent resonant elastic scattering length for a single magnetic ion in the electric dipole approximation is given as a sum of a charge and a magnetic term.

$$f = (f_0 + f_e)(\hat{\varepsilon}'^* \bullet \hat{\varepsilon}) + if_m(\hat{\varepsilon}'^* \times \hat{\varepsilon}) \bullet \hat{m}. \qquad (5.6)$$

Here f_0, f_e and f_m have substituted for $\rho(Q)$, F^0, and F', in the former equations. The much weaker nonresonant magnetic scattering, as well as the linear dichroism term F^2, have been neglected (the latter does not contribute to this scattering geometry). In the first BA, the difference cross section for opposite helicities of CP x-rays for a system of N interfaces with charge and magnetic roughness is given by [38-39]:

$$\Delta_{(L,R)} \frac{d\sigma}{d\Omega} = \frac{4\pi^2 L_x L_y \delta(q_x)\delta(q_y)}{q_z^2}[\cos 2\theta(\hat{k}_i \bullet \hat{m}) + (\hat{k}_f \bullet \hat{m})] \times$$
$$\sum_{i,j}^{N} e^{iq_z(z_i-z_j)}[\Delta\rho^*_{e,i}\Delta\rho_{m,j}e^{-\frac{q_z^2}{2}(\sigma^2_{3,i}+\sigma^2_{m,j})} + \Delta\rho_{e,j}\Delta\rho^*_{m,i}e^{-\frac{q_z^2}{2}(\sigma^2_{e,j}+\sigma^2_{m,i})}] \qquad (5.7)$$

Where $\Delta\rho^*_{e,i} = [n_e^{i+1}(f_0+f_e^*)^{i+1}-n_e^i(f_0+f_e^*)^i]$ and $\Delta\rho^*_{m,j} = n_e^{j+1}(f_m)^{j+1}-n_e^j(f_m)^j$ are charge and magnetic density contrast at interfaces i and j, respectively. Here $f_0 = -Zr_0$ and f_e, f_m are complex, anomalous charge and magnetic scattering lengths; n_e is atomic number density and $\sigma^2_{e,m} = \langle[\delta z_{e,m}(x,y)]^2\rangle$ is the mean squared height fluctuations, assumed Gaussian, about the average position of charge and magnetic interfaces. For simplicity, the cross section above neglects phase retardation and absorption effects; these are included in the fits [26]. For magnetically aligned phases (collinear) the polarization factor involving the local magnetization *direction* is constant throughout the Gd layer thickness at fixed q_z. The *magnitude* of the local magnetization is allowed to vary through the resonant f_m.

Quantitative analysis requires accurate values of complex charge and magnetic anomalous scattering factors at the resonant energy. Their strong energy dependence, which includes solid state (band structure) and excitonic (core-hole) effects calls for their experimental determination on the actual structure under study. Through the optical theorem, the imaginary parts of these factors are related to the absorption coefficient by $f''_{e,m}(E) \propto (e/r_0 n_e hc)\mu_{e,m}(E)$. We measured the energy dependence of the absorption coefficient at 16K in a 100 eV interval around the Gd L_2 edge for opposite helicities of CP x-rays, $\mu^{\pm}(E)$, to obtain edge-step normalized

$f''_{e,m}(\mu_e=[\mu^++\mu^-]/2, \mu_m=\mu^+-\mu^-)$ and used bare-atom scattering factors *away* from resonance for absolute normalization. Real parts were obtained from differential Kramers-Kronig (KK) transforms of imaginary parts.

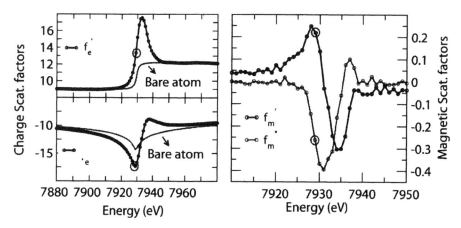

Figure 5-7: Determination of resonant charge and magnetic scattering factors near the Gd L_2 edge. Edge-step normalized charge (μ_e) and magnetic (μ_m) absorption coefficients (left panel) are combined with tabulated bare-atom scattering factors away from resonance to derive $f''_{e,m}(E)$ and KK-related $f'_{e,m}(E)$ (right panel). Values at the resonant energy (7929 eV) used in the magnetic reflectivity measurements shown with circles.

Magnetization density profiles in the Gd layers are described (through variations in f_m) in terms of a few fitting parameters, which are then refined in a nonlinear least-squares fitting of the BA cross section to the data. The most significant finding is summarized in the inset of Fig.5-8. At 300K, best fits indicate that Gd is paramagnetic except for a region 4.1(7)Å in size that remains fully magnetized near the Gd/Fe interface. This magnetization is induced by a strong antiferromagnetic interaction with the magnetically ordered Fe layer, as predicted in mean-field calculations by Camley [40]. This size is a measure of the spatial extent of the AFM interaction at the Gd/Fe interface.

The presence and size of this ordered Gd region were confirmed by T-dependent XMCD measurements. Figure 5-8 shows the XMCD signal at the Gd L_2 edge (left panel) together with its integrated intensity (points, right panel). At 300K, the Gd layers retain $\approx 20\%$ of their saturation magnetization, consistent with the magnetic reflectivity result of $\approx 17\%$ of the layer volume remaining magnetized at 300K. By modeling the XMCD as a superposition of interfacial and "bulk" regions with variable volume fractions and T_c values, we find a 5.2±1.2 Å region remains magnetized at 300K with an estimated $T_c=1050(90)$K. Since this magnetized region is induced by the strong AFM exchange interaction at the Gd/Fe interface, its T_c value quantifies the strength of this interaction; i.e., $J_{AF} \approx J_{Fe} \approx 1000$K. An enhanced T_c of ≈ 800K was previously reported for one monolayer of Gd on a Fe(100) substrate [41].

In summary, this example demonstrates the ability to quantify with high accuracy fundamental parameters characterizing a buried magnetic interface in a layered system. This includes the spatial extent and strength of interfacial exchange coupling and interfacial magnetic roughness. Future effort could be directed towards gaining a better understanding of the interfacial electronic and atomic structure. Spin-dependent x-ray absorption fine structure [5] and XMCD combined with the x-ray standing wave technique [42] should provide further insight into this question.

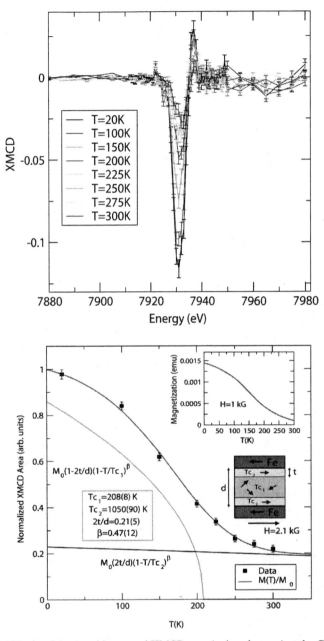

Figure 5-8: XMCD signal (top) and integrated XMCD area (points, bottom) at the Gd L_2 edge. Integrated intensities are fitted as a superposition of interfacial and bulk regions with same saturation magnetization M_0 but variable T_c and volume fractions. Fit (red line) includes a convolution with a Gaussian to account for disorder in the sputtered layers. Top inset shows SQUID magnetization data. *Also see the color plate.*

5.4. SPECTROSCOPY TECHNIQUES

While x-ray resonant scattering contains spectroscopic information through its dependence on the spin polarization of the empty density of states (DOS) at the Fermi level, the retrieval of this information is not always simple. This is particularly true in cases where charge-magnetic interference scattering is measured, and, therefore chemical and magnetic spectroscopic signatures are mixed and need to be deconvolved in order to extract purely magnetic information. Magnetic spectroscopy in the absorption channel, through the x-ray magnetic circular dichroism effect, is, on the other hand, a pure magnetic signal that directly relates to the asymmetry in the DOS at the Fermi level between spin-up and spin-down empty electronic states with a particular orbital character dictated by dipole selection rules that connect the core electron state with the final state. The size of the measured XMCD signal is proportional to the degree of circular polarization in the incident beam, the magnetization of the sample, and the projection of the moment onto the incident photon direction. As such, it can be used to measure the magnetization of a sample as a function of temperature or field. In this manner, XMCD measurements are similar to those taken with a magnetometer. There are two key differences, however, between XMCD and magnetometry. First, XMCD is element specific; therefore the changes in the XMCD signal strength are proportional to the changes in the magnetization of only the particular atomic species excited at the absorption edge where the measurements are taken. Second, the angle between the applied magnetic field and the incident photon direction can be varied. This is useful in measuring magnetic structures in which the magnetic moments are not collinear with the applied field direction. Furthermore, by using a highly focused beam, the XMCD signal can be used to probe the local magnetization of the sample.

In this section, we present a series of examples where the XMCD effect was used to retrieve just such element-specific magnetic information on nanostructured materials. These include magnetic domain mapping in a Fe/SmCo exchange spring magnet, measurement of spin configuration in a Fe patterned array and determination of an inhomogeneous magnetic state in Fe/Gd multilayers.

5.4.1. Magnetic domain mapping of buried nanostructures

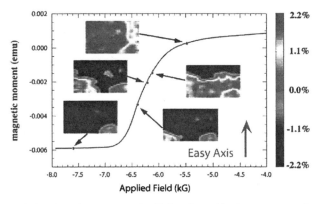

Figure 5-9: Images of the domain structure in Fe/SmCo with the corresponding positions on the magnetization curve. *Also see the color plate.*

Composites of soft and hard magnetic materials have shown a great deal of promise as new high-strength permanent magnets. In these composites, the soft magnet provides a high magnetic saturation, whereas the magnetically hard material provides a high coercive field. Bilayers can be used as model systems to investigate the magnetization-reversal process in these composites [43], where the hard magnetic material is grown epitaxially on a substrate to provide a well-defined magnetization axis, and the soft material is overlaid on top of it. Studies of the spatial magnetic structure in such bilayers, however, have been limited to measurements of the domains in the top soft layer [44]. This is because, the magnetic structure of the buried hard layer is inaccessible to established methods like magnetic force microscopy or the magneto-optical Kerr effect, since these techniques are highly surface sensitive. Thus the structure of the buried layer upon magnetization reversal could not be studied directly using these methods. In this experiment, a polarized x-ray microbeam [13] was used to overcome the limitations of the more conventional techniques. By using ~5 to 12 keV x-rays, the top layers of the structure are penetrated in a nondestructive manner, and the measurement of magnetic domain structure of the buried layer is achieved while an external field is applied.

The experimental setup for this experiment consisted of two parts. First, phase-retarding optics converts the linearly polarized beam from the planar undulator (a device commonly used in third-generation synchrotron facilities to produce high-brightness radiation [45-46]) into a circularly polarized one, and second, focusing optics produces a micron-sized beam. A Kirkpatrick-Baez (KB) mirror pair yielded a focal spot of 9×22 μm^2, with ~10^{10} photons/s. One mirror focuses the beam in the horizontal direction, while the second does it in the vertical direction. It should be noted that, with a state-of-the-art microfocusing optics [47-50] and a dedicated experimental setup, one can achieve a spot size of the order of 100 nm (or less) in this energy range.

X-ray magnetic circular dichroism (XMCD) was used to provide a contrast mechanism sensitive to the orientation of the magnetization. As previously described in Section 5.4, XMCD measures the projection of the magnetic moment onto the incident photon wave vector. Therefore, the orientation of the local magnetic moments can be measured by taking the flipping ratio $(I^+ - I^-)/(I^+ + I^-)$ of the measured intensities for opposite helicities (this ratio is also referred to as the asymmetry ratio).

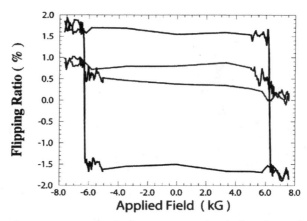

Figure 5-10: Hysteresis measurements for the low-contrast region (red) and rest of sample (blue).

The sample studied was a 200 Å Fe/1600 Å SmCo/200 Å Fe/200 Å Ag layer grown on a MgO substrate. The SmCo was nominally deposited in the Sm_2Co_7 phase, although there are local deviations from the ideal stoichiometry, leading to $SmCo_5$ or $SmCo_3$ phases. Since the sample was grown on a relatively thick substrate, the fluorescence yield from the sample was used to measure the absorption. The fluorescence from the sample is proportional to the x-ray absorption and therefore shows XMCD contrast. Measurements were performed at the Sm L_3 edge, monitoring the L_α fluorescence intensity. First, XMCD spectra were taken as a function of energy with an unfocused beam and the sample fully aligned. The best magnetic contrast was found to be at 6.710 keV, which was the energy then used to obtain all the magnetic structure images. Magnetic domain images were recorded as a function of the externally applied magnetic field. The sample was scanned in two dimensions through the microfocused beam. A magnetic field of up to 8 kG was applied parallel to the axis of easy magnetization.

Figure 5-9 shows a series of 250×500 μm^2 (vertical x horizontal) images [14] for different applied magnetic fields. The relative position of each image along the sample magnetization curve is also indicated. The colors in the images correspond to the measured flipping ratios given by the scale on the right. A red color denotes a region where the local magnetization is antiparallel to the incoming beam and a blue color is where it is parallel.

The images in Fig.5-9 clearly show the magnetic reversal of the domains in the SmCo layer upon increase of the applied field. A large region (> 500 μm) nucleates at the top of the image and grows at the expense of the oppositely oriented domain. The boundary between the two domains is predominantly oriented perpendicular to the direction of magnetization. The direction of the domain wall can be understood from the chemical structure of the SmCo layer. The axis of easy magnetization in SmCo films is given by the c-axis of the Sm_2Co_7 unit cell [51]. Stacking disorders induced by the $SmCo_5$ or $SmCo_3$ phases mentioned earlier will be oriented perpendicular to the easy axis. These stacking disorders may effectively pin the domain walls.

One interesting feature is found at the lower right portion of each image. In this region, very little magnetic contrast was observed for any applied fields. To investigate this further, local hysteresis measurements were performed (shown in Fig.5-10) at the center of this region and at a point where clear domain formation was observed. Figure 5-10 shows that, although the contrast is much smaller than that from the other parts of the sample, there is some change in this region also. The much smaller signal is due to either a local Co deficiency in this region or a misorientation of the epitaxial growth, resulting in a crystal grain whose easy axis is oriented nearly perpendicular to the x-ray beam.

5.4.2. Biquadratic coupling in SmCo/Fe

Physical properties of thin magnetic nanostructures are dominated by exchange interactions between the layers. These interactions in most cases induce collinear coupling of spins. Non-collinear coupling of spins is also allowed through the biquadratic term in the exchange Hamiltonian $H_2 = -j(\mathbf{M}_1 * \mathbf{M}_2)^2$, where $\mathbf{M}_1$ and $\mathbf{M}_2$ are magnetic moments in the layers. However, the biquadratic exchange is typically much smaller than the conventional Heisenberg exchange. Recently, Vlasko-Vlasov and collaborators [44] observed unusual perpendicular coupling of two ferromagnetic layers in direct contact in a now familiar system of SmCo and Fe exchange spring magnets. They deduced noncollinear remanent magnetic configurations based on magneto-optical imaging of the top Fe layer.

In this example, we demonstrate how the application of hard x-rays offers a nondestructive way to simultaneously probe the magnetization in the surface Fe layer and in the buried SmCo layer. Since both Fe and Sm are ferromagnetic, circularly polarized synchrotron radiation was used. The goal of the experiment was to combine element-specific hysteresis loops and magnetic imaging of both the top Fe and the bottom SmCo layers in order to unequivocally show that the Sm and Fe magnetizations were nearly perpendicularly coupled when the hard SmCo hard layer is demagnetized. This perpendicular coupling of magnetic moments is referred to as biquadratic coupling.

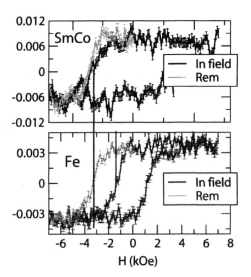

Figure 5-11: Element-specific hysteresis loops measured at the Sm L_3 edge (6.710 keV) (top) and Fe K edge (7.110 keV) (bottom). Measurements were done with the field constantly applied (in-field data) and in remanence (field was turned off during data collection). *Also see the color plate.*

Element-specific remanent hysteresis loops were performed by tuning the energy of the synchrotron radiation to the absorption edges of Fe (7.110 keV) and Sm (6.710 keV). Magnetic contrast was obtained, as in previous examples, by switching the helicity of incident CP photons at each field and measuring the difference in the fluorescence signal. In addition, element-specific imaging of magnetic domains was performed by focusing the circularly polarized x-rays to a spot size of 1 μm by 1 μm. The thickness of both SmCo and Fe layers was 20 nm, and they were grown epitaxially by magnetron sputtering on an MgO substrate. A 20 nm Cr layer was used as a buffer layer between the substrate and the SmCo and a 5 nm Cr layer was a cap. The easy axis was the in-plane c-axis of SmCo.

The experiment involved focusing the beam to a 1x1 μm² spot size using two mirrors in KB geometry. The mirrors had a Pd coating and were each 10 cm long. The sample was mounted on high-resolution stages (0.07 μm step size) between pole pieces of an electromagnet capable of achieving 0.9 T field strength. The asymmetry (flipping) ratio, which is defined as the contrast in absorption coefficient for opposite helicities of incoming x-rays, was measured in fluorescence geometry using two Ge solid-state detectors. Circularly polarized x-rays with opposite helicities were generated by phase retarder optics [52-53] consisting of a single-crystal (111) diamond. Element specific measurements were done at the Fe K edge (7.110 keV) and the Sm L_3 edge (6.710 keV).

First, Sm- and Fe-specific hysteresis loops were acquired by measuring the asymmetry ratio at each applied field with the microbeam. Loops taken in-field and in remanence are shown in Fig. 5-11. The in-field loops show coercive fields of 3.4 kOe and 1.5 kOe for SmCo and Fe layers, respectively. The remanent hysteresis loops were measured by first fully magnetizing the sample at the saturation field of H = +7 kOe and then applying field H_{rem} before turning the applied field to zero. The Sm-remanent hysteresis indicates the nucleation of oppositely oriented domains at the field strength of H_{rem} = −2.7 kOe. Switching occurs at H_{rem} = −3.4 kOe, where the SmCo layer is demagnetized, i.e., broken into equal number of domains with opposite orientation. It is plausible that the strong anisotropy of the hard SmCo layer results in the orientation of domains along the easy axis.

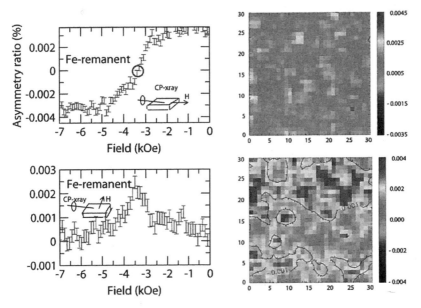

Figure 5-12: Top left: Remanent hysteresis loops performed at the Fe K edge in geometry where field and photon-helicity directions are parallel. At $H_{rem} = -3.4$ kOe, the magnetization is zero. Top right: magnetic imaging over a 30×30 μm² area shows no domains structure. Bottom left: Remanent hysteresis loops at the perpendicular geometry peak at $H_{rem} = -3.4$ kOe, indicating that Fe domains are oriented perpendicular to the easy axis, and thus perpendicular to the Sm domains under the Fe layer. Bottom right: corresponding magnetic imaging clearly shows two large Fe domains, red and green. *Also see the color plate.*

Imaging of magnetic domains was subsequently done by scanning the sample in two dimensions with a submicron step size and measuring the asymmetry ratio at each point. A 30×30 μm² image taken at the peak of the Sm resonance (6.710 keV) and at the remanent field of −3.4 kOe, exhibited no domain structure. This means that the domains are either very large or much smaller than the beam size (1 μm²). The first option can be easily ruled out because the *average* asymmetry ratio corresponds to the zero of the remanent hysteresis loop. A large domain would have resulted in the asymmetry ratio near the extreme parts of the hysteresis loops, either 0.005 or −0.01 as seen in the top part of Fig.5-11. Therefore, the Sm domains are significantly smaller than the probing beam, and, since the microbeam illuminates many domains at each pixel, the average magnetization is zero.

To determine the structure and orientation of the domains at the top Fe layer, hysteresis loops and magnetic imaging (Fig.5-12) were done at the Fe K edge. While the in-field hysteresis shows that the reversal of Fe domains occurs at H= −1.5 kOe, the remanent hysteresis indicates that the switching occurs at a larger (in absolute value) field of $H_{rem} = -2.8$ kOe because of the strong interfacial coupling between the Fe and SmCo layers. The reorientation of Fe domains happens at the same field strength as the magnetically hard SmCo layer. The Fe remanent magnetization becomes zero at $H_{rem} = -3.4$ kOe, where the SmCo underlayer is demagnetized. The zero net magnetization probed with a microbeam could be the consequence of either Fe breaking up into domains oriented along the easy axis that are much smaller than the beam size (similar to Sm domains) or the Fe magnetization is aligned perpendicularly with respect to the applied field and x-ray beam direction. Since XMCD measures the projection of the magnetization along the x-ray helicity (or beam direction), a 90° magnetization direction away from the beam direction would result in the zero asymmetry ratio. The latter scenario could be proved by repeating the measurements in the geometry in which both the sample and magnet are rotated by 90° with respect to photon helicity. In this geometry, the field is still applied along the easy axis, but, if the Fe domains prefer to orient perpendicularly to the easy axis, the projection of the magnetization, and thus the measured asymmetry, would be maximized. This was indeed observed in the Fe remanence loops after rotation: for H_{rem} below –2.5 kOe, the Fe magnetization remained along the easy axis resulting in the zero value of the asymmetry ratio (Fig.5-12 bottom left). Between – 4.5 kOe < H_{rem} < –2.5 kOe, the Fe magnetization rotates towards a direction perpendicular to the easy axis with a peak corresponding to the zero net magnetization (H_{rem} = –3.4 kOe) of the SmCo underlayer. If the H_{rem} field is further decreased, the Fe magnetization direction points again towards the easy axis, resulting in the zero XMCD signal. This conclusion was confirmed by imaging. Two images over the 30 μm × 30 μm area were taken at the Fe K edge resonance and in remanence: one with the field (an easy axis) parallel to the photon helicity (top right of Fig.5-12) and the other perpendicular to the photon helicity (bottom panel of Fig.5-12). The absence of domain structure in the parallel geometry indicates that the domains are oriented perpendicular to the easy axis. The possibility of having domains smaller than the beam size and aligned along the easy axis can be dismissed because the acquired image in the perpendicular geometry clearly showed two large (over 10 μm) domains.

In summary, this example demonstrates the utility of combining spectroscopy and microfocusing techniques in the study of heterogeneous magnetic systems. It also shows that the magnetic field can be used, which, in turn, opens the possibility of studying domain dynamics.

5.4.3. Magnetic reversal in antidot arrays

Antidot (hole) arrays in continuous magnetic films have recently received much attention because of their potential advantages over magnetic dot array systems for data storage [54]. Two advantages are (1) there is no superparamagnetic lower limit to the bit size, and (2) the intrinsic properties of the continuous magnetic film are preserved. Antidot arrays possess unique magnetic properties, such as shape-induced magnetic anisotropy, domain structure, and pinning in laterally confined geometries. Typically, antidot arrays at remanence show three types of domains behaving collectively as a single domain [55]. Domain formation is understood to be mainly the result of the interplay between the intrinsic and shape anisotropy. The coexistence of well-defined domains with individual magnetizations provides an opportunity to study interactions between domains during magnetic switching. Here the XMCD technique is used as a vector magnetometry (VM) to understand the switching mechanism in antidot arrays. The approach is complementary to previous microscopy studies [55]. The results in this example have relevance for future studies of the interlayer coupling under lateral confinement, because the element-specific XMCD technique is ideally suited for heteromagnetic systems such as Gd/Fe multilayers.

For the VM studies, hysteresis loops were measured by recording XMCD signals. Because XMCD is proportional to the projection of the magnetization vector **M** along the photon momentum direction $\mathbf{k}_{ph}$ near resonance energies [i.e., XMCD is proportional to $\mathbf{k}_{ph} \cdot \mathbf{M} = \cos \phi$ as shown in the inset of Fig.5-13(a)], this technique allows element-specific determination of the orientation of the average magnetization. The technique involves collecting hysteresis curves with more than two orthogonal incident photon directions for a given field [56]. Multilayer [Fe (3 nm)/Gd (2 nm)]×8 films were prepared on Si substrates by e-beam deposition. Square-shaped arrays of circular holes with a period of 2 μm and a diameter of 1 μm were manufactured by using standard lithography and liftoff processes. Magneto-optic Kerr effect

(MOKE) hysteresis loops were measured for both unpatterned and patterned films to determine the direction of intrinsic uniaxial magnetic anisotropy. Circularly polarized hard x-rays were produced by a diamond (111) quarter-wave plate operated in Bragg transmission geometry [57]. The XMCD effects were measured in fluorescence around the Fe K absorption edge (7.111 keV) by switching the helicity of the incident radiation. For the VM studies, the sample/electromagnet assembly was rotated with respect to the projected incident photon direction.

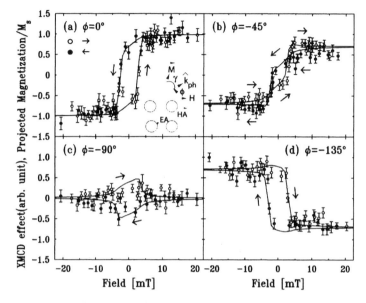

Figure 5-13: XMCD magnetic hysteresis loops (circles) measured at the Fe K edge at room temperature. To obtain vector information on the average magnetization, the incident photon beams were rotated with respect to the positive field direction, by ϕ = (a) 0°, (b) –45°, (c) –90°, and (d) –135°. The inset in (a) shows a schematic of the experimental setup, where ϕ is the angle between the magnetization vector **M** and the incident photon momentum direction $\mathbf{k}_{ph}$, **H** is the applied field, and "EA" and "HA" denote the easy- and hard-axis of the intrinsic anisotropy, respectively. The solid lines represent the calculated hysteresis loops from using micromagnetic simulations.

Figure 5-13 shows XMCD hysteresis loops measured with four different directions of incident x-ray beams. ϕ = (a) 0°, (b) –45°, (c) –90°, and (d) –135° with respect to the field applied in the positive direction. While ϕ = 0° corresponds to the conventional hysteresis loop along the applied field direction, the rotation of the average magnetization of the sample at ϕ = –90° can be described by θ_{avg} = –tan^{-1} (M_{-90}/M_0). This was surprising because many domains were expected to form. Following this relationship, one can determine a counter-clockwise rotation of magnetization from Figs.5-13(a) and 5-13(c) induced by the easy axis orientation of the intrinsic uniaxial anisotropy, as depicted by the inset in Fig.5-13(a). The preferential rotation gives rise to a dramatic asymmetry between the ϕ = –45° and ϕ = –135° loops. Interestingly, ϕ = –45° hysteresis shows three loops whose tie points correspond to the coercive fields. Since XMCD-VM measures a spatially averaged magnetization, numerical micromagnetic simulations have been performed to reconstruct the microscopic domain configuration. The hysteresis loops were calculated by using micromagnetic simulations and were fitted to the experimental data from XMCD-VM by varying the uniaxial anisotropy, exchange stiffness, and saturation magnetization as parameters. The fitted results from the 2-D code are shown as solid lines in Fig.5-13 and are in good agreement with the measured XMCD hysteresis loops. The reconstructed spin configurations with the best-fit parameters clearly showed three main types of domains, as reported previously by Toporov and collaborators [55].

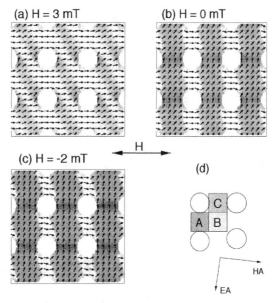

Figure 5-14: The spin configurations of antidot arrays obtained from micromagnetic simulations with a sequence of applied fields of (a) 3 mT, (b) 0 mT, and (c) –2 mT. (d) Schematic of the three characteristic domains labeled A, B, and C in the unit cell.

A sequence of spin configurations reveals that two types of domains rotate coherently while one is pinned (see Fig. 5-14). To understand intuitively the coherent rotations, we have developed a simple phenomenological energy model by employing the Stoner-Wohlfarth single-domain model with an effective shape anisotropy. This model suggests that the interplay between the shape anisotropy and the intrinsic uniaxial anisotropy can explain the coherent rotations of domains, as well as characteristic domain formations in antidot arrays.

This example demonstrates the power of vector magnetometry using XMCD. It may be useful in future applications where one needs to resolve individual contributions to the domain structure in a multicomponent system. The details of this work are given by Lee and collaborators [58].

5.4.4. Inhomogeneous magnetic structures in magnetic multilayers

Inhomogeneous magnetic states, wherein the magnetization direction rotates away from the applied field direction with distance from surfaces and interfaces, are commonly found in a variety of structures. Examples include the spin-flop transitions in giant magnetoresistant (GMR) Fe/Cr multilayers [59], and biquadratic coupling in exchange spring magnet Fe/SmCo bilayers [44]. Understanding the nature of the interactions leading to these magnetic states is important for tailoring the properties of these and other technologically relevant layered magnetic structures.

Artificial Fe/Gd multilayers are an ideal model system to investigate the nature of such interactions. The markedly different bulk Curie temperatures of Fe (1024K) and Gd (293K), together with strong interlayer antiferromagnetic coupling at the Fe/Gd interface, result in inhomogeneous magnetic depth profiles that depend on surface termination, applied field, and temperature [60]. Over a decade ago LePage and Camley [61] predicted that the nucleation site of an inhomogeneous state will either be the surface or the bulk of the multilayer, depending on termination. Termination by the component with the smallest magnetization will lead to surface nucleation wherein the magnetization deviates from the applied field direction only near the surface while the bulk remains field aligned. This inhomogeneous phase has eluded direct experimental detection due to the difficulty in probing surface and bulk states in the same measurement. The challenge is to observe both the existence of a surface-twisted phase and the absence of a bulk twist.

In this example, the penetration depth tunability of x-rays at grazing and larger incidence angles θ_i was exploited to alternately probe surface and bulk magnetic states by XMCD [62]. Figure 5-15 shows Gd and Fe hysteresis loops in an Fe-terminated $[Fe(35\text{Å})/Gd(50\text{Å})]_{15}/Fe(35\text{Å})$ multilayer for selected temperatures below, near and above the ferrimagnetic compensation temperature $T_0 \approx 110K$ at which the Fe and Gd magnetizations cancel. For Gd loops, two sets of data are shown corresponding to surface-enhanced loops at $\theta_i = 0.43°$ (probes ≈ 2 bilayers) and bulk-sensitive loops at $\theta_i = 9.5°$ (probes the whole multilayer). Specular reflectivity data were used to accurately determine the angle used for Gd surface-sensitive loops. These loops are obtained from the asymmetry in the absorption coefficient for opposite x-ray helicities at each applied field, $(\mu^+-\mu^-)/(\mu^++\mu^-)$, at resonant energies that maximize the magnetic contrast [62]. Since XMCD measures the projection of the magnetization along the photon wave vector a "flat" loop indicates aligned magnetic states where the Gd(Fe) magnetization is parallel (antiparallel) to the magnetic field H as dictated by their AFM exchange coupling. A "tilted" loop, however, indicates a reduced projection due to canting of the moments away from H. This canting, which increases with H, can only be driven by a reduction in Zeeman energy, since exchange is already minimized in the aligned geometry. A gain in the net magnetization has to take place for a twist to occur. Since the magnetization of Gd is larger than that of Fe below T_0, this can only be achieved if the Fe sublattice twists more toward the applied field than the Gd sublattice twists away from it, in order to compensate for the increased Zeeman energy of the latter. These different twist angles, however, increase the exchange energy. The competition between this increased exchange energy and the reduction in Zeeman energy determines the magnetic configuration.

At 10K the Gd layers dominate the Zeeman energy and align with H, while Fe is constrained antiparallel by AF exchange. Here a twisted phase would require an applied field outside the experimental range. At 70 and 90 K, "tilted" Gd loops are measured in the top part of the multilayer, while bulk-sensitive Gd loops show less tilting, indicating larger canting of the moments at the surface. The decrease in Gd magnetization with T, as seen from the reduced edge jump, decreases the required field for nucleation of a twist to within the experimental range. At 90K the Gd surface-sensitive XMCD is reduced by 65% at H=600 Oe, while the bulk XMCD decreases only by 20%. Considering the probing depth of ≈ 2 bilayers at $\theta_i=0.43°$ and, given that top and bottom parts of the multilayer are equivalent, the average reduced magnetization m in the inner 11 bilayers can be obtained from $[0.35\times4+11\times m]/15=0.8$. This yields m=0.96; i.e., the interior of the multilayer remains mostly field aligned. The reduction in the bulk Gd loops at 70 and 90K is mainly due to the surface contribution. At 110K, the tilting or twist already propagates throughout the multilayer, as evidenced from the now significantly tilted surface and bulk loops. A correlated reversal in the sign of Gd and Fe loops at this temperature shows that Fe now dominates the Zeeman energy contribution. At 200K the loops are again "flat," with the Fe aligning along the field and Gd antiparallel.

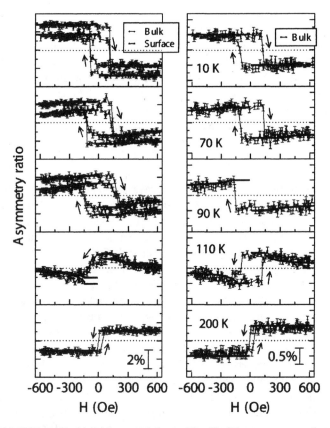

Figure 5-15: Gd (left) and Fe (right) hysteresis loops. The "flat" loops correspond to Gd dominant (10K) and Fe dominant (200K) field-aligned configurations. The "tilted" loops correspond to magnetic twisted configurations where the magnetization deviates from the applied field direction in the surface alone (70, 90K) or also in the bulk (110K). Solid lines are obtained from Landau-Lifshitz calculations of the magnetization profiles. *Also see the color plate.*

Figure 5-16 also shows theoretical calculations of the static magnetization profile. The surface nucleation of the inhomogeneous state is clearly observed. To compare with the experimental data, the calculated magnetization depth profiles were weighted, each element separately, to account for the depth selectivity of our XMCD measurements at the different incidence angles. The results of this averaging are shown by the solid lines on the loops, where the agreement with experiment supports the conclusion of the extent of the penetration depth at nucleation of ≈ 200 Å (2-3 bilayers). The energy barrier for a twist of the minority sublattice (Fe) towards the applied field direction H is decreased at the surface due to the absence of Fe/Gd interlayer exchange coupling at the terminal Fe layers. This results in surface nucleation of the inhomogeneous state, while the increased exchange energy cost in the bulk does not allow the twist to penetrate past the first few Fe/Gd bilayers.

In summary, surface nucleation of a twisted magnetic state when the a Gd/Fe multilayer is terminated by the minority (Fe) component was observed. The surface state penetrates ≈ 200 Å into the bulk due to strong interlayer coupling at Fe/Gd interfaces. These results are the first direct confirmation of the long-ago predicted inhomogeneous magnetic phase in the strongly coupled model system. Furthermore, this method opens a way towards distinguishing surface from bulk states in inhomogeneous magnetic systems.

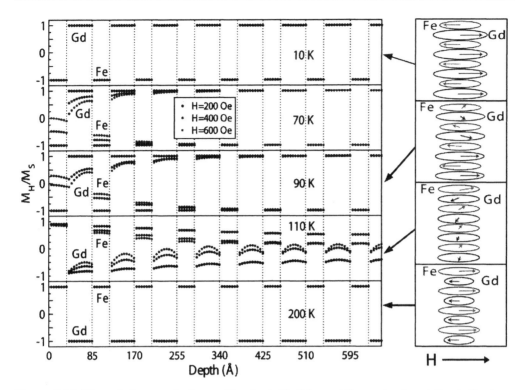

Figure 5-16: Theoretical magnetization profiles for half of the multilayer structure (other half mirror symmetric). Magnetization is normalized to saturation value at each temperature. Schematic diagram (right) represents the magnetization (intralayer averaged) in the upper four bilayers at the different temperatures and H=600 Oe. *Also see the color plate.*

5.5. CONCLUSIONS AND OUTLOOK

In the past several years, the application of synchrotron techniques to the study of magnetic nanostructures has emerged as a viable and complementary tool to the more conventional techniques described in this book. Examples in this chapter are meant to illustrate some of the advantages that diffraction, reflectivity and spectroscopy techniques could offer. We particularly want to emphasize that the penetration ability of hard x-rays enables nondestructive study of buried structures and interfaces. We hope that this feature, coupled with the high brightness of third-generation synchrotron sources, energy selectivity, high momentum resolution and well-defined polarization characteristics will entice practitioners to consider synchrotron radiation techniques for resolving problems in nanomagnetism.

Work at Argonne is supported by the U.S. Department of Energy, Office of Science, under contract W-31-109-ENG-38.

REFERENCES

[1]   P. M. Platzmann and N. Tzoar, Phys. Rev. B **2** (1970) 3556.
[2]   D. Gibbs, D.E. Moncton, K.L. D'Amico et al., Phys. Rev. Lett **55** (1985) 234.

[3]   D. F. McMorrow, D. Gibbs and J. Bohr, *Handbook on the Physics and Chemistry of Rare Earths,* Volume 26, 1 (1999) Elsevier Science.
[4]   G. Schütz, W. Wagner, W. Wilhelm et al., Phys. Rev. Lett **58** (1987) 737.
[5]   G. Schütz, R. Frahm, P. Mautner, et al., Phys. Rev. Lett. **62** (1989) 2620.
[6]   C. T. Chen, F. Sette, Y. Ma, et al., Phys. Rev. B **43** (1990) 7262.
[7]   K. Namikawa, M. Ando, T. Nakajima and H. Kawata, J. Phys. Soc. Jpn. **54** (1985) 4099.
[8]   D. Gibbs, D.R. Harshman, E.D. Isaacs et al., Phys. Rev. Lett. **61** (1988) 1241.
[9]   J. W. Freeland, V. Chakarian, K. Bussmann et al., J. Appl. Phys. **83** (1998) 6290.
[10]   C. S. Nelson, G. Srajer, J.C. Lang, et al., Phys. Rev. B **60** (1999) 12234.
[11]   D. R. Lee, S.K. Sinha , D. Haskel, et al., Phys. Rev. B **68** (2003) 224409.
[12]   D. R. Lee, S.K. Sinha, C.S. Nelson, et al., Phys. Rev. B **68** (2003) 224410.
[13]   J. Pollmann, G. Srajer, D. Haskel, et al., J. Appl. Phys. **89** (2001) 7165.
[14]   J. C. Lang, J. Pollmann, D. Haskel, et al., SPIE Proc. **4499** (2001) 1.
[15]   D. Gibbs, G. Grübel, D.R. Harshman, et al., Phys. Rev. B **45** (1991) 5663.
[16]   C. T. Chen, Y.U. Idzerda, H.-J. Lin, et al., Phys. Rev. Lett. **75** (1995) 152.
[17]   D. Gibbs, J. P. Hill and C. Vettier, Phys. Stat. Sol. (b) **215** (1999) 667.
[18]   D. B. McWhan, J. Synch. Rad **1** (1994) 83.
[19]   G. van Der Laan, B.T. Thole, G.A. Sawatzky, et al., Phys. Rev. B **34** (1986) 6529.
[20]   J. Stöhr, H. A. Padmore, S. Anders et al., Surface Rev. and Lett. **6** (1998) 1297.
[21]   A. Scholl, J. Stöhr, J. Lüning, et al., Science **287** (2000) 1014.
[22]   J. C. Lang, D.R. Lee, D. Haskel, et al., J. Appl. Phys. **95** (2004) 1.
[23]   B. T. Thole, P. Carra, F. Sette and G. van der Laan, Phys. Rev. Lett. **68** (1992) 1943.
[24]   P. Carra , B.T. Thole, M. Altarelli and X. Wang, Phys. Rev. Lett. **70** (1993) 649.
[25]   M. Blume and D. Gibbs, Phys. Rev. B **37** (1988) 1779.
[26]   D. Haskel, G. Srajer, J.C. Lang, et al., Phys. Rev. Lett. **87** (2001) 207201.
[27]   D. Haskel, et al., IEEE Trans. Mag. **40** (2004) 2874.
[28]   M. Blume, J. Appl. Phys. **57** (1985) 3615.
[29]   M. Blume in *Resonant Anomolous X-Ray Scattering – Theory and Practice*, 495 (1994), Elsevier Science.
[30]   J. P. Hannon, G.T. Tramell, M. Blume and D. Gibbs, Phys. Rev. Lett. **61** (1988) 1245.
[31]   J. P. Hill and D.F. McMorrow, Acta Cryst. **A52** (1996) 236.
[32]   J. F. Herbst, Rev. Mod. Phys. **63** (1991) 819.
[33]   R. J. Celotta, J. Unguris, M.H. Kelley, and D.T. Pierce, "Techniques to measure magnetic domain structures," in *Methods in Materials Research: Current Protocols*, edited by E. Kaufmann, Ch. 6b.3 (John Wiley & Sons, New York, 2001).
[34]   C. Sutter, G. Grübel, C. Vettier et al., Phys. Rev B **55** (1997) 954.
[35]   D. D. Awschalom, M.E. Flatté, and N. Samarth, *Scientific American*, p. 68 June (2002).
[36]   R. Schad et al., Europhysics Letters **44** (1998) 379.
[37]   S. Bae, J.H. Judy, W.F. Egeloff et al., J. Appl. Phys. **87** (2000) 6980.
[38]   S. K. Sinha et al., Phys. Rev. B **38** (1988) 2297.
[39]   R. M. Osgood III, S.K. Sinha, J.W. Freeland, et al., J. Magn. Magn. Mater. **198-199** (1999) 698 .
[40]   R. E. Camley, Phys. Rev. B **39** (1989) 12316.
[41]   M. Taborelli et al., Phys. Rev. Lett. **56** (1986) 2869.
[42]   S. K-Kim and J. B. Kortright, Phys. Rev. Lett. **86** (2001) 347.
[43]   E. F. Fullerton, J. S. Jiang, M. Grimsditch et al., Phys. Rev. B **58** (1998) 12193.
[44]   V. K. Vlasko-Vlasov, U. Welp, J.S. Jiang at al., Phys. Rev. Lett. **86** (2001) 4386.
[45]   R. J. Dejus, I.B. Vasserman, S. Sasaki and E.R. Moog, Argonne National Laboratory Report No. ANL/APS/TB-45 (2002).

[46] J. Chavanne and P. Elleaume, *Undulators, Wigglers and Their Applications*" edited by
      H. Onuki and P. Elleaume, Taylor & Francis, London, 2003.
[47] D. H. Bilderback, S.A. Hoffman and D.J. Thiele, Science **263** (1994) 201.
[48] W. Yun, B. Lai, Z. Cai et al., Rev. Sci. Instrum. **70** (1999) 2238.
[49] F. Pfeiffer, C. David, M. Burghamer et al., Science **297** (2002) 230.
[50] H. Takano, Y. Suzuki and A. Takeuchi, Jpn. J. Appl. Phys. **42** (2003) L132.
[51] E. F. Fullerton, J. S. Jiang, C. Rehm at al., Appl. Phys. Lett. **71** (1997) 1579.
[52] K. Hirano, T. Ishikawa, and S. Kikuta, Nucl. Instrum. Methods **A336** (1993) 343.
[53] J. C. Lang, G. Srajer, and R.J. Dejus, Rev. Sci. Instrum. **67** (1996) 62.
[54] R. P. Cowburn, A.O. Adeyeye, and J.A.C. Bland, Appl. Phys. Lett. **70** (1997) 2309.
[55] A.Y. Toporov, R.M. Langford, and A.K. Petford-Long, Appl. Phys. Lett. **77**, (2000) 3063.
[56] V. Chakarian, Y.U. Idzerda, G. Meigs, et al., Appl. Phys. Lett. **66** (1995) 3368.
[57] J. C. Lang and G. Srajer, Rev. Sci. Instrum. **66** (1995) 1540.
[58] D. R. Lee, Y. Choi, C.Y. You, et al., Appl. Phys. Lett. **81** (2002) 4997.
[59] R. W. Wang et al., Phys. Rev. Lett. **72** (1994) 920.
[60] R. W. Wang et al., Phys. Rev. Lett. **72** (1994) 920.
[61] J. G. LePage and E.E. Camley, Phys. Rev. Lett. **65** (1990) 1152.
[62] D. Haskel, G. Srajer, Y. Choi, et al., Phys. Rev. B **67** (2003) 180406.

Spin-resolved photoemission studies
of magnetic films

6.1. INTRODUCTION

Ultra-thin magnetic films, their surfaces and the multilayer structures engineered with them have distinct magnetic properties without counterpart in bulk systems. They truly represent a new class of magnetic materials that is beginning to find technological applications [1]. Investigations of their electronic structures are essential for understanding their magnetic behavior. This article discusses representative applications of such studies, focusing on experiments performed by spin-polarized photoemission spectroscopy (SPPES) with synchrotron radiation. The electronic and magnetic properties of surfaces of clean metal films and the role of quantum size effects in determining the physical properties of magnetic multilayers is illustrated with selected examples. Finally, the last section considers the electronic structure of a few complex transition-metal (TM) oxide systems.

The study of thin films and multilayers recently became a major field of research in magnetism. The interest in these two-dimensional (2D) structures stems from their peculiar magnetic properties that often have no equivalence in bulk materials. These novel magnetic properties, such as indirect exchange coupling between non-adjacent magnetic layers, spin-dependent transport phenomena, and interface and surface magnetic anisotropies, have both technological and fundamental relevance [2]. For practical applications, it is important to note that the properties of magnetic thin films can be partially controlled by the appropriate choice of materials, substrates, growth conditions, thickness, and sequence of various layers. The advanced growth techniques used for generating nanostructured 2D systems have, in practice, produced new magnetic materials that already are applied in information technology. For fundamental research in solid-state magnetism, multilayers and thin films facilitate studies of the effects of structure, chemical composition, and dimensions on magnetic properties.

Recent progress in exploring of magnetic multilayers and thin films reflects a combination of causes, comprising substantial developments in growth techniques, experimental capabilities, and computational tools. Recently, the increasing availability of synchrotron radiation has greatly contributed to the development of spectroscopic methods to study magnetism in thin films. The high-intensity photon-flux available at synchrotron light sources – especially on insertion-device beamlines – has promoted the elaboration of complex spectroscopy techniques, such as spin-resolved photoemission. Moreover, the possibility of controlling the polarization of photons – both circular and linear – has enormously enhanced the scope of magnetic measurements. By introducing polarization control in traditional techniques, such as absorption, reflection, and photoemission, the effects of magnetic dichroism can be easily observed [3]. In this way, synchrotron radiation spectroscopies provide a detailed description of the magnetic and electronic properties of ultra-thin films down to only a few atomic layers. For example, we can determine band dispersions, spin- and orbital-magnetic moments, magnetic anisotropies and susceptibilities, Curie temperatures and critical behaviors at the magnetic phase transitions. Furthermore, by extending spectroscopic methods – which are sensitive to the orientation of magnetic moments – into the field of microscopy allows us to directly observe the structure of magnetic domains in surfaces and thin films.

In this paper I briefly review recent synchrotron-radiation photoemission studies of magnetic systems. Although this field is relatively new, the number of its applications already has far exceeded the possibility of giving a balanced review in the space of a single article. Therefore, instead of attempting a comprehensive review of the subject, the discussion is limited to illustrating particular results that are representative of the entire field, even though not at all exhaustive. These examples are mainly chosen from our original research simply because they are the ones most familiar to me.

After a short review of the technical aspects of spin-polarized photoemission experiments (section 6.2), the applications of SPPES to simple elemental monocrystalline films are illustrated in section 6.3. These examples show the current level of accuracy with which the surface electronic structure of magnetic materials can be probed. Furthermore, they also serve as a useful introduction to general aspects of the spin-resolved photoemission technique. More complex systems are considered in the following sections. The electronic structure of magnetic multilayers is analyzed in section 6.4. Metallic multilayers are obtained by combining in various ways magnetic and non-magnetic thin films. These combinations are extremely interesting: their unusual physical properties are closely linked to the 2-dimensional confinement of their electronic charge. Also in section 6.4, the origin of one of these unusual properties – the indirect exchange interaction coupling magnetic layers separated by non-magnetic ones – is traced back and explained in terms of the properties of the electronic band structure of these 2D systems. These results explain the powerful long-range interactions originating from the magnetic couplings observed in many multilayers. Furthermore, the effects of magnetic coupling directly influence the macroscopic properties of multilayers and are, therefore, of direct technological relevance. In this sense these results are also practically useful because they provide the key parameters for the rational engineering of magnetic multilayers with specific properties.

Finally, in the last section (section 6.5), I consider more complex cases of ultra-thin magnetic films, analyzing in particular the electronic structures of ferromagnetic oxide compounds. For such compounds the situation is less well defined than for the elemental metals, as this area of research's quite new. Finally, I take the opportunity to discuss the expectations and perspectives for new applications of spectroscopic methods to investigate magnetic materials.

6.2. EXPERIMENTAL TECHNIQUE

Photoemission spectroscopy is a well-established experimental technique for studying the electronic structure of solids [4]. This section is mostly limited to brief remarks summarizing the key features of the experimental set up.

In the simplest case, photoemission consists of irradiating a polycrystalline solid sample with UV or soft X-ray light of known properties (photon energy and polarization) to excite the (electron) photoemission process, and then measuring the intensity of the emission as a function of the kinetic energy. As a first approximation, these measurements can be interpreted as directly sampling the electron energy distribution (Density of States) of the solid. However, being a photon-in electron-out technique, photoemission is highly surface-sensitive (the probing depth is limited by the escape depth of electron), and therefore the technique is eminently apt to the study of surfaces and ultra-thin films.

Photoemission is implemented in a more sophisticated way when the sample is a single crystal. In this case, the angular distribution of the photoemitted electrons is highly anisotropic and carries valuable information about the dependence of electronic structure on the crystalline momentum **k** (energy bands of the solid) [5]. In an angle-resolved photoemission experiment, the intensity of photoemission therefore is measured both as a function of emission angle and electron kinetic energy. In this way, the band structures of numerous solid samples were extensively mapped [6].

An additional anisotropy is present in the electronic emission of ferromagnetic materials, resulting from the unbalanced number of electrons with opposite spin inside the solid. Spin-resolved photoemission naturally extends angular-resolved photoemission by further measuring the spin of the ejected electrons [7]. This opens up the possibility of examining the spin-polarized band structures of ferromagnetic materials. This last step is far more difficult than the previous one because the technique to measure the electronic spin is extremely inefficient. For this reason, spin-resolved photoemission is mostly performed at synchrotron light sources which generate sufficiently intense radiation [8].

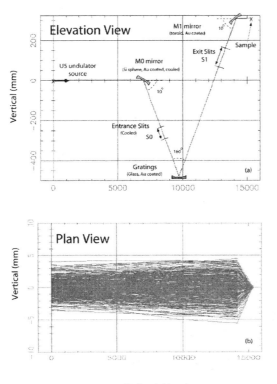

Figure 6-1: Layout of the U5UA 3.5m SGM beamline. (a) Elevation view, showing locations of optical elements. (b) Plan view, showing ray-tracing of the entire optical system.

The U5UA beamline at NSLS, Brookhaven National Laboratory is a typical example of synchrotron beamline used for this kind of experiment [9]. Its operation is based upon a relatively simple 3.5 m spherical grating monochromator (SGM). The beamline is equipped with four gratings for optimum operation in the photon energy range 10-200 eV. The high brightness produced by this light source, especially in the lower half of the energy range, makes it an ideal tool for valence-band spin-polarized photoemission spectroscopy.

The U5 straight section in the NSLS VUV storage ring houses an undulator. The parameters for the undulator (7.5 cm period length, 27 periods, 0.05 T < B < 0.45 T) were chosen to produce high brightness UV radiation in the 10-200 eV photon energy range. Figure 6-1a, and b, respectively, show the schematic layout and the optical ray-tracing simulations of the U5UA beamline. They display the beamline's focusing properties. Fig.6-2 illustrates the ray pattern normal to the beamline axis at the sample's position – the relevant spot size for the photoemission experiment.

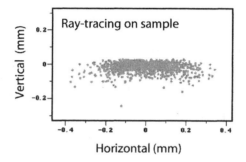

Figure 6-2: Ray-tracing image at the sample position, normal to the beamline axis.

Optically, the beamline consists of a first mirror (M0), located 7m from the center of the undulator source that focuses tangentially (vertically) onto the entrance slit (S0). The SGM proper consists of the entrance slit (S0), the 10 m-radius spherical gratings located 1.0 m downstream of the S0, and a movable exit slit (S1) located 2.527 m downstream of the grating. One of four gratings (groove densities: 300, 600, 1200, 2400 lines/mm) can be used at a time; rotating the grating scans the wavelength. The final optical element of the beamline is a Au-coated glass toroidal mirror (M1), located 2.0 m downstream of S1 that refocuses the monochromator output onto the sample, which lies 1.2 m downstream of M1. The grazing angles of incidence on the M0 and M1 are 5 degrees, so that the beam exiting M1 is level with the M0. The M0 is a back-cooled Au-coated Si sphere (R=32 m); it de-magnifies tangentially (vertical plane) approximately by 4-to-1 at the entrance slit of the SGM. The M0 provides good transmission through small slit openings, but does not overfill the small gratings in the SGM anywhere in their operating angle ranges. Sagittally (horizontal plane), the M0 essentially does no focusing. This is physically acceptable owing to the small opening angle of the U5 undulator radiation, and has many benefits optically. First, the aberration term that causes loss of resolution in the TGM design is astigmatic coma, which depends on the sagittal angular spread at the grating as well as the grating's sagittal radius. By not introducing sagittal angular magnification, this aberration term is quite small (in addition to the large, 10m sagittal radius, which also helps tremendously). Second, by not introducing sagittal focusing until after the SGM monochromator, the M1 refocusing mirror can be made strongly de-magnifying, thereby decreasing the spot size at the sample's position in the end-station. Note that the photoemission spectroscopies carried out at U5UA are not very sensitive to the angular spread of the incoming photon beam.

The spin-resolved photoemission spectrometer is permanently connected to the U5UA beam-line. It consists of a UHV chamber equipped with the standard tools for preparing and characterizing single crystal ultra-thin films and surfaces. The photoemission experiment is carried out with a commercial spherical analyzer (Omicron EA125) coupled to a mini-Mott detector for measuring electron spin polarization [10]. In this device the electron beam is accelerated to high voltage (about 30 keV) to scatter against a target of high-Z material (Th in our case). The (back) scattering asymmetry is measured with symmetrically opposite channeltrons (see Fig.6-3). The measured asymmetry originated from the spin-orbit term in the scattering potential. Analysis of this scattering geometry shows that the spin polarization of the incident electron beam can be obtained from the simple formula $P = (1/S) (I_+ - I_-)/(I_+ + I_-)$ where S, the Sherman factor, at fixed incident electron energy, is a constant characteristic of the spin-polarimeter; it measures its ability to distinguish opposite spins in the direction perpendicular to the scattering plane. I_+ and I_- are the scattering intensities measured in the two opposite channels (left and right, or up and down) and together with the incidence direction determine the scattering plane of the system. The figure of merit used in comparing polarimeters is $S^2 I/I_o$ where I is the sum of intensities collected in the two opposite channels and I_o is the incident intensity. The typical figure of merit of a mini-Mott detector is of the order of 10^{-4}, which, as mentioned above, explains the high intensity required to perform spin resolved photoemission experiments.

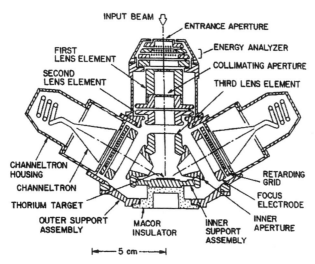

Figure 6-3: schematic drawing of Rice mini-Mott

To illustrate the capability of this kind of apparatus, one can consider the valence band photoemission spectra from an Fe(110) films. Iron is the commonest ferromagnetic element and its electronic structure has been amply studied both experimentally and theoretically [11,12]. The Fe atom has an outer electronic configuration $4s^1 3d^8$; upon forming a solid, the outer electronic states broaden into a continuum forming the $3d$ and $4sp$ bands of the solid.

These band features can be sampled using angular-resolved photoemission. For example, the normal emission spectra from Fe(110) surface measured with linearly polarized 40 eV photon energy light at two incidence angles are shown in Fig.6-4. These curves represent the intensity of emission as a function of the binding energy (the zero energy is the Fermi level E_F). They display the typical qualitative features of valence-band photoemission spectra from a $3d$ metal: the region close to the Fermi level and extending up to about 4 eV binding energy is dominated by the intense emission of the $3d$ states. In this case, at higher binding energy, a broad structure is also visible, reflecting the *sp*-derived states. The two spectra are taken in different conditions of light incidence: $\theta = 35°$ (more *s*-light), and $\theta = 60°$ (more *p*-light). Due to the dipole-selection rules operating in photoemission, these two geometries allow us to sample preferentially states with different orbital symmetry. In the *p*-light condition, photoemission favors the states of Σ_1 symmetry; in the *s*-light spectrum, the Σ_4 states are enhanced. Thus, the incident light-polarization constitutes a powerful tool to distinguish and recognize the orbital symmetry of the various features of a photoemission spectrum. Furthermore, at 40 eV photon energy, the normal emission spectra samples states close to the high-symmetry Γ-point of the bulk band structure. Straightforwardly comparing the positions of the emission peaks with the bulk band structure of Fe (see Fig.6-7 inset) leads to the following interpretation of the spectra in Fig.6-4. The first peak at 0.25 eV binding energy, enhanced with *p*-polarized light, should correspond to the minority Σ_1 band in proximity of the $\Gamma_{2,5}{}^{\downarrow}$ point. The peak at 0.9 eV binding energy should correspond to majority states of both Σ_1 and Σ_4 symmetry, which are degenerate close to the $\Gamma_{1,2}{}^{\uparrow}$ point. The shoulder at about 2.5 eV should represent the exchange-split majority counterpart of the peak close to E_F (i.e.: Σ_1, $\Gamma_{2,5}{}^{\uparrow}$). Finally, at 8.4 eV, the broad feature should originate from the Σ_1 *sp*-derived band at $\Gamma_1{}^{\uparrow\downarrow}$.

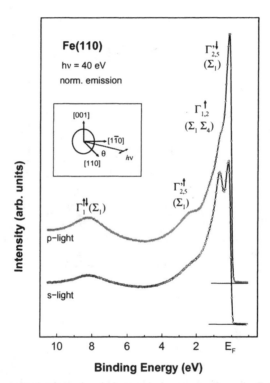

Figure 6-4: Angular resolved valence band photoemission spectra from the Fe(110) surface taken in normal emission at 40 eV photon energy. The light is incident at $60°$ and $35°$ for the upper (close dot, *p*-light) and lower (open dot, *s*-light) spectra respectively.

These spin assignments obtained indirectly from the angular resolved spectra can be directly verified using spin-resolved photoemission. This is shown in Fig.6-5, where the spin-resolved photoemission spectra from the same Fe(110) sample measured with *p*-polarized light is reported. The angular-resolved *p*-spectrum of Fig.6-4 now is decomposed into two components corresponding to electron emission with spin parallel to the *in-plane* [110] direction (majority spin: filled-up-triangles), or opposite to the [110] direction (minority spin: empty-down-triangles).

The spin character of the bands at $\Gamma_{2,5}{}^{\downarrow}$, $\Gamma_{1,2}{}^{\uparrow}$ and $\Gamma_{2,5}{}^{\uparrow}$ are readily identified. Additionally, the inset shows the spin-resolved photoemission spectrum in the binding energy region of the *sp*-derived Σ_1 band. Spin-resolution can separate the *sp*-emission in its two broad, closely spaced exchange-split components corresponding to $\Gamma_1{}^{\uparrow}$ and $\Gamma_1{}^{\downarrow}$ points. Their positions are found at 8.6 and 8.1 eV, respectively.

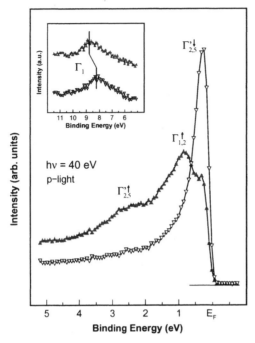

Figure 6-5: Spin-resolved valence band photoemission spectra from Fe(110) taken in the same condition as in Fig.6-4 with *p*-light. Up-triangles: majority spin; down-triangles: minority spin. The region of the *sp*-derived states is shown in the inset.

An extremely important characteristic of a synchrotron light source is its broad photon-energy spectrum. This is very useful in photoemission: by changing the photon energy, different regions of the band structure can be sampled [4,5]. Thus, by changing photon energy, it is possible to directly observe the dispersion of the bands as a function of the crystalline momentum, i.e, to map the bands of the solid. Fig.6-6 depicts the spin-integrated normal emission spectra from Fe(110) measured as a function of the photon energy from 15 to 70 eV. As mentioned, the region close to E_F is dominated by the 3*d* emission and their dispersion is now clearly seen (see marks in Fig.6-6). The measured dispersions of the photoemission features agree well with the calculated Fe-bulk band-structure.

Fig.6-7 shows the corresponding band mapping experiment performed in the spin-resolved mode. The spectra are decomposed into the spin-up (full-up-triangles) and spin-down (empty-down-triangles) components, making the spin characteristics of the various bands fully apparent. From this type of measurement, important quantities can be directly extracted characterizing the magnetic interaction, such as the exchange splitting. Note, for example, how the typical exchange splitting of the 3*d* states is $\Delta_{ex}{}^{\Gamma}{}_{2,5} = 2.5 - 0.3 = 2.2$ eV, while the exchange of the *sp*-states is only $\Delta_{ex}{}^{\Gamma}{}_1 = 8.6 - 8.1 = 0.5$ eV. For this reason, the 3*d* electrons sometimes are called the magnetic electrons, while the *sp* electrons are termed the conduction electrons.

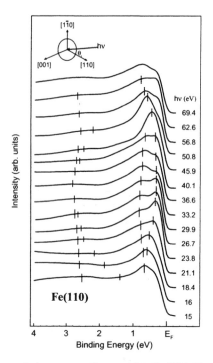

Figure 6-6: Valence band photoemission spectra from a thick Fe(110) film taken in normal emission condition in the photon energy range 16-70 eV. The light incidence is 56°.

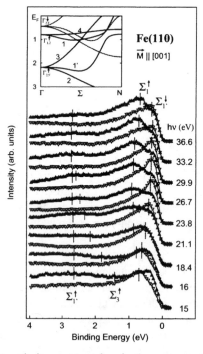

Figure 6-7: Spin-resolved photoemission spectra taken in the same condition as in Fig.6-6, in a smaller photon energy range. The sample is magnetized along the *in-plane* [001] direction. The portion of Fe band structure appropriate for the comparison with the experimental data is reproduced in the inset (continuous lines: majority states; dotted lines: minority states).

6.3. SIMPLE MAGNETIC FILMS

Surfaces and thin films display a different magnetic behavior than do bulk systems. This is partially related to the modification of electronic structure due to the lowered symmetry and coordination number characteristic of the 2D systems. Hence, the study of the magnetic properties of surfaces and thin films and its relation to their underlying electronic structure is an area of intense theoretical and experimental activity. The capability of modern *ab initio* numerical methods, based on local spin density functional theory, allows detailed predictions of the ground-state electronic structure of low dimensional magnetic systems like surfaces and interfaces. On the other hand, synchrotron-radiation-based spin-resolved photoemission with its high surface-sensitivity and the extreme richness in probing the details of the electronic structure is an ideal tool for comparing experimental data with theory.

In this section I discuss the "simplest" 2D ferromagnetic system, i.e., one composed of a single element in the form of a single crystalline ultra-thin film. To give a general picture of applications of the photoemission technique, two different and quite unrelated topics are illustrated in considerable detail. First, I consider a few interesting spectroscopic problems; i.e., problems related to the nature and determination of the surface and interface contributions to the electronic structure of magnetic films. I demonstrate how spin-resolved photoemission constitutes a powerful and indispensable technique to characterize the surface electronic structure of solids. A different aspect of the physics of single elemental films is addressed in the second part of this section. Here, photoemission is applied to investigate the macroscopic magnetic properties of 2D films, in particular, the magnetic reorientation phase transition in ultra-thin Fe(110) films.

6.3.1 Spin-resolved Photoemission as a Spectroscopic Tool: the Electronic Structure of Fe Surfaces

Modifying electronic structure at surfaces is of primary importance for understanding magnetism in ultra-thin films. Therefore, the surface electronic structure of magnetic 3d transition metals has been studied in great detail both experimentally and theoretically. The scope of this traditional line of spectroscopic investigation is too extensive to attempt to summarize it exhaustively. Furthermore, our knowledge still is too fluid to definitively assess the field. Instead, I focus on a few particular but nevertheless important aspects of the surface electronic structure of Fe surfaces. They will show how, despite the tremendous research effort, there are still open questions about the electronic structure of elemental magnetic materials. A close comparison between spin-resolved photoemission data and first-principle calculation may contribute to a fuller understanding of these complex electronic structures.

From the Fe(110) spectra illustrated in the previous section, the impression might be gained that spin-resolved data are not really essential in assessing the spin-polarized electronic structure of magnetic materials. After all, all the photoemission peaks were assigned correctly using only the angular-resolved spectra (Fig.6-4) in comparison with the calculation of the bulk band structure (Fig.6-7 inset). This impression is erroneous. I discuss now two interesting cases in which angular-resolved photoemission alone fails to give the correct answer, and direct measurement of the spin character of the emission with spin-resolved photoemission is mandatory.

Erskine and co-workers extensively studied the electronic structure of various Fe surfaces using high-resolution angular-resolved photoemission [11]. These studies were mostly concerned with establishing the extent of applicability of single-particle approximations to the ferromagnetic $3d$ metals. Ferromagnetism originates from correlation effects in the electronic structure; consequently, it is not taken as granted that a single-particle picture could fully describe these systems. Furthermore, this experimental work constitutes precious source of information on the surface electronic structure of ferromagnets. Indeed, the task of determining, from the photoemission data, the bulk and surface components of the electronic structure is closely related. The comparison between the photoemission data and the bulk band-structure calculation became considerably more reliable once the contribution from surface emissions was clearly identified.

One of the most interesting results of these studies was the discovery of the first example of an exchange split-pair of surface state on the Fe(100) surface [13]. This state is best seen in the angular-resolved spectra taken in normal emission at low photon energy of 16 eV. Its extreme sensitivity to a very small amount of absorbed oxygen clearly indicates its surface nature. Both these facts can be seen in the spectra reported in Fig.6-8, taken from a 30 ML Fe film epitaxially grown on a Pd(100) single crystal. Oxygen exposures as low as 0.075 L are enough to nearly-completely suppress the emission from this state. The comparison with different band-structure calculations all agree in indicating that the peak at 0.3 eV binding energy should reflect the minority component of a Shockley-type surface state; its majority counterpart would then be the other feature located at 2.7 eV binding energy [14]. This was a very natural interpretation of the experimental data. Furthermore, it was satisfying because it was the first time that both components of a surface state from a ferromagnetic surface appeared in a photoemission spectrum, and that fact, in turn, would have allowed the extraction of a measured value of the exchange split energy locally at the surface.

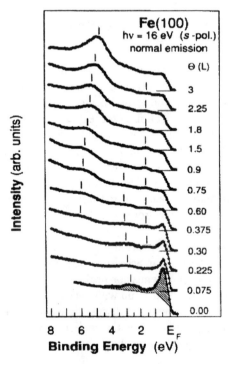

Figure 6-8: Spin-integrated photoemission spectra from a 30 ML Fe(100) film epitaxially grown on Pd(100) as a function of oxygen exposure. The extreme surface sensitivity of the peak close to the Fermi level is clearly apparent.

It came then as a surprise when, a few years later, the spin of this surface emission was directly measured using SPPES and revealed that both surface features have the same spin [15]. The spin-resolved photoemission spectrum from the clean Fe(100) film is shown in Fig.6-9; it clearly demonstrated that both surface sensitive components are majority states. This means that the previous interpretation of these features as an exchange split pair of surface states was incorrect. Unfortunately, reinterpretation of this data is quite complex and would demand undertaking a set of special calculations with very accurate simulation of the surface potential of the Fe(100) surface. Up to now, these calculations have not been made.

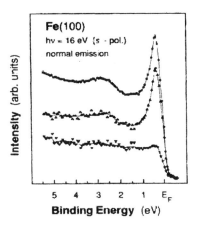

Figure 6-9: Spin-resolved photoemission spectrum from the same clean Fe(100) film shown in Fig.6-8. Up-triangle: majority; down-triangle: minority; circle spin-summed spectrum.

A second analogous case, and one for which the full theoretical analysis is available, is that of the Fe(110) surface [16]. Here, according to band- structure calculations, a favorable way to observe surface states should be to probe the electronic states along the high-symmetry Γ_{bar} - S_{bar} line of the two-dimensional surface Brillouin zone [17]. Along this line lies a distinct symmetry gap, filled with a surface state extending nearly all of the way along it. Particularly, its minority component should have a maximum binding energy around the middle of the zone and disperse upward, crossing E_F at about three quarters of the Γ_{bar} - S_{bar} line (see the calculations reported in Fig.6-12).

The Γ_{bar} - S_{bar} line can be accessed by properly choosing the geometry of the angular-resolved photoemission experiment, as shown in Fig.6-10 where the electronic states originating from the Γ_{bar} - S_{bar} line are sampled by scanning the emission angle. In the off-normal spectra given in Fig.6-10(a), we see that a new feature appears in the spectra, around the middle of the zone (spectra from 16^0-22^0), at about 0.3 eV binding energy. This peak disperses toward E_F increasing k_{parr} and eventually crosses it at about three quarters of the length of the Γ_{bar} - S_{bar} line. Furthermore this feature is sensitive to the surface cleanliness as can be deduced from its conspicuous sensitivity to oxygen (Fig.6-10(b)). The two-dimensional character of this state is proved in Fig.6-10(c). Here, this state does not show any appreciable dispersion when the photon energy (i.e., k_{perp}) is varied between 21 to 57 eV, as should be the case for a two-dimensional surface state. The binding energy vs. k_{parr} for this state is plotted in Fig.6-10(d). The experimental points are plotted against the minority symmetry projected bulk gap, as calculated in ref.17. In summary, as in the preceding case, this state satisfies all the ordinary conditions usually ascribed to surface states: 1) it is sensitive to surface contamination, 2) it behaves as a 2-dimensional electronic state, and, 3) it lies in a (minority) symmetry gap. From angular-resolved spectra we might conclude that this is a *minority* surface state.

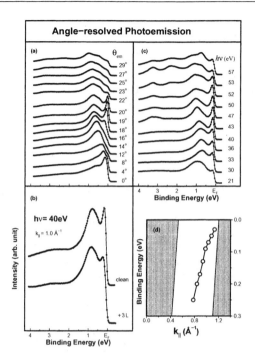

Figure 6-10: Photoemission spectra of a 50 ML Fe(110) film. Electronic states with parallel wave vectors along the symmetry line Γ_{bar} - S_{bar} of the two-dimensional surface Brillouin zone (SBZ) are selected in the chosen geometry. (a) Spectra taken at 40 eV photon energy as a function of the emission angle, (b) at 19° off normal, as a function of oxygen exposure, (c) k_{par} = 1.0 A^{-1} at the Fermi level, as a function of photon energy, (d) Energy dispersion of the state near to E_F as a function of k_{par} as extracted from the spectra in (a). The experimental point are superimposed to the calculated minority bulk gap, projected onto the 2-dimensional SBZ.

Once more, the spin-resolved spectra for this state reveals a different situation. The spin-resolved spectra corresponding to the angular-resolved spectra of Fig.6-10 are reported in Fig.6-11. Opposite to expectations, the surface peak near to E_F is unambiguously found to be of *majority* spin. Sharp contradicting with what would have been concluded without the direct spin analysis, no *minority* state is found in the spectra near E_F.

To understand the paradox generated by the spin-resolved spectra, *first-principle* slab calculations were made, and are shown in Fig.6-12 along the appropriate Γ_{bar} - S_{bar} high-symmetry line.

First, we notice that there is no majority gap in Fe(110) along the Γ_{bar} - S_{bar} line close to the Fermi level, so that no real majority surface state can be found in this energy region. To explain our surface-sensitive majority feature we have than to look for surface resonances: this is done in the calculations by selecting states with a considerable charge density in the first two atomic layers. These surface and subsurface states are marked with gray-dark circles in Fig.6-12. This type of calculation helps to clarify the experimental results. Along the Γ_{bar} - S_{bar} line, we now see that at about three quarters of way along the Γ_{bar} - S_{bar} line and crossing the Fermi level, there is a majority surface resonance. Clearly, this is the electronic state corresponding to the surface-sensitive feature observed in the experiment.

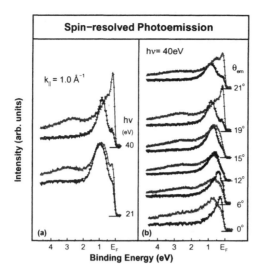

Figure 6-11: Spin-resolved photoemission spectra of a 50 ML Fe(110) film. Up-triangles: spin-up, down-triangles: spin-down. Electronic states with parallel wave vectors along the symmetry line Γ_{bar} - S_{bar} of the 2-dimensional SBZ are selected in the chosen geometry. The remanent magnetization is along the [001] direction which is aligned to the axis of the spin-polarimeter. (a) off normal (k_{par} = 1.0 A^{-1}), as a function of the photon energy, (b) at 40 eV photon energy, as a function of photoelectron emission angle.

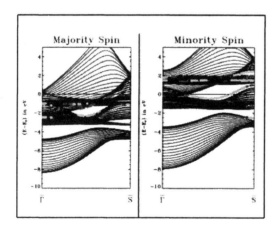

Figure 6-12: Band structure calculation for a 23 layer Fe(110) film along the Γ_{bar} - S_{bar} symmetry line: Light circles represent states where more than 50 % of the charge density distribution is confined in the first atomic layer and the vacuum. Slightly darker circles represent states with more than 60 % of the charge density distribution confined in the first two atomic layers and the vacuum. This second type of state is marked with an even darker circle if, additionally, its charge distribution is such that the difference in the two layers is smaller than 20 %.

In summary, extending the analysis from only pure surface states, i.e., those strictly confined into symmetry gaps, to surface resonances, i.e., electronic states whose charge density is considerably spread on both surface and subsurface layers is necessary to understand the experimental results from the Fe(110) surface.

6.3.2 Spin-resolved Photoemission as a Tool to Investigate Magnetic Properties: The Magnetic Reorientation Transition in Fe(110) films

In this section I consider another aspect of the spin-resolved photoemission technique; showing how this technique can be used to obtain information on the macroscopic magnetic behavior of ultra-thin films. A typical magnetic phenomenon of 2-dimentional systems is related to the different magnetic anisotropy of a thin film with respect to its bulk counterpart. In general the total free energy of a ferromagnet depends on the orientation of the magnetization with respect to the crystallographic axis; thus originates a magnetic anisotropy. Magnetic anisotropies are of great technological importance; they define easy and hard axes for the magnetization of materials. Indeed, the prospect of effectively manipulating magnetic anisotropies is a major drive in applied research. Switching magnetic transitions occupy a special place in this field, the direction of the magnetization being very relevant for applications such as data storage, where stability of the magnetic configuration is mandatory. Although the basic mechanisms of magnetic anisotropy are relatively well understood, the detailed analysis of their electronic origin in specific systems still challenges theoretical computations. Magnetic anisotropy energies are typically extremely small (a few $\mu eV/atom$); only recently have first-principle calculations approached the level of accuracy required to treat them correctly. Therefore, it seems appropriate to provide experimental data for comparison with theory.

In ultra-thin magnetic films, the interplay between exchange interaction, (long-range) dipolar interaction and on-site magnetocrystalline anisotropy generates a variety of interesting new magnetic phenomena with no counterpart in three-dimensional systems, such as perpendicular magnetic anisotropy and magnetic reorientation transitions. Accordingly, the study of magnetic properties of thin films has become an important testing ground to understand the basic interactions and how they are affected by such factors as composition and microscopic structure. This is particularly true for magnetic reorientation transitions that can be triggered by different factors, such as temperature and/or the film's thickness and/or chemical composition. For example, morphology-dependent magnetic anisotropy has been studied in systems such as Fe/Ag(100) and Co/Cu [18]; in alloy systems like $Fe_{1-x}Co_x/Cu(100)$ composition-driven spin reorientation transition has also been observed [19]; additionally, surface states also were reported to be responsible for the magnetic anisotropy in the Fe/W(100) system [20].

I illustrate the case of reorientation transition in epitaxial Fe(110) ultra-thin films grown on a W(110) single crystal. In these films, an *in-plane* to *in-plane* reorientation of magnetization occurs as a function of the film's thickness. This dependence on thickness usually is explained by the competition between the bulk Fe anisotropy term, which favors the bulk's easy axis [001] direction, and the two-dimensional surface and interface anisotropy terms, which favor the [110] direction [21].

This view seems to be consistent with the fact that by capping the clean Fe surface with a noble metal overlayer, thereby strongly modifying the surface properties of the Fe layer, the surface anisotropy term can be modified, effectively altering this equilibrium [22]. However, recently it also was found that the elastic properties of the Fe(110) film vary considerably with its thickness. A change in the elastic characteristic of the material could have profound effect on the magnetic anisotropy due to the strong magneto-elastic anisotropy term. In other words, the origin of the observed reorientation transition in the Fe(110) films might be due to this magneto-elastic term rather than to a modification of the balance between surface and bulk anisotropy [23]. Spin-resolved photoemission can advantageously used to study both aspects of the problem.

First, I show the dependence of magnetization on the thickness of the noble metal overlayer, and then confirm that indeed the surface contribution is relevant here [24]. Second, I discuss Fe(110) films alloyed with small percentages of V, Co, or Ni. In this condition, the lattice constant of the film is considerably changed and a strong elastic stress is induced in the Fe films. These second type of experiments clearly demonstrate that the effect of the magneto-elastic anisotropy is extremely pronounced, forcing the conclusion that the complete picture of the magnetic reorientation transition in Fe(110) films needs to account for both aspects of magnetic anisotropy.

Fig.6-13 shows the appearance of the Fe(110) magnetic reorientation transition, monitored with valence-band photoemission. The spin-resolved spectra are taken along an Fe wedge of variable film thickness (indicated in the Fig.6-13) in a range corresponding to the magnetic transition. In the left panel, the spectra are decomposed using the spin-polarization measured along the *in-plane* [110]; direction on the right panel, along the *in-plane* [100] direction. On the thin side of the Fe wedge (bottom spectra), the spin polarization is present only on the left panel, i.e., all the magnetization is along the [110] direction. Increasing the Fe film's thickness (moving upward in Fig.6-13) a critical thickness t_r is reached between 80 and 86A. Above this value, the situation is reversed; the spin-polarization is now only in the left panel, i.e., the easy magnetization axis has switched from the [110] direction (thin films) to the [100] direction (thick films).

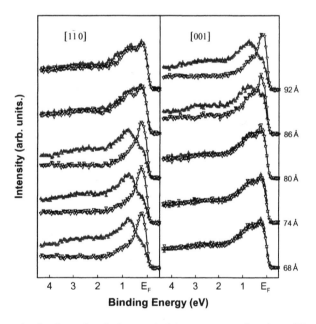

Figure 6-13: Spin-resolved valence band photoemission spectra as a function of Fe thickness along the two *in-plane* [110] and [100] high-symmetry directions of the Fe(110) surface. The spectra are measured at 40 eV photon energy and in normal emission. The filled-up and empty-down triangles represent majority- and minority-spin components respectively.

In the first model discussed, the reorientation transition in Fe(110) rests on the balance between the bulk and the surface contributions to the magneto-crystalline anisotropy. It should then be possible to modify this balance by modifying the Fe surface, possibly by covering it with overlayers of noble metal. The choice of noble metals seems convenient to minimize the interaction between the two materials, as is well illustrated in Fig.6-14. Here, the valence-band photoemission spectra during Ag (left panel) and Au (right panel) depositions are shown as a function of the noble-metal film's thickness. The bottom spectra are the Fe spectra, characterized by the typical $3d$ emission that extends about 3 eV below the Fermi level. By depositing the noble metal, the Ag $4d$ and Au $5d$ states appear. In both cases, the d-d interaction between the noble metal and the Fe is expected to be quite low because the d states of the noble metal are well separated from the Fe $3d$. Furthermore, a stronger interaction would be expected for Au-Fe rather than Ag-Fe because the Ag d-bands are further separated from the Fe $3d$-bands than the Au $5d$-band is from Fe $3d$. In this sense, Au deposition should have a bigger effect on the anisotropy of Fe than Ag deposition.

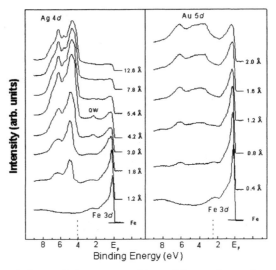

Figure 6-14: Valence band photoemission spectra of Ag and Au overlayers deposited on a thick Fe(110) film as a function of Ag thickness (left panel) and of Au thickness (right panel).

Figure 6-15 proves that this is indeed the case. These data were obtained by depositing on Fe wedges encompassing the magnetic transition, Ag or Au uniform overlayers. For each overlayers' thickness, spin-resolved spectra similar to the one shown in Fig.6-13 were measured and the corresponding Fe reorientation thickness t_r extracted. In both, magnetic reorientation occurs at lower thickness compared to clean Fe(110). This finding seems reasonable because the effect of the weakly interacting Ag or Au overlayers is mainly geometric; the presence of the overlayers makes the Fe top-most layer more bulk-like, thus reducing its contribution to surface anisotropy. Therefore, the balanced thickness between the anisotrophy of the bulk and surface becomes smaller.

However, although the effect of the two noble metal is qualitatively the same, i.e. to anticipate the reorientation transition at thinner Fe film thicknesses, quantitatively there is a considerable difference between Ag/Fe(110) and Au/Fe(110). With Ag, the transition thickness decreases quite slowly with coverage. Moreover, the effect saturates above the first two Ag monolayers. Apparently Ag-growth on Fe(110) is close to the ideal layer-by-layer growth at room temperature. This circumstance is consistent with the presence of intense Ag sp-derived quantum well states in the photoemission valence-band spectra of Ag(111)/Fe(110) (indicated by QW in Fig.6-14).

This saturation effect, reached at only a few monolayers of Ag, is also in agreement with theoretical considerations. According to first-principle calculations the magnetic moment of surface Fe atom is 2.65 μ_B and is reduced to 2.17 μ_B with a single Ag monolayer; indeed, 2.17 μ_B is very close to the bulk Fe moment (2.2 μ_B).

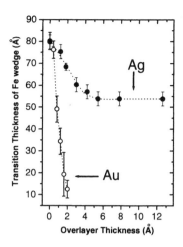

Figure 6-15: Fe reorientation thickness t_r as a function of the noble metal overlayer thickness.

For Au coverage, t_r decreases much more drastically than Ag. This behavior can be partly attributed to the stronger Au-Fe electronic interaction compared with that of Ag-Fe. The maximum of Au $5d$ band is located at ~2 eV binding energy, which is much smaller than that of Ag $4d$ band (~4 eV). However, we did not observe any saturation with Au, suggesting that some other mechanism is operating that is not present in the Ag case. Possibly Au intermixes with Fe at the interface; indeed, Au is known to form various stable bulk alloys with Fe. Furthermore, recent STM observations demonstrated the formation of Au-Fe surface alloys for submonolayer depositions of Au on Fe(100) at temperature above 370C [25]. Therefore, the formation of an Fe-Au interface alloy possibly is the main reason for the strong modification the Fe(110) surface anisotropy by Au.

From these overlayer studies, we concluded that modification of the surface region constitutes an effective way to shift the critical thickness t_r. Electronically, the deposition of noble metal overlayers results in the least possible alteration of the Fe surface. The fact that even in this case t_r varies so considerably clearly demonstrates that an equilibrium between a surface and a "bulk" contributions is realized and plays a key role in determining the reorientation transition of Fe(110) thin films.

Nevertheless, in view of the considerable variation of the magneto-elastic constants experimentally observed during the growth of Fe(110) ultra-thin films, [23] it is worth considering whether magneto-elastic anisotropy could also play a part in the reorientation. Note that these magneto-elastic terms vary with thickness and can, therefore, induce a reorientation transition as a function of thickness.

To investigate this aspect, the lattice constant of the Fe film would need to be varied in a controlled way and its effect on the reorientation transition monitored. Fortunately, this situation can be approximated quite closely by alloying small percentages of V, Co, or Ni in the Fe(110) films. If the percentage of doping is small, the lattice structure of the Fe(110) film remains bcc and the main effect only is an expansion of the lattice to accommodate the foreign atoms, as Fig.6-16 shows for FeNi alloys. Here, the LEED patterns from the clean W(110) and Fe(110) surfaces are reported in the top panels, while the other panels show the LEED pattern from the $Fe_x Ni_{1-x}$(110) alloy films for various Fe concentrations, x. On these pictures, the spots from the W surface are superimposed as white dots for comparison. This reveals that the large *in-plane* lattice mismatch of ~ 10% between the W(110) and the Fe(110) is greatly reduced by adding a small amount of Ni. This figure then demonstrates that the FeNi alloy lattice is expanded more than the one of pure Fe. In the bottom right panel, this expansion is quantified.

Analogous results can be obtained by substituting Co for Ni. However, substitution with V does not appreciably change the in-plane lattice constant because V is antiferromagnetically coupled to the Fe. To determine the effect of these expansions on the reorientation transition, we prepared wedges of these alloys and measured the critical thickness analogously to what was shown above for the pure Fe films. The results are summarized in Fig.6-17. On left panel are the changes in the *in-plane* lattice mismatch for V, Co, and Ni in the Fe(110) films; in right panel the corresponding Fe critical reorientation thicknesses t_r are plotted.

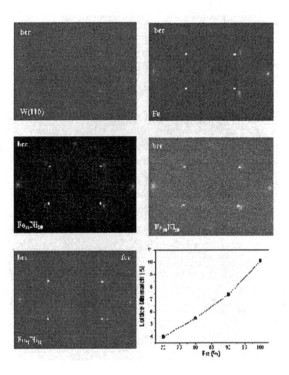

Figure 6-16: LEED patterns from W(110) and Fe(110) surfaces (top panels) and from the surfaces of thin films of $Fe_xNi_{1-x}(110)$ alloys. All films are 50ML thick. The W spots are superimposed as white spots to all the other LEED patterns for reference. On the bottom right, the measured in plane lattice mismatch is plotted vs. the Fe concentration.

The main finding from the results in Fig.6-17 is that the changes in the critical reorientation thickness t_r directly mirrors the corresponding induced lattice strain in the films. At about 30% Ni concentration, the film undergoes a structural bcc → fcc phase transition; this critical thickness is captured in the bottom-left panel of the LEED (Fig.6-16) where both [110]-bcc and [111]-fcc spots are present. Thus, for these high concentrations the stress built up in the bcc phase is so large that it induces a structural transition. Indeed, the case of Ni is quite extreme. From Fig.6-17, we see that corresponding to this considerable increment in stress caused by Ni doping, the Fe reorientation thickness increases tremendously with Ni concentration. Unfortunately, we could not follow this process for Fe films above about 400A thick because it took too long to evaporate them with e-beam evaporators to obtain clean surfaces. Accordingly, we could not explore the behavior of the Fe reorientation thickness close to the structural transition. Nevertheless, these results clearly indicate the effectiveness of magneto-elastic anisotropy in modifying the behavior of the Fe(110) films. These results are important in directly demonstrating the impact of the magneto-elastic contribution in the reorientation transition of ultra-thin films.

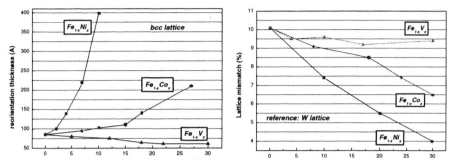

Figure 6-17: left panel: in-plane lattice mismatch for small substitution of V, Co and Ni in Fe(110) films; right panel: corresponding Fe critical reorientation thickness t_r.

In conclusion the appropriate models to describe the magnetic anisotropies in the Fe(110) films, and possibly more generally in other ultra-thin films, should consider both the effect of the magneto-elastic contribution as well as that of surface anisotropy.

6.4. METALLIC MULTILAYERS

Metallic multilayers are artificially engineered structures prepared by alternate superposition of layers of magnetic and non-magnetic (spacers) materials. These systems present interesting physical properties that do not correspond to the simple superposition of the properties of the constituent materials. Among these are the electronic transport properties that are greatly relevant in technological application. For example, the electric current flowing in a magnetic multilayer was found to be strongly affected by the magnetization state of the multilayer, thus establishing a useful connection between magnetic and electric properties in these systems [26].

In multilayers with antiparallel alignment of the magnetization, the electrical resistivity depends on the applied magnetic field, as qualitatively shown in Fig.6-18; the relative resistance is maximum at remanence and decreases with increasing magnetic field, being strongly suppressed in a high magnetic field. At high enough fields, the intrinsic antiparallel magnetic coupling of the multilayer is overcome, forcing the reordering and parallel magnetization of all the magnetic layers. In this condition, the flow of current is greatly facilitated. The variation of relative resistance is quite strong in some cases (for example, in Fe/Cr at low temperatures, values of $\Delta R/R = 220$ % were measured). This is particularly remarkable in comparison with the ordinary magneto-resistance effects of alloys (for example, in $Fe_{0.2}Ni_{0.8}$, $\Delta R/R = 2$ %). For this reason, the magneto-resistance effect in metallic multilayers is known as the giant magneto resistance (GMR) effect. Magnetic field sensors based on the GMR effect in commercial films of a few nanometers thick are routinely employed in industrial applications.

Research on magnetic multilayers gained considerable momentum in the eighties when strong coupling was discovered between the magnetization directions of adjacent magnetic layers. The ground state of the ensemble of layers is such that the system spontaneously orders. The coupling interaction forces the magnetization of each magnetic layer in a super-ordered magnetic structure extending along the entire multilayer. Most commonly, the adjacent magnetic layers are aligned parallel or antiparallel. Furthermore, it soon became clear that this coupling is far too strong to be simply explained by dipolar interaction between the magnetic layers. Its origin must instead be connected to an electronic interaction mediated by the layers of non-magnetic materials.

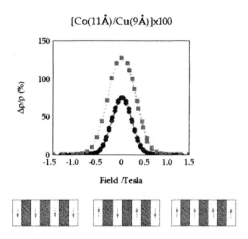

Figure 6-18: Relative variation of the electrical resistance in an anti-parallel coupled metallic multi-player vs. the applied external magnetic field.

The first observation of indirect magnetic coupling was in Fe/Cr multilayers [27]. It was shown that two thin (~ 10 ML) layers of Fe are magnetically coupled when separated by a non magnetic layer of Cr. The magnetic coupling, the parallel or anti-parallel magnetization directions of adjacent magnetic layers in the absence of any external magnetic field– i.e., in the ground state of the multilayer system– crucially depends on the thickness of the intervening Cr film. Two important characteristic of this type of coupling should be mentioned. First, the coupling survives up to as many as tens of Cr monolayers. This means that the magnetic coupling is a *long-range* phenomenon that cannot be accounted for by the ordinary short-range exchange interaction of magnetic materials. In other words, the magnetic interaction is an indirect one and must necessarily be mediated by the spacer material. Second, the sign of this indirect interaction changes periodically as a function of the thickness of the Cr layer. The indirect magnetic interaction is, then, an oscillatory interaction.

These two properties of the Fe/Cr magnetic multilayer are almost universal and have been observed for numerous combinations of magnetic and non-magnetic materials. These studies further revealed that the oscillation period is essentially a characteristic of the non-magnetic spacer materials and their crystallographic orientation in the multilayer system.

In this section, I give examples of photoemission studies from magnetic multilayers, and, in particular, discuss how spin-resolved photoemission helps in clarifying the electronic origin of this electronic interaction, called indirect exchange magnetic coupling.

6.4.1 The Electronic Origin of Indirect Exchange Coupling

From the properties of the magnetic coupling mentioned above, it might be supposed that the origin of the indirect exchange coupling is strictly connected to the electronic structure of the intervening spacer materials. Indeed, recent theoretical and experimental studies clearly established the close link between the indirect magnetic coupling and the finite size effect on the electronic states of the layered structure [28-30].

The fundamental difference between the electronic structure of a bulk material rather than as a thin film is the change in the system's dimensions and boundary conditions. Fig.6-19 qualitatively visualizes the ideal case of a perfect crystalline film composed by n layers. The film has a finite number of monoatomic layers (6 and 8 ML in Fig.6-19), and its electronic levels can be thought of by analogy with the problem of a particle in a one-dimensional potential box. The substrate and the vacuum interfaces provide the boundary conditions (i.e., the potential height of the walls). The localization of the electrons inside the film is the origin of the individuality of the levels. The number of states increases with the film's thickness while their energy separation decreases. The electronic structure converges to that of the bulk material with increasing film thickness.

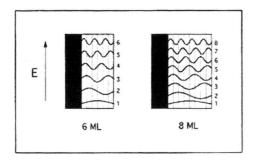

Figure 6-19: Schematic representation of the electronic states of a thin film grown on a substrate material.

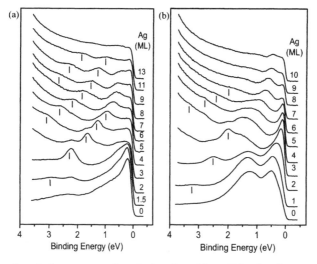

Figure 6-20: Normal emission spectra taken during deposition of the first few atomic layers of Ag on Fe(110) and Ni(111) substrata. The formation of discreet Ag QW states is indicated by the marks.

Angular-resolved photoemission can be used to study this quantization. As an example, Fig.6-20 shows the normal emission spectra from Ag(111) overlayers grown on Fe(110) and Ni(111) substrates. There are various peaks that regularly disperse toward lower binding energies: these are the signatures of the discrete quantum well (QW) states of the Ag sp bands. In the Fe case, the peaks are more intense representing real QW states of Ag; those on Ni are less intense because they are QW resonances (see below).

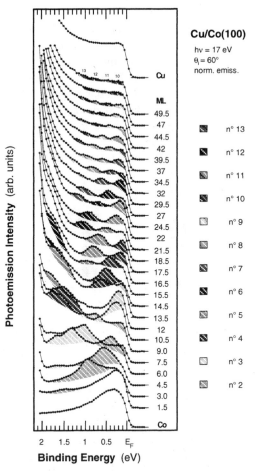

Figure 6-21: Normal emission spectra from Cu overlayers grown on a Co(100) substrata. The spectra are measured at 17 eV photon energy.

As described for a film on a substrate, or part of a multilayer, the boundary conditions are determined by the film's interfaces with the adjacent materials. It is the detailed situation at these interfaces that determines the form of the potential well and, therefore, the electronic structure of the thin film.

There are two important limiting cases. The first happens when the electronic structure of the substrate material contains electronic states of the same orbital symmetry and energy as one of the overlayer material. In this case, the levels of the films are less discrete. The electronic states of the film, even for one of a few atomic layers, easily hybridize with the continuum of states of the substrate and consequently, delocalize in the semi-infinite system (substrate/overlayer) by forming resonance states with finite band-widths. In the opposite case, the substrate material does not have corresponding states of symmetry and degeneracy with one of those of the overlayer film, i.e., the substrate material presents a symmetry gap corresponding to the energy region of the states of the film. In this case no hybridization occurs, and well-developed QW states are observed during the growth of the overlayer material. Fig.6-21, an example of such a favorable case, plots the normal emission spectra collected during the evaporation of Cu overlayer on a Co(100) substrate. The discreetness of the electronic structure is apparent up to considerable thicknesses of the Cu films (about 50 ML). In this case, the QW levels originate from electronic states of Δ_1 symmetry of Cu. They can be grouped in families starting at about 1.5 eV binding energy, and dispersing toward E_F as the film's thickness increases. Indeed, the energy of the bulk Cu state toward which they are all tending is above the Fermi level. From this picture, we see that the QW states cross the Fermi level at very regular intervals of the thickness of the Cu film (period ~ 6 ML). This fact is extremely important because this periodicity of 6 ML coincides with the period of oscillation observed in the indirect exchange coupling in the Cu/Co(100) magnetic multilayers [31].

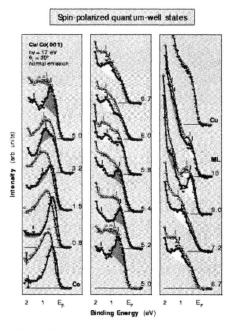

Figure 6-22: Spin-resolved Cu overlayers grown on a Co(100) substrata taken in normal emission condition at 17 eV photon energy. Up-triangles: majority spin; down-triangles: minority spin. The minority character of the Cu QW states is clearly visible.

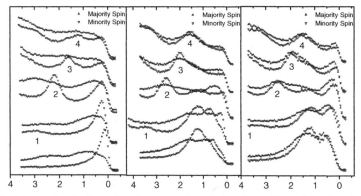

Figure 6-23: Spin resolved valence-band spectra taken during deposition of the first four monolayers of Ag on Fe(110) and Fe_xNi_{1-x}(111) substrata. Up-triangles: majority spin; down-triangles: minority spin.

To understand the details of the oscillating magnetic coupling, these considerations must be added to the fact that in a magnetic material the band structure is different for the two spins due to the exchange interaction. This means that the boundary conditions (the position of the symmetry gaps) will be different for the two-spin direction.

Therefore, when a film of non-magnetic spacer material is grown in direct contact with a ferromagnetic substrate, the boundary conditions are different for the two spins and a magnetic character is induced in the two-dimensional states of the film.

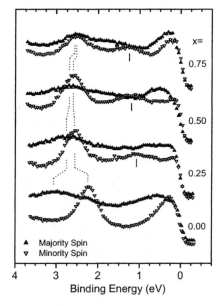

Figure 6-24: Spin resolved valence-band spectra for 1 monolayer Ag deposited on Fe and FeNi alloys.

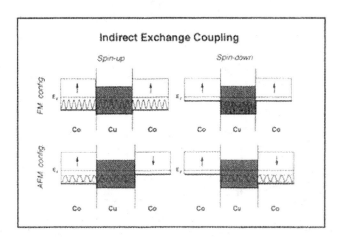

Figure 6-25: Schematic of one-dimensional potential wells (thick black lines) which describe the indirect exchange coupling between two magnetic Co layers mediated by a non-magnetic Cu layer. The top left (right) panel represent the potential seen by spin-up (-down) electrons in the case of parallel magnetization alignment in the two ferromagnetic layers. The bottom left (right) panel represents the potential for spin-up (-down) electrons in the case of antiparallel alignment. Only in the parallel alignment case and only for spin-down electrons are localized levels formed.

For example, for the case of Cu on Co, the Co substrate in the region close to the Fermi level presents a minority gap of Δ_1 symmetry but not a majority one (the majority is shifted at ~1.5 eV lower binding energy by the exchange interaction). When a Cu layer is grown on the Co substrate, the Cu states of Δ_1 symmetry and majority spin find in the Co layers available corresponding states, and so they hybridize and delocalize along the entire multilayer thereby loosing their discreet character. The Cu Δ_1 minority states instead do not find corresponding states in the Co layers and so remain localized. In other words, both discrete and bulk states coexist in the same magnetic multilayer with opposite spin character. The electronic structure of a paramagnetic Cu film in contact with a Co ferromagnet thus acquires a strong dependence on the orientation of the spin.

This situation can be seen in Fig.6-22 that shows the spin-resolved spectra from the QW states of Cu/Co(100). The QW states are mostly in the minority channel. Similar observations were made for Ag overlayers grown on Fe(110) and Ni(111). The spin-resolved spectra for the Ag QW states are shown in Fig.6-23: the minority character of this Ag QW is apparent.

Furthermore, Fig.6-24 compared the same QW state for 1 ML Ag grown on Fe and various FeNi alloys. The exchange splitting of Ni bands is much smaller than the one of Fe, and after adding Ni, the boundary conditions vary and transform the well-defined minority QW state present on Fe into a QW resonance with comparable components of majority and minority spin.

We are now able to understand the discreet electronic structure of the spacer materials. The situation for the Co/Cu/Co(110) multilayers is schematically represented in Fig.6-25 for the parallel and anti-parallel magnetic configurations. On the top row is represented the case of parallel alignment; the boundary conditions are, therefore, symmetric; the majority (left) and minority (right) potential wells are marked by thick lines. The majority states of Cu see a small potential well and consequently, form resonance states that delocalize along the entire multilayers. On the other hand, the minority Cu states remain confined due to the deep minority well and thus develop sharp QW states. In the anti-parallel case (bottom row) the boundary conditions are opposite on the two interfaces of the Cu film and both majority and minority states delocalize in the opposite Co layer. In this case, no real QW state and associated discreet electronic structure is formed. Pure QW states exist only for parallel magnetization.

To qualitatively assess the oscillatory behavior of the magnetic coupling with the thickness of the non-magnetic spacer material, we then consider the total energy of the system as a function of the material's thickness. Assuming a hypothetical permanent anti-parallel aligned multilayer, the total energy increases continuously by increasing the thickness of the non-magnetic material. In an hypothetical parallel aligned multilayer, however, the total energy increases in discrete steps with the spacer's thickness, corresponding to the filling of the discrete QW levels. In practice, the system always moves towards the configuration of minimum energy between these two cases. The periodic crossing the Fermi level of the QW states then modifies the relative energy of the two configurations, alternatively favoring the parallel or anti-parallel magnetization configuration. In other words, the observed periodic oscillation of the density of states at the Fermi level due to the individual levels directly determines the period, phase, and intensity of indirect coupling.

At present, the magnetic coupling in noble metal multilayers is well understood [32]. Systematic spectroscopic studies show complete series of quantified and spin-polarized states for many different systems. These investigations have contributed to developing an accurate quantitative description of the indirect interaction, which finds very satisfactory agreement with the corresponding theoretical calculations [33]. Problems still exist in the detailed description of more complex cases in which the interaction is mediated by transition metals with partially filled $3d$ shells (as in the case of Fe/Cr multilayers). In these cases, important contributions can be expected from spin-resolved photoemission studies in identifying the relevant branches of the Fermi surface and the electronic states which determine the oscillation periods [34].

6.5. COMPOUND MATERIALS

The examples described in the previous sections validate the importance of spectroscopic techniques for studying ultra-thin magnetic films. They allow to identify the microscopic mechanisms that determine the film's macroscopic magnetic behavior. However, the range of magnetic phenomena is enormous and the application of spectroscopy in this field is still quite new. For example, apart from multilayers, other materials of great technological interest display strong links between electric transport phenomena and magnetic properties. Unfortunately, these systems are complex, such as colossal magneto-resistance (CMR) oxides or ternary Heusler alloys. Experimental information on their electronic structure therefore is extremely scarce. The main difficulty lies in obtaining high-quality thin films of these alloys and compounds. The difficulty is even greater in preparing compositionally and structurally well-defined surfaces of these materials, a necessary step for undertaking surface-sensitive photoemission experiments.

In this section, I illustrate this sort of problem, but to limit the scope of discussion, I focus on a particular goal of the current research in magnetic thin films by considering materials useful for preparing a magnetic tunnel junction (MTJ) [35]. An MTJ is a magneto-electronic device in which a tunneling current between two magnetic materials separated by a ultra-thin film of insulating material (the tunneling barrier) is governed by the layers' relative magnetization. Should such a device become available, it would constitute a key element in producing magnetic memories. Hence, much research is currently devoted to this subject. Here, my consideration is limited to idealized MTJs, looking at the problem only from aspects of the electronic structure, and neglecting practical considerations in constructing the device.

In its simplest form, an MTJ is a three-layered structure in which two ferromagnetic metallic layers are separated by a very thin film (typically < 5 monolayers) of highly insulating material. This insulating layer act as a barrier between the metallic layers and, assuming that the layer is perfectly continuous, any conduction through the junction must happen by quantum tunneling. In a ferromagnet, there is an imbalance between spin-up and spin-down electrons at the Fermi level, so that quantum tunneling depends on the relative alignment of magnetization of the two ferromagnets. The relative difference in conductivity between parallel and anti-parallel configurations is essentially proportional to the number of electrical carriers available for hopping between the two ferromagnets (i.e., the density of states at the Fermi level) and to their spin-polarization. In other words. the magneto-resistivity of this device will be closely linked to the character of the electronic structure close to the Fermi level of the ferromagnetic materials. This problem is highly suited to spin-resolved photoemission investigations.

In principle, to realize an MTJ we might imagine using a strong $3d$ ferromagnet, like Co or Ni, in which the majority of the spin $3d$ bands are completely filled and only minority electrons are left at the Fermi level. Indeed, early work followed this direction until very small measured values of magneto-resistance forced researchers to realize that this approach does not work in practice [36]. In a simple $3d$ ferromagnet, it is the localized $3d$ electrons that carry the magnetic information, while the electrical conductivity is mostly due to the delocalized sp-derived bands.

As discussed previously, the *sp* bands have a much lower exchange split and, therefore, are present in nearly the same numbers with spin-up and spin-down. When two 3*d* ferromagnets are separated by an oxide barrier, the tunneling current is carried by these weakly polarized *sp* electrons; therefore, it is quite insensitive to the magnetizations of the ferromagnets.

To practically generate an efficient MTJ, we must have a system in which the 3*d* electrons, the spin-polarized electrons, are also the ones conducting electricity. This is possible only by considering a more complex electronic structure than the one of elemental 3*d* ferromagnets. In particular, it would be desirable to have a material in which the 3*d* TM ions interact strongly with other atomic species so as to loose their *sp* electrons in the bond, leaving there only the polarized 3*d* electrons close to the Fermi level. In principle, this situation can be attained with appropriate compound materials; typically, in transition metal-oxide systems. In simple terms, oxygen is extremely electronegative and, in bonding with the 3*d* metal atoms, acquires charge from the metal atoms, most likely from the *sp* electrons which can easily hybridize with the O 2*p* levels. In this way, the *sp* electrons of the TM are effectively transferred to the oxygen 2*p* bands and consequently, move toward higher binding energies away from the Fermi level. The density of states close to the Fermi level of a TM oxide contains mostly 3*d* electrons.

Not unexpectedly, using a TM-oxide to replace the simple metal in an MTJ introduces quite serious disadvantages. First, most magnetic TM-oxides order antiferromagnetically (due to the super-exchange interaction via the oxygen atoms [37]) and therefore, are not appropriate for MTJ applications. Furthermore, many are good insulators, which prevents their use for MTJ. But even considering the few cases of ferromagnetic metallic TM-oxides, the 3*d* electrons are much poorer electrical carriers compared with *sp* electrons. This means that these materials are poor metals characterized by low conductivity values. To compensate for this serious problem, an effective MTJ can be built from TM-oxides systems only if the spin polarization of the 3*d* electrons left at the Fermi level is very high; ideally, 100 %. This situation is realized in particular systems called half-metallic ferromagnets. These have only one type of spin present at the Fermi level (and consequently, with a gap across E_F for the other spin).

The next section will explore various attempts at constructing such a system using TM metal oxides.

6.5.1 Magnetic Transition Metal Oxides

Considering metallic magnetic TM-oxides with high spin polarization at the Fermi level, Fe_3O_4, the common magnetite material, would seem to be an excellent candidate for MTJ applications. Fe_3O_4 is a metallic oxide (albeit with a small conductivity) that orders ferrimagnetically and has a Curie temperature (T_c=848 K) well above room temperature. Furthermore, its Fe atoms are in two different oxidation states, Fe^{3+} with atomic configuration 3d^5 and Fe^{2+} with atomic configuration 3d^6. The five *d*-electrons in the Fe^{3+} ions are aligned parallel, while on Fe^{2+} sites five electrons have parallel spin and the sixth is aligned anti-parallel. Thus, the electron with lower binding energy is the anti-parallel one from the Fe^{2+}. Accordingly, we can expect Fe_3O_4 to display a highly negative spin polarization close to the Fermi level.

Until recently, spectroscopic studies of TMO were limited to bulk properties: the problem of preparing thin films of TMO with well-defined stoichiometry (and even more so, their surfaces) seemed unsolvable. However, recent work stabilized a few of these complex structures in the form of ultra-thin films by epitaxial growth on appropriate crystalline substrates. Particularly, thin films of Fe oxides were grown on different substrates, using different techniques: FeO was reported to grow on Pt(111) [38], and on Fe(110) [39]; Fe_3O_4 was grown on Pt(111) [40] and on Fe(110) [41] as well as on MgO(100) and NiO [42]; and, thin films of α- Fe_2O_3 and γ- Fe_2O_3 were stabilized on Al_2O_3(0001) and MgO(100), respectively [43].

This progress is remarkable, considering the complex structures of TMO and opens new possibilities for investigating their electronic properties. This is particularly true for ferromagnetic TMO, which, as a bulk material, was essentially out of reach of the most valuable magnetic spectroscopies such as spin-resolved photoemission and spin-resolved inverse photoemission.

We recently grew *in-situ* epitaxial Fe oxide films starting from the clean Fe(110) surface [44]. This was done by simply exposing the clean Fe(110) surface to oxygen while annealing the film (see, Fig.6-26). The oxidation reaction proceeds through two distinct phases. At low oxygen levels, a paramagnetic FeO(111) layer is formed, followed by a stable Fe_3O_4(111) layer obtained at high oxygen doses (>600 L). In the resultant Fe_3O_4(111)/Fe(110) bilayer, the magnetizations of the two materials are coupled anti-parallel. The LEED picture in Fig.6-26 demonstrates the existence of these two different phases.

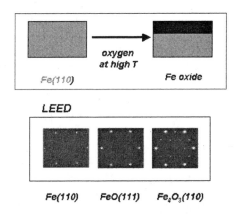

Figure 6-26: Well defined Fe oxides films can be formed by exposing the clean Fe(110) surface to oxygen atmosphere (p ~ 5 10^{-7} torr) while maintaining the sample to 250 °C. The LEED patterns during the oxidation process are shown in the bottom panel. Left: pure Fe(110); middle: after 300 L oxygen exposure; right: after 300 L oxygen exposure.

The Fe(110) surface displays a rectangular LEED pattern (bottom left panel). After a first exposure to 300 L of oxygen, the LEED is modified into an hexagonal pattern (middle panel), characteristic of the formation of an FeO(111) surface. Additional exposures to 300L of oxygen, modify the pattern a second time by adding a 2x2 reconstruction to the hexagonal pattern (right panel). This second hexagonal LEED pattern is the one expected for an unreconstructed Fe_3O_4(111) surface [38]. Furthermore, independent absorption experiments with circular polarized light unequivocally show that the oxide films obtained at high oxygen exposures are Fe_3O_4 (111) [45].

The presence of two distinct phases can be clearly distinguished in the photoemission data. Representative valence band spectra are depicted in Fig.6-27 as a function of oxygen exposure. The spin-summed valence-band photoemission spectra are shown on the left panel, while the corresponding spin-resolved spectra are on the right. The Fe(110) spectrum (top-left) shows the 3*d* emission, extended over the first 3 eV below the Fermi level with a small feature at about 8 eV, due to the $4s$ states. The first oxidation stage is characterized by the development of the O $2p$-derived emission at about 6 eV below E_F, accompanied by a smooth decrease in the intensity of the Fe $3d$ states. The electronic structure is markedly modified above 600 L (second stage), as denoted by the appearance of a new feature in the spectra at about 2 eV binding energy.

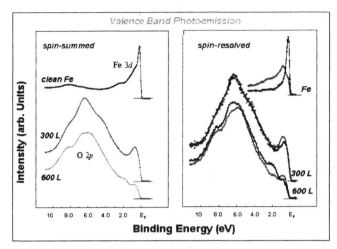

Figure 6-27: Spin-summed (left-panel) and spin-resolved (right-panel) valence band photoemission spectra taken in normal emission with 40 eV photon energy from the clean Fe surface (top spectra) and after exposure to 300 L and 600 L oxygen.

The two reaction stages are even more apparent in the spin-resolved spectra that contain information about the magnetic ordering and the spin character of the electronic states. The Fe spectrum (top-right) displays various features, which, as shown above, can be interpreted in terms of the Fe bulk bands. Initially, exposure to oxygen effectively reduces the spin polarization. Essentially, the entire spectrum is not polarized at 300 L. The small polarization near E_F is probably due to a residual contribution from the Fe metal underneath. The first Fe oxide is then not ferromagnetic, at least not on its surface. Surprisingly, an appreciable polarization reappears in the spin-resolved spectra above 600 L. In this second stage, the oxide overlayer, or at least its surface, has regained long-range magnetic order.

It is interesting to note that the reappearance of a spin-polarization is accompanied by a change of its sign: i.e., the Fe metal and the thick oxide overlayer have opposite magnetization. In other words, there is an anti-parallel magnetic coupling in the $Fe_3O_4(111)/Fe(110)$ bilayer.

Unfortunately, these results are not very promising for preparing Fe_3O_4 based MTJs. Firstly, the valence band data from the surface of the oxide films on Fe(110) are quite different from the ones expected from a simple bulk-terminated $Fe_3O_4(111)$ surface. Various groups have measured spin-integrated valence band spectra from bulk Fe_3O_4 single crystals, cleaved *in-situ* to expose the [110], or the [100] surface [46]. Despite the different surface orientations, the intensity at the Fermi level in all these photoemission spectra is consistently lower than in our films. A larger intensity at E_F seems to be an intrinsic feature of the films prepared on the Fe(110) surface and possibly indicates the formation of an oxygen-deficient $Fe_3O_4(111)$ surface. Further, the lack of pure $Fe_3O_4(111)$ stoichiometry on the surface of these films is apparent from the small polarization (16%) observed close to E_F. Alvarado et al. studying bulk $Fe_3O_4(111)$ by spin-polarized threshold spectroscopy found a polarization between 40 to 60% for the first state near E_F [47].

Note that the reduced polarization is confined only at the surface region. Absorption measurements shows that the bulk magnetization of the oxide overlayer is as high as 80% of the value for pure Fe_3O_4 [45] suggesting that the small surface polarization could partially be due to unequivalent terraces on the $Fe_3O_4(111)$ surface. Along the [111] direction, the Fe_3O_4 crystal consists of pure oxygen planes separated by pure Fe^{2+} or Fe^{3+} planes. Therefore, if different terraces of Fe atoms (and/or of oxygen planes in contact with different Fe planes) are exposed on the surface, the polarization measured by photoemission is effectively reduced due to the ferrimagnetic ordering of Fe_3O_4. This view is supported by recent STM images showing that on the (111) surface of bulk Fe_3O_4 samples two unequivalent Fe planes are present [48].

These considerations suggest that it would be worthwhile to look with photoemission on different crystallographic surfaces of Fe_3O_4 to see if there is an increase in spin-polarization. To this end, we considered thin films of $Fe_3O_4(100)$. Good quality films can be grown on MgO (100) substrates as was recently shown [42]. However, preparing this system involves exposure to atomic oxygen that cannot be performed *in-situ* in our experimental set-up. Therefore, these films were grown in a separate UHV apparatus and were shortly exposed to air during their transfer in the spin-resolved spectrometer. The damaged surface was subsequently "repaired" by a cleaning process, consisting of repeated annealing (in cycles of 5-10 minutes) to 250 ^{0}C in an oxygen atmosphere ($\sim 10^{-8}$ torr).

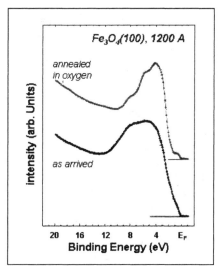

Figure 6-28: Normal emission valence band spectra from a film of $Fe_3O_4(100)$ epitaxially grown on an MgO(100) single crystal substrate.

Fig.6-28 compares the photoemission spectra recorded from the Fe_3O_4 film on MgO before ('as arrived') and after the annealing procedure. The initial spectrum is rather broad and structureless. In contrast, the annealed spectrum is very similar to the one measured from the (100) cleaved surface of Fe_3O_4 bulk single-crystal, suggesting that a good surface had formed. In particular, the spectra from the annealed Fe_3O_4 film and the cleaved surface exhibit the same intensity-ratio between the oxygen $2p$ structure and the Fermi energy;

Fig.6-29 shows the spin-polarized PES data taken from these films. The valence band photoemission spectrum in the first 12 eV binding energy region is depicted on the left side. As expected from a TM oxide system, the UV photoemission spectrum is dominated by the oxygen

$2p$-derived states extending from 2 to about 10 eV BE. These states are heavily hybridized with the Fe $3d$ states and form a complex series of multiplet lines, the most intense of which are directly detectable in the spectrum. A detailed analysis of these features, using resonant photoemission at the Fe $3p$ threshold, has been published [49]. The spin resolution shows that the various features have distinct degree of spin-polarization.

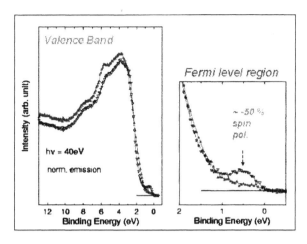

Figure 6-29: Spin-resolved valence band photoemission spectra from an $Fe_3O_4(100)$ film. On the right panel, the region close to the Fermi level is expanded to evidence the high negative spin polatization.

An expanded view of the region close to the top of the valence band is shown on the right side of Fig.6-29. Here, the spin polarization is much higher than that measured on Fe-oxide grown on the Fe(110) film. However, it is still smaller than the 100% spin-polarization suggested by the atomic considerations reported above.

Those consideration were somewhat too simple. Bagus analyzed quantitatively the localized atomic model and predicted a spin-polarization of about -67% [50]; this is the case because, basically, the photoelectron energy is determined by the final state energy of the solid left behind through the energy conservation law, and the energy of the final states of the solid depends on the total angular momentum quantum number. S. but is independent of its z-component S_z. Hence, for the Fe^{2+} state, if the emission at the Fermi level corresponds to the S=5/2 final state of the solid left behind, then its measured spin-polarization should be -2/3, as calculated by Bagus, since both minority and majority spin photoelectrons can combine with the S=5/2 solid to recover S=2 of the initial state. This situation is characteristic of the "localized" electronic states that preserve the coherence (correlation) among electrons.

In other words, data showing less than 100% spin polarization at E_F probably simply indicate the localized nature of Fe $3d$ electrons in the magnetite. This conclusion should also hold for transport property and spin-injection. Maybe we cannot expect 100% spin polarization for the so-called half-metallic samples if the number of d electrons is more than 5 and the electrons retain their localized character.

These considerations suggest looking at metallic TMO ferromagnets based on TM atoms in the first half of the d row, i.e., with less than 5 d electrons. One possibility would than to consider CMR materials, the complex oxides of Mn derived from the parent compound $LaMnO_3$ by partial substitution of La with Ca or Sr. The physical properties of these compounds are exceptional in many ways. For example, they display combined electrical- and magnetic-phase transitions such as insulator/paramagnet → metal/ferromagnet. Clearly, there is a very close relationship between transport and magnetic properties in these materials.

Zener gave a basic explanation of these complex magneto-transport phenomena many years ago, introducing the concept of double-exchange interaction [51]. Here, it is convenient to recall that in these compounds Mn is in the two oxidation states of Mn^{4+} with $3d^3$ atomic configuration, and Mn^{3+} with configuration $3d^4$. All the electrons on the Mn sites therefore are aligned parallel. Consequently, a 100% spin-polarization can be expected at the Fermi level.

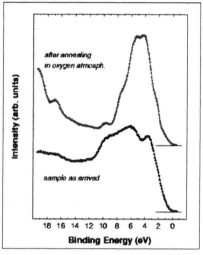

Figure 6-30: Valence band photoemission spectra from a 1900A film of $La_{0.7}Sr_{0.3}MnO_3$ epitaxially grown on a $SrTiO_3(100)$ single crystal substrate using pulsed deposition technique.

Once again thin films of this type of sample are very difficult to grow due to the many elements present. Therefore,the $La_{0.7}Sr_{0.3}MnO_3$ sample was grown in a separate chamber and briefly exposed to air in the transfer to the spin-polarized spectrometer, followed by cycles of mild annealing in oxygen atmosphere to restore a good surface [52]. Fig.6-30 compares the photoemission spectra taken before and after the "cure".

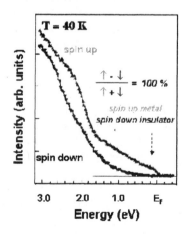

Figure 6-31: Spin-resolved photoemission spectrum from a film of $La_{0.7}Sr_{0.3}MnO_3$ in the region close to the Fermi level. The spectra is measured at 40 eV photon energy with the semple maintained at T= 40 K.

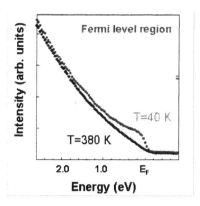

Figure 6-32: Comparison between photoemission spectra from $La_{0.7}Sr_{0.3}MnO_3$ in the region close to the Fermi level at to different temperatures. At T= 40 K the system is ferromagnetic (see Fig.6-31) and a conductor. At T= 380 K the system is a paramagnetic insulator.

The remarkable spin-resolved spectrum in the region close to the Fermi level is reported in Fig. 6-31. At the top of the valence band of this compound, the spin polarization is nearly 100%. Only spin-up electrons are present, while the intensity of the spin down ones is totally suppressed at E_F: this is the characteristic property of a half-metallic system.

Unfortunately, this compound is not suitable for technological application [53]. The Curie temperature of this system is barely above room temperature, so that close to ambient temperature the film will be very poorly magnetized. Furthermore, the magnetic-phase transition is linked to the metal insulator one, implying that the $LaSrMnO_3$ also will be nearly an insulating material at room temperatures. The metal insulator phase transition can be easily detected in photoemission (Fig.6-32).

6.6. Future and Perspectives

The examples described above demonstrate how using spectroscopic techniques allows to indentify the microscopic mechanisms that determine the magnetic behavior of thin films. However, the range of magnetic phenomena is enormous and it is certain that the application of spectroscopy in this field is in its infancy.

Indeed, there are important aspects of thin film magnetism that are practically unexplored microscopically. Although our comprehension of the microscopic aspects of magnetism, for example, the close relationship between microscopic structure and magnetic properties, has greatly improved recently, new advances can be expected: first, because new techniques are emerging which allow to investigate new aspects of magnetism, and second, because many developed methods are being constantly improved.

Considering synchrotron-radiation-based techniques, we can mention the recent advances in magnetic microscopy [3]. Here, dichroism effects in photoemission or in absorption are exploited to offer magnetic contrast; the result is the direct mapping of the element-resolved structure of magnetic domains. This has facilitated investigations of magnetic coupling at the interface between a ferromagnet and an antiferromagnet, providing important experimental information to understand exchanged-biased multilayers.

Furthermore, two different types of microscopies, PEEM (Photo-Emission Electron Microscopy), and LEEM (Low Energy Electron Microscopy), often can be combined at synchrotron sources to generate topographic-and magnetic-images of materials. There is a strong connection between the geometrical disposition of a material and its superimposed magnetic microstructures. Magnetic phenomena are characterized by long-range order, while domain size typically is of the order of microns, which is easily compatible the spatial resolution of these techniques. Also, crystal defects and terrace boundaries have long been recognized to strongly influence magnetic anisotropy, acting either as pinning boundaries to direct the motions of magnetic domains or as nucleation centers for domain formation. The ability to compare magnetic- with topographic-images from the same specimen allows quantitative assessments of this aspect of the problem. The large field of view of LEEM and PEEM will be essential to this task because of the long range of magnetic order (up to several microns).

Another emerging technique is related to the study of the dynamics of magnetic phenomena; in particular, the dynamics of magnetization reversal. Synchrotron radiation is an intrinsically pulsed light-source; in the more modern facilities, the pulse length is as short as 100 ps. Then, pump-probe experiments can be set up using dichroism effects, to probe the dynamic response down to at least the nanosecond time-scale [54]. This is a very interesting scale because, with current magnetic memories operating at GHz rate, their proper function depends critically upon these dynamical processes [55].

As discussed, the very high photon emissions available at third-generation synchrotron light sources have greatly extended the capabilities of more traditional techniques. As a dramatic example of enhanced sensitivity reached with X-rays dichroism techniques in absorption spectroscopy, I mention that element-specific orbital and spin moments have been measured for individual atomic chains or even for isolated clusters containing only three atoms [56].

Finally, I believe that spectroscopic studies have not yet contributed to their fullest potential in the field of magnetic anisotropies or of critical phenomena. Essentially, this is because variations in the magnetization states of a system are accompanied by very modest variation of the energy of the electronic states. However, the continuous improvements in the resolution of the instruments (both the light sources and the electron analyzer) are such that we are now very close to the required performance, and it is extremely likely that soon high-resolution spectroscopy studies will significantly advance the understanding of these microscopic aspects of magnetic phenomena.

REFERENCES

[1] B. Heinrich, J.A.C. Bland (Eds.), Ultrathin Magnetic Structures I,II, Springer-Verlag (1994)

[2] F. J. Himpsel, J. E. Ortega, G. J. Mankey and R. F. Willis, Advances in Phys. **47**, 511 (1998)

[3] J. B. Kortright, D. D. Awschalom, J. Stöhr, S. D. Bader, Y. U. Idzerda, I. K. Schuller, S. S. P. Parkin, and H. C. Siegmann, *J. Mag. Mag. Mat.* **207**, (1999) 7.

[4] S. Huefner, Photoelectron Spectroscopy, Springer-Verlag (1995)

[5] E.W. Plummer, W. Eberhardt, Adv. Chem. Phys. **49**, (1982) 533.

[6] T. C. Chang, K. H. Frank, H. J. Himpsel, U. Karlsson, R. C. Leckey, W. D. Schneider, "Electronic Structure of Solids: Photoemission Spectra and Related Data", Landolt-Boernstein, New Series, Vol. III/23a, Springer-Verlag (1989).

[7] E. Kisker, C. Carbone in: S. Kevan (Ed.), Angle Resolved Photoemission, Elsevier, Amsterdam (1992)

[8] P. D. Johnson, Annu. Rev. Mater. Sci. **25**, (1995) 455.

[9] E. Vescovo, H.-J. Kim, Q.-Y. Dong, G. Nintzel, D. Carlson, S. L. Hulbert, N. V. Smith, Synchr. Rad. News **12**, (1999) 10.

[10] C. G. Burnett, T. J. Monroe, F. B. Dunning, Rev. Sci. Instrum. **65**, 1893 (1994)

[11] A. M. Turner, J.L.Erskine, Phys. Rev. **B25**, (1982) 1983, ibid. Phys. Rev. **B29**, (1984) 2986.

[12] W. Pepperhoff, M. Acet, "Constitution and Magnetism of Fe and its Alloys", Springer-Verlag (1995)

[13] A. M. Turner, Yu-Jeng Chang and J. L. Erskine, Phys. Rev. Lett. **48**, (1982) 348.

[14] P. J. Berlowitz, J. W. He and D. W. Goodman, Surf. Sci. **231**, (1990) 315; T.J. Yates, ibid. **232**, (1990) 113.

[15] E. Vescovo, O. Rader and C. Carbone, Phys. Rev. **B47**, (1993) 13051.

[16] H.-J. Kim, E. Vescovo, S. Heinze, S. Bluegel, Surface Science **478**, (2001) 193.

[17] D. G. Dempsey, L. Kleinman, E. Caruthers, Phy. Rev. **B13**, (1976) 1498.

[18] D. M. Schaller, D. E. Brugler, C. M. Schmidt, F. Meisinger, and H.-J. Guentherodt, Phys. Rev. **B59**, (1999) 14516.

[19] W. Weber, C. H. Back, A. Bischof, Ch. Wuersch, and R. Allenspach, Phys. Rev. Lett. **76**, (1996) 1940.

[20] A. Dittschar, M. Zharnikov, W. Kuch, M.-T. Lin, C. M. Schneider, and J. Krischner, Phys. Rev. **B57**, (1998) R3209.

[21] H. J. Elmers and U. Gradmann, Appl. Phys. **A51**, (1990) 255.

[22] H. J. Elmers, T. Furubayashi, M. Albrecht and U. Gradmann, J. Appl. Phys. **70**, (1991) 5764.

[23] D. Sander, A. Enders, J. Kirschner, J. Magn. Magn. Mater. **200**, (1999) 439.

[24] I.-G.Baek, HG.Lee, H.-J.Kim,and E.Vescovo Phys. Rev. **B 67**, (2003) 75401.

[25] M. M. J. Bischoff, T Yamada, A.J. Quinn, R.G.P. van der Kraan, and H. van Kempen, Phys. rev. Lett. **87**, (2001) 246102.

[26] see for example G.A. Prinz, Physics Today, **48**, (1995) 59.

[27] P. Grunberg, R. Schreiber, Y. Pang, M. B. Brodsky and H. Sowers, Phys. Rev. Lett. **57**, (1986) 2442.

[28] J. Ortega and F. J. Himpsel, Phys. Rev. Lett. **69**, 844 (1992), Phys. Rev. **B47**, (1994) 16441.

[29] D. M. Edwards, J. Mathon, R. B. Muniz, M. S. Phan, Phys. Rev. Lett. **67**, (1991) 493; M. D. Stiles, Phys. Rev. **B48**, (1993) 7238; L. Nordstrom, P. Lang, R. Zeller and P. H. Dederichs, Europhys. Lett. **29**, (1995) 395.

[30] K. Garrison, Y. Chang, P.D. Johnson, Phys. Rev. Lett. **71**, (1993) 2801; C. Carbone, E. Vescovo, O. Rader, W. Gudat, W. Eberhardt, Phys. Rev. Lett. **71**, (1993) 2805

[31] S. S. P. Parkin, Phys. Rev. Lett. 67, 3598 (1991); Appl. Phys. Lett. **61**, (1992) 1358

[32] P. Bruno and C. Chappert, Phys. Rev. Lett. **67**, (1991) 1602, 2592.

[33] for a recent review of the photoemission work in this field see: T.C. Chang, Surf. Sci. Rep. **39**, (2000) 181.

[34] D. Li, J. Pearson, S. Bader, E. Vescovo, D.-J. Huang, P. D. Johnson, B. Heinrich, Phys. Rev. Lett. **78**, (1997) 1154.

[35] S. A. Wolf, D.D. Awschalom, R.A. Buhrman, J. M. Daughton, S. von Molnar, M. L. Roukes, A. Y. Chtchelkanova, D. M. Treger, Science **294**, (2001) 1488.

[36] M. B. Stearns, J. Magn. Magn. Mat. **5**, (1977) 167.

[37] C. Kittel, Introduction to Solid State Physics, John Wiley and Sons (1986)

[38] A. Barbieri, W. Weiss, M. A. Van Hove, G. A. Somorjai, Surface Science **302**, (1994) 259; C. S. Fadley, M. A. Van Hove, Z. Hussain, A. P. Kaduwela, J. Electr. Spectr. Relat. Phenom. **75**, (1995) 273.

[39] D. Cappus, M. Hassel, E. Neuhaus, M. Heber, F. Rohr, and H.-J. Freund, Surf. Sci. **337**, (1994) 268 ; K. Koike and T. Furukawa, Phys. Rev. Lett. **77**, (1996) 3921

[40] W. Weiss, A. Barbieri, M. A. Van Hove, G. A. Samorjai, Phys. Rev. Lett. **71**, (1993) 1848.

[41] V. S. Smentkowsy and J. T. Yates, Surf. Sci. **232**, (1990) 113.

[42] Y. J. Kim, Y. Gao, S. A. Chambers, Surf. Sci. **371**, (1997) 358.

[43] Y. Gao, Y. J. Kim, S. Thevuthasan, S. A. Chambers, P. Lubitz, J. Appl. Phys. **81**, (1997) 3253.

[44] H.-J. Kim, J.-H. Park and E. Vescovo, Phys. Rev. **B61**, (2000) 15288.

[45] H.-J. Kim, J.-H. Park and E. Vescovo, Phys. Rev. **B61**, (2000) 15284.

[46] A. Chainani, T. Yokoya, T. Morimoto, T. Takahashi, S. Todo, Phys. Rev. **B51**, (1995) 17976; J.-H. Park, L. H. Tjeng, J. W. Allen, P. Metcalf, C. T. Chen, Phys. Rev. **B55**, (1997) 1.

[47] S. F. Alvarado, W. Eib, F. Meier, D. T. Pierce, K. Sattler, H. C. Siegmann, and J. P. Remeika, Phys. Rev. Lett. **34**, (1975) 319; S. F. Alvarado, M. Erbudak, and P. Munz Phys. Rev. **B14**, (1976) 2740.

[48] A. R. Lennie, N. G. Condon, F. M. Leibsle, P. W. Murray, G. Thornton, D. J. Vaughan, Phys. Rev. **B53**, (1996) 10244; N. G. Condon, F. M. Leibsle, T. Parker, A. R. Lennie, D. J. Vaughan, G. Thornton, ibid., **B55**, (1997) 15885.

[49] R. J. Lad and V.E. Henrich, Phys. Rev. **B39**, (1989) 13478.

[50] S. F. Alvarado, and P.S. Bagus, Phys. Lett. **67A**, (1978) 397; P. S. Bagus, J. L. Freepuf, D. E. Eastman, Phys. Rev. **B15**, (1977) 3661.

[51] C. Zener, Phys. Rev. **82**, (1951) 403.

[52] J. H. Park, E. Vescovo, H.-J. Kim, C. Kwon, R. Ramesh and T. Venkatesan, Nature **392**, (1998) 794.

[53] J. H. Park, E. Vescovo, H.-J. Kim, C. Kwon, R. Ramesh and T. Venkatesan, Phys. Rev. Lett. **81**, (1998) 1953.

[54] M. Bonfim, G. Ghiringhelli, F. Montaigne, S. Pizzini, N. B. Brookes, F. Petroff, J. Vogel, J. Camarero, and A. Fontaine, *Phys. Rev. Lett.*, vol. **86**, 2002, pp. 3646-3649.

[55] W. Bailey, P. Kabos, F. Mancoff, and S. Russek, IEEE Trans. Magn. **37**, (2001) 1749.

[56] P. Gambardella et al., Science **300**, (2003) 1130.

Electron Scattering

CHAPTER 7

Magnetic phase imaging with transmission electron microscopy

7.1. INTRODUCTION

Intensive research has been carried out to understand magnetic structure of domain configuration, dynamic behavior, and coupling between magnetic building-blocks of magnetic bulk materials, thin films and artificially structured assemblies at the nanometric scale, reflecting closely the trend of the ever-decreasing bit-size of the magnetic constituents in recording and storage media as well as the rapid advances in spintronics devices [1,2]. This effort stimulates the continued development of magnetic imaging methods at different length-scales to bridge the gap between magnetic structure and properties. Widely used imaging techniques include magneto-optic Faraday and Kerr microscopy [3,4], x-ray magnetic circular dichroism [5,6], neutron topography [7,8], magnetic force microscopy [9,10] and electron-optical method [11]. Among them, the electron-optical method, or electron microscopy, represents the widest family of the techniques for magnetic imaging. Today, the most advanced electron microscopy techniques can provide extremely high spatial resolution of the order of 1 nm and display high contrast and sensitivity to small changes in the local magnetization of a sample. The magnetic images observed in various modes of electron microscopy are often explained by the classical particle-like picture of electrons through the Lorentz force acting upon an electron in a magnetic field, i.e., the trajectory displacements of the electrons yield magnetic contrast revealing local magnetization. Nevertheless, in many cases correct interpretations have to be based on the quantum wave-like picture of electrons. It is the phase change, or phase shift, of the electron wave passing through the vector potential generated by the magnetic fields in the sample that gives rise to a detectable magnetic signals.

The electron-optical family can be mainly divided into two branches: transmission electron microscopy (TEM), and scanning electron microscopy (SEM). In the scanning electron microscope a finely focused electron beam is scanned across the sample and the secondary or backscattered electrons thus generated from the sample surface are collected by a detector. There are two different types of magnetic contrast in SEM, depending on the Lorentz force acting on either secondary electrons (type-I contrast) or backscattered electrons (type-II contrast). The secondary electron contrast is generated from the magnetic stray field existing above the surface of a sample, when the low energy (0-50 eV) secondary electrons ejected from the sample are deflected by the field [12-14]. The magnetic contrast increases with decreasing electron energy and is very sensitive to the magnetization at the probe position on the sample surface with respect of the detector. The backscattered electron contrast, on the other hand, is produced when high energy incident electrons (typically >30 kV) deflected by the magnetic flux density towards or away from the sample surface depending on local magnetization as they travel through it, give rise to different backscattering yields [15-17]. This method does not require the presence of stray fields, and therefore, is suitable for studying soft magnetic materials with low anisotropy. The domain contrast is usually low depending strongly on the angle of incidence; it varies typically from tenths of a percent for conventional SEM energies of 30 kV to one percent in a 200 kV instrument. The spatial resolution generally is no better than half of a micron. However, with high incident electron energy, a probing depth up to 15 μm may be achieved, allowing investigation of domain structure in the bulk.

A more powerful scanning technique that offers high spatial resolution and quantitative information of surface magnetization is scanning electron microscopy with polarization analysis [18-20] wherein spin-polarized secondary electrons excited by the probe are transported from the sample surface to a spin-polarized analyzer. The magnetization distribution of the sample surface is thus obtained point-by-point. The primary energy is chosen to obtain high resolution (>5 kV) and high secondary-electron yield (<50 kV). The crucial part of the instrument is the spin detector that usually has very low efficiency; therefore, long acquisition times are required to obtain a reasonable signal to noise ratio and statistics. Details of the technique and its applications have been given in Chapter 8. A similar technique utilizing electron-spin is the spin polarized low energy electron microscopy (SPLEEM) that adopts a parallel, rather than scanning acquisition mode [21-23]. It offers very high spatial resolution and efficient detection and is particularly attractive for simultaneously studying surface growth and magnetic structure. Readers are referred to Chapter 9 by Ernst Bauer for details.

Transmission electron microscopy (TEM), the other branch of electron microscopy, is the focus of this chapter. TEM has a long tradition of use in studying magnetic materials [24]. For example, Livingston in 1958 [25-26] explored the size, shape, and distribution of ferromagnetic Co precipitates in relation to their magnetic characteristics. In all the early studies, TEM was used merely as a tool for observing microstructure, with little work correlating magnetic properties with magnetic domain configurations. In early 1960s, Lorentz microscopy was developed and became the first and the most widely used TEM method to reveal magnetic structure [27-30]. Lorentz TEM includes Fresnel mode and Foucault mode. While the former gives rise to contrast along domain walls, the latter displays the domains themselves. The two modes are often considered as qualitative since they either involve defocusing the image-forming lens of the microscope or displacing aperture of the lens to generate magnetic contrast. The exact amount of the defocus and the accurate positions of the aperture often are difficult to measure and control.

Since the 1980s, other techniques were developed based on retrieving the phase of the exit electron wave to obtain quantitatively magnetization information, including differential phase contrast microscopy [31-34], off-axis electron holography [35-37], and TEM-based phase microscopy methods [38-40]. This development is partly due to the advances in instrumentation, especially that of the coherent electron source. These methods have been outstandingly successful for direct observations of magnetic structure in ferro-magnetic materials. The differential phase microscopy is an in-focus imaging technique using the scanning mode in TEM. A quadrant-split detector is placed in the diffraction plane of the microscope. The difference in the signals from different parts of the detector, which are proportional to the magnetic deflection of the beam, is analyzed to reveal the magnetic components of the sample [31]. In principle, the technique is insensitive to non-magnetic information since the symmetrically scattered signals are expected to be canceled out in the differential procedure. On the other hand, electron holography offers the opportunity to map local magnetization with a spatial resolution down to a few nanometers and sensitivity of about the order of $\pi/100$ [37]. Its main advantage is the quantitative capability allowed by phase-shift retrieval that provides information on the electrostatic and magnetostatic potential throughout the sample, as well as on the fringing field in the regions above and below the sample. In off-axis holography, an electron hologram is recorded with a coherent electron beam and an electron biprism, and then magnetic information is extracted by an optical (laser) or electronic (digital) image reconstruction process. While the technique has played a crucial role in nanoscale characterization, its applications are often hampered by the limited area of view (<0.5 μm in width), and the requirement for special hardware including a field-emission source and a bi-prism. The limitation may be lessened by the recent development of phase-retrieval method based on the Transport-of-Intensity (TIE) approach, originally proposed for light-optics [41]. We give detailed descriptions of the TEM-based technique, discussing its advantages and drawbacks in this chapter.

In the following, we first briefly review the image formation theory in TEM, then describe the magnetic imaging techniques using TEM, including Fresnel and Foucault methods, in-situ Lorentz microscopy, electron holography, and the phase retrieval method based on TIE. We discuss the electrostatic and magnetic phase shifts, and the separation of electrostatic potential and magnetostatic potential. We then focus on characterization of magnetic structure at nanoscale. We present theory and structure models for magnetic domains and domain-walls and compared them with experimental observations of $Nd_2Fe_{14}B$ hard magnets and artificially structured magnetic arrays of Ni, Co, and Permalloy films. Image simulation of magnetic contrast and induction distribution of magnetic nanoparticles with various geometries is also presented. Finally, we brief summarize the chapter on magnetic phase imaging and induction mapping.

7.2. IMAGE FORMATION IN TEM

To understand how to extract magnetic information from TEM experiments on thin samples, it is necessary to recall the main concepts of image formation. Only by knowing how the image recorded either on the CCD camera, or on the photographic plate, is actually formed, and what are the sources of the contrast features observed, is it possible to correlate the observations to the magnetic properties of the material.

In the electron microscope, certain physical properties of the object under study are transferred to an image that encodes the information in the form of contrast, or intensity modulation. To extract this information, we have to examine step by step the processes leading to the formation of the features of the images, and understand how to correlate the two. Several approaches describe the image formation process inside an electron-optical system. One of the most instructive, and most easily exploitable for theoretical simulations, is the approach reviewed by Lenz [42], which assimilates the microscope to an information channel. A source signal (electron beam) is propagated through the channel (electron microscope), modified by the channel structure (specimen, lenses and devices), and finally detected (photographic film or CCD camera).

7.2.1. Image-formation theory

The electron wave, coming from the electron source and traveling towards the specimen, passes through the upper part of the microscope column, containing electromagnetic fields. By properly tuning the lenses and devices, the electron beam can be treated as an ideal plane or spherical wave. Since the main ideas underlying image formation do not strictly depend on the wave geometry, we consider the simplest case of a plane illumination.

After passing through the specimen, the wave is modified in amplitude and phase, and can be described as a two-dimensional signal. The *input signal* $S_0(\mathbf{r}_0)$ contains the information we are interested in. The vector $\mathbf{r}_0$ represents a point in the object plane. The electron wave after the propagation in the lower part of the column, is called *output signal* $S_1(\mathbf{r}_1)$, ($\mathbf{r}_1$ is the vector representing a point in the image plane). Its intensity corresponds to the image recorded, which contains, in the form of contrast modulation, the information about the object investigated. To retrieve this information, we must establish a biunivocal connection between the output and input signal. In this way, each image feature can be interpreted correctly as the representation of some physical property of the object. For this purpose, some assumptions are necessary.

If the transmission system is free of noise, the output signal will depend only on the input signal. For the electron microscope, this means the instrument is operated in an ideal environment free of vibrations with a good quality data recording system.

Moreover, the connection between signals is simpler when a linear channel is considered. Thus, if S_1 is the response to S_0, and S_1' is the response to S_0', then $\alpha S_1 + \beta S_1'$ is the response to the signal $\alpha S_0 + \beta S_0'$, for arbitrary α and β. In electron microscopy, this linearity follows directly from the Schrödinger equation describing the electron waves. Under these assumptions, we can establish an image formation theory to describe a unique biunivocal relationship between the signals.

One function that can be used to describe the relation between input and output signals in a linear system is its *impulsive response* $G(\mathbf{r}_1, \mathbf{r}_0)$. It describes the response of the channel to a short pulse signal $S_0(\mathbf{r}_0) = \delta(\mathbf{r}_0 - \mathbf{r}_0')$. The properties of the Dirac-delta distribution allow us to decompose each arbitrary signal in terms of elementary pulses

$$S_0(\mathbf{r}_0) = \iint S_0(\mathbf{r}_0')\delta(\mathbf{r}_0 - \mathbf{r}_0')d\mathbf{r}_0' \tag{7.1}$$

Since we have assumed that G is the response of the linear system to a δ-signal, the output signal S_1 of an arbitrary input signal will be

$$S_1(\mathbf{r}_1) = \iint S_0(\mathbf{r}_0)G(\mathbf{r}_1, \mathbf{r}_0)d\mathbf{r}_0 \tag{7.2}$$

Formally, the connection between S_0 and S_1 is established, but the impulsive response function is still unknown. To find an expression for G, another important assumption is necessary. In the imaging system, a point-like signal is transformed in a disk of a certain radius due to the instrumentation limits. This can only be improved, and never avoided completely (it is linked to the point resolution of the microscope). Nevertheless, it is important that the image-disk shape does not depend on the position in the image plane. In other words, if we have two object points in $\mathbf{r}_0'$ and $\mathbf{r}_0''$, their image-disks can be displaced, but not modified. The desirable property of an imaging system in which all object points at $\mathbf{r}_0$ would produce an image disk of equal shape around the point $\mathbf{r}_1 = M\mathbf{r}_0$ (M is the magnification) in the image plane is called *isoplanacy* (see Fig.7-1). This property can be expressed by saying that the impulsive response is a function not of two separate vectors $\mathbf{r}_0$ and $\mathbf{r}_1$ but only of the difference: $G(\mathbf{r}_1, \mathbf{r}_0) = G(\mathbf{r}_1 - M\mathbf{r}_0)$.

The condition of isoplanacy is not precisely satisfied in optical and electron optical imaging systems. If the system has aberrations depending on $\mathbf{r}_0$ such as distortion, third-order

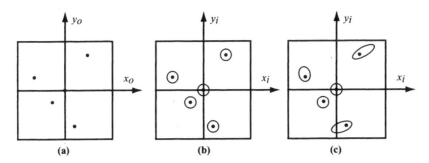

Figure 7-1: Condition of isoplanacy. (a) point sources in the object plane (input signal); (b) image-disks of the points (output signal, isoplanacy condition); (c) aberration-distorted image-disks (output signal, without isoplanacy). The distortion of the image-disks in (c) increases with the distance from the optical axis.

astigmatism, or coma, the isoplanacy condition is violated, i.e. the image-disk of an off-axis point looks different from an axis point. On the contrary, aberrations depending on the initial direction of an electron trajectory, such as spherical aberration, defocusing, axial astigmatism and axial coma, does not affect isoplanacy.

Under the isoplanacy condition, we can write eq.(7.1) as

$$S_1(\mathbf{r}_1) = \iint S_0(\mathbf{r}_0) G(\mathbf{r}_1 - M\mathbf{r}_0) d\mathbf{r}_0 \tag{7.3}$$

which becomes a convolution integral. Exploiting the convolution theorem, we can Fourier-transform the signals, and obtain a more manageable form of the input-output connection: $S_1(\mathbf{k}) = S_0(\mathbf{k}) \, T(\mathbf{k})$, where $S_0(\mathbf{k})$ and $S_1(\mathbf{k})$ are the signals spectrums (Fourier transforms), and the new function $T(\mathbf{k})$, the Fourier transform of the impulsive response, is called *transfer function*:

$$T(\mathbf{k}) = \iint G(\mathbf{r}) e^{-2\pi i \mathbf{k} \cdot \mathbf{r}} d\mathbf{r} \tag{7.4}$$

It follows that when the input spectrum is multiplied by the microscope transfer function, then it is inverse Fourier transformed and eventually becomes the output signal, whose intensity is the final image. Using the concept of the transfer function, the linear relation between the input and output signals can be described by the diagram in Fig.7-2.

If the transfer function, or the impulsive response of a system is known, the relation between S_0 and S_1 is uniquely defined, and one can calculate S_1 if S_0 is known, and vice versa. If, on the other hand, the relation between S_1 and S_0 were known empirically by taking a great number of micrographs of different objects with known properties, one would be able to determine the transfer function $T(\mathbf{k})$.

7.2.2. The transfer function

The transfer function, without loss of generality, can be written as

$$T(\mathbf{k}) = \frac{1}{M} B(\mathbf{k}) e^{-\frac{2i\pi}{\lambda} W(\mathbf{k})} e^{-g(\mathbf{k})} \tag{7.5}$$

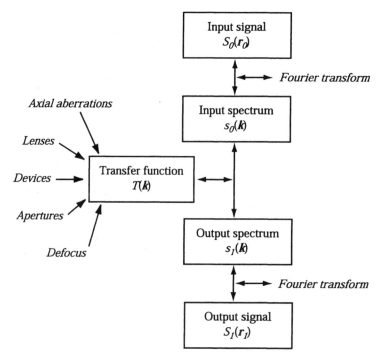

Figure 7-2: Scheme describing the image formation process for a linear system satisfying the isoplanatic condition. A biunivocal relation is established between input and output signal.

where the amplitude part, $B(\mathbf{k})$, includes the apertures intercepting part of beam, the magnification is reduced to a scaling factor M, the phase part $W(\mathbf{k})$ includes all the phase-shifting effects due to aberrations, defocusing and additional devices which may be present, like phase- plates or biprisms, and the damping part $g(\mathbf{k})$ is mostly due to the beam divergence.

The simplest component of $T(\mathbf{k})$ is the aperture function $B(\mathbf{k})$. The microscope may include many apertures intercepting part of the signal spectrum. Each aperture can be totally opaque, or partially transmissive, and with an arbitrary shape. A generic aperture can be described by means of step-like functions, and may also include phase-shifting apertures.

More difficult is to include the phase-shifting effects in $W(\mathbf{k})$. Aberrations can be interpreted as perturbative corrections of high order to the incident wave. Up to the fourth-order we have

$$W(\mathbf{k}) = \frac{C_s}{4}\lambda^4 k^4 + \frac{\Delta_z}{2}\lambda^2 k^2 - \frac{C_A}{2}\left(k_y^2 - k_x^2\right)\lambda^2 \qquad (7.6)$$

where C_s is the spherical aberration coefficient, C_A is the axial astigmatism (the off-axis astigmatism cannot be included within the isoplanacy condition), and Δ_z is the defocus parameter (see [43]).

Higher-order aberrations may be important in some particular cases. However, the main effects in our imaging set-up are due to defocus, associated with the Fresnel technique, and apertures, associated with the Foucault technique (see sections 3.2 and 3.4 for details). In fact, the spatial frequencies involved in the phase objects under investigation are so small that the terms of order higher than k^2 are not influential. Therefore, not only higher-order aberrations, but even C_s and C_A can be neglected safely, considering also that modern microscopes can be equipped with correctors for astigmatism and, more recently, with spherical aberration correctors [44-46].

The damping envelope $\exp[-g(\mathbf{k})]$ can be defined from

$$g(\mathbf{k}) = 2(\pi\theta_0\Delta_z)^2 k^2 \qquad (7.7)$$

where the beam divergence angle θ_0 is the angular half width of the normalized Gaussian intensity distribution describing the electron source. Small values of θ_0 describe more coherent electron sources. In fact, for $\theta_0=0$ the electron source can be considered as a delta function.

Once the transfer function is known, the relationship between input and output signal is fully established. What remains now is to specify the physical nature of the signals in transmission electron microscopy. The information is transferred through the electron microscope by means of electrons whose propagation in space can be described by a linear wave equation such as Schrödinger's (the condition of linearity is exactly fulfilled in this case). An obvious definition would be to identify the input signal with the object electron wave $\psi_o(\mathbf{r})$, and the output signal with the electron wave in the image plane $\psi_i(\mathbf{r})$.

7.2.3. The Schrödinger equation

The wave equation describing the motion of electrons inside electromagnetic fields can be deducted from the general Dirac equation (for a full treatment of the image formation theory by Dirac formalism, see ref.[47]), neglecting spin and fixing the reference frame of the laboratory [48]. The scalar equation obtained under these approximation, is a scalar wave equation, which is the relativistically corrected version of the time-independent Schrödinger equation:

$$\frac{1}{2m_e}(-i\hbar\nabla + e\mathbf{A})^2\psi(\mathbf{r}) = e[U^* + \gamma W(\mathbf{r})]\psi(\mathbf{r}) \qquad (7.8)$$

where m_e is the electron rest mass, $\mathbf{A}$ is the vector potential, $V(\mathbf{r})$ is the electrostatic potential, γ is the relativistic Lorentz factor, which in this case can be written as

$$\gamma = 1 + \frac{eU}{m_e c^2} \qquad (7.9)$$

and

$$U^* = U\left(1 + \frac{eU}{2m_e c^2}\right) = \frac{U}{2}(1+\gamma) \qquad (7.10)$$

is the relativistic corrected accelerating potential.

To find a convenient solution of eq.(7.8), describing the motion of electrons within the microscope, we must resort to an approximation, called *Eikonal approximation*, and treat the problem semi-classically. The solution we are looking for should be a wave function with an amplitude, a phase, and monochromatic ($E=\hbar\omega=eU+m_e c^2$) temporal dependence

$$\Psi(\mathbf{r},t) = a(\mathbf{r})e^{\frac{i}{\hbar}[S(\mathbf{r})-Et]} \qquad (7.11)$$

Introducing this wave function into the time-dependent form of eq.(7.8), two coupled equations for the phase $S(\mathbf{r})$ and the amplitude $a(\mathbf{r})$ are obtained

$$\nabla \cdot \left[a^2(\nabla S + e\mathbf{A})\right] = 0 \qquad (7.12)$$

$$(\nabla S + e\mathbf{A})^2 = g^2 + \hbar^2\frac{\nabla^2 a}{a} \qquad (7.13)$$

where $g^2(\mathbf{r})=2m_e[eU^*+\gamma V(\mathbf{r})]$ represents the classical kinetic momentum, relativistically corrected, related to the electron wavelength through the De Broglie relation $g=h/\lambda$.

The Eikonal approximation consists in neglecting the amplitude variation term of eq.(7.13),

$$\left|\frac{\nabla^2 a}{a}\right| << \frac{g^2}{\hbar^2} \tag{7.14}$$

to have it in a more manageable form

$$(\nabla S + e\mathbf{A})^2 = g^2 \tag{7.15}$$

The Eikonal approximation is generally applicable when the amplitude in spatial variation is small compared to the electron wavelength. Typically, for microscopic fields far from singularities, the approximation holds well. By contrast, in the proximity of borders, where diffraction phenomena appear, or near the beam focus, where rapid variations of amplitude are present, the approximation fails (Fig.7-3).

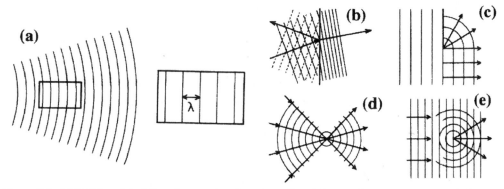

Figure 7-3: (a) Validity of the Eikonal approximation: the radius of the wavefront curvature must be larger than the wavelength. On the right: cases in which the Eikonal approximation breaks down: (b) reflection and refraction at surfaces, (c) diffraction at sharp edges, (d) high intensity in foci, and (e) scattering in atomic fields.

From Eikonal equation (7.15) an important variational principle can be deducted. In fact, it is possible to demonstrate that given two points in space P_1 and P_2, the integral

$$\int_{P_1}^{P_2} g ds - e\mathbf{A} \cdot d\mathbf{r} \tag{7.16}$$

has extremal values over the trajectory tangent to the vector $\mathbf{g} = \nabla S + e\mathbf{A}$.

This variational principle is equivalent to the Least Action principle of classical mechanics, from which all the Newtonian laws of motion can be derived: the vector $\mathbf{p}=\mathbf{g}-e\mathbf{A}$ is the canonical momentum and the field lines of $\mathbf{g}$ are equivalent to the classic motion trajectories. Recalling also the Fermat principle of light optics, an electron-optical refraction index given by

$$N = \sqrt{2m_e e[U^* + \gamma V(\mathbf{r})]} - e\mathbf{A} \cdot \frac{d\mathbf{r}}{ds} \tag{7.17}$$

can be introduced to establish a correspondence between light and electron optics.

However, different from the optical case, due to the presence of the vector potential, the electron trajectories cannot always be considered perpendicular to the wave front. Since the Schrödinger equation is gauge-invariant, and an arbitrary choice of the vector potential make it impossible to define uniquely the wavefront of electrons. This is neither a paradox, nor a problem, since observable quantities are not influenced by gauge choices. As the wavefront, or the absolute phase, is not a physical observable, a vague definition generally is acceptable.

7.2.4. Beam-object interaction

While in light optics the image contrast comes mainly from the absorption of photons by the specimen observed, in electron optics the situation is rather different. The high-energy of the beam allows us to neglect the absorption if the object is in the form of a thin film, and the main effect of the interaction is a phase modulation of the electron wave induced by the electromagnetic fields in the specimen, as shown in Fig.7-4.

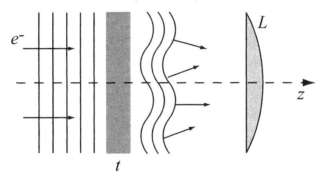

Figure 7-4: Modulation of a high-energy electron wave in presence of a phase object.

For high-energy beams (U $\geq$ 30 keV), the elastic forward scattering is predominant, and we can safely assume that the electron trajectories are not sensibly affected. Roughly, we can say that electrons travel straight through the specimen, undergoing only some small phase variations in their motion. However, as the electron absorption from the specimen is not completely negligible, we will keep an amplitude term in the description of the object wave. This is particularly useful in presence of thick regions of the specimen, where the amplitude has to be put to zero and the phase information is unavoidably lost.

Considering a monochromatic plane wave $\psi_K(z)$=exp($2\pi iKz$) (where $\hbar^2K^2=2m_eU^*$) impinging on the specimen, we can express the exit wave transmitted through the object as

$$\psi(\mathbf{r}) = \psi_K(z)\psi_o(x,y) = e^{2\pi iKz}a(x,y)e^{i\varphi(x,y)} \tag{7.18}$$

The object wave $\psi_o(x,y)$ represents the input signal of the microscope interpreted as an information channel (see section 7.2.1).

Inserting eq.(7.18) into the Schrödinger equation, following the concepts introduced in the previous section, and ignoring some terms which are negligible when the A_z component of the vector potential is present, we obtain from eq.(7.17) the solution for the phase shift in the high-energy approximation

$$\varphi(x,y) = \frac{\pi\gamma}{\lambda U^*} \int V(x,y,z)dz - \frac{e}{\hbar} \int A_z(x,y,z)dz \qquad (7.19)$$

where the electron wavelength is

$$\lambda = \frac{h}{\sqrt{2em_e U^*}} \qquad (7.20)$$

The physical (magnetic and electrostatic) information on the object is therefore encoded in a phase modulation of the electron wave. As the image intensity is proportional to the square modulus of the wave amplitude, it becomes necessary to find a way to transform the phase modulation into amplitude modulation, which is then recorded in some medium (photographic plate, imaging plate, or CCD camera). For this purpose, several phase-contrast and phase-retrieval techniques were developed; some are reviewed next.

7.3. MAGNETIC IMAGING BY TEM

In this section, we discuss magnetic imaging techniques using the transmission electron microscope. We first describe the microscope's basic components, the accessories of the instrument, and the major requirement for characterizing magnetic materials. We then review various imaging techniques ranging from the Lorentz method to phase microscopy. The pros and cons of the two phase-microscopy techniques, off-axis electron holography and the method based on TIE equation are discussed.

7.3.1. The electron microscope

The cross-section of a typical modern electron microscope in shown in Fig.7-5, based on the lens layout and detector assignment of the JEOL JEM3000F electron microscope at Brookhaven National Laboratory [49]. The instrument is a 300kV transmission electron microscope with scanning capabilities. It has a field emission source that generates a highly coherent electron beam with a brightness of the order of 10^9A/cm²str. Its first and second condenser lenses are designed to control the illumination, while the condenser mini-lens is to provide a strong pre-field to form a small probe.

The objective lens is the image formation lens which determines the image quality and the spatial resolution of the microscope. Nevertheless, since for magnetic imaging, the sample has to be in a magnetic-field-free environment, in most situations the objective lens is either switched off, or under weakly excited conditions, resulting in reduced magnification and spatial resolution. The goniometer provides 5 degrees of freedom for the sample (2 tilt and 3 positional). The intermediate and projector lenses, in imaging mode, serve as magnifying lenses. Traditionally, photographic films were used for recording images and diffraction patterns. Today, films are increasingly replaced by CCD cameras, or imaging plates for quantitative data acquisition.

The microscope has several critical attachments. The EDS (labeled as 1: energy dispersive x-ray spectroscopy) detector is designed for detecting electron-excited x-rays for chemical analysis. The BEI (2: back-scattered electron imaging) detector is for imaging using back-scattered electrons in scanning mode. The biprism (3) is biased to split the incident electron beam for

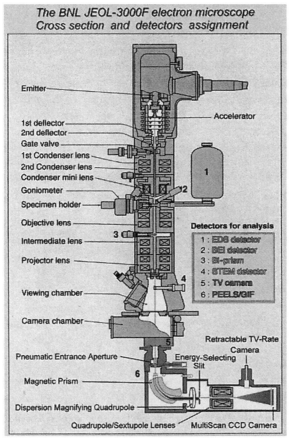

The BNL JEOL-3000F electron microscope Cross section and detectors assignment

Emitter

1st deflector
2nd deflector
Gate valve
1st Condenser lens
2nd Condenser lens
Condenser mini lens
Goniometer
Specimen holder

Objective lens
Intermediate lens
Projector lens

Viewing chamber

Camera chamber

Pneumatic Entrance Aperture
Magnetic Prism
Dispersion Magnifying Quadrupole
Quadrupole/Sextupole Lenses

Accelerator

Detectors for analysis
1 : EDS detector
2 : BEI detector
3 : Bi-prism
4 : STEM detector
5 : TV camera
6 : PEELS/GIF

Retractable TV-Rate Camera
Energy-Selecting Slit
MultiScan CCD Camera

Figure 7-5: Schematic view of the JEM3000F field-emission transmission electron microscope at Brookhaven National Laboratory.

off-axis electron holography, while the STEM detector (4) is for imaging in the scanning transmission mode. With a special designed annular dark-field (ADF) detector, the microscope can offer atomic-number sensitive (Z-contrast) imaging. Location 5 is nowadays often accommodates two cameras (TV-rate and CCD cameras) with a retractable mechanism. The TV camera provides analog signals; it is fast and suitable for in-situ experiments. The CCD camera records the data digitally, thus is suitable for quantitative imaging, but is slow in data transfer partially due to its large pixel size (up to 4096×4096 pixels). The CCD camera is extremely useful in performing real-time fast Fourier Transform (diffractogram) to analyze the observed images.

The most complex attachment is the Gatan Image Filter (GIF) system with Parallel Electron Energy-Loss Spectroscopy (PEELS) capabilities (labeled 6 in Fig.7-5). The system itself is an electron microscope with multiple quadrupole and sextupole lenses. It is powerful for energy filtering to eliminate unwanted electrons and to perform spectroscopy imaging and thickness mapping of the sample. It also provides additional 20× magnification that is useful for magnetic imaging when objective lens is turned off.

The entire microscope is an enclosed system and is evacuated to the range of $10^{-7} \sim 10^{-10}$ torr with the highest vacuum in the emitter and accelerator areas, and the lowest in the region of the sample holder and camera chamber. One of the keys for successful quantitative magnetic imaging is to ensure that the field at the sample area, which typically ranges from 200 G to 3 T depending on the lens setting, is well calibrated. There are two independent methods to measure the fields: 1) using the known magnetic response of a well-calibrated magnetic sensor (or sample), and 2) using a Hall probe. The significant advantage of the former is that it does not require the special designed TEM holder with a Hall-probe, or taking the microscope apart to place a commercial Hall-probe for the measurements [50]. The calibration curve of the magnetic field in the sample area of the JEM3000F microscope at Brookhaven was carefully calibrated. The field as a function of the objective lens (OL) potential with the two methods is shown in Fig.7-6. The circles represent the measurements using a calibrated $Nd_2Fe_{14}B$ sample and the squares are the measurements using a conventional Hall-probe (BELL610). The low-field regime suitable for magnetic imaging is plotted on the right. The linear fitting using $B=B_0+kU_{obj}$ for both methods gives $B_0=(0.15\pm0.12)$ kG and $k=(15.4\pm0.3)$ kG/V for the $Nd_2Fe_{14}B$ sensor and $B_0=(0.17\pm0.01)$ kG and $k=(13.56\pm0.03)$ kG/V for the Hall-probe.

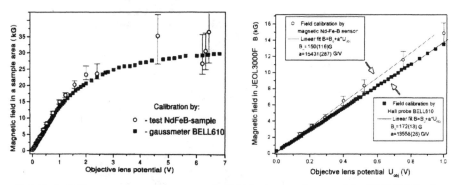

Figure 7-6: Left: Calibration curve of the magnetic field in the sample area of the JEM3000F microscope as a function of the OL potential measured by two independent methods (see text for details). Right: Low-field part of the calibration curve suitable for magnetic imaging and in-situ magnetization experiments.

7.3.2. Fresnel technique

The out-of-focus, or Fresnel, technique is the most known and utilized among the standard phase-contrast methods in Lorentz microscopy [51]. The basic idea of this technique is to observe the specimen in a plane different from the image plane to recover information about the phase shift induced by the object. A schematic view is depicted in Fig.7-7.

Following the image formation theory presented earlier, the effect of defocusing the image can be described by means of the phase term in the transfer function of the microscope: $T(\mathbf{k})=\exp(-i\pi\Delta_z\lambda k^2)$, where Δ_z is the distance between the image plane and the observation plane.

It is interesting in this case to find the impulse response function $G(\mathbf{r})$, which is the inverse Fourier transform of the transfer function. Since $T(\mathbf{k})$ is a complex Gaussian function, it is straightforward to find that also $G(\mathbf{r})$ has a similar form:

$$G(\mathbf{r}) = \frac{1}{\lambda\Delta_z}\exp\left[i\pi\frac{x^2+y^2}{\lambda\Delta_z}\right] \qquad (7.21)$$

apart from an unessential constant phase term.

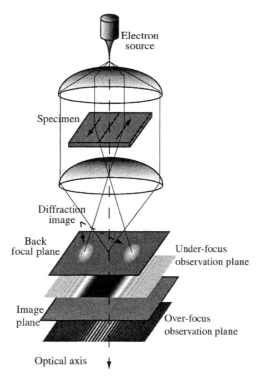

Electron
source

Specimen

Diffraction
image

Back
focal plane

Under-focus
observation plane

Image
plane

Over-focus
observation plane

Optical axis

Figure 7-7: Schematic view of the experimental setup for Fresnel (out-of-focus) observations. The magnetic specimen is imaged in some plane located over or under the image plane. Typical feature of this technique is the presence of Fresnel diffraction fringes.

The convolution theorem states that multiplying the input spectrum by the transfer function is equivalent convoluting the input signal with the impulse response function. For a generic input signal $\exp[i\varphi(x,y)]$ we obtain as a final image (square amplitude of the output signal)

$$I(x,y) = \frac{1}{\lambda^2 \Delta_z^2} \left| \iint \exp[i\varphi(x,y)] \exp\left[i\pi \frac{(\xi-x)^2 + (\eta-y)^2}{\lambda \Delta_z} \right] d\xi d\eta \right|^2 \qquad (7.22)$$

This convolution integral is well known in optics, and is called *Kirchhoff-Fresnel integral*. It describes the propagation of the phase modulated wavefront $\exp[i\varphi(x,y)]$ in free space in the paraxial approximation.

7.3.3. Defocus calibration

As it will become clearer in the following, an accurate knowledge of the defocus parameter is essential for obtaining quantitative results when employing the Fresnel technique. Despite the basic need for a controlled defocus, it is very difficult to obtain the values with high accuracy. Modern microscopes are not well calibrated in the low/medium magnification regimes that are used for magnetic imaging. The defocus values shown on the microscope control-screen are often inaccurate, especially at large defocus values, by up to *two or three* orders of magnitude ! This is mainly because the calibration is only performed for high-resolution imaging when the objective lens is on. The defocus curve, in general, is extrapolated from the defocus parameter of the order of a few nanometers. It is therefore necessary, before attempting any experiment by Fresnel technique, and even more so for TIE, to undertake preliminary experiments to obtain a calibration curve for the magnification range suitable for magnetic imaging.

Several methods were proposed to accomplish this task (see [52] and references therein). Here, we just review the most elementary one, namely the analysis of the Fresnel diffraction fringes by an opaque edge.

Following once again the image-formation scheme, we can model an opaque straight edge as a one-dimensional amplitude object, completely absorbing by one side and transparent on the other. In this case, the object wave can be simply expressed as

$$\psi_o(x) = H(x) = \begin{cases} 0 & x < 0 \\ 1 & x > 0 \end{cases} \tag{7.23}$$

i.e. a unit-step function (also called Heaviside function). In this case, it is instructive to follow the image formation scheme illustrated in section 2 from the beginning to the end. In fact, the diffraction from an opaque edge is one of the few examples in optics that can be solved analytically.

The starting point is to calculate the object spectrum $s_o(k)$ (see Fig.7-2; also note that we are trating a one-dimensional object, therefore the wave vector $\mathbf{k}$ becomes a scalar k), which we leave unevaluated as the generalized Fourier transform of the unit step function involves distribution theory, which is outside of the scope of this chapter

$$s_o(k) = \int_0^{+\infty} e^{-2\pi i k \xi} d\xi \tag{7.24}$$

Then, the object spectrum is multiplied by the transfer function

$$T(k) = e^{-\sigma k^2} \tag{7.25}$$

where we introduced a complex parameter σ which includes both the defocus and the beam divergence terms

$$\sigma = i\pi\lambda\Delta_z + 2(\pi\theta_0\Delta_z)^2 \tag{7.26}$$

The image spectrum will then be

$$s_i(k) = e^{-\sigma q^2} \int_0^{+\infty} e^{-2\pi i k \xi} d\xi \tag{7.27}$$

which can be inverse-Fourier transformed into real space obtaining the image wave

$$\psi_i(x) = \int_{-\infty}^{+\infty} dk\, e^{2\pi i k x} e^{-\sigma q^2} \int_0^{+\infty} e^{-2\pi i k \xi} d\xi \tag{7.28}$$

Changing the integration order, we can finally obtain

$$\psi_i(x) = \int_0^{+\infty} d\xi \int_0^{+\infty} e^{2\pi i k (x-\xi)} e^{-\sigma k^2} dk = \sqrt{\frac{\pi}{\sigma}} \int_0^{+\infty} d\xi \exp\left[-\frac{\pi^2(x-\xi)^2}{\sigma}\right] = \frac{1}{2}\left[1 + \mathrm{Erf}\left(\frac{\pi x}{\sigma}\right)\right] \tag{7.29}$$

and, for the image intensity we just take the square modulus of the image wave.

The same solution can be expressed, rather than by an error function of complex argument, by the Fresnel C and S functions. The only advantage of this would be that we can express directly the image intensity in explicit form. However, this can be done only for a perfectly coherent beam, therefore, we keep the given solution as is.

An example of defocus calibration is shown in Fig.7-8. The procedure is simple: from an out-of-focus image Fig.7-8(d) taken at suitable magnification, we extract a line scan. Then a series of simulations for different defocus parameter is calculated, and compared one-by-one with the experimental line scan of the Fresnel image (Fig.7-8(a-c)). Judging either by visual inspection, or more precisely by minimizing the χ-square of the simulation vs. the experiment, we can extract the correct value of the defocus. It is important to emphasize that even if the beam divergence of the beam is not well known, the calibration is not dramatically affected. In fact, as seen in the right side of Fig.7-8(a-c), the main effect of the coherence is to damp the highest order diffraction fringes, without affecting the spacing. Therefore, in this case, a good coherence will only slightly improves the statistics, but does not have a dramatic effect. The accuracy of this elementary method for defocus calibration is about 5%. In the case shown in Fig.7-8, the measurement yields a defocus value of 21 ± 1mm. Although a 5% uncertainty is not considered as highly accurate, it is a much better than the two or three orders of magnitude of uncertainty in the value obtained from a microscope console.

Performing a series of experiments under different electron optical conditions, we may finally arrive at a calibration curve. A small part of it, the one suitable for magnetic imaging by TEM (where the defocus ranges from 1 to 50 mm) is shown in (e).

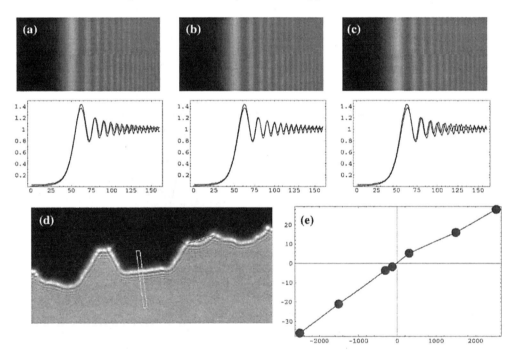

Figure 7-8: Comparison between simulated (above) and experimental (below) images (top row) and line scans (middle row) of Fresnel diffraction fringes. Defocus value: (a) 20 mm; (b) 21 mm; (c) 22 mm. (d) The low-magnification image of the sample. The boxed area indicates the region where the line scan was extracted. The result from the calibration procedure is the curve shown in (e), where the real defocus (line) vs. the defocus step (dots) in nm of the microscope is displayed. The microscope was operated at 297 kV, with a magnification of 2000×. The thickness of the spot is indicative of the 5% accuracy of the method.

7.3.4. Foucault technique

With the Foucault technique, the image contrast is provided by intercepting part of the diffraction pattern, usually near the forward direction due to the Lorentz diffraction. Positioning an opaque mask in the focal plane, where the input spectrum is formed, masks the information associated with special k vectors. The signal is then propagated to the image plane, deprived of some components, thus creating a visible effect in the final image.

The main advantage of the Foucault technique is to maintain the focus condition that allows us to correlate the object under observation with the image features. This technique is widely employed in observing magnetic domains [53,54] and sometimes used also in the observation of other phase objects such as p-n junctions [55] or superconducting vortices [56]. In fact, by intercepting the diffraction spot correspondent to a certain magnetization direction, all the regions of the specimen having such magnetization are clearly revealed as dark regions. In Fig.7-9, a simple experimental set-up for Foucaul observations is sketched.

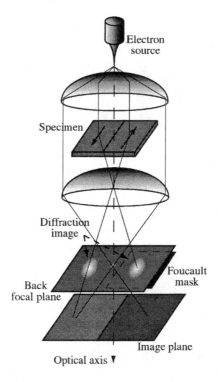

Figure 7-9: Schematic view of the experimental setup for Foucault observations. The magnetic specimen is imaged in-focus and the contrast is provided by intercepting part of the diffraction beam with a mask.

In recent studies [57-59], Chapman successfully applied a new phase-contrast imaging technique, the so-called Coherent Foucault method, to the observation of magnetic domains in electron microscopy. With this method the images of the domains are obtained by inserting a transparent film acting as phase plate into the back focal plane of the objective lens. Modern technology enables the film to be prepared by accurately controlling the material as well as its thickness to obtain a phase shift equal to π which is the value giving the best results [60]. Despite the promising results obtained by Chapman, and the theoretical possibility of applying the technique to observe superconducting fluxons [61], the method has not been widely used, mainly due to the additional hardware and the challenging experimental setup required.

7.3.5. In-situ magnetization

An advantage of TEM magnetic imaging is that microstructure and magnetic structure of a sample can be observed simultaneously, thus enabling us to directly examine the interaction between magnetic domain and structural defects to understand, for example, whether grain boundary pinning or reversal nucleation controls the coercivity strength of $Nd_2Fe_{14}B$ hard magnets, which are currently the most powerful permanent magnets with a wide range of practical applications from generators to motors and computer devices [62,63]. By in-situ magnetization and imaging we can directly view magnetic structural evolution, thus, clarify magnetic dynamic behaviors. Figure 7-10 shows an example of domain nucleation at a grain boundary in sintered Nd-Fe-B. The consecutive Lorentz images were captured from video (originally recorded at 30 frame/sec) showing in-situ irreversible remagnetization under a decreasing magnetic field from 4000 Gauss (image #1) to 500 Gauss (image #8). The images (#1-3) are Fresnel images showing difference in domain density of two neighboring grains due to the crystal anisotropy. With the decrease of the field, the domain density increases and reaches to an equilibrium equal-spacing domain-configuration in the right grain. The images (#4-8) are Focault images revealing the nucleation process. The enlarged images of the same domain area for the images # 5-8 are shown underneath, where the black and white arrows denote domains with opposite magnetization and GB indicates the grain boundary position. It is interesting to note that nucleation of the reversal domain at the grain boundaries is the dominant mechanism in the demagnetization process in the sintered magnet. The rapid process takes place usually at the grain boundary, where the nucleation of negative (reversal) domains occurs via several sudden splitting of positive domains [53]. The nucleation transforms a "positive" domain into a positive-negative-positive domain configuration, enclosing a pair of newly generated 180°-domain walls, followed by lateral expansion of the newly created domains. This nucleation process is essentially irreversible, and may be considered as a cascade-like discharge of over-saturation of magnetic charge (or poles) at the boundary. The formation of reversal domains through this process reduces the magnetostatic energy of the boundary and is energetically favorable process.

The magnetic field generated by the objective lens in a TEM are parallel to the electron beam. Nevertheless, with TEM we only observe the in-specimen-plane (perpendicular to the beam direction) component of the magnetization. Thus, it would be desirable to develop a specimen holder that can produce a well controlled in-plane magnetic field. Figure 7-11 shows a modified double-tilt holder for the JEM3000F that can provide a pure in-plane magnetic field of up to 25 G. It consists of a single solenoid, wherein the specimen is embedded in the geometric center. The relative low maximum field created at the sample is attributed to the limited space caused by the narrow gap of the high resolution pole piece of the microscope. Larger fields (up to 120 G) are generated when a magnetic core is used in the solenoid, however, this produces an unwanted remnant field and hysteresis problems. The applied field can be accurately calibrated by measuring the beam deflection in diffraction mode to ensure the high quality of in-situ magnetization experiments [64].

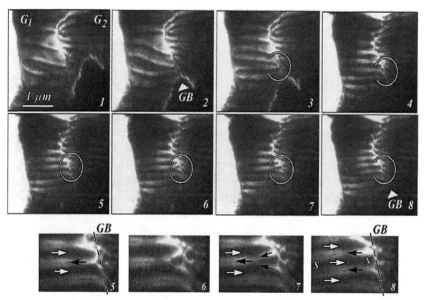

Figure 7-10: Successive Lorentz images excerpted from a video recorded at 30 frames per second showing *in-situ* irreversible remagnetization in a sintered Nd-Fe-B magnet under a decreasing magnetic field. The process takes place by the nucleation of new domains at the grain boundary (marked with arrow heads in image #2 and #8) via splitting of single domains of the same magnetization towards the grain interior, as shown in the enlarged images below (#5-8). The circle indicates the nucleation site at the grain boundary (GB) and the black and white arrows in the bottom panel denote domains with opposite magnetization.

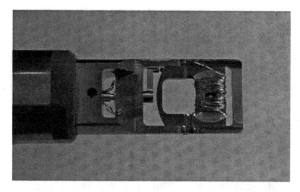

Figure 7-11: Single solenoid attachment that on a high-resolution double-tilt specimen holder allowing for the application of precise and well-calibrated in-specimen-plane magnetic fields.

An alternative way to apply a magnetic field in a TEM is to utilize the field of the objective lens through tilting the sample to project an in-plane magnetic component. Figure 7-12 gives an example of in-situ magnetization with a 3mm disk consisting of a periodic 2-D array of more than 10,000 Co-elements with a size and spacing of about 6µm [65]. To vary the in-plane magnetic fields we tilted the sample to obtain $H = H_o \sin\theta$ (where $-30° < \theta < 30°$). The set of Lorentz Fresnel images shown in Fig.7-12 corresponds a complete cycle of an in-situ re-magnetization process under a fixed lens excitation $H_o = 160$ Oe. To clearly illustrate the dynamic behavior of the evolution of the magnetic structure, only one element is shown, with inverted contrast for better visibility. The twelve images of the same element on the top panel

(Fig.7-12a) were recorded with applied fields starting at H = 70 Oe, and ending at H = -70 Oe. The corresponding value of H is indicated above or below of each image and its position on the magnetization (M) - field (H) curve is marked by the black dot in the upper part of the hysteresis loop. The reverse process is shown in the bottom set of twelve images (Fig.7-12b) from H = -70 Oe to H = 70 Oe, also with the corresponding half hysteresis loop on the right. In general, the structural evolution corresponding to the whole cycle of the re-magnetization process of the 6μm Co elements can be divided into four steps, according to the sign and amplitude of the applied fields: 1) H=70→14 Oe, a process of coarsening and deformation of the magnetic ripple structure (the local magnetization directions, marked by small arrows in Fig.7-12, are derived using the phase retrieval method to be discussed in Section 7.3.6); 2) H=14 → 0 Oe, a process dominated by the nucleation and rapid expansion of the reverse domains with motion of the domain walls and spin-rotation; 3) H=0 → -28 Oe, a process of expulsion of non-favorite domains towards the edges of the Co-element; and, 4) H=-41 → -70 Oe, a process of domain-wall annihilation at element edge and completion of the single domain structure. Here, we use Fig.7-12(a) as an example; the reverse process (Fig.7-12(b)) is very similar.

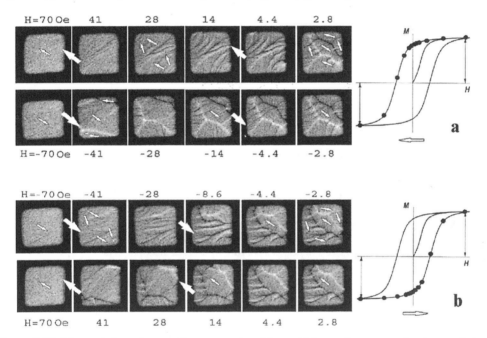

Figure 7-12: The evolution of magnetic structure in a 25-nm thickness 6μm across Co-element during magnetization and demagnetization. The sign and amplitude of the applied magnetic field are indicated above or below each image. The large and small arrows show the direction of applied field and local magnetization, respectively. The complete cycle of the corresponding hysteresis loop is shown on the right. The shape distortion of the square element at H=±70Oe is due to the projection of the sample at the large tilt.

The occurrence of these steps for individual Co elements is not necessarily identical. The switching fields for domain nucleation and annihilation for each element can differ significantly due to the pre-history and the existence of the structural defects of the element as well as the fringing fields including those from neighboring elements. Since these Co-elements are polycrystalline with random crystallographic orientations, the wavy-line ripple contrast is the dominant feature in the Fresnel images. This is largely because of the weak magneto-crystalline anisotropy effects determined by the crystal symmetry (a mixture of cubic and hexagonal), and the high demagnetizing effects determined by the element shape, or its aspect ratio. Unlike the 180° domain walls observed in highly anisotropic crystals, such as $Nd_2Fe_{14}B$, whose anti-parallel magnetization can be easily defined, the magnetization direction, not mention the magnitude, associated with faint ripple-contrast is hard to determine.

As we noted above, the magnetic structure and ripple contrast in polycrystalline films are quite complex. We have to emphasize that the above conclusions on domain evolution were reached from our detailed knowledge of local magnetic structure. Such structure can be revealed by phase microscopy methods rather than conventional Lorentz microscopy since the latter (both Frensel and Foucault) is limited to visualization of anti-parallel domains and their domain-wall position.

Figure 7-13 demonstrates the complexity of the magnetic structure associated with the Co-element shown in Fig.7-12. Figure 7-13(c) is phase image reconstructed from two slightly defocused images (one is shown in Fig.7-13(a)) using the so-called the Transport-of-intensity (TIE) algorithm (outlined in the following section 7.3.6). Based on the phase image we can derive a vector map which reveals detailed local projected induction distribution. The analysis allowed us to model the "zig-zag" ripple magnetization (Fig.7-13(b)) widely existing in the magnetic Co-islands [65].

7.3.6. Transport of Intensity

The Transport-of-Intensity (TIE) technique, as demonstrated in Fig.7-13, is strictly connected to the Fresnel technique. TIE was first proposed by Teague in 1983 for light optics [41]; and then developed for electron microscope by Paganin and Nugent [66]. The technique is now beginning to find its collocation among phase retrieval methods. As there is not yet a great deal of data available, it can be considered still a technique under development. New research results are opening interesting perspective [67]. As it will be shown in this section, and afterwards by direct comparison with electron holography, this technique has several advantages. In particular, the experimental setup is not demanding as that for holography.

The TIE equation relates the intensity change of the electron wave along the direction of propagation to the gradient of the phase in a plane normal to the propagation direction. The equation is expressed as

$$\nabla \cdot [I(\mathbf{r},0)\nabla \varphi(\mathbf{r})] = -\frac{2\pi}{\lambda} \frac{\partial I(\mathbf{r},z)}{\partial z}\bigg|_{z=0} \tag{7.30}$$

where $\varphi(\mathbf{r})$ is the phase of the image wave, λ the wavelength of the incident electrons, I the image intensity (function of $\mathbf{r}$ and of the defocus parameter), and ∇ the 2-dimensional nabla operator acting on $\mathbf{r}$. The equation can be derived either by following the original concepts of Paganin and Nugent [66], or, as done by De Graef [43], following the image formation scheme outlined in section 2.

Equation (7.30) is put to practical use by recognizing that the intensity derivative can be approximated by a finite difference of images taken at vanishingly small defocus distances, i.e.

$$\frac{\partial I(\mathbf{r},z)}{\partial z}\bigg|_{z=0} = \lim_{\Delta z \to 0} \frac{I(\mathbf{r},\Delta_z) - I(\mathbf{r},-\Delta_z)}{2\Delta_z} \tag{7.31}$$

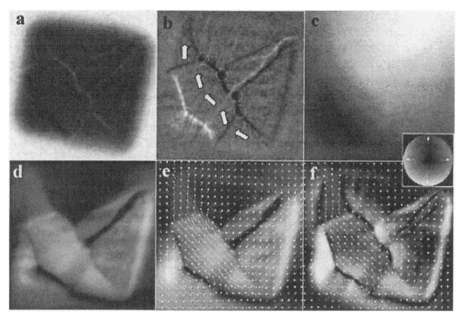

Figure 7-13: Analysis of Fresnel ripple contrast in Co-element by a phase microscopy approach (for details, see section 3.6): (a) experimental Fresnel image, (b) enhanced defocused image with "zig-zag" ripple magnetization (marked by arrows), (c) recovered phase map, (d) projected in-plane induction. The color background represents the direction and amplitude of the induction using the color wheel (inset) as a reference. (e) projected magnetization with a vector map, and (f) as (e) but after minimizing the thickness effect. *Also see the color plate.*

This approximation is valid if the defocus value is small enough compared to the scale of the object under investigation and of the intensity variation. A well defined method to insure that the defocus is "small enough" is not easy to establish because it depends on many parameters and on the experimental setup. A crude criterion, of general applicability, is to defocus the image to the smallest amount necessary to generate visible Fresnel contrast. The presence of diffraction fringes, though, means that the chosen value is too large, and would spoil the phase reconstruction. In fact, the small defocus limit corresponds to a geometrical-optics regime, where no interference or diffraction phenomena should be present. However, TIE remains fundamentally a wave-optical technique, even if the working regime resembles geometrical optics. The intensity recorded in the out-of-focus plane can be considered as a near-field Fresnel diffraction image (where near-field is not referred to the electron wavelength, but to the length scale of the Fresnel diffraction regime, which extends further beyond the small defocus limit), and it has indeed an interferometric fingerprint. Following the TIE derivation in [43], it can be shown that the out-of-focus image is an interference between the object wave and its Laplacian, which carries phase information. More details on the analysis of the small defocus limit which bounds the applicability of the TIE technique will be given in [68].

Once the intensity derivative is available, by differentiating the under- and over-focus images, the electron phase can be retrieved formally by the inverse of eq.(7.30), or

$$\varphi(\mathbf{r}) = -\frac{\pi}{\lambda \Delta_z} \nabla^{-2} \nabla \cdot \left[\frac{1}{I(\mathbf{r},0)} \nabla \{ \nabla^{-2} [I(\mathbf{r},\Delta_z) - I(\mathbf{r},-\Delta_z)] \} \right] \qquad (7.32)$$

where ∇^{-2} is the 2D inverse Laplacian operator. While eq.(7.32) appears complicated, it may readily be solved by Fourier techniques requiring input of only 3 images, i.e., the in-, over- and under-focus images. In principle, the in-focus image is not required, as the image derivative can be also approximated by the difference between over- and in-focus images. However, it can be used for intensity normalization to reduce the diffraction effect, and it helps to minimize the beam divergence effects, thus expanding the range of applicability of the technique also to non perfectly coherent illuminations.

The phase sensitivity of the technique is mainly limited by the visibility of Fresnel contrast. It is generally assumed that the lowest detectable contrast in a generic image is around 3%, where the contrast is defined as $(I_{max}-I_{min})/(I_{max}+I_{min})$. From this point of view, images should be recorded at the highest possible defocus, in order to have a higher signal/noise ratio, and higher contrast. On the other hand, the small defocus limit imposes an upper limit on the value of the defocus parameter, which, if crossed, undermine the derivation of the TIE equation which becomes incorrect. All considered, it is possible, for a given object, establish a range of possible defocus values where TIE equation is valid, and where the contrast is high enough to retrieve some information. Considering a phase object which varies in space in some length scale L (L may represent the periodicity of the object, if the object is periodic, or the half-widht of a Gaussian, if the object is Gaussian-like, or its decay length, if the object is exponential, and so forth), we can write the defocus limit as $\pi\lambda z/L^2 << 1$, thus establishing a maximum possible defocus for a given object $z_{max}=L^2/\pi\lambda$. Note that the *generalized defocus parameter* $\delta=\pi\lambda z/L^2$ is related to the *Fresnel number* which characterizes Fresnel diffraction [69].

The visibility requirement, which still depends on the scale length L (Fresnel contrast is linked to the phase second derivative), allows us to determine a z_{min}. For a generic object, a contrast higher than 3% is obtained for $z>z_{min}=0.03 \ L^2/(\pi\lambda\varphi_{max})$, where φ_{max} is the maximum phase shift induced on the electron beam. The previous expression is not of general validity and, depending on the particular object under investigation, it is possible to find deviations. For a generic object of a given spatial extent L, the visibility requirement becomes also a phase detection limit. Considering both the visibility and the small defocus requirements, and including also the influence of noise, it is possible to establish a working region for the application of the TIE procedure where the phase detection limit is univocally determined as a function of the experimental conditions. See again [68] for a more accurate analysis.

Figure 7-14 shows an application of the TIE method for assessing the magnetization of an α-Fe particle in $Nd_2Fe_{14}B$ magnet [70]. Figure7-14(a) and (b) are the under-focus and over-focus image of the particle, respectively, and Fig.7-14(c) is their difference image, or z-gradient. Based on eq.(7.32) the phase of the wave function was reconstructed (Fig.7-14(d)). The in-plane component of the magnetization B_x and B_y was then calculated by a 2-D gradient operation, as shown in a gray-scale map (Fig.7-14(e-f)) where the white contrast corresponds to positive x and y directions (x is the horizontal and y the vertical direction of the image), black to the opposite directions. Here, we assume a constant thickness, t, across the entire area. The determination of thickness is described in section 7.4.1. A closure configuration of four magnetic domains within the Fe particle can be revealed when Fig.7-14(e) and (f) are combined. An induction map B(x,y) with color codes, such as those shown in Fig.7-13((e-f), can also be generated [71].

The solution by Fourier techniques of the TIE equation raises several questions about boundary conditions. Whenever fast Fourier transform (FFT) is involved, periodic boundary conditions are assumed. However, a TEM image is hardly symmetric; thus, the phase shift reconstructed

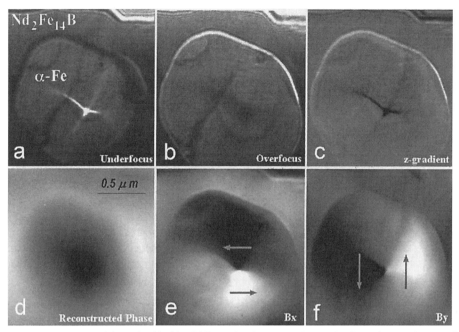

Figure 7-14: An α-Fe particle embedded in $Nd_2Fe_{14}B$ hard magnet. (a) Under-focus image, (b) over-focus image, (c) z-gradient image, (d) reconstructed phase retrieved using TIE method, (e) phase gradient along y direction, and, (f) phase gradient along x direction.

from the original image is strongly affected near the image boundaries. A thorough investigation of the effect of periodic boundary condition on the TIE-scheme for phase retrieval was made in [71]. The main result of this analysis led to the development of a *symmetrized solution* for phase retrieval by TIE. The concept is to reconstruct an artificial image created by a simple symmetrization rule as explained in Fig.7-15, rather than use the original image directly.

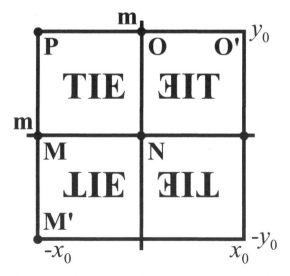

Figure 7-15: Scheme of the symmetrization procedure for a more accurate phase reconstruction. The top-left quadrant is the original.

7.3.7. Electron holography

Electron holography was invented by Gabor [72] as a spherical-aberration correction technique to improve the microscope resolution. A hologram is a superposition of the object wave, coming from the specimen and containing the physical information, and a coherent reference plane wave, coming from outside the specimen. The image is, in this case, an interference pattern containing the information on phase and amplitude of the object wave.

Off-axis electron holography requires a coherent electron source, such as provided by a field-emission electron gun and a biprism wire, which is an electron-optical device situated below the sample. The sample is positioned off the optic axis of the microscope so that incident plane-wave illumination passes partially through the vacuum (reference wave) and partially through the sample (object wave). The biprism is oriented such that the reference and object waves pass on opposite sides of the wire, being brought together by a positive bias applied to the biprism [73,74]. The resulting interference pattern (hologram) is recorded on a CCD camera, and the complex image wave (amplitude and phase) is mathematically reconstructed from the hologram intensity through the knowledge of the interfering reference wave, i.e., a plane-wave vacuum reference.

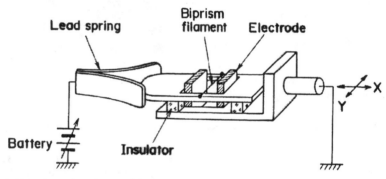

Figure 7-16: Side view of an electron biprism.

The electrostatic biprism, shown in Fig.7-16, is basically a biased conducting thin wire with a diameter of the order of 1μm (platinum or quartz coated with gold). In modern TEM, the biprism is a retractable assembly located approximately in the selected area aperture plan of the microscope. The wire is freely rotatable by ±90° and its positive or negative potential has a convergent or divergent effect on the incoming electron wave. If we suppose that the electrostatic field generated by the biprism wire of radius r_f is equivalent to the field of a cylindrical condenser of radius R equal to the distance between the wire and the earthed plates, the deflection induced on a electron traveling at a distance x_0 from the optic axis is given by

$$\alpha = \frac{\pi e V_B}{2E \log(r_f / R)} \text{Sign}(x_0) \tag{7.33}$$

Consequently, all the trajectories on one side of the biprism are deflected at the same angle, independently of the initial distance and incoming angle. From a wave-optics point of view, the biprism is characterized by the following transmission function in real space [75]:

$$T_B = \begin{cases} \exp[iCV_B|x|] & |x| \geq r_f \\ 0 & |x| < r_f \end{cases} \tag{7.34}$$

where the constant C is related to the energy of the incident electrons. A plane wave impinging on a biassed biprism, is therefore modified by introducing two symmetric linear terms on the two wavefront halfs passing through opposite sides of the filament.

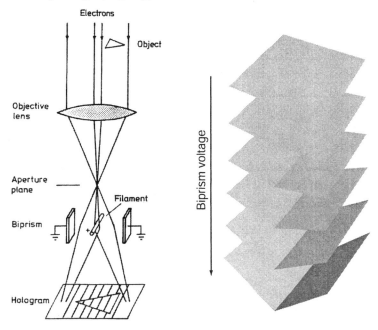

Figure 7-17: Left: schematic view of the formation of an electron hologram inside a microscope. Right: effect of the biprism on a plane electron wave, as a function of voltage.

In the ideal case of no aberrations, the electron beam overlaps, creating an interference pattern referred to the object plane (see Fig.7-17):

$$\psi(\mathbf{r}) = \psi_o\left(\mathbf{r} - \frac{\mathbf{D}}{2}\right)e^{i\pi\mathbf{F}\cdot\mathbf{r}} + \psi_o\left(\mathbf{r} + \frac{\mathbf{D}}{2}\right)e^{-i\pi\mathbf{F}\cdot\mathbf{r}} \tag{7.35}$$

where $\mathbf{D}$ represents a vector connecting the interference points and $\mathbf{F}$ is the carrier spatial frequency of the fringes (the fringe spacing is then $1/F$). $\psi_o(\mathbf{r}) = \exp[i\varphi(\mathbf{r})]$ is the object wave. The reference wave is in general the same object wave but at a different point $\mathbf{r} + \mathbf{D}/2$ of the image plane. Only if the object under investigation has no long-range fields can the reference wave be considered as a plane wave.

The hologram intensity recorded on the film is then

$$I(\mathbf{r}) = \left|\psi_o\left(\mathbf{r} - \frac{\mathbf{D}}{2}\right)e^{i\pi\mathbf{F}\cdot\mathbf{r}} + \psi_o\left(\mathbf{r} + \frac{\mathbf{D}}{2}\right)e^{-i\pi\mathbf{F}\cdot\mathbf{r}}\right|^2 = 2 + 2\cos(\Delta\varphi + 2\pi\mathbf{F}\cdot\mathbf{r}) \tag{7.36}$$

where

$$\Delta\varphi = \varphi\left(\mathbf{r} - \frac{\mathbf{D}}{2}\right) + \varphi\left(\mathbf{r} + \frac{\mathbf{D}}{2}\right) \tag{7.37}$$

This implies that the hologram records information relative to a fictious phase object $\exp[i\Delta\varphi]$, rather than the object phase itself. As a general guideline, during the experiments it is necessary to take particular care to have the reference wave as plane as possible. First of all, the reference wave should travel through vacuum regions, which means that the object in holography should sit close to a specimen edge. But, even in this case, in the presence of long-range fringing field protruding from the object, the retrieved phase will be affected. This is the problem of the *perturbed reference wave* which is as yet unfortunately unavoidable. No valid algorithm has been developed for deconvoluting the perturbed phase to get the unspoiled phase. For magnetic specimens, however, the dipolar nature of the magnetic field implies that fringing fields decay more rapidly than the $1/r^2$ Coulomb fields; therefore, from this point of view, the perturbed reference wave does not dramatically limit phase retrieval as may happen in electric-field observations.

An experimental hologram must be reconstructured to retrieve the phase of the electron wave. While during the early stages of electron holography the hologram was reconstructed by optical methods in a laser optical bench, today the situation is rather different. All modern microscopes are equipped with CCD cameras that allow digitally record images and holograms. Therefore, reconstructing the hologram to extract the information encoded in the interference pattern is now performed via digital techniques, in particular the FFT algorithm. So far, we have shown how the phase information is coded; here we briefly recall how to decode it.

Considering a generic electron wave, with amplitude and phase, as our object

$$\psi_o(\mathbf{r}) = a(\mathbf{r})e^{i\varphi(\mathbf{r})} \tag{7.38}$$

and for the sake of simplicity assuming that we record the hologram in focus and with an ideal electron-optical system (no aberrations or apertures), then the object wave is transferred without changes onto the image plane, where it interferes with the ideal plane-reference-wave (we also neglect the perturbed reference wave problem) $\psi_r(\mathbf{r})$=exp[$2\pi i \mathbf{F}.\mathbf{r}$]. Therefore, the recorded hologram can be written (see also eq.(7.36))

$$I(\mathbf{r}) = \left| a(\mathbf{r})e^{i\varphi(\mathbf{r})} + e^{2\pi \mathbf{F}\cdot\mathbf{r}} \right|^2 = 1 + a(\mathbf{r})^2 + 2a(\mathbf{r})\cos[\varphi(\mathbf{r}) + 2\pi\mathbf{F}\cdot\mathbf{r}] \tag{7.39}$$

Performing a Fourier transform of the hologram intensity, we obtain

$$I(\mathbf{k}) = \Im[I(\mathbf{r})] = \delta(\mathbf{k}) + \Im[a(\mathbf{r})^2] + \delta(\mathbf{k}+\mathbf{F})*\Im[a(\mathbf{r})e^{i\varphi(\mathbf{r})}] + \delta(\mathbf{k}-\mathbf{F})*\Im[a(\mathbf{r})e^{-i\varphi(\mathbf{r})}] \tag{7.40}$$

where * is the convolution operator. The hologram spectrum $I(\mathbf{k})$ is composed of three parts: the central autocorrelation signal, and the two sidebands, symmetrically located with respect to the origin .

If we now shift the spectrum, and define a new origin in the center of the sideband, then apply a circular mask to delete all the unwanted information (autocorrelation and second sideband), we are left only with, where $B(\mathbf{k})$ is an appropriate mask for the object (sharp, soft, exponential, Lorentzian, depending on the sideband features). Performing now an inverse Fourier transform, we can finally retrieve the wave object, in amplitude and phase:

$$\Im^{-1}[\Im[a(\mathbf{r})e^{\pm i\varphi(\mathbf{r})}]B(\mathbf{k})] = a(\mathbf{r})e^{\pm i\varphi(\mathbf{r})} * \Im^{-1}[B(\mathbf{k})] \approx a(\mathbf{r})e^{\pm i\varphi(\mathbf{r})} * \delta(\mathbf{r}) = a(\mathbf{r})e^{\pm i\varphi(\mathbf{r})} \tag{7.41}$$

where the inverse Fourier transform of the mask has been considered as a delta function due to the fact that a large circular aperture in Fourier space corresponds to a narrow circular aperture in real space, due to the duality between the two spaces. The convolution of a function with a very narrow Gaussian-type function is very similar to the function itself. The reconstruction procedure is summarized in Fig. 7-18, where the amplitude and phase of the electron wave are retrieved to measure the electrostatic potential of a (001) twist grain boundary in $Bi_2Sr_2CaCu_2O_8$ high-temperature superconductor [76]. Since the standard holography modes of the JEM3000F do not provide optimized holographic fringe contrast and area of view, we routinely use the free-lens mode for our experiments. For example, a reconstructed image wave with 3 Å resolution requires ~1 Å fringes, which gives only ~5-10% fringe contrast in the standard holography mode. Considerably better quality holograms were obtained by adjusting the three intermediate lenses and projector lens of the microscope. Phase and amplitude reconstruction from holograms was performed in the scripting language of DigitalMicrograph (used for image acquisition) with code written in-house, thereby conferring flexibility to customize application-specific data analysis.

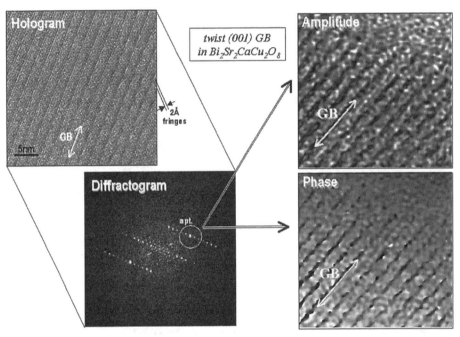

Figure 7-18: Scheme of the digital reconstruction for electron holograms. The recorded hologram is first Fourier-transformed to a diffractogram. Then the selected sideband, carrying the information about the object wave, is centered and selected by a circular aperture. Finally, after an inverse Fourier transform, the amplitude and phase of the wave function are retrieved.

7.3.8. Holography vs. TIE

Off-axis electron holography is a relatively established technique. To validate our TIE results, we made numerous comparisons on the same samples using both techniques. Figure 7-19 shows a square array of magnetic Permalloy elements ranging from 100-500 nm. Figure 7-19(a) is an in-focus Lorentz image of the array which mainly gives mass-thickness, or electrostatic contrast. The retrieved phase image and the phase contour map of the same area using the TIE method are shown in (b) and (c), respectively. Figure 7-19(d) is the induction map where color represents the amplitude and direction of the induction. It clearly indicates that the three small elements (B, C, and D) have a closure structure with 90° domain walls, i.e., a stable vortex state, while the large element A has a triple-domain configuration with nearly anti-parallel magnetization. In-situ experiments and calculations show that both the vortex state and triple-domain configuration are in low-energy states; the adoption of the final state depends on the element's aspect ratio (its thickness vs in-plan dimension). Fig.7-19 (e) is phase profile of the line scan across element B, as marked in Fig.7-19(b), similar to the schematics of the total phase shown in Fig.7-23. The reconstructed vector map of element B superimposed on the color code is shown in Fig.7-19(g). For comparison, we include analysis of using off-axis electron holography (Fig.7-19(g-j)) with the maximum area of view in our experimental setting that included only element B. Fig.7-19(g-h) shows the wrapped and unwrapped phase image of the element B, respectively. The reconstructed phase profile and the reconstructed induction distribution are shown in Fig.7-19(i) and (j). The results from the TIE agree well with those from holography, except that the data retrieved from the hologram are noisy. The TIE procedure imposes a low-pass filter that suppresses high-frequency noise in the images.

As shown in Fig.7-19, although the results from both methods agree well, each technique has its own advantages and drawbacks, as stated in the introduction. In general, TIE is more versatile in length scale. It is easy and fast to perform, and does not require special hardware. The speed of the process promises real-time phase-retrieval that is extremely important for studying magnetic dynamics. On the other hand, holography is quantitatively more robust, especially when a good quality reference wave from vacuum is available. Its main limitation is the small area of view (<0.5 μm) in the direction of the biprism wire, and the requirement for a nearby vacuum region for reference wave. Quantitative analysis for TIE can be very demanding; its quality strongly depends on the contrast difference at the periphery of the image [67,71], as well as the corrections made for the image shift, rotation, and change of magnification of the defocused image pairs used in the analysis.

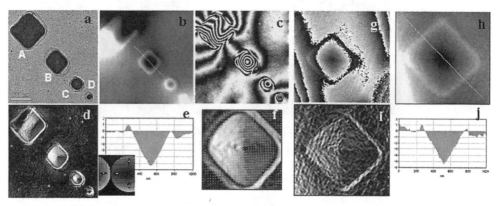

Figure 7-19: A lithographically patterned $Ni_{82}Fe_{18}$ Permalloy array. (a) In-focus Lorentz image of the array, (b-f) analysis based on the TIE method, (b) phase image, (c) phase contour map (cosine of the phase), (d) induction map, color represents the amplitude and direction of the induction using the color vector wheel (inset) as a reference, (e) phase profile of the line scan across the element B in (b), and (f) vector map of the element B, superimposed on the color map. (g-j) analysis of element B based on off-axis electron holography, (g) wrapped (h) unwrapped phase image, (i) color induction map, and (j) phase profile of the line scan across the element B in (h). *Also see the color plate.*

7.4. ELECTROSTATIC AND MAGNETIC PHASE SHIFTS

After reviewing the main TEM techniques for phase contrast and phase retrieval, it is necessary to describe more in detail the nature of the signal encoded in the electron wave. In the next section it will be shown how the local distribution of the magnetic field within nanoparticles and domains results in a phase shift. Aim of this short section is to clarify that i) part of the encoded signal is non-magnetic, and this contribution must be properly taken into account and ii) the signal is related to the *projected* magnetic field in the specimen, and a particular care is necessary when the magnetic structures under investigation has not a constant thickness (the most frequent case while dealing with nanoparticles).

Examining the expression for the phase shift based on the Aharonov-Bohm effect [77], eq.(7.19), we may note that it is composed of two parts, electrostatic and magnetic:

$$\varphi(x,y) = \varphi_e + \varphi_m = C_E \int V(x,y,z)dz - C_B \int A_z(x,y,z)dz \qquad (7.42)$$

where $C_E = \pi\gamma/\lambda U^*$ is the electrostatic constant [78], which depends on the beam accelerating potential ($C_E = 6.53$ V^{-1}μm^{-1} for 300 keV electrons), and $C_B = \pi/\phi_0$ is the magnetic constant, where ϕ_0 is the flux quantum $h/2e = 2.07 \times 10^3$ T nm^2, which does not depend on the electron energy. As we are interested in studying magnetic materials, the electrostatic part is here treated as an unwanted perturbation with respect to the magnetic signal we want to retrieve. However, even if no long-range electric fields are present, the electrostatic contribution cannot be neglected because of the mean inner potential of the material.

7.4.1. The mean inner potential

When the specimen is not charged, or biassed by applying an external potential, the only remaining electrostatic contribution is the averaged crystal potential of each atom. It is well known that the mean inner potential is related to the work function, as well as to the Fermi level of the material. The crystal potential of a material can be seen as the infinite sum of the contributions from each atomic site, each adding a $1/r$ potential from the nucleus, together with the potential from the electron cloud. In low magnification the resolution is not high enough to get information on the atomic-scale potentials. Therefore, the only contribution resulting in a measurable phase shift is the mean inner potential. In a very rough approximation, V_0 can be seen as the zero-order term of the Fourier-series expansion of the crystal potential. It can be estimated by known material parameters for each sample, such as scattering factors and unit cell volume

$$V_0 = \frac{1}{\Omega} \int_\Omega V(\mathbf{r})d^3\mathbf{r} \qquad (7.43)$$

where Ω is the unit cell volume of a crystalline material, or the whole body volume in a disordered solid. In principle then, the mean inner potential can be calculated for any material in which the crystal potential is known. However, the crystal potential is hardly ever known. Therefore, the mean inner potential must be estimated only in a very approximate form, and in general is a parameter which must be experimentally determined. The simplest approximation often used is the non-binding approximation, in which the atoms of the material are considered neutral. Assuming this treatment, the mean inner potential can be written as

$$V_0 = \frac{h^2}{2\pi m_e e\Omega} \sum_j f_j(0) \qquad (7.44)$$

where $f_j(0)$ are the atomic scattering amplitudes for electrons scattered in the forward direction.

Measuring mean inner potentials represents a challenge, in spite of the apparent simplicity of the problem. The importance of the mean inner potential determination for magnetic imaging by TEM is evident considering that the phase shift is always and unavoidably composed also of the electrostatic term. Especially for magnetic particle at the nanoscopic scale, the electrostatic contribution to the phase shift can mask completely the magnetic signal in which we are interested. Therefore, an accurate knowledge of the mean inner potential and thickness profile of the specimen is needed to separate the two contributions.

To evaluate the electrostatic phase shift φ_e, we integrate the mean inner potential over the film thickness. In general, we can write φ_e as

$$\varphi_e = C_E V_0 t(x, y) \tag{7.45}$$

where $t(x,y)$ is the projected thickness in the beam direction. For a flat specimen, $t(x,y)$ is reduced to the specimen constant thickness t, or to its effective thickness $t\cos\alpha$ if the specimen is tilted by an angle α with respect to the electron beam. In general, for specimens of non-constant thickness, such as spherical nanoparticles, or wedge samples, there is no way to further simplify the electrostatic contribution. The thickness information is therefore very important in disentangling the magnetic signal.

The projected thickness distribution of a sample can be retrieved by energy filtered imaging. Two images, a zero-loss filtered image collected with a narrow energy selecting slit width (e.g., 3 eV), and an unfiltered image, aligned by a standard cross correlation method from the same area, are needed to derive a thickness map according to [79]

$$t(x, y) = \lambda \log \frac{I(x, y)}{I_0(x, y)} \tag{7.46}$$

where λ is the inelastic mean free path (MFP) of the material and $I(x,y)$ and $I_0(x,y)$ are the unfiltered and filtered image intensities, respectively. If the MFP for the sample is known, a 2-D thickness map can then be obtained.

Another option for thickness mapping, widely employed in experiments when EELS setup (or an energy filter) is not available, is the specimen rotation, followed by image subtraction. The method is, in principle, very simple, but it suffers from the need of image centering and alignment which often results in induced artifacts. Considering a generic magnetic specimen, we may easily realize that the electrostatic contribution is not dependent on the beam direction while the magnetic part is. This means that recording two images after a 180 degree flip of the specimen along an axis perpendicular to the optical axis z, we record two images which contain $\varphi_e + \varphi_m$ and $\varphi_e - \varphi_m$. The separation between the two contributions is therefore obtained simply subtracting and adding the two retrieved phases. This method is only apparently simple, and relies on several assumptions which, unfortunately, are not always reasonable. The specimen magnetic field can be altered by the flipping procedure due to the remnant field of the objective lens and of the stray fields around. If the specimen holder does not allow a 180 degree flip, and the specimen needs to be extracted from the holder, we may also induce temperature variations, chemical reactions, and other effects. Moreover, as mentioned, it is often very challenging to perform the flip procedure ending up in exactly the same position under the beam. Alignment procedures are therefore very important in a correct separation, and especially for objects with very little contrast and very few contrast features to exploit as reference positions, this often results in a residual mixing between the two components.

7.4.2. Magnetic induction mapping

After the phase has been retrieved from experimental data, either through the use of electron holography or TIE, the magnetic signal is often displayed as *magnetic induction maps*. These maps are typically derived from the phase gradient [80]:

$$\nabla \varphi = \frac{\pi}{\phi_0}(\mathbf{B} \times \mathbf{n})t \qquad (7.47)$$

where $\mathbf{B}$ is the magnetic induction, $\mathbf{n}$ a unit vector parallel to the incident electron beam, and t the foil thickness. The magnetic induction components derived from this equation are (for a right handed cartesian reference frame with $\mathbf{n}=-\mathbf{e}_z$)

$$(B_x, B_y) = \frac{\phi_0}{\pi t}\left(-\frac{\partial \varphi}{\partial y}, \frac{\partial \varphi}{\partial x}\right) \qquad (7.48)$$

It is important to point out the conditions under which this equation may be used. The reconstructed phase φ is usually the total phase, which consists of both electrostatic (φ_e) and magnetic (φ_m) contributions. The gradient operation would give meaningless results on the electrostatic component of the phase, implying that φ_e must be constant. However, this is only possible if the sample thickness t is constant over the region of interest. Furthermore, since the electron-optical phase shift represents the integral of the vector potential along the electron trajectory, the phase bears only a simple relation to the magnetic induction components inside the particle if there is no fringing field. This means that eq.(7.48) may only be used if there are no fringing fields above or below the particle or foil *and* if the magnetic structure has constant thickness.

Consider the following numerical example [81]: a magnetic ring is constructed by taking 40 isosceles triangles with top angle $\pi/20$, height h, and magnetization $\mathbf{m}$, and subtracting from the resulting plate a slightly smaller plate, constructed from triangular plates of height $h-t$ and magnetization $-\mathbf{m}$. This ring has a square cross-section. Next, we compute the gradient of the phase for two different magnetization configurations: the vortex state (circular magnetization), and the uniform state. Figure 7-20(a-d) shows the resulting phase (a), the cosine of the phase (b), and the magnetic induction components B_x (c) and B_y (d). White represents an induction component along the positive $\mathbf{e}_{x,y}$ directions. The computed magnetic induction map for the vortex state agrees well with the input state. Equation (7.47) is applicable in this case since there are no fringing fields and the thickness is constant. For the uniformly magnetized state, with $\mathbf{m}=\mathbf{e}_x$, the phase (Fig.7-20) is not constant outside the ring, due to the fringing field which is taken into account in the simulations. The gradient operation on this phase function results in the magnetic induction maps (g) and (h). For B_x the agreement with the input phase is good only near the top and bottom of the ring, whereas the B_y component should vanish completely.

An experimental example of a quantitative induction mapping which does not suffer from the presence of fringing field or thickness variations is shown Fig.7-21. A polycrystalline Co island is imaged by the Fresnel technique. Black and white lines in correspondence of domain walls are clearly visible, together with four spots (two bright and two dark) where magnetization vortices are present. The magnetic element is in a multidomain configuration, with a very little total magnetization and fringing fields. Here we can safely assume that the contribution to the phase shift given by the external field surrounding the element is negligible. Therefore, assuming also that the magnetization is constant through the element thickness, we can extract a quantitative induction map of the element provided that we know the thickness and the defocus value with sufficient accuracy.

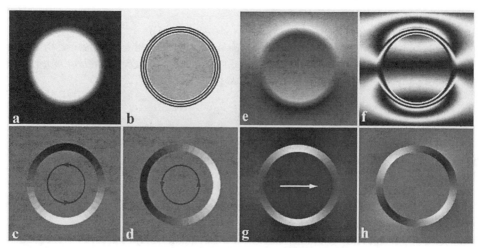

Figure 7-20: Phase simulation and magnetic induction components (following eq.(7.47)) for a double ring consisting of 40 isosceles triangles. The center ring has the opposite magnetization pattern, resulting in zero magnetization inside the ring. The counterclockwise vortex state is shown in the top half (a-d), and the uniformly magnetized state in the bottom half (e-h).

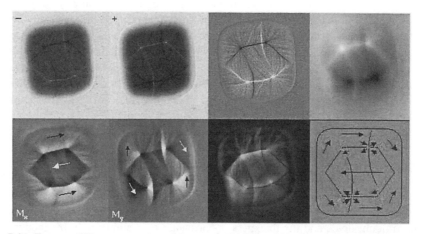

Figure 7-21: Top row, TIE reconstruction of the magnetic phase shift: (a) overfocus image, (b) underfocus image, (c) z-gradient of the intensity, (d) retrieved phase shift.. Second row, retrieved local projected induction maps: (e,f) Foucault images showing the main features of the magnetic field, (g) color map of the local induction, (f) scheme of the magnetic configuration of the magnetic element. *Also see the color plate.*

7.5. CHARACTERIZATION OF NANOSCALE MAGNETIC STRUCTURES

Characterization of nanoscale magnetic structures and properties of bulk materials including those with colossal magnetoresistance (CMR) effects [82] as well as artificially patterned films is an area of immense importance to technology and fundamental science. For example, the most powerful permanent magnet to date is $Nd_2Fe_{14}B$, whose energy product is still far below the theoretical predication, suggesting there is plenty of room for improvement with tailored magnetic structure. On the other hand, with the tremendous efforts on nanoscience, lithographically patterned magnetic films have been developed for various applications, including recording and spin-valve devices, and magnetic sensors. Optimizing the film properties critically depends on our understanding of magnetic structures that are sensitive to the specific geometry of elements when they diminish in size. Fig.7-22 shows patterned Co-film under a magnetic field of 28 Oe. The projected induction distribution (Fig.7-22(d)) derived from the defocused Lorentz image (Fig.7-22(a)) using TIE algorithm reveals fringing field and interaction among the four islands. The challenge we face is the quantification of the phase microscopy via comparing the experimental observation with the theoretical calculation and modeling.

In this section, we first describe the modeling and measurements of the magnetic domain wall and its width using $Nd_2Fe_{14}B$ bulk magnet as examples. We then focus on magnetic particles with common shapes and their associated magnetic structure. The magnetic configuration and induction distributions of these particles with various well-defined geometries are then compared with micromagnetic simulation.

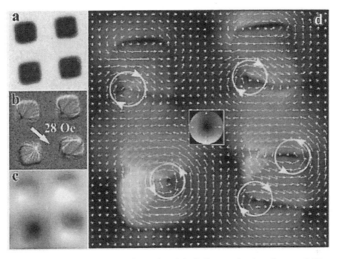

Figure 7-22: Patterned Co-islands of 6x6 μm^2. (a) defocused experimental Fresnel image, (b) z-gradient of the intensity, (c) reconstructed phase image, and, (d) projected magnetic-induction vector map quantitatively displayed both by the arrows and color-vector notation (inset). The circles in (d) highlight a magnetization curling around local vortices. *Also see the color plate.*

7.5.1. MAGNETIC DOMAINS

A fairly superficial interpretation of magnetic domains imaging in TEM does not require a deep mathematical or physical background. Considering a magnetic element of a certain size, the electrostatic contribution will be proportional to the projected local thickness of the element (see Fig.7-23). Each magnetic domain present in the specimen will induce a linear phase shift (a phase ramp) on the incident electron wave, proportional to the magnetic flux enclosed in the domain. A series of 180° domains, for example, will produce a sawtooth-like phase shift, as shown in Fig.7-23.

Referring to the TEM techniques that we have described, we can correlate the phase shift with electron holography and TIE, while Foucault and Fresnel techniques are related to phase derivatives. For example, the derivative of a sawtooth function is a square-wave, which clearly resembles the Foucault contrast (dark where the spot corresponding to a certain magnetization was intercepted, and bright elsewhere). The second derivative is closer to the typical features of Fresnel technique, where the contrast is in correspondence with the domain walls.

In Fig. 7-23 we use the classical scheme of interpretation, where magnetic contrast is linked to Lorentz forces and the electrons are deflected at a certain angle. Unfortunately, an interpretation based solely on deflections is often misleading, and cannot be employed, in particular when weak phase objects are involved, without incurring in fundamental problems as suggested by H. Lichte [83]. Nevertheless, the classical scheme may be useful for a first qualitative evaluation of the magnetic structure of the specimen, and does not require any knowledge of wave-optics or quantum mechanics. An example is shown in Fig.7-24.

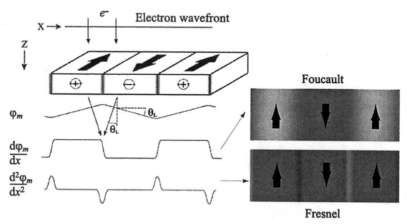

Figure 7-23: Schematics of electron deflection and phase shift due to scattering by an infinite array of magnetic stripe domains. Profiles of the phase shift, and of its first and second derivatives, are sketched on the left. Image simulations corresponding to the Fresnel and Foucault techniques are shown on the right.

Figure 7-24 (a) is a Lorentz image showing the configuration of magnetic domains from a $Nd_2Fe_{14}B$ single crystal with its surface normal near the [001] easy-axis. The maze pattern corresponds to domains with their magnetization directions anti-parallel, pointing either into, or out of, the paper. The domain width changes considerably with the area thickness, as evident in the image where the bright and dark background intensities correspond to thin and thick areas, respectively. A similar dependence of the domain width on areal thickness was also observed for (100) oriented crystals which consist of strips of parallel domains separated by Bloch-walls that lie in the projected [001] direction, as shown in Fig.7-24 (b,c) for sintered polycrystalline $Nd_2Fe_{14}B$. When the samples were heated above the Curie temperature and then cooled down to room temperature, the domains in the thin regions split (Fig.7-24(c)), while the domains in thicker regions remained unchanged.

The elementary scheme for image interpretation of magnetic domains based on Lorentz forces and deflections has to be handled with care. In fact, especially when fringing fields are present, such as near the specimen edge the image features may differ quite dramatically from the simple behavior sketched in Fig.7-23 and employed in the interpretation of the experimental results in Fig.7-24. To better analyze more realistic cases, and have a way to interpret more accurately situations as in Fig.7-24(b,c), we will now derive a complete description for an array of 180° domains which terminate at a specimen edge and show how the phase shift and corresponding phase contrast images are affected by the presence of fringing fields.

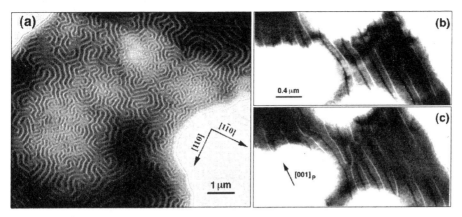

Figure 7-24: (a) Magnetic domains from a (001) single crystal of $Nd_2Fe_{14}B$ magnetized and viewed along the *c*-axis. (b-c) Magnetic domains from the same area in a sintered bulk $Nd_2Fe_{14}B$ polycrystal, magnetized with a large [001] component projected into the viewing plane (marked as [001]$_p$). Note, the domains in the thin region (center area of (b)) split after the sample was heated above T_c (312 °C), and then cooled back to 25 °C (c).

7.5.1.1. Model for magnetic domains

Let us consider a thin specimen of thickness $t=2d$, lying on the (x,y) plane and containing an array of 180° magnetic domains of width w, each alternatively oriented along the positive or negative direction on the x-axis. The specimen is considered semi-infinite, which means that there is an abrupt termination along the y-axis at $x=0$. The setup is sketched in Fig.7-25.

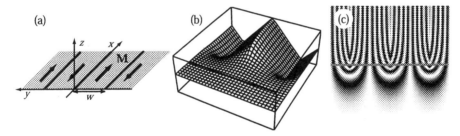

Figure 7-25: (a) scheme of an array of 180 degree magnetic domains and coordinate system employed in the mathematical description; (b) phase simulation of the domains near the specimen edge; (c) cosine map of the same phase shift, showing a bending of the fringes in close proximity to the edge.

The magnetization can be expressed as

$$\mathbf{M(r)} = \frac{N_f \phi_0}{\mu_0 tw}(1,0,0)H(x)Q(y)U(z) \tag{7.49}$$

where N_f is the number of flux quanta trapped inside the domain (not necessarily an integer number, as the flux quantization does not apply here). The function $H(x)$ is simply a unit-step function, equal to one for $x>0$ and to zero for $x<0$, the function $Q(y)$, representing a square wave of width w, is given by

$$Q_w(y) = \begin{cases} 1 & (2n)w \leq y \leq (2n+1)w \\ -1 & (2n+1)w < y < (2n+2)w \end{cases} \tag{7.50}$$

for $n=\ldots-2,-1,0,1,2\ldots$, and the function $U(z)$, describing the specimen thickness, is defined as

$$U(z) = \begin{cases} 1 & |z| \leq d \\ 0 & |z| > d \end{cases}$$

(7.51)

The expression linking magnetization and magnetic vector potential,

$$\mathbf{A(r)} = \frac{\mu_0}{4\pi}\int \mathbf{M(r')}\times\frac{\mathbf{r}-\mathbf{r'}}{|\mathbf{r}-\mathbf{r'}|^3}d^3\mathbf{r'}$$

(7.52)

represents an invaluable resource for calculating magnetic configurations starting from a known magnetization. In fact, exploiting the convolution theorem and the linearity of the vector product operation, equation (7.52) can be written in 3D-Fourier space as

$$\mathbf{A(k)} = \frac{\mu_0}{4\pi}\mathbf{M(k)}\times\Im\left[\frac{\mathbf{r}}{r^3}\right] = \frac{i\mu_0}{k^2}\mathbf{M(k)}\times\mathbf{k}$$

(7.53)

Hence, the calculation of the vector potential is reduced to a vector product if the Fourier transform of the magnetization is computable.

As the three functions in eq.(7.49), $H(x)$, $Q(y)$, $U(z)$, depend on different variables, we can express the Fourier transform of the magnetization as the product of the transforms of these functions, namely

$$\mathbf{M(k)} = \frac{N_f\phi_0}{\mu_0 tw}(1,0,0)H(k_x)Q(k_y)U(k_z)$$

(7.54)

with

$$H(k_x) = \pi\delta(k_x) + \frac{1}{ik_x}$$

(7.55)

$$Q_w(k_y) = \frac{4\pi i}{wk_y}\sum_{n,odd}\delta\left(k_y + \frac{\pi n}{w}\right)$$

(7.56)

$$U(k_z) = \frac{2}{k_z}\sin(dk_z)$$

(7.57)

Considering eq.(7.53), thus performing the cross product between $(1,0,0)$ and $\mathbf{k}=(k_x,k_y,k_z)$, we can directly write the expression for the vector potential

$$\mathbf{A(k)} = \frac{8\pi N_f\phi_0}{w^2 t}\frac{(0,-k_z,k_y)}{k_y k_z k^2}\left[\pi\delta(k_x) + \frac{1}{ik_k}\sum_{n,odd}\delta\left(k_y + \frac{\pi n}{w}\right)\right]\sin(dk_z)$$

(7.58)

Extracting the z-component of the vector potential, integrating along the z-axis and going back to real space, we obtain the phase shift in an analytical form after a lengthy computation:

$$\varphi_m = \frac{N_f}{2\pi}\mathrm{Re}\left[2H(x)e^{\pi\frac{iy}{w}}\phi_{1/2}^2\left(e^{2\pi\frac{iy}{w}}\right) - S(x)e^{\pi\frac{iy-|x|}{w}}\phi_{1/2}^2\left(e^{2\pi\frac{iy-|x|}{w}}\right)\right]$$

(7.59)

where and $S(x)$ is the Sign function. The generalized Φ_u^v function (a generalization of the Riemann Zeta and Polylogarithmic functions) is also called Lerch function [84]. This expression for the phase shift of a semi-infinite array of stripe domains was first reported in [85].

The phase shift (in arbitrary units) corresponding to a region enclosing three domains is shown in Fig.7-25(b). The simulated holographic fringes, Fig.7-25(c), are curved near the specimen edge, indicating a strong demagnetizing effect. Moreover, inside the specimen (for $x>0$) the fringes form a sharp angle, while in the vacuum they connect more smoothly. This effect is mainly due to the zero-width model assumed for the domain walls.

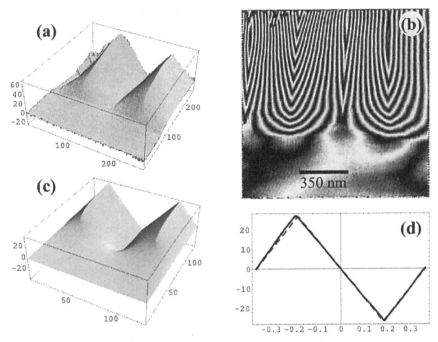

Figure 7-26: (a) A three-dimensional representation of the phase shift; (b) experimental holographic cosine map; (c) phase simulation; (d) matching between experimental and simulated line scan. In (a,c,d) the phase shift is measured in radians. In (b) each fringe (black or white line) corresponds to a phase shift of π.

7.5.1.2. Measurement of domain wall widths

To validate the above theoretical description, we performed a series of holographic experiments on NdFeB thin film, searching for a region of the specimen where two or more 180° domains were close to the edge. Moreover, as a model for an array of domains forming an angle with the specimen edge has not yet been developed, we choose to extract the phase from the domain walls which were as perpendicular as possible to the edge. The result of one experiment is shown in Fig.7-26, where the retrieved phase shift is displayed both as a surface plot (a) and as a cosine map (b). The phase simulation in (c), calculated based on the model presented, shows excellent agreement with the experiment.

After retrieving also a thickness map of a wide region around the domain wall (not shown here) we also were able to determine with great accuracy the magnetic field in the domain. The good match between the simulated and experimental line scan allowed us to determine that, in absence of an applied field, the magnetization of the material is 1.2 T. The obtained value agrees well with the known value of the remnant field for $Nd_2Fe_{14}B$.

Comparing Fig.7-26(c) with Fig.7-25(c), several differences prompted us to further deepen and develop the theoretical description. While the visual comparison between the two cosine maps is rather satisfactory, nevertheless a more detailed examination shows that in the experiment the fringes start closing *before* reaching the specimen edge. By contrast, the simulated map shows that across the edge there is an abrupt change in the smoothness of the fringes: before the edge the fringes appear to form an acute angle over the position of the wall width, while just outside the edge they connect more smoothly. Moreover, examining the region just outside the specimen edge, we notice that the simulation shows a fringing field protruding to the vacuum. The experiment, instead, shows a rather flat phase shift in the vacuum.

To interpret correctly those differences, we should consider that the theoretical description assumes a uniform magnetization in each domain. The magnetization direction never changes from the bulk to the edge. This assumption is most likely incorrect, because the demagnetizing field at the edge modifies the magnetization topography in close proximity to the material-vacuum interface. Possibly when the magnetization is close to the edge due to the presence of the surface through the demagnetizing field, it starts rotating in order to align along the edge, rather than remaining perpendicular to it. In other words, the presence of the specimen edge may induce a transition from a 180° domain wall in the bulk, and a variable angle which continuously decreases to zero at the edge. From the demagnetization point of view, this transition would dramatically reduce the fringing field and the magnetic energy associated with the domain termination. The theoretical descriptions, therefore, must be improved to cover more complicated magnetization topographies.

Several modeling techniques were developed to estimate the wall width, which is connected to important material parameters such as anisotropy constant and exchange coupling. In the following we show how TEM can be very useful in extracting physical quantitative information from the region of transition between two magnetization states. A transition that occurs on a length scale of a few nanometers is within the realm of nanoscience.

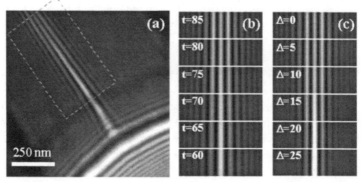

Figure 7-27: (a) Fresnel image of a domain wall near the specimen edge; (b) effect of the thickness on the Fresnel diffraction fringes: from top to bottom the specimen thickness decreases from 85 down to 60 nm keeping Δ=0; (c) effect of the domain wall width enlargement in proximity of the specimen edge: from top to bottom the wall width increases from 0 nm to 25 nm keeping *t*=85 nm.

First, we examine in more detail the effect of the specimen edge on the topography of magnetization close to the vacuum-sample interface. We mentioned previously that the demagnetizing effect near the specimen edge may induce a continuous decrease of the domain wall angle (from 180° to zero at the exact location of the edge). From another perspective, this can be also interpreted as an enlargement of the domain wall width, due to the magnetization divergence. A Fresnel image of a domain wall in close proximity to the specimen edge is shown in Fig.7-27. The effect of the edge is evident: the diffraction fringes appear to change in contrast, periodicity, and spatial extent while going from the bulk to the edge. At the domain termination, the wall fringes, running parallel to the wall and the edge fringes, running parallel to edge, overlap and interfere creating complicated contrast features. More towards the vacuum, the primary bright edge fringe, shows a higher contrast with respect to the wall.

There are two possible interpretations, which may account for the observed contrast variations around the edge. First, the specimen thickness may not be perfectly constant throughout the observed area. This would induce in the phase shift that depends on the distance from the edge because the magnetic flux is directly proportional to the thickness, for a given constant magnetization. This, in turns, accounts for the change in periodicity observed in the experimental image. A series of simulations for different thicknesses is shown in Fig. 27(b), where the specimen thickness decreases from 80 nm (top) down to 60 nm (bottom). The simulations show a change in fringe periodicity and contrast that appears to be in good agreement with the experimental image. However, the contrast of the primary fringe in the center of each simulation does not increase as much as the experimental one. Therefore, we conclude that the specimen is likely to be a wedge, with a jump of around 60 nm at the vacuum interface; however, this alone is not enough to fully account for the observed contrast.

The second effect considered, shown in the simulations series Fig.7-27(c), is the enlargement of the wall width. The domain wall changes in this case from 0 nm (top) to 30 nm (bottom). We note that the overall appearance is not very different from the thickness effect in Fig.7-27(b). However, there are two main differences: first, the primary fringe appears to be brighter and brighter, becoming similar to the experimental one; second, the periodicity of the fringes is not affected. To better show the dependence of the Fresnel diffraction contrast on the spatial extent of the phase shift, connected to the wall width, another series of simulations was computed, and

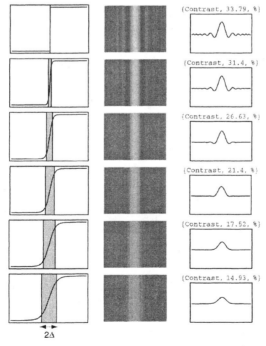

Figure 7-28: Effect of the Bloch wall width enlargement on the Fresnel diffraction contrast. First column: phase line scan across the wall and definition of the wall width. Second column: out-of-focus image simulation. Third column: line scan of the image contrast.

the results are shown in Fig.7-28. Starting from a zero-width wall, where the phase shift is a step function having an amplitude which depends on magnetic field and thickness, we increase the wall width assuming that the phase shift enlarges as $\tanh(x/\Delta)$, which is typical of Bloch walls.

It can be concluded that the variations in the Fresnel diffraction contrast near the specimen edge can be ascribed to a combination of thickness and wall enlargement. The first has been confirmed by a direct measurement of thickness by electron energy-loss spectroscopy (EELS), showing that indeed the thickness of the thin sample linearly increases from its edge to its interior. The second is the main reason why holography experiments focused on measuring of domain wall widths tend to give results larger than expected, as shown in the following.

An example of wall width measurement by holography is summarized in Fig.7-29. An electron hologram was recorded at the highest possible resolution, in a low magnification mode (with the objective lens off), for our microscope. Under these conditions we can achieve a subnanometer resolution, adequate to extract information from the domain wall. Then, from the reconstructed phase shift, shown in Fig.7-29(a) and (b), a line scan was extracted and compared with phase simulations computed for different wall-width values. Three such comparisons are shown in (c-e), where the noisy line is the experiment and the continuous line is the simulation. Then, from a numerical χ-square analysis, the plot in Fig.7-29(f) was obtained. The χ-square shows a minimum around 13 nm; therefore, we take this value as the result from electron holography.

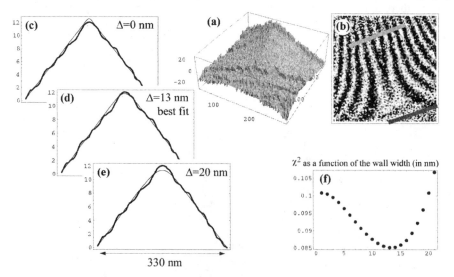

Figure 7-29: (a) Phase shift of a domain wall displayed as 3D plot; (b) cosine map of the same phase shift; the upper thick line locates the area from which the line scan was extracted; the lower thick line represents the specimen edge; (c-e) three different simulations (regular line) compared with the experimental line scan extracted from (a); (f) chi-square analysis of the data and location of the best fit value around 13 nm.

In general, the wall width is related to the material properties by $\Delta = c\sqrt{A/K}$, where A is the exchange coupling and K is the anisotropy constant of the material. The numerical factor c actually depends on the assumed definition, and can be either 1, or 2 or π [54]. An estimate of the wall width for NdFeB would result in a value of the order of 5 nm. The result by holography is of the same order of magnitude, confirming once again that this technique can be fruitfully employed for the nanoscale characterization of materials, but is also an indication of the edge effect that tends to enlarge the domain wall width near the interface with the vacuum.

7.5.1.3. Measurement of magnetic potential in bulk magnets

The advantage of phase retrieval for Lorentz applications is that the image wave is, in a very good approximation, equal to the sample exit wave. The spherical aberration of the lenses and transfer function of the microscope can be neglected since the scattering angles of the incident electrons from the sample in magnetic imaging is about two orders of magnitude smaller than the Bragg angles. Although phase retrieval from ion milling-thinned bulk samples with non-uniform thickness and specimen geometry is difficult, it is possible if the results are cross-checked with different methods.

We map the distribution of magnetic induction in a $Nd_2Fe_{14}B$ permanent magnet first by electron holography. The standard low-magnification holography mode for our JEOL 3000F suffers somewhat poorer performance compared to the standard high-magnification modes. Typically, we obtain holographic fringe contrast around 10% or lower for bi-prism voltages around 60 V, which gives 19.1 nm interference fringes and 4.2 μm hologram width. Efforts to maximize the fringe contrast using free-lens control allowed us to record high-quality holograms and to retrieve induction maps at low magnifications. Fig.7-30(a) is an electron hologram recorded from a sintered $Nd_2Fe_{14}B$ permanent magnet, while (b) and (c) are, respectively, the reconstructed amplitude and cosine of the phase image. The hologram shown in (a) was recorded with fringe contrast (in vacuum) about 16% and 60.2 V biprism voltage giving 10.8 nm fringes over a 1.46 μm field-of-view on the CCD camera.

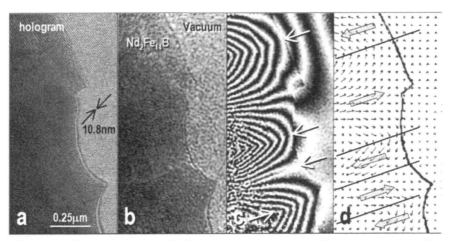

Figure 7-30: (a) Electron hologram of $Nd_2Fe_{14}B$ permanent magnet; (b) reconstructed amplitude; (c) cosine of reconstructed phase. Arrows indicate presence of domain boundaries in the material. (d) Calculated vector-map of projected induction. Thick arrows indicate trend of induction lines suggesting 180°-domain configuration [86].

Domain boundaries (marked by arrows in Fig.7-30(c)) appear as discontinuities in the gradient of the phase image. Fig.7-30(d) shows a vector map of the calculated phase-gradient that represents the projected magnetic induction, neglecting contributions from sample thickness and electrostatic potentials in the sample. (Even accounting for the contributions due to the sample thickness and electrostatic potentials, one must be aware that the component of magnetic field parallel to the beam does not contribute to the final induction map, so that interpretation must be performed with caution). Fig.7-30(d) shows qualitatively the 180° domain configuration (as thick arrows) present in the sample. Details on the magnetic domain, domain wall, and their dynamics can be found in [87,88].

We applied the TIE formalism to the study of the same $Nd_2Fe_{14}B$ sample in conjunction with the off-axis holography method. Fig.7-31(a,b) are, respectively, over- and under-focused (Lorentz) images from the same region shown in Fig.7-30. The nominal defocus (not calibrated) was $\Delta_z=16$ nm and the magnetic domain-wall structure is clearly visible as lines of contrast in the images that reverse contrast on either side of the zero-defocus. Figure 7-31(c) is the cosine of the phase image (similar to Fig. 30(c)) retrieved with the TIE formalism of eq.(32). Figure 31(d) shows a vector map of the calculated phase-gradient, which, as with the holography results, represents approximately the projected magnetic induction, i.e., the 180°-domain configuration (as thick arrows) present in the sample. The results obtained with holography, Figs. 30(c,d), and with the TIE formalism, Figs. 31(c,d), agree favorably and illustrate the potential of the TIE formalism to retrieve quantitative phase information under conditions less restrictive than that of holography experiments.

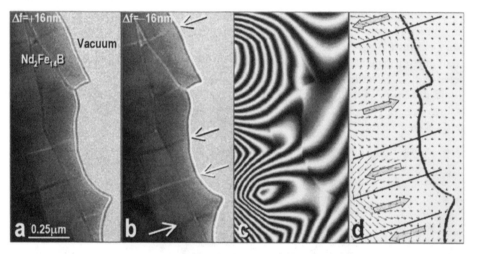

Figure 7-31: (a) Over-focused and (b) under-focused Lorentz images from same region of the $Nd_2Fe_{14}B$ permanent magnet shown in Fig.7-30. As in Fig.7-30c, arrows indicate presence of domain boundaries in material. (c) cosine of phase image retrieved with the TIE formalism. (d) Projected induction vector-map. The trend of induction lines suggests a 180°-domain configuration in good agreement with results obtained with electron holography.

7.5.2. MAGNETIC NANOPARTICLES

Magnetic nanoparticles on thin films with well-defined geometries are ideal for quantitative studies to understand nanomagnetism. Their assemblies can be used to investigate magnetic interactions with others and spin dynamics and switching behaviors as a function of size, shape of the particles and the distance from their neighbors. Figure 7-32 shows a Lorentz image of Permalloy ($Ni_{20}Fe_{80}$) arrays fabricated using an in-house UHV film deposition system with an artificially structured mask. The small magnetic disks are about 1 μm in diameter. The white and black contrast in the magnetic elements corresponds to opposite magnetization directions. The induction maps derived using phase retrieval techniques can be validated by magnetic structural modeling and the observed low-energy state, including fringing fields, can be compared with micromagnetics simulations.

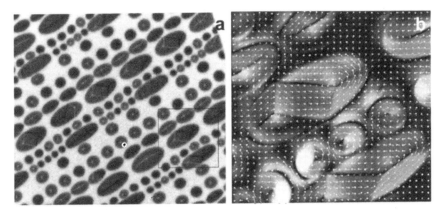

Figure 7-32: Lorentz image and projected induction distribution (boxed area) of the artificially patterned $Ni_{20}Fe_{80}$ Permalloy. The diameter of the small dots are about $1\,\mu m$. *Also see the color plate.*

7.5.2.1. Model for magnetic nanoparticles

To cover all the possible geometries involved in the experiments, we can start with the three basic shapes: rectangular, cylindrical, and spherical. As a first approximation, we will consider uniformly magnetized particles.

In general, we can write the magnetization vector as $M=M_0\,\hat{m}$ for **r** inside the particle, and zero outside introducing the dimensionless shape function $D(\mathbf{r})$ (also called characteristic function) representing the region of space bounded by the particle surface

$$\mathbf{M}(\mathbf{r}) = M_0\,\hat{\mathbf{m}}\,D(\mathbf{r}) \tag{7.60}$$

Recalling the basic concepts of the Fourier-space treatment of magnetic structures outlined in section 7.5.1.1, namely that once the Fourier transform of the magnetization is known, the vector potential can be derived by a simple vector product, we compute the Fourier trasform of eq.(7.60)

$$\mathbf{M}(\mathbf{k}) = M_0\,\hat{\mathbf{m}}\,\int d^3\mathbf{r}D(\mathbf{k})e^{-i\mathbf{k}\cdot\mathbf{r}} = M_0\,\hat{\mathbf{m}}\,D(\mathbf{k}) \tag{7.61}$$

where $D(\mathbf{k})$ is the Fourier transform of the shape function, often called shape amplitude, or shape transform.

From equations (7.53) and (7.61) we can calculate directly the vector potential in Fourier space

$$\mathbf{A}(\mathbf{k}) = -\frac{iB_0}{k^2}D(\mathbf{k})(\hat{\mathbf{m}}\times\mathbf{k}) \tag{7.62}$$

where $\mu_0 M_0 = B_0$ is the magnetic induction corresponding to the magnetization M_0.

From the knowledge of the vector potential, one can readily calculate the magnetic induction, as $\mathbf{B}=\nabla\times\mathbf{A}$ and the phase shift as a line-integral along the electron trajectory. Let us first concentrate on the magnetic induction, in order to establish a correspondence between well-known results of electromagnetic theory and our approach.

As any differential operator in real space is a reciprocal vector in Fourier space, the nabla operator becomes $\nabla \rightarrow i\,\mathbf{k}$. Therefore, the Curl is translated into a vector product as follows:

$$\mathbf{B(k)} = i\mathbf{k} \times \mathbf{A(k)} = \frac{B_0}{k^2} D(\mathbf{k})(\mathbf{k} \times \hat{\mathbf{m}} \times \mathbf{k}) \tag{7.63}$$

which, exploiting the vector identity $\mathbf{k} \times \hat{\mathbf{m}} \times \mathbf{k} = \hat{\mathbf{m}} k^2 - \mathbf{k}(\mathbf{k} \cdot \hat{\mathbf{m}})$, can be also written, after an inverse Fourier transform, as the sum of the induction proportional to the magnetization and the demagnetizing field:

$$\varphi_m(\mathbf{k}) = \frac{i\pi B_0}{\phi_0} \frac{D(k_x, k_y, 0)}{k_\perp^2} \left(\hat{\mathbf{m}} \times \mathbf{k}\right)\Bigg|_z \tag{7.64}$$

The integral in eq.(7.64), namely the demagnetizing field $\mathbf{H}$, evaluated in the inner part of a spherical particle yields $\mathbf{H} = -1/3\, B_0\, \mathbf{m}$. The factor $1/3$ is exactly the demagnetizing factor for a sphere, for which the relationship $\mathbf{B} = 2/3\,\mu_0\mathbf{M}$ holds. It has been thus demonstrated how the vector potential calculated from eq.(7.58) takes into account also the demagnetizing field extending in the vacuum surrounding the particle. This field depends on the geometry of the uniformly magnetized particle.

The magnetic component of the phase shift φ_m can be calculated from the knowledge of the vector potential. After integrating along the z-axis in Fourier space, we obtain

$$\varphi_m(\mathbf{k}) = \frac{i\pi B_0 V_R}{\phi_0} \frac{k_y \cos\beta - k_x \sin\beta}{k_\perp^2} \mathrm{sinc}(L_x k_x)\mathrm{sinc}(L_y k_y) \tag{7.65}$$

where .

Equation (7.65) suggests that to calculate the phase shift of a uniformly magnetized nanoparticle, all we need to know is the shape amplitude, along with the direction and intensity of the magnetization.

7.5.2.2. Basic shapes for nanoparticles

Let us now recall the main results obtained for the phase shift of magnetic nanoparticles of different shapes [89]. An analytical expression for the phase shift, to compare directly with experiments, can be obtained in real space only for the three basic shapes: rectangular, cylindrical, and spherical. As shown in earlier, in Fourier space the analytical calculation can be performed whenever the shape function of the particle is known in Fourier space. In all the other cases, a numerical approach can be used to evaluate the phase shift of all the possible particle shapes. In fact, the shape function can be always expressed as a three-dimensional matrix made of 1s and 0s, to mimic the particle topography. It has been shown that this numerical approach is reliable when compared to analytical calculations [90].

For the rectangular geometry, sketched in Fig.7-33, the phase shift can be expressed in Fourier space by

$$\varphi_m(\mathbf{k}) = \frac{i\pi B_0 V_R}{\phi_0} \frac{k_y \cos\beta - k_x \sin\beta}{k_\perp^2} \mathrm{sinc}(L_x k_x)\mathrm{sinc}(L_y k_y) \tag{7.66}$$

where V_R is the particle volume, L_x and L_y are the half-dimensions along x and y, and β is the magnetization angle. The inverse Fourier transform can be calculated analytically, but the expression is rather cumbersome and will not be discussed here.

An example of the magnetic phase shift, calculated from eq.(7.66) by means of an inverse FFT, is shown in Fig.7-33(b) for the following set of parameters: L_x=60 nm, L_y=100 nm, d=10 nm, β=300° and B_0=1.6 T.

As the particle surfaces are flat, the electrostatic contribution φ_e (see eq.(7.45)) reduces to a constant term inside the particle, and zero outside. This means that the only effect of the thickness is a phase jump across the particle borders, which is responsible for the discontinuity of the holographic contour fringes displayed in Fig.7-33(c), where the particle shape was slightly smoothed to mimic the real specimen edges usually encountered in experiments.

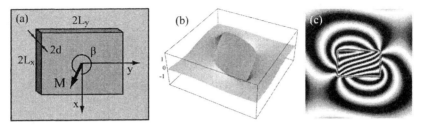

Figure 7-33: (a) Rectangular nanoparticle, coordinate system and main parameters involved; (b) purely magnetic phase shift displayed as a surface plot; and (c) holographic contour map (12×amplified) of the total phase shift $\varphi_e + \varphi_m$.

A cylindrical particle is not very different from the rectangular case, as the electrostatic contribution is still represented by a phase discontinuity on the particle edge. The calculation of the magnetic contribution yields

$$\varphi_m(\mathbf{k}) = \frac{2i\pi^2 B_0 R t}{\phi_0} \frac{k_y \cos\beta - k_x \sin\beta}{k_\perp^3} J_1(k_\perp R) \qquad (7.67)$$

where $J_1(x)$ is the Bessel function of first order, R is the particle radius and t the particle thickness.

The magnetic phase shift for a sphere, calculated from eq.(7.67) using an inverse FFT, is shown in Fig.7-34(b) for the parameters: R=50 nm, d=10 nm, β=300° and B_0=1.6 T. The effect of the electrostatic phase shift (a mean inner potential V_0=10 V is assumed) is shown in Fig.7-34(c), and, as before, it reveals itself only near the particle edges.

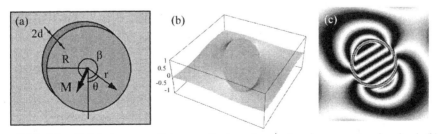

Figure 7-34: (a) cylindrical nanoparticle, coordinate system and main parameters involved; (b) purely magnetic phase shift displayed as a surface plot; (c) holographic contour map (16×amplified) of the total phase shift $\varphi_e + \varphi_m$.

The spherical geometry, shown in Fig.7-35(a), is rather different from the previous configurations. The basic difference is in the electrostatic contribution that has a dramatic effect on the total phase shift as there are no flat surfaces on a sphere. Moreover, as the ratio between the two contributions depends on the sphere radius, as will be shown in the following section, for very small particles under 50 nm in radius, the magnetic signal is masked by the predominant electrostatic one. This is even more striking when the sphere is in the vortex state, as the two contributions are not only similar in amplitude, but in shape as well.

The phase shift is then

$$\varphi_m(\mathbf{k}) = \frac{4i\pi^2 B_0 R^2}{\phi_0} \frac{k_y \cos\beta - k_x \sin\beta}{k_\perp^3} j_1(k_\perp R) \qquad (7.68)$$

where $j_1(x)=[\mathrm{sinc}(x)\text{-}\cos(x)]/x$ is the spherical Bessel function of first order, and R is the particle radius.

The magnetic phase shift, calculated from eq.(7.68) by means of an inverse FFT, is shown in Fig. 7-35(b) for the following set of parameters: $R=50$ nm, $\beta=300°$, and $B_0=1.6$ T. Once again, the electrostatic component φ_e is added in Fig.7-35(c). Unlike the geometries considered previously, the effect of the electrostatic potential in a spherical particle changes dramatically the projected potential configuration.

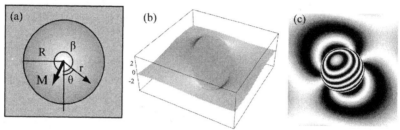

Figure 7-35: (a) Spherical nanoparticle with main parameters used in the calculations; (b) purely magnetic phase shift displayed as a surface plot; (c) holographic contour map (3×amplified) of the total phase shift $\varphi_e + \varphi_m$.

7.5.2.3. The vortex state

Magnetic structures with circular magnetizations are frequently found in real materials due to the fact that the demagnetizing energy is minimized as no **H** field is associated to a closure-domain configuration (also referred to as *vortex state*). The stability of the vortex state depends mainly on geometry, size and magnetic properties of the material (anisotropy and exchange constants). It was proven by Brown [91] that for any particle shape a critical size can be established where the single-domain state is energetically favourable due to the predominance of the exchange interaction over magnetostatic and crystal anisotropy. This issue was then further analyzed by Aharoni [92]. Therefore, vortex states will be found only in nanoparticles larger than the Brown's limit, which is, very roughly, a few hundred Angstroms. The critical size is in general inversely proportional to the exchange constant of the material, i.e. smaller for high-exchange materials like Co and larger for low-exchange materials like permalloy.

In order to describe circular domains, we have to extend the approach employed insofar. While the basic eq.(7.58) remains valid, the magnetization unit vector is no longer a constant, as it is now flowing circularly inside the particle. Therefore, the considerations employed for a uniform magnetization are not valid, and we have to generalize the approach.

Let us consider a circular flux line carrying an arbitrary non-integer number of magnetic flux quanta $N_f\phi_0$. If we obtain a solution for the vector potential, and then for the phase shift, we can use it to reproduce a realistic magnetic field topography of the circular domain.

The magnetization corresponding to a closed circular flux line of radius R, as shown in Fig.7-36(a), written in cylindrical coordinates is

$$\mathbf{M(r)} = \frac{N_f\phi_0}{\mu_0}(-\sin\theta,\cos\theta,0)\delta(r-R)\delta(z) \tag{7.69}$$

where the unit vector $(-\sin\theta,\cos\theta,0)$, describing a counter-clockwise flux flow, has the role previously assumed by $\mathbf{m}$. Its Fourier transform is directly obtained as

$$\mathbf{M(k)} = \frac{2\pi i N_f\phi_0 R}{\mu_0}\frac{J_1(k_\perp R)}{k_\perp}(-k_y,k_x,0) \tag{7.70}$$

We can now calculate the vector potential, by means of eq.(59), and integrate along the electron trajectory to obtain the phase shift in Fourier space

$$\varphi_m(\mathbf{k}) = 2\pi^2 N_f R\frac{J_1(k_\perp R)}{k_\perp} \tag{7.71}$$

This expression can be inverted to real space:

$$\varphi_m(\mathbf{r}) = \pi N_f R\int_0^\infty J_1(k_\perp R)J_0(k_\perp r)dk_\perp = \begin{cases} -\pi N_f & r < R \\ 0 & r > R \end{cases} \tag{7.72}$$

which is a cylinder of radius R, and height $-\pi N_f$ as shown in Fig.7-36(b).

This result can be employed to describe the phase shift of a cylindrical particle of uniform vertical magnetization, embedded in a medium of opposite magnetization, as shown in Fig. 7-36(c) [93]. A thorough treatment of similar structures, including a deep analysis of their energetics and stability can be found in [94-96].

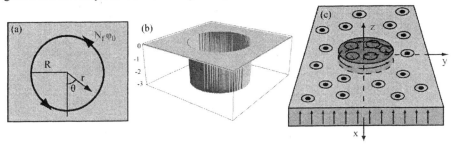

Figure 7-36: (a) Circular flux line carrying N_f flux quanta; (b) resulting magnetic phase shift; (c) physical situation represented by the model.

7.5.2.4. Domain wall widths

We can extend our result to a realistic circular domain structure by describing the flux distribution across a Bloch domain wall with a suitable function $\phi(r)$. For instance, it is suggested in [93] that a suitable wall description is provided by

$$\cos\theta(r) = \tanh\left(\frac{r-R}{\Delta}\right) \tag{7.73}$$

where $\theta(r)$ represents the angular change of the magnetization unit vector across the Bloch wall (from 0° to 180°), and Δ is equal to $\sqrt{A/K}$ (see 7.5.1.2). Following the considerations found in [86], we can define the wall width as $W=\pi\Delta$.

Resorting to numerical procedures, we can evaluate the FT of the function

$$\phi(r) = \sin\theta(r) = \mathrm{sech}\left(\frac{r}{\Delta}\right) \tag{7.74}$$

and compute the convolution $\varphi_{tot}(\mathbf{k})=\varphi(\mathbf{k})\phi(\mathbf{k})$ to include the effect of the wall-width on the phase shift. The plots of the phase shift corresponding to the three different values $\Delta=1,5,10$ nm are shown in Fig.7-37.

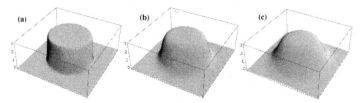

Figure 7-37: Magnetic phase shift of a circular domain calculated for $N_f=1$ and for different values of the wall width: (a) 1 nm, (b) 5 nm, (c) 10 nm.

The sensitivity of the phase shift to domain wall structures may open a door to direct determination of various physical properties of magnetic materials through measuring the domain wall width. An example of wall width measurement by holography was given in section 7.5.1.2.

7.5.2.5. Vortex state of nanoparticles

Using the same formalism, we can generalize the previous results to a circularly magnetized disk of radius R and thickness t, for which the phase shift in real space is

$$\varphi_m(r) = \frac{B_0 t}{\phi_0}(R-r) \quad \text{for} \quad r < R \tag{7.75}$$

to a circularly magnetized sphere of radius R

$$\varphi_m(r) = \frac{\pi B_0 t}{\phi_0}\left[R\arcsin\sqrt{1-\frac{r^2}{R^2}} - r\sqrt{1-\frac{r^2}{R^2}}\right] \quad \text{for} \quad r < R \tag{7.76}$$

and, extending the approach, to a non-circular closed flux line on a rectangular domain of dimensions $(2L_x,2L_y)$, in which the geometry induces the formation of 90° domain walls (with the assumption of zero wall width):

$$\varphi_m(x,y) = \frac{\pi B_0 t}{\phi_0}\,\mathrm{Min}\left\{\begin{array}{l}L_x-|x|\\L_y-|y|\end{array}\right\} \quad \text{for} \quad |x| < L_x\,;|y| < L_y \tag{7.77}$$

Since any closure domain structure has no fringing fields, in the previous three expressions the phase shift outside the particle is identically zero. The situations are depicted in Fig.7-38.

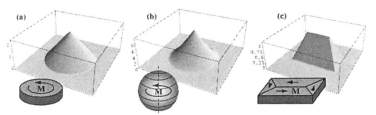

Figure 7-38: Phase shift for magnetic particles in vortex state: (a) cylindrical, (b) spherical, and (c) rectangular geometry. The magnetization is also sketched for each plot.

Figure 7-39(a) shows an experimental out-of-focus image of Ni nano dot-array, with the diameter of each dot smaller than 40 nm. The retrieved phase contour of the dots near the array edge is displayed in Fig.7-39(b) using the TIE method. The sample was prepared with an exploratory home-made system based on TEM lithography, or patterning. The pattern was first written by the electron beam in TEM, the beam-exposed sites were processed, and then subjected to magnetic film deposition. The TEM-based method has a potential to have better resolution than SEM-based lithography due to the negligible beam boarding effect in the sample.

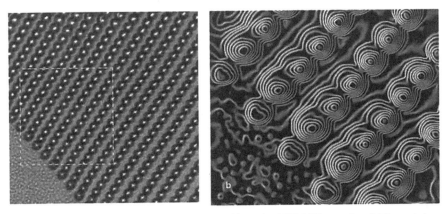

Figure 7-39: Ni dot array (40 nm in diameter) prepared by using TEM lithography. (a) Lorentz image and (b) phase contour map. *Also see the color plate.*

The most challenging problem regarding magnetic phase retrieval at nanoscale is that when the particle dimensions become smaller and smaller, the electrostatic contribution to the phase shift tends to overwhelm the magnetic signal. Therefore, it is important to compare the electrostatic and magnetic contributions to the phase shift in a circularly magnetized spherical particle, as this can give useful hints on the ability of TEM to retrieve a magnetic signal from nanoparticles. Both contributions to the phase shift strongly depend on the particle radius. If we choose a reasonably high accelerating voltage (300 kV), and an embedding medium with a mean inner potential not very different from that of the magnetic particle (e.g., V_0=10 V), we can plot φ as a function of the particle radius R (Fig.7-40(a)). We can define the characteristic radius R_c=$4C_E V_0/\pi C_B B_0$=34 nm, for which the electrostatic and magnetic contributions are equal [89]. This is roughly the case for the Ni dot-array shown in Fig.7-39. For smaller R, the electrostatic contribution predominates, and overwhelms the magnetic phase, which reaches the limit of detectability (here assumed equal to $\pi/20$) around R=7 nm, as displayed in Fig.7-40(b).

For a spherical particle of radius R=$R_c/2$=17 nm, the magnetic signal can be considered as a perturbation in relation to the predominant electrostatic phase shift shown in Fig.7-40(e). This poses a serious limitation to the robustness of TEM for magnetic observations of small objects. However, by carefully choosing the experimental set-up and specimen geometry, it is possible to reach the now inaccessible region below R_c.

The assumed limit of detectability $\pi/20$ is very dependent on the experimental conditions, and on the phase retrieval technique employed. As electron holography is generally claimed to be able to retrieve phase shifts as small as $\pi/100$ [97], in principle there is no lower limit for extracting the magnetic signal for nanoparticles with TEM, as $\pi/100$ corresponds to a particle radius smaller than 3 nm, very close to the atomic scale. Certainly, the electrostatic contribution should be precisely taken into account first, otherwise the real limit for magnetic observation in TEM remains R_c. A thorough analysis of the problems involved in separating the magnetic and electrostatic components by in-situ magnetization reversal was given by Dunin-Borkowski et al. [98]. The same author succeeded in retrieving the magnetic phase from nanocrystals in magnetotactic bacteria [99] and from Co particles of a few nanometers in size [100], thus establishing the current status-of-the-art for these kind of observations.

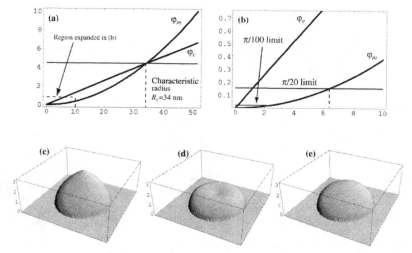

Figure 7-40: (a) Comparison between the electrostatic and magnetic components of the phase shift in a spherical particle. The region around $R=0$ is expanded in (b) to emphasize the detectability limits. (c) total phase shift, clockwise magnetization; (d) total phase shift, counterclockwise magnetization; (e) electrostatic component of the phase shift for a spherical particle of radius $R=R_c/2=17$ nm.

7.5.2.6. Arrays of nanoparticles

An advantage of having a Fourier representation of the phase shift is its straightforward extension to arrays of nanoparticles. Assuming that the specimen is made of a regular array of dots, or disks, or bars (respectively spherical, cylindrical, or rectangular geometry), each element is located by a standard Bravais lattice vector $\mathbf{r}_j$. The total phase shift for the array composed of N elements, each of them having a phase shift φ_j, can be written in real space as

$$\varphi_{\text{tot}}(\mathbf{r}) = \sum_{j=1}^{N} \varphi_j(\mathbf{r} - \mathbf{r}_j) \tag{7.78}$$

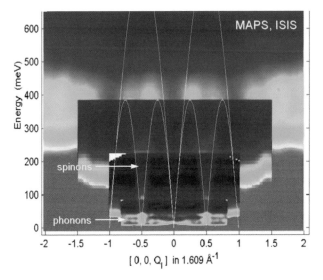

Figure 1-1

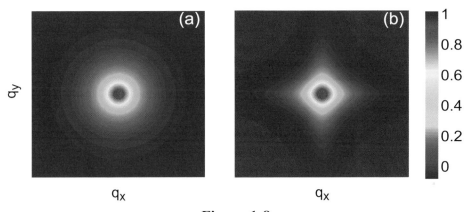

Figure 1-8

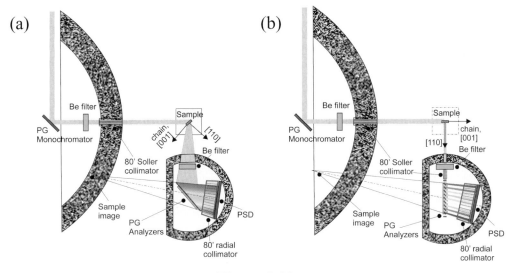

Figure 1-13

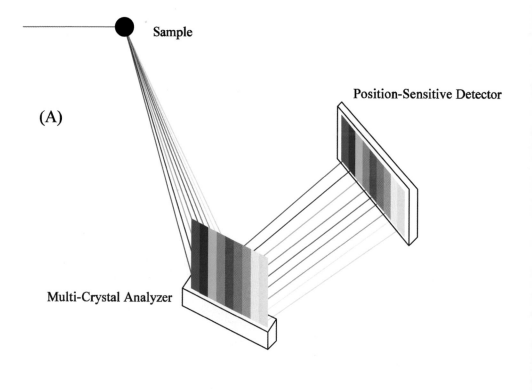

(A) Sample

Position-Sensitive Detector

Multi-Crystal Analyzer

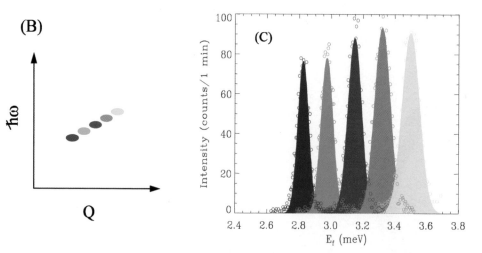

(B)

$\hbar\omega$

Q

(C)

Intensity (counts/1 min)

E_f (meV)

Figure 1-14

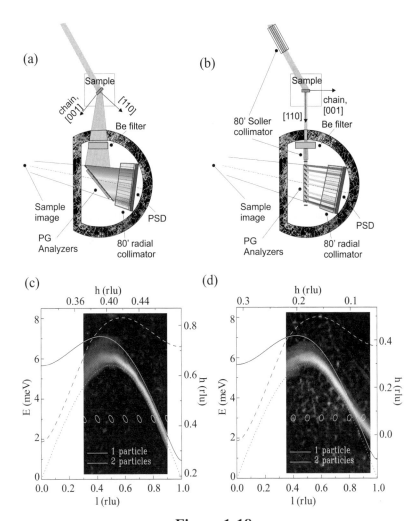

Figure 1-18

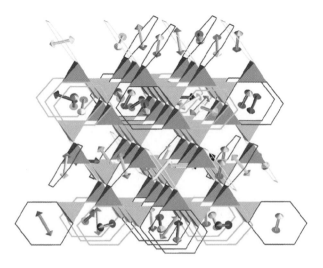

Figure 1-19

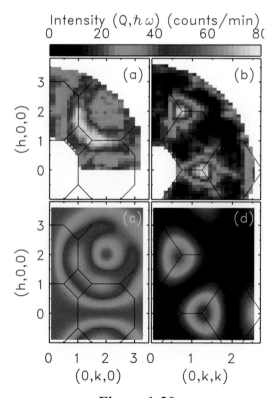

Figure 1-20

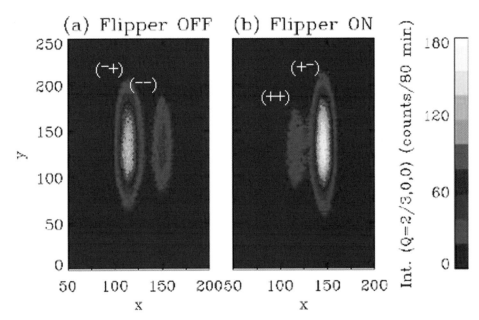

Figure 1-24

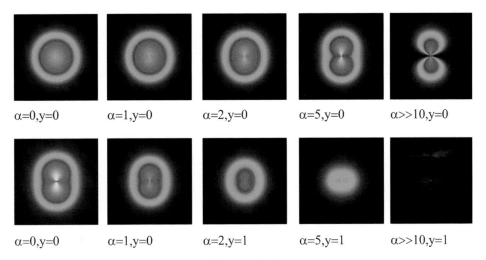

Figure 2-4

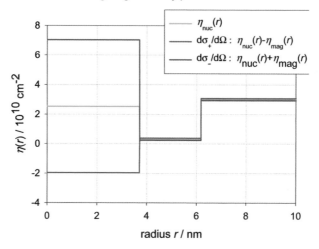

Figure 2-6

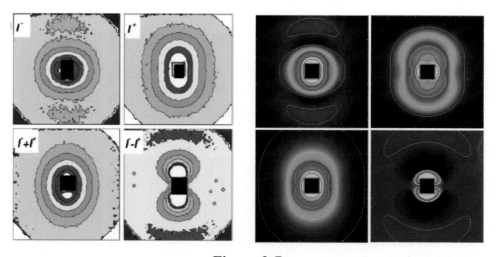

Figure 2-7

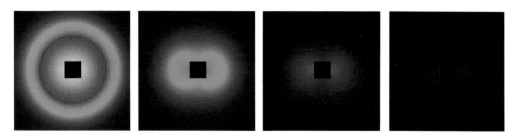

Figure 2-18

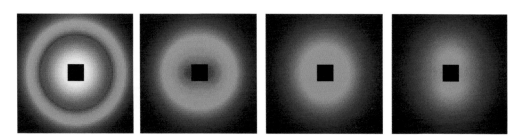

Figure 2-19

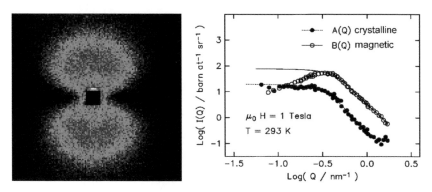

Figure 2-20

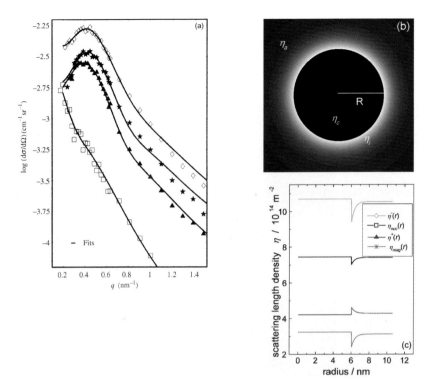

Figure 2-25

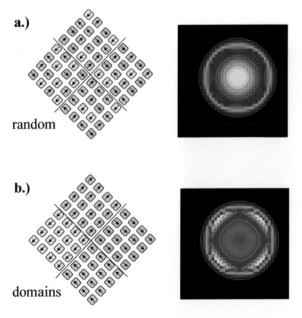

Figure 2-30

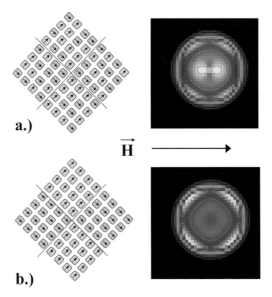

a.)

b.)

$\vec{H}$ ⟶

Figure 2-31

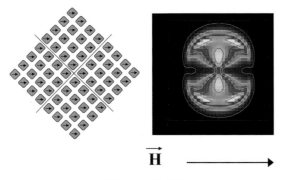

$\vec{H}$ ⟶

Figure 2-32

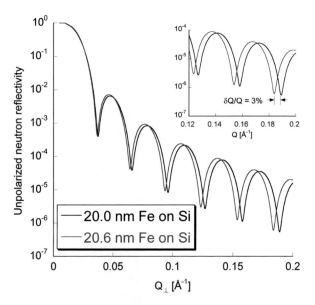

Figure 3-6

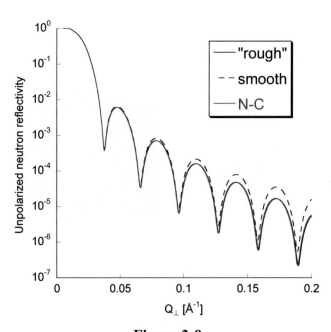

Figure 3-9

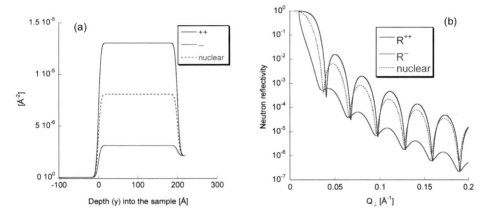

Figure 3-11

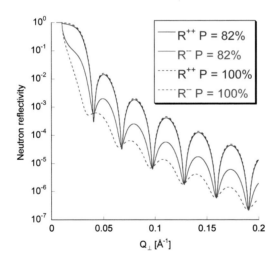

Figure 3-12

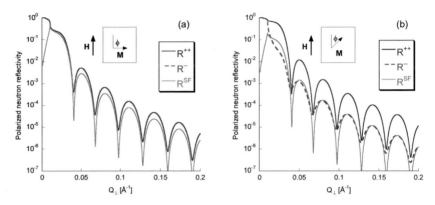

Figure 3-15

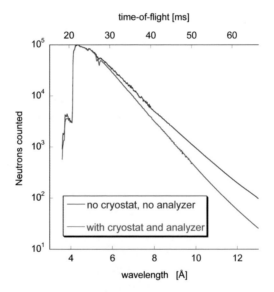

Figure 3-19

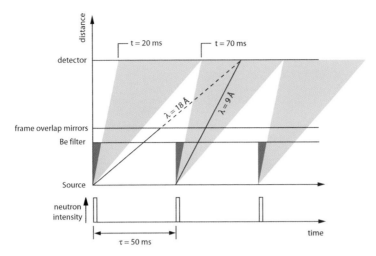

Figure 3-20

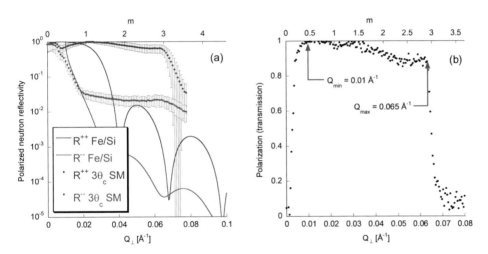

Figure 3-21

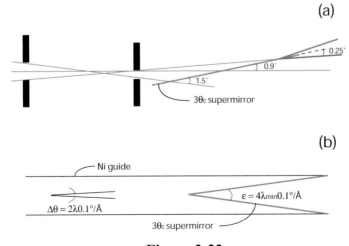

Figure 3-22

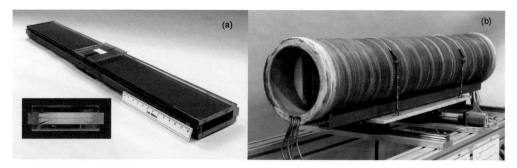

Figure 3-23

Figure 3-27

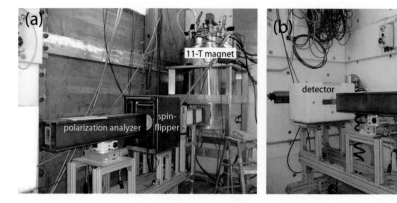

Figure 3-31

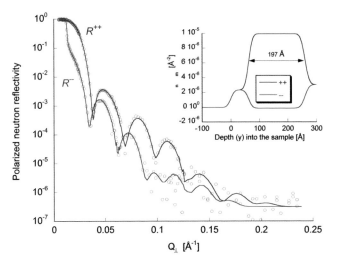

Figure 3-34

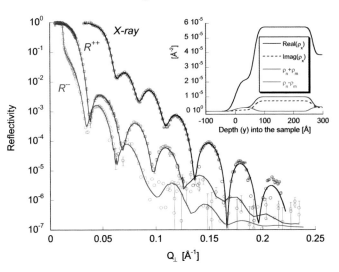

Figure 3-35

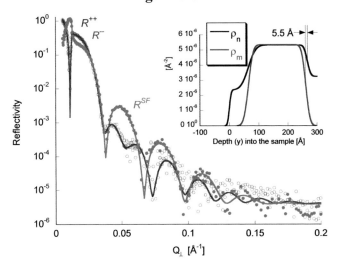

Figure 3-37

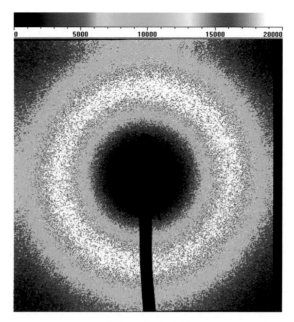

Figure 4-21

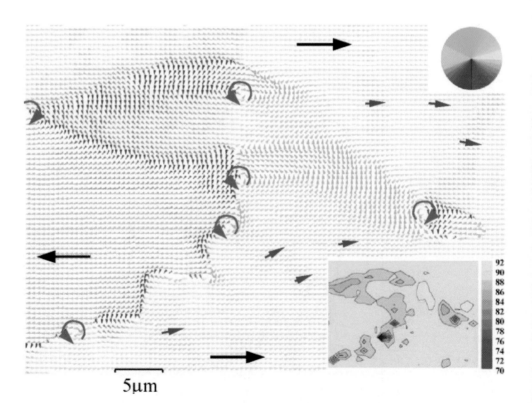

5µm

Figure 4-26

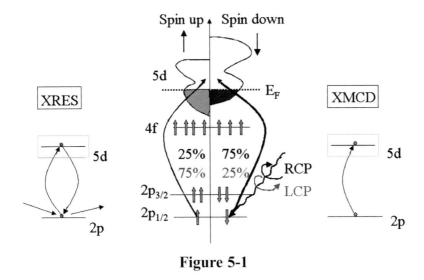

Figure 5-1

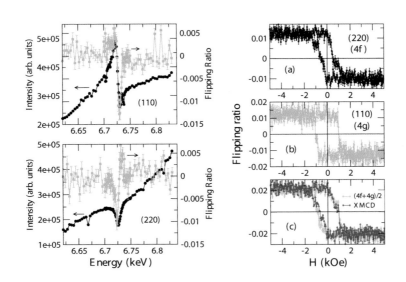

Figure 5-2

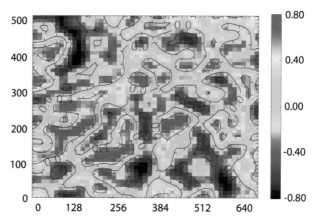

Figure 5-5

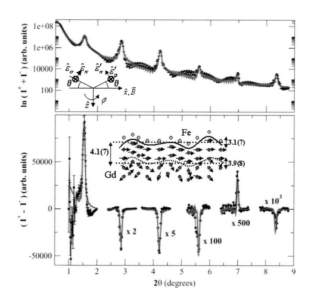

Figure 5-6

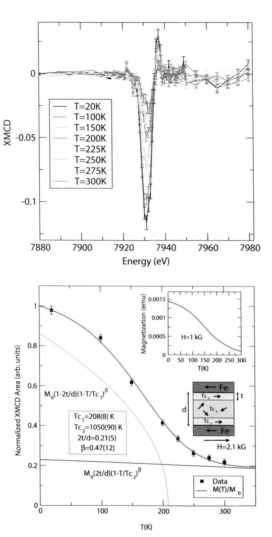

Figure 5-8

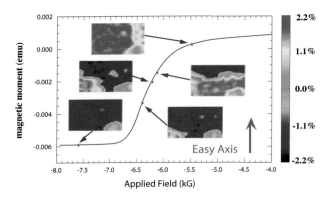

Figure 5-9

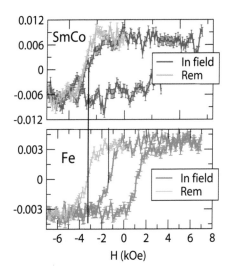

Figure 5-11

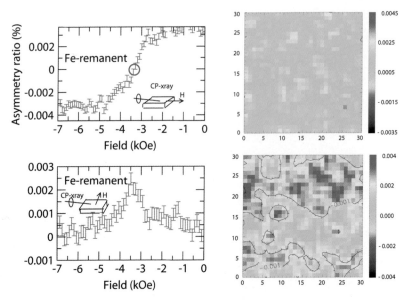

Figure 5-12

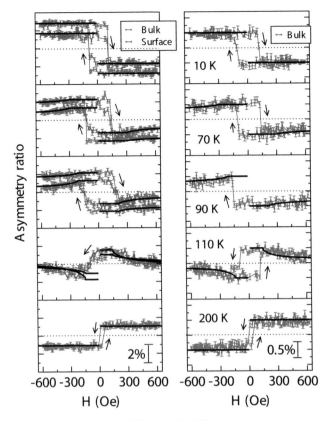

Figure 5-15

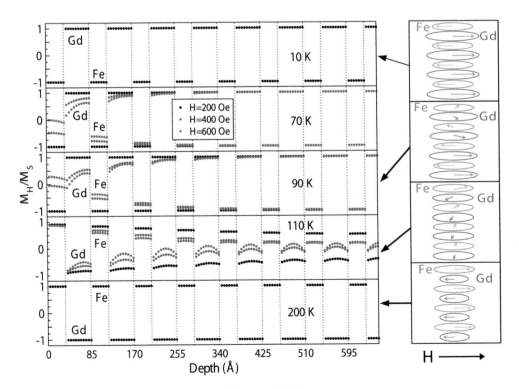

Figure 5-16

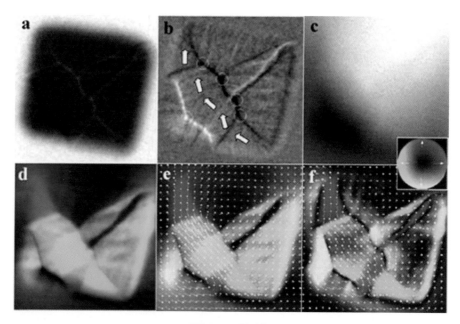

Figure 7-13

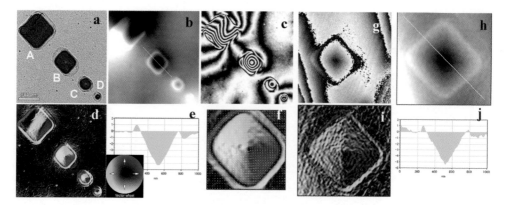

Figure 7-19

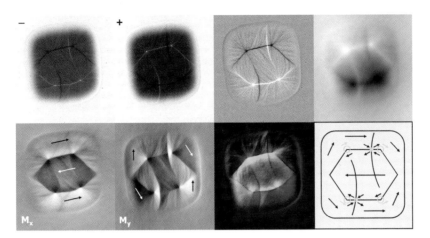

Figure 7-21

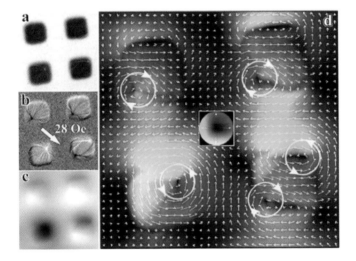

Figure 7-22

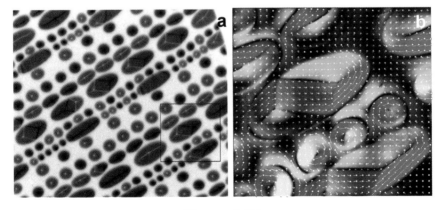

Figure 7-32

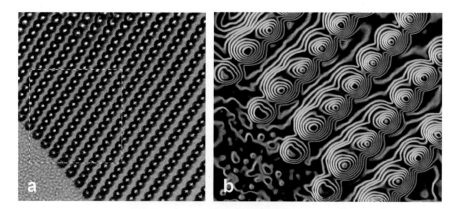

Figure 7-39

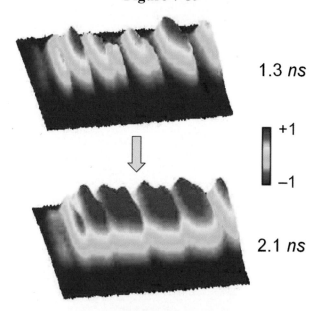

1.3 *ns*

+1

−1

2.1 *ns*

Figure 13-8

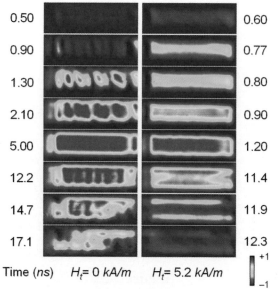

| Time (*ns*) | H_t= 0 kA/m | H_t= 5.2 kA/m |

Figure 13-10

which in Fourier space becomes

$$\varphi_{tot}(\mathbf{k}) = \sum_{j=1}^{N} \varphi_j(\mathbf{k}) e^{-i \mathbf{r}_j \cdot \mathbf{k}} \tag{7.79}$$

i.e. a Fourier series composed of N terms.

A simple example is shown in Fig.7-41: an array composed of four rectangular elements with different aspect ratios, each element having the same magnetization $\mathbf{m}$ unit vector, oriented at an angle $\beta=45°$. While the electrostatic component is still only responsible for the phase discontinuity at each element edges, as shown in Fig.7-41(b), the magnetic phase shift is now giving information about the interaction between the elements. Phase contour lines go from one element to another, revealing that the local field inside one rectangle is influenced by the nearby elements.

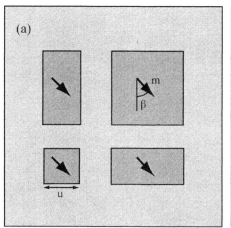 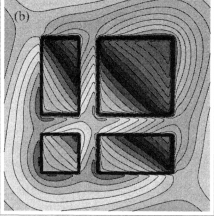

Figure 7-41: (a) Array of four rectangular elements with variable aspect ratio (1:1, 1:2, 2:1, 2:2). The element unit size u in the figure is 50 nm; (b) contour line plot of the total phase shift $\varphi_e + \varphi_m$, where each contour line represents a phase shift of $\pi/4$.

Moreover, we can extract some information on the effect of the demagnetizing field. In fact, the contour lines within the rectangular elements appear not to be oriented in a direction exactly parallel to the magnetization unit vector. They appear to be tilted at an angle which depends on the aspect ratio of the element. This is the combined effect of the projection of the magnetic field along the beam direction, typical of any TEM experiment, and of the demagnetizing field generated by each element of the array.

7.5.2.7. Nanoparticles of arbitrary shapes

Let us now consider the case of a polyhedral particle. There are two major reasons for studying polyhedral particles. First, the current interest in the properties of nanoparticles includes not only particles which have solidified or were otherwise formed into a polyhedral shape, but also artificially designed ones, such as patterned arrays of disks, and plates. A clear understanding of the magnetic behavior of such particles requires a knowledge of the effect of shape on the magnetization state at the nanoscale, and, perhaps more importantly, an analysis of the type of information that can be extracted from electron-optical observations on such particles. An experimental example of magnetic structure imaging in circular, square, triangular and, pentagonal Co and $Ni_{80}Fe_{20}$ films can be found in [101]. Second, the polyhedral shape can be analyzed analytically, using a formalism developed by Komskra [102]. Explicit expressions for the shape amplitude can be used to obtain an analytical expression for the electron-optical phase shift. This in turn, allows a detailed study of the types of images that can be obtained from either Lorentz microscopy or electron holography.

The formalism for describing particles of arbitrary shapes was outlined in [103]. We recall here the main results of the analysis, and provide a few examples of the current capabilities of the phase computation scheme.

The shape amplitude $D(\mathbf{k})$ of a polyhedral particle with E edges and F faces is given by [101] (using $\mathbf{k}=2\pi\mathbf{q}$):

$$D(\mathbf{k}) = -\frac{1}{k^2}\sum_{f=1}^{F}\frac{\mathbf{k}\cdot\mathbf{n}_f}{k^2-(\mathbf{k}\cdot\mathbf{n}_f)^2}\sum_{e=1}^{E_f}L_{fe}\mathbf{k}\cdot\mathbf{n}_{fe}\,\mathrm{sinc}\left(\frac{L_{fe}}{2}\mathbf{k}\cdot\mathbf{t}_{fe}\right)\exp\left(i\mathbf{k}\cdot\boldsymbol{\xi}_{fe}^{C}\right) \qquad (7.80)$$

This equation is only valid if the second denominator is non-zero. If $\mathbf{k}=\pm k\mathbf{n}_f$ (in other words, if $\mathbf{k}$ is parallel to any one of the face normals), then the contribution of that particular face (or faces) must be replaced by

$$D_f(\mathbf{k}) = i\frac{\mathbf{k}\cdot\mathbf{n}_f}{k^2}P_f\exp\left(-id_f\mathbf{k}\cdot\mathbf{n}_f\right) \qquad (7.81)$$

where P_f is the surface area of the face f, and d_f the distance between the origin and the face f. In the origin of Fourier space, the shape amplitude is equal to the particle volume, i.e. $D(\mathbf{0})=V$. The symbols in eq.(7.80) are defined as (see also Fig.7-41): $\boldsymbol{\xi}_{fe}^{c}$: coordinate vectors of the center of the edge e of face f; $\mathbf{n}_f$: unit outward normal to face f; $\mathbf{L}_{fe}$: length of the eth edge of the fth face; $\mathbf{t}_{fe}$: unit vector along the eth edge of the fth face, defined by

$$\mathbf{t}_{fe} = \frac{\mathbf{n}_f\times\mathbf{N}_{fe}}{|\mathbf{n}_f\times\mathbf{N}_{fe}|} \qquad (7.82)$$

where $\mathbf{N}_{fe}$ is the unit outward normal *on* the face which has the edge e in common with the face f; $\mathbf{n}_{fe}$: unit outward normal *in* the face f on the edge e defined by $\mathbf{n}_{fe}=\mathbf{t}_{fe}\times\mathbf{n}_f$.

The input parameters needed to complete this computation for an arbitrary polyhedron are the N_v vertex coordinates $\boldsymbol{\xi}_v$ and a list of which vertices make up each face (counterclockwise when looking towards the polyhedron center). All other quantities can be computed from these parameters.

A simple example of computation is shown in Fig.7-42, where the phase shift is calculated for two plates of hexagonal and triangular shape. While the triangular is just a curiosity (but it also can be used as a template for creating more complicated configurations), the hexagonal shape is very important in general. In fact, a uniformly magnetized hexagonal plate can be the basis for a 3D micromagnetic simulation, representing a cell in the grid. By the Fourier approach, it is possible to calculate not only the phase shift, but also the total magnetic induction starting from a known magnetization. If we allow a numerical approach, then non-uniform magnetizations can also be studied. This will be shown in the next section.

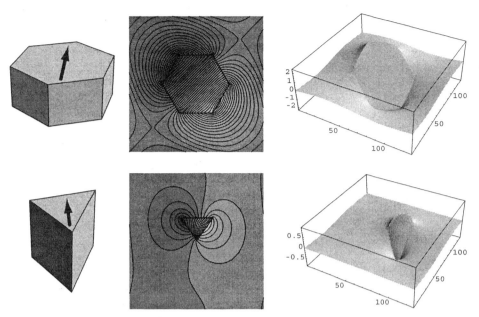

Figure 7-42: Magnetic phase shift for an hexagonal (upper row) and a triangular (lower row) plate. From left to right: shape outline with magnetization direction, phase contour plot (each contour line represents a phase variation of $\pi/25$) and phase surface plot. The phase is measured in radians, and is obtained considering a magnetization equivalent to 2 T in the material.

7.5.2.8. Micromagnetic simulations

In section 7.4 it was demonstrated that whenever fringing fields are present and/or the specimen has not constant thickness, it is not possible to retrieve a quantitative local magnetization map by means of a simple gradient operation on the retrieved phase. The maximum obtainable information in such cases is the projected local induction, which not necessarily is the information we are interested in. The magnetic induction **B** is, in fact, composed of two parts: the magnetization **M** and the demagnetizing field **H** generated by **M** (see also section 7.5.2.1) following the relation $\mathbf{B}=\mu_0(\mathbf{M}+\mathbf{H})$. It will be shown in this section how to make a further step: disentangle the demagnetizing field from the magnetization, and obtain a truly quantitative magnetization map by TEM.

The basic idea is to apply a statistical fitting procedure to the results of the experiment. Assuming that the phase shift has been retrieved from a magnetic structure by means of holography or TIE, we can calculate, starting from several magnetization states, the corresponding demagnetizing fields and then the total phase shifts. Finally, by comparing each of them with the experimental phase, we can extract the best-fit state, which will then represent the magnetization map of the structure. We will give an experimental example of this procedure in the following.

The theoretical framework developed in the previous sections, suitable for uniform or vortex states, can be extended to cover the case of a generic magnetization state. However, the results are only semi-analytical, and numerical analysis is required to obtain results in real space. Considering again eq.(7.59), namely the connection between magnetization and vector potential, we may note that it is not necessary to assume a particular state. The relationship is rather general, and can be used to calculate the phase shift for an arbitrary shape *and* magnetization state. If (m_x, m_y) are the two in-plane components of the magnetization state (two numerical arrays in this case), and $(\tilde{m}_x, \tilde{m}_y)$ are their numerical FFT, from eq.(7.59) we can directly evaluate the z-component of the vector potential in Fourier space:

$$A_z = -\frac{2i\mu_0 M_0}{k_z k^2}\left(\tilde{m}_x k_y - \tilde{m}_y k_x\right)\sin\left(k_z \frac{t}{2}\right) \quad (7.83)$$

where t is the thickness of the magnetic element and M_0 its magnetization. We are assuming that the specimen has flat surfaces, and that the magnetization is constant through the thickness (in other words, we are examining a 2D situation). Integrating eq.(7.83) along the optical axis z we can calculate the phase shift which, in Fourier space, turns out to be

$$\varphi_m = -\frac{i\pi\mu_0 M_0 t}{\phi_0}\frac{\tilde{m}_x k_y - \tilde{m}_y k_x}{k_x^2 + k_y^2} \quad (7.84)$$

The phase shift eq.(7.84) includes the effect of the demagnetizing field **H** generated by **M**, as demonstrated in section 7.5.2.1.

As input for the phase simulation series, we can either take an arbitrary magnetization distribution, suitably chosen as a possible candidate for generating the observed magnetic induction, or employ the results of micromagnetic simulations computed with slightly different parameters, such as relaxation constant, anisotropy coefficient, temperature, initial state. While the first choice does not require any additional computational effort, it is rather random, and it leads to a huge number of possible candidates for the match. On the other hand, the micromagnetic approach appears to be more physical, and has the side advantage of justifying from physical principles the choice of a particular state for the magnetic structure under investigation.

However, achieving quantitative agreement between micromagnetic output and retrieved phase remains problematic [98]. In fact, both the sample and the modeled data must refer to a sample with exactly the same geometry, history and experimental conditions. The second requirement may be relaxed if a match of experiment and simulation is attempted on a sample in a well defined state. For a magnetic element the easiest approach is to match the ground state (lowest energy) configuration. Further difficulties can arise if multiple structures are situated in close proximity as they can both influence the configuration of the ground state and distort the reference wave in the electron holography experiments. This complicates both the micromagnetic simulations and the reconstruction of electron holography data.

To test this approach leading to quantitative magnetization mapping, we prepared thin film samples of Permalloy elements with rectangular geometry on 50 nm thick silicon nitride membrane using electron beam patterning of poly(methyl methacrylate) (PMMA) photoresist followed by lift-off of a sputter deposited film. The Permalloy film was deposited in a field-free environment and, to prevent charging under the electron beam, the sample was coated with a 2 nm amorphous carbon layer. The composition of the Permalloy film was measured to be Ni 88% Fe 12% by EELS, with estimated accuracy of 10% based on inaccuracies of background subtraction and inelastic cross-section calculation [79]. For the thickness measurement of the sample, we calculated the inelastic mean free path of permalloy $Ni_{88}Fe_{12}$ to be 119 nm for 300 kV incident beam energy and 10 mrad collection semi-angle using the empirical formula by Malis [104]. After the subtraction of the substrate and coating, the measured thickness turned out to be (45±5) nm, which is in good agreement with the thickness expected from in-situ measurements by a crystal film thickness monitor.

A set of two holograms, one is shown in Fig.7-43(a), was acquired for phase retrieval with sample flipped between the acquisitions. The difference and sum of the phases from the two holograms, shown in Fig.7-43(c,d), allowed us to quantitatively separate the magnetic and electrostatic contributions to the phase shifts (see section 7.4.1). The sample was studied at room temperature with the main objective lens off. From the electrostatic component of the phase shift, and the measured thickness, we estimate the mean inner potential of the Permalloy to be $V_0 = (26 \pm 3)$ V.

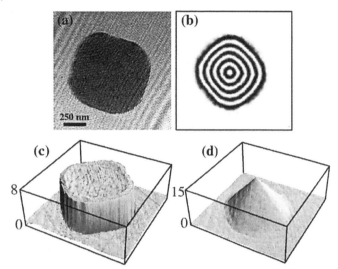

Figure 7-43: (a) Electron hologram of the magnetic element; (b) contour map (2× amplified) of the total phase shift; (c) electrostatic and (d) magnetic contributions to the phase shift (in radians) retrieved from the experiment.

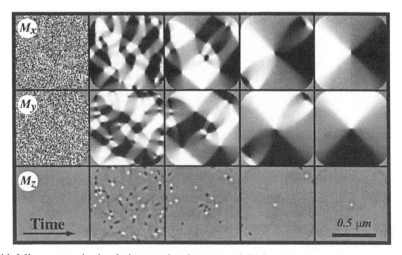

Figure 7-44: Micromagnetic simulations on the element: an initial random mangetization (left) evolves to a vortex state (right). The element size is 830×855 nm.

Micromagnetic simulations based on Landau-Lifsitz-Gilbert (LLG) equations [105], describing the dynamical evolution of the magnetization under a set of external constrains (temperature, applied field, material parameters, etc.) were performed on the element starting from the initial random configuration shown in the first column of Fig.7-44. To achieve the best possible match with experiment we used the in-focus image of the element as a template to define the shape in the micromagnetic simulation. The magnetization then evolves in a few nanoseconds to the final vortex state shown in the last column of Fig.7-44, passing through intermediate unstable states.

By eq.(7.84) it is possible to evaluate the magnetic phase shift for each step of the micromagnetic simulation. The resulting phase shift, obtained by an inverse FFT and corresponding to the last four states in Fig.7-44 are shown in Fig.7-45. The phase variations visible outside the element, emphasized in the 32× amplified contour map displayed in third column, can be ascribed to residual fringing fields: part of the magnetic field leaks to the vacuum outside, and this is a clear indication of the presence of stray fields surrounding the element.

As the match between the phase shift reported in Fig.7-45(d) and the experimental one displayed in Fig.7-43(b) is rather good, we can clearly state that the final state from micromagnetic, last column on the right in Fig.7-44, is the quantitative magnetization map of the magnetic element [106]. It should be emphasized that this approach allows us to obtain some information on the third component of the magnetization as well, one which is not accessible by the local projected induction map calculated by phase gradients. It was therefore possible to detect the presence of the central vortex with vertical magnetization in the magnetic structure we examined.

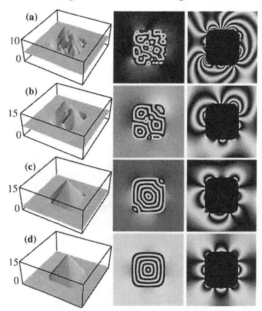

Figure 7-45: Phase shift (in radians) from micromagnetic simulations (last four states of Fig.7-44) displayed as a surface plot (first column), 2× (second column) and 32× (third column) amplified contour maps.

7.6. SUMMARY AND FUTURE PERSPECTIVES

Electrons, photons, and neutrons are three major and complementary probes used in characterizing materials. The strength of electron microscopy, however, is its ability to provide an extremely small bright probe as well as a suite of arsenal for nanoscale analysis, ranging from atomic imaging, nano-probe diffraction to x-ray and energy-loss spectroscopy. Very often, several techniques have to be used to understand the magnetic structure and physical properties of a sample. The magnetic information extracted from the sample has to be complemented by the knowledge of electronic structure, crystal symmetry, chemical composition, and atomic arrangement of the same sample. The current worldwide focus on nanoscience has promised to put electron microscopy techniques on a par with synchrotron x-ray and neutron methods for structural investigations.

In this chapter, we reviewed various TEM–based techniques for magnetic imaging and induction mapping. We discussed the image formation theory, the interaction of electrons with the sample, and the propagation of the electron wave through the microscope column to the detector. The concepts we provided are not limited to magnetic imaging, but are versatile with a wide range of applications for non-magnetic materials. We also offered the underlying physical principles of magnetic phase retrieval techniques coupled with extensive mathematical descriptions. We included the newly developed mathematic expressions for nanoscaled particles with basic and arbitrary geometries to explicate the observed magnetization and induction distribution. The image-simulation algorithms we presented can be used to compare directly with experiments.

Looking ahead into future magnetic imaging, the conventional Lorentz Frensnel microscopy, occasionally supplemented by the Foucault method, will still be the most widely used technique. It is easy to implement and can provide useful, but qualitative, information on domain wall and domain configuration. Information on the direction and amplitude of local magnetization and induction including fringing fields can be acquired using phase microscopy. Although off-axis holography is quantitative, it is technically demanding. In addition to the experiment set-up, it requires a high-quality reference wave next to the area of interest that may not always be available and its spatial resolution is limited by the spacing of the biprism carrier fringes. A new holographic phase shift technique acquiring a series of holograms has been developed to overcome the problem [107]. The TIE-based method is very promising, besides which, its versatility in length scale also opens a door to real-time phase retrieval for revealing magnetic dynamics. Nevertheless, care must be taken in dealing with the artifacts induced in its image reconstruction procedure, especially when Fourier transforms are used. For images with highly uneven contrast and intensities at their periphery, non-FFT method may be used to solve the TIE equation. The use of iterations based defocused images using maximum likelihood algorithms has been explored and demonstrated great potential [108].

The recent development on aberration correctors in electron microscopes is also noteworthy. It pledges new generation of microscopes with a large pole-piece gap for in-situ magnetization experiments while improving the spatial resolution of the objective lens (the so-called Lorentz lens with a long focal-length) with negligible field in the sample area for magnetic imaging. For current instrumentation, the expected minimum size for phase retrieval, which in terms of retrieving the magnetic signal from nanoparticles using electron holography, can be the order of 10 nm in particle size, has not yet been reached. In many areas of magnetic phase imaging, theory is ahead of experiment, especially for small objects. Experiments are in progress to push observations to the scale where transmission electron microscopy is the only technique available for the characterizing magnetic structure and properties of individual magnetic building blocks.

ACKNOWLEDGEMENTS

The authors would like to thank their collaborators V.V. Volkov, M.A. Schofield, M. De Graef, M. Malac, J. Lau, M. Freeman, and Z. Li for their contribution and useful discussions. The work at BNL was supported by Division of Materials Sciences, U.S. Department of Energy, under contract No. DE-AC02-98CH10886.

REFERENCES

[1]   R.P. Cowburn and M.E. Welland, *Science* **287** (2000) 1466.
[2]   J.B. Wedding, M. Li, and G.-C. Wang, *J. Magn. Magn. Mater.* **204** (1999) 79.
[3]   M.J. Freiser, *IEEE Trans*, **MAG-4** (1968) 152.
[4]   S.D. Bader, *J. Magn. Magn. Mat.* **100** (1991) 440.
[5]   J. Stohr et al. *Science* **259** (1993) 658.
[6]   C.T. Chen et al. *Phys. Rev. Lett.*, **75** (1995) 152.
[7]   M. Schlenker and J. Baruchel, *J. Appl. Phys.* **49** (1978) 1996.
[8]   G.E. Bacon, *Neutron Diffraction*, 3rd edition, Clarendon Press, Oxford (1975).
[9]   J.J. Saenz et al, *J. Appl. Phys.* **62** (1987) 4293.
[10]   P. Grutter, D. Rugar and H.J. Mamin, *Ultramicroscopy* **47** (1992) 393.
[11]   see for example, "*Magnetic imaging and its applications to materials*", Marc De Graef and Y. Zhu eds, Academic Press, (2001).
[12]   J.R. Banbury, W.C. Nixon, *J. Sci. Instrum.* **44** (1967) 889.
[13]   D.C. Joy and J.P. Jakubovics *J. Phys. D* **2** (1969) 1367.
[14]   G.A. Wardly, *J. Appl. Phys*, **42** (1971) 376.
[15]   J. Philibert and R. Tixier, *Micron* **1** (1969) 174.
[16]   D.J. Fathers, J.P. Jakubovics, and D.C. Joy, *Phil. Mag.* **27** (1973) 765.
[17]   D.J. fathers, J.P. Jakubovics, D.C. Joy, D.E. Newbury and H. Yakowitz, *Phys. Status Solidi* **A 20**, (1973) 535; **A22** (1974) 609.
[18]   K. Koike and K. Hayakawa, *Appl. Phys. Lett.* **45** (1984) 585.
[19]   J. Unguris, D.T. Pierce, A. Galejs and R.J. Celotta, *Phys. Rev. Lett.* **49** (1982) 72.
[20]   M. R. Scheifein, J. Unguris and M.H. Kelley, P.T. Pierce and R.J. Celotta, *Rev. Sci. Instr.* **61** (1990) 2501.
[21]   E. Bauer and W. Telieps, *Emission and low energy reflection electron microscopy*" in *Surface and interface characterization by electron optical methods*, A. Howie and U. Valdre Ed., Plenum, New York, 1988, pp. 195-233.
[22]   H. Poppa, E. Bauer and H. Pinkvos, *MRS Bull.* **20** (1995) 38.
[23]   T. Duden and E. Bauer, *Phys. Rev. Lett.* **77** (1996) 2308.
[24]   P.B. Hirsch, A. Howie, R.B. Nigholson, D.W. Pashley and M.J. Whelan, *Electron Microscopy of Thin Crystals*, Lodon, Butterworths, 1965
[25]   J.D. Livingston and J.J. Becker, *Trans. Amer. Inst. Met. Engrs.*, **212** (1958) 316.
[26]   J.D. Livingston, Trans. *Amer. Inst. Met. Engrs.*, **215** (1959) 566.
[27]   M.E. Hale, H.W. Fuller and H. Rubinstein, *J. Appl. Phys.* **30** (1959) 789.
[28]   H.W. Fuller and M.E. Hale, *J. Appl. Phys.* **31** (1960) 238.
[29]   H. Boersch, H. Raith and D. Wohlleben, *Z. Phys.* **159** (1960) 388.
[30]   J.P. Jakubovics: *Lorentz microscopy*, in *Handbook of Microscopy*, Vol.1, S. Amelinckx et al. Ed., VCH, Weinheim (1997) 505.
[31]   J.N. Chapman, P.E. Batson, E.M. Waddell and R.P. Ferrier, *Ultramicroscopy* **3** (1978) 203.
[32]   E.M. Waddell and J. N. Chapman, *Optik* **54** (1979) 83.

[33]   J. N. Chapman, *J. Phys. D: Appl. Phys.* **17** (1984) 623.
[34]   N. Chapman, I.R. McFadyen and S. McVitie, *IEEE Trans. Magn.* **26** (1990) 1506.
[35]   A. Tonomura et al., *Phys. Rev. Lett.* **44** (1980) 1430.
[36]   A. Tonomura, *Rev. Mod. Phys.* **59** (1987) 639; *J. Magn. Magn. Mat.* **31** (1983) 963.
[37]   *Introduction to electron holography*, E. Volkl, L.F. Allard and D.C. Joy eds, Kluwer Academic/Plenum Publishers, New York 1999.
[38]   D. Paganin and K.A. Nugent, *Phys. Rev. Lett.*, **80** (1998) 2586.
[39]   V.V. Volkov, Y. Zhu and M. De Graef, *Micron* **33** (2002) 411.
[40]   V.V. Volkov and Y. Zhu, *Phys. Rev. Lett.* **91** (2003) 0439041.
[41]   M.R. Teague, *J. Opt. soc. Am.* **73** (1983) 1434.
[42]   F.A. Lenz, in *Electron microscopy in material science* **2**, Proc. Int. School of El. Micr. Erice (Ed. U. Valdrè, Academic Press, NewYork, 1971).
[43]   M. De Graef, in *Magnetic Microscopy and its Applications to Magnetic Materials*, M. De Graef and Y. Zhu Eds., chapter 2 (Academic Press, 2001) pp. 27-67.
[44]   M. Haider, *et al.*, *Electron microscopy image enhanced. Nature*, **392** (1998)768.
[45]   P.E. Batson, N. Dellby and O.L. Krivanek, *Nature,* **418** (2002) 617.
[46]   C.L. Jia, M. Lentzen, K. Urban, (Science) **299** (2003) 870.
[47]   R. Jagannathan, *Phys. Rev. A* **42** (1990) 6674.
[48]   P.W. Hawks and E. Kasper, *Wave Optics*, vol. 3 (Ed. Academic Press, 1994).
[49]   M.A. Schofield, Y. Zhu, L. Wu, V.V. Volkov and M. Malac, *JEOL News: Electron Optics Instrumentation*, **36E** (2001) 2.
[50]   V.V. Volkov, D.C. Crew, Y. Zhu and L.H. Lewis, *Rev. Sci. Instruments* **73** (2002) 2298.
[51]   J. Chapman, *J. Phys. D* **17** (1984) 623.
[52]   P.F. Fazzini, P.G. Merli and G. Pozzi, proc. of 15th International Crogress on Electron Microscopy, vol.1, p.327, Durban, South Africa, Sept. 1-6, 2002.
[53]   V.V. Volkov, and Y. Zhu, *J. Magn. Magn. Mater.* **214** (2000) 204.
[54]   A. Hubert and R. Schafer, *Magnetic Domains,* Springer-Verlag, 1998.
[55]   P.G. Merli, G.F. Missiroli and G. Pozzi, *Phys. Stat. Sol.* **30** (1975) 699.
[56]   T. Yoshida, M. Beleggia, J. Endo, K. Harada, H. Kasai, T. Matsuda, G. Pozzi and A. Tonomura. *J. Appl. Phys.* **85** (1999) 4096.
[57]   M. Beleggia, G. Pozzi, K. Harada, H. Kasai, T. Matsuda, T. Yoshida and A. Tonomura. *Microscopy & Microanalysis* **3(2)** (1997) 509.
[58]   J.N. Chapman, A.B. Johnston and L.J. Heydeman, *J. Appl. Phys* **76** (1994) 5349.
[59]   J.N. Chapman, A.B. Johnston, L.J. Heydeman, S.Mc Vitie, W.A.P. Nicholson and B. Bormans, *IEEE Trans. Magn.* **MAG-30** (1994) 4479.
[60]   A.B. Johnston, J.N. Chapman, *J. Microsc.* **179** (1995) 119.
[61]   M. Beleggia, S. Fanesi, R. Patti and G. Pozzi. *Materials Characterization* **42** (1999) 209.
[62]   J.F. Herbst, *Rev. Mod. Phys.*, **63** (1991) 819.
[63]   E. Burzo, *Rep. Pro. Phys.*, **61** (1998) 1099.
[64]   J. W. Lau, M. A. Schofield, Y. Zhu, and G. F. Neumark, *Microscopy & Microanalysis,*. **9(2)** (2003) 130.
[65]   V.V. Volkov, Y. Zhu and M. Malac, *Phil. Mag. A*, **84** (2004) 2607.
[66]   D. Paganin and K. Nugent, *Phys.Rev.Lett.* **80** (1998) 2586.
[67]   V.V. Volkov, Y. Zhu and M. De Graef, *Micron* **33** (2002) 411.
[68]   M. Beleggia, M.A. Schofield, V.V. Volkov and Y. Zhu, D01: 10.1016/ j.ultramic.2004.08.004, *Ultramicroscopy* (2004) in press.
[69]   M. Borne and E. Wolf, *Principles of optics*, 7th ed., Cambridge University Press (1999).
[70]   V.V. Volkov, M.A. Schofield and Y. Zhu, *Mod. Phys. Lett. B* **17** (2003) 791.
[71]   Y. Zhu, V.V. Volkov and M. DeGraef, *J. Electron Microsc.* **50** (2001) 447.

[72] D. Gabor, *Nature* **161** (1948) 777.
[73] G. Möllenstedt and H. Düker, *Naturwissenschaften* **42** (1955) 41.
[74] A. Tonomura, *Advances in Physics* **41** (1992) 59.
[75] G. Matteucci, G.F. Missiroli and G. Pozzi in *Advances in imaging and electron physics* **99** (1998) 171 (P.W.Hawkes Ed., Academic Press, 1998).
[76] M.A. Schofield, L. Wu and Y. Zhu, *Phys. Rev. B*, **67** (2003) 224512.
[77] Y. Aharonov and D. Bohm, *Phys. Rev.* **115** (1959) 485.
[78] G.F. Missiroli, G. Pozzi and U. Valdre', *J. Phys. E: Sci. Instrum.* **14** (1981) 649.
[79] R. F. Egerton, *"Electron Energy-Loss Spectroscopy in the Electron Microscope"*, Plenum, New York, 1986.
[80] J. Chapman, *Mat. Sci. Eng. B* **3** (1989) 355.
[81] M. Beleggia, S. Tandon, Y. Zhu and M. De Graef, *Phil. Mag. B* **83** (2003) 1143.
[82] Y. Murakami, J.H. Yoo, D. Shindo, T. Atou, and M. Kikuchi, *Nature*, **423** (2003) 965.
[83] H. Lichte, private communication
[84] I.S.Gradshtein and I.M. Ryzhik, *Table of Integrals, Series and Products,* Academic Press, San Diego, 1980.
[85] M. Beleggia, P.F. Fazzini and G. Pozzi, *Ultramicroscopy* **96** (2003) 96.
[86] M.A. Schofield and Y. Zhu, Proc. of *the 7th International Symposium on Advanced Physical Fields: Fabrication and characterization of nano-structured materials,* Nov.12-15, 2001, Tsukuba, Japan, p41-44.
[87] M.R. Mc Cartney and Y. Zhu, *Appl. Phys. Lett.* **72** (1998) 1380.
[88] Y. Zhu and M. Mc Cartney, *J. Appl. Phys.* **84** (1998) 3267.
[89] M. Beleggia and Y. Zhu. *Philos. Mag. B* **83** (2003) 1043.
[90] S. Tandon, M. Beleggia, Y. Zhu and M. De Graef, *J. Magn. Magn. Mater.* **271** (2004) g9.
[91] W.F. Brown, Jr., *Ann. N.Y. Acad. Sci.* **147** (1969) 463.
[92] A. Aharoni, *J. Appl. Phys.* **90** (2001) 4645.
[93] M. Mansuripur, *The Physical Principles of Magneto-optical Recording*, Cambridge University Press, 1995.
[94] A.A. Thiele, *Bell Syst. Tech. Journal* **48** (1969) 3287.
[95] A.A. Thiele, A.H. Bobeck, E. Della Torre and U.F. Gianola, *Bell Syst. Tech. Journal* **50** (1971) 711.
[96] A.A. Thiele, *Bell Syst. Tech. Journal* **50** (1971) 725.
[97] A. Tonomura, *Electron Holography* (Ed. Springer, Berlin, 1993).
[98] R.E. Dunin-Borkowski, M.R. McCartney, D.J. Smith and S. Parkin, *Ultramicroscopy* **74** (1998) 61.
[99] R.E. Dunin-Borkowski, M.R. McCartney, R.B. Frankel, D.A. Bazylinski, M. Posfai and P.R. Buseck, *Science* **282** (1998) 1868.
[100] S.L. Tripp, R.E. Dunin-Borkowski and A. Wei, *Angew. Chem. Int. Ed.* **42** (2003) 5591.
[101] K.J. Kirk, S. McVitie, J.N. Chapman and C.D.W. Wilkinson, *J. Appl. Phys.* **89** (2001) 7174.
[102] J. Komskra, *Optik,* **80** (1987) 171.
[103] M. Beleggia, Y. Zhu, S. Tandon and M. De Graef. *Philos. Mag. B* **83** (2003) 1143.
[104] T. Malis, S.C. Cheng, R.F. Egerton, *J. Electron Microsc. Tech.* **8** (1988) 193.
[105] Y. Nakatani, Y. Uesaka, N. Hayashi, *Jap. J. Appl. Phys.* **28** (1989) 2485.
[106] M. Beleggia, M.A. Schofield, Y. Zhu, M. Malac, Z. Liu and M. Freeman, *Appl. Phys. Lett.* **83** (2003) 1435.
[107] F.R. Chen, unpublished.
[108] K. Yamamoto, I. Kawagiri, T. Tanji, M. Hibino and T. Hirayama, *J. Electr. Microsc.* **49** (2000) 31.

CHAPTER 8

Spin-polarized scanning
electron microscopy

8.1. INTRODUCTION

This chapter describes and reviews a domain-imaging technique called spin-polarized scanning electron microscopy (spin-SEM) or scanning electron microscopy with polarization analysis (SEMPA) [1-4]. One of its premier characteristics is its very high surface sensitivity, making the technique ideal for investigating phenomena related to magnetism in ultrathin films and small structures. Already before the development of this technique, magnetic contrast could be observed in scanning electron microscopy because the path of the electrons is deflected by the Lorentz force whose origin is either the stray field above or the magnetic field within the specimen. Here however, domain contrast is only superimposed on the topographic image and generally very small. Because of this low contrast, spin-SEM has essentially replaced the older, established contrast mechanisms in scanning electron microscopy. Interestingly, the technique was originally developed to observe magnetic domain patterns and domain walls in bulk ferromagnets with high resolution under "realistic" conditions, i.e., without the need to thin the specimens like in transmission electron microscopy. Spin-SEM has been applied to enhance our understanding of domains in amorphous as well as crystalline materials, for fundamental studies on the origin of ferromagnetism in two-dimensional systems as well as for technologically relevant problems such as the investigation of the transition region between written bits or the performance of write heads in magnetic data storage.

The chapter is organized as a short tutorial on several relevant quantities in ferromagnetism rather than a mere illustration of the strength of the technique. After a description of the principle and technical implementation of the technique, domain patterns and domain formation in bulk are compared with domains in thin films and small structures. A similar comparison is then made for the domain walls. Studies on magnetic anisotropy and magnetization reversal are presented next, followed by a section on phase transitions. Finally, the exchange coupling between ferromagnetic layers across a metallic spacer is illustrated by one specific example.

8.2. TECHNIQUE

8.2.1. Principle

Spin-polarized scanning electron microscopy is an offspring of standard scanning electron microscopy, the difference being that the spin polarization of the secondary electrons can be used to form a magnetic image of the sample surface. That secondary electrons emitted from a ferromagnetic specimen are spin-polarized was suspected since the early days of spin-polarized electron spectroscopy, and proven in 1976 in the experiment by Chrobok and Hofmann [5]. Two years later DiStefano proposed that spin-polarized secondary electrons could be used for magnetic domain viewing [6]. Interestingly, the idea for this highly successful magnetic imaging tool came as a byproduct in the search for a beam-addressable memory device based on an electron gun and a spin detector. While it took only a few more years to build the first working experimental spin-polarized scanning electron microscope, the memory device was never realized. In 1984, Koike and Hayakawa [7] combined an electron gun having a beam diameter of 10 µm with a spin detector to visualize magnetic domains on Fe(001) single crystals. Shortly afterwards, Unguris et al. [8] modified an ultrahigh-vacuum SEM by attaching a home-built spin analyzer to image magnetic domain patterns. Since these days, two acronyms exist for the same method: Spin-SEM was chosen by Koike et al., whereas Unguris et al. coined the term SEMPA—scanning electron microscopy with polarization analysis. Currently there are about ten systems in use at laboratories on three continents. These systems mainly differ in the type of spin detector used for the polarization analysis and in the magnetization components they are able to determine. In addition, each of these systems has further specific tools attached, for example to anneal or cool the sample during imaging, to apply magnetic fields, to perform reflection high-energy electron diffraction (RHEED) for structural characterization, or to investigate surface morphology by scanning tunneling microscopy.

The principle of spin-SEM or SEMPA as well as a photograph of its experimental realization are shown in Fig.8-1. The system consists of an SEM equipped with a spin-polarization detector. As in SEM, a focused beam of high-energetic unpolarized electrons scans along a specimen surface. These primary electrons scatter at the electrons close to the surface of the sample in various ways. The predominant mechanism is inelastic scattering: the primary electron transfers some of its energy to an electron of the sample. As in most cases the incoming electron loses only a small amount of its energy, this process occurs repeatedly until the electron has essentially lost its entire energy and a cascade of excited low-energetic electrons has been created. A considerable number of these secondary electrons travels back to the surface, eventually undergoes additional elastic and inelastic scattering events, and finally might exit the sample if the energy is still sufficient to overcome the vacuum level. The number of these electrons depends on the local curvature of the surface and the local work function. Hence an image of the sample topography and its chemistry is obtained by recording the number of these electrons for each position of the incoming beam. For ferromagnetic samples, a net spin density is present in the sample that directly relates the magnetization M to the spin imbalance of up and down spin electrons: $M = -\mu_b (N\uparrow - N\downarrow)$, where $N\uparrow$ ($N\downarrow$) is the number of spins per unit volume aligned parallel (antiparallel) to the magnetization, and μ_b is the Bohr magneton. The minus sign ensures that the spin of the negatively charged electron is antiparallel with respect to its magnetic moment. Because of this spin imbalance in ferromagnets, the emitted secondary electrons are spin-polarized, with a spin polarization P that can be conveniently defined as $P = (N\uparrow - N\downarrow)/(N\uparrow + N\downarrow)$. Thus, by measuring the spin polarization along a certain direction in space, a map of the magnetization component in this direction is obtained.

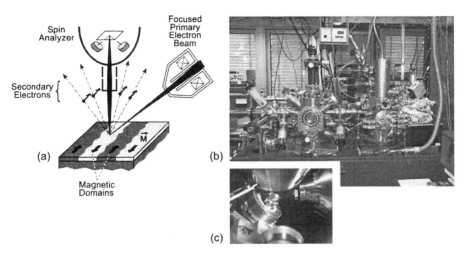

Figure 8-1: (a) The principle of spin-SEM: An unpolarized electron beam focused by electromagnetic lenses scans along a ferromagnetic surface, thereby exciting secondary electrons. Those electrons whose energy is higher than the vacuum level are emitted into vacuum, collected by an electron optics that transfers the electrons with low energy to the spin analyzer. (b) The spin-SEM setup at the IBM Zurich Research Laboratory: The spin-SEM chamber is at the right-hand side, the spin analyzer (not visible) at the rear with a Faraday cage for high-voltage protection. The preparation and analysis chamber is positioned in the center, the introduction chamber with air lock is to the left. (c) A close up of the sample region with the objective lens of the primary electron beam at the top and secondary electron transport optics to the right. The sample is tilted by 45°.

While this simplified picture describes the essentials, the reality of course is more complex. The spin polarization of the secondary electrons is strongly dependent on the secondary electron energy, see Fig.8-2. The polarization at energies above typically 10 eV is close to the one calculated from the spin imbalance of the bands near the Fermi level. For the $3d$ transition metals Fe, Co, and Ni, these values are 28%, 19%, and 5%, which reasonably agree with the measured values. At lower energies, the spin polarization increases and peaks close to the vacuum level with a value that typically is 2 to 3 times larger, reaching $P = 50\%$ in the case of Fe. This enhancement at very low energies is attributed to preferential inelastic scattering of $\downarrow$-spin electrons, which leads to a higher escape probability for $\uparrow$-spin electrons. As the intensity of secondary electrons is also highest at the lowest energies, an efficient instrument for magnetic imaging collects the abundance of electrons from a relatively large energy window (typically 0 to 10 eV) exploiting their high spin polarization. It is this advantageous combination of high intensity and polarization enhancement that allows us to image magnetic patterns in ultrathin films that are only a single atomic layer thick.

An important characteristic of spin-SEM is the complete insensitivity of the magnetic pattern to spurious effects stemming from fluctuations in the incoming beam current and from varying secondary electron emission at different specimen topographies. The reason is that the spin polarization as defined above is a normalized quantity: therefore changes in the number of emitted electrons cancel out. The total number N of electrons emitted, $N = (N\uparrow + N\downarrow)$, is a direct map of the secondary electron yield. Thus when the spin polarization is measured, the topographic map comes for free.

An illustration of the reconstruction of topographic and magnetic images is shown in Fig.8-3 for an epitaxial ultrathin Fe film grown on Cu(001). This film has a thickness of only 3 monolayers (ML), and its preferred magnetization direction is perpendicular to the surface plane. Two individual images are measured in the spin analyzer that is sensitive to this magnetization direction, the one for $N\uparrow$ and the one for $N\downarrow$. Both images display topographic features such as scratches and defects on the surface, but superimposed is a contrast that reverses between the two images. From these two images, the maps of the sum and of the normalized difference are calculated, see Fig.8-3(c) and (d). As expected, the defects visible in the topographic map do not show up in the magnetic image, and no magnetic contrast remains superimposed on the topography map. Thus, spin-SEM not only provides topography and magnetic information simultaneously, but completely separates the two. On the other hand, in this example a correlation of structure and magnetism is clearly identified: Most domains actually pin at scratches visible in the topography, as can be concluded straightforwardly from Fig.8-3(a) and (b).

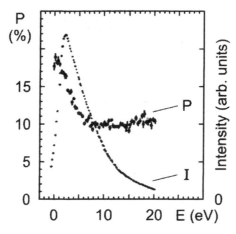

Figure 8-2: Energy dependence of the intensity I and spin polarization P of secondary electrons emitted from an Fe(001) single crystal covered by 20 L O_2. At energies above ~8 eV, the polarization is approximately constant. At low energies, a large enhancement of the spin polarization is observed. Similarly, the intensity peaks at low energy. The polarization values are lower than those from clean Fe because of the O_2 coverage. Adapted from Ref. [9].

These images exemplify another very prominent feature of the technique: its high surface sensitivity. Only a few techniques exist that are able to image magnetic domains in films of thicknesses down to a few atomic layers. Spin-SEM not only has the sensitivity to detect the polarization signal in very thin magnetic layers, it is in addition essentially "blind" to the bulk magnet below the top surface. The reason is the short probing depth of the technique. The probing depth is a convolution of the attenuation length of the incoming primary electron and the escape depth of the outgoing secondary electron. The attenuation length for inelastic scattering of an incoming primary electron of 5 keV energy is only about 4 nm [10], and reduces to ~2.5 nm for our incidence angle of 45°. The secondary electrons can also undergo scattering on their way to the surface before escaping into vacuum, and hence elastic and inelastic scattering lengths have to be considered. Experimentally it is found that for metals the escape depth is limited to a few nanometers [11], in particular if empty states are available close to the Fermi level [12,13] and secondary electrons up to several tens of eV are collected. This means that it is the short attenuation length of the incoming primary electron and the abundance of elastic scattering that make SEM techniques so surface-sensitive [14]. For metallic materials, the probing depth amounts to 1 to 2 nm. A typical example of how this probing depth can be determined is shown in Fig.8-4. A ferromagnetic film of Fe is covered by a shallow wedge of Ag. The spin polarization of the secondary electrons leaving the Fe specimen is attenuated upon Ag coverage, weakening the domain contrast appreciably. The quantitative determination of this attenuation leads to an accurate measurement of the probing depth. In the case of an Ag overlayer, the probing depth is 1.1 nm [15].

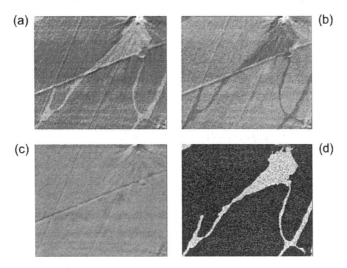

Figure 8-3: Image acquisition in spin-SEM illustrated by a perpendicularly magnetized 3-ML fcc-Fe/Cu(001) film. The spin detector signals $N\uparrow$ and $N\downarrow$ are reproduced in (a) and (b), respectively. From these, the total number of electrons, $N\uparrow + N\downarrow$, is calculated and corresponds to the topographic image (c). The normalized difference, $(N\uparrow - N\downarrow) / (N\uparrow + N\downarrow)$, is the spin polarization, which is proportional to the perpendicular magnetization component (d). A magnetic pattern with up (white) and down (black) domains is obtained. Magnetic domains are observed to pin at structural defects, but topography and magnetization are completely separated in (c) and (d). The simultaneously acquired in-plane polarization component (not shown) is homogeneously grey because the film is fully magnetized out-of-plane. Beam parameters: primary energy 2 keV, beam current <1 nA. The data-acquisition time is 20 ms per pixel; image area: 140 μm × 124 μm.

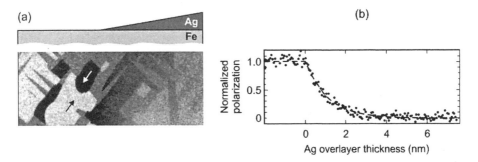

Figure 8-4: (a) Attenuation of spin polarization in an Fe(001) single crystal covered by an Ag wedge. The start of the wedge is indicated in the figure, and the Ag thickness increases linearly up to 2.6 nm at the right-hand edge of the image. The arrows indicate the magnetization directions. (b) Linescan along an Ag wedge after the Fe specimen has been brought to a single-domain state. This simplifies the interpretation of the result by avoiding different contrast from domains with different magnetization directions. Adapted from Ref. [15].

This high surface sensitivity has several direct consequences. It means that even in a bulk sample, only the surface is imaged. This fact led, for instance, to the discovery that a domain wall in a bulk ferromagnet, such as Fe(001), which is of Bloch wall type, is terminated as a Néel wall at the top surface [16, 17], as will be shown in Section 8.4.1. It also means that domain imaging by spin-SEM requires clean, uncovered surfaces. Therefore, an ultrahigh-vacuum environment is necessary.

Clean magnetic surfaces can be made in various ways. For fundamental studies in magnetism, *in-situ* grown epitaxial films and structures are the obvious choice. For technologically relevant samples, usually a mild ion-bombardment procedure is sufficient to remove a few atomic layers at the surface to reveal a magnetic top surface layer. Because the spin polarization of the secondary electrons is essentially the same regardless of whether a specimen is single-crystalline or amorphous, the technique can also be applied in cases in which some structural damage might result from the milling process. The third method to achieve a magnetic top layer on a contaminated sample is inspired by the well-known colloid decoration technique [18,19]: A small amount of a ferromagnet is deposited on top of the magnetic specimen that has too small a spin polarization in the near-surface region to be imaged successfully. Typically, a few atomic layers of Fe are deposited *in-situ*. The thin overlayer assumes a domain configuration that is identical to that of the underlying specimen in the near-surface region, either by exchange coupling or through dipolar fields. This "marker" technique has been proposed for enhancing the spin-polarization signal [20], but can also be used to image the stray field of patterned magnetic structures [21]. An example of stray-field imaging is given in Fig.8-5(a), which shows a patterned track in an experimental hard disk. In the nonmagnetic region to the left and right of the track, the stray field from the magnetic bits is sufficiently large to magnetize the thin Fe marker film on a length scale extending beyond a micrometer. The decoration technique extends the use of spin-SEM to areas of research that are seemingly unrelated to thin-film magnetism, e.g., to the geophysically relevant problem of magnetic self-reversal in rocks [22]. An example is given in Fig.8-5(b), which shows the domain pattern in a titanomagnetite rock sample after decoration with 2.5 nm of Fe. Without Fe adlayer, the oxide sample has an excessively small spin polarization and, in addition, tends to charge because it is poorly conducting. The limits for this technical trick clearly are related to the overlayer film thickness. It must be thin enough so that the magnetic dipolar energy of the overlayer remains small compared to that of the sample being investigated, to avoid a modification of the domain pattern in the sample. There is no lower thickness limit, on the other hand, in contrast to speculations in the original work [20]. In particular, it is not even necessary that the overlayer be a ferromagnet below its Curie temperature. Indeed we have observed that a paramagnetic Fe film having a thickness of less than two atomic layers becomes polarized in the exchange field of the specimen under study [23].

Spin-SEM stands out among the several magnetic microscopy techniques because of its high lateral resolution, which is determined primarily by the probing electron-beam diameter at beam currents that are high enough to form a magnetic image with a sufficient signal-to-noise ratio. The magnetic resolution routinely is below 50 nm, i.e., it is superior to that of most other domain-imaging methods such as classical optical microscopies (magneto-optical Kerr microscopy [24] or the colloid technique), which are diffraction-limited to several hundred nanometers. A similar resolution is obtained with other electron microscopy techniques, such as photoelectron emission microscopy [25] or spin-polarized low-energy electron microscopy [26], and scanning probe techniques [27]. The only routinely available technique with an even better resolution is transmission electron microscopy (Lorentz microscopy, electron holography) [28], which, however, requires sophisticated sample preparation, and spin-polarized STM with its atomic scale resolution [29]. The published best resolution obtained in spin-SEM is below 10 nm, and has been achieved by increasing the beam current, by reducing the distance between the SEM objective lens and the sample, and by improving the aberration of the electron optical lens system [30]. Figure 8-6 illustrates the high resolution capability of the technique. A domain wall in the permanent magnet $SmCo_5$ has been imaged. From differential-phase contrast transmission electron microscopy, the wall width has been deduced to be 2.6 nm [31], compatible with the large anisotropy of the material. The experimental line profile gives an experimental transition width of <10 nm. The fit to the data in Fig.8-6 is a convolution of the intrinsic domain-wall width and the finite beam diameter. From this fit, it was concluded that the spin-SEM tool has a resolution of 5 nm, which is the best published to date.

Figure 8-5: Two examples of the overlayer technique: (a) A patterned track in an experimental hard disk. The left image shows the topography: the two magnetic tracks appear bright, the nonmagnetic material dark. A dust particle is attached at the track. The right image gives the magnetic contrast of the written bits in the center track, the left track is uniformly magnetized. The Fe marker film deposited over the entire sample enhances the polarization contrast on the track and is magnetized in the stray field of the bits on the sides of the track. The bit size is huge compared with today's standards; image size: 10 μm × 10 μm. (b) Magnetic domains in a titanomagnetite ore grain imaged after covering the specimen with 2.5 nm of Fe. The irregular stripe pattern is indicative of a large uniaxial, but locally varying anisotropy caused by stress in the sample. The grain is surrounded by grains of different composition and hence different domain patterns. Image size: 25 μm × 25 μm; adapted from Ref. [22].

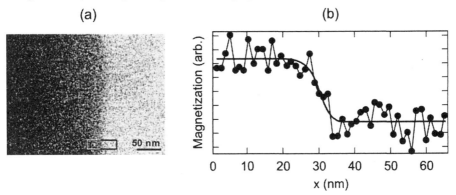

Figure 8-6: (a) Magnetic domain image of $SmCo_5$. (b) Experimental data of the magnetization profile in the rectangular area shown close to the bottom of (a) averaged along the vertical direction; the smooth line is the simulation assuming a spin-SEM resolution of 5 nm. Adapted from Ref. [30], used with permission.

However, despite these advanced design changes, the resolution in magnetic imaging is still worse than in standard topographic imaging because of the low efficiency of the spin analyzers. In our system, the primary electron-beam diameter can be made as small as 5 nm, whereas the best magnetic resolution is not better than ~20 nm. Resolution is limited by line-frequency electromagnetic interferences. In standard SEM operation, synchronization with the line frequency is done routinely. In spin-SEM, data-acquisition times are much larger than in standard SEM operation, and hence electromagnetic noise and mechanical drift are more strongly disturbing factors. Moreover, the microscope is routinely operated at beam voltages that are too low for optimum resolution. The primary energy typically ranges from 1 to 5 keV compared with values of 20 to 30 keV in standard SEM. The reason for this is the secondary-electron yield: For the 3*d* transition-metal ferromagnets, it has a broad maximum at an energy of ~1 keV but drops at higher energies. Hence a compromise is made between high lateral resolution and low noise in the magnetic image.

8.2.2. Instrumentation

A considerable design and construction effort is needed to set up a spin-SEM tool. It requires a high-brightness electron gun producing a narrow beam of primary electrons and an efficient spin analyzer. Most of the tools currently in use around the world are based on a commercially available SEM, to which a spin analyzer is attached. As discussed above, the short probing depth of the technique means that an ultrahigh vacuum environment is a necessity. Standard electron microscopes, however, operate in the 10^{-6} mbar range. Therefore extending an SEM to a spin-SEM not only requires a spin analyzer but generally also means that the entire vacuum and possibly also the pumping systems have to be exchanged, so that the base pressure of the tool is reduced to the low 10^{-10} mbar range.

The electron source should have a high brightness because spin detectors are inherently inefficient. Typically, thermionic LaB$_6$ cathodes [32] or W field-emitters are used, with either cold [3,4] or thermally assisted emission [33]. The best lateral resolution is achieved with field emitters, whereas for long-term stability thermionic cathodes are advantageous. The electron-optical column contains the same electromagnetic elements as in a standard SEM, but in general the distance between the end of the objective lens and the sample is kept larger, typically 10 to 15 mm. The reason is two-fold: First, the magnetic stray fields from the magnetic objective pole piece should be negligible at sample position. For most specimens, a field of less than 50 A/m is acceptable. Second, also the electron optics for transferring the secondary electrons to the spin analyzer has to be positioned close to the specimen. Typical beam currents reaching the sample are on the order of 1 nA. A detailed discussion of the tradeoffs between primary-electron energy, beam diameter, and electron-optical aberrations can be found in Ref. [33].

At the heart of the experimental setup is the electron spin analyzer. The determination of the spin of an electron is an atomic or surface physics experiment in itself. Various types of spin detectors exist, but not all of them are suited for use as a polarimeter in the spin-SEM technique. A comprehensive review is found in Ref. [32]. The principle behind most types is to use the spin-orbit interaction to transform a spin asymmetry into a spatial asymmetry. Spin-SEM tools exist with Mott detectors [34], low-energy electron diffraction (LEED) detectors [35], and the low-energy diffuse scattering (LEDS) detector [36]. In a Mott detector, the electrons are accelerated to high energies and scatter at a thin target foil. Because of spin-orbit coupling, electrons with spin ↑ or ↓ with respect to the scattering plane are deflected into different locations and can then be counted in a pair of particle detectors. Mott scattering is only efficient at high energies, so typically 50 to 100 kV are used as acceleration voltage. The target foil needs to be a material with high atomic number to have a large spin-orbit coupling, and usually Au is used. The foil is kept as thin as possible to avoid multiple scattering. Typical thicknesses are on the order of 100 nm. The spin polarization is related to the number of electrons N_L and N_R counted in the detectors at the left and the right by $P = 1/S\ (N_L - N_R)/(N_L + N_R)$, where S is the Sherman function, which is determined solely by the scattering conditions. The figure of merit of such a spin detector has been defined as $F = S^2 N/N_0$, where N/N_0 is the fraction of collected electrons in the detector [34]. Although the Sherman function is comparatively large, $S \sim 0.25$, N/N_0 is disappointingly small: Only about 1 to 5 out of 1000 electrons scatter into the detector pair. Thus, F is on the order of 10^{-4}: A Mott detector is notoriously inefficient because most high energetic electrons traverse the target foil without a scattering event.

Both the LEED and LEDS detectors have been designed to address this low figure of merit. The same physical principle is used, but scattering at much smaller energy is employed (on the order of 100 eV). Therefore, N/N_0 increases to some extent, but, on the other hand, S is reduced. The result is that the figure of merit for all three detector types is essentially the same, differing by not more than about a factor of two [32]. Therefore it is at least as important that the maximum number of secondary electrons be collected and transmitted to the spin detector, and that the spin detector be insensitive to spurious instrumental asymmetries. Two efficient electron transport optics are described in [32] and [33]. Asymmetries that are unrelated to spin polarization in the specimen result from changes of the electron-beam position and from angular variations. Our high-energy Mott detector has a spherically focusing electrical field distribution [37], which is a simple and effective remedy to minimize these unwanted effects. In low-energy-type spin detectors instrumental asymmetries are larger, but can largely be compensated by sophisticated "descan" schemes [32].

The inherently low figure of merit of all spin analyzers makes the technique a permanent struggle for more electrons: Domain imaging by spin-SEM is "slow", with typical acquisition times of 5 to 50 ms per pixel, leading to image-acquisition times of several minutes. This fact must be considered as the main experimental constraint of the technique because dynamic magnetization processes cannot be observed.

An appealing feature of the technique is the possibility to determine not only the magnetization component along a given axis but also the magnetization vector. Two detector pairs are generally mounted in orthogonal planes and allow two magnetization directions to be measured simultaneously. Depending on the actual arrangement, these directions can be the two in-plane components or, as in our system, one in-plane and the out-of-plane component. The third magnetization component is determined by rotating the sample. Alternative methods to determine all three magnetization components include an electrostatic "switch yard" [32] to deflect the electrons into a second spin detector, or a spin rotator [38] to turn the spin in a controlled manner into a direction perpendicular to the scattering plane of the spin detector. Hence, a vectorial map of the magnetization can be measured, which is crucial for the interpretation of complicated domain patterns.

8.3. MAGNETIC DOMAINS

8.3.1. Bulk domains

Even though magnetic domains have been introduced as a concept by Weiss in 1907 [39], they escaped experimental observation for almost half a century. Only in 1949 did Williams *et al.* [40] report evidence of regular regions of different magnetization directions in Fe single crystals. These structures could be satisfactorily explained by energy considerations based on the fundamental macroscopic magnetic quantities, such as magnetic anisotropy, exchange, and magnetic dipolar energy.

The textbook example of a bulk domain structure is the pattern observed in Fe(001) single crystals. This pattern has been investigated with all available techniques, so no new information is gained by imaging Fe(001) also by spin-SEM. Nevertheless it is instructive to do so in view of the completely different domain patterns observed in ultrathin films, which will be discussed in subsequent sections. The domain pattern in Fe(001) was presented in Fig.8-4 in a different context. It is characterized by the fact that the magnetization direction is confined to four specific directions—the four equivalent [100] directions within the plane. Straight domain walls are typical for bulk materials: By keeping the length to a minimum, the energy cost of the domain wall is reduced. Although the bulk domains extend throughout the thickness of the single crystal, spin-SEM maps the topmost end of the domains at the surface, a region ~1 nm thick. From the spiky appearance of the domains in Fig.8-4, the experienced observer concludes that this particular Fe single crystal is slightly miscut from the (001) direction. We will encounter such miscut surfaces again in Section 8.5.2, when magnetic step anisotropies are discussed.

8.3.2. Domains in ultrathin films

Ultrathin magnetic films behave completely different from bulk specimens. Defect-free films with in-plane magnetization are single-domain in the entire sample area, except for small closure domains at the edge [41,42]. An infinitely extended, homogeneously magnetized film with thickness approaching zero is stray-field-free, therefore the energy cannot be further reduced by splitting the magnetic state into several domains. However, magnetic domains are observed at structural defects, and they can be induced by demagnetizing the sample with alternating magnetic field cycles [42]. Domain patterns of such multi-domain states are shown in Fig.8-7 for various epitaxial thin films. The striking difference to a bulk domain state is the irregular shape of the domains and their boundaries. In contrast to the three-dimensional case, the magnetization component across the wall is neither constant nor is the divergence of the magnetization vanishing. While such a "charged" wall is prohibitively expensive in magnetostatic energy in a thick specimen, it comes almost at no cost in the ultrathin limit. As the film becomes thinner and thinner, the magnetostatic energy contributes less and less to the wall energy.

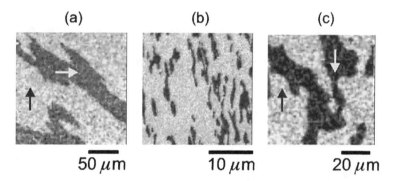

Figure 8-7: Spin-SEM images of three in-plane magnetized epitaxial films: (a) 17 ML fcc-Co/Cu(001), (b) 7 ML fcc-Fe/Cu(001), (c) 14 ML hcp-Co/Pt(111). The out-of-plane magnetization component vanishes for all these films.

Ultrathin films with perpendicular magnetization are generally observed to decay into a multi-domain pattern in the as-grown state, even without demagnetization [43], see Fig.8-8. Again, as in the case of in-plane magnetized films, irregular domain shapes are observed. From energy considerations, one would also expect that a perpendicularly magnetized ultrathin film is single domain in its ground state [44], evolving into a multi-domain state only at larger thicknesses [45]. The irregular walls are compatible with a standard, divergence-free Bloch wall, in analogy to perpendicularly magnetized thick specimens, such as garnets or ferrite platelets [46]. It has to be mentioned, however, that the wall profile in a perpendicularly magnetized ultrathin film typically has a calculated thickness of <5 nm and has not yet been resolved by spin-SEM.

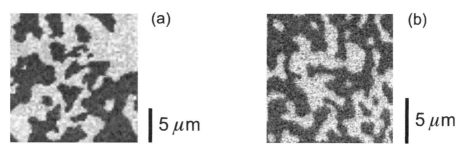

Figure 8-8: Spin-SEM images of two epitaxial films with perpendicular magnetization: (a) a 3-ML Co film on Au(111), and (b) a 18-ML Ni film on a Cu(001) buffer on Si(001). Irregularly shaped domains of varying size have magnetization "up" (white) or "down" (black). The in-plane magnetization components vanish.

8.3.3. Domains in patterned structures

The rapid progress in the patterning of magnetic structures is motivated by applications in data storage and possible future magnetic memories. Read and write heads are laterally confined to small and ever shrinking areas; patterned media [47] have been proposed as a possibility to push the superparamagnetic limit, which threatens future progress in magnetic storage, further out; and small magnetic structures are the key element in nonvolatile magnetic random-access memories [48]. From a fundamental point of view, the investigation of laterally confined magnetic structures is essentially a study of the influence of edges on the magnetic domain state.

Spin-SEM has contributed to this field for more than a decade. The complex domain pattern in the soft magnetic film of recording heads was investigated by Mitsuoka *et al.* in 1987 [49]. A first attempt to observed domains in ultrathin, patterned structures in the Fe/Ag(001) system came shortly afterwards [41]. Spin-SEM clearly is a technique that is very well suited for the inspection of magnetic domain states in ultrathin micro- or nanostructures, where high sensitivity and high lateral resolution are crucial.

Small structures are produced most frequently by lithographical means [50]. While this approach allows feature sizes down to less than 20 nm, it is restricted to films that are protected against ambient atmosphere. For studying the intrinsic magnetic properties of ultrathin uncovered structures, a different *in-situ* approach is used: The patterns are produced by evaporation through shadow masks [41] or—to have more flexibility in terms of the shape of the structures—by a nanostencil approach [51].

The scientific questions these ultrathin small elements pose are manifold, and many of them are still unsolved. The current understanding of small magnetic structures is in a state quite similar to what has been known about infinitely extended ultrathin films 15 years ago. At that time, conflicting results were reported for nominally identical systems, mainly because film quality varied strongly. Today, similarly conflicting results are obtained for microstructures. The classic example is Co/Cu(001): Not even the basic question on the nature of the equilibrium domain state in small, thin elements can be answered unambiguously [52,53]. For thick Fe elements on GaAs, the remanent state changes from single-domain to multi-domain when the lateral size shrinks to values below 50 µm. This is attributed to the predominance of the magnetostatic energy over anisotropy, if the ratio of edge length to element area increases [54]. Domain formation in small elements clearly is a rather complex process.

What is the equilibrium domain state of a small element with in-plane magnetization? As we have seen in the preceding section, the extended film is single-domain, unless domains are introduced by demagnetization in an external field. It has recently been argued that the same holds for a small element: Regardless of its size, shape, and magnetic history, the element has been found to be magnetically uniform [52]. However, we do not find this uniformity for all length scales and thicknesses. Figure 8-9 illustrates Co/Cu(001) square elements with 1 μm lateral size, imaged after ac-field demagnetization. Whereas the thinner elements indeed are a single-domain, the thicker ones form flux-closure patterns to reduce magnetostatic energy. A recent, detailed investigation of magnetization reversal in Co/Cu(001) elements of somewhat bigger size revealed that the single-domain state prevails only after the sample has been saturated for the first time [53]. The as-grown state was multi-domain. Moreover, it has been calculated that the sharpness of the element edges is important for the formation of domains.

(a) 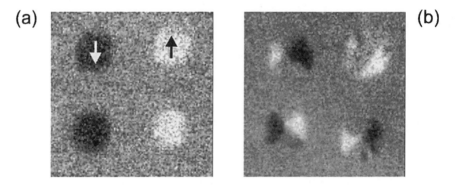 (b)

Figure 8-9: Spin-SEM images of Co/Cu(001) square magnetic elements with 1 μm lateral size, measured in remanence after ac demagnetization. The arrows indicate the magnetization direction which is measured, the in-plane component perpendicular to this direction appears grey. (a) At a thickness of 6 ML, all elements are single-domain. (b) At a thickness of 17 ML, three out of four elements display flux-closed domain patterns.

The formation of flux-closed domain patterns is hindered if the lateral size of the element is reduced. Figure 8-10 shows an array of 300-nm-diameter Co/Cu(001) dots and a thickness of 15 nm. All dots are uniformly magnetized, as-grown as well as after demagnetization. Only in rare cases can a multi-domain state be induced. Domain formation in even the most regular magnetic patterns is more complex than anticipated: It depends on both lateral size and element thickness. Clearly more work is needed, in particular also to determine the limits of ferromagnetic stability in small ultrathin elements. We have observed, for instance, that the magnetization in 300-nm dots that are only few atomic layers thick begins to fluctuate on the time scale of the spin-SEM experiment. Recent speculations that ultrasmall elements containing only a few atoms may provide a stable magnetic bit for "nanorecording" [52] hence are neither supported by experimental evidence nor expected from thermodynamic considerations.

In bulk ferromagnets, the overall shape determines the domain pattern to a large extent because it is the magnetostatic energy that has to be minimized. The square or circular elements described above are designed to be highly symmetric, and therefore the in-plane shape anisotropy is not expected to be important, except for a possible small configurational contribution [55]. Changing the shape of the elements is an easy way to test the influence of magnetostatics on domain patterns and magnetization reversal. Various shapes have been tested. In the ultrathin limit, the shape is found to be unimportant for both the domain state [52] and the average reversal field [53]. As soon as the thickness is increased, the magnetic dipolar energy increases to a point at which shape matters again. This is illustrated in Figs.8-11 and 8-12 by elements of completely different shape. In Fig.8-11, Co bars with a length of 100 µm and a width of 1.8 µm are seen to decay into a narrow sequence of antiparallel domains with the magnetization pointing along the width of the bar. As the easy axis is parallel to the width, this directly proves that shape anisotropy is negligible compared with the crystalline and step anisotropies. Figure 8-12 shows a complex structure consisting of 20-µm-sized squares linked by a narrow 1.5-µm wire. This element has been subjected to a magnetic field that is large enough to switch the magnetization direction of the large squares, but the wire remains magnetized along its original direction. This example clearly shows that size and shape can matter in magnetization reversal. Agreement exists in the literature about the reversal mode: In all small elements with lateral sizes above 100 nm, magnetization reversal is dominated by domain nucleation. Coherent rotation is not observed at this length scale, even if the thickness is reduced to a few atomic layers.

(a)   (b)   (c)

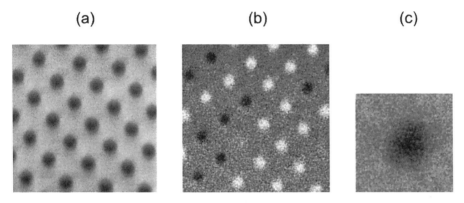

Figure 8-10: Regular array of circular dots of Co on stepped Cu(001) with a diameter of 300 nm and a thickness of 15 nm. (a) Map of secondary electron yield showing the Co dots. The image displays chemical rather than topographical contrast. The Co dots appear dark, the Cu substrate bright because of the lower work function of Cu. (b) In-plane magnetization component; (c) enlarged view of an individual dot. All dots are in a single-domain state.

Figure 8-11: Patterned epitaxial Co structures on stepped Cu(001) having a length of 100 µm, a width of 1.8 µm, a bar-to-bar separation of 0.8 µm, and a thickness of 2 nm. (a) Map of secondary electron yield showing the Co bars in dark. (b) Magnetization component along the width of the bars. The sample has been ac-demagnetized before the image was taken. A long sequence of narrow antiparallel domains is observed, mainly in the left-hand part of the bars.

These examples have illustrated isolated magnetic elements: Their surroundings consist of nonmagnetic material. "Negative" patterning is also possible and is interesting for the investigation of domain nucleation and pinning behavior. The first approach used lithography to create a mesh of square holes in a 40-nm-thick $Ni_{80}Fe_{20}$ film [56]. This "antidot" lattice is a controlled array of holes in the film, and considerably affects the magnetization reversal. The use of scanning Kerr microscopy required that antidotes have a large size of 15 µm because smaller antidots introduced noise in the images due to optical diffraction. Spin-SEM is insensitive to this type of noise, and hence smaller antidot arrays can be investigated. We have fabricated an epitaxial Co film on stepped Cu(001) containing an antidot array without lithography steps by an *in-situ* nanostencil technique. Figure 8-13 shows the spin-SEM image of such an array with an antidot diameter of 300 nm and a spacing of 3.8 μm. The domain pattern is completely determined by the presence of the antidots: Only three out of 25 dots do not nucleate or pin a domain. The as-grown film without defect array is in a single-domain state as discussed in Section 8.3.2. The magnetic anisotropy is locally dominated by dipolar contributions, as was found for the larger-scale dot array studied in Ref. [56]. It is remarkable that the antidot lattice can easily be identified with high accuracy, even though the topographic variations in a thin film are minimal. As in Figs. 8-10 to 8-12, the contrast in the "topographic" image is a chemical contrast originating in the different work functions of Co and Cu, which lead to a difference in secondary electron yield.

(a)   **(b)**

Figure 8-12: Patterned epitaxial structure of 5 nm Co on a Cu(001) buffer on Si(001). Two squares of 20 μm side length are connected via a 1.5 μm wide wire. (a) In-plane magnetization component parallel to the wire; (b) the corresponding map of the chemical contrast. The element has been subjected to a magnetic field large enough to revert the magnetization in the squares. The wire magnetization is unaffected. Note the head-to-head walls near both ends of the wire.

(a)   **(b)**

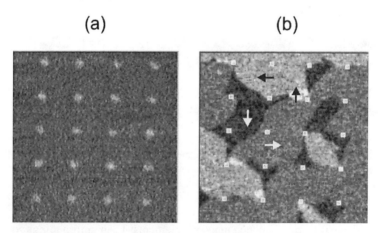

Figure 8-13: Co antidot array fabricated *in-situ* by the nanostencil technique on a stepped Cu(001) substrate. (a) Map of secondary electron yield showing the position of the antidots where Co is missing (bright). (b) In-plane magnetization distribution; the different grey levels represent the four magnetization directions (arrows), which lie along the four [110] directions. No magnetic field has been applied prior to image acquisition. The position of the antidot lattice as determined from (a) is indicated by the white squares. Image size: 18 μm × 18 μm.

A "top-down" approach with lithography or nanostencil masks is not the only way to create defects in a controlled manner. Nature's capability for self-organization leads to vicinal surfaces that magnetically act as a sequence of one-dimensional defect lines. Some of these aspects will be discussed in Section 8.5.2. However, even more complex geometrical shapes can be produced by self-organization. An example is presented in Fig.8-14, which shows two hexagonal pits in Pt(111), created by repeated cycles of ion bombardment at elevated temperatures and annealing during oxygen exposure. By depositing 3.7 ML Co on this substrate, a continuous epitaxial film is fabricated with localized deep hexagonal holes. The magnetic image shows that the preferred state is an arrangement of perpendicular domains, as is also observed in a continuous film without such a defect. A comparison of topographic and magnetic image reveals directly that if two opposite domains touch each other close to the hole, the domain wall is anchored at the corners of the hexagon. Moreover, the pit itself is seen to be nonmagnetic, even though some Co has certainly reached its bottom. This last point nicely illustrates that spin-SEM is capable of determining, in the same scan, the topography and the magnetism even at largely different heights and curvatures, and that image sharpness is maintained thanks to the large depth of focus of the high-energy electron-beam technique. This particular advantage has been exploited, for instance, to determine the side-plane domain distribution in thin-film recording heads [57] or the magnetization distribution in tips used for magnetic force microscopy [58].

(a) 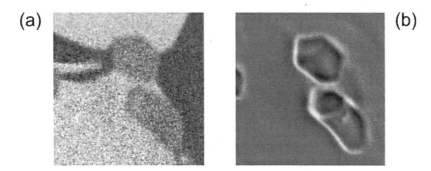 (b)

Figure 8-14: The magnetic influence of hexagonal pits in 3.7 ML Co on Pt(111). (a) Perpendicular magnetization component and (b) topography. The domain walls bend to pin at the corners of the irregular hexagons. Image area: 5.6 μm × 6.0 μm.

8.4. DOMAIN WALLS

8.4.1. Surface termination of a bulk domain wall

At the interfaces between the magnetic domains the magnetization must change its direction. These domain walls have been the subject of intensive research efforts, both theoretical and experimental. In the simplest models, the domain wall width and energy are described by two material parameters only, namely, the exchange and the anisotropy. The prototypical magnetic domain wall separates two domains of opposite magnetization direction and hence is called a 180° domain wall. In the gradual transition from the magnetization direction in one domain to the opposite one in the adjacent domain, the spins within the wall point into directions that are not the energetically preferred easy magnetization directions. Therefore, to minimize this energy contribution, magnetic anisotropy tries to keep the domain wall thin. Exchange, on the other hand, tries to keep the angle between adjacent spins as small as possible. The interplay between these two energy contributions then leads to a finite domain wall width.

In an infinitely extended ferromagnet, the domain wall would be a Bloch wall, in which the magnetization rotates within the plane of the wall. If such a Bloch wall is undisturbed at a surface, the magnetization would point out of the surface plane. This configuration no longer is the lowest energy state because the magnetic stray field is large. Indeed it has been proven by spin-SEM that the domain wall at the surface is not of Bloch type. The magnetization at the surface rotates entirely within the plane, thereby changing the character of the domain wall to a Néel wall [16]. This transition of a bulk Bloch wall to a surface Néel wall has important consequences: First, the domain wall width determined at the surface is by no means the one expected for the bulk wall [17]. Second, the symmetry of the Bloch wall is broken as the surface is approached, leading to a lateral displacement of the Néel wall with respect to the Bloch wall center. Because the surface termination of the Bloch wall can turn over in two opposite directions, there is an offset between Néel wall segments pointing into opposite directions, as can be observed in spin-SEM. An example illustrating these aspects in an Fe single crystal is shown in Fig.8-15. Micromagnetic simulations of a Bloch wall in the bulk terminated at the surface have been performed for various materials and are fully consistent with the observations [59]. They prove that the Néel surface termination of a Bloch wall is a general feature of bulk domain walls in low-anisotropy materials. Remarkably, a highly surface-sensitive magnetic imaging technique was needed to conclude that the model of the most common bulk domain wall can no longer be maintained.

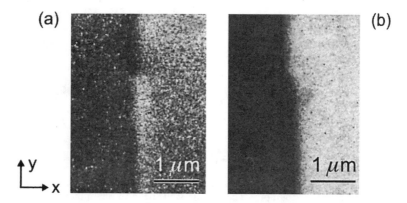

Figure 8-15: In-plane magnetization components (a) M_x and (b) M_y at the surface of Fe(001). The Néel surface domain wall runs vertically through the image. Two laterally displaced wall segments can be distinguished. A magnetic singularity exists near the center, where these two segments join. From Ref. [59], used with permission.

8.4.2. Domain walls in ultrathin films

In the preceding section we have seen that a Bloch wall is generally encountered in a bulk ferromagnet, and that it will be terminated at the surface by a Néel wall with in-plane magnetization rotation. However, this complex wall type is no longer the preferred one if the sample thickness is reduced to length scales that approach the width of a domain wall. Then the Néel cap evolves into a complete Néel wall, with the magnetization rotating entirely within the plane for an in-plane magnetized specimen to avoid an excess of magnetostatic energy. With typical domain wall widths of 100 nm in low-anisotropy ferromagnets, the transition between a Néel wall and a Bloch wall occurs at thicknesses of several hundred nanometers, going through more complex domain wall types such as crosstie walls [60]. Many experimental studies were devoted to the investigation of these Néel walls, but almost all were performed on films with thicknesses of more than 10 nm, mainly because of limitations in the sensitivity of the imaging technique employed. If the film thickness becomes smaller than the exchange length, the domain walls should become truly one-dimensional because the magnetization across the film thickness will be constant. Typical exchange lengths are on the order of a few nanometers, so it can be expected that ultrathin films a few atomic layers thick are model systems to investigate the fine structure of a pure, one-dimensional Néel wall.

Figure 8-16 shows a domain wall profile across a 180° Néel wall in an 5.5-ML-thick epitaxial fcc-Co film grown on Cu(001) [61]. The entire wall consists of a core area limited to a scale of ~200 nm and long tails on both sides of the core extending to distances of more than 1 μm. In the core, the magnetization rotates rapidly, whereas the tail region is characterized by small, gradual changes of the magnetization direction. A fit of the wall profile with the standard tanh function reveals the discrepancy between experimental observation and the Néel wall model which neglects magnetostatic dipole-dipole interaction. The large extensions of the wall tails clearly are not described by the tanh function. A micromagnetic simulation taking into account the magnetostatic energy leads to improved agreement but systematic deviations between experimental profile and simulation still are present. It has been suggested that the discrepancy between experiment and the one-dimensional simulations could be an indication that even the conceptually simplest case of a Néel wall in an ultrathin film is a two-dimensional magnetic object rather than a one-dimensional one [61].

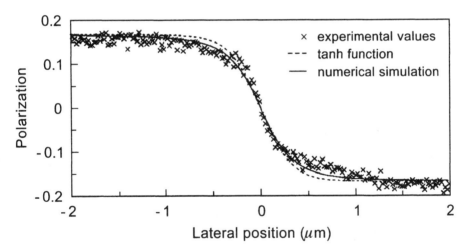

Figure 8-16: Domain wall profile across a 180° Néel wall in a 5.5-ML-thick Co/Cu(001) film. The experimental values are compared with fits to the tanh function generally employed and to the micromagnetic numerical simulation. Note the discrepancy between experiment and tanh fit in the tail region and the improved agreement with the simulation. From Ref. [61], used with permission.

8.4.3 Domain walls in confined geometries

The width of a domain wall, be it of Bloch or Néel type, is solely determined by intrinsic material parameters. In the simplest case, the domain wall width w is given by the exchange A and anisotropy K alone: $w \sim (A/K)^{1/2}$. Only very recently has it been suggested that the width of a wall can be controlled by the geometry of the specimen, provided the dimensions are rather small, i.e. approaching the length scales of the undisturbed domain wall width [62]. Future magneto-resistive sensors are likely to exploit ballistic electron transport [63], and in such devices the domain wall resistance might become relevant. Therefore it is important to control the domain wall properties independently of the material parameters, which are optimized for other aspects. An example of such a confined magnetic domain wall is shown in Fig.8-17. A small circular Co dot of 10 nm thickness has been fabricated by molecular beam epitaxy through a stencil mask as described in Section 8.3.3. The dot is so thick that in rare cases it decays into a multi-domain state. From a linescan across the 180° domain wall, the width of the wall can be determined to be about 50 nm. This value is roughly a factor of four smaller than the corresponding wall width in the unconfined Co film, which nominally has the same material parameters such as magnetization, anisotropy, and exchange. The wall width is reduced due to the modified dipolar contributions at the element edges and decreases when the lateral size of the element shrinks. The wall can be further reduced by introducing a notch into the element, again because of significant dipolar energy in the edge region of these constrictions. A systematic study investigating the width of Néel walls in such constrictions has been done very recently [64].

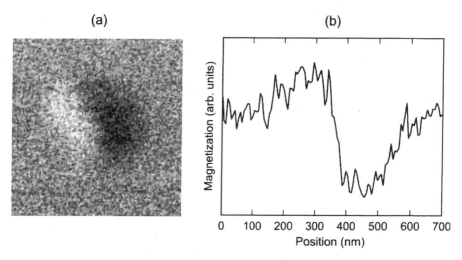

Figure 8-17: (a) Magnetization image of an individual epitaxial 300-nm-diameter Co dot on Cu(001), with a thickness of 10 nm, fabricated and measured *in-situ* by spin-SEM. The dot decays into a multi-domain state containing two antiparallel regions with a 180° domain wall in between. (b) Linescan across the wall taken in the center of the dot along the horizontal axis. The wall width is about 50 nm.

This section on domain walls illustrates an important aspect of spin-SEM. For a quantitative comparison of experimentally determined domain-wall profiles and model calculations, it is important that the magnetization is directly accessible. The most widely used alternative technique with high lateral resolution, magnetic force microscopy [27], is sensitive to the magnetic fields rather than to the magnetization, which renders the interpretation of wall profiles difficult. In addition, the ability of spin-SEM to measure all three magnetization components allows the identification of fine structure within wall profiles.

8.5. MAGNETIC ANISOTROPIES

The concept of magnetic anisotropy is fundamental for a correct description of the magnetization directions in a ferromagnet without an applied magnetic field. Phenomenologically such "easy" magnetization directions are derived from the series expansion of the anisotropy energy, which reflects all symmetry operations of the crystal lattice. The quantum mechanical origin of magnetic anisotropy is the spin-orbit coupling of the electrons: The magnetization feels the environment and symmetry of the crystal lattice through the orbital motion of the electrons. Therefore, any change in structure will affect the magnetic anisotropy. Alternatively, if the symmetry of a ferromagnet is broken, the anisotropies are modified as well, reflecting this change. Néel [65] pointed out that symmetry breaking at a surface can create an additional anisotropy term that is compatible with the changed symmetry, even for perfectly flat surfaces. Imperfections, such as defects or steps locally change the symmetry, again affecting magnetic anisotropies.

Most experiments investigating magnetic anisotropies are performed by macroscopically averaging techniques such as torque magnetometry, ferromagnetic resonance or Brillouin light scattering [66], or through the analysis of hysteresis loop shapes. In specific cases, however, spatial resolution helps to visualize the effects of changing easy magnetization directions or to correlate effects of structure and magnetic anisotropy. This section summarizes a few of these experiments.

8.5.1. Magnetization reorientation

In various magnetic systems the magnetization direction has been found to rotate or jump from one high-symmetry axis to a different one. In three-dimensional systems, magnetic insulators such as $Sm_{1-x}Dy_xFeO_3$ exhibit such a magnetization reorientation upon a temperature change, caused by a temperature-dependent lattice distortion. This reorientation can be described as a phase transition of first or second order, depending on whether the magnetization changes direction abruptly or continuously [67]. In ultrathin films, magnetization reorientation can take place by a different mechanism, namely, the interplay between surface anisotropy and shape anisotropy. While shape anisotropy favors in-plane magnetization due to magnetostatic interaction, surface anisotropy can stabilize perpendicular magnetization. If surface anisotropy wins at very small thicknesses, shape anisotropy will predominate at large thicknesses. The thickness range in which these two contributions balance each other has been the subject of intensive research in the past decade. Fundamental questions addressed were a comparison of the reorientation transition with the film thickness on the one hand and with temperature on the other hand, the microscopic nature of the transition manifested in its domain patterns, and the order of the phase transition.

Spin-SEM has both the surface sensitivity to determine the magnetization directions in these ultrathin films and the capability to determine the magnetic microstructure during reorientation. The first experiment studying the reorientation transition microscopically [68] is reproduced in Fig.8-18. Films of fcc-Fe have been grown on Cu(001) in the shape of a shallow wedge, and the magnetization direction has been determined locally across the wedge. The start of the wedge can easily be identified with the help of the standard "topographic" image, because the work functions of Fe and Cu are different, resulting in a chemical contrast in the secondary electron yield. Films thinner than 2.3 ML are paramagnetic at the experimental temperature, whereas those with thickness between 2.3 ML and 5.3 ML are perpendicularly magnetized at a temperature of 175 K. Above this thickness, the magnetization lies completely in the plane. The reorientation is rather abrupt: It takes place within a thickness interval of less than 0.2 ML.

The thickness at which reorientation occurs varies with temperature because surface anisotropy and—to a lesser extent—magnetostatic energy are temperature-dependent. This means that reorientation can also be induced in a perpendicularly magnetized film if the temperature increases, provided the film thickness is close to the reorientation thickness. The temperature dependence of the magnetization pattern is shown in Fig.8-19. In an Fe/Cu(001) wedge, the in-plane domain front is observed to expand to smaller thicknesses if the temperature is increased. The transition spreads over a relatively large temperature interval, starting at around 80 K for this particular thickness and extending beyond 120 K.

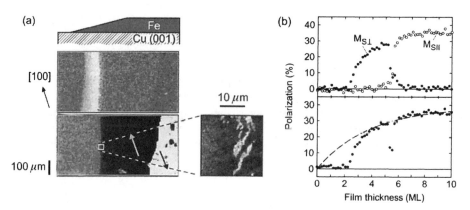

Figure 8-18: Magnetization reorientation in an fcc-Fe wedge grown on Cu(001). (a) The schematic shows the wedge position, with the Fe thickness extending from 0 to 10 ML. The upper and lower spin-SEM images show the out-of-plane and in-plane magnetization components which have been determined simultaneously. A sharp transition from out-of-plane to in-plane magnetization occurs at 5.3 ML. The domain structure at reorientation thickness is identified in the close-up. The temperature of the specimen was 175 K. (b) Linescan across the wedge of (a), with the out-of-plane (filled circles) and in-plane (empty circles) magnetization component (top figure), and the total magnetization (bottom figure). The expected increase of polarization with film thickness according to the small probing depth is given by the dashed line. Adapted from Ref. [68].

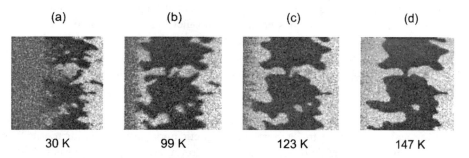

Figure 8-19: Magnetization reorientation in an fcc-Fe wedge grown on Cu(001). The wedge thickness increases linearly from 6.5 ML at the left to 7.0 ML at the right edge. The in-plane magnetization component is shown. The grey part of the images indicates that the magnetization is out-of-plane. The reorientation is characterized by the front of the in-plane magnetization pattern, which moves to the left with increasing temperature. The temperature is indicated below each image. Image size: 20 μm × 20 μm.

Several theoretical studies have investigated the magnetization reorientation and in particular its consequences on the domain pattern. It has been predicted that a stripe domain phase should evolve at the reorientation thickness or temperature [69], with orientational long-range order but positional disorder, in analogy to a smectic liquid crystal [70]. Indeed, a striped phase with micrometer-sized domains has been identified.

One of the important parameters of the stripe phase is the domain width D. It can be approximated by an analytical formula that relates the film thickness d, domain wall energy σ_w, and saturation magnetization M_s [71]. The mean domain width is predicted to decrease upon approaching reorientation thickness according to $D = 0.955 \, d \, \exp[\sigma_w/(4 \, M_s^2 \, d)]$. To keep the perpendicular magnetization even though the surface anisotropy decreases, the magnetostatic energy has to be reduced as well. To some extent this can be achieved by reducing the domain size. Speckmann *et al.* have investigated this prediction in detail in the Co/Au(111) system [72]. Figure 8-20 shows their statistical analysis of many domain images in a wedge-type sample. Good agreement with the theoretical prediction [71] is found: The number of perpendicular domains per unit length increases rapidly with increasing film thickness.

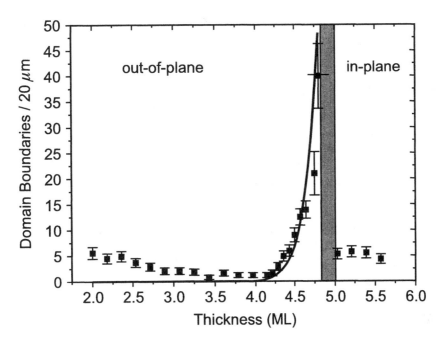

Figure 8-20: Number of domain boundaries per unit length vs. film thickness for an annealed Co wedge on Au(111), deduced from a statistical analysis of magnetic domain images. The shaded region gives the thickness range in which the magnetization reorientation takes place. Experimental data (filled squares) are compared with a fit to the formula given in Ref. [71]. Adapted from Ref. [72], used with permission.

Since these experiments, stripe domain phases have been investigated in even more detail, and geometrical as well as thermodynamic aspects remain the focus of research, because fundamental properties of two-dimensional systems going beyond the topic of ultrathin ferromagnets can be investigated. A more detailed account of these results is presented in Section 8.7.

8.5.2. Magnetic step anisotropies

The understanding of the magnetization reorientation presented in the preceding section relies on the concept of surface anisotropy. It is the surface or interface, i.e., a two-dimensional object, which leads to a uniaxial anisotropy perpendicular to the film. In this section it will be illustrated that also one-dimensional objects contribute to magnetic anisotropy. More specifically, step edges can lead to magnetic anisotropy contributions that influence easy magnetization directions, change domain patterns, or act as pinning sites for domains.

The effects of steps can be investigated in a controlled manner on vicinal surfaces. These surfaces are slightly miscut, typically up to a few degrees, from a high-symmetry crystallographic orientation. The first spin-SEM experiments on vicinal surfaces were performed by Berger *et al.* [73] on Co films on vicinal Cu(1 1 13). This surface deviates from the (001) orientation by 6.2°, with the step edges running along the [1 -1 0] direction. With this choice of step direction, the easy magnetization axes of the corresponding low-index surface, Co/Cu(001), are oriented exactly parallel and perpendicular to the step edges. It was directly proven by spin-SEM on Co/Cu(1 1 13) that the magnetization direction in 5–9-ML-thin films is parallel to the step edges, see Fig.8-21. No domains with magnetization perpendicular to the steps could be induced, not even after applying magnetic field pulses perpendicular to the edges. Despite the step anisotropy, the domain structure of Co/Cu(1 1 13) is very similar to that of Co/Cu(001), demonstrating that for this system the domain patterns are not determined by the uniaxial step anisotropy.

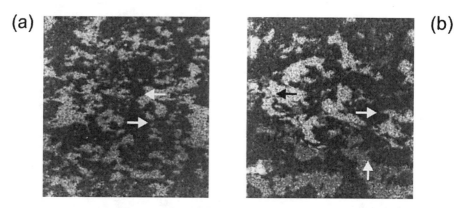

Figure 8-21: Domain patterns in ultrathin Co films after demagnetization by an ac field. (a) 5 ML Co on Cu(1 1 13); image size: 1 mm × 1 mm; (b) 9 ML Co on Cu(001); image size: 500 μm × 500 μm. The relation of the grey levels and the magnetization direction is indicated by the arrows: On the stepped substrate, the Co film is uniaxial, whereas on the flat substrate, the film is biaxial. Adapted from Ref. [73], used with permission.

The situation is different in the Co/Pt(111) system. For ultrathin Co layers, the magnetization is out-of-plane. Co/Pt layers grown on MgO(111) miscut by 0.5° to 1.5° exhibit an anisotropic shape of the domains as observed by optical microscopy, with the actual form depending on the miscut and step arrangement [74]. Moreover, also domain-wall propagation is highly anisotropic and correlated with the miscut. The steps themselves cannot be resolved in optical microscopy.

It is not clear at this point why the Co/Cu and the Co/Pt systems behave so differently. One possible explanation could be that in perpendicularly magnetized films it is easier to pin domains at steps because demagnetizing fields at structural inhomogeneities are in general larger. However, also thicker, 10-ML Co/Pt(111) films with in-plane magnetization behave anisotropically: The domains are elongated and pin preferentially at steps, see Fig.8-22. This substrate is not a vicinal surface with regular step distances but rather shows large, flat (111) terraces separated by step-bunching regions. The topographic image acquired by spin-SEM is able to resolve these bunches because the yield of secondary electrons depends both on the angle of the incoming electron beam and on the emission angle, which at edges are different from those on a flat terrace [75]. The magnetic image shows in-plane magnetized domains elongated along the steps. In Fig.8-22(c) the topographic image is superimposed on the magnetic one. Clearly domain wall positions are correlated with the step edges: The step anisotropy is the crucial factor in determining the pinning strength of magnetic domains and therefore influences the shape of the domains. It is not known at present whether this difference in the behavior of Co/Cu(001) and Co/Pt(111) is caused by the strength of the step anisotropy or by structural factors such as modified growth in the presence of steps. However, the examples nicely illustrate that the capability to view and correlate both magnetic and structural information is decisive to extract detailed information on how and where magnetic domains pin.

(a)   (b)   (c)

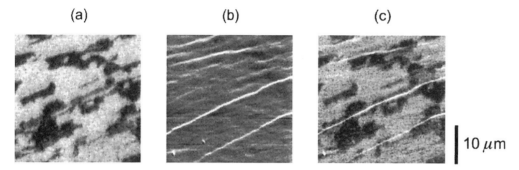

10 μm

Figure 8-22: (a) In-plane magnetic pattern and (b) topography of a 10-ML Co film grown on a miscut Pt(111) substrate. The step bunches are visible as bright narrow lines in the topography. (c) The superposition of magnetic and topographic images shows that most domains are pinned at the steps.

Nonmagnetic adsorbates are known to affect magnetic anisotropies on flat surfaces. This effect cannot be explained by the Néel model of surface anisotropy, because the breaking of the symmetry at the surface is unchanged. Here the hybridization of electron wave functions between adsorbate and film atoms becomes important. Again, this two-dimensional effect is present also in one dimension, namely, at step edges. As the density of step edges can be controlled, the effects related to the steps can be separated from those occurring on flat surfaces. This approach has allowed us to discover anisotropy oscillations upon covering a stepped Co/Cu(001) film with small amounts of Cu [76]. Their magnitude can be sufficient to switch the magnetization direction by 90° within the plane. Now the easy axis is perpendicular to the steps. Microscopically it could be shown that the Cu atoms diffuse on the surface until they attach at a Co step edge [77]. Hence the environment of the Co step atoms changes locally, modifying the relevant electron wave function at the steps by hybridization. Figure 8-23 exemplifies this effect. A Cu wedge has been evaporated onto the Co/Cu(001) film. On the uncovered part, the magnetization is aligned with the step edges. Right at the beginning of the wedge, the spin-SEM image shows that the magnetization has rotated by 90° and stays perpendicular to the steps up to a Cu coverage of ~1.3 ML. At this thickness, a second switch happens spontaneously with external field, reestablishing the original magnetization [78]. It is remarkable that minute amounts of Cu are sufficient to induce this effect. On average, the spins of 500 Co atoms will switch direction for each Cu atom attaching at a step.

The examples presented in this section show that spin-SEM can be successfully used to elucidate the physics and mechanisms behind the concept of magnetic step anisotropy. In principle these experiments do not necessarily require sophisticated domain imaging. It is sufficient to measure hysteresis loops by the magneto-optical Kerr effect. However, for the last example given above, spin-SEM turned out to be crucial: only by a technique that is able to determine the direction of the magnetization without ambiguity could it be proved that the magnetization switches discontinuously and does not rotate, as has been inferred from magneto-optical experiments [79].

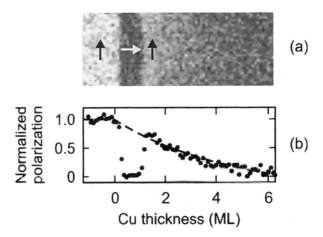

Figure 8-23: (a) Spin-SEM image of a 7-ML Co film on Cu(001) miscut by 1.6° covered by a Cu wedge ranging from 0 to 6.5 ML. The magnetization direction parallel to the steps is measured, hence the component perpendicular to the steps appears in grey. The sharp transitions prove that the magnetization switches direction discontinuously. (b) Linescan along the wedge through image (a) shows the switched range between 0.1 and 1.3 ML Cu coverage and the decaying spin polarization at larger coverage caused by the small probing depth of the technique as described in Section 8.2.1. Image size: 150 μm × 60 μm.

8.6. MAGNETIZATION REVERSAL

On a microscopic scale, magnetization reversal is largely governed by inhomogeneities in the sample: A perfect crystal should behave like a single-domain particle in which magnetization reversal can only occur by coherent rotation [80]. The less perfect reality is that coherent rotation is a rare incident, and that magnetization reversal generally occurs by nucleation of reversed domains and subsequent propagation of domain walls. These microscopic events determine the macroscopic properties of the ferromagnet, such as remanence and coercive field. To understand these quantities, domain structures and their response to a magnetic field in the presence of structural imperfections have to be investigated.

For an electron spectroscopy technique such as spin-SEM, the application of magnetic fields during imaging appears impossible because low-energetic electrons are deflected by magnetic fields owing to the Lorentz force, and because the spin will precess in the presence of an external field. Techniques of choice for these studies are those alternatives that do not rely on charged particles, such as magneto-optical techniques [24]. The investigation of magnetization reversal in ultrathin films on a very small lateral scale, however, is not easily accessible by any technique. Therefore it is worthwhile to devise ways to employ spin-SEM for these studies, despite the intrinsic technical problems of electrons traveling in magnetic fields. In this section, it will be shown that under favorable circumstances spin-SEM can be used to visualize magnetic domains in applied fields [58,81,82].

Many interesting ultrathin-film systems are magnetically very soft and can be reversed by small fields. A typical example is fcc-Fe/Cu(001). A 3-ML-thick film has a perpendicular magnetization and can be reversed by fields of less than 1 kA/m. In this case, the complete reversal sequence has been studied by spin-SEM, starting with a single-domain remanent state and ending with the reversed saturated state [58], see Fig.8-24. The disturbing effect of spin precession is absent in this configuration because the magnetic field and the spins are aligned. Various characteristic reversal features have been observed: Nucleation at extended defects in the sample, domain wall propagation, pinning and subsequent unpinning, until finally the "hard" magnetic centers are removed by the saturation field. Even though reversal on a microscopic scale is rather complex, the macroscopic hysteresis loop is close to being square, i.e., it resembles the simplest possible reversal mode of coherent rotation or of a reversal completely dominated by propagation [83].

(a)   (b)

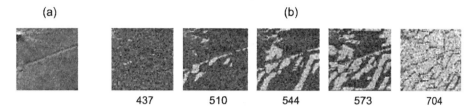

| | 437 | 510 | 544 | 573 | 704 |

Figure 8-24: Magnetization reversal in a perpendicularly magnetized 3-ML fcc-Fe/Cu(001) film at room temperature. (a) Topographic map showing a defect at the upper left-hand corner and a scratch diagonally across the sample. (b) Sequence of spin-SEM images from single-domain remanence to reversed single-domain saturation. The applied external field during imaging is given in units of A/m. Both nucleation and wall propagation can be identified. Image size: 100 μm × 100 μm.

A thorough investigation of the effects of in-plane magnetic fields on the performance and limits of spin-SEM has been done by Iwasaki *et al.* [81] and Steierl *et al.* [82]. Three effects have to be taken into account: First, the incoming electron beam is deflected. This mainly leads to a shift of the overall image that can be compensated by reshifting the image electronically. Typical values depend on the beam energy but are reported to be as large as 250 μm for a magnetic field of 1 kA/m [81]. Second, the path of the secondary electrons is affected by the Lorentz force, which means that the count rate at the spin detector is reduced. Third, the electron spin precesses in magnetic fields that are not aligned with the spin axis. In principle, this effect can be compensated by a Wien filter [34], but this has not yet been done. A highly successful approach to suppress these deteriorating effects has recently been reported [82]. Its main idea is that the transit time of the electrons in the large magnetic field has to be reduced. This means that on the one hand the accelerating voltage to transfer the secondary electrons away from the sample is increased from typical electric field values of 10^4 V/m to as much as 10^7 V/m. On the other hand, the magnetic field is confined to a distance of typically 100 μm from the sample surface.

Figure 8-25 reports the magnetization reversal in a micrometer-sized $Ni_{80}Fe_{20}$ element. The spin-SEM tool used for this study has been optimized for domain imaging with in-plane magnetic fields, implementing the idea described above [82]. The center domain shrinks with increasing applied field of opposite direction, and the flux closure domains at the shorter edges of the rectangle enlarge and become distorted, as expected from the external field direction, which is inclined with respect to the rectangle edges. Using this approach neither an appreciable loss of resolution nor a distortion of the images can be observed.

These experiments show that in magnetically soft thin films and small patterns spin-SEM is able to image the magnetization reversal on a microscopic scale. It is important to keep the traveling time of the secondary electrons in the region of large magnetic fields brief, in particular if the magnetization component perpendicular to the field direction is investigated. Careful design of combined magnetic coils and transport lenses is key to increase the range of applicable fields, so that technologically relevant samples such as write heads or sensors can be investigated in applied fields.

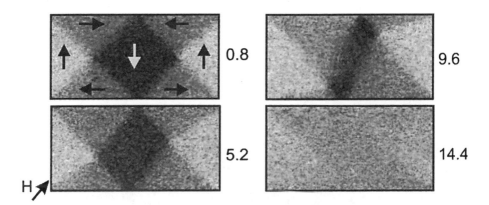

Figure 8-25: Switching process in 50-nm-thick $Ni_{80}Fe_{20}$ rectangles with lateral dimensions of 4 μm × 2 μm, imaged by spin-SEM. The in-plane magnetization component along the page is displayed. The magnetic field was increased during imaging along the direction of the arrow. The field values are given at each image in kA/m. Adapted from Ref. [82], used with permission.

8.7. PHASE TRANSITIONS AND TRANSFORMATIONS

One of the most fundamental properties of a ferromagnet is the existence of a critical temperature, the Curie temperature T_C. At this temperature, the spontaneous magnetization vanishes and long-range order of the spins no longer persists. The approach to T_C is described by the theories of phase transitions, with power-law behavior and the corresponding critical exponents. Phase transitions have been investigated for decades in diverse fields in physics, ranging from magnetism to binary alloys or quark–hadron interactions. Recently it has even been suggested that money-market movements follow the same physical laws [84]. This section summarizes how spin-SEM has been applied to contribute to a deeper understanding of important aspects of phase transitions. It is not surprising that these new insights come from investigations in which ultrathin, two-dimensional magnets are investigated, and in which local, microscopic results rather than spatially averaged data are required.

Experimentally experiments to investigate the ferromagnetic-to-paramagnetic phase transition in ultrathin films are challenging. Because the spontaneous magnetization drops rapidly upon approaching T_C, the spin polarization is small, and temporal fluctuations might become important. High lateral resolution is required, and temperature control and stability are important. The first spin-SEM experiment done with variable temperatures concentrated on the investigation of the aspects of a two-dimensional ferromagnet close to T_C [85]. In the in-plane magnetized Co/Cu(001) model system, a magnetic domain pattern was induced by an external magnetic field and then imaged while approaching T_C, with the temperature stabilized to better than 1 K. By comparing the decay of the spin polarization with the macroscopic remanent magnetization determined by the magneto-optical Kerr effect, the correspondence of microscopic and macroscopic quantities could be established: Spontaneous and remanent magnetization coincide.

Perpendicularly magnetized ferromagnets add an additional degree of complexity. The balance of exchange and long-range dipole-dipole interaction favors a state with alternating domains even for "bulk" ferromagnets [46,86]. The extrapolation to ultrathin films is not straightforward because the magnetostatic energy scales with film thickness. It was Yafet and Gyorgy who calculated in their pioneering paper that also in the ultrathin limit an inhomogeneous magnetization distribution is the equilibrium state, provided that anisotropy and magnetostatic energy are almost the same [69]. This requirement is not necessarily met at the ferromagnetic-to-paramagnetic phase transition, so that the search for this predicted stripe domain at first concentrated on experimental situations in which anisotropy or magnetostatic energy can be tuned by means of the film thickness or temperature. Indeed, the stripe-domain phase evolves during reorientation of the magnetization from perpendicular to parallel to the surface [68], as described in Section 8.5.1. Experimentally it was found that the stripe-domain phase is not limited to the narrow range of parameters predicted in Ref. [69]. In 2 to 3 ML of Fe/Cu(001), i.e., about 2 ML below the reorientation thickness, first indications of meandering elongated domains were found, even though the ratio of crystalline to shape anisotropy is about 3. For such ratios a stripe width exceeding the sample size by several orders of magnitude would be expected, and hence experimentally a single domain state should prevail [69].

Recently the evolution of the stripe-domain phase has been imaged in a detail by spin-SEM in Fe/Cu(001) films only ~2 ML thick [87,88]. These films remain perpendicularly magnetized up to T_C and hence do not undergo spin reorientation. The observed domain patterns upon approaching the paramagnetic phase have been determined by recording the domain patterns in films of slightly different thicknesses at room temperature. As T_C is thickness-dependent, the films thus have a different effective temperature. These results have been verified and compared with experiments on films of constant thickness, in which the temperature was varied. The rich variety of stripe patterns is shown in Fig.8-26. Far below T_C, a stripe phase is observed that meanders around a preferential orientation. Four types of dislocations have been identified experimentally, confirming the predictions of Ref. [89]. At higher effective temperatures (or, equivalently, smaller thickness), orientational melting occurs, and the domain phase transforms into a higher-symmetry labyrinthine phase. In this phase, the stripes have broken up and mainly terminate in square corners. Such a "tetragonal liquid" has been predicted on a fourfold-symmetric substrate lattice [89,90]. Surprisingly, at even higher effective temperatures, the less symmetric stripe phase reoccurs with a reduced stripe width before the paramagnetic phase sets in. Portmann et al. [88] have elucidated the mechanism of the transformations microscopically. A transverse instability drives the stripe-to-labyrinthine transition. The labyrinthine phase converts into the reentrant stripe phase by straightening knee-bend domains. This process is quite complex, going through repetitive stages of cutting or detaching short segments and reshaping to arrive at smoother stripes.

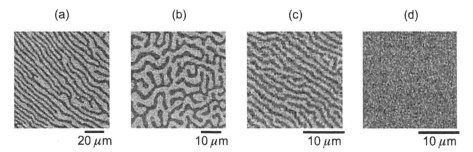

Figure 8-26: The four domain phases in perpendicularly magnetized ultrathin fcc-Fe films grown on Cu(001), measured at room temperature by spin-SEM. The thickness d of the film slightly decreases from (a) to (d), corresponding to a gradual approach to the Curie temperature. (a) Stripe phase at $d = 2.1$ ML; (b) labyrinthine phase at $d = 1.91$ ML; (c) reentrant stripe phase at $d = 1.84$ ML; (d) contrastless paramagnetic phase at $d = 1.73$ ML. Note that the image size varies as indicated by the scale bars. Adapted from Ref. [88], used with permission.

While some of these stripe-phase patterns have already been observed also in thick films, these experiments have discovered new and unexpected features that are intrinsic to truly two-dimensional systems. New insight into a very fundamental problem in phase transitions and transformations has been gained, thanks to a microscopic imaging technique with high sensitivity and lateral resolution.

8.8. EXCHANGE COUPLING

The discovery of antiparallel exchange coupling [91] and shortly thereafter of oscillatory exchange coupling [92] between two ferromagnetic layers separated by a nonmagnetic metallic spacer material has led to a wealth of investigations to resolve the fundamental issues of this phenomenon. Whether the parallel or the antiparallel alignment of the magnetizations is preferred depends on the thickness and nature of the intervening spacer material as well as on the structure and quality of the interfaces between film and spacer. Almost any metallic element of the periodic table is able to couple the two magnetic films [93]. The physical origin of this oscillatory exchange was soon traced back to the quantum confinement of wave function of the majority spin holes in the spacer material [94]. In this model, oscillation periods are directly related to the calipers of the Fermi surface, whereas oscillation amplitudes and phases are also related to the ferromagnetic material. Thus, the periods depend on the crystallographic orientation of the spacer and can be quite complex because more than one period can exist. One of the key ingredients to test and refine the theoretical models was the accurate and precise quantitative determination of oscillation periods. This required the preparation of numerous samples with varying spacer thicknesses but otherwise identical structural perfection. Alternatively, techniques with lateral resolution were of great help because they were able to characterize the magnetic coupling locally on a wedge-type sample: The intervening spacer was grown as a wedge, whereas the magnetic layers were of constant thickness.

The textbook example that set new standards in structural perfection and magnetic characterization in this field was the Fe/Cr/Fe system. The almost identical lattice constants of Fe and Cr make the growth of epitaxial high-quality Fe/Cr/Fe superlattices possible. Unguris *et al.* [95] applied spin-SEM to visualize the variations from parallel to antiparallel magnetic coupling along a chromium wedge in an Fe/Cr/Fe(001) structure. The bottom magnetic layer is a bulk Fe whisker, the highest-quality single crystal known with atomically flat terraces extending over 1 μm. First, the domain state in the bare Fe whisker is imaged, see Fig.8-27(a). Two antiparallel domains magnetized in-plane can be identified. After growth of the Cr wedge and subsequent deposition of a 1-nm-thick top Fe layer, the magnetization pattern in the top layer is determined, see Fig.8-27(b). A regular sequence of oppositely magnetized domains is visible, corresponding to parallel and antiparallel coupling to the bottom Fe whisker, mediated by the Cr spacer layer. The coupling extends up to Cr thicknesses of 80 atomic layers, with a short oscillation period close to 2 ML. Phase slips without reversals take place at 25, 44, and 64 ML, indicating that the oscillation period is incommensurate with the atomic layer spacing. A detailed analysis of these data yields a period of 2.105 ± 0.005 ML [96]. This precise determination of the period was only possible because the Cr thickness of the wedge was measured with equally high precision during growth of the Cr wedge using the spin-SEM tool operated in a different mode: Instead of collecting the polarized low-energy secondary electrons, the reflected high-energy elastically diffracted electrons were monitored. These electrons are known to be an extremely sensitive means of checking the epitaxial quality and determining film thicknesses with atomic-layer precision. The combination of the two techniques was the important step in testing the models for the exchange coupling oscillations with unprecedented accuracy.

This example impressively illustrates that spin-SEM provides a direct map of the sign of the exchange coupling, i.e., whether the two ferromagnets are magnetized parallel or antiparallel to each other. This allows a very small antiparallel coupling to be distinguished from a parallel coupling, a realm that is not easily accessible with the commonly employed techniques of recording the switching field in magnetic hysteresis loops [92]. On the other hand, a determination of the coupling strength is not directly amenable to the technique. Other techniques such as Brillouin light scattering or the magneto-optical Kerr effect serve that purpose and have been employed accordingly.

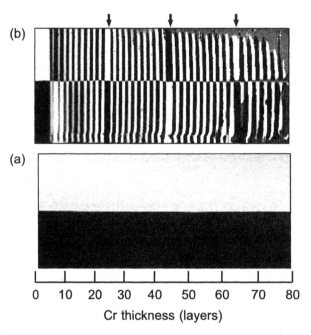

Figure 8-27: Magnetic patterns revealing the exchange oscillations in epitaxial Fe/Cr/Fe(001). (a) Two-domain state with oppositely oriented magnetizations in the Fe(001) substrate. (b) Alternating magnetization directions in the top Fe film illustrating the exchange coupling oscillations across the Cr wedge. Phase slips at 25, 44, and 64 ML are marked by arrows. The Cr thickness is given below the images. Adapted from Ref. [97], used with permission.

8.9. CONCLUDING REMARKS

In this chapter, an overview on spin-polarized scanning electron microscopy has been given. Spin-SEM or SEMPA has made many important contributions to the understanding of ferromagnetism on a microscopic scale. The impact of the technique mainly stems from a combination of different features that make it a unique tool for the investigation of magnetic phenomena in ultrathin films and small structures: It offers not only the high lateral resolution of an electron-beam technique, but also high surface sensitivity, complete separation of topographic or chemical information from magnetic information, and the capability to determine the direction of the magnetization vector. I have chosen not to illustrate these characteristics individually but rather by means of a short tour through some fundamental aspects of ferromagnetism. Such an approach admittedly must remain incomplete but emphasizes the insights into the characterization of magnetic materials on a microscopic scale that spin-SEM has provided. We have started by reviewing the formation of domains and domain walls and have compared bulk magnets with magnetic films and small structures. Magnetic anisotropy in various flavors has been reviewed next, relying on the strength of the spin-SEM technique to determine magnetization directions unequivocally. After a short account of new technical developments in adapting spin-SEM to incorporate magnetic fields and its application to magnetization reversal, phase transitions and exchange coupling have been discussed using a specific example in each case.

I am aware that some topics could only be touched upon. Too many beautiful examples exist that have to be presented for an exhaustive treatment of all aspects, and therefore a subjective choice has been made. I had to omit, for instance, the contributions to the study of exchange bias at the interface between ferromagnets and antiferromagnets [98]. Likewise, investigations of technological relevance such as the direct observation of the transition zone between bits in storage media have been omitted [33] as well as studies relating temperature-dependent domain formation and structural roughness [99], or studies of precessional magnetization reversal [100].

Since its first implementation two decades ago, spin-SEM has matured from a pure research tool to a versatile technique. As long as the trend to ever smaller length scales persists in magnetism, it will make its contributions in both fundamental and applied topics.

ACKNOWLEDGMENTS

The magnetism project at the IBM Zurich Research Laboratory would not have been possible without the valuable contributions of many people. I am particularly grateful to my colleague Andi Bischof for his major contributions to the design and construction of the experimental set-up as well as for keeping the system running. The experimental data presented in this overview were taken in collaboration with Samy Boukari, Urs Dürig, Jürgen Fassbender, Peter Grütter, Maja Haag, Pierre-Olivier Jubert, Gebhard Marx, Michele Stampanoni, and Wolfgang Weber.

REFERENCES

[1]   K. Koike, H. Matsuyama, and K. Hayakawa, *Scanning Microsc.* **1** (1987) 241.
[2]   G.G. Hembree, J. Unguris, R.J. Celotta, and D.T. Pierce, *Scanning Microsc.* **1** (1987) 229.
[3]   H.P. Oepen and J. Kirschner, *Scanning Microsc.* **5** (1991) 1.
[4]   R. Allenspach, *J. Magn. Magn. Mater.* **129** (1994) 160.
[5]   G. Chrobok and M. Hofmann, *Phys. Lett.* **57A** (1976) 257.
[6]   T.H. DiStefano, *IBM Techn. Discl. Bull.* **20** (1978) 4212.
[7]   K. Koike and K. Hayakawa, *Jpn. J. Appl. Phys.* **23** (1984) L187.
[8]   J. Unguris, G.G. Hembree, R.J. Celotta, and D.T. Pierce, *J. Microscopy* **139** (1985) RP1.
[9]   R. Allenspach, M. Taborelli, and M. Landolt, *Phys. Rev. Lett.* **55** (1985) 2599.
[10]   M. P. Seah and W. A. Dench, *Surf. Interface Anal.* **1** (1979) 2.
[11]   D.L. Abraham and H. Hopster, *Phys. Rev. Lett.* **58** (1987) 1352.
[12]   H. C. Siegmann, *J. Phys. Cond. Mat.* **4** (1992) 8395; G. Schönhense and H.C. Siegmann, *Ann. Phys. (Leipzig)* **2** (1993) 465.
[13]   H.-J. Drouhin, *Phys. Rev. B* **62** (2000) 556.
[14]   L. Reimer, *Scanning Electron Microscopy, Springer Series in Optical Sciences, Vol. 45, 2nd Edition* (Springer, Berlin, Heidelberg, 1998).

[15]   G. Marx, P.-O. Jubert, A. Bischof, and R. Allenspach, *Appl. Phys. Lett.* **83** (2003) 2925.

[16]   H.P. Oepen and J. Kirschner, *Phys. Rev. Lett.* **62** (1989) 819.

[17]   M.R. Scheinfein, J. Unguris, R.J. Celotta, and D.T. Pierce, *Phys. Rev. Lett.* **63** (1989) 668.

[18]   L. von Hámos and P.A. Thiessen, *Z. Phys.* **71** (1931) 442.

[19]   F. Bitter, *Phys. Rev.* **38** (1931) 1903.

[20]   T. VanZandt, R. Browning, and M. Landolt, *J. Appl. Phys.* **69** (1991) 1564.

[21]   H.P. Oepen, G. Steierl, and J. Kirschner, *J. Vac. Sci. Technol. B* **20** (2002) 2535.

[22]   M. Haag and R. Allenspach, *Geophys. Res. Lett.* **20** (1993) 1943.

[23]   C.H. Back, W. Weber, A. Bischof, D. Pescia, and R. Allenspach, *Phys. Rev. B* **52** (1995) R13114.

[24]   Chapter 13 in this book.

[25]   J. Stöhr and S. Anders, *IBM J. Res. Develop.* **44** (2000) 535.

[26]   Chapter 9 in this book.

[27]   Chapter 11 in this book.

[28]   Chapter 7 in this book.

[29]   Chapter 10 in this book.

[30]   T.Kohashi and K.Koike, *Jpn. J. Appl. Phys.* **40** (2001) L1264.

[31]   G.R. Morrison, H. Gong, J.N. Chapman and V. Hrnciar, *J. Appl. Phys.* **64** (1988) 1338.

[32]   M.R. Scheinfein, J. Unguris, M.H. Kelley, D.T. Pierce, and R.J. Celotta, *Rev. Sci. Instrum.* **61** (1990) 2501.

[33]   H. Matsuyama and K. Koike, *J. Electron Microsc.* **43** (1994) 157.

[34]   J. Kessler, *Polarized Electrons, 2nd Edition* (Springer, Berlin, Heidelberg, 1985).

[35]   J. Kirschner, *J. Appl. Phys. A* **36** (1985) 121.

[36]   M.R. Scheinfein, D.T. Pierce, J. Unguris, J.J. McClelland, R.J. Celotta, and M.H. Kelley, *Rev. Sci. Instrum.* **60** (1989) 1.

[37]   M. Landolt, unpublished; a sketch of the spherical arrangement is shown in M. Landolt, R. Allenspach, and D. Mauri, *J. Appl. Phys.* **57** (1985) 3626.

[38]   T. Kohashi, H. Matsuyama, K. Koike, and H. Miyamoto, *J. Magn. Soc. Jpn.* **18** (1984) S1, 7.

[39]   P. Weiss, *J. Phys. (France)* **6** (1907) 661.

[40]   H.J. Williams, R.M. Bozorth, and W. Shockley, *Phys. Rev.* **75** (1949) 155.

[41]   J.L. Robins, R.J. Celotta, J. Unguris, D.T. Pierce, B.T. Jonker, and G.A. Prinz, *Appl. Phys. Lett.* **52** (1988) 1918.

[42]   H.P. Oepen, *J. Magn. Magn. Mater.* **93** (1991) 116.

[43]   R. Allenspach, M. Stampanoni, and A. Bischof, *Phys. Rev. Lett.* **65** (1990) 3344.

[44]   C. Kittel, *Phys. Rev.* **70** (1946) 965.

[45]   R. Allenspach and M. Stampanoni, *Mat. Res. Soc. Symp. Proc.* **231** (1992) 17.

[46]   C. Kooy and U. Enz, *Philips Res. Rep.* **15** (1960) 7.

[47]   R.M.H. New, R.F.W. Pease, and R.L. White, *J. Vac. Sci. Technol. B* **13** (1995) 1089.

[48]   S.S.P. Parkin, K.P. Roche, M.G. Samant, P.M. Rice, R.B. Beyers, R.E. Scheuerlein, E.J. O'Sullivan, S.L. Brown, J. Bucchigano, D.W. Abraham, Yu Lu, M. Rooks, P.L. Trouilloud, R.A. Wanner, and W.J. Gallagher, *J. Appl. Phys.* **85** (1999) 5828.

[49]   K. Mitsuoka, S. Sudo, S. Narishige, K. Hanazono, Y. Sugita, K. Koike, H. Matsuyama, and K. Hayakawa, *IEEE Trans. Magn.* **23** (1987) 2155.

[50]   S.Y. Chou, P.R. Krauss, and L. Kong, *J. Appl. Phys.* **79** (1994) 6101.

[51]   R. Lüthi, R.R. Schlittler, J. Brugger, P. Vettiger, M.E. Welland, and J.K. Gimzewski, *Appl. Phys. Lett.* **75** (1999) 1314.

[52]   C. Stamm, F. Marty, A. Vaterlaus, V. Weich, S. Egger, U. Maier, U. Ramsperger, H. Fuhrmann, and D. Pescia, *Science* **282** (1998) 449.

[53]   H.P. Oepen, W. Lutzke, and J. Kirschner, *J. Magn. Magn. Mater.* **251** (2002) 169.

[54]   E. Gu, E. Ahmad, S.J. Gray, C. Daboo, J.A.C. Bland, L.M. Brown, M. Rührig,
       A.J. Gibbon, and J.N. Chapman, *Phys. Rev. Lett.* **78** (1997) 1158.

[55]   R.P. Cowburn, A.O. Adeyeye, and M.E. Welland, *Phys. Rev. Lett.* **81** (1998) 5414.

[56]   R.P. Cowburn, A.O. Adeyeye, and J.A.C. Bland, *Appl. Phys. Lett.* **70** (1997) 2309.

[57]   H. Matsuyama, K. Koike, T. Kobayashi, R. Nakatani, and K. Yamamoto, *Appl. Phys.
       Lett.* **57** (1990) 2028.

[58]   R. Allenspach, *IBM J. Res. Develop.* **44** (2000) 553.

[59]   M.R. Scheinfein, J. Unguris, J.L. Blue, K.J. Coakley, D.T. Pierce, R.J. Celotta,
       and P.J. Ryan, *Phys.Rev.B* **43** (1991) 3395.

[60]   S.Methfessel, S. Middelhock, and H. Thomas, *J. Appl. Phys.* **31** (1960) 302S.

[61]   A. Berger and H.P. Oepen, *Phys. Rev. B* **45** (1992) 12596.

[62]   P. Bruno, *Phys. Rev. Lett.* **83** (1999) 2425.

[63]   N. Garcia, M. Munoz, and Y.-W. Zhao, *Phys. Rev. Lett.* **82** (1999) 2923.

[64]   P.-O. Jubert, R. Allenspach, and A. Bischof, *Phys. Rev. B* **69** (2004) 220410(R).

[65]   L. Néel, *J. Phys. Rad.* **15** (1954) 376.

[66]   B. Hillebrands, Chapter in this book.

[67]   L.M. Levinson, M. Luban, and S. Shtrikman, *Phys. Rev.* **187** (1969) 715.

[68]   R. Allenspach and A. Bischof, *Phys. Rev. Lett.* **69** (1992) 3385.

[69]   Y. Yafet and E.M Gyorgy, *Phys. Rev. B* **38** (1988) 9145.

[70]   A.B. Kashuba and V.L. Pokrovsky, *Phys. Rev. Lett.* **70** (1993) 3155.

[71]   B. Kaplan and G.A. Gehring, *J. Magn. Magn. Mater.* **128** (1993) 111.

[72]   M. Speckmann, H.P. Oepen, and J. Kirschner, *Phys. Rev. Lett.* **75** (1995) 2035.

[73]   A. Berger, U. Linke, and H.P. Oepen, *Phys. Rev. Lett.* **68** (1992) 839.

[74]   P. Haibach, M. Huth, and H. Adrian, *Phys. Rev. Lett.* **84** (2000) 1312.

[75]   Y. Homma, M. Tomita, and T. Hayashi, *Surf. Sci.* **258** (1991) 147.

[76]   W. Weber, A. Bischof, R. Allenspach, Ch. Würsch, C.H. Back, and D. Pescia, *Phys. Rev.
       Lett.* **76** (1996) 3424.

[77]   R. Allenspach, A. Bischof, and U. Dürig, *Surf. Sci.* **381** (1997) L573.

[78]   W. Weber, C.H. Back, A. Bischof, D. Pescia, and R. Allenspach, *Nature* **374**
       (1995) 788.

[79]   M.E. Buckley, F.O. Schumann, and J.A.C. Bland, *Phys. Rev. B* **52** (1995) 6596.

[80]   W. Brown, *Rev. Mod. Phys.* **17** (1945) 15.

[81]   Y. Iwasaki, K. Bessho, J. Kondis, H. Ohmori, and H. Hopster, *Appl. Surf. Sci.* **113**
       (1997) 155.

[82]   G. Steierl, G. Liu, D. Iogorov, and J. Kirschner, *Rev. Sci. Instrum.* **73** (2002) 4264.

[83]   J. Ferré, V. Grolier, P. Meyer, S. Lemerle, A. Maziewski, E. Stefanowicz, S.V.
       Tarasenko, V.V. Tarasenko, M. Kisielewski, and D. Renard, *Phys. Rev. B* **55**
       (1997) 15092.

[84]   V. Plerou, P. Gopikrishnan, and H.E. Stanley, *Nature* **421** (2003) 130.

[85]   D. Kerkmann, D. Pescia, and R. Allenspach, *Phys. Rev. Lett.* **68** (1992) 686.

[86]   T. Garel and S. Doniach, *Phys. Rev. B* **26** (1982) 325.

[87]   A. Vaterlaus, C. Stamm, U. Maier, M.G. Pini, P. Politi, and D. Pescia, *Phys. Rev. Lett.* **84**
       (2000) 2247.

[88]   O. Portmann, A. Vaterlaus, and D. Pescia, *Nature* **422** (2003) 701.

[89]   Ar. Abanov, V. Kalatsky, V.L. Pokrovsky, and W.M. Saslow, *Phys. Rev. B* **51**
       (1995) 1023.

[90]   I. Booth, A.B. MacIsaac, J.P. Whitehead, and K. De'Bell, *Phys. Rev. Lett.* **75**
       (1995) 950.

[91] P. Grünberg, R. Schreiber, Y. Pang, M.N. Brodsky, and H. Sower, *Phys. Rev. Lett.* **57** (1996) 2442.

[92] S.S.P. Parkin, N. More, and K.P. Roche, *Phys. Rev. Lett.* **64** (1990) 2304.

[93] S.S.P. Parkin, *Phys. Rev. Lett.* **67** (1991) 3598.

[94] D.M. Edwards, J. Mathon, R.B. Muniz, and M.S. Phan, *Phys. Rev. Lett.* **67** (1991) 493.

[95] J. Unguris, R.J. Celotta, and D.T. Pierce, *Phys. Rev. Lett.* **67** (1991) 140.

[96] D.T. Pierce, J.A. Stroscio, J. Unguris, and R.J. Celotta, *Phys. Rev. B* **49** (1994) 14564.

[97] J. Unguris, D.T. Pierce, R.J. Celotta, and J.A. Stroscio, *Proceedings of the NATO Advanced Research Workshop*, Corsica, France, June 15-19, 1992, in *Magnetism and Structure in Systems of Reduced Dimension*, ed. by R.F.C. Farrow et al. (Plenum Press, New York, 1993).

[98] H. Matsuyama, C. Haginoya, and K. Koike, *Phys. Rev. Lett.* **85** (2000) 646.

[99] H. Hopster, *Phys. Rev. Lett.* **83** (1999) 1227.

[100] C.H. Back, R. Allenspach, W. Weber, S.P.P. Parkin, D. Weller, E.L. Garwin, and H.C. Siegmann, *Science* **285** (1999) 864.

CHAPTER 9

Spin-polarized low energy electron microscopy (SPLEEM)

9.1. INTRODUCTION

Probing the spin of matter with slow electrons is based on their strong spin-spin interaction in contrast to fast electrons that experience a strong spin-orbit interaction, the basis of the Mott detector. In addition to the strong spin-spin interaction slow electrons have a short inelastic mean free path in solids. Therefore their penetration depth into solids is small and the spin can be probed only in reflection experiments. This has been done for many years in spin-polarized low energy electron diffraction (SPLEED) [1]. In SPLEED electrons with energies between several 10 eV and about 100 eV are used. At lower energies the spin-spin interaction is stronger, the inelastic mean free path becomes spin-dependent and in many materials the back-scattering cross section is higher. This allows probing the spin not only by diffraction but also by imaging in a low energy electron microscope (LEEM) provided that it is equipped with a source of spin-polarized electrons. This type of microscope uses normal incidence and reflection of the electrons. The resolution of modern LEEM instruments is limited by lens aberrations to 5 – 10 nm, that of SPLEEM instruments to somewhat larger values because of the image subtraction necessary in order to obtain pure magnetic contrast, which reduces the signal/noise ratio.

The power of LEEM and in particular of SPLEEM is due to the fact that in single crystals the backscattered intensity is focused into a few diffracted beams, in particular into the specular beam, which generally is used for imaging. This is also true for epitaxial layers and for the specularly scattered beam ("00 beam") in polycrystalline films with strong preferred crystal orientation ("fiber texture"). In amorphous materials and polycrystalline films without preferred crystal orientation the backscattered electrons are distributed over a wide angular range of which only a small fraction can be used for imaging because of the lens aberrations. Imaging is still possible as illustrated in Fig. 9-1, though with much longer image acquisition time. Therefore SPLEEM is most useful for the study of the magnetization in the topmost layers of single crystal surfaces, epitaxial layers and layers with strong fiber texture. For amorphous and polycrystalline materials without preferred crystal orientation X-ray magnetic circular dichroism photo emission electron microscopy (XMCDPEEM) is preferable, although this method sees mainly the in-plane component of the magnetization. If all three components are of interest, then scanning electron microscopy with polarization analysis (SEMPA) is preferable. All three methods have presently comparable resolution (several 10 nm).

Figure 9-1: Typical LEEM image of 20 nm thick Co elements. The brightness of the elements indicates a preferred orientation of the crystallites with their surface approximately parallel to the film surface. Electron energy 5.1 eV. Diameter of field of view 10 μm.

9.2. PHYSICAL BASIS

As already mentioned in the introduction, magnetic contrast has two causes: i) the spin-spin or exchange interaction between the spin-polarized incident electron and the spin-polarized electrons in the magnetic material, ii) the different inelastic mean free paths of electrons with spin parallel and anti-parallel to the spins in the material. The details of these causes have been reviewed elsewhere [2,3] in more detail. Therefore they are only briefly discussed here.

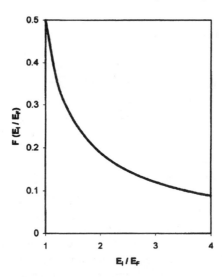

Figure 9-2: Energy dependence of the exchange potential relative to its value at the Fermi energy.

The scattering potentials of atoms for slow electrons depend upon the relative orientation of the spin of the incident electron and of the unpaired electrons of the atom. This dependence is weak because most of the scattering is determined by the nucleus and the filled shells that is spin-independent. It decreases rapidly with increasing energy because of the energy dependence of the exchange potential, which is shown in Fig. 9-2 for the free electron gas. It is large at the Fermi level E_F, but at 1 eV above the vacuum level it is only ½ of the value at the Fermi level, at 5 eV only 1/3 and at 20 eV only 1/5 of it in Fe and Co. Spin-dependent scattering potentials cause spin-dependent back-scattered intensities for electrons with spin parallel and anti-parallel to the magnetization **M** in ferromagnetic materials. Therefore the lowest energies give the strongest magnetic contrast.

The second reason for using very slow electrons is the difference in the inelastic mean free paths $l_\uparrow$, $l_\downarrow$ of electrons with spin parallel and anti-parallel to that of the unpaired electrons in the material. Calculated values of the energy dependence of $l_\uparrow$ and $l_\downarrow$ for Fe are shown in Fig. 9-3 [4]. This difference also contributes to the magnetic contrast. In Fe, for example, $l_\uparrow / l_\downarrow = 3.2$ and 2.0 at 1 eV and 5 eV, respectively, above the vacuum level while at 20 eV the inelastic mean free paths are practically identical. It should be noted that in materials with a high density of unoccupied states above the vacuum level the inelastic mean free paths l are much shorter than those of the so-called universal curve. In Fe and Co with their high density of unoccupied spin down states l is in the 1 monolayer range compared to the 10 monolayer range of the universal curve. This, together with band gaps above the vacuum level, gives SPLEEM a very high surface sensitivity.

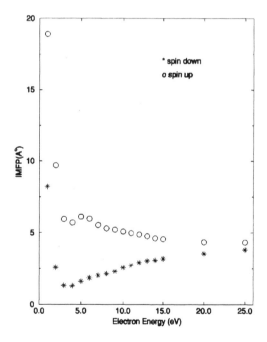

Figure 9-3: Inelastic mean free paths of majority spin and minority spin electrons in Fe. From [4] with permission.

While magnetic contrast optimization favors very low energies, resolution considerations suggest the use of higher energies. The resolution is determined primarily by the chromatic aberration of the objective lens, which is a cathode (immersion) lens, secondarily by its spherical aberration. In the low energy range considered here the aberrations of the homogeneous field in front of the specimen dominate. Fig. 9-4 shows how the two aberrations together with the diffraction at the angle-limiting aperture determine the resolution in the idealized case of a homogeneous filed of length 3 mm, a start energy of 2.5 eV at the specimen, an energy spread

ΔE of 0.25 eV and an acceleration voltage of 25 kV. The chromatic aberration is approximately proportional to ΔE/E. Thus, at fixed ΔE, the resolution can be improved by a factor of 10 by going from 1 eV to 10 eV. However, in Fe, Co and Ni and their alloys, the backscattered intensity decreases considerably when this is done and so does the magnetic contrast. Therefore, energies of a few eV are generally used in SPLEEM. In practice ΔE is larger than 0.25 eV, the acceleration voltage is smaller than 25 kV and the approximation of a homogeneous field in front of the specimen gives only a lower limit of the resolution. In SPLEEM it is therefore presently in the 10 nm range.

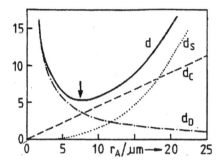

Figure 9-4: Resolution limitation by the homogeneous field part of a cathode lens at low energies resulting from diffraction at the angle-limiting aperture, chromatic and spherical aberration.

9.3. INSTRUMENTATION AND METHODIC

A SPLEEM instrument is a LEEM instrument with a spin-polarized electron source. Presently there are worldwide only two home-built SPLEEM instruments and a commercial one that is equipped with both a conventional and a spin-polarized electron source. As a consequence, the contributions of SPLEEM to the understanding of the magnetic microstructure of materials are limited up to now, which is reflected in the fact that the method is only briefly mentioned [5] or not mentioned at all [6] in recent reviews of magnetic imaging. The results reported below have been obtained with the first two instruments and only these will be sketched briefly. The first one is the original LEEM instrument [7] in which the field emission source was replaced by a spin-polarized electron source [8]. This source consisted of a GaAs cathode, activated with Cs and O_2 to negative electron affinity, and an electrostatic 90° sector field. It allowed only imaging of in-plane **M** distributions. It has been modified considerably since then, in particular by replacing the complete illumination system with a "spin manipulator" [9]. Except for a combined electrostatic-magnetic 90° sector field and a spin rotator lens, all lenses in the illumination system are electrostatic, while those in the imaging column are magnetic. The specimen chamber that houses the specimen manipulator and the objective lens is connected to the illumination and imaging columns by a 60° magnetic sector field. The lenses in the imaging column allow switching between imaging and diffraction. In addition to the lenses and sectors the system contains two apertures, a field-limiting aperture in the illumination system that has to be introduced in the diffraction mode and an angle-limiting aperture ("contrast aperture") that has to be introduced for imaging. Other electron-optical components are deflectors for beam alignment and stigmators for astigmatism correction. The electron source is at a potential of –15 kV, the specimen, which is part of the electrostatic objective lens, is at –15 kV + E/e so that the electrons arrive at the specimen with the energy E after having had 15 keV energy in the rest of the microscope. The specimen can be exposed to vapors from several evaporators before and during imaging, allowing in situ growth studies. As no magnetic shielding is used, the complete system is surrounded by large Helmholtz coils that compensate DC and (actively) AC fields. The instrument is shown in Fig. 9-5.

Figure 9-5: Present configuration of the first SPLEEM. The spin manipulator is on the left side, the imaging column on the right side, the 60° beam separator and the specimen chamber, with several evaporators mounted, in the center. On the right side of the specimen chamber is the preparation chamber and the airlock facility.

The second SPLEEM instrument whose schematic is shown in Fig. 9-6 [10] is completely electrostatic except for the magnetic fields needed in the spin manipulator and in the beam separator. It was designed as a flange-on-instrument. Therefore the specimen is at ground potential while all lenses are at +5 kV except for the first electrode of the objective that determines the field strength at the specimen. A high field strength at the specimen is important for good resolution so that this electrode is at +15 kV. The use of 45° double deflectors in the illumination and imaging beams allow a compact design. Magnetic shielding makes Helmholtz coils unnecessary. The other components of the instrument, in particular the spin manipulator, are similar to those of the first one.

In the spin manipulator the negative electron affinity GaAs cathode is illuminated with circular polarized light from a diode laser, which excites photoemission of spin-polarized electrons with the polarization **P** perpendicular to the surface. When the helicity of the light is switched by a Pockels cell or some other device, the direction of **P** inverts. While polarization degrees $P = |\mathbf{P}|$ of up to 50% can be achieved with suitably prepared GaAs sources and up to 80% with more complex cathodes [11] the present instruments operate usually with about 20%. When the 90° sector field is operated purely electrostatic, the direction of **P** is not influenced and the electrons enter the polarization rotator with **P** perpendicular to **v**, the propagation direction of the electrons along the optical axis. Exciting the magnetic rotator lens allows to rotate **P** around the optical axis. Finally, if the sector field is operated purely magnetic, **P** follows **v** so that it points along the optical axis. Combining electrostatic and magnetic deflection with rotation allows alignment of **P** along any direction in space. Usually two preferred direction in the surface of the specimen and the specimen normal are chosen to determine the three components of **M**. Fig. 9-7 sketches the design of this part of a SPLEEM instrument.

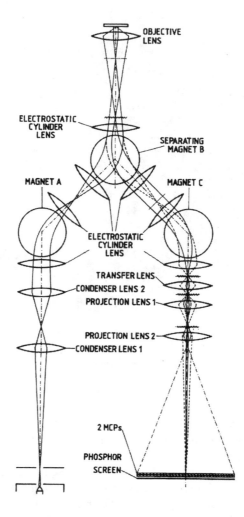

Figure 9-6: Schematic of the flange-on-SPLEEM. From [10]. The spin manipulator replaces the emitter shown schematically at the lower left.

Provided that **M** and **P** do not change during the interaction between the spin-polarized incident electrons and the electron in the ferromagnet - an assumption to be discussed below - the magnetic contrast between regions with different **M** directions is proportional to **P·M**. It is expressed by the so-called exchange asymmetry $A_{ex} = (1/P)(I_\uparrow - I_\downarrow)/(I_\uparrow + I_\downarrow)$ where $I_\uparrow$ and $I_\downarrow$ are the local intensities obtained by inverting the direction of **P**. In the difference image the **P**-independent nonmagnetic structural contrast is cancelled out, in the sum image the **P**-dependent magnetic contrast. This allows easy correlation between magnetic and structural features. The images are acquired from the channel plate-fluorescent screen combination with a CCD camera and are subtracted and added pixel by pixel. Because of the low P and the small contribution of the magnetic contrast to the total contrast usually the contrast in the asymmetry images has to be enhanced. This is done as follows. First it should be noted that A_{ex} might either be positive or negative. Negative values cannot be displayed, therefore the zero in the A_{ex} images is shifted up by half of the dynamic range N of the camera. What is finally displayed is $A_{ex}^* = N/2 + C (I_\uparrow - I_\downarrow)/(I_\uparrow + I_\downarrow)$ where C may be as large as 10 or even larger when the contrast is very weak. The resolution is then limited by noise [12].

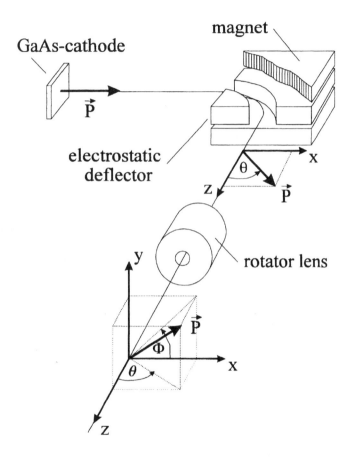

Figure 9-7: Schematic of the spin manipulator. From [10].

The assumption that **M** and **P** are constant during the interaction process is not strictly correct whenever **M** and **P** are not collinear. In this case **P** precesses around **M** and rotates into **M** and vica versa. Recent transmission experiments of slow spin-polarized electrons through thin ferromagnetic films [13] allow to estimate the possible influence of these effects in SPLEEM studies. The specific precession angles ε of **P** of electrons 7 eV above E_F in Fe and Co are 33°/nm and 19°/nm, respectively, the spin-averaged inelastic mean free paths λ are about 0.4 nm [4]. In a simple model ε has to be averaged over the penetration depth with exp(-2x/λ). For films more than a few monolayers thick this gives values of only about 7° and 4° for Fe and Co, respectively. They depend but little upon energy up to about 14 eV above E_F. For the rotation of **P** into **M** even smaller angles can be estimated from the transmission experiments ($\approx$ 3°). The influence of **P** on **M** is negligible at the low current densities used in SPLEEM so that no domain switching as observed in experiments with high current densities [14,15] can occur. The precession and rotation effects are cancelled in magnetic domain images by the fact that they are difference images taken with opposite **P**.

9.4. APPLICATIONS

9.4.1. General comments

The main application of SPLEEM is the in situ study of the change of the magnetic domain structure of ultrathin epitaxial films with thickness, temperature and applied magnetic field normal to the surface. This can be done efficiently because image acquisition times are short so that at least one of the magnetization components can be monitored continuously during growth, temperature or field change. However, externally prepared samples can be studied as well after proper cleaning, for example by sputtering in the preparation chamber. In fact, one of the first SPLEEM images was obtained from a sputter-cleaned Co (0001) single crystal surface (see Fig. 9-8 [8]). Thin films that have been prepared in another system and have been coated with a protective coating also pose no problems for SPLEEM after the coating has been sputtered off [16]. Inasmuch as most of the limited work done to date is in situ work, only examples of in situ studies will be given. They may be divided into three groups: studies of single layers, double layers and trilayers ("sandwiches").

Figure 9-8: SPLEEM image of the in-plane magnetization component on a Co(0001) surface. Electron energy 2 eV. From [8] with permission.

9.4.2. Single layers

The main subjects of the single layer studies have been the connection between structure and magnetism and the spin reorientation transition (SRT) with increasing thickness or temperature. The systems that have been studied are Co on W surfaces and on Au(111), Fe-Co alloys on Au(111) and Fe on Cu(100). Before discussing these subjects, an application that is unique to SPLEEM, will be briefly mentioned. It is based on the fact that the electrons used in SPLEEM have a well-defined energy E, momentum **k** and wavelength $\lambda = 2\pi/k$. In the ferromagnetic solid the E(**k**) relationship is spin-dependent because of the exchange splitting. In a thin film with parallel reflecting boundaries and thickness t the electron wave produces at normal incidence standing waves in the film whenever $n\lambda/2 = t + \varphi$, where φ represents the finite penetration depth of the wave beyond the boundaries ("Quantum size effect"). This causes extrema in the reflected intensity, similar to the Fabry-Perot etalon in optics. Because of the exchange splitting the energies at which this condition is fulfilled are different for spin-up and spin-down electrons, a measurement of the reflectivity oscillations as a function of thickness and energy allows the determination of the exchange-split excited state band structure. This has been demonstrated for thin Fe films on W(110) using special preparation procedures which produce films that are atomically flat and single domain over several μm and in which regions with different thickness coexist [17]. Fig. 9-9 shows an example of the reflectivity oscillations. The band structure can be deduced from them with such accuracy that a decision between various band structure calculations can be made [18].

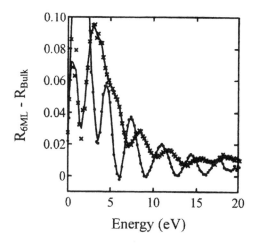

Figure 9-9: Spin-dependent quantum size oscillations in the reflected intensity from a 6 monolayer thick Fe film region as a function of energy. The intensity was measured from a 1 μm² region in the spin-up and spin-down images. From [18] with permission.

The most detailed SPLEEM study of a single layer to date is that of Fe on Cu(100) [19-22]. The system Fe/Cu had previously been the subject of numerous laterally averaging studies because of its complex evolution of structure and magnetism with increasing film thickness. From these studies it was concluded that up to 4 monolayers (ML) Fe grows in a tetragonal distorted face-centered (fct) structure. This phase is ferromagnetic and shows perpendicular magnetization. Between 4 ML and about 11 ML the Fe film has fcc structure with an interlayer spacing of 0.178 nm, except for the topmost layers whose distance is 0.187 nm. The topmost layers (1-2 ML) remain ferromagnetic with perpendicular magnetization while the lower layers are antiferromagnetic due to their reduced interatomic distances. Finally, above 11 ML the Fe film converts into the bcc structure with in-plane magnetization. This information came from studies with a wide variety of different experimental techniques.

SPLEEM not only complements these studies with lateral resolution but allows also combining structural and magnetic study in one experiment. The structural information comes from the LEED pattern and from the LEEM image ($I_\uparrow + I_\downarrow$) that is determined by the interference effects at the surface and, therefore, shows the intensity oscillations that occur during monolayer-by-monolayer growth (Fig. 9-10 [19]). The magnetic images (a) – (g), taken at the points of the LEEM intensity oscillations marked by arrows, show that the domain size increases initially with thickness and decreases near 4 ML again. This has been attributed to corresponding changes of the Curie temperature. Quasi-continuous image acquisition and low deposition rate was used to determine the exchange asymmetry in small thickness increments and with high accuracy (Fig.9-11 [19]. With the assumption that the asymmetry is proportional to the magnetization M – which is correct under certain conditions (for references see [2]) – the thickness dependence of M could be fitted with the two-dimensional Ising model, similar to the results of the laterally averaging studies.

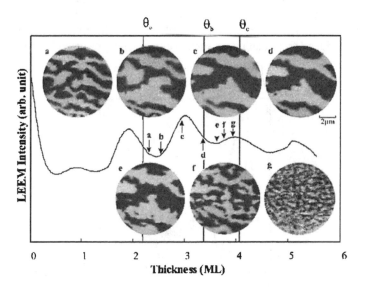

Figure 9-10: SPLEEM images of the out-of-plane magnetization taken during the growth of Fe on Cu(100) at the deposition times marked at he LEEM intensity oscillation curve. Room temperature, deposition rate 0.067 ML/min, electron energy 1.8 eV. From [19] with permission.

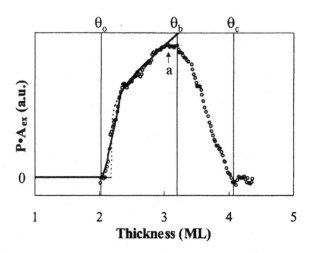

Figure 9-11: Exchange asymmetry as a function of Fe film thickness on Cu(100). The solid and dashed curves are fits to the data of $\theta M(T,\theta)$ assuming the usual thickness (θ) and temperature (T) dependence of M in the two-dimensional Ising model without (-----) and with (——) thickness broadening. From [19] with permission.

It should be noted that the theoretical models, such as the Ising or the Heisenberg model, of the thickness dependence of the Curie temperature assume thickness-independent interatomic distances, an assumption that is frequently not fulfilled in the first several monolayers. The system Fe/Cu(100) is a good example as shown by a recent high resolution STM study [23,24]. In this study it was found that between 2 and 4 ML Fe did not grow in a fct structure but consisted of a nanoscale mixture of fcc regions and regions of stripe-shaped nanotwins with a bcc-like structure. The fraction of the bcc-like structure changed with thickness in a manner similar to the asymmetry shown in Fig. 9-11. This suggests that the thickness dependence of the magnetization is determined by the bcc-like fraction of the film, which is believed do be ferromagnetic on the basis of its interatomic distances in contrast to the more densily packed fcc regions, and not by finite size effects at constant interatomic distances. The observation that the coverages θ_b and θ_c in Figs. 9-10 and 9-11 depend upon growth rate [19,20] may also be connected with the fraction of the bcc-like phase. Obviously, the much-studied system Fe/Cu(100) still poses some questions.

The strong exchange asymmetry and large domain size of 3 ML thick Fe films on Cu(100) make them ideal for demonstrating the capability of SPLEEM to obtain magnetic information from submicron regions. One example is the determination of the hysteresis curve in fields perpendicular to the film [21]. Another one is the demonstration of the transition from a bubble-domain phase via a stripe phase to a inverse bubble phase upon magnetization reversal on a submicroscopic level [22]. The study of the system Ni/Cu(100) showed no magnetic signal up to 5 ML above which suddenly very large domains (several 10 μm diameter) with in-plane magnetization appeared [25].

Another extensively studied system is the system Co/Au(111) because of its interesting SRT. The first microscopic investigation with scanning electron microscopy with polarization analysis (SEMPA) [26] came to the conclusion that in the SRT the spins rotated within the domains from perpendicular to in-plane orientation. Later SEMPA studies [27-29] showed that the SRT was more complex: it occurred via a decay of the domain size in a narrow thickness range. This thickness range was studied in more detail with SPLEEM, which allowed much finer thickness steps because of its fast image acquisition capability. Fig. 9-12 [30,31] shows some images of this transition that occurs between 4.1 and 4.6 ML. It is evident that the in-plane and out-of-plane domains are initially (e.g. at 4.3 ML) not randomly distributed but are intimately connected with each other. Superposition of in-plane and out-of-plane images leads to the conclusion that the SRT begins with broadening of the Bloch domain walls of the large out of-plane domains, which leads to the coexistence of in-plane and out-of-plane domains. This domain wall broadening and domain coexistence has also been found in theoretical work [32-34]. In the later stages, once the in-plane magnetization becomes dominant (e.g. at 4.45 ML), the vectorial addition of in-plane and out-of-plane images leads to a fine-grained wrinkled magnetization distribution, similar to the one described below in the system Co/W(110)), in which the spins are tilted within the domains.

While there is no evidence that the SRT in Co/Au(111) is caused by a structural transition but is rather solely a result of the competition between shape anisotropy and surface anisotropy, the SRT in Fe/Au(111) is clearly connected with a structural change. It has been observed at low temperatures with lateral averaging methods and the structure has been studied with STM and diffraction methods. SPLEEM at room temperature did not allow studying the SRT because the Curie temperature of the out-of-plane magnetized phase is below room temperature. Therefore up to 30% Co was alloyed to the Fe, which increased T_C sufficiently so that (out-of-plane) magnetic domains could be observed starting at 1.2 ML [35]. The SRT took place between about 2 and 2.75 ML in a more complicated manner than in the other systems. The out-of-plane magnetization phase developed a well–defined stripe pattern with increasing thickness (Fig.9-13 [35]) in which the spins were tilted ("canted phase") in such a manner that their in-plane components formed an antiferromagnetic pattern. During the SRT not only the tilt increased up to 35° – not to 90° (in-plane)!(Fig.9-14), but also the direction of the stripes rotated by 15°–20°, depending upon Co concentration. The subsequent in-plane magnetization phase had initially a large out-of-plane spin component that rapidly decreased with thickness so that at 3.5 ML the magnetization was completely in-plane (Fig.9-14). In contrast to Co on Au(111) the domain walls do not broaden during the transition within the resolution limit, which is noise limited. The domain size increases linearly with thickness below the SRT and decreases exponentially above the SRT, as observed also in other SRTs such as in the system Co/Au(111) in agreement with theory.

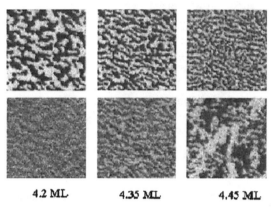

Figure 9-12: Selected images of the spin reorientation transition of Co on Au(111). Top row: out-of-plane magnetization component, bottom row: in-plane magnetization component. Field of view: 7×7 μm^2, electron energy 1.2 eV. From [31] with permission.

The exchange asymmetry A_{ex} of the stripe phase decreased approximately linearly with decreasing Co concentration, which was attributed to a corresponding decrease of T_C. At constant Co concentration it decreased during the SRT to a lower value while the in-plane phase showed a large A_{ex} from the very beginning of its appearance (Fig. 9-14 [35]. The strong magnetic order at room temperature already between 1 and 2 ML is due to the large interatomic distances in the initially formed pseudomorphic layer that produces a "high-spin state" whose spin magnetic moment is about 15% larger than that of Fe in the bcc bulk [36]. The in-plane magnetization phase beyond the SRT has the bcc structure and the bulk spin magnetic moment. The fact that A_{ex} is larger in this phase is simply due to the higher Curie temperature and to the larger thickness that contributes to the signal.

Two other film-substrate pairs that have been studied with SPLEEM, though in less detail are the systems Co/W(110) and Co/W(111). Co films on W(110) are strained in one direction (W[1-10]) to fit the substrate and floating in the perpendicular direction (W[001]). This causes a pronounced uniaxial magnetic anisotropy. SPLEEM corrected the previously held picture of a pure in-plane magnetization. The easy switching between the three magnetization components allowed following the evolution of in-plane and out-of-plane components. At the smallest thickness, 3 ML, at which sufficient magnetic contrast could be obtained at room temperature so that the tilt angle could be determined, **M** was tilted about 35° out of plane [37]. The tilt decreased in an apparently oscillatory manner to zero at about 10 ML. The out-of-plane component had the same sign over relative large areas and changed sign preferentially at monatomic substrate steps, while the in-plane component formed very large domains, essentially uninfluenced by steps (Fig. 9-15 [37]).

This resulted in a wrinkled magnetization quite different from the tilted magnetization in the $Fe_{1-x}Co_x$ films discussed above. Thickness fluctuations of +/-1 ML, as imaged in Fig. 9-17c with the quantum size effect, had no influence on the magnetization distribution. In the system Co/W(110) the first monolayer has already a hexagonal close packing – though somewhat strained to fit the substrate in one direction – and maintains it during further growth with rapidly decreasing strain.

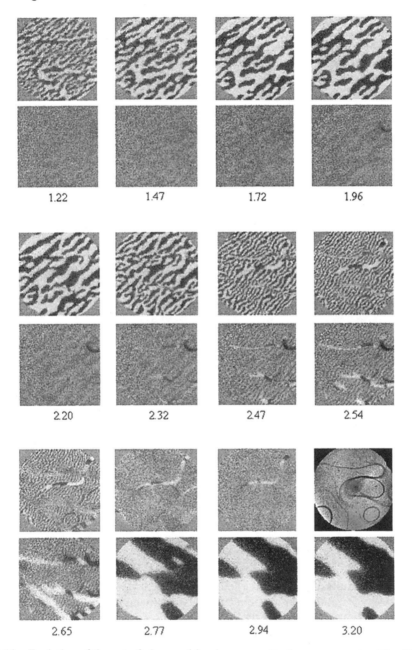

Figure 9-13: Evolution of the out-of-plane and in-plane magnetization components of $Fe_{0.7}Co_{0.3}$ alloy films on Au(111) with increasing thickness. The snake-like features are from step bunches that have a smaller or larger thickness than the flat regions because of the small grazing angle of incidence of the Fe and Co vapor. Field of view: 10×10 μm^2, electron energy 2.5 eV. From [35] with permission.

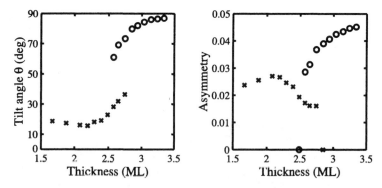

Figure 9-14: Tilt angle against the surface normal of the magnetization (left) and exchange asymmetry (right) of the stripe domains (×) and in the large in-plane domains (o) as a function of $Fe_{0.8}Co_{0.2}$ film thickness on Au(111). Points on the abscissa indicate absence of the corresponding domains. Asymmetry not corrected for the degree of the polarization of the incident beam. From [35] with permission.

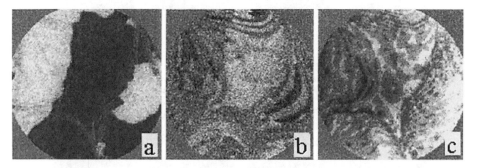

Figure 9-15: In-plane (a), out-of-plane (b) magnetization component images and structural image (c) of a 5 monolayer thick Co film on W(110). The quantum size contrast in (c) shows 4, 5 and 6 monolayer thick regions. Electron energy 1.5 eV (a,b) and 3 eV (c). Field of view: $6{\times}6\ \mu m^2$. From [30] with permission.

In contrast the system Co/W(111) undergoes considerable temperature-dependent structural changes within the first ten monolayers. Furthermore, the threefold symmetry of the substrate does not induce a strong magnetic anisotropy. At room temperature and up to about 450 K thick pseudomorphic (ps) Co layers – at least 12 ML, the thickest layer studied - can be grown. 3 ps ML may be considered as a highly corrugated and laterally slightly compressed Co(0001) layer [38]. Magnetic contrast – purely in-plane - appears first at about 7.5 ML, that is at about 2.5 (0001) Co ML, which is slightly higher than on W(110) where contrast appears first at about 2 ML. On a well-oriented surface small domains with three orientations are observed, on a somewhat misoriented surface large domains with only one orientation, which is attributed to a step-induced anisotropy.

Summarizing the results of single layer studies, the usefulness of SPLEEM in providing complementary magnetic, morphologic (from LEEM) and structural (from LEED) information is evident. Other important aspects are the high magnetic sensitivity, the access to all three spin components and the fast image acquisition. Making use of these features, the few SPLEEM studies made to date have already revealed a fascinating variety of phenomena in ultrathin ferromagnetic films.

9.4.3. Double layers

There are many interesting layer combinations such as Cr/Fe, which has been studied extensively with SEMPA, but SPLEEM has been used up to now only to obtain a microscopic understanding of the thickness dependence of the spin reorientation upon the thickness of nonmagnetic overlayers. Laterally averaging studies had shown that nonmagnetic overlayers such as Cu or Au cause a transition from in-plane to out-of-plane magnetization in thin Co films that is most pronounced at about one ML. SPLEEM is well-suited for the study of the thickness dependence because both in-plane and out-of-plane components of the magnetization can be followed quasi-continuously during the growth of the overlayer. Co films on W(110), which had already an out-of-plane component before overlayer deposition as discussed above, were used. Fig. 9-16 [39] illustrates the effect of 1.5 ML auf Au on the in-and out-of-plane images of a 6 ML thick Co film. From the relative A_{ex} values the tilt angle can be determined. The overlayer increases the out-of-plane tilt of **M** from $10°$ to $30°$ without changing the domain structure. In thinner films this tilt can be changed more, for example from $25°$ to $70°$ but no complete perpendicular magnetization could be achieved and the previously reported maximum at about 1.5 ML was only weakly pronounced. With Cu overlayers the tilt enhancement was smaller but the maximum around 1 ML was clearly evident. By varying the deposition temperature conditions could be found, which produced with increasing average thickness regions of bare, monolayer- and trilayer-covered surface that were large enough for local evaluation of A_{ex}. Fig. 9-17 [39] shows the result. Below 1 ML regions with bare surface, in which the **M** tilt angle is about $12°$, coexist with monolayer-covered regions with a tilt angle of $25°$, above 1 ML monolayer- and trilayer-covered regions with tilt angles of about $15°$ and $10°$, respectively. This shows that the tilt angle is very sensitive to the thickness distribution, which cannot be seen in laterally averaging measurements, and may vary accordingly with growth conditions.

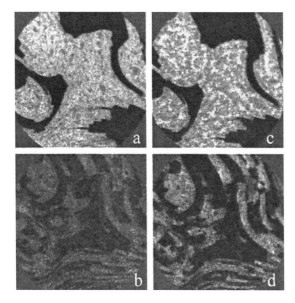

Figure 9-16: Out-of-plane (a,c) and in-plane (b,d) magnetization component images of a 6 monolayer thick Co film on W(110) before (a,b) and after (c,d) deposition of 1.5 monolayers of Au. Note that the contrast in (b) has been enhanced by a factor of 1.5 over that in (d) in order to make the domain structure more visible. Electron energy 1.2 eV. Field of view $\approx 7{\times}7$ µm². From [39] with permission.

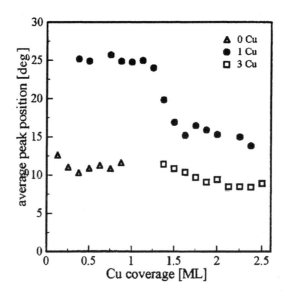

Figure 9-17: Local tilt angle of the magnetization against the surface in 0, 1 and 3 monolayer thick regions of a Cu film on a 5 monolayer thick Co film on W(110). From [39] with permission.

9.4.4. Trilayers (sandwiches)

Trilayers or sandwiches are of particular interest in connection with the giant magnetoresistance (GMR) in multilayers and its practical applications. GMR can occur when the magnetization in alternating layers of a multiplayer is antiparallel, that is when the exchange coupling between them is antiferromagnetic (AF). Some of the largest GMR effects have been obtained in sputtered Co/Cu multilayers but in Co/Cu multilayers deposited by molecular beam epitaxy, in particular in films with (111) orientation, the GMR and AF were much weaker. Therefore, SPLEEM was used with the goal to obtain a microscopic understanding of the causes of the weak GMR in these films, in particular of the influence of the microstructure on the AF coupling. Co/Cu/Co and for comparison Co/Au/Co sandwiches were prepared in situ on W(110) and the domain structure of the Co overlayer with was compared with that of the Co underlayer as a function of Cu and overlayer thickness. The domain structures of the system Co/Cu/Co confirmed the AF coupling for about 4.3 ML Cu found earlier in laterally averaging studies but were very sensitive to the deposition temperature of the Cu spacer layer. For example, at a deposition temperature of about 400 K the domain structure in the top Co layer was very fine-grained without any relation to that of the bottom Co layer. This was attributed to the fine-scale roughness of the resulting top Cu/Co interface which leads to corresponding regions with AF and ferromagnetic (F) coupling [40].

When Au was used as a spacer layer, growth at about 400 K also completely destroyed any correlation with the domains in the bottom layer but produced rather large domains with **M** perpendicular to **M** in the bottom layer, indicative of biquadratic coupling. This type of coupling, which was assumed to occur only in the transition region between AF and F coupling was very evident at all spacer thicknesses. An example is shown in Fig.9-18 [41]. The 6 ML thick Au film was grown slightly above room temperature on a 7 ML thick Co layer that still has some out-of-plane component of **M** as seen in the image with **P** perpendicular to the film. This weak **M** component produced via F coupling a pronounced out-of-plane **M** domain pattern up to 4

ML (center row), which disappeared in the SRT region to be replaced by a pure in-plane **M** domain pattern (bottom row). **M** has now two equally strong components, one with **M** parallel to **M** in the bottom layer (F coupling), the other with **M** perpendicular to it and a completely different domain distribution. Vectorial addition of the A_{ex} values pixel by pixel produces a domain structure in which **M** is rotated $\pm 45°$ with respect to the orientation in the bottom layer, the classical picture of the 90° coupling.

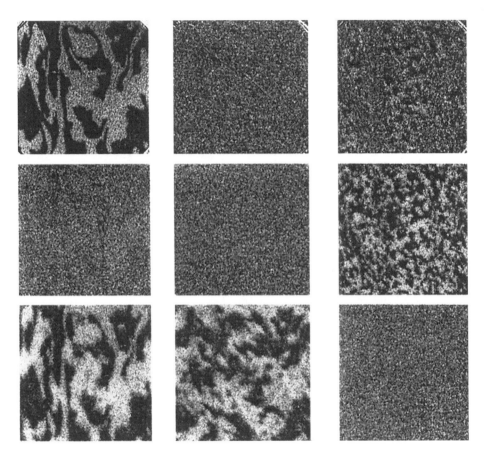

Figure 9-18: Magnetization component images of a Co/Au/Co sandwich on W(110) with ferromagnetic interlayer coupling. Left and center: parallel to easy axis W[1-10] and hard axis W[001] of bottom layer, right perpendicular to the film. The thickness of the top Co layer is 0, 3 and 7 monolayers from top to bottom. Electron energy 1.2 eV. Field of view: 6×6 μm². From [41] with permission.

In these trilayer studies one important aspect of SPLEEM is used: the short inelastic mean free path of electrons in the 1 eV range in ferromagnetic materials. While the domain structure of the bottom layer is still visible trough metals with filled d shells such as Cu or Au, a few mono-layers of Co or Fe on top of Cu or Au completely suppress the contribution of the bottom layer. Of course the other advantages of SPLEEM mentioned earlier are used too.

9.5. SUMMARY

The introduction, the discussion of the physical basis and of the instrumental aspects of SPLEEM, together with the application examples make it clear that SPLEEM is not a routine method for microscopic imaging but rather a research tool. In amorphous and polycrystalline samples without preferred orientation other methods such as SEMPA or XMCDPEEM are more useful. However, in samples, which confine the backscattered electrons into a well-defined diffracted specular beam, SPLEEM has the advantage of giving also structural and morphological information via LEEM and LEED, an important feature in view of the close connection between structure and magnetism.

REFERENCES

[1]   J. Kirschner, *Polarized Electrons at Surfaces*. Berlin, Springer 1985
[2]   E. Bauer, *SPLEEM. In Magnetic Microscopy of Nanostructures,* Hopster and H. P. Oepen, eds. Berlin: Springer, 2004
[3]   E. Bauer, *Polarized Electrons in Low Energy Electron Microscopy*. In *Spin-Polarized Electrons in Solids: Fundamentals and Applications to Spintronics,* K. Hricovini, G. Lampel, J. Peretti, C. Richter, eds. Berlin: Springer, in print
[4]   J. Hong, D.L. Mills, *Phys. Rev. B* **62** (2000) 5589.
[5]   A. Hubert, R. Schäfer, *Magnetic Domains*. Berlin: Springer (1998) 11.
[6]   R.J. Celotta, J. Unguris, D.T. Pierce, *Magnetic Domain Imaging of Spintronic Devices*. In *Magnetic Interactions and Spin Transport Electronics,* eds. S. Wolf, Y. Idzerda. Hingham, MA: Kluwer Academic/Plenum Publishers (2002) 341.
[7]   W. Telieps, E. Bauer, *Ultramicroscopy* **17** (1985) 51.
[8]   M.S. Altman, H. Pinkvos, J. Hurst, H. Poppa, G. Marx, E. Bauer, *MRS Symp. Proc.* **232** (1991) 125.
[9]   T. Duden, E. Bauer, *Rev. Sci. Instrum.* **66** (1995) 2861.
[10] K. Grzelakowski, T. Duden, E. Bauer, H. Poppa, S. Chiang, *IEEE Trans. Magnetics* **30** (1994) 4500.
[11] D.T. Pierce, *"Spin Polarized Electron Sources"*. In *Atomic, Molecular, and Optical Physics (Experimental Methods in the Physical Sciences, Vol. 29A)*, F.B. Dunning and R.G. Hulet, eds. 1995. p. 1-38
[12] T. Duden, E. Bauer, *Surf. Rev. Lett.* **6** (1998) 1213.
[13] W. Weber, S. Riesen, H.C. Siegmann, *Science* **291** (2001) 1015.
[14] C.H. Back, R. Allenspach, W. Weber, S.S.P. Parkin, D. Weller, E.L. Garwin, H.C. Siegmann, *Science* **285** (1999) 864.
[15] E.B. Myers, D.C. Ralph, J.A. Katine, R.N. Louie, R.A. Buhrman, *Science* **285** (1999) 867.
[16] E.D. Tober, G. Witte, H. Poppa, *J. Vac. Sci. Technol.* **A18** (2000) 1845.
[17] R. Zdyb, E. Bauer, *Surf. Rev. Lett.* **9** (2002) 1485.
[18] R. Zdyb, E. Bauer, *Phys. Rev. Lett.* **88** (2002) 166403.
[19] K.L. Man, M.S. Altman, H. Poppa, *Surf. Sc.* **480** (2001) 163.
[20] K.L. Man, W.L. Ling, Y. Paik Silena, H. Poppa, M.S. Altman, Z.Q. Qui, *Phys. Rev. B* **65** (2001) 024409.
[21] Poppa H., Tober E.D., Schmid A.K., *J. Appl. Phys.* **91** (2002) 6932.
[22] Schmid A.K., Bartelt N.C., Poppa H., (2002) unpublished
[23] A. Biedermann, R. Tscheliessnig, M. Schmid, P. Varga, *Phys. Rev. Lett.* **87** (2001) 086103.

[24] A. Biedermann, R. Tscheliessnig, M. Schmid, P. Varga, *Appl Phys. A* **78** (2003) 807.

[25] R. Ramchal, A.K. Schmid, M. Farle, H. Poppa, *Phys. Rev. B* **68** (2002) 054418.

[26] R. Allenspach, M. Stampanoni, A. Bischof, *Phys. Rev. Lett.* **65** (1990) 3344.

[27] M. Speckmann, H.P. Oepen, H. Ibach, *Phys. Rev. Lett.* **75** (1995) 2035.

[28] H.P. Oepen, Y.T. Millev, J. Kirschner., *Appl. Phys.* **81** (1997) 5044.

[29] H.P. Oepen, M. Speckmann, Y. Millev, J. Kirschner., *Phys. Rev. B* **55b** (1977) 2752.

[30] T. Duden. *Entwicklung und Anwendung der Polarisationsmanipulation in der Niederenergie-Elektronenmikroskopie.* Ph.D. Thesis. TU Clausthal 1996

[31] T. Duden, E. Bauer, *MRS Symp. Proc.* **475** (1997) 283.

[32] Y. Yafet, E.M. Gyorgy, *Phys. Rev. B* **38** (1988) 9145.

[33] E.Y. Vedmedenko, H.P. Oepen, A. Ghazali, J.-C.S. Lévy, J. Kirschner, *Phys. Rev. Lett.* **84** (2000) 5884.

[34] E.Y. Vedmedenko, H.P. Oepen, J. Kirschner, *Phys. Rev. B* **66** (2002) 214401.

[35] Zdyb R., E. Bauer, *Phys. Rev. B* **67** (2003) 134420.

[36] P. Ohresser, N.B. Brookes, S. Padovani, F. Scheurer, H. Bulou, *Phys. Rev. B* **64** (2001) 104429.

[37] T. Duden, E. Bauer, *Phys. Rev. Lett.* **77** (1996) 2308.

[38] K.L. Man, R. Zdyb, S.F. Huang, T.C. Leung, C.T. Chan, E. Bauer, M.S. Altman, *Phys. Rev. B* **67** (2003) 184402.

[39] T. Duden, E. Bauer, *Phys. Rev. B* **59** (1999) 468.

[40] T. Duden, E. Bauer, *J. Magn. Magn. Mater.* **191** (1999) 301.

[41] T. Duden, E. Bauer, *Phys. Rev. B* **59** (1999) 474.

Proximal Probe

CHAPTER 10

Spin-polarized scanning tunneling microscopy

10.1. HISTORICAL BACKGROUND

Although first surface science experiments go back to the 1930s, in the following years this branch of solid state physics was a playground of very few physicists and chemists only, mainly because of the rather specialized equipment needed for the preparation of clean surfaces and their analysis. Only in the 1960s commercial vacuum components and surface analysis tools, as, e.g., the ion getter pump and low energy electron diffraction (LEED) optics, became available, leading to an increasing popularity of surface science and a boosting number of publications in this field [1]. The lack of lateral and vertical real space resolution, however, led to a rather unrealistic, idealized view: mostly, surfaces were considered as being perfect without any step edges, dislocations, adatoms or impurities. This made the interpretation of data difficult, in particular in cases, where unavoidable inhomogeneities play an essential role for the understanding of physical and chemical properties of surfaces.

With the advent of scanning tunneling microscopy (STM) in 1982 the situation changed dramatically [2,3]. The invention by G. Binnig and H. Rohrer allowed the real-space imaging of surfaces with lateral and vertical atomic resolution up to a larger lateral scale of about $1\mu m$ thereby bridging the limit of optical microscopy. Only now it was realized that surfaces, which so far have been considered as highly ordered, contained a high percentage of defect sites or even looked like "the surface of the moon" [4]. Soon, improved surface preparation procedures led to much better surface qualities.

By making use of different spectroscopic modes of the STM it became also possible to measure electronic properties with unprecedented lateral resolution and to correlate the results with the topography. In 1988, Pierce considered the possibility to make the STM sensitive for the spin of the tunneling electron by employing spin-sensitive tip materials [5]. This project became the more important since the rapid increase in storage density of magnetic data storage devices, as, e.g., computer hard drives, desperately required magnetically sensitive high spatial resolution imaging tools for an improved understanding of nanometer-sized domain and domain wall structures.

In this chapter we will summarize the important steps that lead to the successful development of spin-polarized scanning tunneling microscopy (SP-STM) and a few milestones which have recently been reached.

10.2. MEASUREMENT PRINCIPLE

10.2.1. Planar junctions

The working principle of spin-polarized scanning tunneling microscopy (SP-STM) rests on the conservation of the electron spin during the tunneling process. In the absence of inelastic processes like Stoner or magnon excitations—a condition which is almost perfectly fulfilled for tunneling across a vacuum barrier—the spin of an electron that tunnels from an occupied state that belongs to one magnetic electrode into an empty state of another magnetic electrode must be preserved. Effectively, as already proposed by Julliere [6] and schematically represented in Fig.10-1, the total tunnel current may be considered as the sum of two independent currents which flow in separate channels, i.e., the spin-up and spin-down or the minority-spin and majority-spin channel.

If both electrodes, which for simplicity may exhibit identical electronic properties, are magnetized parallel, the Fermi level of both electrodes is dominated by electrons of the same spin orientation, leading to a high tunneling current (left panel of Fig.10-1). In case of an antiparallel magnetization, however, the density of states of the two electrodes is dominated by opposite spin orientations. E.g., the right panel of Fig.10-1 shows a situation where the Fermi level of the left (source) electrode is dominated by spin-down electrons, while mainly spin-up electrons are found in the right (drain) electrode. As a results, less empty states with an adequate spin orientation are available and the tunneling current in Fig.10-1(b) is lower than in a parallel configuration [Fig.10-1(a)].

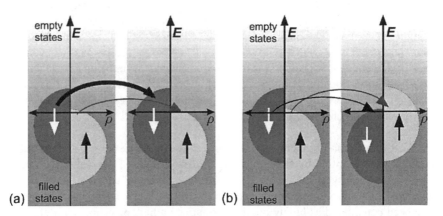

Figure 10-1: Principle of spin-polarized tunneling between two magnetic electrodes which are (a) parallel or (b) antiparallel magnetized. Since the spin is conserved during elastic tunneling events the tunnel current and the conductance depend on the relative magnetization orientation of the electrodes.

Spin-polarized electron tunneling in planar tunnel junctions was established in the early 1970s in pioneering experiments by Tedrow, Meservey, and Fulde, who used the well-defined Zeeman-splitting of a superconducting electrode as a reference that allows the determination of the Fermi level spin polarization of ferromagnets [7-9]. In 1975, Julliere reported on the first observation of spin-polarized tunneling between two ferromagnetic electrodes [6]. The geometry of a planar tunnel junction is schematically represented in Fig.10-2(a). Two ferromagnetic films are separated by an insulating barrier which typically consists of Al_2O_3. While one electrode is magnetically soft and can easily be reversed by an external field, the other one is magnetically hard. For applications the hard electrode is usually exchange-biased to an antiferromagnetic pinning layer.

As a magnetic field is applied opposite to electrodes which were previously aligned by a sufficiently strong pulse of the external magnetic field, the soft electrode soon changes its magnetization direction. This results in an increased junction resistance (see Fig.10-2(b)). Further increasing the external field eventually switches the hard electrode, too, and the resistance returns to its original, low value. As long as the field is strong enough to switch both electrodes the resistance for every field cycle adopts twice a small and twice a large value, corresponding to the antiparallel and parallel relative orientation. The magnetoresistance may be determined very accurately by small loop cycles, i.e., cycles of the external field which switch the soft electrode only, and detecting the resistance variation by lock-in technique. As we will see below, this technique has also been applied for spin-polarized STM.

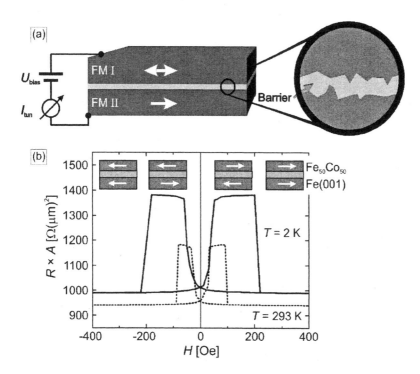

Figure 10-2: (a) Scheme of a planar magnetic tunnel junction. Inset: the insulating barrier is not perfectly flat but exhibits a slight thickness fluctuation which leads to an inhomogeneous current density. (b) Plot of the tunnel magnetoresistance (MR) of a $Fe(211)$-Al_2O_3-$Fe_{50}Co_{50}$ junction measured at $T = 2$ K (full line) and $T = 293$ K (hatched line). The temperature increase reduces the (MR) by a factor of 2 (with courtesy of Y. Suzuki [10]).

10.2.2. Spin-polarized tunneling with the STM

In this part of the chapter we will describe different operational modes of SP-STM using (anti)ferromagnetic probe tips. Other concepts, as the generation of spin-polarized electrons by the illumination of semiconducting GaAs with helical light or by exposing superconducting materials to an external field, have also been proposed. However, the latter two methods have not yet been successfully applied for imaging magnetic domains on surfaces. Therefore, they will only briefly be described in Sec. 10.3.2.

Magnetic probe tips are used for SP-STM in three modes of operation, which have been discussed recently from a theoretical point of view by Wortmann *et al.* [11]:

    a) constant current mode;
    b) spectroscopy of the local magnetoresistance dI/dm;
    c) spectroscopy of the local differential conductance dI/dU.

We shall start with the total tunnel current between a magnetic surface and a spin-polarized tip, which—according to Wortmann *et al.* [11]—can be described by the sum of a spin-averaged and a spin-dependent term

$$I(\vec{r}_\mathrm{T},U,\theta) = I_0(\vec{r}_\mathrm{T},U) + I_\mathrm{SP}(\vec{r}_\mathrm{T},U,\theta)$$
$$= \frac{4\pi^3 C^2 \hbar^3 e}{\kappa^2 m^2}[n_\mathrm{T}\tilde{n}_S(\vec{r}_\mathrm{T},U) + \vec{m}_\mathrm{T}\tilde{\vec{m}}_S(\vec{r}_\mathrm{T},U)] \tag{10.1}$$

where n_T is the non–spin-polarized local density of states (LDOS) at the tip apex, $\tilde{n}_s$ is the energy-integrated LDOS of the sample and $\tilde{\vec{m}}_S$ is the corresponding energy-integrated spin-polarized LDOS

$$\tilde{\vec{m}}_S(\vec{r}_\mathrm{T},U) = \int \vec{m}_S(\vec{r}_\mathrm{T},E)\mathrm{d}E \tag{10.2}$$

with

$$\vec{m}_S = \sum \delta(E_\mu - E)\Psi_\mu^{S*}(\vec{r}_\mathrm{T})\sigma\Psi_\mu^{S*}(\vec{r}_\mathrm{T}). \tag{10.3}$$

σ is Pauli's spin matrix and

$$\Psi_\mu^S = \begin{pmatrix} \Psi_{\mu\uparrow}^S \\ \Psi_{\mu\downarrow}^S \end{pmatrix}. \tag{10.4}$$

It has been shown that the simulated lateral variation of the tunneling current at constant tip-sample distance qualitatively corresponds to the height variation which is experimentally found in constant-current images [12-14]

$$I_\mathrm{SP}(\vec{r}_\mathrm{T}) \propto z(\vec{r}_\mathrm{T}) \propto n_\mathrm{T}\tilde{n}_S(\vec{r}_\mathrm{T}) + \vec{m}_\mathrm{T}\tilde{\vec{m}}_S(\vec{r}_\mathrm{T}) \cdot \cos\theta(\vec{r}_\mathrm{T}) - C. \tag{10.5}$$

Here, Θ describes the angle between the (static) tip magnetization direction and the (local) magnetization direction of the sample.

Nevertheless, the constant-current mode can only be applied in some limited cases. This is related to the integral in Eq.(10.2) which is performed over all energies between the Fermi level E_F and eU, i.e., the electron charge e multiplied with the applied bias voltage. As soon as the spin-polarization reverses sign, I_SP decreases. This can be avoided by measuring the differential conductance dI/dU, which for spin-polarized STM can be written as

$$\frac{\mathrm{d}I}{\mathrm{d}U}(\vec{r}_\mathrm{T},U) \propto n_\mathrm{T}\tilde{n}_S(\vec{r}_\mathrm{T},E_\mathrm{F}+eU) + \vec{m}_\mathrm{T}\tilde{\vec{m}}_S(\vec{r}_\mathrm{T},E_\mathrm{F}+eU)] \tag{10.6}$$

and which is no longer proportional to the energy-integrated spin polarization but to the spin polarization within a narrow energy window around $E_F + eU$. In practice, the differential conductance dI/dU is measured by the superposition of a modulation to the applied sample bias voltage and detection of the resulting current modulation by lock-in technique. Then, the width of the energy window is given by the amplitude of the modulation voltage. Ultimately, it is determined by the thermal broadening.

As long as electronically homogeneous surfaces are considered, differential conductance maps reflect the magnetization pattern and any signal variation must originate from the second, spin-dependent term of Eq.(10.6). The situation becomes more complicated, however, if the magnetic domain structure of electronically heterogeneous surfaces is to be imaged. However, as we will demonstrate in Sec. 10.4.2 of this chapter, a detailed comparison between spin-averaged and spin-resolved measurements allows the clear identification of magnetic contrasts. An alternative method measures the local magnetoresistance dI/dm between the sample and the STM tip by dynamic modulation of the tip magnetization direction at a frequency far above the cutoff frequency of the STM feedback circuit and detecting the resulting variation of the tunnel current by lock-in technique. This measurement scheme was first proposed by Johnson and Clarke [15] and later accomplished by Wulfhekel and Kirschner [16]. By taking the derivative of Eq.(10.1) we find that

$$ dI / d\vec{m}_t \propto \widetilde{\vec{m}}_s \, , \tag{10.7} $$

i.e., the signal is proportional to the energy-integrated spin-polarized LDOS. In contrast to the differential conductance mode, a non-zero signal in the dI/dm mode is only obtained if a local magnetization exists. It has to be emphasized, however, that the interpretation of chemical heterogeneous surfaces may still be difficult: Since the sign and the magnitude of the material-specific spin-polarization may vary, it cannot directly be identified with the domain structure. If, for example, the surface consists of two chemical species with opposite spin polarization, the dI/dm map may be interpreted as a complicated domain pattern although it simply reflects the chemical order.

10.3. EXPERIMENTAL

Before we consider the experimental conditions, which have to be fulfilled in order to allow the successful operation of a SP-STM, we shall remind that spin-polarized electron tunneling in planar tunnel junctions is a well-established technique since the mid-1970s. As already mentioned above pioneering experiments on the magnetic field dependence of the tunnel current between a ferromagnetic and a superconducting electrode being separated by an insulating Al_2O_3 barrier were performed by Tedrow, Fulde and Meservey [7-9]. The existence of a spin dependent contribution was clearly proven. Soon Julliere replaced the superconducting electrode by a second ferromagnetic layer. Meanwhile, spin-polarized electron tunneling in planar ferromagnet-insulator-ferromagnet junctions is widely applied in magnetic field sensors.

If the further reduction of the coordination, which necessarily occurs at the tip apex, does not principally affect the spin imbalance at the Fermi level E_F which exists in any (anti) ferromagnet, the concept shall be transferable to the STM. There are, however, two important differences between experiments with planar junctions and those performed with a STM. Firstly, planar junctions consist of a single piece of solid material as both electrodes are rigidly connected by an Al_2O_3 barrier. Therefore, mechanical vibrations cannot influence the tunnel current. In contrast, the vacuum barrier in a STM experiment does not restrict the relative movement of the electrodes. Therefore, special care has to be taken in order to make the instrument insensitive against mechanical vibration. Secondly, the electrodes in planar tunnel junctions consist of relatively thick, in-plane magnetized, continuous films. This makes them rather insensitive against dipolar interactions. Only a detailed quantitative analysis reveals weak dipolar effects [17]. Since SP-STM is considered as an instrument for the analysis of nanostructures, which may have extremely low anisotropies, the sample may be very sensitive against the tip's stray field. This problem is serious in the STM geometry as the tip is necessarily very close to the sample surface to be investigated, in particular, if the magnetization points along the tip axis. A solution of this problem is based on the use of probe tips made from antiferromagnetic materials, the alternating magnetic moments of which cancel eachother.

10.3.1. STM design—requirements

In order to be suited for spin-resolved studies the STM setup must fulfill certain requirements. First of all a tip exchange mechanism is absolutely necessary. Since stray fields may distort the intrinsic domain structure of the sample, a magnetic clamping mechanism for interconnecting the tip with the piezo scanner tube or with the microscope's body shall be avoided. Several mechanical clamping mechanisms have been developed; some are commercially available. As we will describe in the following section bulk tips as well as thin film tips have to be annealed during the cleaning and preparation process. Furthermore, it was found that the mechanical stability of the magnetic thin film is improved by flashing the W tips before deposition [18]. This thermal flash cleaning requires a temperature $T > 2000$ K. Therefore, only few high-melting materials, as, e.g., W, Mo, and Ta are suitable to be used for the design of the tip holder and shuttles which are needed for moving the tip between different positions during the preparation procedure [19]. Secondly, the STM body itself shall be non-magnetic. Magnetic parts may cause substantial field strength at the sample position. In the past the machinable ceramics MARCOR® was mainly used [19]. Alternative materials with a better thermal conductance for low temperature applications may be non-magnetic alloys like bronze (CuSn) or light elements as Be.

In recent years it was experienced that low-temperature operation with its improved mechanical stability—often considered to be necessary for SP-STM—is not essential [12, 20]. The stability of a well-designed STM operated at room temperature is absolutely sufficient for spin-resolved experiments. In principle, SP-STM experiments may be performed at any temperature up to the respective material-specific ordering temperature, i.e., the Curie- or Néel-temperature.

10.3.2. Tip design

A key issue of SP-STM is the preparation of spin-sensitive tips. In principle, any effect that produces a spin-polarized density of electron states at or close to E_F is suitable. So far three different classes of materials have been proposed:

  a) Magnetic materials;
  b) Semiconductors with a significant spin-orbit splitting;
  c) Superconductors in an external magnetic field.

Obviously, the latter material is not suited for the investigation of ferromagnetic surfaces as the magnetic field strength that is required to obtain a sufficient spin-splitting of the superconductor [9] certainly exceeds the saturation field of most magnetic samples. It may, however, be used for imaging of antiferromagnetic surfaces which are insensitive against an external field up to field strengths equivalent to the exchange field. Although superconducting electrodes played an important role in the development of planar tunnel junctions they have not yet successfully been used in SP-STM experiments.

Semiconducting GaAs has often been considered as the ideal material for SP-STM tips [5]. GaAs is not ferromagnetic and therefore produces no stray field which may destroy the sample's domain structure. The band gap of GaAs is located at the gamma bar point of the surface Brillouin zone. Due to significant spin-orbit coupling the degeneracy at the bottom of the conduction band is lifted. The spin-orbit splitting between the $p_{3/2}$- and the $p_{1/2}$-level amounts to 0.34 eV. It is sufficiently large to allow the selective excitation of spin-polarized electrons from the valence band into the lower lying $p_{3/2}$ band edge. Due to the optical selection rules, irradiation with helically polarized light with an energy just above the threshold for excitation into the $p_{3/2}$–level but below the $p_{1/2}$–level leads to the photoemission of electrons with a spin-polarization of 50% [21]. This principle is routinely used as a source of spin-polarized electrons [22].

In spite of the fact that this technique is relatively simple and well established we are only aware of very few successful spin-polarized tunneling experiments with the STM involving GaAs tips [23-26]. Recent luminescence experiments performed with Ni tips on GaAs(110) surfaces [27] suggest that the problem may be related to the special electronic properties of GaAs step edges, which necessarily exist around the atom which acts as the tip apex. It was found that close to a step edge not only the intensity of recombination luminescence decreases by about three orders of magnitude, but also the polarization of the luminescent light is reduced by a factor of 6. This was explained by a reduction of either the spin injection efficiency, or of the spin relaxation lifetime and attributed to the metallic nature of the step edge caused by mid-gap states of the (111) surface [27]. It is a straightforward conclusion that the spin relaxation lifetime may also be drastically reduced at the very end of the tip.

Yet, the only class of materials, which has successfully been used for tips in SP-STM experiments, are ferro- and antiferromagnets. Basically, three concepts were followed:

    a) Amorphous ferromagnets;
    b) Ferromagnetic thin films;
    c) Antiferromagnets.

In any case the intrinsic exchange splitting and the resulting spin imbalance of the magnetic material is used [5, 28]. The main problem of ferromagnetic materials is that their use is necessarily associated with the existence of a stray field. As already mentioned above the stray field of the tip may interact with the sample's domain structure, especially, if the tip is magnetized along its rotational axis. So far, two concepts were used in order to avoid a dipolar interaction between a ferromagnetic tip and the sample surface, i.e., the use of low-saturation and low-coercivity amorphous CoFeSiB [16, 29] and thin film tips [18].

Figure 10-3 schematically shows two experimental setups to obtain (a) out-of-plane and (b) in-plane sensitivity by using a pointed or a ring-shaped CoFeSiB tip, respectively. Although the saturation magnetization of the amorphous ferromagnet CoFeSiB amounts to 0.5 T only, the flux guiding shape of a conventional, pointed STM tip still exhibits a field strength of several

10 mT at the position of the sample surface [29]. As a result this method may be useful for the investigation of perpendicular domains on surfaces of bulk-like samples or antiferromagnets, which are rather robust against external fields, but is probably not suited for the investigation of thin films. Ring-shaped CoFeSiB tips as shown in the photographic image of Fig.10-4 have successfully been used for in-plane sensitive measurements even of magnetically soft samples as Fe(001) whiskers [30]. Although the ring geometry leads to a perfect flux closure in most cases with no stray field emanating from the ring, a small influence to soft Fe whiskers was found occasionally [30].

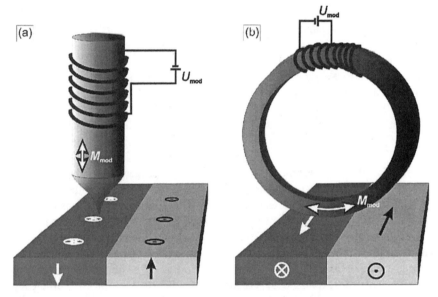

Figure 10-3: Schematic experimental setup for local magnetoresistance measurements with the STM using (a) out-of plane pointed and (b) in-plane sensitive ring-shaped "tips".

Figure 10-4: Photography of an in-plane sensitive ring-shaped CoFeSiB "tip" (with courtesy of W. Wulfhekel [30]).

Alternatively, coating a non-magnetic wire with a very thin layer of magnetic material can reduce the stray field of a tip. Polycrystalline W wires have successfully been coated with a large variety of materials. Typically the wires are ac-etched (etching voltage: 5 V) in a solution of 8 g NaOH per 100 g H_2O. Without any further treatment the non-magnetic tip is transferred into the UHV chamber via a load lock. The W tips are cleaned in the UHV system by flashing to $T > 2200$ K, subsequently coated with a thin film of magnetic material and finally annealed at

600 K. Without flashing the W wire prior to film deposition the mechanical stability of the films was significantly worse. It turned out that the final magnetization axis of a thin film tip can quite accurately be anticipated when the easy axis of an equivalent film on W(110) is known [18]. E.g., 10 ML Fe/W(110) exhibit an in-plane magnetization and 7 ML thick Gd or Gd$_7$Fe films on *flat W(110)* are perpendicularly magnetized. Consistent results were found with coated *W tips*. The observed similarity is a result of the flashing procedure: While freshly etched W tips exhibit a tip radius of about 20-50 nm, the tip was found to be rather blunt with a tip diameter of about 1 μm after the flashing procedure [see Fig.10-5(a)]. Probably, the tip apex melts and upon forms a densely-packed, (110)-terminated W surace. As the schematic drawing of Fig.10-5(c) demonstrates, the remaining curvature of the film is very small and can basically be neglected.

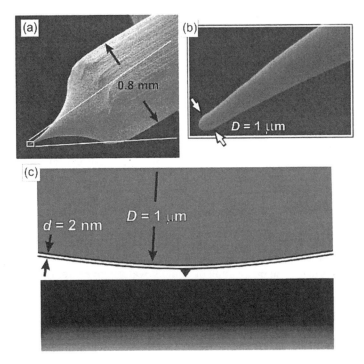

Figure 10-5: (a) Scanning electron micrograph of a W tip after a high-temperature flash at T > 1800 K. While the overall tip shape with a relatively high aspect ratio is maintained, the tip apex becomes very blunt. Melting may cause this. (b) Typically, very thin ferromagnetic films are evaporated onto the tip apex. As can be seen in the schematic drawing a flat plane with a small cluster can approximate the tip structure.

In the following we want to discuss, why a thin film tip exhibits a significantly reduced stray field: if the film is much thinner than the tip-sample distance, opposite "magnetic charges" being located at the tip-film interface and the film-vacuum surface compensate, resulting in a vanishing stray field at the sample position. With decreasing tip-sample distance the stray field at the position of the sample surface increases. Perpendicularly magnetized thin film tips used so far were made from 7±1 ML Gd or Gd$_7$Fe. They were successfully applied to study the domain structure of high-coercitive Fe/W(110) films [31-33] and antiferromagnetic surfaces [20, 34]. Since the film thickness is comparable with the tip-sample distance, the remaining field strength may modify the domain structure of magnetically soft samples. Even magnetically hard samples as the above mentioned Fe/W(110) films may be affected if they are investigated close to the saturation field. In this situation even the small additional stray field of a ferromagnetic thin film tip is sufficient to push the system out of a metastable multi-domain state into the stable single domain state [35].

Similar to the above mentioned ring geometry used with CoFeSiB tips, in-plane sensitive Fe-coated thin film tips shall exhibit no stray field, as no magnetic charges exist on the tip surface. Recent experiments performed on 8 nm high Fe islands with a magnetic vortex pattern revealed, however, that the vortex core is affected by the stray field even for Fe coatings as thin as 3 ML [36]. This may be caused by the fact that a STM tip, which is well suited for the imaging of stepped surfaces, cannot be absolutely flat. As schematically represented in Fig.10-5(c), the tip apex is probably formed by a small cluster, which leads to a stray field at the sample position. It remains to be investigated whether the same argument has to be applied to ring-shaped CoFeSiB tips.

In principal, a stray field can completely be avoided by using antiferromagnetic tips. Since adjacent moments compensate each other the effective magnetization of antiferromagnets is zero. Indeed, no hint of a significant field was found in recent experiments [35]. Since, however, the symmetry is broken, the tip apex probably exhibits a different spin structure than the bulk material. Uncompensated magnetic moments may lead to an—yet unknown—extremely small remaining field at the sample position.

10.4. RESULTS

In the following we will present results which have been obtained with the three different types of magnetic tips discussed in Sec. 10.3.2 of this chapter. The sample systems cover a wide range of magnetic materials including surfaces of bulk crystals, thin films, nanoparticles, superparamagnetic islands, and antiferromagnets, thereby demonstrating the broad applicability of SP-STM and its spectroscopic modes of operation.

Throughout this section, magnetic domain wall profiles will be compared with micromagnetic continuum theory [37]. Although this theory disregards the atomic structure of matter, it has been proven to allow the precise description of domain walls with a width down to a few nanometers. It is based on the fact that a static domain and domain wall configuration requires, that the sum of all forces must be zero at any position of the sample. The course of the magnetization angle Θ with respect to the easy magnetization direction within an (uncharged) bulk Bloch wall is described by the following function:

$$\cos \varphi(x) = \tanh\left(\frac{x - x_0}{\sqrt{A/K}}\right), \tag{10.8}$$

where x_0 is the domain wall center. The material specific parameters A and K are the exchange stiffness, which describes the force which acts on a spin in order to tilt it by a certain angle with respect to an adjacent spin, and the anisotropy energy density, respectively. The square root of the ratio of A and K is the so-called magnetic exchange length. SP-STM is sensitive to the projection of the sample magnetization onto the tip magnetization direction. For an arbitrary angle Θ between the easy axis of the sample and the tip magnetization, the SP-STM signal variation across a domain wall can be written as

$$y_{sp}(x) = \cos(\varphi(x)) = y_0 + y_{sp} \cdot \cos\left\{\arccos\left[\tanh\left(\frac{x - x_0}{A/K}\right)\right] + \phi\right\}, \tag{10.9}$$

with y_{sp} the spin-dependent signal strength and y_0 a non-spin-dependent offset.

10.4.1. The local magnetoresistance mode

As mentioned above the local magnetoresistance mode of SP-STM is capable to image out-of-plane as well as in-plane magnetic domains by choosing an appropriate tip geometry. In the ring geometry the magnetic flux is closed within the "tip," leading to a virtually stray field free situation which even allows the imaging of extremely soft (low coercivity) sample surfaces [30].

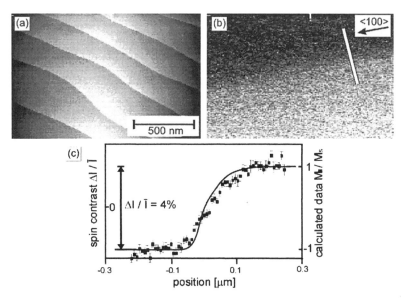

Figure 10-6: (a) Topography and (b) local tunneling magnetoresistance dI/dm as measured with a ring-shaped magnetic tip on a Fe(001) whisker. (c) Line section of the dI/dm signal. The data are fitted with a micromagnetic domain wall profile. Data with courtesy of W. Wulfhekel [30].

Figure 10-6 shows the topography (a) and the local magnetoresistance dI/dm map (b) of the central region of a Fe(001) whisker [30]. Due to the elongated whisker geometry the sample typically adopts the so-called Landau magnetization pattern, which reduces magnetic charges at the sample rim. The topography image shows atomically flat terraces, which are separated by mono-atomic step edges. The dI/dm map reveals that the magnetoresistance of the scanned region is not homogeneous but exhibits two different levels, high (bright) and low (dark). No cross talk between the topographic and the magnetic signal is visible. The change of magnetoresistance is caused by a domain wall, which separates two domains being magnetized in opposite in-plane directions. Schlickum et al. [30] have drawn a line section across the domain wall, which is shown in Fig.10-6(c). The measured magnetic contrast amounts to ~ 4 %. Obviously, the domain wall extends over a width of about 100 nm. The experimental data of Fig.10-6(c) were fitted by micromagnetic theory.

In contrast to the ring, which exhibits practically no stray field, out-of-plane sensitivity requires a wire geometry which necessarily causes a stray field at the sample location underneath the tip apex [29]. Although magnetically soft and thin film samples probably cannot be studied with this technique, the surface domain structure of bulk-like samples can be imaged because it is stabilized by the underlying bulk domains [38]. For example, Fig.10-7 shows dI/dm maps of the Co(0001) surface. Bright and dark branches of the typical fractal surface domain pattern can be recognized in the overview image of Fig.10-7(a). One branch is magnified in Fig.10-7(b).

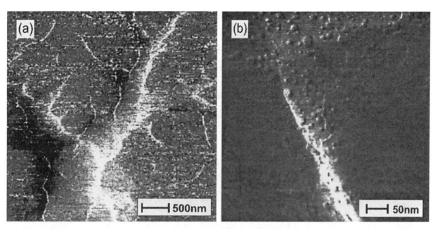

Figure 10-7: (a) Overview of the fractal domain structure of Co(0001) and (b) zoom to a single branch. The local tunneling magnetoresistance dI/dm maps were measured with an out-of-plane sensitive CoFeSiB tip. Data with courtesy of W. Wulfhekel [38].

Images of the simultaneously recorded topographic and magnetic dI/dm signal of even higher magnification are shown in Fig.10-8(a) and (b), respectively. These data prove that no correlation between the topographic and the magnetic dI/dm signal is observed. A quantitative analysis revealed that the domain wall, which separates the two domains in Fig.10-8(b), has a width of 1.1 nm only, i.e., it is one order of magnitude narrower than 180° domain walls in bulk Co. This apparent discrepancy was interpreted by small angle domain walls (20°) which are frequently found between closure domains on Co(0001) surfaces. It is supported by the observation that the contrast of Fig.10-8(b) amounts to only 20% of the total contrast in Fig.10-7.

Figure 10-8: High spatial resolution images of (a) the Co(0001) topography and (b) the perpendicular component of the local tunneling magnetoresistance dI/dm. The measurement was performed at the branch shown in the previous figure. Data with courtesy of W. Wulfhekel [38].

10.4.2. The differential conductance mode

10.4.2.1. Domain structure of mesoscopic islands

The domain structure of ferromagnetic particles with lateral dimensions of several µm down to approximately 100 nm—mostly prepared by lithographic techniques—has frequently been investigated with different techniques in the past. The role of the competing energy terms, i.e., exchange, stray field, and anisotropy, are well understood [37]. Large particles typically exhibit a multi domain state in which the magnetization is parallel to the particle rim. The resulting reduction of stray field energy of the particle overcompensates the energy being necessary to form domain walls which separate different domains. Since the remanent domain configuration is not reproducible in the course of successive hysteresis cycles, multi-domain particles are not suitable for magnetic recording applications.

As the particle becomes smaller the energy gain due to closure domains decreases and more and more domain walls are omitted. This may lead to the so-called vortex or Landau state: the magnetization is parallel to the particle edges thereby avoiding any stray field. Depending on the particle geometry this may be accomplished by a four domain state (rectangular particle) or by the continuous rotation of the magnetization (spherical disk). Eventually, as the particle size approaches the exchange length $(A/K)^{1/2}$ the domain wall energy even of a two domain particle exceeds the stray field energy of a single domain particle of the same size. Therefore, small particles adopt a single domain state [37].

The latter two domain configurations, i.e., the vortex and the single domain particle, are considered as potential building blocks for future non-volatile magnetic data storage devices [39]. While the single-domain state is conceptually easy, the usefulness of the vortex state depends on the diameter of the perpendicularly magnetized central region, the so-called vortex core. The existence of the vortex core, which is a direct consequence of the requirement of a continuous magnetization configuration, has been predicted already in 1964 [40]. The diameter of a vortex core in Fe was estimated to be smaller than 10 nm. Only in year 2000 the very existence of vortex cores was proven experimentally by magnetic force microscopy (MFM) [41, 42]. Due to the limited resolution of this and other techniques, however, neither the diameter nor the internal spin structure of the vortex core could be explored. Currently, SP-STM is the only technique capable to resolve the details of magnetic vortex cores [36].

Figure 10-9 shows a rendered perspective representation of a topographic STM image (scan range: 1.5 µm × 1.5 µm) of numerous Fe islands grown on W(110). The typical lateral dimensions are about 400 nm × 200 nm and the average island height amounts to 8 nm. Between the islands the substrate is covered with a single Fe wetting layer. Fe islands in this size regime shall exhibit a magnetic vortex pattern as indicated by the arrows. The spin configuration in the vortex core with its perpendicular component is schematically represented in the inset. Although the spin-averaged electronic structure of the Fe(110) surface is homogeneous, scanning tunneling spectroscopy (STS) using in-plane sensitive, stray field free Cr-coated probe tips shows a strong spatial variation of the spectral intensity over wide sample bias regions.

This variation is due to spin-polarized vacuum tunneling between the STM tip and the magnetic sample surface and may be used for the imaging of surface regions which are magnetized along different directions, i.e., magnetic domains. With the particular tip used in Fig.10-10 only a small dI/dU contrast is found at positive bias voltages $U > 0.5$ eV. At U = +0.56 V all spectra cross. Consequently, a dI/dU map (inset) shows no systematic contrast. At –0.43 V <

$U < +0.56$ V the dI/dU signal in the lower right corner is higher than in the upper left corner of the dI/dU map. The highest asymmetry of about 0.5 is obtained at $U = -0.18$ V. The dI/dU peaks are probably caused by a minority d-like surface resonance, which is well known from spin-resolved photoemission experiments. At $U = -0.43$ V the dI/dU contrast vanishes and inverts at even lower bias voltage. At $U = 0.70$ V the dI/dU spectrum measured in the lower left corner of the dI/dU map exhibits a local minimum while the spectrum in the opposite corner exhibits a shoulder. This results in a very high asymmetry of about 70%. The dI/dU spectra and maps of Fig.10-10 were acquired be measuring a full spectrum at every pixel which is not only very time consuming but also unnecessary if the domain configuration is to be imaged. For this purpose it is sufficient, to perform the measurement at one particular bias voltage which exhibits a high asymmetry. As the asymmetry does not only depend on sample but also on tip magnetic properties the exact bias voltage which gives a high asymmetry is not always the same but varies between different tips/measurements [43]. Usually, a suitable voltage is identified by a trial and error procedure.

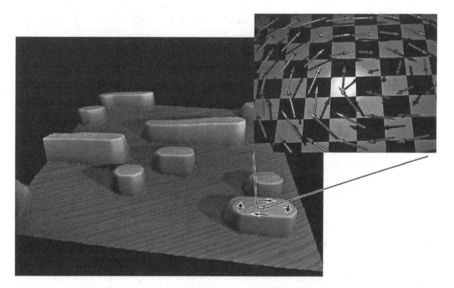

Figure 10-9: Rendered perspective STM image of Fe islands on W(110) which are expected to form a magnetic vortex structure. The image size is 1.5 μm × 1.5 μm. The islands which are elongated along the [001] direction have an average height of 8 nm. The inset shows a schematic representation of a vortex core. Far away from the vortex core the magnetization continuously curls around the center with the orientation in the surface plane. In the center of the core the magnetization is perpendicular to the plane (highlighted).

For example, Fig.10-11 shows the topography (a) and the spin-resolved dI/dU signal (b) of a single, about 8 nm thick Fe island on W(110). The simultaneously measured in-plane sensitive magnetic dI/dU map of Fig.10-11(b) shows the expected vortex pattern. In the upper and lower part of the island the dI/dU signal is diminished and enhanced with respect to a spin-averaged measurement, respectively, indicating the collinear magnetic orientation of tip and sample. In contrast, an intermediate dI/dU signal is found at the left and right side of the Fe island. Here, the sample magnetization is orthogonal to the magnetization of the tip, which results in a vanishing magnetic contribution to the tunnel current.

As mentioned above it is the unique strength of SP-STM to allow for the comparison of the spin structure of the vortex core with theoretical predictions. Fig.10-12 shows high spatial resolution images of the vortex core region as measured with different Cr-coated tips. By varying the thickness of the Cr coating we are able to specifically prepare probe tips with different easy magnetization axis, i.e., with either in-plane (relatively thick Cr films with $\theta > 200$ ML) or out-of-plane sensitivity (thin Cr films with $\theta < 50$ ML). While the in-plane sensitive dI/dU signal (left column) exhibits the typical Landau pattern the out-of-plane image (right column) shows a homogeneous dI/dU signal for the entire island except for a small bright spot approximately located in the island center. This spot is caused by the perpendicular orientation of the magnetization in the vortex core as already indicated in the inset of Fig.10-9.

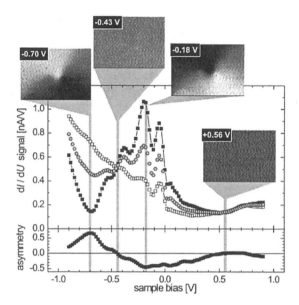

Figure 10-10: Spin-resolved dI/dU spectra measured on a Fe island on W(110) using a Cr coated probe tip. Although the spin-averaged electronic structure of the Fe(110) surface is homogeneous, a strong spatial variation of the dI/dU signal can be recognized in wide bias voltage regions. This is due to spin-polarized tunneling between the tip and the sample. The lower panel reveals that asymmetries as high as 70% are obtained. As can be seen in the dI/dU maps (insets), the magnetic domain structure can be imaged at values of the sample bias where the asymmetry is high. Data with courtesy of A. Wachowiak and J. Wiebe [36].

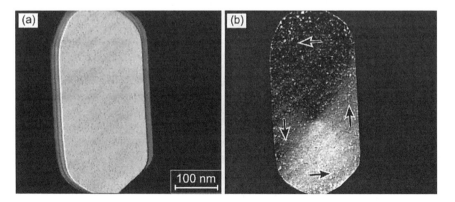

Figure 10-11: (a) Topography and (b) spin-resolved dI/dU map of a single 8 nm thick Fe island on W(110). By keeping the magnetization parallel to the island edges (arrows) a stray field can be avoided. Note, that the in-plane magnetization direction cannot be determined absolutely and the curl may have the opposite sense. Data with courtesy of A. Wachowiak and J. Wiebe [36].

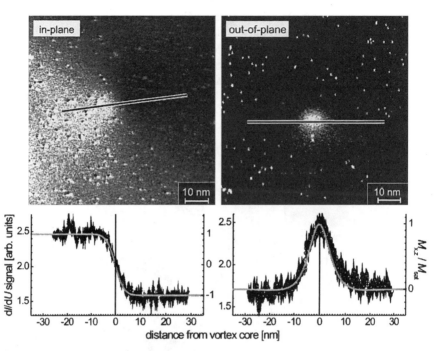

Figure 5-12: In-plane and out-of-plane component in the vicinity of the magnetic vortex core. The lower panel shows averaged sections of the experimental data (black data) drawn along the lines. The sections were numerically simulated by micromagnetic theory (gray lines). Data with courtesy of A. Wachowiak and J. Wiebe [36].

In the lower panels of this figure we have plotted line sections drawn along the lines in the experimental dI/dU maps (black points) across the vortex core. The in-plane as well as the out-of-plane signals show a vortex core width of about 10 nm. The magnetization distribution of a vortex core was simulated by micromagnetic calculations employing the widely used OOMMF software [44]. For the simulation the topography of the island shown in Fig.10-11 was divided into cuboids with lateral dimensions of 1 nm × 1 nm and a height of 8 nm. The simulation was started in a "perfect" vortex state, i.e., without any perpendicular component even in the vortex core. Upon relaxation the simulated profiles (gray lines) are in excellent agreement with the experimental data.

10.4.2.2 Size-dependent reorientation transition of nanometer-scale islands

As the vertical size of Fe islands on W(110) is reduced to two atomic layers only, a perpendicular anisotropy is obtained. As already described by Weber *et al.* [45], room temperature deposition of Fe onto W(110) leads to Fe islands with a local coverage $\Theta_{loc} = 2$ ML (DL islands) which are surrounded by a single, perfectly closed and thermodynamically stable Fe wetting layer. From a micromagnetic point of view, this sample system is particularly interesting as the surrounding monolayer exhibits an in-plane easy axis. Therefore, the sample consists of nanostructures with spatially switching anisotropies. Exactly this case has recently been discussed theoretically by Elmers [46].

Figure 10-13 shows the coverage dependent topography and magnetic dI/dU maps of Fe/W(110). Since the data were measured with a 7 ML Gd tip which is magnetized along the tip axis the dI/dU maps are sensitive to the magnetization component along the surface normal. Just after completion of the first Fe layer ($\Theta = 1.22$ ML) the surface topographic image (top) shows double-layer islands, typically 10-30 nm long and 5-10 nm wide, which have nucleated on atomically flat terraces. The spin-sensitive dI/dU map, which shows a high dI/dU signal for some islands (bright) and a low dI/dU signal for others (dark), reveals that these islands are perpendicularly magnetized. Since no external field was applied to the surface (the sample is in the magnetic virgin state), the number of bright and dark islands is approximately equal. In such a configuration the stray filed is minimized. A similar behavior has also been found for narrow Fe DL nanowires where adjacent stripes couple antiparallel [31, 32].

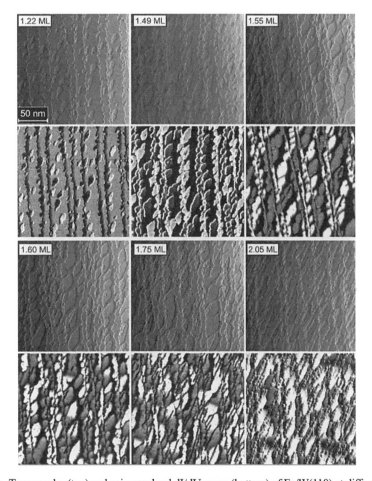

Figure 10-13: Topography (top) and spin-resolved dI/dU maps (bottom) of Fe/W(110) at different coverages.

At low coverage all islands are single domain particles. With increasing total coverage the size of the islands increases. At $\Theta = 1.60$ ML a domain wall is found on few islands. Obviously, the two domain state is energetically more favorable than a single domain state as the reduction of the stray field energy overcompensates the additional domain wall energy. As the coverage is increased further, more and more domain walls appear until at $\Theta = 2.05$ ML the domain structure resembles a higly disordered version of the stripe domain phase found on more homogeneous samples. Figure 10-14 shows (a) the topography and (b) the magnetic dI/dU signal of Fe DL islands separated by a closed monolayer. In this case the average Fe coverage amounts to

$\Theta = 1.28$ ML. Within the center of the box in Fig.10-14(b) an island with a constriction can be seen. It may originate from the nearby nucleation of two individual Fe DL islands in an early stage of the preparation process, which coalesced as the growth proceeded, thereby forming a single island with a constriction.

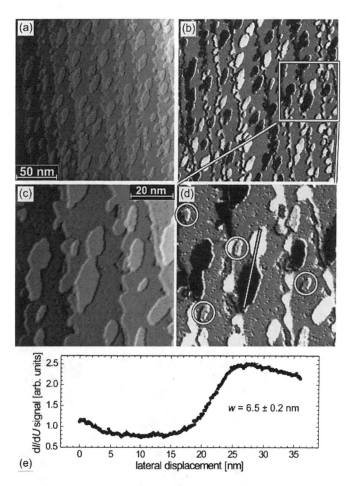

Figure 10-14: Overview and zoom-in of the topography (a,c) and the spin-resolved dI/dU signal (b,d) of 1.28 ML Fe/W(110), respectively. A high and low dI/dU signal represent a magnetization pointing up or down. The magnetic structure is governed by dipolar coupling preferentially leading to an antiparallel orientation of adjacent islands. The island in the middle of (c) exhibits a constriction where a domain wall is located. The line section drawn across the domain wall reveals a domain wall width of 6.5 ± 0.2 nm. Circles denote islands which are neither dark nor bright.

The region around this particular island is shown at higher magnification in Fig.10-14(c) and (d). It can clearly be recognized that a high dI/dU signal is found in the upper part of the island while a much lower value is found in the other part. The domain wall is located at the position where the constriction becomes narrowest which allows the domain wall energy to be minimized. Fig.10-14(e) shows a line section of the dI/dU signal measured along the line in (d). The y-scale indicates that at the position of the domain wall the dI/dU signal changes by a factor of three that corresponds to an effective spin polarization of the junction of about 50%. Again, the wall profile has been fitted by micromagnetic theory with Eq.(10.9) resulting in a domain wall width w = 6.5 ± 0.2 nm.

Close inspection of Fig.10-14(d) reveals that some Fe DL islands are neither white nor black but instead exhibit an intermediate dI/dU signal. Circles have marked four of these islands. Obviously, these islands are rather small suggesting that a size-dependent effect is responsible for the variation of the dI/dU signal. In order to check this assumption the strength of the dI/dU signal of individual islands has been analyzed and plotted versus the width of each particular island along the $[1\bar{1}0]$ direction, i.e., the short island axis. The result for about 140 islands is shown in Fig.10-15. The error bar represents the standard deviation over the island area. Three different size regimes can be recognized: (I) Large islands exhibit either a high or a low value of the dI/dU signal. Intermediate signal strength has never been observed for islands with a width of more than approximately 3 nm. (II) For islands, which exhibit a width between 1.8 and 3.2 nm a strong variation of the dI/dU signal, is observed. (III) Finally, Fe DL islands with a width below 1.8 nm show an intermediate dI/dU signal.

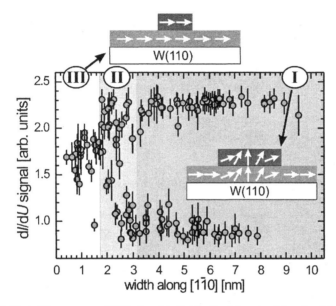

Figure 10-15: Plot of the average dI/dU signal individually determined for about 140 islands. Large islands with a width of more than 3 nm (region III) exhibit a high or low dI/dU signal. For an island width below 2 nm only an intermediate dI/dU signal was found (region I). The transition occurs in region II.

This behavior has been analytically described by a one-dimensional model by Kubetzka *et al* [47]. The model is based on spatially switching anisotropies as introduced by Elmers [46]. Qualitatively, it can be understood by the sample's complicated nanostructure which is governed by the close proximity of regions with different anisotropies: while the closed ML exhibits an in-plane easy axis it is perpendicular for the DL. For simplicity discontinuous changes of the anisotropy between regions which are covered by a Fe ML or DL islands were assumed.

As long as the DL island is sufficiently large the local magnetization rotates by 90° from in-plane to out-of-plane at the boundary between the closed Fe ML and the DL islands. This situation is schematically represented in the right inset of Fig.10-15 (region I). Since the corresponding material parameters of the Fe ML on W(110) allow the rotation to take place on a much narrower scale than in the DL the lateral reorientation transition mainly occurs in the ML region around the DL island. As the DL islands become smaller and smaller the energy, which is gained by turning the magnetization into the easy magnetization direction of the DL, decreases, until it is smaller than the energy that has to be paid for the 90°-domain wall that surrounds the

DL island. Then it is energetically favorable to keep the magnetization of the DL in-plane, in spite of the fact that the local anisotropy favors a perpendicular magnetization direction. This situation is shown in the upper inset of Fig.10-15.

10.4.2.3. Superparamagnetic islands

So far we have discussed results obtained on ferromagnetic surfaces. In the latter example long-range order was only possible because the Fe islands are interconnected by a continuous monatomic Fe layer which wets the W(110) substrate. As soon as individual magnetic objects of comparable size, which are separated by a non-magnetic substrate, are considered, thermally induced magnetic reorientation processes shall become relevant. The rate of thermally induced magnetization reversals was calculated by Néel [48] and Brown [49] under the assumption of coherent rotation, i.e., at any time—even during the reversal—the magnetic moments of the entire particle remain magnetized in the same direction, behaving like a single spin. Under these requirements the switching rate is described by the so-called Néel-Brown law

$$v = v_0 \cdot \exp\left(\frac{E_b}{k_B T}\right), \tag{10.10}$$

with v_0 the so-called attempt frequency, the energy barrier E_b that separates two degenerate magnetization states (up and down), the Boltzmann constant $k_B = 1.38 \times 10^{-23}$ J/K, and the sample temperature T.

Due to the lack of adequate experimental techniques with sufficient sensitivity most experimental investigations were performed on ensembles of supposedly identical particles in the past. Any real sample is, however, not strictly monodisperse but exhibits a certain size and shape variation. Therefore, the experimental averaging process may conceal relevant information on the dependence of physical properties on these parameters. As shown above SP-STM allows the investigation of individual particles and the direct correlation between structural, electronic, and magnetic properties. This may be utilized for studying whether the superparamagnetic switching rate is affected by other parameters than the particle size. Obviously, the use of ferromagnetic tips shall be avoided when superparamagnetic particles are to be investigated as the tip's stay field may easily influence the island's intrinsic properties. In contrast, antiferromagnetic probe tips are virtually free of any stray field and highly recommended for the study of superparamagnetic islands.

Epitaxial superparamagnetic particles can be prepared by the evaporation of about 0.1-0.3 ML Fe on Mo(110). The topography of 0.25 ML Fe on Mo(110) is shown in Fig.10-16(a). Mo is non-magnetic and therefore cannot couple adjacent islands by direct exchange. As can be seen in Fig.10-16(b) which was measured with an out-of-plane sensitive Cr tip at $T = 13 \pm 1$ K most of the Fe islands are too large (area $a > 40$ nm^2) and therefore magnetically stable on the time scale of imaging, i.e., several seconds. These islands exhibit either a high or a low dI/dU signal representing islands being magnetized (anti) parallel with respect to the tip. Some islands, however, change their dI/dU signal during imaging [marked by arrows in Fig.10-16(b)], indicating that they are magnetically unstable. For example, the upper marked island had a low dI/dU signal (dark) in the early phase of the scanning process, then reversed its magnetization direction resulting in a high dI/dU signal (bright), but eventually returned into the original state (dark). The time-dependent behavior of superparamagnetic islands can be investigated by imaging the surface repetitively and comparing consecutive scans.

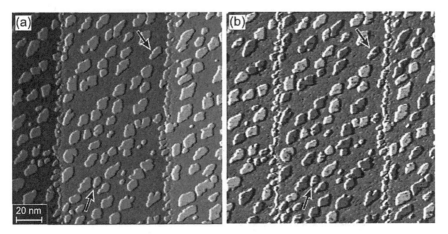

Figure 10-16: (a) Topography and (b) spin-resolved dI/dU signal of perpendicularly Fe islands on Mo(110) (overall coverage 0.25 ML) measured with a Cr-coated tip. While most of the islands are magnetically stable and exhibit a dI/dU signal which is either high (bright) or low (dark), some change it during the scanning process (arrows). These islands reverse their magnetization direction due to thermal excitations.

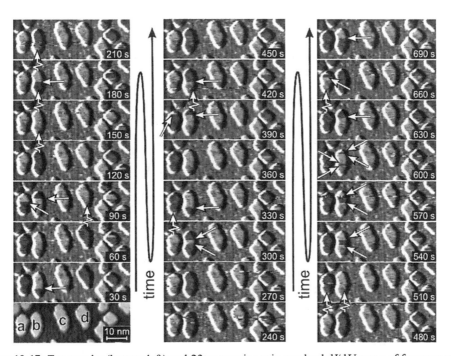

Figure 10-17: Topography (bottom left) and 23 successive spin-resolved dI/dU maps of four superparamagnetic Fe islands on Mo(110) labeled "a"-"d". Switching events which were directly observed are marked by straight arrows. Broken arrows indicate switching events which must have occurred when the tip did not scan across the islands.

Figure 10-17 shows the topography (bottom left) and 25 consecutive dI/dU maps of the same Fe islands labeled "a"-"d". The data were recorded with a 30 s increment. 21 individual magnetization reversals, which are marked by straight arrows, were observed directly. Additionally, from the fact that occasionally the dI/dU signal of single islands changes between successive images, we can conclude that 9 magnetization reversals must have happened when the tip scanned a line that does not cross the reversed island (broken arrows). With the exception of one event all reversals occurred to islands "a" and "b". This is understandable as islands "a" and "b" are smaller than "c" and "d" and therefore—according to Eq.(10.10)—exhibit a lower anisotropy barrier. Furthermore, a closer inspection reveals that islands "a" and "b" are magnetized antiparallel at about 80 % of the observation time. This experimental finding is inconsistent with our expectation for independent particles, for which the probability of a parallel and antiparallel relative orientation is equal. Seemingly, an interaction forces the particles to be antiparallel. Most probably, the dipolar interaction is responsible for this behavior.

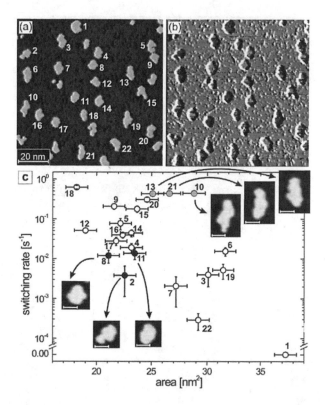

Figure 10-18: (a) Topography and (b) magnetic dI/dU signal of numerous numbered Fe islands on a Mo(110) substrate. (c) Plot of the switching rate versus the area of individual islands. The scatter of the switching rate points to a shape-dependent crossover from coherent rotation of compact Fe islands towards a non-uniform intermediate magnetization pattern, i.e., the nucleation and expansion of reversed domains in elongated islands. The insets show the topography of selected Fe islands (scale bar: 5 nm).

Since the analysis and understanding of interacting superparamagnetic particles is much more complicated than of particles which can be regarded as isolated, the inter-particle distance was increased by reducing the amount of Fe evaporated onto the Mo(110) substrate. Fig.10-18(a) shows the topography of 0.09 ML Fe/Mo(110). In the scan range of 100×100 nm^2 22 numbered islands can be recognized. With the exception of islands 5 and 9 their distance is sufficiently large to be considered as non-interacting. As can be recognized in the spin-resolved dI/dU map of Fig.10-18(b) many islands switch on a time scale just above the line frequency. In total, within 5 hours more than 50 successive images were recorded on the same sample location. By making use of the three different time scales of the scanning process, i.e., the time increment between successive pixels (~ 1 ms), between successive scan lines (~ 1 s), and successive images (~ 5 min), a wide range of superparamagnetic switching frequency could be detected. In Fig.10-18(c) the switching rate of all 22 islands is plotted versus the island area. According to the Néel-Brown law (Eq.10.10) the switching rate shall increase exponentially with decreasing island volume, which, due to the mono-atomic height of all islands, is proportional to the island area. Note, that the y-axis is logarithmic. Therefore, the apparent decay shall be linear. In contrast, a strong scatter of the data points is observed in Fig.10-18(c). As indicated by the six insets, this scatter is related to the island shape, which has not been considered in the Néel-Brown law.

One may ask: which time resolution can be achieved with SP-STM? Several groups have demonstrated that STM images can be taken with video frequency (~25 Hz). We do not see any principle obstacle why the same frequency cannot be reached in spin-resolved experiments. In a static mode, i.e., while keeping the tip at a fixed position above the island to be investigated, much higher frequencies up to the band width of the current preamplifier can be detected, which typically amounts to 100 kHz–1 MHz. It will be impossible, however, to investigate the magnetization dynamics within a single reversal process, as it occurs on the time scale of several hundred pikoseconds (ps). Under typical STM operation conditions the tunnel current amounts to 1–10 nA and the average waiting time between two successive tunneling electrons is 10–100 ps, i.e., comparable with the reversal time.

10.4.3. The constant current mode

So far we have described results, which were measured with two different spectroscopic modes of SP-STM. However, magnetic imaging is also possible using a magnetic tip in the constant current mode. In fact, first spin-polarized vacuum tunneling experiments were performed in the constant-current mode using a ferromagnetic CrO_2 tip on a Cr(001) surface [50]. Below the Cr Néel temperature $T_N = 311$ K the (001) planes oriented parallel to the surface are ordered ferromagnetically. Since adjacent planes couple antiferromagnetic, however, the magnetization direction of the next terrace, which is separated by a single-atomic step, must be reversed.

Therefore, the so-called topological antiferromagnetic order of Cr(001) is formed [51]. It can clearly be recognized in Fig.10-19. At any step edge the surface magnetization direction reverses leading to A and B terraces [see topographic image Fig.10-19(a)]. In fact, the spin-resolved dI/dU map of Fig.10-19(b) confirms the expected behavior. While an enhanced dI/dU signal is found on terraces of type A, a reduced dI/dU signal is found on terraces of type B. This difference is caused by spin-polarized tunneling between the in-plane sensitive Fe tip and the magnetic sample surface. Interestingly, a domain wall is formed between the two screw dislocations marked by an arrow in Fig.10-19(a). A detailed analysis of the dislocation-induced domain structure of Cr(001) can be found in Ref. [34].

Figure 10-19(c) shows line sections drawn along the boxes in (a) and (b) analyzing the simultaneously measured topography (top panel) and dI/dU signal (bottom panel). As expected, corresponding with the positions of the step edges, the dI/dU signal exhibits four sudden jumps. A close inspection of the topographic signal reveals that the apparent step height deviates by 0.07 Å from the expected value of 1.44 Å. This is caused by the net spin polarization of the total tunneling current. At constant tip-sample separation it led to a tunneling current, which depends on the relative magnetic orientation of tip and sample. Since, however, the STM is operated in the constant-current mode, the tip has to get closer to the surface wherever tip and sample are antiparallel, and *vice versa*, resulting in different apparent heights of A-B and B-A steps. The polarization-induced apparent step height variation usually amounts to several tenth of an Å only. Even if topographic features as, e.g., step edges, were exactly known, the tiny magnetic contribution would impede a clear separation of topographic and magnetic effects. Only if the surface topography is extremely simple, SP-STM constant-current images may be useful for imaging magnetic structures. This condition is fulfilled for atomically flat, homogeneous surface planes without any step edges. Then the topographic contribution reduces to a constant or periodic term and any additional variation of the tip-sample distance is exclusively caused by spin-dependent effects.

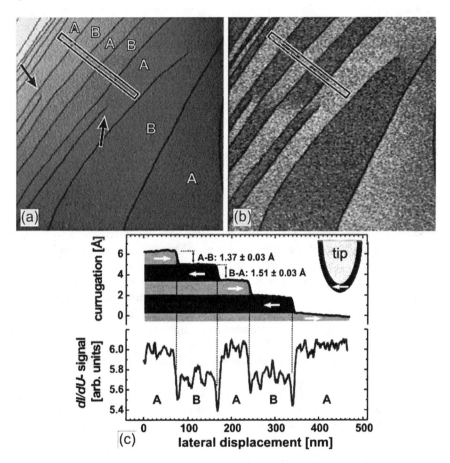

Figure 10-19: (a) Topography and (b) spin-resolved dI/dU map of a Cr(001) surface measured at T = 300 K. Adjacent terraces, which are separated by single-atomic steps, are alternatingly magnetized in opposite directions. (c) Apparent step height modulations can be found in constant current scan lines (top panel) corresponding to the modulation of the spin-resolved dI/dU signal. The measurement parameters were I = 0.18 nA and U = -60 mV. Data with courtesy of M. Kleiber and R. Ravlić [20].

By making use of constant-current SP-STM Heinze *et al.* [51, 52] revealed the antiferromagnetic coupling of nearest-neighbor atoms in a Mn monolayer on W(110) [Fig.10-20]. While the chemical atomic order—irrespective of the magnetic orientation of the particular Mn atom—was imaged when a non-magnetic tip was employed, the use of a magnetic tip resulted in an additional spin-dependent variation of the tip-sample distance thereby revealing the magnetic unit cell.

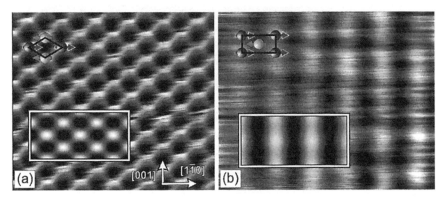

Figure 10-20: Atomic resolution constant-current STM images measured on the antiferromagnetic Mn monolayer on W(110) using (a) a non-magnetic W tip and (b) a Fe-coated tip (tunneling parameters for both images: I = 40 nA, U = -3 mV). With a spin-averaging W tip all Mn atoms have the same apparent height irrespective of their magnetic orientation. This leads to a STM image with the periodicity of the chemical unit cell. In contrast, the Fe tip is sensitive to the spin of the tunneling electrons and the periodicity of the antiferromagnetic c(2×2) unit cell shows up in (b). Data with courtesy of S. Heinze, X. Nie, and S. Blügel [51].

Another example for the successful application of the SP-STM constant-current mode to the Mn_3N_2 surface by Yang *et al.* [14] is shown in Fig.10-21. In the bulk Mn_3N_2 exhibits a face-centered tetragonal (fct) rocksalt-type structure and the Mn moments of (001) planes are ferromagnetically ordered with the magnetization pointing in the [100] direction. Adjacent (001) planes couple antiferromagnetic. The bulk Néel temperature of Mn_3N_2 amounts to 925 K. Since every third (001) layer along the c direction has all N sites vacant the bulk unit cell has a length of c = 12.13 Å and contains six atomic layers.

Due to lattice relaxation effects N-vacancy atomic rows slightly protrude from the surface plane resulting in a one dimensional stripe pattern with a periodicity of $c/2$ = 6.07 Å between adjacent rows and a corrugation of ~0.07 Å. As long as a non-magnetic tip is used all stripes have the same height (cf. Fig.10-1 in Ref. [14]). As described in Sec. 10.2.2 of this chapter the use of a magnetic tip results in a spin-dependent contribution to the tunneling current, which—if the STM is operated in the constant-current mode—is transferred into a tip height variation.

Figure 10-21(a) shows a constant-current image of Mn_3N_2(010) measured with a Mn-coated probe tip. Two domains D1 and D2 with different c-axis orientation relative to the underlying MgO(001) substrate can be recognized. In Fig.10-21(b) Yang *et al.* [14] have plotted area-averaged line sections drawn within the boxes in D1 and D2, perpendicular to the atomic rows, respectively. Both sections reveal an alternating variation of the apparent row height leading to a periodicity of c = 12.13 Å. Obviously, the variation is about a factor of 2 larger in D1 than in D2. This was attributed to the fact that—according to Eq.(10.5)—the magnetic contribution to the tunnel current scales with $\cos(\theta)$, where θ is the angle between the magnetization directions of tip and sample. Since, as schematically represented in Fig.10-20(c), θ is much smaller in D1 than in D2—the magnetization directions of which are rotated by 90° with respect to each other—, the spin-dependent variation is larger above D1. The relative magnetic orientation of the tip with respect to the surface moments is shown in the inset of Fig.10-21(c).

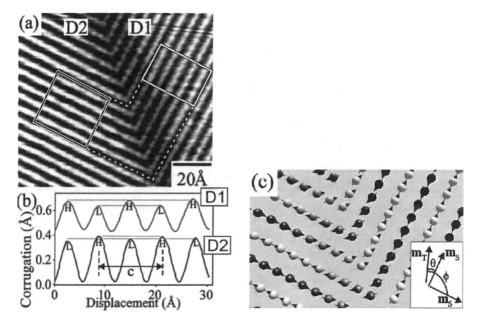

Figure 10-21: (a) Room temperature SP-STM image of a $Mn_3N_2(010)$ film grown on a MgO(001) substrate. The data were acquired using a Mn-coated W tip at U = -0.6 V and I = 0.8 nA. The two-fold symmetry of the substrate results in two domains D2 and D1 with different orientation of the Mn_3N_2 c-axis relative to the substrate. (b) Two area averaged line profiles drawn perpendicular to the lines corresponding to the regions inside the black (D1) and white (D2) rectangles in (a). (c) Schematic model of the imaged spin configuration. The inset shows the moments of tip (m_t) and the sample (m_s) for the two different domains and the angles between them. Each ball represents a magnetic atom. Data with courtesy of A. R. Smith [14].

10.5. CONCLUSIONS

Spin-polarized scanning tunneling microscopy and spectroscopy is capable to image the surface domain structure of single crystalline surfaces and thin films with unprecedented spatial resolution down to atomic scales. In this chapter three modes of operation have been described, i.e., the differential conductance or dI/dU mode, the local magnetoresistance dI/dm mode, and the constant current mode. The former two modes allow the clear separation and correlation of topographic and magnetic contributions. In contrast, topographic and magnetic effects cannot be separated in the constant-current mode. Therefore, this mode of operation is restricted to extremely simple surface topographies, as, e.g., atomically flat terraces.

Beside this "classical" field of application, i.e., the imaging of static domain structures of magnetic nanostructures, spin-polarized STM can also be applied for the investigation of unstable magnetic cofigurations, as the switching behavior of superparamagnetic particles. Since the magnetic bahavior of indvidual particles can directly be correlated with their topography, spin-polarized STM offers the unique possibility to unravel hidden links between structural and magnetic properties.

ACKNOWLEDGEMENTS

We would like to acknowledge contributions of and discussions with G. Bihlmayer, S. Blügel, M. Getzlaff, S. Heinze, M. Kleiber, M. Morgenstern, X. Nie, R. Ravlić, A. Wachowiak, J. Wiebe.

REFERENCES

[1] C. B. Duke, J. Vac. Sci. Technol. **21** (2003) S34.
[2] G. Binnig and H. Rohrer, Helv. Phys. Acta **55** (1982) 726.
[3] G. Binnig and H. Rohrer, Rev. Mod. Phys. **59** (1987) 615.
[4] M. Lagally, J. Vac. Sci. Technol. **21** (2003) S54.
[5] D. T. Pierce, Phys. Scr. **38** (1988) 291.
[6] M. Julliere, Phys. Lett. **54** (1975) 225.
[7] P. M. Tedrow, R. Meservey, and P. Fulde, Phys. Rev. Lett. **25** (1970) 1270.
[8] P. M. Tedrow and R. Meservey, Phys. Rev. B **7** (1973) 318.
[9] R. Meservey, P. M. Tedrow, and R. C. Bruno, Phys. Rev. B **11** (1975) 4224.
[10] S. Yuasa, T. Sato, E. Tamura, Y. Suzuki, H. Yamamori, K. Ando, and T. Katayama, Europhys. Lett. **52** (2000) 344.
[11] D. Wortmann, S. Heinze, P. Kurz, G. Bihlmayer, and S. Blügel, Phys. Rev. Lett. **86** (2001) 4132.
[12] R. Wiesendanger, H. J. Güntherodt, G. Güntherodt, R. J. Gambino, and R. Ruf, Phys. Rev. Lett. **65** (1990) 247.
[13] S. Heinze, S. Blügel. R. Pascal, M. Bode, and R. Wiesendanger, Phys. Rev. B **58** (1998) 16432 .
[14] H. Yang, A. R. Smith, M. Prikhodko, and W. R. L. Lambrecht, Phys. Rev. Lett. **89** (2001) 226102.
[15] M. Johnson and J. Clarke, J. Appl. Phys. **67** (1990) 6141.
[16] W. Wulfhekel and J. Kirschner, Appl. Phys. Lett. **75** (1999) 1944.
[17] T. Miyazaki and N. Tezuka, J. Magn. Magn. Mater. **139** (1995) L231.
[18] M. Bode, O. Pietzsch, A. Kubetzka, and R. Wiesendanger, J. Electr. Spectr. Rel. Phenom. **114-116** (2001) 1055.
[19] O. Pietzsch, A. Kubetzka, D. Haude, M. Bode, and R. Wiesendanger, Rev. Sci. Instr. **71** (2000) 424.
[20] M. Kleiber, R. Ravlić, M. Bode, and R. Wiesendanger, Phys. Rev. Lett. **85** (2001) 4606.
[21] D. T. Pierce and F. Meier, Phys. Rev. B **13** (1976) 5484.
[22] D. T. Pierce, R. J. Celotta, G. C. Wang, W. N. Unertl, A. Galejs, C. E. Kuyatt, and S. Mielczarek, Rev. Sci. Instr. **51** (1980) 478.
[23] M. W. J. Prins, R. Jansen, and H. van Kempen, Phys. Rev. B **53** (1996) 8105.
[24] Y. Suzuki, W. Nabhan, and K. Tanaka, Appl. Phys. Lett. **71** (1997) 3153.
[25] W. Nabhan, Y. Suzuki, R. Shinohara, K. Yamaguchi, and E. Tamura, Appl. Surf. Sci. **144-145** (1999) 570.
[26] R. Jansen, R. Schad, and H. van Kempen, J. Magn. Magn. Mater. **198-199** (1999) 668.
[27] V. P. LaBella, D. W. Bullock, Z. Ding, C. Emery, A. Venkatesan, W. F. Oliver, G. J. Salamo, P. M. Thibado, and M. Mortazavi, Science **292** (2001)1518.
[28] R. Wiesendanger, D. Bürgler, G. Tarrach, T. Schaub, U. Hartmann, H. J. Güntherodt, I. V. Svets, and J. M. D. Coey, Appl. Phys. A **53** (1991) 349.
[29] W. Wulfhekel, R. Hertel, H. F. Ding, G. Steierl, and J. Kirschner, J. Magn. Magn. Mater. **249** (2002) 368.

[30] U. Schlickum, W. Wulfhekel, and J. Kirschner, Appl. Phys. Lett. **83** (2003) 2016.

[31] O. Pietzsch, A. Kubetzka, M. Bode, and R. Wiesendanger, Phys. Rev. Lett. **84** (2000) 5212.

[32] O. Pietzsch, A. Kubetzka, M. Bode, and R. Wiesendanger, Science **292** (2001) 2053.

[33] A. Kubetzka, O. Pietzsch, M. Bode, and R. Wiesendanger, Phys. Rev. B **67** (2003) 020401.

[34] R. Ravlić, M. Bode, A. Kubetzka, and R. Wiesendanger, Phys. Rev. B **67** (2003) 174411.

[35] A. Kubetzka, M. Bode, O. Pietzsch, and R. Wiesendanger, Phys. Rev. Lett. **88** (2002) 057201.

[36] A. Wachowiak, J. Wiebe, M. Bode, O. Pietzsch, M. Morgenstern, and R. Wiesendanger, Science **298** (2002) 577.

[37] A. Hubert and R. Schäfer, Magnetic Domains", Springer (1998).

[38] H. F. Ding, W. Wulfhekel, and J. Kirschner, Europhys. Lett. **57** (2002) 100.

[39] K. Bussmann, G. A. Prinz, S.-F. Cheng, D. Wang, Appl. Phys. Lett. **75** (1999) 2476.

[40] E. Feldkeller and H. Thomas, Phys. Kondens. Mater **4** (1965) 8.

[41] J. Raabe et al., J. Appl. Phys. **88** (2000) 4437.

[42] T. Shinjo, T. Okuno, R. Hassdorf, K. Shigeto, and T. Ono, Science **289** (2000) 930.

[43] M. Bode, Rep. Prog. Phys. **66** (2003) 523.

[44] OOMMF, Object Oriented Micromagnetic Framework, version 1.2 alpha2 (http://math.nist.gov/oommf).

[45] N. Weber, K. Wagner, H. J. Elmers, J. Hauschild, and U. Gradmann, Phys. Rev. B **55** (1997) 14121.

[46] H. J. Elmers, J. Magn. Magn. Mater. **185** (1998) 274.

[47] A. Kubetzka, O. Pietzsch, M. Bode, and R. Wiesendanger, Phys. Rev. B **63** (2001) 140407.

[48] M. L. Néel, Ann. Géophys. **5** (1949) 99.

[49] W. F. Brown, Phys. Rev. **130** (1963) 1677.

[50] S. Blügel, D. Pescia, and P. H. Dederichs, Phys. Rev. B **39** (1989) 1392.

[51] S. Heinze, M. Bode, A. Kubetzka, O. Pietzsch, X. Nie, S. Blügel and R. Wiesendanger, Science **288** (2000) 1805.

[52] M. Bode, S. Heinze, A. Kubetzka, O. Pietzsch, M. Hennefarth, M. Getzlaff, R. Wiesendanger, X. Nie, G. Bihlmayer, and S. Blügel, Phys. Rev. B **66** (2002) 014425.

CHAPTER 11

Magnetic force microscopy

11.1. INTRODUCTION

A general feature of Scanning Probe Microscopes (SPM) is the use of a sharp probe properly prepared to localize a specific interaction between the probe and a surface under investigation. The probe is generally raster-scanned over the surface and the interaction is measured and displayed as a function of the probe position. The measured interactions can range from current in the case of the Scanning Tunneling Microscope (STM) to force in the case of the Atomic Force Microscope (AFM) [1] and evanescent optical excitations in the case of the Scanning Nearfield Optical Microscope (SNOM).

The Magnetic Force Microscope (MFM) is a specific type of SPM based on the AFM where the sharp probe is magnetic and interacts with the magnetic fields generated by the specimen. The magnetic tip is located at the end of a diving board-shaped cantilever and interacts with fields, resulting in a flexing of the cantilever. This flexing is detected and its magnitude is a measure of the fields. It is fundamentally the same process as attaching a small permanent magnet to your finger and investigating some arrangement of similar magnets glued to a surface. Some existing reviews of MFM techniques provide other views of the technology [2-9].

The MFM has become a widespread, easy-to-use-tool that can provide high-resolution (sub-100 nm) images of the micromagnetic structures of a variety of materials. One notable advantage of the MFM over other high resolution techniques is that it is possible to image magnetic structures in air and in the presence of a protective overcoat. Thus, samples such as magnetic hard disk surfaces can be imaged quickly with very high resolution. In general, sample preparation for the MFM is much less demanding than for other techniques such as Lorentz Microscopy or Scanning Electron Microscopy with Polarization Analysis (SEMPA) while yielding similar resolution. Note that while it is possible to operate in a vacuum or at low temperatures, these conditions are not required. Disadvantages of MFM include that it is (i) an indirect probe of the sample magnetization; it is sensitive to the external fields of the sample and does not provide a direct probe of the sample magnetization and (ii) that the interactions are quite difficult to quantify.

The MFM is a magnetic imaging technique that is sensitive to the spatial derivatives of the magnetic fields generated by a sample. These fields do not depend on the sample magnetization directly but result from the divergence of the magnetization ($-\nabla \cdot \mathbf{M}$ for the bulk and $\mathbf{M} \cdot \hat{\mathbf{n}}$ for the surface). Put another way, the "conventional" MFM is a magnetic charge imaging device. This results in contrast over features such as domain walls or when the magnetization directly intersects the surface of a specimen.

Figure 11-1: This Figure shows a photograph of Kaczer's permalloy probe microscope. It used an oscillating permalloy probe (1) whose magnetic state depended on the stray field from the sample (2). The magnetization of the permalloy probe was detected with a simple induction coil. (3) specimen movement control and (4) leveling tripod.

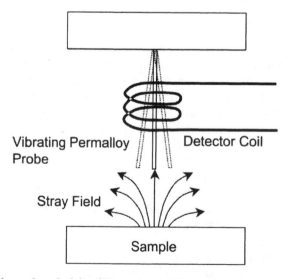

Figure 11-2: This shows the principle of Kaczer's permalloy probe microscope. The permalloy probe is oscillated in a detector coil. As the probe is scanned over the stray field from the sample, the flux through the permally probe changes which in turn changes the voltage induced in the detector coil.

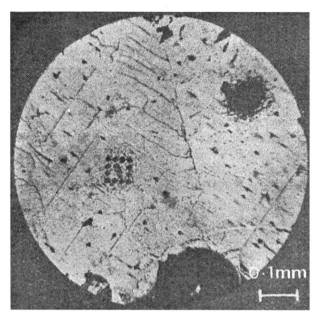

Figure 11-3: Colloid (Bitter) pattern on the (100) surface of Silicon-Iron crystal [11].

Figure 11-4: Probe image of the same region as Fig.11-3. Note the square array of Point defects visible in both here and in Figs.11-3.

Although SPM exploded with Binnig and Rohrer's Nobel prize winning development of the STM [10] we must note that some of the first magnetic domain observations were made with a scanning probe microscope in 1955 and 1956 by Jan Kaczer [11,12]. Kaczer's microscope (shown in Fig.11-1) inductively measured the magnetic state of a small permalloy (a NiFe alloy of approximately 80% Ni) probe. As the probe was scanned over a sample surface, the magnetically soft permalloy probe's magnetic state changed in response to the localized field from the sample. This change in state was detected with a coil (Fig.11-2). The resulting images correlated well with colloid images of the same surface, clearly mapping domain walls and a defect in the sample. This is apparent in the Bitter pattern image of domains in Fig.11-3. Figure 11-4 shows the corresponding permalloy probe image. A square array of pits are visible in both images, allowing easy registration. Perhaps because of the relative simplicity of colloid techniques and the high resolution of Lorentz electron microscopy, more than thirty years passed before another scanning probe microscope used for magnetic imaging appeared in the literature. Work on the AFM [1], brought new technology and ideas to the problem and what we now think of as the MFM [13,14] followed.

11.1.1. Magnetic forces and force gradients

As mentioned above, the MFM is a type of scanning probe microscope with a magnetic tip sensitive to the magnetic field gradients external to a sample as opposed to being directly sensitive to the sample magnetization. The conventional starting point for understanding MFM operation is to calculate the force, $\mathbf{F}^t$, between a MFM tip having a moment $\mathbf{m}^t$ and the fields $\mathbf{H}^s$ from a sample. In general MFM tip moments involve extended geometries and calculating the total force involves integrating the following expression over all of the dipole moments in the MFM tip [15]:

$$d\mathbf{F}^t = -\nabla_r \left[\mathbf{M}^t(\mathbf{r}') \cdot \mathbf{H}^s(\mathbf{r} + \mathbf{r}') \right] dV^t. \tag{11.1}$$

This expression almost always requires numerical evaluation. The geometry used to integrate this expression is shown in Fig.11-5.

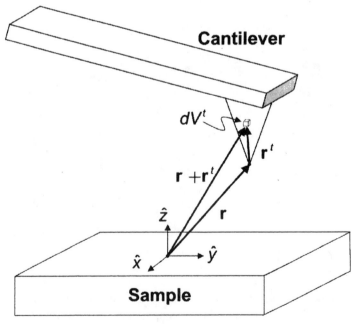

Figure 11-5: This Figure shows the geometry for integrating eq.(11.1).

Depending on the details of the experimental of theoretical situation, it is sometime more convenient to work in terms of the force acting on the sample $\mathbf{F}^s$ rather than on the tip. Newton's third law provides a simple relationship between these two forces [16,17]:

$$d\mathbf{F}^t = -d\mathbf{F}^s.$$ (11.2)

Simply put, the magnetic force acting on the tip is equal and opposite the force acting on the sample. Reciprocity gives us the force acting on the sample; similar in form to that acting on the tip. It is given by

$$d\mathbf{F}^t = -\nabla_r \left[\mathbf{M}^s(\mathbf{r}^s) \cdot \mathbf{H}^t(-\mathbf{r} + \mathbf{r}^s) \right] dV^s.$$ (11.3)

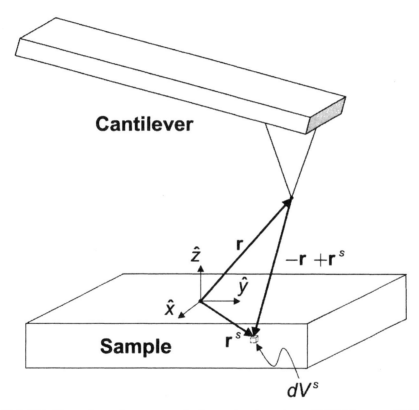

Figure 11-6: This Figure shows the geometry for integrating the reciprocal MFM force, eq.(11.3).

In this expression, $\mathbf{M}^s$ is the sample magnetization and $\mathbf{H}^t$ is the magnetic field from the tip. The geometry associated with this expression is shown in Fig.11-6. This interpretation is particularly useful where there is some knowledge of the magnetic field from the tip. As will be discussed in section 11.5.1, there are a number of techniques for quantifying $\mathbf{H}^t$. With the field quantified, interpretation of MFM response is simplified; one can then concentrate solely on the sample magnetization.

Many interpretations start with a non-extended tip and sample that does not require integration. If the tip is treated like a point magnetic dipole of orientation and magnitude $\mathbf{m}^t = \mathbf{M}^t dV^t$, in the presence of a magnetic field $\mathbf{H}^s$, the force acting on the dipole is given by the gradient of the magnetic energy [18]

$$\mathbf{F}' = -\nabla_r \left[\mathbf{m}'(\mathbf{r}') \cdot \mathbf{H}^s(\mathbf{r} + \mathbf{r}') \right]. \tag{11.4}$$

Similarly, eq.(11.4) can be rewritten in terms of a sample point dipole $\mathbf{m}^s = \mathbf{M}^s d V^s$ and the magnetic field from the tip as

$$\mathbf{F}' = -\nabla_r \left[\mathbf{m}^s(\mathbf{r}^s) \cdot \mathbf{H}'(-\mathbf{r} + \mathbf{r}') \right]. \tag{11.5}$$

The force acting on the tip is a three dimensional one. However, AFM cantilevers are usually compliant along one dimension and relatively stiff along the others. This means that the cantilever will only deflect in a particular direction, $\hat{\mathbf{d}}$ (for detection axis). From eq.(11.4), the component of the force acting along this direction is

$$F' = \hat{\mathbf{d}} \cdot \left[\nabla (\mathbf{m}' \cdot \mathbf{H}^s) \right]. \tag{11.6}$$

The force gradient is given by

$$F'' = \hat{\mathbf{d}} \cdot (\nabla F'). \tag{11.7}$$

Typically, is assumed collinear with the z-axis and eq.(11.6) simplifies further to

$$F'_z = \frac{\partial}{\partial z} (\mathbf{m}' \cdot \mathbf{H}^s) \tag{11.8}$$

and eq.(11.7) becomes

$$F'_z = \frac{\partial^2}{\partial z^2} (\mathbf{m}' \cdot \mathbf{H}^s). \tag{11.9}$$

Equations (11.8) and (11.9) can be expanded to give

$$F'_z = H_x^s \frac{\partial m'_x}{\partial z} + m'_x \frac{\partial H_x^s}{\partial z} + H_y^s \frac{\partial m'_y}{\partial z} + m'_y \frac{\partial H_y^s}{\partial z} + H_z^s \frac{\partial m'_z}{\partial z} + m'_z \frac{\partial H_z^s}{\partial z} \tag{11.10}$$

for the force and

$$\begin{aligned} F'_z = & H_x^s \frac{\partial^2 m'_x}{\partial z^2} + m'_x \frac{\partial^2 H_x^s}{\partial z^2} + 2 \frac{\partial m'_x}{\partial z} \frac{\partial H_x^s}{\partial z} + \\ & H_y^s \frac{\partial^2 m'_y}{\partial z^2} + m'_y \frac{\partial^2 H_y^s}{\partial z^2} + 2 \frac{\partial m'_y}{\partial z} \frac{\partial H_y^s}{\partial z} + \\ & H_z^s \frac{\partial^2 m'_z}{\partial z^2} + m'_z \frac{\partial^2 H_z^s}{\partial z^2} + 2 \frac{\partial m'_z}{\partial z} \frac{\partial H_z^s}{\partial z} \end{aligned} \tag{11.11}$$

for the force gradient.

If the magnetic moment of the tip is a fixed point dipole these reduce to

$$F'_z = m'_x \frac{\partial H_x^s}{\partial z} + m'_y \frac{\partial H_y^s}{\partial z} + m'_z \frac{\partial H_z^s}{\partial z}. \tag{11.12}$$

$$F'_z = m'_x \frac{\partial^2 H_x^s}{\partial z^2} + m'_y \frac{\partial^2 H_y^s}{\partial z^2} + m'_z \frac{\partial^2 H_z^s}{\partial z^2}. \tag{11.13}$$

Equations (11.12) and (11.13) are notable in that they imply that the response of the MFM can be adjusted to respond to various components of the magnetic field. If, for example, the tip is magnetized along the z-axis, then $m_x=0$, $m_y=0$, and eqs.(11.12) and (11.13) further simplify to

$$F_z^t = m_z^t \frac{\partial H_z^s}{\partial z}$$

(11.14)

and

$$F_z'^t = m_z^t \frac{\partial^2 H_z^s}{\partial z^2}.$$

(11.15)

Reciprocity allows us to equivalently express eq.(11.10) though (11.15) in terms of the force and force gradient acting on the sample by exchanging the "t" and "s" superscripts and adding a minus sign to the entire right hand side of the equation (see eq.(11.2)). This results in, for example, eq.(11.15) being rewritten as

$$F'^t = -m_z^s \frac{\partial^2 H_z^t}{\partial z^2}.$$

(11.16)

11.1.2. Measuring the force or the force gradient with a cantilever

The force and the force gradient describe the interactions between the magnetic tip at the end of the cantilever and the sample. The cantilever acts as a mechanical transducer where, as we discuss in more detail below, mechanical parameters, including dc deflection, ac amplitude, the in-phase and/or quadrature amplitude are used to form an image as it is raster scanned over the sample surface. The quantities all depend on the force or force gradient. To interpret the response of the MFM requires some means of extracting the force or the force gradient. There are many details associated with this; they break down into two steps. (i) The first is measuring and quantifying the motion of the cantilever. This is typically done continuously during AFM or MFM imaging. The measured quantities, say phase and amplitude are used to form images and control signals that are used to operate the instrument. (ii) The second is quantifying the sensitivity of this motion to a force or force gradient. For example, if in step (i) we measure a deflection, we still need a spring constant to convert that deflection into a quantitative value of the force (see eq.(11.17) below)

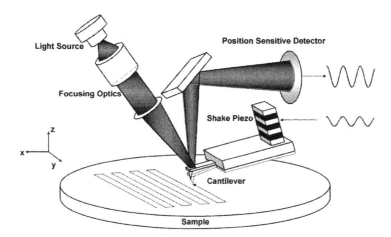

Figure 11-7: This shows the basic components of an AFM or MFM. Most AFMs use an optical lever technique where light from an intense source is focused onto a flexible, reflective cantilever. The cantilever deflection moves tilts the angle of the optical beam which in turn is converted to an electrical signal by the position sensitive detector (PSD). The sample is raster scanned relative to the cantilever (in the x-y plane in this Figure). AC techniques often use a "shake" piezo to excite oscillations in the cantilever.

The basic detection scheme for these follows that of a "generic" AFM as shown in Fig.11-7, which illustrates the commonly used optical lever technique [19]. In this implementation, light from an intense source, typically a gas laser, laser diode or superluminescent diode [20] is focused onto a small, flexible cantilever. The angle of the reflected light is measured by a position sensitive detector. As the cantilever deflects, the angle of the reflected light changes. The resulting motion of the spot on the surface of the position sensitive detector is converted into an electrical signal. Despite its apparent simplicity, the optical lever method has very high sensitivity, competing favorably with other methods such as interferometry [21], tunneling detection [22], and better than piezo resistive methods [23]. It turns out that the theoretical sensitivity and noise immunity of an optical lever detector is identical to that of an interferometer [24].

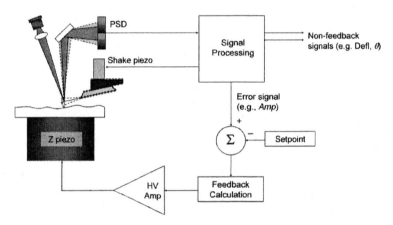

Figure 11-8: Most AFM imaging modes involve a feedback loop that regulate the tip-sample distance. The signal processing block generates an Error signal. In the case of AC mode imaging, this is generally the cantilever amplitude. Other, non-feedback signals such as the deflection, phase or perhaps tip-sample currents can be passed on to the display system to form an image. The Setpoint is subtracted from the Error signal. This is then input into a feedback calculation. The output of this calculation is then used to control the tip-sample separation through a high voltage amplifier and a piezo that modulates the tip-sample separation. The job of the feedback calculation is to do this in such a manner as to keep the Error signal at the sepoint value (that is, to zero the input into the Feedback calculation. The output of the feedback calculation is then a representation of the sample topography.

To create an image, a force microscope is usually operated in a feedback loop (see Fig.11-8) where one of the parameters, such as the ac amplitude or deflection, is maintained at a constant value (referred to as the "setpoint") by varying the height of the sample relative to the base of the cantilever. This allows the tip–sample interactions to be maintained at a roughly constant value. In the case of AC mode imaging, this is generally the cantilever amplitude. Other, non-feedback signals such as the deflection, phase or perhaps tip-sample currents can be passed on to the display system to form an image. The job of the feedback calculation is to do this in such a manner as to keep the error signal at the sepoint value (that is, to zero the input into the Feedback calculation.

11.2. SENSING FORCES AND FORCE GRADIENTS

11.2.1. DC Forces

Conceptually, the simplest imaging mode involves monitoring the cantilever deflection as a function of position. The cantilever is deflected by a force according to the simple relationship

$$\Delta z = -\frac{F'(z)}{k}, \tag{11.17}$$

where Δz is the cantilever deflection, k is the cantilever spring constant and $F'(z)$ is the interaction force acting on the cantilever tip. When the deflection is used as the "error signal" in the feedback loop controlling the sample height, the AFM is operating in "contact mode".

Although not the first or the most popular method of operation, dc detection of magnetic forces has been used to form images of a variety of magnetic microstructures. One example is given by Giles et al. [25]. This work is rather novel in that the magnetic bits written onto the disk were imaged under water. They used a method to first image the topography of their hard disk sample and then, using that as a reference, to image the long range magnetic forces at a predetermined distance above the sample surface. This two pass technique, pioneered by Martin and Wickramasinghe [26] as well as Hosaka [27] for AC imaging will be discussed more completely below. In another example, Gomez et al. [22] exploited the tunneling current to stabilize the cantilever tip above the surface while simultaneously measuring magnetic forces by monitoring the cantilever deflection. More recently, Hug and co-workers used dc imaging in a vacuum MFM to quantify the response of their MFM [28]. In their case, operating the MFM in dc mode simplified the interpretation of the response considerably.

The sensitivity of the dc method is limited by noise sources in the cantilever deflection system. A well-designed detection system should be limited by fundamental physical processes such as Brownian motion and shot noise in the detector [29]. In practice, most AFMs and MFMs have some vulnerability to low frequency noise. This is often characterized as $1/f$ noise and arises from thermal expansion and contraction of the microscope components caused by variations in the ambient conditions or internal heating of the electronics and building and/or acoustic vibrations that can excite unwanted modes in the mechanics of the instrument. Because these low frequency noise sources are often difficult or impossible to control, experimentalists often resort to making measurements at higher frequencies.

11.2.2. AC detection methods

With the exceptions mentioned previously, MFM imaging is regularly performed in a mode where the cantilever is oscillated at or near its resonant frequency [13]. This allows the use of a variety of ac techniques to extract the signal from the noise and moves the measurement away from the ubiquitous $1/f$ or low frequency noise present in all practical experimental environments [13].

The AC detection modes require a detailed understanding of the equations of motion of the three dimensional cantilever (see Fig.11-7). The rewards, however, are well worth the extra effort. Two examples of the added benefit are magnetic dissipation imaging [30] and eddy current microscopy [31,32].

Magnetic interactions, primarily through the force gradient, shift the resonant frequency of the cantilever. This shift in resonance can be measured in a number of ways, three of which will be described below. In addition, information on non-conservative interactions between the oscillating tip and sample can be extracted.

Typically, cantilever motion is simplified into a one dimensional problem. In this simplified case, the equation of motion for the transverse (perpendicular to the specimen surface) position w of the cantilever is a fourth order, time dependent partial differential equation given by

$$EI \frac{\partial^4 w(x,t)}{\partial x^4} + \mu \frac{\partial^2 w(x,t)}{\partial t^2} = F_d(x,t) \tag{11.18}$$

where $w(x,t)$ is the transverse displacement of the cantilever beam, E is the elastic or Young's modulus of the cantilever, I is the moment of inertia, μ is the mass per unit length and $F_d(x,t)$ is the force acting on the cantilever per unit length. Solutions to this equation are not trivial [33-37] and a considerably simpler approach is to approximate the extended cantilever as a simple point mass on a spring [38,39]. Following Marion and Hornyak [40], the differential equation for the motion is that of the more familiar damped harmonic oscillator expression [40],

$$m\ddot{z} = -kz + F'(z) - b\dot{z}, \tag{11.19}$$

where $z=(L,t)$ if the tip of the cantilever is at position $x=L$, k is the spring constant, $F'(z)$ is again the force acting on the tip, m is the effective mass of the cantilever, ω_0 is the angular resonant frequency, and b is the viscous damping. For the case when $F'(z)$ is independent of z or constant, then the zero point is shifted and the oscillatory motion is identical to the oscillator without $F'(z)$, i.e., just like a mass and spring system on a frictionless horizontal surface and when suspended in a gravitational field. If, however, the force gradient, $F''(z)$, is non zero then we have an effective spring constant given by $k-F''(z)$.

If we drive the system with an external force $F_d\cos(\omega t)$, then eq.(11.19) has a steady state solution of the form

$$z = A(\omega)\cos[\omega t + \varphi(\omega)] \tag{11.20}$$

where $A(\omega)$ and the phase $\varphi(\omega)$ are given by

$$A(\omega) = \frac{\dfrac{F_d}{k - F''(z)}}{\left[\left(\dfrac{\omega_0^2}{\omega^2} - 1 \right)^2 + \dfrac{\omega^2}{\omega_0^2 Q^2} \right]} \tag{11.21}$$

and

$$\tan\varphi(\omega) = Q(\omega/\omega_0 - \omega_0/\omega). \tag{11.22}$$

In these expressions ω_0 is the resonant frequency determined by both the spring constant and the force gradient, $\omega_0^2 = (k-F')/m$, and $Q = m\omega_0/b$ is the quality factor. As the expression for the resonant frequency shows, attractive interactions lower the resonant frequency while repulsive interactions raise the resonant frequency.

The expressions for $A(\omega)$ and $\varphi(\omega)$ are plotted in the vicinity of the resonant frequency in Fig.11-9. This Figure also shows the effect of the cantilever interacting with both attractive and repulsive force gradients. These small force gradients shift the amplitude and phase curves as shown by the dashed lines in Fig.11-9.

There are two effects of the damping which need to be discussed here. The first, is the frequency with the maximum response of the cantilever, ω_{max}, is no longer ω_0 but is found to be reduced to

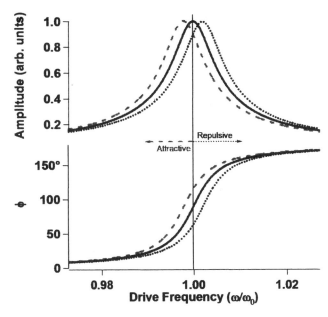

Figure 11-9: The amplitude and phase of a simple harmonic oscillator is plotted in this figure. The dashed lines show the effects of a an attractive force gradient that lowers the resonant frequency while the dotted lines show the effect of a repulsive force gradient that shifts the resonant frequency to a higher value.

$$\omega_{max}^2 = \left[\left(\frac{k - F''}{m}\right)^2 - \frac{\omega_0^2}{2Q^2}\right]^{1/2}. \tag{11.23}$$

Damping has a direct effect on the response speed of the cantilever sensor. In addition to the steady state solution SHO2 above, there is a transient solution to the equation of motion given by

$$z_t = A_t e^{-\omega_0 t / 2Q} \sin(\omega_0 t + \varphi_t). \tag{11.24}$$

One main feature of this equation is it reaches $1/e$ of it's initial amplitude after $2Q$ cycles. Put another way, this transient has a relaxation time $\tau = 2Q/\omega_0$. This implies a higher Q requires the AFM or MFM to scan more slowly for the motion to relax into the steady state solution. For typical MFM cantilevers operated in ambient conditions the Q is ~150 and with a resonant frequency of ~70 kHz, the time constant is τ~0.7 ms. This is, in general, sufficiently small to raster at a scan line frequency of 1 Hz or so. This acquisition speed translates into a time of roughly 9 minutes for a 256×256 pixels image. If the same cantilever was operated in a vacuum environment, however, its Q would approach something like 10^4, and the time constant would become τ~45 ms, increasing time to acquire a 256×256 image to something on the order of 5.5 hours.

The response time of a cantilever can be modified with so-called "Q-control" [41]. With this technique, the phase of the cantilever oscillation is monitored and used as a feedback signal on the drive amplitude of the oscillation piezo. In this, it is similar to the FM-AFM [42]. Q-control uses the oscillation actuator to simulate an increase or decrease in the damping forces acting on the cantilever. This changes the effective time constant of the cantilever, allowing faster response, but lower sensitivity or slower response and higher sensitivity detection. Put another way, Q-control allows one to change the natural averaging time of the cantilever.

There are three main AC detection modes we will cover: "slope detection", phase detection and "phase-locked" (also referred to as "FM detection"). The physics behind these modes are all interrelated and depend heavily on the above description of a simple harmonic oscillator.

11.2.2.1. Slope Detection

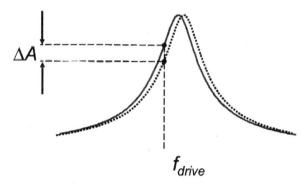

Figure 11-10: In this detection mode, a force gradient acting between the tip and the sample shifts the resonant frequency (in this particular case, to a higher value). Because the cantilever is being driven off of its resonant frequency, the amplitude changes (in this example, it decreases).

The "slope detection" technique was initially used by Martin and Wickramasinghe [26]. The idea is illustrated in Fig.11-10. Changes in the resonant frequency are measured by measuring the cantilever oscillation amplitude with fixed drive at a frequency where it varies as a function of the resonant frequency. Figure 11-10 illustrates the idea. The original SHO cantilever response is plotted as a solid line in the Figure. The dashed line shows the increased resonant curve when the cantilever is interacting with a repulsive force gradient. If the cantilever is driven above (below) its resonant frequency, the amplitude increases (decreases) if the resonant frequency gets larger. With this mode of operation, the sensitivity is optimized by operating on the steepest part of the amplitude curve; for a high Q oscillator, this steepest point is located at $\omega_d \sim \omega_0(1 \pm 1/\sqrt{8}Q)$ on either side of the resonance frequency [13]. The relationship between changes in the cantilever amplitude and changes in the force gradient δF this frequency is given by

$$\delta A = \frac{\partial A}{\partial F'} \delta F'' = \frac{2QA_0}{3\sqrt{3}(k - F')} \delta F'' \qquad (11.25)$$

where A_0 is the peak amplitude at the resonant frequency.

In the original implementation of this approach the RMS amplitude of the cantilever response was used for the signal. This has the advantage that the electronics required are relatively simple, only requiring detection of the RMS amplitude of the cantilever. This technique, however, is more vulnerable to low frequency noise since RMS detection is not frequency selective. In fact, even with the single-frequency single-phase techniques described in the next section, this approach's sensitivity to the drive amplitude A_0 makes it sensitive to any noise in the cantilever drive-detector path which can produce such amplitude variations such as the amplitude noise from the laser, interference effects and building and acoustic vibrations. More general difficulties are first the data are complicated by a mixing of the conservative and dissipative interactions, i.e., the amplitude of oscillation is affected by both changes in the resonant frequency δF and second the quality factor Q and since it is an amplitude measurement, there is a relaxation time associated with changes in the frequency [42].

11.2.2.2. Phase detection

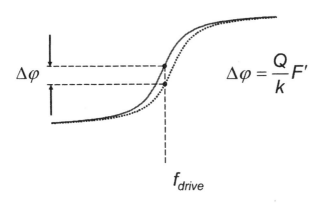

Figure 11-11: This Figure shows the idea behind a phase sensitive measurement of the force gradient. In this example, a repulsive force gradient shifts the resonant frequency to a higher value while the system is being driven at a constant frequency. This causes the measured phase to decrease.

Just as the amplitude changes with the presence of force gradients, the phase angle between the drive and cantilever response will also change as shown in eq.(11.25). If the cantilever is driven at a constant frequency and the resonant frequency shifts in response to a force gradient, there is a shift in the cantilever phase given by

$$\delta\phi = \frac{Q}{k}\delta F''$$

(11.26)

as illustrated in Fig.11-11. The phase thus provides a simple, direct interpretation of the image signal in terms of the force gradient, the quality factor and the spring constant of the cantilever but not the amplitude thereby making it relatively immune to laser noise, interference effects and variations in sample damping as well as building and acoustic vibrations, i.e., the main noise sources in amplitude detection.

For understanding the approach to detection of the phase shift, we begin with some basic features of a driven harmonic oscillator. First, the phase angle between the drive and the oscillator response vary as a function of frequency from 0 radians or in-phase to π radians or out-of-phase as the drive frequency is changed from zero to infinity. The major portion of this phase shift occurs in the frequency range defined by $2Q\omega_0$ with a maximum rate of change at its free resonance frequency ω_0 (at ω_0 the phase shift is $\pi/2$). Thus, as in the example shown, when driven at the no-signal-present ω_0, the addition of a repulsive (attractive) force gradient shifts the resonant frequency to a higher (lower) value which results in a decreased (increased) phase relative to the drive phase.

The amplitude and phase of a signal at a given frequency can be extracted from a phase sensitive detector (PSD) such as a lockin amplifier. A PSD or lockin is commonly used to extract either the magnitude or the phase or both when a small signal is buried in noise. In principle, a lockin signal channel detects a single phase of a specific frequency; if two input signal channels, separated by a phase shift of $\pi/2$ are sensed then both the amplitude and the phase of the input signal can be determined as shown below. The process below is that used in a digital controller since this simplifies the analysis.

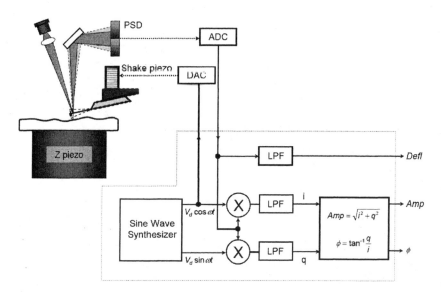

Figure 11-12: The output of the PSD can be processed to yield the dc deflection, amplitude and phase of an oscillating cantilever. First of all, the cantilever is excited with the sinusoidal output from a direct digital synthesizer (DDS). This device generates two sinusoidal waveforms 90° out of phase with each other, designated $\cos\omega t$ and $\sin\omega t$. One of these signals, $\cos\omega t$ is used to excite the cantilever with a "shake" piezo. This in turn excites the cantilever to oscillate, the motion of which is detected by the PSD. The analog to digital converter (ADC) converts this into a digital signal again. This digital signal is multiplied by both the $\cos\omega t$ and $\sin\omega t$ components and then low-pass filtered to get the in-phase (i) and quadrature (q) components of the cantilever motion. These components are then used to calculate the amplitude and phase of the cantilever motion as described in the text. The low frequency deflection can also be extracted from the digitized PSD signal by simply low pass filtering. The same functionality can be obtained with analog electronics, though digital electronics have quite a few advantages. In this figure, everything within the dashed box is a digital calculation.

The components in Fig.11-12 that are within the dashed boundary are all digital computations. The digital computer communicates with the analog world through a Digital-to-Analog Convertor (DAC) and an Analog-to-Digital Convertor (ADC). The DAC is used to drive the cantilever "shake" piezo at a particular frequency ω,

$$V_d \cos(\omega t) \tag{11.27}$$

This in turn causes the cantilever to oscillate at this frequency with some amplitude and at some phase φ relative to the drive. This oscillation is measured by the position sensitive detector (PSD) which outputs a response to the ADC,

$$V_{resp} \cos[\omega t + \varphi(\omega)] \tag{11.28}$$

where V_{resp} is the cantilever detector voltage magnitude. Equation 11.28 is similar to 11.20 with the exception that the amplitude (units of length) have been replaced with voltages. The voltage can be calibrated in terms of amplitude, typically using an amplitude-distance curve while oscillating above a surface.

To determine the phase angle φ, first the response voltage eq.(11.28) is multiplied by the shake piezo drive voltage, resulting in

$$V_{resp}\cos(\omega t + \varphi) \cdot V_d \cos(\omega t) = \frac{1}{2}V_{resp}V_d[\cos(2\omega t + \varphi) + \cos\varphi] \qquad (11.29)$$

This expression contains both a dc and a 2ω ac component. If this signal is low-pass filtered, the resulting dc component contains a term proportional to the cosine of the phase,

$$i = \frac{1}{2}V_{resp}V_d \cos\varphi; \qquad (11.30)$$

i is referred to as the "in-phase" component. The "quadrature component" is obtained in a manner similar to the in-phase component, except that the cantilever response is first multiplied by $V_d\sin(\omega t)$ rather than eq.(11.26). Similar algebra yields

$$q = -\frac{1}{2}V_{resp}V_d \sin\varphi \qquad (11.31)$$

for this component. These two quantities, i and q, can be used to extract the cantilever amplitude A and phase φ using the following expressions:

$$A = \sqrt{i^2 + q^2} \qquad (11.32)$$

$$\varphi = \arctan\left(\frac{q}{i}\right) \qquad (11.33)$$

11.2.2.3. Phase locked loop (or FM mode)

This imaging mode uses the phase signal from the cantilever as an error signal to a feedback loop that controls the drive frequency. The image is then a plot of cantilever resonance frequency [42-45]. The resonant frequency depends on the force gradient acting (or the conservative interactions) between the tip and the sample according to the relationship

$$F''(z) = k\left[1 - \left(\frac{\omega_0'(z)}{\omega_0}\right)^2\right]. \qquad (11.34)$$

For small values of the force gradient, this expression simplifies to

$$\delta\omega_0 \approx -\frac{\omega_0 \delta F''}{2k}. \qquad (11.35)$$

This mode is similar to phase imaging, except that the phase signal is used as an error signal and the drive frequency is varied to keep the phase constant, typically at a value of 90° or resonance (See Fig.11-13).

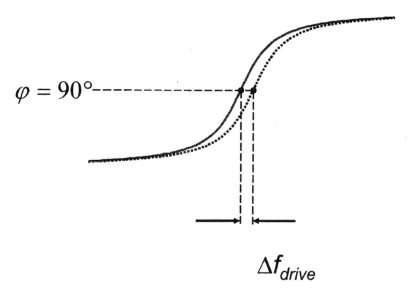

$$\varphi = 90°$$

$$\Delta f_{drive}$$

Figure 11-13: This Figure shows the idea behind the "phase-locked or "FM" detection mode. A feedback loop maintains the phase at a preset value (typically 90 degrees) by varying the drive frequency. In this example, a repulsive force gradient shifts the resonant frequency to a higher value while the system is being held at a constant phase of 90 degrees. This causes the drive frequency to increase.

11.3. CHARACTERIZING THE PROBE

The mechanical and magnetic properties of the cantilevers and tips are important to both understand and optimize for MFM imaging. In the following section we will discuss some of the issues with sensitizing the MFM cantilever to magnetic fields and to calibrating their sensitivity.

11.3.1. Cantilevers and coatings

All efforts to quantify MFM images require a detailed knowledge of both the mechanical and magnetic properties of the MFM cantilever. Determining possible invasive effects of the imaging process on the specimen's magnetic structure, i.e., the effects of the tip generated fields on the specimen require similar information. Initially, AFM and MFM cantilevers were laboriously manufactured by hand. For example, in the AFM described by Binnig, Quate and Gerber[1] the cantilever was cut from a 25 μm thick gold foil and a diamond shard was attached to the end for a tip; this is similar to the styli used by surface profilometers, the immediate predecessors of AFMs. Similarly, the first MFM cantilevers were manufactured from electrochemically etched Nickel wires [2].

Binnig, Quate and Gerber suggested using existing microfabrication techniques to make smaller, higher resonant frequency cantilevers which have the added benefit of being manufactured in large batches [46]. At the present, cantilevers are commercially available in wafers with between 200-400 chips, each having 1-6 cantilevers per chip with resonant frequencies up to or even above 1 MHz [47]. Etched Si cantilevers can have tips with a radius on the order of 2-10 nm and for MFM, a magnetic coating, typically a CoCr alloy, on top of this will increase the radius to something typically on the order of 30-50 nm. As most researchers use commercially available tips we will focus on these in the following discussions.

11.3.2. Cantilever spring constants

Any procedure to quantify the fields or field gradients in a MFM image requires both a determination of the magnetic fields or magnetic moments of the tips, using one of the above techniques, and a value for the cantilever spring constant k. There has been a great deal of work on determination of cantilever spring constants as it is of considerable interest to other applications of force microscopy; for a complete review of the state of the art, we refer the reader to the review by Sader [48]. Here we will only briefly cover three of the most successful methods; (i) the added mass method where known masses are attached to the cantilever and the change in resonant frequency is measured, (ii) the thermal noise method where Brownian motion of the cantilever is used to measure k via the equipartition theorem and (iii) the "Sader" method, where the plan view and the experimentally measured resonant frequency and quality factor. Q, are used to quantify k.

In the added mass calibration technique, the spring constant of the cantilever is calibrated by measuring the shift of the resonant frequency in response to test masses attached to the cantilever [49]. Taking the resonant frequency of the first oscillatory mode in a cantilever to be $\omega_0^2 \sim k/m_c$, where ω_0 is the angular resonant frequency, k is the spring constant at the tip of the cantilever and m_c is the effective mass of the cantilever [1]. If a test mass m_t is attached to the end of the cantilever, the resonant frequency will shift according to

$$\omega_0^2 = \frac{k}{m_t + m_c}.$$

(11.36)

Equation (11.36) can be rearranged to give

$$m_t = \frac{k}{\omega_0^2} - m_c$$

(11.37)

Thus, if several test masses are placed on the cantilever and a plot of m_t vs. $1/\omega_0^2$ is made, one should get a straight line whose slope is the spring constant k and whose y-intercept is the effective mass of the cantilever m_c. Although this technique is somewhat labor intensive it can, in principle, provide an accurate measurement of k.

For the thermal noise calibration method, the spring constant of the cantilever is calibrated by measuring the Brownian fluctuations on the lever [50]. The equi-partition theorem applied to the first flexural mode in the cantilever relates the mean squared motion of the cantilever $\langle \Delta x^2 \rangle$ and the spring constant k to the absolute temperature T and Boltzmann's constant k_B by the relationship

$$\frac{1}{2}k\langle \Delta x^2 \rangle = \frac{1}{2}k_B T.$$

(11.38)

To accurately determine a value for the mean squared amplitude, it is necessary to calibrate the Inverse Optical Lever Sensitivity S (with units of nm/V) of the cantilever which is done with a force curve against the surface (this quantity is the main source of error in this procedure). The expression for the for the spring constant then becomes

$$k = \frac{k_B T}{\langle \Delta V^2 \rangle S^2}.$$

(11.39)

In this expression, $\langle \Delta V^2 \rangle$ is the mean squared voltage fluctuations due to movement in the first flexural mode of the cantilever [51].

Sader and coworkers have worked extensively on quantifying the spring constant of cantilevers and in their most recent efforts have developed a simple procedure to determine k. In brief, their technique models the response of the cantilever with inclusion of the air hydrodynamic damping. The significant advantage of their approach is only the required information for a determination of spring constant are the cantilever resonant frequency ω_f the Q_f of the cantilever in in the fluid (often air) and the plan dimensions of the cantilever: its width b and length L. In their approach the spring constant is given by the real and imaginary components of the hydrodynamic function $k = 0.1906 \, \rho_f b^2 L \, Q_f \Gamma_i(\omega_f) \, \omega_f^2$ [52], where ρ_f is the density of the fluid (taken as $\rho_f = 51.18$ kg/m^3 for air) and $\Gamma_i(\omega_f)$ is the imaginary component of the hydrodynamic function (see Ref. [52] for a discussion of and analytical expression for the hydrodynamic function). Note ω_f and Q_f may be measured with software available on commercial microscopes and that both b and L are easily measured in a plan view of a cantilever with an optical microscope. Thus all required information is easily measured (neither the cantilever thickness nor the mass density are required; both difficult parameters to determine). The values determined by The Sader method procedure compare quite favorably with both the thermal noise and the added mass techniques. An on-line calibration of torsional and normal spring constants using the Sader method is available at http://www.ampc.ms.unimelb.edu.au/afm/calibration.html.

11.4. SEPARATING MAGNETIC FORCES FROM OTHERS

As we said, a MFM image is made by plotting the magnetic effect on one or more of the cantilever parameters as a function of position over the sample. The key words in the above sentence are "magnetic effect." Although simple in principle, we must separate magnetic interactions from any other forces which may be present. These include van der Waals, short-range repulsive, and capillary wetting interactions between the tip and sample. These forces typically fall off rapidly above the surface, becoming negligible at a distance of angstroms to a few nanometers above the surface. On the other hand, something that generally differentiates magnetic forces from others is their range; magnetic forces, though often weaker than the others mentioned when close to the surface, fall off much more slowly and are therefore still present hundreds of nanometers above the surface. Early schemes at measuring magnetic forces made use of a variety of methods to control the tip-sample separation and to keep the tip in a range where magnetic forces dominated. Unfortunately, these techniques were difficult to implement and were often less than stable, especially in cases where the sample was rough [3,53].

11.4.1. Electric field biasing

Magnetic interactions between the tip and sample can be repulsive and attractive. This complicates conventional feedback loops. In the case of a conductive tip and sample, the overall interaction can be made attractive by applying a potential between the tip and sample [2]. Then, either dc deflection, ac amplitude or phase feedback, could be used with a fixed force gradient for the feed back zero point. The variations in this zero would then be a measure of additional force gradients, either positive or negative, attributed to the similarly long ranged magnetic interactions.

11.4.2. Using short range interactions as a reference

The topography of a surface is the natural reference point for interpreting magnetic domain structure. This topography can also be used as a reference point for positioning the magnetic force probe at a controlled distance above it. One of the first attempts at this exploited the tunneling current to stabilize the cantilever tip above the surface while simultaneously measuring magnetic forces by monitoring the cantilever deflection [22].

Electric and magnetic fields, although typically quite weak, are usually longer ranged than most other tip-sample interaction forces. This different distance dependence can be exploited to allow the near simultaneous measurement of both the sample topography and the long ranged interactions. The first reference to appear in the literature that used this difference in the length scales was the pioneering paper by Martin and Wickramasinghe on ac AFM imaging [26]. In this work, they demonstrated a technique to obtain both the topographic profile of the surface and longer ranged which they hypothesized were van der Waals in nature. Their goal was to characterize the material properties of different regions of a sample. The particular sample they investigated was a silicon wafer partially covered with photoresist. Although there was no magnetic contrast in this sample, it did help point the way for them, as well as others, to separate strong short ranged topographic and weaker but longer ranged magnetic and electrostatic forces. "Section V: Materials Sensing and Imaging" of their paper describes the invention. They slowly (~1 kHz) oscillated the cantilever towards and away from the surface. When the amplitude decreased, the feedback loop would pull the sample away from the surface. A peak detector then measured this turn-around point which provided information about the sample topography. At the same time, another feedback loop measured the resonant frequency of the cantilever away from the surface, providing a measure of the long range forces (in this case, van der Waals). In what is generally considered the first paper describing MFM [13], the same group used the techniques described in their original paper [26] to image a recording head.

Hosaka et al. [27] implemented a similar technique and what was probably the most immediate predecessor to what would become the most popular MFM imaging mode, commercialized by Digital Instruments under the trademarked name of "LiftMode" [25]. The idea was to separate the short range topographical from the long range magnetic interactions by making two separate measurements or passes over the same region of a specimen. The first pass was a measurement of the surface position (essentially of the short ranged, topographical forces). The data from this first pass provided a template to follow on a second pass with the tip withdrawn a predefined distance, i.e., the height information obtained from the first pass provides a path for a second pass but with a constant additional offset from the surface. This second pass is too far from the surface for sensitivity to the short ranged interactions but is close enough to measure the long ranged magnetic interactions. The idea is illustrated in Fig.11-14. The cantilever first interacts with the surface with a relatively large interaction (labeled 1 in the Figure from Hosaka et al. [27]) thus sensing the large short ranged interactions which unambiguously measures the sample topography at a given x-y location. The cantilever was then lifted a pre-programmed height above this level (generally 5-500 nm), with the topographic feedback disabled. The amplitude and phase of the cantilever at this position was then measured. This put the cantilever at a height where magnetic forces are dominant and can be measured without being perturbed by topographic interactions.

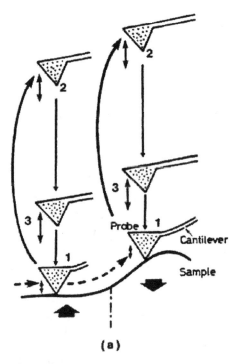

(a)

Figure 11-14: This figure from Hosaka [27] illustrates the sequential probe control they used to first locate the surface (1) and then to raise the oscillating cantilever a predetermined height above the surface to measure the long-ranged magnetic force gradients (2) and (3).

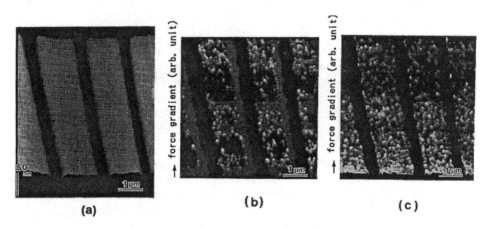

(a)   **(b)**   **(c)**

Figure 11-15: This Figure shows the resulting topographic image (a) and force gradient images at a separation of 50 nm (b) and 200 nm (c).

It also obviously provided an image of the sample topography along with the magnetic image. Since the topography interacts strongly with the magnetization through the divergence term, as well as internal grain structure, this can be quite useful in interpretation of the magnetic data. The results presented by Hosaka et al. are shown in Fig.11-15. Another example of this is shown in Figure 11-16. In this Figure, the image on the left shows the topography of a textured hard disk and the image on the right shows the magnetic image made at a z-offset of 50 nm.

First Pass   **Second Pass**

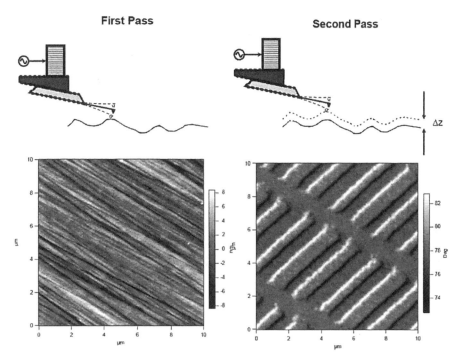

Figure 11-16: In this Two-pass method, the topography is first measured (left). The topography is then used as a reference for a second scan where the probe is positioned a controlled distance above the surface (in this case Δz=50 nm). This allows imaging of the relatively weak but longer ranged magnetic interactions. In this case, the cantilever phase is plotted on the right (see eq.(11.26)). Note that the transitions visible in the MFM image are not particularly smooth. This "zig-zagging" can lead to noise in the playback process.

11.4.3. Fourier methods in MFM

Schönenberger and Alvarado[54] introduced an analytical theory for interpreting the response of an MFM based around a Fourier analysis of the response. Their analysis assumed that the MFM tip and sample magnetization was completely stable. By using a two dimensional Fourier Transform of the sample magnetization they were able to gain a great deal of insight into MFM response that was later verified numerous times. These include:

(i)   The observation that the ultimate resolution of the MFM was limited by both the tip radius and tip sample separation. They went on to predict the ultimate resolution being somewhere near 10 nm, a value that is currently accepted as a practical limit in most cases [55], even with nanofabricated tips [56] (see Sec. 11.6).

(ii) For large scale magnetic structures, the tip behaves like a dipole interacting with the sample and hence the MFM force is proportional to the field derivative. This is the assumption implicit in our initial analysis in Sec. 11.2.

(iii) for intermediate magnetic structures, only a single pole of the MFM tip interacted significantly with the sample, leading to a force that depended on the field directly rather than its derivative. In one example, this interpretation was used to quantitatively extract the magnetization of a thin film sample [57]. One implication of this is that a small magnetic particle, used as MFM tips on the end of a non-magnetic cantilever (see Ref. [58]) will act as a bandpass filter, producing images with reduced information at both high and low spatial wavelengths.

(iv) The definition of a "Magnetic Force Derivative Response" function. This was a transfer function in wavelength space of the MFM to a sample. This transfer function and variations of it have proven useful for example, in image "reconstruction" [59,60], interpreting MFM sample structures [61-63], calibrating MFM tips [56], and reciprocity [16,17].

11.5. IMAGING STABILITY

Hartmann proposed some basic conditions for stable MFM imaging [64] based on the ratio of the anisotropy field of the tip (sample) $H_k^{t(s)}$ to the magnetization of the sample (tip) $M_s^{s(t)}$:

$$H_k^t / M_s^s \geq 1 \tag{11.40}$$

and

$$H_k^s / M_s^t \geq 1. \tag{11.41}$$

In practice, these conditions are usually only fulfilled with magnetically hard tips and samples.

A beautiful example of the tip magnetic field perturbing the sample is shown in Fig.11-17. The first panel in this Figure shows an undisturbed Faraday effect image of stripe domains in a garnet film with perpendicular anisotropy. MFM images of the same sample, while showing stripe domains, show an asymmetry in the domain size. In particular, the domains that are magnetized anti-parallel (repulsively, bright in the Figure) magnetized (bright) domains appear smaller in area than the domains parallel to the tip magnetization. The discrepancy between the Faraday and MFM images is elegantly explained by the third panel which is a Faraday image of the sample with an MFM tip placed close to the surface. The field from the tip has locally distorted the domain structure of the Garnet sample, temporarily expanding the attractive domain and shrinking the repulsive [65].

The magnetization of the tip can also be a source of confusion for interpreting MFM images. One example comes from dissipation images of recording media with relatively large stray fields where the tip magnetization switched in response to the sample fields [67]. To help quantify this, Babcock demonstrated a method for measuring the hysteretic properties of MFM tips by imaging a current carrying conductor in the presence of an applied field [66].

11.5.1. Imaging in an applied field

A powerful tool for understanding the dynamics of a magnetic sample is an applied field. A number of researchers have applied the field to a sample external to the MFM and used the MFM to image the resulting remanant state. This includes a study by Lederman [68], who examined the thermal stability of arrays of particles that were subjected to both externally applied fields andtemperature variations. Others have applied the field to the sample in-situ [69-78, 177]. Fig.11-18 shows a series of permalloy bars with different aspect ratios and sizes imaged in the presence of an applied field. Part of a study that combined magnetooptical Kerr magnetization measurements and MFM of (110) Fe particles with differing shape anisotropies is shown in Fig.11-19 [79]. Finally, Liu[80] has studied magnetic dissipation in the presence of an applied field.

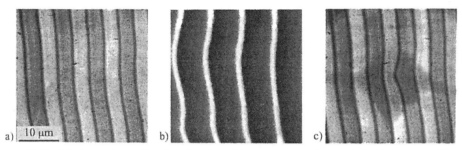

Figure 11-17: (a) shows the undisturbed Faraday effect image of stripe domains in a garnet film with perpendicular anisotropy. (b) the same sample observed with an MFM. The repulsively magnetized (bright) domains appear smaller in area than the attractive. The discrepancy between (a) and (b) is explained by (c), a Faraday image where an MFM cantilever (visible as a dark diamond shaped shadow) has been placed near the sample surface. The field from the tip has locally distorted the domain structure of the Garnet sample, temporarily expanding the attractive domain and shrinking the repulsive [65].

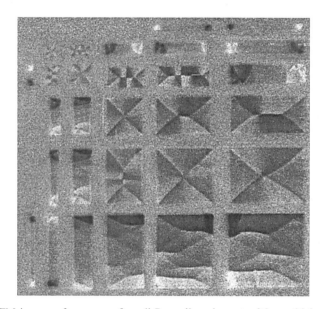

Figure 11-18: MFM images of an array of small Permalloy elements, 26 nm thick on Si, showing the transition from solenoidal to non-solenoidal configuration with increasing aspect ratio. A 40 Oe field was applied parallel to the vertical axis during imaging. The smallest bar dimensions are 0.25 µm and the largest are 4 µm [175].

One way to simplify interpretation in an applied field is to use a tip with a coercivity that is very different from that of the sample, either much lower or much higher. Then, when the contrast is changing, one can be relatively confident of the origin of the change. For this reason, superparamagnetic tips [81,82] or very high coercivity tips [83] can be useful for applied field imaging.

One example of an application of magnetically soft tips is the "Fluxgate MFM" [84]. In this imaging mode, a relatively soft MFM tip [81,82] is used to image a magnetically hard sample. The tip and the sample are subjected to an externally applied magnetic field that includes both a dc component and a small AC dither. The DC component is varied to cancel the DC component from the sample surface, in a manner quite similar to a conventional fluxgate magnetometer.

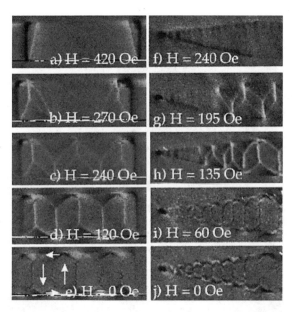

Figure 11-19: Field dependent MFM images on (a–e) rectangular and (f-j) needle-shaped (110) Fe particles imaged in a longitudinal applied field. The white arrows show the direction of the domain magnetizations [79].

11.5.2. Quantifying tip magnetic fields

The magnetic field from the MFM tip is important both for determining the mechanical response of the cantilever in the reciprocal interpretation of the tip-sample interactions (see Equations Force 6-8) as the tip is scanned over a surface and for understanding how the tip field may alter the magnetization of the specimen.

MFM tip field perturbations of the sample magnetization have been observed in many situations [46,85-88]. Because these perturbations originate from the tip magnetic field, they necessarily have a tip-sample distance dependence. This has historically complicated the interpretation of MFM images of soft samples. One notable experiment was reported by Garcia et al. [89], where they included tip-induced perturbations of a soft permalloy sample in their image simulations. They found that including the perturbations, which in turn required quantitative values for the tip field, allowed a quantitative simulation of the forces above their samples and a convincing reconstruction of the MFM images.

There have been a number of experimental approaches to characterizing MFM tips. Arguably the most accurate measurements have used electron microscopy techniques. For example, electron holography can directly image and quantify the fields from tips as a function of distance from the tip apex [90-92]. An example from Streblechenko is shown in Figs.11-20 and 11-21. They used electron holography to quantify the magnetic field from an MFM tip. Figure 11-21 shows the axial (z) field of the tip as a function of z [178].

Other electron microscopy studies have included Foucault-mode Lorentz microscopy [2] and differential phase contrast Lorentz microscopy [93]. For the average practicioner of MFM, however, these techniques are complicated, time consuming and require a great deal of expertise in electron microscopy.

A somewhat easier approach is the use of a miniature magnetic field sensor to directly image the fields. Sensors have included a 2D electron gas magnetic Hall effect sensor [94] and hard drive read head sensor [95]. Unfortunately, to date, these sensors are much larger than the tip being studied, resulting in the need for complex analysis to extract information about the tip.

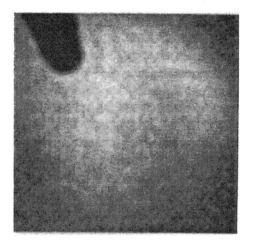 

Figure 11-20: The electron amplitude (left) and phase (right) near an MFM tip visible as a dark shadow on the upper left corner of the left image.

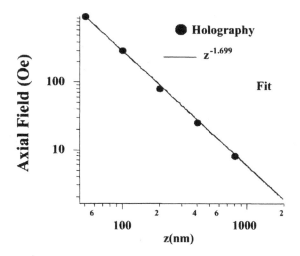

Figure 11-21: This shows the axial field along the axis from the MFM tip in Fig.11-17. The red markers are from the holography measurements and the blue curve is a power law fit.

The most popular approach has been the use of a reference sample. A number of researchers have imaged the bit transitions in hard disks and compared the images to models of bit transitions [2,96,97]; the use of modeled fields from a bit transition is a limitation. Other reference samples have included lithographically defined current strips to create known fields [98], a variant of this which involves the balancing of the electrostatic and magnetostatic forces to estimate a MFM tip moment [99], the magnetosomes from magneto-tactic bacteria [100], and recently the use of paramagnetic films [101].

Of the above, the lithographically defined current strip, is arguably one of the best and easiest to implement. Lithography can produce a current path from which magnetic fields and field gradients can be calculated accurately. Interestingly enough, the current strip was also one of the first techniques [98]. An extension of this technique has been to use the imaging of the fields in a current loop with the addition of a uniform applied magnetic field to measure the magnetic switching properties or coercive fields of the tips [66]. Given its sensitivity to the magnetic fields generated by currents, the MFM has been suggested as a diagnostic tool for integrated circuits [102-107]. Note that because the current is adjustable, a current strip is also useful for a determination of the minimum detectable MFM signal [108].

For the most complete modeling of MFM tips, one must also include the position of the monopole or dipole moments with respect to the tip apex. This was the approach taken in two of the above efforts with a reference sample [101,109]. It is important to note, although these two works used different reference samples, a current loop and a paramagnetic film, they had similar conclusions regarding the magnitude and position of the moments in commercial MFM tips.

11.6. MFM RESOLUTION

It is generally accepted that the resolution one can expect is in the 30-100 nm range [110]. The main limitations to improving the resolution are probe geometry and working distance [54]. This is very similar to the challenges faced by reducing the size of inductive magnetic recording systems [111]. If the magnetization of both the sample and tip are stable (see section) For example, a resolution of 10 nm has been claimed while imaging a strong, hard ferromagnetic (rare earth) sample [55]. Domain boundaries and other micromagnetic features in soft materials on the other hand often appear to be a sizeable fraction of a micron.

11.6.1. Thin film coatings

The most ubiquitous technique for higher resolution cantilevers has combined silicon microfabricated cantilevers and tips with thin magnetic films [46,93,97]. This approach allows the considerable control of the tip geometry in AFM cantilevers to be combined with the good control of magnetic and mechanical properties of thin magnetic films. It allows a large variety of magnetic and mechanical properties to be selected for both magnetic and topographic imaging purposes. In addition to providing better repeatability, thin film coatings also minimize the volume of magnetic material which both improves the spatial resolution and reduces irreversible magnetic modification of the sample magnetization.

11.6.2. Nanofabricated tips: electron beam deposited, lithographically defined, focused ion-beam milled and carbon nanotubes

Electron beam deposition is a technique that has been used to deposit carbonaceous materials in a spatially controlled manner onto a variety of substrates. These tips can then be coated with a magnetic film for use in an MFM [112,113]. Another direction that has been explored is the use of carbon nanotubes as sharp probes [114,115]. Again, this requires the deposition of a thin magnetic film to sensitize the probe to magnetic forces.

In another variation, Bauer et al. [116] began with a sputter coated MFM cantilever, deposited a cap of material at the tip using electron beam deposition and then removed the remainder of the unprotected magnetic material. This left a well defined "particle" of magnetic material behind. This approach had the advantage of beginning with a well-controlled sputtered thin film with relatively well understood and controllable magnetic properties. The resulting particle was also protected from the usual wear and tear of the cantilever tip during AFM imaging. However, it is also a relatively time consuming process and, as with most other EBD techniques, more difficult to batch process.

Focused ion beam (FIB) technology provides a means of directly shaping nanometer scale objects. FIB milling has been used previously to improve MFM imaging. Kikukawa et al. [45] were probably the first to apply this technique to MFM tips. They first defined the tip with the FIB and then deposited a magnetic thin film. Khizroev et al. [117] modified tips coated with a 10 nm thick film of high magnetic anisotropy CoCrPt resulting in extremely low aspect ratio tips. They used these tips were developed for the quantification of magnetic fields from write heads. Folks et al. [118] obtained images of data tracks possessing 50 nm bit lengths with tips modified by FIB in such a way as to be sensitive to in-plane components of the sample stray magnetic field. Phillips et al. [119] modified Co thin films deposited on a Sil cantilever using a FIB mill. The majority of the Co thin film was etched away, leaving behind thin bars with extremely high aspect ratios.

Finally, it is also possible to directly deposit magnetic material with an electron beam, thus growing a ferromagnetic spike tip [120]. This technique has the disadvantage of poor materials and magnetic properties control. Nevertheless, it demonstrates the direct and localized growth of a controlled geometry ferromagnetic tip.

It is important to discuss three topics intimately related to MFM probe characterization; these are MFM resolution, effects of the tip fields on the specimen magnetic state, and, in light of these, modifications to tips to control or alter these. Starting with resolution, here we mean the ability to separate the fields from two objects and not the sensitivity or minimum detectable signal. It is becoming standard practice to define the resolution of an MFM tip by the minimum detectable wavelength written into a standard sample [56]. In the absence of a standard sample with a wavelength pattern, the use of a "smallest feature observable" has been another technique to define the minimum resolution [55]. This area of MFM has become somewhat of a research area itself and the most extensive reviews of this subject have been by the University of Twente group [110].

Because the tip of the MFM is coated with a ferromagnetic material, interactions between the tip and a ferromagnetic sample can cause perturbations in the magnetic structure of both. Whether these perturbations, always present to some extent, are significant depends on the particular situation. As well as an unwanted perturbation, the tip field has been used to to write localized domains [121] and to measure the localized coercivity of media [122]. Some of the first observations of samples strongly perturbed by the tip field include Göddenheinrich [123], Hartmann [124], Mamin [85], and Abraham [86] who used the localized perturbation to interpret MFM contrast.

The changes induced by the MFM tip can be reversible in which case one can investigate the localized susceptibility by imaging with two opposite magnetization directions of the MFM tip [87,88]. This idea was extended to image the susceptibility and magnetostatic charge by subtracting and adding images taken with the same MFM tip again with the magnetization reversed [87,88,125,126]. In principle, reversing the sign of the interaction to isolate tip-induced effects is very similar to the technique that Martin and Wickramasinghe [13] applied to extract the "real" magnetic field pattern of a recording head; instead of reversing the tip magnetization, they simply reversed the dc current driving the recording.

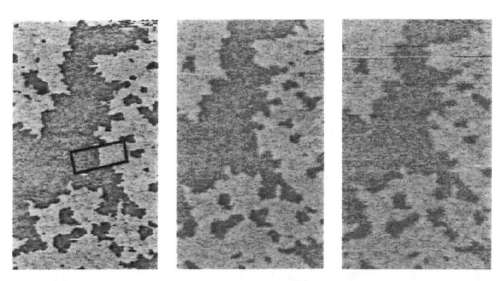

Figure 11-22: A series of three MFM images made over a 13.9 Å Co thin film with lift heights of 5, 60 and 100 nm. The images are 9×15 μm². All three images show clear examples of domain contrast. The existence of contrast in the center of domains, far away from the domain walls implies that the contrast originates from nonlinear (perturbative) tip-sample interactions [127].

An important investigation of the role of tip perturbations in MFM contrast is that of Belliard (see Fig.11-22) [127], who found the "domain contrast" seen on a great number of MFM images requires the inclusion of nonlinear tip-sample interactions [86]. The conclusion to be drawn from all the studies and the last in particular is, at a minimum, one must record images with opposing tip magnetizations if quantitative information on the micromagnetic state is required.

One means of reducing the tip-sample perturbations is to use a MFM tip with a smaller moment which takes us into the subject of modifications to tips. Tip modifications can both reduce the invasive nature of MFM and also improve spatial resolution; there are also tips special to specific purposes. Non-commercial coating of AFM tips tips are are the easiest to implement [46,93,97]. This approach allows the control of the tip geometry in AFM cantilevers to be combined with the good control of the magnetic and mechanical properties of thin magnetic films. It allows a large variety of magnetic and mechanical properties to be selected for both magnetic and topographic imaging purposes. In addition to providing better repeatability, thin film coatings also minimize the volume of magnetic material which both improves the spatial resolution and reduces irreversible magnetic modification of the sample magnetization. The more sophisticated techniques mentioned above (electron beam deposition, lithography, focused ion-beam milling and carbon nanotubes) also have the benefit of reducing the tip-sample perturbations, typically at the cost of poorer signal to noise.

11.7. MODELING MFM RESPONSE

Practical interpretation of an oscillating cantilever in terms of the tip-sample interactions is a difficult problem. For typical experimental oscillation amplitudes, the force and force gradient can vary significantly over the range of the oscillation. If this is the case, the harmonic approximation discussed above is no longer valid [128]. In particular, this is almost always the case for the commonly used topographic imaging mode referred to as amplitude modulation or "tapping mode" microscopy. For MFM, the force and force gradients are significantly more long ranged than most topographic forces and the harmonic approximation may be more valid [129-131].

Most theoretical discussions of MFM response start with a model based on a magnetic monopole or dipole tip that has a linear interaction with the sample, i.e., the tip fields do not perturb the magnetic structure of the sample or vice-versa. These assumptions are useful for simple image interpretation but in practice they are insufficient for a detailed or quantitative understanding of MFM response [2]. In reality, the tip structure is complicated, both geometrically and magnetically. Given specific knowledge of the tip shape, the MFM response can be obtained by performing an integration over the tip volume [132]. More elegantly, in the absence of perturbations, the Green's function of the tip can be calculated by imaging a point-like sample [59,60,63]. Other computational techniques that have proved useful are based on reciprocity [16,17].

The reality behind the linear response assumption is that many, if not all MFM images involve some nonlinear and dissipative interactions; the magnetic state of the tip or sample may change during the imaging process. This can come about because of either mutual magnetic interactions, an applied field, or, possibly mechanical forces between the tip and sample. Although generally it is unwanted, in some cases the tip perturbation effects on a specimen can be used as a probe of the micromagnetic properties of the sample. The perturbations lead to magnetic dissipation, domain contrast and magnetization reversals, all active topics of research. Other active areas of work include imaging in applied fields, at variable temperature, combining the MFM with other microscopies and improving the spatial and time resolution.

11.7.1. Sample magnetization modeling

Because the MFM is sensitive to the sample fields, it is only indirectly dependent on the sample magnetization. A great deal of effort has gone into interpreting MFM results in terms of the sample magnetization [2,100,133-136]. Although progress has been made, one should not underestimate the difficulties in quantifying the micromagnetic structure of a sample using the MFM [137]. To date, all "quantitative" measurements of a sample's micromagnetic structure have relied on reasonable but ultimately un-testable (at least using only the conventional MFM) assumptions about the sample micromagnetics.

11.8. DISSIPATIVE INTERACTIONS

Monitoring the energy or power dissipated by the cantilever allows one to quantify the irreversible magnetic perturbations. One caution with the technique is that the magnetic structure of the MFM tip can be perturbed as easily as that of the sample, leading to problems with the interpretation. Figure 11-23 show a conventional MFM image and a dissipation image, respectively, of a magnetic hard disk with written bits [138]. There is a strong correlation between the domain walls imaged with the MFM and the dissipation signal.

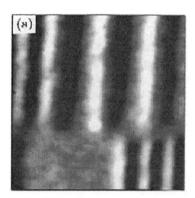
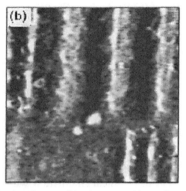

Figure 11-23: A standard MFM image (a) and a dissipation image (b) taken over bits written in a CoPtCr longitudinal magnetic recording medium. The image is 10×10 μm^2. There are correlations between the standard MFM image and the dissipation features. There are also a number of high dissipation features visible in image (b) that have no obvious correlation in the standard MFM image [138].

Cleveland et al. related the power dissipated by the tip-sample interactions in terms of the sin of the phase angle between the cantilever excitation and the response. A modification of the expression derived by Cleveland et al. for the out of phase component of the cantilever response q yields

$$q = A_{sp} \sin\varphi = \frac{\omega A_{sp}^2}{\omega_0 A_0} + \frac{2QP_{ts}}{k_c A_0}.$$ (11.42)

In this expression, A_{sp} is the "set-point" or operating amplitude of the cantilever, φ is the phase referenced to the excitation of the cantilever base (usually accomplished with a piezo crystal), Q is the native quality factor of the cantilever, typically measured at a reference point, A_0 is the peak amplitude of the cantilever measured at the same point, ω is the angular drive frequency, ω_0 is the resonant frequency of the cantilever and k_c is the cantilever spring constant. The dissipated power in this expression is simply $P_{ts} = \langle F_{ts}\dot{z}\rangle$, where F_{ts} is the instantaneous force acting between the tip and sample, $\dot{z}$ is the time derivative of the cantilever deflection and $\langle\ \rangle$ denotes a time averaging of the enclosed quantities. Rewriting eq.(11.42) in terms of these quantities gives

$$q = \frac{\omega A_{sp}^2}{\omega_0 A_0} + \frac{2Q\langle F_{ts}\dot{z}\rangle}{k_c A_0}.$$ (11.43)

The first term in this expression depends on the native cantilever dynamics and a second that depends on the dissipative tip-sample interactions. During imaging, the feedback loop keeps the first term in this expression constant (or nearly so). Then any changes in the quadrature component are directly related to dissipative tip-sample interactions, $\langle F_{ts}\dot{z}\rangle$.

Using the virial theorem, San Paulo and Garcia [139] derived a similar result for the in-phase component. This was slightly more complicated, given by a slight modification of eq.(11.4) in Ref. [139],

$$i = A_{sp} \cos\varphi = \frac{2Q}{k_c A_0}\left[\frac{\langle F_{ts}\rangle^2}{k_c} - \langle F_{ts}z\rangle + \frac{1}{2}k_c A_{sp}\left(1-\frac{\omega^2}{\omega_0^2}\right)\right].$$ (11.44)

Following this, if we assume that the cantilever deflection is always significantly smaller than the amplitude, eq.(11.3) can be written as

$$i \approx \frac{QA_{sp}^2}{A_0}\left(1 - \frac{\omega^2}{\omega_0^2}\right) - \frac{2Q\langle F_{ts}z\rangle}{kA_0} \, . \tag{11.45}$$

This equation has some obvious similarities to eq.(11.43). The first term remains constant for AM-AFM (where the cantilever is driven at a fixed frequency and the feedback loop keeps the amplitude constant). The second term in this expression contains a time averaged term that depends on the elastic forces acting between the tip and the sample.

Equations (11.43) and (11.45) have a nice symmetry. Both have first terms that are constant during AM-AFM. The second term in eq.(11.43) contains the time averaged dissipative forces with a pre-factor. The second term in eq.(11.45) contains a term that depends on the conservative time averaged tip-sample forces with the same pre-factor. Although there is still a great deal of work to be done to extract quantitative materials parameters (indeed, it may not be possible), we can at least separate the conservative and dissipative interactions using these relationships.

11.9. HIGH FREQUENCY MFM (HF-MFM)

The resonant frequencies of most MFM cantilevers are well below 1 MHz. However, there are many magnetic phenomena that have much faster time scales. By using the relatively rapid reaction time of the magnetic material in or on the MFM tip, it is possible to get a glimpse at higher speed magnetic behavior.

11.9.1. DC HF-MFM

A pioneering study on magnetic recording heads [140] revealed that the MFM could be sensitive to high frequency fields. In this early work, the recording head was excited with a high frequency current. The resulting images contained information about both the dc and high frequency magnetic fields response of the recording head.

11.9.2. AC HF-MFM

The MFM can also be used to observe the field from the write head, while operating at radio frequencies up to 50 MHz using an unmodulated high frequency (HF) excitation of the recording head [140], and recently, past 1 GHz using a modulated HF excitation field [141]. The idea is based around using the non-linear response of the magnetic MFM tip to demodulate the high frequency signal. If a point dipole MFM tip has a frequency dependent susceptibility χ in addition to a fixed moment m_0, the total moment is given by $m=m_0+\chi H(f)$. The recording head is driven with a time dependent current, resulting in a time dependent magnetic field $H=H_0+H_1 f(t)$. H_0 is the low frequency (dc) component of the head field and H_1 is the magnitude of the high frequency component. If the waveform is 100% modulated, the time dependent portion takes the form f(t)=(1+cosκt)cosΩt. In this expression, Ω is the high carrier frequency and κ is the modulation frequency. The z-component of the force from Equation Force01 in this case is given by

$$F_z = m_z \frac{\partial H_z}{\partial z} = \chi H \frac{\partial H_z}{\partial z} \, . \tag{11.46}$$

Assuming that the magnetic field derivative $\partial H_z/\partial z$ has the same time-dependence as the magnetic field and some algebra gives the force between the tip and the recording head:

$$F = F_{dc} + \chi(\Omega)H_1 \frac{\partial H_1}{\partial z}\cos \kappa t + \frac{1}{4}\chi(\Omega)H_1 \frac{\partial H_1}{\partial z}\cos 2\kappa t . \qquad (11.47)$$

This expression has a component at the modulation frequency and at twice this value. This process is illustrated in Fig.11-24. Both of these terms are dependent on the high frequency susceptibility of the tip and the high frequency field being applied by the recording head. If the modulation frequency of the recording head is adjusted to match the resonant frequency of the MFM cantilever, the response is given by the first term in eq.(11.47). The response over a recording head is shown in Fig.11-24. In this Figure, the cantilever amplitude is plotted as a function of the modulation frequency κ.

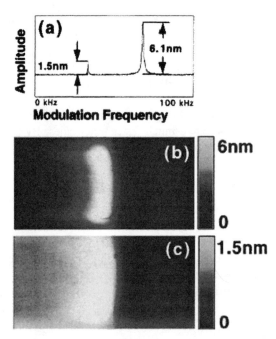

Figure 11-24: A recording head imaged while being operated at 100 MHz. The drive current was 10 mA rms. (a) shows the cantilever amplitude as a function of the modulation frequency. The relative magnitudes of the two peaks agrees well with eq.(11.47). (b) Image made with the modulation frequency set to the cantilever resonant frequency (roughly 75 kHz) and the carrier frequency at 100 MHz. (c) Image made with the modulation frequency set to half the cantilever resonant frequency (roughly 75/2 kHz) and the carrier frequency at 100 MHz. The response with this modulation is much less localized.

11.10. COMBINING MFM WITH OTHER TECHNIQUES

Many microscopies profit from combining techniques and MFM is no exception. In the next two sections, we'll look at two areas where there has been work combining MFM first with optical microscopy and second with low temperature measurements.

11.10.1. Combining MFM with Kerr microscopy

It is often the situation where one would like to explore a specimen for a specific feature such as a domain wall. This process can be rather time consuming, however, as MFM scans are limited to maximum areas on the order of 100 μm on a side and each scan taking several minutes. One can immediately see the difficulties looking for a specific feature in a sample on the order of 1 cm on a side.

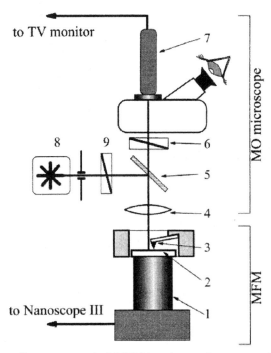

Figure 11-25: Design of a magnetooptical MFM based around a commercial AFM/MFM: (1) MFM scanner, (2) sample, (3) MFM tip, (4) objective, (5) semitransparent mirror, (6) first polarizer, (7) CCD camera, (8) lamp, (9) second polarizer.

One technique developed to predetermine a region of interest is to combine the MFM with an optical Kerr microscope [142-144]. These two magnetic imaging techniques are quite complimentary to each other as Kerr microscopy is fast, has a large field of view, and it is directly sensitive to the sample magnetization whereas the MFM has both high spatial resolution and sensitivity comparatively speaking. In materials with magneto-optical contrast, combining the techniques can be powerful. One example is the work of Pokhil [142] where a reflectance Kerr microscope was combined with an MFM to study Co/Pd multilayers. The experimental setup is shown in Fig.11-25. Briefly, a Kerr microscope was adapted to a commercial top view MFM system. The Kerr microscope was used to first locate regions of interest in large domains and to then position the MFM cantilever at the boundaries. The MFM was then used to observe the real time nucleation of domains and displacement of domain walls in an external magnetic field near the coercive point (see Fig.11-26). Other applications of combined magneto-optical microscopes and MFMs include studies of high anisotropic materials and UHV investigations [143].

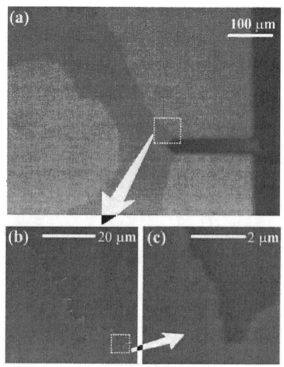

Figure 11-26: Images of the domains in 8 nm thick Co/Pd multilayer film obtained with the magneto-optic/ magnetic force microscope: (a) MOKE image of the domains showing location of the MFM cantilever; (b) MFM image of the area outlined in (a); (c) high resolution MFM image of the domain boundary in the area outlined in (b). Note that the gray scale contrast is reversed in the MO and MFM images.

11.10.2. Vacuum and low temperature MFM

Although it is experimentally challenging, there is a large body of work involved with MFM at low temperatures. We will not go into great depth here, but note that work includes a number of studies of vortices in both high and low temperature superconductors [166-171], domain walls in magnetite as the material went through its Verwey transition [172], and magnetoresistive films of interest to magnetic recording [173]. Recently, Volodin et al., have demonstrated enhanced sensitivity operation of a low temperature MFM using higher order oscillation modes of a piezo-resistive cantilever [168,169,174].

11.11. OTHER APPLICATIONS OF MFM

Finally, we'll look at three examples where MFM has been applied to the study of different sorts of materials; magnetic recording media, biological and geological magnetite and magnetostrictive materials.

11.11.1. Magnetic recording studies

Magnetic force microscopy has been intimately connected with magnetic recording applications since the beginning of the field [2,145-147]. The majority of MFMs in the world today are employed by the data storage industry. One of the reasons is that as data densities meet and exceed the Gb/in^2 range, the sub-100 nm resolution of the MFM becomes a requirement for useful microscopy. Rapid imaging in ambient conditions on samples that have protective overcoats also means the MFM can be used for failure analysis. Applications that have appeared in the literature include characterizing media noise with the MFM as well as write and read media and head testing [148-152].

As discussed above, the MFM tip can act as a localized magnetic field "applicator". In addition to probing micromagnetic structures, the field has also been used as a local magnetic flux source for testing the sensitivity of a read sensor in a head. The basic idea is as follows. The MFM is operated in the usual manner, imaging both topography ad magnetic structures. At the same time, because the tip is oscillating, it is applying a varying magnetic field to the sample. If the sample is a magnetic sensor, standard techniques can be used to measure the response of the sensor, both in response to the dc field or at the oscillation frequency of the MFM cantilever [153-156].

11.11.2. Magnetite

Magnetite, commonly known as lodestone, is a ferrimagnetic oxide (Fe_3O_4) that occurs naturally. The micromagnetic behavior of magnetite is of great interest to geophysicists because it is thought to carry a record of the geomagnetic field in the past. For example, the remanent magnetization of geological magnetite samples have been used to establish ancient reversals in the earth's magnetic field. This geomagnetic record is thought to be carried mostly by particles that are smaller than 50um, with just a few to a single magnetic domain [157]. MFMs have been used to study the micromagnetic structure of domain walls in magnetite, including particles [158,159] and larger single crystal samples [87,88,96,160].

A great deal of the particulate magnetite in the earth's crust that carries a good record of the paleomagnetic field was likely produced by biological organisms. Although there may be other purposes, magnetite produced by these organisms is often linked to navigation; it is used as a sensor for detecting the earth's magnetic field. Examples of this include magnetotactic bacteria [161]. Magnetotactic bacteria mineralize intracellular magnetosomes, which are membrane-enclosed, single magnetic domain particles of magnetite, Fe_3O_4. The particles are characterized by a narrow size distribution and species-specific crystalline habit. Magnetosomes are arranged in one or more linear chains along the symmetry axis of the cell, which constitutes a permanent magnetic dipole in the cell. The torque exerted by the ambient magnetic field on the permanent cellular dipole causes the bacterium to be oriented and to migrate along the magnetic field lines, a phenomenon referred to as magnetotaxis. Because they provide more stable magnetic moment per volume, magnetic particles produced biologically tend to be in the single domain size range. Single domain particles have also been imaged with MFM [100,162]. Larger animals have also been shown to have a sensitivity to the geomagnetic field.

A recent example is the rainbow trout, where the MFM was used to conclusively identify the location of magnetic navigational receptors (see Fig.11-27) [163]. MFM images that show the response of a putative single magnetic particle (within trout tissue) in the presence of an applied field. The "smoking gun" for identifying the navigational magnetic particle(s) in the trout tissue was the behavior of the particle in an applied magnetic field. Fig.11-27 shows the MFM contrast as an applied field was used to first magnetize and then flip the magnetization of a particle.

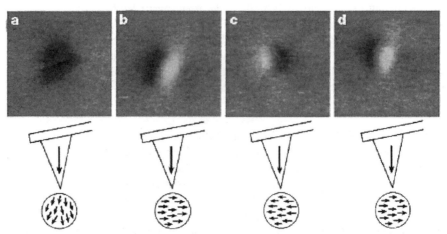

Figure 11-27: MFM images that show the response of a putative single magnetic particle (within trout tissue) in the presence of an applied field. The applied field was used to first magnetize the particle (b) and then to flip the particle magnetization (c) and finally to flip it back one more time (d). The cartoons below illustrate the hypothetical magnetizations of the MFM tip and particle in each of the four cases. A cantilever tip with a very high coercivity (>5,000 Oe) was chosen for this experiment to simplify the interpretation.

11.11.3. Magnetostrictive materials

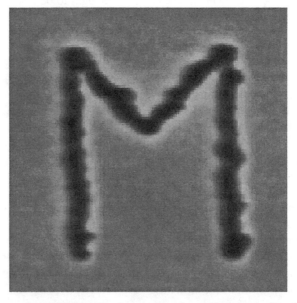

Figure 11-28: A 7.5 μm square MFM image of a magnetic domain in an amorphous $Tb_{17}Fe_{66}Co_6Gd_{11}$ alloy film of thickness 540 Å. This domain was formed by applying a local stress with a non-magnetic scanning probe tip. The letter is approximately 5.5 μm square and the width of the line comprising the letter is on average 300 nm. The rough perimeter of the domain is indicative of a large number of domain wall pinning sites and a large domain wall pinning energy [176].

The MFM has been used to study magnetostrictive effects at a sub-micron level. Initial studies on magnetostrictive terfenol samples were made by Lord. In their work, the stress caused by the domains resulted in large topographical features in the shape of the domain patterns, visible in the tapping mode height image. Recently, Schmidt et al. [164,165] have written domains on a terfenol surface by locally exerting a force with the tip. Figure 11-28 shows an MFM image of a domain written in a terfenol surface in this manner. An uncoated cantilever, operating in contact mode exerted a 20 nN force on the surface. This nucleated a reverse domain (in this case in the shape of an 'M') which could be subsequently characterized with the MFM. This work also implies that the localized coercivity studies of Babcock [122], may be more complicated that first thought, possibly involving both magnetostatic and magnetostrictive interactions.

11.12. CONCLUSIONS

Over the past few years the magnetic force microscope has evolved from a purely research tool to one of the most widely used micromagnetic imaging techniques. It is a very adaptable instrument, lending itself to imaging a very large variety of samples, often without any special preparation. The main obstacle to even more widespread use at this point seems to be interpretational issues. As with a confocal light microscope or a scanning electron microscope, it is now possible for non-specialists to routinely use the MFM.

REFERENCES

[1]   G. Binnig, C.F. Quate, and C. Gerber, *Phys. Rev. Lett.* **56** (1986) 930.
[2]   D. Rugar, H.J. Mamin, P. Guethner et al., *J. Appl. Phys.* **68** (1990) 1169.
[3]   P. Grutter, D. Rugar, and H.J. Mamin, *Ultramicroscopy* **47** (1992) 393.
[4]   U. Hartmann, T. Goddenhenrich, and C. Heiden, *J. Magn. Magn. Mater.* **101** (1991) 263.
[5]   K. Babcock, M. Dugas, S. Manalis et al., presented at the Evolution of Thin Film and Surface Structure and Morphology. Symposium, Boston, MA, USA, 1995 (unpublished).
[6]   S. Porthun, L. Abelmann, and C. Lodder, *J. Magn. Magn. Mater.* **182** (1998) 238.
[7]   R. Proksch, *Curr. Opin. Solid St. M.* **4** (2), (1999) 231.
[8]   E. Dan Dahlberg and R. Proksch, *J. Magn. Magn. Mater.* **200** (1999) 720.
[9]   U. Hartmann, *Ann. Rev. Mater. Sci.* **29** (1999) 53.
[10]   G. Binnig and H. Rohrer, *Helv. Phys. Acta* **55** (1982) 726.
[11]   J. Kaczer and R. Gemperle, *Czech. J. Phys.* **6** (1956) 173.
[12]   J. Kaczer, *Czech. J. Phys.* **5** (1955) 239.
[13]   Y. Martin and H.K. Wickramasinghe, *Appl. Phys. Lett.* **50** (1987) 1455.
[14]   J.J. Saenz and N. Garcia, *J. Appl. Phys.* **63** (1988) 2947.
[15]   P. Grütter, H. J. Mamin, and D. Rugar, in *Springer Series on Surface Sciences* (Springer, Berlin, Heidelberg, 1992), Vol. 28.
[16]   C.D. Wright and E.W. Hill, *Appl. Phys. Lett.* **67** (1995) 433.
[17]   C.D. Wright and E.W. Hill, *IEEE T. Magn.* **32** (1996) 4144.
[18]   J.D. Jackson, *Classical Electrodynamics*, 2nd Ed. (John Wiley & Sons, 1975).
[19]   G. Meyer and N.M. Amer, *Appl. Phys. Lett.* **53** (1988) 1045.
[20]   A. Garcia-Valenzuela and J. Villatoro, *J. Appl. Phys.* **84** (1998) 58.
[21]   D. Rugar, H.J. Mamin, and P. Guethner, *Appl. Phys. Lett.* **55** (1989) 2588.
[22]   E.R. Burke, R.D. Gomez, A.A. Adly et al., *IEEE T. Magn.* **28** (1992) 3135.
[23]   T. Itoh and T. Suga, *Jpn. J. Appl. Phys.* **33** (1994) 334.

[24]   C.A.J. Putman, B.G. Degrooth, N.F. Vanhulst et al., *Ultramicroscopy* **42** (1992) 1509.
[25]   R. Giles, J.P. Cleveland, S. Manne et al., *Appl. Phys. Lett.* **63** (1993) 617.
[26]   Y. Martin, C.C. Williams, and H.K. Wickramasinghe, *J. Appl. Phys.* **61** (1987) 4723.
[27]   S. Hosaka, A. Kikukawa, Y. Honda et al., *Jpn. J. Appl. Phys.* **31** (1992) L904.
[28]   H.J. Hug, B. Stiefel, P.J.A. van Schendel et al., *J. Appl. Phys.* **83** (1998) 5609.
[29]   D. Sarid, *Scanning Force Microscopy* (Oxford University Press, 1990).
[30]   P. Grutter, Y. Liu, P. LeBlanc et al., *Appl. Phys. Lett.* **71** (1997) 279.
[31]   M.A. Lantz, S.P. Jarvis, and H. Tokumoto, *Appl. Phys. Lett.* **78** (2001) 383.
[32]   B. Hoffmann, R. Houbertz, and U. Hartmann, *Appl. Phys. A-Mater.* **66** (1998) S409.
[33]   H.-J. Butt and M. Jaschke, *Nanotechnology* **6** (1), (1995) 1.
[34]   J.E. Sader, *J. Appl. Phys.* **84** (1998) 64.
[35]   R.W. Stark, T. Drobek, and W.M. Heckl, *Appl. Phys. Lett.* **74** (1999) 3296.
[36]   R.W. Stark, T. Drobek, and W.M. Heckl, *Ultramicroscopy* **86** (2001) 207.
[37]   R.W. Stark and W. M. Heckl, *Surf. Sci.* **457** (2000) 219.
[38]   J. Chen, R. K. Workman, D. Sarid et al., *Nanotechnology* **5** (1994) 199.
[39]   D. Sarid, T.G. Ruskell, R.K. Workman et al., *J. Vac. Sci. Technol. B* **14** (1996) 864.
[40]   W. Marion and W. F. Hornyak, *Physics for Scientists and Engineers* (Saunders
        College Publishing, Philadelphia, 1982).
[41]   B. Anczykowski, J.P. Cleveland, D. Kruger et al., *Appl. Phys. A-Mater.* **66** (1998)
        S885.
[42]   T.R. Albrecht, P. Grutter, D. Horne et al., *J. Appl. Phys.* **69** (1991) 668.
[43]   W.A. Ducker, R.F. Cook, and D.R. Clarke, *J. Appl. Phys.* **67** (1990) 4045.
[44]   U. Durig, H.R. Steinauer, and N. Blanc, *J. Appl. Phys.* **82** (1997) 3641.
[45]   A. Kikukawa, S. Hosaka, Y. Honda et al., *Appl. Phys. Lett.* **61** (1992) 2607.
[46]   P. Grutter, D. Rugar, H.J. Mamin et al., *Appl. Phys. Lett.* **57** (1990) 1820.
[47]   Missing reference
[48]   J.E. Sader, *Calibration of Atomic Force Microscope Cantilevers*, in: Encyclopedia of
        Surface and Colloid Science, 846 (2002).
[49]   J.P. Cleveland, S. Manne, D. Bocek et al., *Rev. Sci. Instrum.* **64** (1993) 403.
[50]   J.L. Hutter and J. Bechhoefer, *Rev. Sci. Instrum.* **64** (1993) 1868.
[51]   D.A. Walters, J.P. Cleveland, N.H. Thomson et al., *Rev. Sci. Instrum.* **67** (1996) 3583.
[52]   J.E. Sader, J.W.M. Chon, and P. Mulvaney, *Rev. Sci. Instrum.* **70** (1999) 3967.
[53]   C. Schonenberger, S.F. Alvarado, S.E. Lambert et al., *J. Appl. Phys.* **67** (1990) 7278.
[54]   C. Schonenberger and S.F. Alvarado, *Z. Phys. B-Cond. Mat.* **80** (1990) 373.
[55]   P. Grutter, Th. Jung, H. Heinzelmann et al., *J. Appl. Phys.* **67** (1990) 1437.
[56]   S. Porthun, L. Abelmann, S.J.L. Vellekoop et al., *Appl. Phys. A-Mater.* **66** (1998)
        S1185.
[57]   G. Bochi, H.J. Hug, D.I. Paul et al., *Phys. Rev. Lett.* **75** (1995) 1839.
[58]   U. Hartmann, T. Goddenhenrich, H. Lemke et al., *IEEE T. Magn.* **26** (1990) 1512.
[59]   T. Chang, M. Lagerquist, J.-G. Zhu et al., *IEEE T. Magn.* **28** (1992) 3138.
[60]   J.-G. Zhu, X. Lin, R.C. Shi et al., *J. Appl. Phys.* **83** (1998) 6223.
[61]   Y. Honda, N. Inaba, F. Tomiyama et al., *Jpn. J. Appl. Phys.* **34** (1995) L987.
[62]   B. Vellekoop, L. Abelmann, S. Porthun et al., *J. Magn. Magn. Mater.* **190** (1998) 148.
[63]   S.J.L. Vellekoop, L. Abelmann, S. Porthun et al., *J. Magn. Magn. Mater.* **193** (1999)
        474.
[64]   U. Hartmann, *J. Appl. Phys.* **64** (1988) 1561.
[65]   A. Hubert and R. Schafer, *Magnetic Domains* (Springer, Berlin, 1998), p. 84.
[66]   K.L. Babcock, V.B. Elings, J. Shi et al., *Appl. Phys. Lett.* **69** (1996) 705.
[67]   R. Proksch, K. Babcock, and J. Cleveland, *Appl. Phys. Lett.* **74** (1999) 419.
[68]   M. Lederman, G.A. Gibson, and S. Schultz, *J. Appl. Phys.* **73** (1993) 6961.

[69]   R.D. Gomez, I. D. Mayergoyz, and E. R. Burke, *IEEE T. Magn.* **31** (1995) 3346.
[70]   R.D. Gomez, M. C. Shih, R.M.H. New et al., *J. Appl. Phys.* **80** (1996) 342.
[71]   K.L. Babcock, L. Folks, R.C. Woodward et al., *J. Appl. Phys.* **81** (1997) 4438.
[72]   J. Shi, D.D. Awschalom, P.M. Petroff et al., *J. Appl. Phys.* **81** (1997) 4331.
[73]   B. Walsh, S. Austvold, and R. Proksch, *J. Appl. Phys.* **84** (1998) 5709.
[74]   R.D. Gomez, T.V. Luu, A.O. Pak et al., *J. Appl. Phys.* **85** (1999) 4598.
[75]   S. Foss, C. Merton, R. Proksch et al., *J. Magn. Magn. Mater.* **190** (1998) 60.
[76]   G.A. Gibson and S. Schultz, *J. Appl. Phys.* **73** (1993) 4516.
[77]   G.A. Gibson, J.F. Smyth, S. Schultz et al., *IEEE T. Magn.* **27** (1991) 5187.
[78]   G. Gubbiotti, L. Albini, G. Carlotti et al., *J. Appl. Phys.* **87** (2000) 5633.
[79]   J. Yu, U. Rudiger, A. D. Kent et al., *Phys. Rev. B* **60** (1999) 7352.
[80]   Y. Liu and P. Grutter, *J. Appl. Phys.* **83** (1998) 7333.
[81]   P.F. Hopkins, J. Moreland, S.S. Malhotra et al., *J. Appl. Phys.* **79** (1996) 6448.
[82]   S.H. Liou, S.S. Malhotra, J. Moreland et al., *Appl. Phys. Lett.* **70** (1997) 135.
[83]   S.H. Liou and Y.D. Yao, *J. Magn. Magn. Mater.* **190** (1998) 130.
[84]   R. Proksch, G. D. Skidmore, E. D. Dahlberg et al., *Appl. Phys. Lett.* **69** (1996) 2599.
[85]   H.J. Mamin, D. Rugar, J.E. Stern et al., *Appl. Phys. Lett.* **55** (1989) 318.
[86]   D.W. Abraham and F.A. McDonald, *Appl. Phys. Lett.* **56** (1990) 1181.
[87]   S. Foss, R. Proksch, E.D. Dahlberg et al., *Appl. Phys. Lett.* **69** (1996) 3426.
[88]   S. Foss, E.D. Dahlberg, R. Proksch et al., *J. Appl. Phys.* **81** (1997) 5032.
[89]   J.M. Garcia, A. Thiaville, J. Miltat et al., *Appl. Phys. Lett.* **79** (2001) 656.
[90]   G. Matteucci, M. Muccini, and U. Hartmann, *Appl. Phys. Lett.* **62** (1993) 1839.
[91]   D.G. Streblechenko, M.R. Scheinfein, M. Mankos et al., *IEEE T. Magn.* **32** (1996) 4124.
[92]   B.G. Frost, N.F. van Hulst, E. Lunedel et al., *Appl. Phys. Lett.* **68** (1996) 1865.
[93]   P. Grutter, D. Rugar, H.J. Mamin et al., *J. Appl. Phys.* **69** (1991) 5883.
[94]   A. Thiaville, L. Belliard, D. Majer et al., *J. Appl. Phys.* **82** (1997) 3182.
[95]   K. Sueoka, K. Inagami, T. Imamura et al., presented at the Conference Proceedings. 10th Anniversary. IMTC/94. Advanced Technologies in I & M. 1994 IEEE Instrumentation and Measurement Technology Conference, Hamamatsu, Japan, 1994 (unpublished).
[96]   R.B. Proksch, S. Foss, and E.D. Dahlberg, *IEEE T. Magn.* **30** (1994) 4467.
[97]   K. Babcock, V. Elings, M. Dugas et al., *IEEE T. Magn.* **30** (1994) 4503.
[98]   T. Goddenhenrich, H. Lemke, M. Muck et al., *Appl. Phys. Lett.* **57** (1990) 2612.
[99]   R.D. Gomez, A.O. Oak, A.J. Anderson et al., *J. Appl. Phys.* **83** (1998) 6226.
[100]   R.B. Proksch, T.E. Schaffer, B.M. Moskowitz et al., *Appl. Phys. Lett.* **66** (1995) 2582.
[101]   J. Vergara, P. Eames, C. Merton et al., *Appl. Phys. Lett.* **84** (2004) 1156.
[102]   A.N. Campbell, E.I. Cole, Jr., B.A. Dodd et al., presented at the 31st Annual Proceedings. Reliability Physics 1993 (Cat. No.93CH3194-8), Atlanta, GA, USA, 1993 (unpublished).
[103]   A.N. Campbell, E.I. Cole, Jr., B.A. Dodd et al., *Microelectron. Eng.* **24** (1994) 11.
[104]   R. Yongsunthon, E.D. Williams, A. Stanishevsky et al., presented at the Electrically Based Microstructural Characterization III. Symposium (Materials Research Society Symposium Proceedings Vol.699), Boston, MA, USA, 2002 (unpublished).
[105]   R. Yongsunthon, A. Stanishevsky, P.J. Rous et al., presented at the Spatially Resolved Characterization of Local Phenomena in Materials and Nanostructures. Symposium (Mater. Res. Soc. Symposium Proceedings Vol.738), Boston, MA, USA, 2003 (unpublished).

[106]   F. Kral, G. Kostorz, N. Ari et al., *Czech. J. Phys.* **49** (1999) 1567.
[107]   R. Yongsunthon, A. Stanishevsky, J. McCoy et al., *Appl. Phys. Lett.* **78** (2001) 2661.
[108]   L. S. Kong and S. Y. Chou, *J. Appl. Phys.* **81** (1997) 5026.
[109]   J. Lohau, S. Kirsch, A. Carl et al., *J. Appl. Phys.* **86** (1999) 3410.
[110]   L. Abelmann, S. Porthun, M. Haast et al., *J. Magn. Magn. Mater.* **190** (1998) 135.
[111]   H. Neal Bertram, *Theory of Magnetic Recording* (Cambridge University Press, Cambridge, 1994).
[112]   M. Ruhrig, S. Porthun, and J.C. Lodder, *Rev. Sci. Instrum.* **65** (1994) 3224.
[113]   M. Ruhrig, S. Porthun, J.C. Lodder et al., *J. Appl. Phys.* **79** (1996) 2913. S.Y. Chou, M.S. Wei, and P.B. Fischer, *IEEE T. Magn.* **30** (1994) 4485.
[114]   Takayuki Arie, Hidehiro Nishijima, Seiji Akita et al., *J. Vac. Sci. Technol. B*, **18** (2000) 104.
[115]   N. Yoshida, M. Yasutake, T. Arie et al., *Jpn. J. Appl. Phys.*, **41** (7B), (2002) 5013.
[116]   P. Bauer, H.P. Bochem, P. Leinenbach et al., *Scanning* **18** (1996) 374.
[117]   S. Khizroev, J.A. Bain, and D. Litvinov, *Nanotechnology* **13**, (2002) 619.
[118]   L. Folks, M. E. Best, P. M. Rice et al., *Appl. Phys. Lett.* **76** (2000) 909.
[119]   G. N. Phillips, M. Siekman, L. Abelmann et al., *Appl. Phys. Lett.* **81** (2002) 865.
[120]   Y.M. Lau, P.C. Chee, J.T.L. Thong et al., *J. Vac. Sci. Technol. A* **20** (2002) 1295.
[121]   T. Goddenhenrich, U. Hartmann, and C. Heiden, *Ultramicroscopy* **42-44** (1992) 256.
[122]   K. Babcock, S. Manalis, V. Elings et al., *J. Appl. Phys.* **79** (1996) 6440.
[123]   T. Goddenhenrich, U. Hartmann, M. Anders et al., *J. Microsc.* **152** (1988) 527.
[124]   U. Hartmann, *Phys. Status Solidi A* **108** (1988) 387.
[125]   E. Zueco, W. Rave, R. Schafer et al., *J. Magn. Magn. Mater.* **190** (1998) 42.
[126]   A. Hubert, W. Rave, and S. L. Tomlinson, *Phys. Status Solidi B* **204** (1997) 817.
[127]   L. Belliard, A. Thiaville, S. Lemerle et al., *J. Appl. Phys.* **81** (1997) 3849.
[128]   H. Holscher, W. Allers, U.D. Schwarz et al., *Phys. Rev. Lett.* **83** (1999) 4780. 4783
[129]   U. Durig, *Appl. Phys. Lett.* **76** (2000) 1203.
[130]   U. Durig, *New J. Physics* **2** (2000) ??.
[131]   T.E. Schaffer, M. Radmacher, and R. Proksch, *J. Appl. Phys.* **94** (2003) 6525.
[132]   U. Hartmann and C. Heiden, *J. of Microscopy* **152** (1988) 281.
[133]   A. Wadas, P. Grutter, and H.-J. Guntherodt, *J. Vac. Sci. Technol. A* **8** (1990) 416.
[134]   A. Wadas, P. Grutter, and H.-J. Guntherodt, *J. Appl. Phys.* **67** (1990) 3462.
[135]   H. J. Hug, B. Stiefel, A. Moser et al., *J. Appl. Phys.* **79** (1996) 5609.
[136]   H.J. Hug, B. Stiefel, P.J.A. Van Schendel et al., *J. Appl. Phys.* **83** (1998) 5609.
[137]   P. Grutter and R. Allenspach, *Geophys. J. Int.* **116** (1994) 502.
[138]   P. Grutter, Y. Liu, P. LeBlanc et al., *Appl. Phys. Lett.* **71** (1997) 279.
[139]   Alvaro San Paulo and Ricardo Garcia, *Phys. Rev. B* **64** (2001) 193411.
[140]   K. Wago, K. Sueoka, and F. Sai, *IEEE T. Magn.* **27** (1991) 5178.
[141]   R. Proksch, G. Skidmore, and S. Austvold, *J. Appl. Phys.* **81** (1997) 4522.
[142]   T.G. Pokhil and R.B. Proksch, *J. Appl. Phys.* **81** (1997) 3846.
[143]   W. Rave, E. Zueco, R. Schaefer et al., *J. Magn. Magn. Mater.* **177-181** (1998) 1474.
[144]   D. Peterka, A. Enders, G. Haas et al., *Rev. Sci. Instrum.* **74** (2003) 2744.
[145]   P. Grutter, E. Meyer, H. Heinzelmann et al., *J. Appl. Phys.* **63** (1988) 2947.
[146]   P. Grutter, E. Meyer, H. Heinzelmann et al., *J. Vac. Sci. Technol. A* **6** (1988) 279.
[147]   U. Hartmann, *Phys. Status Solidi A* **115** (1989) 285.
[148]   P. Rice, B. Hallett, and J. Moreland, *IEEE T. Magn.* **30** (1994) 4248.
[149]   P. Rice and J.R. Hoinville, *IEEE T. Magn* **32** (1996) 3563.
[150]   J. H. Chen, X. D. Lin, J. G. Zhu et al., *IEEE T. Magn.* **33** (1997) 3040.
[151]   Y.J. Chen, W.Y. Cheung, I.H. Wilson et al., *Appl. Phys. Lett.* **72** (1998) 2472.

[152] Xing Song, Qixu Chen, C. Leu et al., *IEEE T. Magn.* **34** (1998) 1549.
[153] G. A. Gibson, S. Schultz, T. Carr et al., *IEEE T. Magn.* **28** (1992) 2310.
[154] M. Todorovic and S. Schultz, *J. Appl. Phys.* **83** (1998) 6229.
[155] K. Sueoka, K. Wago, and F. Sai, *IEEE T. Magn.* **28** (1992) 2307.
[156] C.X. Qian, H.C. Tong, F.H. Liu et al., *IEEE T. Magn.* **35** (1999) 2625.
[157] D. J. Dunlop, *Rep. Prog. Phys.* **53** (1990) 707.
[158] W. Williams, V. Hoffmann, F. Heider et al., *Geophys. J. Int.* **111** (1992) 417.
[159] T.G. Pokhil and B.M. Moskowitz, *J. Appl. Phys.* **79** (1996) 6064.
[160] S. Foss, B.M. Moskowitz, R. Proksch et al., *J. Geophys. Res.* **103** (1998) 30551.
[161] R. Blakemore, *Science* **190** (1975) 377.
[162] H. Suzuki, T. Tanaka, T. Sadaki et al., *Jpn. J. Appl. Phys.* **37** (1998) L1343.
[163] C.E. Diebel, R. Proksch, C.R. Green et al., *Nature* **406** (2000) 299.
[164] J. Schmidt, S. Foss, G.D. Skidmore et al., *J. Magn. Magn. Mater.* **190** (1998) 108.
[165] J. Schmidt, R. Tickle, G.D. Skidmore et al., *J. Magn. Magn. Mater.* **190** (1998) 98.
[166] A.P. Volodin and M.V. Marchevsky, *Ultramicroscopy* **42-44** (1992) 757.
[167] A. Moser, H.J. Hug, B. Stiefel et al., *J. Magn. Magn. Mater.* **190** (1998) 114.
[168] A. Volodin, K. Temst, C.V. Haesendonck et al., *Physica B* **284-288** (2000) 815.
[169] A. Volodin, K. Temst, C. Van Haesendonck et al., *Physica C* **332** (2000) 156.
[170] U.H. Pi, D.H. Kim, Z.G. Khim et al., *J. Low Temp. Phys.* **131** (2003) 993.
[171] C.C. Chen, Q. Lu, C. Yuan et al., presented at the Proceedings of the 10th Anniversary HTS Workshop on Physics, Materials and Applications, Houston, TX, USA, 1996 (unpublished).
[172] K. Moloni, B.M. Moskowitz, and E.D. Dahlberg, *Geophys. Res. Lett.* **23** (1996) 2851.
[173] C.C. Chen, Q. Lu, and A. de Lozanne, *Appl. Phys. A* **66** (1998) S1181.
[174] A. Volodin, K. Temst, C. Van Haesendonck et al., *Rev. Sci. Instrum* **71** (2000) 4468.
[175] R.D. Gomez, T.V. Luu, A.O. Pak et al., *J. Appl. Phys.* **85** (1999) 6163.
[176] J. Schmidt, S. Foss, G.D. Skidmore et al., *J. Magn. Magn. Mater.* **190** (1998) 108.
[177] R. Proksch, E. Runge, P.K. Hansma et al., *J. Appl. Phys.* **78** (1995) 3303.
[178] M. Scheinfein, Private communication.

Light Scattering

CHAPTER 12

Scanning near-field magneto-optic microscopy

12.1. INTRODUCTION

Optical microscopy has ever since the development of the first microscopes at the beginning of the 17th century been an important tool for the investigation of small structures in nature. Unfortunately, the resolving power cannot be increased endlessly. E. Abbe [1,2] discovered 130 years ago a fundamental limit of optical resolution for coherently illuminated objects in a microscope which is directly related to the wavelength of the light (an introduction to the theory of optical resolution will be given in Chapt. 12.2.1.2). Some years later, in 1896, Lord Rayleigh found a similar limit for incoherently illuminated objects [3]. The limit of optical resolution could, therefore, only be lowered by using a smaller wavelength of the illuminating source. This principle is optimized in scanning electron microscopes which use high-energy electron beams as a source of illumination. According to quantum theory, an electron has also a wave nature with a wavelength λ given by the de Broglie relation.[1] For a 30 keV electron beam, the wavelength results to $\lambda \cong 0.1$ Å yielding atomic resolution in transmission electron microscopy (*TEM*). For more details see Chapter 7 of this book.

An alternative way of circumventing the classical limit of optical resolution was proposed in 1928 by E. H. Synge [4]. He suggested to introduce in close distance to the sample surface an aperture with a size much smaller than the wavelength of the light. By scanning the aperture across the sample, the optical resolution should only be limited by the size of the aperture and not by the wavelength of the light. As aperture, he proposed a pin hole in a thin metallic film. To produce the hole, he suggested to spread small transparent particles on a glass slide and to deposit a metallic film on top. The aperture is then created by planing down the film until the transparent particles are exposed. However, it took another 40 years until E. Ash and G. Nicholls succeeded to demonstrate with microwaves the validity of Synge's concept. They achieved a resolution of $\lambda/60$ using a wavelength of $\lambda = 3$ cm [5]. By using 3 cm microwaves, they had not to bother with the problem of a precise distance control.

To attain a resolution far beyond that of a classical optical microscope, i.e. in the visible light region, the aperture should be less than 100 nm. Hence, a very accurate distance-control mechanism is crucial because the tip has to be kept at a distance from the sample surface of the order of the size of the aperture. The demonstration of a scanning tunneling microscope (*STM*) by G. Binnig and H. Rohrer [6,7] in 1982 initiated the development of precise scanning and distance-control devices based on piezoelectric transducers. As a consequence, a variety of novel scanning-probe microscopes were accomplished including the scanning near-field optical microscope (*SNOM*[2]) realized independently by A. Lewis [8] and D. Pohl [9].

In the meantime, *SNOM* has become a widely used technique in surface science. The reason for that is not so much the lateral resolution, which is still much less than in atomic force microscopy or in *STM* due to the complicated dependence between topography and optical information. It is the appealing fact that topographic and optical information can be obtained simultaneously and virtually any optical method can be combined with *SNOM*. Recent examples include investigations of single-molecule fluorescence [10-12], surface plasmon effects [13,14], femtosecond time-resolved microscopy [15], Raman spectroscopy [16], second-harmonic generation [17], and magneto-optic effects [18-21]. In addition, *SNOM* bears a considerable potential for applications such as high-density optical [22] and magneto-optic [18,23] storage devices, investigations of biological specimens in liquid [24], and near-field optical lithography [25-29].

The lateral resolution which can be obtained with metal-coated single-mode glass fiber tips is limited to about 12 nm due to the skin depth of light in a metal [30,25] but has not yet been achieved experimentally. A higher lateral resolution of about 6 nm is accomplished by using a tetrahedral tip in a combined *SNOM/STM* set-up [31,32] or by apertureless probes (apertureless *SNOM*, see Chapter 12.2.3), where an atomically sharp tip acts as a local scattering center in the vicinity of a focused light beam [33]. Features of 1 nm size have been claimed to be observed by this method [34].

However, in order for *SNOM* to become a technique which is competitive with other scanning-probe-microscopy methods it is crucial that the lateral resolution of the optical information is improved to better than 20 nm. In addition, other key problems, such as topographic cross-talk in the optical images and tip-shape effects, have to be solved.

In this article, a thorough survey is given on how to characterize magnetic materials by using polarization-sensitive *SNOM*. By means of the magneto-optic Faraday and Kerr effect (see Chapter 12.4.1 for an explanation), magnetic domains can be imaged and magnetic properties can be measured optically. In order to determine magneto-optic effects, the change of polarization of the light has to be determined. This requires the usage of a polarization-sensitive *SNOM* which allows to determine local magnetic properties and to image magnetic domains with sub-micron resolution. The fundamental advantage of this method in comparison with magnetic force microscopy (*MFM*) is that *SNOM* does not require the use of a magnetic tip. Therefore, domain walls are not influenced while scanning across the surface. However, the lateral resolution in *SNOM* is still lower than in *MFM*. Furthermore, the investigation of opaque samples, such as thin metallic films, requires working in reflection mode using the Kerr effect which is technically much more demanding than working in transmission mode. Besides giving an introduction to *SNOM* and to the theory of magneto-optics, emphasis is put in this article also on the problems that must be solved in order to develop a user-friendly, high-resolution Kerr microscope with a topographic and magneto-optic lateral resolution of considerably less than 100 nm.

For a complete review of *SNOM* see the books of D. Pohl and D. Courjon [35], M. Paesler and P. Moyer [36], M. Ohtsu [37], and M. Ohtsu and H. Hori [38] and the review article by D. Courjon and C. Bainier [39].

12.2. PRINCIPLE OF *SNOM*

In this chapter, an introduction to the principle of *SNOM* is given. We will first discuss the classical limit of optical resolution as developed by Lord Rayleigh and E. Abbe. Furthermore, the connection between optical resolution and Heisenberg's uncertainty principle is explained. In the second part, the principle of *SNOM* using an aperture is described while in the third part apertureless *SNOM* will be discussed.

12.2.1. Classical limit of optical resolution

12.2.1.1. Rayleigh's criterion

In classical optics, the resolution of an optical instrument such as a microscope or a telescope is generally limited by diffraction. In a simple approach, the determination of the resolution can be reduced to the limiting effect of the aperture of the objective lens. It will act as a circular hole diffracting the incoming light as shown in Fig.12-1(a).

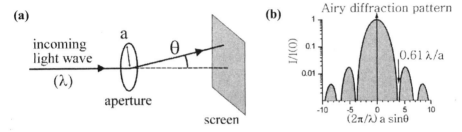

Figure 12-1: (a) Diffraction by a circular aperture and (b) cross section through the resulting Airy diffraction pattern.

The resulting diffraction pattern is the so-called Airy disk [40] which consists of concentric consecutive intensity maximums determined by the first Bessel function J_1:

$$I(\theta) = I(0)\left[\frac{2 J_1(k\, a \sin\theta)}{k\, a \sin\theta}\right]^2, \quad k = 2\pi/\lambda. \tag{12.1}$$

where θ is the diffraction angle, $k = 2\pi/\lambda$ is the wave vector, λ is the wavelength of the light, and a is the radius of the aperture. A cross section in a logarithmic scale is shown in Fig.12-1(b). The distance of the first minimum from the center line depends on the size of the aperture and the wavelength of the light:

$$\sin\theta = 0.61\frac{\lambda}{a}, \tag{12.2}$$

In order to derive the lateral resolution of the instrument, Rayleigh's criterion is applied as plotted in Fig.12-2. It simply states that two objects are just about resolvable when the Airy disks are overlapping such that the central maximum of one disk coincides with the first minimum of the second disk. This yields an angular separation θ of the two objects according to Eq.(12.2).

Rayleigh's criterion is valid for incoherently illuminated objects or luminescent objects without phase relation such as, e.g., the stars in the sky. Hence, it can be used to determine the resolution of a telescope.

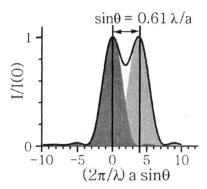

Fig. 12-2: Rayleigh's criterion of optical resolution.

12.2.1.2. Theory of microscopy

In principle, any microscope can be reduced to an objective lens and an ocular lens. In reality, of course much more lenses are necessary to deal with imperfections of the imaging system such as chromatic aberration and spherical aberration. The ocular lens acts as a magnifying glass by increasing the angle of perception in the eye. The magnification v_{ocular} is given by

$$v_{ocular} = \frac{d_0}{f_{ocular}} \tag{12.3}$$

where f_{ocular} is the focus length of the ocular lens and $d_0 = 250$ mm is the average minimal distance for an object in order to be imaged sharply on the retina of a human eye. The objective acts as a collecting lens and creates an enlarged image of the object in the image plane, which coincides with the focal plane of the ocular lens as outlined in Fig. 12-3.

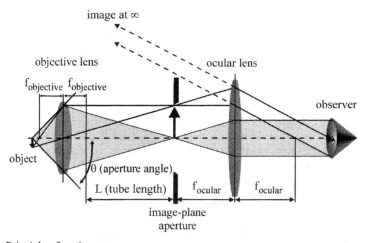

Figure 12-3: Principle of a microscope.

The objective lens enlarges the object according to the well-know lens equation:

$$v_{objective} = \frac{L}{f_{objective}}, \tag{12.4}$$

where $f_{objective}$ is the focus length of the objective lens and $L = 160$ mm is a typical tube length.

The total magnification is given by the product of the magnification of ocular and objective, i.e. of Eq. (12.3) and (12.4).:

$$V_{total} = V_{objective} \cdot V_{ocular} = \frac{d_0 L}{f_{objective} f_{ocular}} \qquad (12.5)$$

However, the resolution of a microscope is solely determined by the objective lens because the ocular lens only magnifies the image produced by the objective lens. The numerical aperture *N.A.* of the microscope is defined by the opening angle and the index of refraction *n* on the object side of the object lens:

$$N.A. = n \sin \theta \qquad (12.6)$$

In order to understand the theory of optical resolution [41], let us just focus on the objective lens of a microscope as plotted in Fig.12-4. Let *n* be the index of refraction on the object side and *n'* inside the microscope, i.e. on the image side. The resolution of the microscope is defined by the minimum distance *Y* for two point objects *P* and *Q* in order to be resolved in the image plane as *P'* and *Q'*. Point *P* lies on the optical axis and *Q* is very close to it. Due to the small distance between *P* and *Q* and the large magnification, the angles θ' and φ are very small, i.e. θ', φ << 1. The limiting ray of point *P* reaching the image *P'* is defined by the circular aperture with radius *a'* in the focal plane. This defines the numerical aperture and the opening angle θ at the object plane. The corresponding angle at the image plane θ' is then given by

$$\tan \theta' = \frac{a'}{D'} \cong \sin \theta'. \qquad (12.7)$$

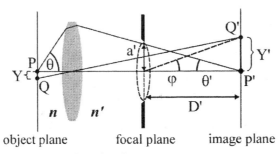

object plane   focal plane   image plane

Figure 12-4: Theory of optical resolution of a microscope.

The rays of the two point sources *P* and *Q* are diffracted at the aperture in the focal plane. They will be imaged as an Airy disk in the image plane as explained before. According to Rayleigh's criterion, the minimum distance *Y'* for the image *P'* and *Q'* to still be resolved is given by the angle φ in accordance with Eq. (12.2):

$$Y' = D' \tan \varphi \cong D' \sin \varphi = 0.61 D' \frac{\lambda'}{a'} = 0.61 \frac{\lambda}{n' \sin \theta'} . \qquad (12.8)$$

In Eq. (12.8) we took into account that the wavelength λ' inside the microscope is given by λ' = λ/n', where λ is the vacuum wavelength. To derive a relation between object separation *Y* and image separation *Y'*, we need Abbe's sine condition [41]:

$$n Y \sin \theta = n' Y' \sin \theta'. \qquad (12.9)$$

Combining Eqs. (12.8) and (12.9), we end up with the relation for the optical resolution of a microscope:

$$Y = 0.61 \frac{\lambda}{n \sin \theta} \qquad (12.10)$$

The prefactor 0.61 in Eqs. (12.8) and (12.10) is valid for incoherently illuminated or incoherently radiating objects. For coherently illuminated objects, Abbe found a slightly smaller value [40, 41]:

$$Y = 0.5 \frac{\lambda}{n \sin \theta} \; . \qquad (12.11)$$

12.2.1.3. Optical diffraction limit and Heisenberg's uncertainty principle

Before explaining what evanescent waves are and before explaining the principle of *SNOM*, it is illustrative to discuss the relation between the limit of optical resolution and Heisenberg's uncertainty principle. W. Heisenberg himself used the example of an optical experiment to illustrate the uncertainty principle [42]. This idea was further developed by N. Bohr [43]. A nice description of this gedanken experiment is given in a paper by J. M.. Vigoureux and D. Courjon [44].

The idea of Heisenberg's gedanken experiment was to consider a microscope which uses gamma rays in order to observe the position of an electron. Let the wavelength of the gamma rays in vacuum be λ and let the direction of propagation be along the z axis with wave vector $\vec{k}_\gamma$. Furthermore, the index of refraction in the object space is n. As shown in the previous section, the resolution of an optical microscope is given by Eq.(12.10), where Y is the smallest distance between two points that can be resolved. Since in quantum theory light is not a neutral observer, it will transfer momentum to the electron by means of the Compton effect [45]. After interaction, only those photons within an acceptance angle of 2θ will be collected by the microscope as depicted in Fig.12-5(a).

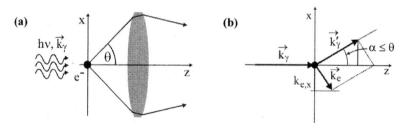

Figure 12-5: Heisenberg's microscope: (a) geometry, (b) momentum distribution.

The electron receives a recoil momentum along the x axis $\hbar \vec{k}_{e,x}$ ($2\pi\hbar = h$ is the Planck constant) according to the law of conservation of momentum as depicted in Fig.12-5(b):

$$\hbar k_{e,x} = \hbar k'_\gamma \sin \alpha = \frac{h \, n}{\lambda'} \sin \alpha \, , \qquad (12.12)$$

where $\lambda' \geq \lambda$ is the red-shifted wavelength according to the Compton effect. The uncertainty $\Delta p_{e,x}$ in the electron momentum along the x axis is defined by the maximum acceptance angle of the microscope:

$$\alpha_{max} = \theta \Rightarrow \Delta p_{e,x} = 2\hbar k_{e,x}'^{max} = 2\hbar k_{\gamma}' \sin \theta = \frac{2h\,n}{\lambda'} \sin \theta. \tag{12.13}$$

Multiplying with the lateral resolution of the microscope $\Delta x = Y$ according to Eq. (12.11), we derive the Heisenberg uncertainty principle:

$$\Delta x \cdot \Delta p_{e,x} = 0.5 \frac{\lambda'}{n \sin \theta} \cdot \frac{2h\,n}{\lambda'} \sin \theta = h. \tag{12.14}$$

This gedanken experiment shows the close connection between optical resolution limit and quantum mechanical uncertainty principle.

12.2.1.4. Evanescent waves

In this section we will give a short introduction to evanescent waves as they will be crucial for understanding the superior resolution of near-field optics. All electromagnetic waves have to fulfill the dispersion relation in vacuum (c is the speed of light in vacuum):

$$\left|\vec{k}\right| = k = \sqrt{k_x^2 + k_y^2 + k_z^2} = \frac{\omega}{c}. \tag{12.15}$$

For propagating waves, all components of the wave vector must be real:

$$k_x, k_y, k_z \in \mathbb{R}. \tag{12.16}$$

Examples of propagating electromagnetic waves are plane waves,

$$\vec{E}(\vec{r},t) = \vec{E}_0 e^{i(\vec{k}\cdot\vec{r}-\omega t)}, \tag{12.17}$$

and spherical waves,

$$\vec{E}(\vec{r},t) = \vec{E}_0 \frac{e^{i(kr-\omega t)}}{r}, \tag{12.18}$$

where $\vec{E}(\vec{r},t)$ is the electric field vector of the electromagnetic wave. The magnetic field vector has a similar form and is related to $\vec{E}(\vec{r},t)$ according to the Maxwell relations [46]. Propagating waves can transport energy over long distances. The energy which is transmitted through a certain area at a fixed time will be preserved as shown in Fig.12-6 for plane and spherical waves. As a consequence, the electric field vector will decrease less or equal to $1/r$.

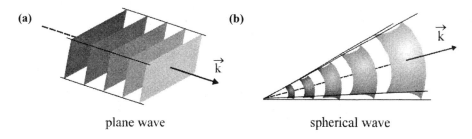

Figure 12-6: Energy flow for (a) plane wave and (b) spherical wave.

For evanescent waves, the energy transport through a certain area at a fixed time will not be conserved, it will decrease. This implies that $\vec{E}(\vec{r},t)$ has to decrease faster than $1/r$. A classical example for evanescent waves is the electric field distribution of a Hertz dipole which consists of two equal charges of opposite sign at an oscillating distance d as shown in Fig.12-7.

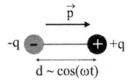

Figure 12-7: Hertz dipole

The dipole moment $\vec{p}$ is defined as

$$|\vec{p}(t)| = q\,d = q\,d_0\cos(\omega t) = p_0\cos(\omega t), \tag{12.19}$$

The general solution for the electric field is

$$E_r \propto 2\left(\frac{[p]}{r^3} + \frac{[\dot{p}]}{c\,r^2}\right)\cos\theta \tag{12.20}$$

for the radial component, where

$$[p] = p_0\cos[\omega(t - r/c)] \tag{12.21}$$

is the retarded dipole moment, and

$$E_\theta \propto \left(\frac{[p]}{r^3} + \frac{[\dot{p}]}{c\,r^2} + \frac{[\ddot{p}]}{c^2\,r}\right)\sin\theta \tag{12.22}$$

for the angular component in spherical coordinates (z axis parallel to the dipole moment $\vec{p}$). The angular component of the magnetic field is given by

$$H_\varphi \propto \left(\frac{[\dot{p}]}{c\,r^2} + \frac{[\ddot{p}]}{c^2\,r}\right)\sin\theta \tag{12.23}$$

In the far field, i.e. at a distance $r \gg c|p/\dot{p}|$ and $r \gg c|\dot{p}/\ddot{p}|$ which implies $r \gg c/\omega = \lambda/2\pi$, the electric field becomes equal to that of a spherical wave:

$$E_\theta = c\,H_\varphi \propto p_0\sin\theta\,\frac{e^{i\omega(t - r/c)}}{r}, \tag{12.24}$$

$$E_r = 0 . \tag{12.25}$$

In the near-field, however, there are many other components as is obvious from Eqs. (12.20)-(12.23) which are decreasing faster than $1/r$.

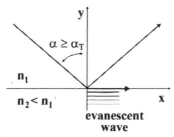

Figure 12-8: Total internal reflection.

Another example for evanescent waves is the total internal reflection of light as plotted in Fig.12-8. Although 100 % of the light intensity is reflected, an evanescent wave exists, which travels along the interface (i.e. in x direction) and whose amplitude is exponentially damped perpendicular to the interface (i.e. along the negative y axis):

$$E_{evan} = E_{0,evan} e^{ky \sinh \beta''} e^{-ikx \cosh \beta''} . \qquad (12.26)$$

The first term on the right side of Eq.(12.26) is an exponentially damped amplitude along the $-y$ direction while the second term is a propagating wave along the x direction. The angle β'' is part of the complex angle of the diffracted beam according the Snellius law in the case of $\alpha \geq \alpha_T$, where α_T is the critical angle for total internal reflection:

$$n_1 \sin \alpha = n_2 \sin \tilde{\beta}, \quad \tilde{\beta} = \pi/2 + i\beta'' . \qquad (12.27)$$

In this case, the wave vector $\vec{k}$ has a complex component,

$$\tilde{k}_y = ik_y \sinh \beta'' , \qquad (12.28)$$

in y direction producing an exponentially damped amplitude. For the dispersion relation Eq. (12.16) to be fulfilled, the x component k_x must become larger than the absolute value k of the wave vector ($k_z = 0$):

$$\left(\frac{\omega}{c}\right)^2 = \left|\vec{k}\right|^2 = k^2 = k_x^2 + \tilde{k}_y^2 = k_x^2 - k_y^2 \sinh^2 \beta'' . \qquad (12.29)$$

$$\Rightarrow \left|k_x\right| > k = \frac{\omega}{c} . \qquad (12.30)$$

As we will see in the following sections, such a component of the wave vector being larger than the absolute value will lead to an enhanced resolution beyond the classical diffraction limit. This is explained in Fig.12-9 in more detail. Evanescent waves with one component, say k_x, larger than $k = \omega/c$ can yield a superior resolution $\Delta x < \lambda/2$ without violating Heisenberg's uncertainty principle. This is not true for propagating waves.

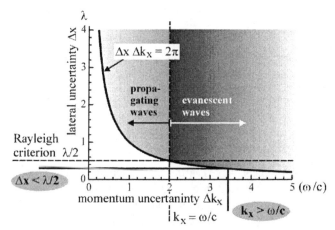

Figure 12-9: Classical limit of optical resolution and superior resolution in near-field optics. The hyperbolic function (thick solid line) represents the ultimate limit given by Heisenberg's uncertainty principle. Evanescent waves with $k_x > \omega/c$ provide a possibility for superior resolution $\Delta x < \lambda/2$ without violating Heisenberg's uncertainty principle.

12.2.2. Aperture SNOM

The basic principle of near-field detection using an nanocollector is illustrated in Fig.12-10(a). A similar description is found in a review article by D. Courjon and C. Bainier [39]. The sample surface is illuminated by a light source. The electric field vector of the source will generate local currents in a surface layer of thickness of the penetration depth of the light according to Ohm's law:

$$\vec{J}_{local}(\vec{r}, t) = \ddot{\sigma}(\vec{r}, \omega)\vec{E}_{light}(\vec{r}, t), \tag{12.31}$$

where $\ddot{\sigma}$ is the optical conductivity tensor. The excited local currents can be considered as due to dipoles $\vec{P}_{local}$ being stimulated by the light source because of the polarizability, described by the tensor $\ddot{\alpha}$, of the material:

$$\vec{p}_{local}(\vec{r}, t) = \ddot{\alpha}_{local}(\vec{r}, \omega)\vec{E}_{light}(\vec{r}, t) \tag{12.32}$$

The local dipole moments create local currents due to the well-known relation between the polarizability $\ddot{\alpha}$, the dielectric function $\ddot{\varepsilon}$ and the conductivity tensor $\ddot{\sigma}$:

$$\ddot{\varepsilon} = 1 + 4\pi\ddot{\alpha} = 1 - i\frac{4\pi}{\omega}\ddot{\sigma} \ \ (\text{cgs}), \qquad \ddot{\varepsilon} = 1 + \frac{1}{\varepsilon_0}\ddot{\alpha} = 1 - i\frac{1}{\varepsilon_0\omega}\ddot{\sigma} \ \ (\text{SI}). \tag{12.33}$$

Because the local dipoles are oscillating at the frequency of the illuminating light, they will emit themselves electromagnetic waves according to the theory of Hertz (see Chapt. 12.2.1.4). The evanescent waves will die out very rapidly leaving only far-field components which build up the reflected wave. If, however, a small collector is brought in close distance to the sample surface, it will be excited by the evanescent waves yielding information on the local electric-field distribution. Now, the collector acts again as a Hertz dipole and will radiate electromagnetic waves with an intensity which is proportional to the exciting local electric field. A small part of the radiated waves are propagating and can be detected using classical optics. As a consequence, the lateral resolution is only determined by the size of the collector and not by the wavelength of the illuminating light. By scanning the collector across the sample surface, the local optical properties can be measured with a resolution given by the size of the collector.

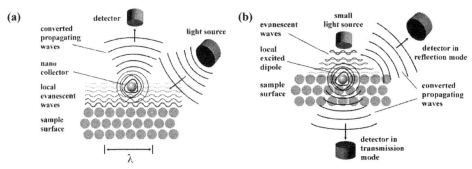

Fig. 12-10: principle of near-field detection: (a) collection mode, (b) illumination mode.

Of course, the concept is still valid if, instead of a nanoscopic collector, a very small light source is used as shown in Fig.12-10(b). If the light source is brought in close distance to the sample surface, the local dipoles will be excited by the evanescent waves of the light source and will radiate again a small fraction as propagating waves, which can be detected either in transmission or reflection. The resolution is again determined by the small size of the light source as it only illuminates a small volume of the surface.

12.2.2.1. Simple theory of aperture SNOM

The superior resolution of *SNOM* as discussed above can be demonstrated by a simple mathematical model proposed by J. M. Vigoureux and D. Courjon [44]. The object is described as an illuminated slit of infinite length in y direction acting as a line source at position x_0:

$$f(x) = \delta(x - x_0),$$ (12.34)

Applying the Fourier transformation we get

$$\hat{f}(k) = \int_{-\infty}^{\infty} \delta(x - x_0) e^{ikx} dx = e^{ikx_0}.$$ (12.35)

For $x_0 = 0$, this corresponds to the unit function. Looking at this object with a conventional microscope, the spatial frequencies are limited by the finite aperture angle 2θ of the microscope limiting the spatial frequencies k_x in x direction to:

$$|k_x| \in [0, k_{max}], \quad k_{max} = k \sin \theta = \frac{\omega}{c} \sin \theta.$$ (12.36)

This reduces in the Fourier transformation the integral from $\pm\infty$ to $\pm k_{max}$. The image that can be constructed by the microscope is then given by

$$f(x) = \int_{-k_{max}}^{k_{max}} e^{ikx_0} e^{-ikx} dk = \int_{-k_{max}}^{k_{max}} e^{ik(x_0-x)} dk = 2 \frac{\sin\left[\frac{\omega}{c} \sin \theta (x - x_0)\right]}{(x - x_0)}.$$ (12.37)

Although the width of the source is infinitely small, its image will have a finite width given by the *sinc* function:

$$\Delta x = \frac{\pi c}{\omega \sin \theta} = \frac{\lambda}{2 \sin \theta}. \tag{12.38}$$

This result corresponds to the Rayleigh criterion of Eq. (12.2), i.e., the image looks as if it would be coming from a source of width Δx.

For an object of arbitrary form $f(x-x_0)$, the image as seen through the microscope is found in the same way:

$$\begin{aligned} g(x) &= \int_{-k_{max}}^{k_{max}} \hat{f}(k) e^{-ikx} dk = \int_{-\infty}^{\infty} \left[\Theta\left(k + \frac{\omega}{c}\sin\theta\right) - \Theta\left(k - \frac{\omega}{c}\sin\theta\right) \right] \hat{f}(k) e^{-ikx} dk \\ &= \int_{-\infty}^{\infty} dk \left[\Theta\left(k + \frac{\omega}{c}\sin\theta\right) - \Theta\left(k - \frac{\omega}{c}\sin\theta\right) \right] \int_{-\infty}^{\infty} dx' f(x') e^{ikx'} e^{-ikx} \\ &= \int_{-\infty}^{\infty} dx' f(x') \int_{-\infty}^{\infty} dk \left[\Theta\left(k + \frac{\omega}{c}\sin\theta\right) - \Theta\left(k - \frac{\omega}{c}\sin\theta\right) \right] e^{ik(x'-x)} = \int_{-\infty}^{\infty} dx' f(x') \int_{-k_{max}}^{k_{max}} e^{-ik(x-x')} dk \\ &= 2\int_{-\infty}^{\infty} f(x') \frac{\sin\left[(x-x')\frac{\omega}{c}\sin\theta\right]}{x-x'} dx' = 2\int_{-\infty}^{\infty} f(x-x') \frac{\sin\left(x'\frac{\omega}{c}\sin\theta\right)}{x'} dx', \end{aligned} \tag{12.39}$$

where $\Theta(k-k_0)$ denotes the Heaviside or step function[3] and $\hat{f}(k)$ is the Fourier transform of $f(x)$. If L is the width of the object, then it follows from Eq.(12.39) that the width Δx of the image $g(x)$ is given by the larger of the two quantities L and $\lambda/(2\sin\theta)$, i.e., the width cannot be smaller than $\lambda/(2\sin\theta)$. This ultimately limits the resolution of the microscope.

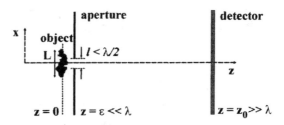

Fig. 12-11: Simple model for near-field detection.

In order to show the superior resolution of near-field optics, we will now place a small aperture of width $l < \lambda/2$ in close distance $\varepsilon \ll \lambda$ to the object with size $L > l$ as shown in Fig.12-11. The detector is placed far away at $z = z_0 \gg \lambda$ where only the far field is detected. The object at $z = 0$ is supposed to emit a wave (neglecting the time dependence)

$$E_1(x, z = 0) = E_0(x, z = 0)[\Theta(x + L/2) - \Theta(x - L/2)] \tag{12.40}$$

Without aperture, the electric field falling onto the detector at $z = z_0$ is given by [$\hat{E}_1(k,z)$ being the Fourier transformation of $E_1(x,z)$]

$$\begin{aligned} E_{det}(x, z = z_0) &= \int_{-\infty}^{\infty} dk_x \hat{E}_1(k_x, z = 0) e^{-ik_z z_0} e^{-ik_x x} \\ &= \int_{-\omega/c}^{\omega/c} dk_x \hat{E}_1(k_x, z = 0) e^{-i\sqrt{k^2 - k_x^2} z_0} e^{-ik_x x}, \end{aligned} \tag{12.41}$$

where we took into account that the wave vector is given by $k^2 = k_x^2 + k_z^2$. The y component is neglected because we assume for simplicity an infinite width of the object and the aperture in y direction.

Next, we consider the aperture. At the position $z = \varepsilon$, the electric field evolves to

$$E_2(x, z = \varepsilon) = \int_{-\infty}^{\infty} dk_x \hat{E}_1(k_x, z = 0) e^{-i\sqrt{k^2 - k_x^2} \varepsilon} e^{-ik_x x} \tag{12.42}$$

The electric field just behind the aperture can be written as

$$E_3(x, z = \varepsilon) = E_2(x, z = \varepsilon)[\Theta(x + l/2) - \Theta(x - l/2)] \tag{12.43}$$

The electric field that reaches the detector at $z = z_0$ is then

$$E_{det}(x, z = z_0) = \int_{-\omega/c}^{\omega/c} dk_x \hat{E}_3(k_x, z = \varepsilon) e^{-i\sqrt{k^2 - k_x^2}(z_0 - \varepsilon)} e^{-ik_x x} \tag{12.44}$$

$$= \int_{-\omega/c}^{\omega/c} dk_x \int_{-\infty}^{\infty} dx' E_2(x', z = \varepsilon)[\Theta(x' + l/2) - \Theta(x' - l/2)] e^{ik_x x'} e^{-i\sqrt{k^2 - k_x^2}(z_0 - \varepsilon)} e^{-ik_x x}$$

$$= \int_{-\omega/c}^{\omega/c} dk_x \int_{-\infty}^{\infty} dx' \int_{-\infty}^{\infty} dk_x' \hat{E}_1(k_x', z = 0) e^{-i\sqrt{k^2 - k_x'^2} \varepsilon} e^{-ik_x' x'} [\Theta(x' + l/2) - \Theta(x' - l/2)] e^{ik_x x'}$$
$$\times e^{-i\sqrt{k^2 - k_x^2}(z_0 - \varepsilon)} e^{-ik_x x}$$

$$= \int_{-\omega/c}^{\omega/c} dk_x e^{-i\sqrt{k^2 - k_x^2}(z_0 - \varepsilon)} e^{-ik_x x} \int_{-\infty}^{\infty} dk_x' \hat{E}_1(k_x', z = 0) e^{-i\sqrt{k^2 - k_x'^2} \varepsilon}$$
$$\times \int_{-\infty}^{\infty} dx' [\Theta(x' + l/2) - \Theta(x' - l/2)] e^{i(k_x - k_x')x'}$$

$$= \int_{-\omega/c}^{\omega/c} dk_x e^{-i\sqrt{k^2 - k_x^2}(z_0 - \varepsilon)} e^{-ik_x x} \int_{-\infty}^{\infty} dk_x' \hat{E}_1(k_x', z = 0) e^{-i\sqrt{k^2 - k_x'^2} \varepsilon} \int_{-l/2}^{l/2} dx' e^{i(k_x - k_x')x'} \quad .$$

Executing the last integral we end up with

$$E_{det}(x, z = z_0) \tag{12.45}$$
$$= 2 \int_{-\omega/c}^{\omega/c} dk_x e^{-ik_x x} e^{-i\sqrt{k^2 - k_x^2}(z_0 - \varepsilon)} \int_{-\infty}^{\infty} dk_x' \hat{E}_1(k_x', z = 0) e^{-i\sqrt{k^2 - k_x'^2} \varepsilon} \frac{\sin[(k_x - k_x')l/2]}{k_x - k_x'} \cdot$$

To evaluate this equation and Eq. (12.41), we consider only one spatial frequency K:

$$\hat{E}_1(k_x, z = 0) = E_1 \delta(k_x - K) \ . \tag{12.46}$$

Inserting this into Eq. (12.41), we obtain the electric field falling onto the detector without aperture:

$$E_{det}(x, z = z_0) = \int_{-\omega/c}^{\omega/c} dk_x E_1 \delta(k_x - K) e^{-i\sqrt{k^2 - k_x^2} z_0} e^{-ik_x x} \tag{12.47}$$

$$= \begin{cases} E_1 e^{-i\sqrt{k^2 - K^2} z_0} e^{-iKx}, & |K| < \omega/c \\ 0, & |K| > \omega/c \end{cases}$$

Without aperture, the microscope can only detect spatial frequencies $|K| < \omega/c$ consistent with the classical limit of resolution.

Inserting Eq.(12.46) into (12.45), the electric field falling onto the detector with an aperture in between is obtained as

$$E_{det}(x, z = z_0) = 2 \int_{-\omega/c}^{\omega/c} dk_x e^{-ik_x x} e^{-i\sqrt{k^2 - k_x^2}(z_0 - \varepsilon)} \tag{12.48}$$

$$\times \int_{-\infty}^{\infty} dk_x' E_1 \delta(k_x' - K) e^{-i\sqrt{k^2 - k_x'^2}\varepsilon} \frac{\sin\left[(k_x - k_x')l/2\right]}{k_x - k_x'}$$

$$= 2E_1 e^{-i\sqrt{k^2 - K^2}\varepsilon} \int_{-\omega/c}^{\omega/c} dk_x e^{-ik_x x} e^{-i\sqrt{k^2 - k_x^2}(z_0 - \varepsilon)} \frac{\sin\left[(k_x - K)l/2\right]}{k_x - K} \quad .$$

It is obvious that Eq.(12.48) is unequal to zero even for spatial frequencies $K > \omega/c$! Figure 12-12 summarizes the result. In the classical case [Fig.12-12(a)], the microscope detects the spatial frequencies in the center of the Airy pattern. As described in the previous sections, the resolution is, therefore, limited by the Rayleigh criterion. By introducing an aperture in close distance to the object, the spatial frequencies are shifted as showing in Fig.12-12(b). Although the width of the frequency spectrum is not larger as in (a), higher spatial frequencies can be detected due to the shift in the spectrum. Hence, information on sub-wavelength details of the object can now be detected. This explains the superior resolution of near-field optics. From Eq. (12.48) it is apparent that the efficiency of the near-field microscope depends on two factors:

(i)   the size of the aperture l. It determines the width of the sinc function, i.e., the smaller the aperture the broader the sinc function giving a larger contribution to the integral for higher spatial frequencies K.

(ii)   The aperture-object distance ε. A smaller distance leads to a smaller argument $-i\sqrt{k^2 - K^2}\varepsilon$ of the exponential damping factor in Eq.(12.48) allowing to detect higher spatial frequencies.

In summary, building a near-field optical microscope with high lateral resolution requires to position a nanoscale aperture as close as possible to the sample surface.

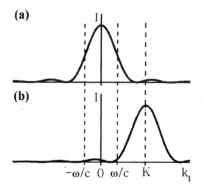

Figure 12-12: Spatial-frequency spectrum. Non-shaded region depicts the spatial frequencies detectable by a microscope with $N.A. = 1$. (a) Classical case without aperture, (b) shifted spectrum due to the insertion of a small aperture.

12.2.2.2. Experimental configurations of aperture SNOM

The basic principle of near-field optics has been discussed in the previous sections. Placing an aperture close to an object can be done in different ways creating a variety of configurations used in *SNOM* which will be briefly reviewed in this section. Most of the commonly used configurations are sketched in Fig.12-13. In illumination mode, as plotted in Fig.12-13(a) and (b), the aperture serves as a nanoscopic light source. Usually, a laser is focused into a single-mode fiber at the end of which the aperture has been formed as described in Chapt. 12.3.1. For the magneto-optic measurements, on which we are focusing in this article, we used this configuration. The converted evanescent waves can be detected either in transmission mode or in reflection mode depending on the opaqueness of the sample. Figure 12-13(c) and (d) describe the collection mode, where the aperture serves as a nanocollector. This mode is often used in fluorescence measurements or single molecule experiments, it is the most simplest set-up. The frustrated total reflection mode (e) is used for plasmon studies because it allows to excite surface plasmons. The last configuration to be mentioned is the shared-aperture mode (f) [47, 48]. Here, the tip aperture serves as light source and collector simultaneously. This mode is used, e.g., in the Sagnac *SNOM* (see Chapt. 12.4.2.1). It has the advantage of a very high lateral resolution but at the trade off of a strongly reduced transmission because the light has to penetrate the aperture twice.

12.2.3. Apertureless SNOM

In 1994, H. K. Wickramasinghe and coworkers proposed another near-field optical microscope [33,49]. It is based on measuring the modulation of the scattered electric field from the end of a sharp tip which is usually an atomic-force-microscope (*AFM*) tip. A simple sketch is given in Fig.12-14. A laser beam is focused on the backside of the sample. Most of the light is reflected from the backside. But, part of the light is scattered by a sharp tip in close distance to the back surface of the sample. The reflected and the scattered light are analyzed by means of an optical interferometer. The physical idea behind this concept is that a sharp tip locally enhances the electric field. The high resolution is due to local dependence of the scattered electric field at the tip. It reaches the highest intensity where the curvature is the largest, i.e. at the apex. The scattered light interacts on his way back with the features at the sample surface yielding information on the sample of the order of the tip size.

The difficulty of this approach is how to detect the very small intensity of the scattered amplitude. By using an *AFM*, the scattered amplitude is modulated by the resonance frequency used for distance control. Therefore, a lock-in technique may be applied. The scattered electric field is phase shifted relative to the reflected beam [49] allowing to detect it in a sensitive differential phase optical interferometer [50]. The scattered field from the tip apex is on top of a spurious background of light scattered from tip shank and cantilever. By using confocal optical illumination and detection, the tip-interaction region is restricted to 100 nm of the tip end. Further reduction of the background noise is achieved by using appropriate modulation amplitudes and frequencies: A vertical modulation of $f_{vert} = 1$ kHz in conjunction with a lateral modulation at $f_{lat} = 3$ kHz, both with an amplitude which is approximately the tip radius. The detection is done at the sum frequency $f_{vert} + f_{lat}$. With this technique, the authors claim to achieve a resolution of 1 nm on dispersed oil droplets on mica [34]. However, to our knowledge only topographic features have been optically imaged with such a high resolution by this method leaving it an open question whether the high resolution originates from topographic cross-talk due to the tip-sample interaction or from a true optical effect due to the local nature of the scattered electric field.

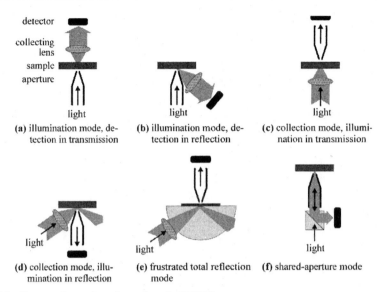

Figure 12-13: Commonly used configurations in *SNOM*.

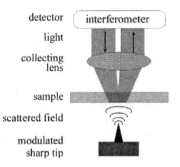

Figure 12-14: Principle of apertureless *SNOM*. The light is reflected from the backside of the sample and analyzed by an interferometer. The electric field scattered from the sharp tip perturbs the interference pattern.

12.3. Technical realization of SNOM

12.3.1. Fabrication of apertures

There are many ways of making suitable apertures for *SNOM*. We will focus in this section on apertures made of single-mode fibers as they are the most widely used apertures and we exclusively worked with them in all our experiments.

12.3.1.1. Apertures made by etching

The most simplest way to make apertures is by etching single-mode fibers. The easiest way to do this is by meniscus etching [51-54]. A thorough review is given in the books by M. Ohtsu [37] and M. Ohtsu and H. Hori [38]. In meniscus etching, a single-mode fiber without protecting buffer is immersed in 40% HF acid covered with a surface layer of an organic solution (e.g. silcon oil, toluene, bromodecane) as outlined in Fig.12-15. Due to surface tension between the fiber and the fluid, a meniscus is not only formed at the surface but also at the interface between the HF acid and the organic solution. The procedure is self-terminating as the height of the meniscus formed at the interface is reduced while the fiber diameter decreases. As a consequence, the organic surface layer moves down protecting the fiber from further etching. The whole process lasts about one hour. The cone angle can be varied from 9-40° with an apex diameter as small as 60 nm [37]. The advantage of this method is, apart from its simplicity and speed, a fairly good reproducibility of up to 80% [37] and the robustness of the tips. In order to fabricate an aperture of only a few ten to hundred nanometer, the etched fiber must be coated by a thin metallic layer as described in Chapt. 12.3.1.3. To fabricate a probe with an apex diameter less than 10 nm, a more complicated selective-etching technique has been developed [55, 56]. The throughput of etched and pulled fiber tips (see next paragraph) are comparable.

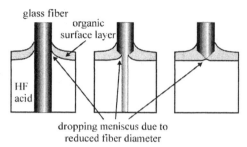

Figure 12-15: Meniscus etching [71].

12.3.1.2. Apertures made by pulling

Another way of producing sharp fiber tips with a slightly better reproducibility than etching is by pulling the single-mode fibers as first described by E. Betzig et al. [30,25]. For this purpose, the authors had modified a commercial micropipette puller from Sutter Instruments [57]. After removing the protecting buffer, a single-mode fiber is drawn while heating it locally by a CO_2 laser. A sketch of the pulling process is shown in Fig.12-16(a). This results in highly uniformly tapered tips with apex radii of less than 100 nm at cone angles between 20-40° [30,37] as shown in Fig.12-16(b). By varying the pulling parameters, the length of the tapered section as well as the cone angle can be adjusted [58]. The advantage of this method is a good reproducibility and a quick manufacture. A disadvantage is the mechanical stress induced in the tip by the pulling process. This renders the tips rather fragile. The throughput of pulled fiber tips is of the order of 10^{-6}-10^{-4} [59].

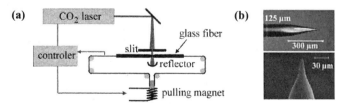

Figure 12-16: (a) Modified micropipette puller. (b) Thermally pulled single-mode fiber tip [60].

12.3.1.3. Forming an aperture by metallic coating

The fiber tips described in the previous two chapters can be used as apertures for *SNOM*.

However, because for etched tips the cladding is removed first, the fiber looses its wave-guide properties before the light reaches the outermost tip region. The same is true for pulled fibers. The core is tapered as well causing the light to leak into the cladding prior to reaching the apex. In order to reduce this effect, the fiber tips are coated under high-vacuum conditions with a metallic layer of high reflectivity such as aluminum, silver or gold as outlined in Fig.12-17(a). Generally, chromium is used as a buffer layer because of its large sticking coefficient onto glass. The thickness of the metallic coating is chosen such that it well exceeds the penetration depth of light, typically 50-100 nm. If the fiber tip is kept at an angle slightly larger than 90° during deposition of the metal film, the very end of the tip will not be coated leaving a circular aperture of a few hundred to less than one hundred nanometer as shown in Fig.12-17(b).

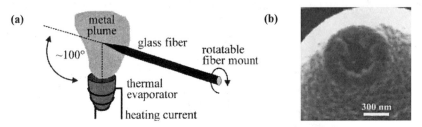

Figure 12-17: (a) Aperture forming by metallic coating. (b) Thermally pulled single-mode fiber coated with a 100 nm Al film. The circular aperture of less than 300 nm is clearly seen in the center of the tip [60].

To achieve an homogeneous coating, the fiber tips are rotated along their axis. The diameter of the apertures can be made as small as 20 nm [25]. Ultimately, the size of the aperture is limited by the finite penetration depth of light which amounts to 10-20 nm for aluminum and noble metals in the visible part of the spectrum. A shortcoming of apertures made by metallic coating onto etched or pulled glass fibers is the relatively low damage threshold with respect to laser intensity. For cw laser intensities of only a few mW coupled into the fiber, the tips heat up over the melting point of glass due to the large absorption in the apex region of the tip [60]. For pulsed laser, the damage threshold is much higher.

12.3.1.4. Apertures made by nanostructuring

A more sophisticated way to prepare small apertures is by nanostructuring using standard lithographic procedures [61-67]. The advantage of this method is an almost perfect reproducibility and a high throughput due to conventional wafer technology as well as a much higher damage threshold than for etched or pulled fibers. In addition, the nanostructures are very robust. The disadvantage is the need for expensive equipment in order to fabricate such apertures limiting the manufacturing to a few groups worldwide. It is also very difficult to buy them commercially. Another method, to make nanostructured apertures is by focused-ion-beam milling (*FIB*) [68-70] which we will not discuss in this article. *FIB* is mainly used to optimize pulled or etched fibers by flattening the apex region or removing excessive metal coating yielding very well defined apertures. The disadvantage of the method is again the high cost of a *FIB* equipment.

In some of the experiments discussed later, nanostructured apertures from the group of E. Oesterschulze [64-66] were used. They consist of hollow metal pyramids with a small aperture of 100-300 nm diameter grown on a silicon waver as shown in Fig.12-18(a). The base length of the pyramid is 20 µm. As a metal aluminum is used protected with a thin SiO$_2$ layer. Originally, the pyramids were grown on *AFM* cantilevers in order to combine

AFM with *SNOM* [64]. In this configuration, it is very difficult to focus the laser onto the back of the pyramid in order to illuminate the aperture. Therefore, we developed a method to place the pyramids onto the flat end of a single-mode fiber in order to produce a new type of *SNOM* tips [60]. For this purpose, the pyramids have later been placed on round silicon pads of the same size as the diameter of a single-mode fiber as shown in Fig.12-18(b) [71].

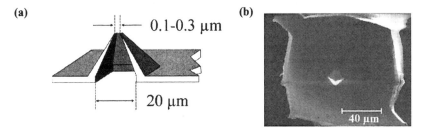

Figure 12-18: (a) Sketch of a cross-section through a hollow metal pyramid. (b) Silicon pad with hollow metal pyramid in the center.

Figure 12-19: Silicon pad with hollow metal pyramid positioned on single-mode fiber.

In order to achieve a high throughput, the nanoscopic aperture of the metal pyramid has to be very precisely positioned onto the core of the fiber which has only a diameter of approximately 3 μm. For this purpose, a mounting stage with a resolution of a few nanometer has been developed [60]. The fiber is then positioned at a few micrometer distance under a pad which is still attached to the waver. The correct position with respect to the aperture and the fiber core is controlled by coupling laser light into the fiber and monitoring the intensity transmitted through the aperture. At the optimum position, the pad is glued to the fiber by means of a special epoxy resin that hardens by ultraviolet light [60,71]. Thereafter, the fiber with the pad can be broken off from the wafer. Figure 12-19 gives an example of such a tip.

12.3.2. Shear-force distance control

Apart from fabricating well-defined apertures, measuring in the optical near-field requires a precise control of the aperture-sample separation on a nanometer scale. The aperture has to be kept at a fixed distance in order for the optical signal not to be influenced by the spatial dependence of the evanescent waves. There are numerous ways to control the distance but the method most widely used is the shear-force distance control first proposed by E. Betzig et al. [72]. The fiber tip is excited to oscillate parallel to the sample surface at an amplitude of a few nanometer. While approaching the surface, the oscillation will be damped due to shear forces between sample surface and tip. The damping starts at a distance of about 10 nm in ambient atmosphere. A typical approach curve is plotted in Fig.12-20. Up to now, it is not clear what the physical origin of the shear force is. There have been made a lot of speculative suggestions in literature which will not be discussed here.

However, the shear-force distance control even works in ultra-high vacuum on surfaces where the water film has been removed. In this case, the damping starts at a distance of about 1-2 nm [73]. This clearly proofs two things: (i) the damping is influenced by adsorbates such as water, (ii) part of the shear force originates from the surface of the solid. A simple tapping mode, as proposed in literature [74], is definitely not correct.

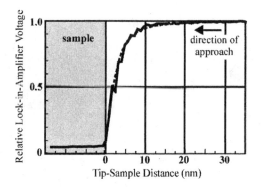

Figure 12-20: Approach curve (solid line) for shear-force distance control. The dashed line is a fit according to a simple damped-oscillator model as described in the text [60].

The approach curve can be qualitatively simulated by a simple damped-oscillator model:

$$\frac{\partial^2 x}{\partial t^2} + 2\delta \frac{\partial x}{\partial t} + \omega_0^2 x = A_0 \sin(\omega t). \tag{12.49}$$

In order to account for the tip-sample interaction, we assume a damping constant

$$\delta = \delta_0 + \delta(z), \qquad \delta(z) = \frac{\gamma}{z}. \tag{12.50}$$

Eq.(12.49) is solved by

$$x = A_S(\omega, z)\cos(\omega t + \varphi). \tag{12.51}$$

This yields for large distance z a resonance frequency ω_R which is shifted with respect to the free ($\delta = 0$) resonance frequency ω_0:

$$\omega_R^2 = \omega_0^2 - 2\delta_0^2. \tag{12.52}$$

For the amplitude, we get

$$A_S(\omega = \omega_R, z) = \frac{A_0}{\sqrt{4\delta_0^4 + 4\left(\delta_0 + \frac{\gamma}{z}\right)^2 (\omega_0^2 - 2\delta_0^2)}}. \tag{12.53}$$

The result is plotted in Fig.12-20 as the dashed line. Obviously, this simple model describes very well the behavior of the tip approach. Nevertheless, it gives no answer on the nature of the shear-force interaction. The region below 10 nm is best suited for distance control as the change in shear-force amplitude is monotonically decreasing. The dependence is not exponential as in *STM* limiting the vertical sensitivity to a few Ångstrom.

12.3.2.1. Detection of the shear-force signal

The most commonly used method to detect the shear-force signal is by using a quartz tuning crystal in order to measure the oscillation of the tip as proposed by K. Karrai and R. D. Grober [75,76]. Figure 12-21 sketches the set-up used in our experiments.

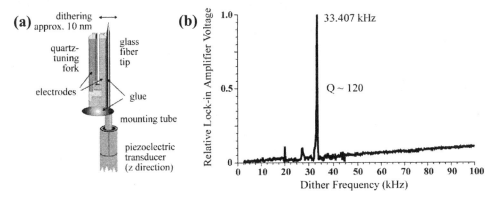

Figure 12-21: (a) Shear-force detection scheme using a quartz tuning fork. (b) Typical resonance curve [60].

As detector serves a quartz tuning fork with a resonance frequency of nominally 32.768 kHz which is commercially used in watches and, thus, very cheap. It is glued to the fiber tip mounted in a tube. The tube is connected to a piezoelectric transducer enabling the tip-sample approach as described later. The fiber is brought to resonance by being excited either by the piezoelectric transducer itself or through a small piece of piezoelectric ceramic that has been glued to the tuning fork [77]. The first method has the disadvantage that the dither amplitude and the phase change when approaching the sample because of the non-linear behavior of the piezoelectric transducer. This requires a recalibration of the resonance before a stable tip-sample contact can be established [60]. The second method does not have this shortcoming. However, the dither amplitude depends in the second case quite sensitively on how the piece of piezoelectric ceramic is glued to the tuning fork. A third method uses the tuning fork itself to excite the fiber. For this purpose, the two electrodes that cover both arms of the tuning fork are separated in such a way that one arm serves to excite the fiber while the other arm is used to detect the shear-force signal [78]. This method leads to a very stable detection without drawback for the excitation.

12.3.2.2. Experimental set-up of a SNOM in transmission and reflection mode

The *SNOM* shown in Fig.12-22, which has been used for the measurements described in this article, is a home-built device. A remarkable feature is the miniaturized cylindrical scanner head with only 40 mm length and 8 mm diameter. The small size makes the microscope almost insensitive to external mechanical distortions or to thermal drift effects and allows easy integration of the *SNOM* into all kinds of experimental set-ups. For reflection mode, the *SNOM* head is placed through a hole in a spherical reflector in such a way that the sample is positioned slightly below the focus [79]. For transmission mode, the spherical reflector simply has to be removed. The miniaturization of the *SNOM* head was achieved by the employment of a commercial piezoelectric transducer [80], used as the z drive, which is a linear motor with a driving range of 5 mm and a sub-nm resolution. The tip-sample distance is kept constant by a shear-force feedback loop using a quartz tuning fork as sensor for the damping of the lateral motion of the tip as previously described. Scanning the sample is done by means of the outer piezoelectric tube. The scan range amounts to 20×20 µm². Using the inner piezoelectric tube only for the *z* direction has the advantage that the aperture is not moved relative to collecting lens and detector ensuring steady conditions even while scanning.

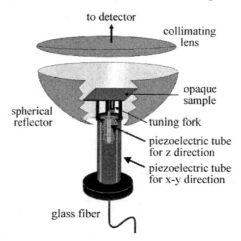

Figure 12-22: Sketch of the SNOM set-up in reflection mode.

12.4. MAGNETIC IMAGING BY MAGNETO-OPTICS

Magnetic imaging is a very important tool for understanding the dynamics of magnetic properties especially in thin magnetic films, in recording media, and in nanostructured materials. In addition to measuring magnetic properties such as magnetization, understanding the dynamics of the magnetization-reversal process is crucial for improving recording media. The reversal process can be easily studied by measuring hysteresis curves, i.e., the change of magnetization as a function of applied magnetic field. However, this does not yield direct information about the local dynamics of magnetic moments such as nucleation process or domain-wall movement. For this purpose, imaging of magnetic domains is the key method. In this book, several contributions deal with magnetic imaging: spin-polarized low energy electron microscopy (chapter 9), advanced transmission electron microscopy (chapter 7), Brillouin light scattering (chapter 14), magnetic force microscopy (chapter 11), and time-resolved magneto-optic microscopy (chapter 13). The advantage of optical methods as opposed to electron-based techniques or techniques involving a magnetic tip (*MFM*) is the insensitivity to external magnetic fields allowing to study the dynamics even in high applied fields.

In this chapter, we will first give a short introduction to the theory of magneto-optics and then explain how magneto-optics can be implemented in *SNOM*.

12.4.1. Theory of magneto-optics

In a general but not very specific way *magneto-optics* is defined as the interaction of electromagnetic radiation with magnetized matter or with material located in a magnetic field. Hence, magneto-optics is a fairly large area in physics. Looking at the electromagnetic spectrum as outlined in Fig.12-23, we can find a variety of magneto-optic effects which are related to a particular part of the spectrum.

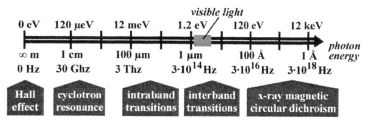

Figure 12-23: Magneto-optic effects and their relation to the electromagnetic spectrum.

The *dc limit* of magneto-optics is the well-known *Hall effect* which describes the appearance of a transverse voltage when an electric current flows through a material which is exposed to a magnetic field, $\vec{B}$. It is caused by the Lorentz force, $\vec{F}_L = q\vec{v} \times \vec{B}$ leading to a pile-up of electric charge which is equivalent to an electric voltage, the Hall voltage. In the *microwave* region, magneto-optic effects appear in form of the *cyclotron resonance*. When exposed to a constant magnetic field, free carriers (e.g. electrons or holes) will be forced on a circular trajectory in a plane perpendicular to the magnetic field. When an additional electromagnetic microwave is interacting with the carriers, the time-dependent electric field of the wave will accelerate the carriers. If the frequency of the wave is at resonance, the carriers are accelerated in phase on their circular trajectory and will strongly absorb the microwaves. From the magnetic field and the frequency of the microwaves the effective mass of the free carriers averaged over the trajectory can be determined. The reason that this effect occurs in the microwave regime is because *dc* magnetic fields can not be made much larger than 20 T. In the *far and near infrared* region, magneto-optic effects due to *intraband* transitions are prevalent. Intraband transitions are transitions of free carriers within the conduction band. There are two contribution to the magneto-optic signal which have different frequency dependence. The first term is related to the classical Drude conductivity. The second term is nonclassical and is due to spin-orbit scattering. Both contributions are decreasing with frequency. The Drude-like term decreases with the third power and the spin-orbit term with the first power of the frequency. Therefore, intraband transitions do usually not play a role in magneto-optic spectra above a photon energy of approximately 1 eV. In the *near infrared*, the *visible*, and the *near ultraviolet* part of the electromagnetic spectrum, *interband* transitions are dominating. These are excitations from the valence band to the conduction band or, in conducting materials such as metals, also from the conduction band to empty higher-level states. If the electronic transition is dipole-allowed, i.e. it fulfils the dipole selection rules, and there exists a large spin polarization in the ground state or a large exchange splitting in the final state, substantial magneto-optic effects are to be expected. In the *far ultraviolet* and the *x-ray* regime, transitions from core levels to the conduction band occur generating *x-ray magnetic circular dichroism* (*XMCD*). If the final state is exchange split, an asymmetry will happen between left (*lcp*) and right (*rcp*) circularly polarized light. This is due to the fact that the selection rules are different for *rcp* and *lcp* light leading to slightly shifted transition energies and amplitudes which produce the *XMCD* effect.

Besides Faraday and Kerr effect, there exists other magneto-optic effects. In this article we will, however, concentrate on these two effects as they are by far those that are used the most.

12.4.1.1. Faraday-effect

The oldest magneto-optic effect, the Faraday effect, is named after his discoverer M. Faraday [81]. It is also described as *circular magnetic dichroism* because it originates from a splitting of the index of refraction for left (*lcp*) and right circularly polarized (*rcp*) light. When describing magneto-optic effects, it is important to define the relation between the applied magnetic field $\vec{B}$ (or in the case of magnetic materials the direction of the magnetization $\vec{M}$) and the direction of the light defined by the wave vector $\vec{k}$. In this context we will often talk about Faraday *geometry* when we think of a set-up which measures the Faraday effect.

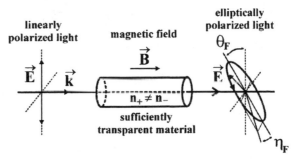

Figure 12-24: Description of the Faraday effect.

Linearly polarized light is transmitted through an optically isotropic, transparent material as shown in Fig.12-24. This can be, e.g., an insulator or semiconductor with a band gap larger than the photon energy of the light so that it is transparent. But even in strongly absorbing materials such as metals, light can travel a few nanometer as the penetration depth of light is given by the inverse value of the absorption constant K which is about to 10^6 cm^{-1} for metals in the visible [82]. Therefore, in thin films, which are only a few nanometer thick, it is very easy to transmit light in order to measure a Faraday effect. A magnetic field is applied to the material parallel to the light direction. As we will explain later, this will lead to a splitting of the index of refraction n into n_+ and $n_- \neq n_+$ initiating a magnetic-field-induced anisotropy in the material. The index "+" and "-" denotes *rcp* and *lcp* light, respectively. After transmission, the light is in general *elliptically polarized*. The rotation θ_F of the major axis of the ellipse with respect to the polarization direction of the incoming light is called *Faraday rotation*. The tangent of the *Faraday ellipticity* η_F is defined as the ratio between minor and major axis. A nonzero ellipticity arises only if the material has a finite absorption. The Faraday rotation and ellipticity are not independent of each other. They can be combined to a *complex Faraday rotation*

$$\tilde{\theta}_F = \theta_F - i\,\eta_F. \tag{12.54}$$

The complex Faraday rotation defines a function in the complex plane which is continuous. As a consequence, the real and imaginary part are related by the Kramers-Kronig relation [83].

Faraday himself had already described the dependence of the effect on the experimental parameters. He noticed a linear dependence of θ_F on the strength of the applied magnetic field $\vec{B}$ and on the length d of the material. The proportionality constant is called *Verdet constant V*. However, V is not a constant since it depends on the photon energy and in magnetically ordered materials it depends on the magnetization rather than on the applied magnetic field. Wiedemann and Verdet gave an empirical law for θ_F, which also includes a directional dependence between the applied magnetic field and the wave vector of the light:

$$\theta_F = V\,B\,d\cos\alpha. \tag{12.55}$$

Here, α is the angle between $\vec{B}$ and $\vec{k}$. Faraday already noticed that there is no connection between the sign of θ_F and the type of magnetism of the material such as para- or diamagnetism.

In addition, he gave a sign convention for the Faraday rotation by observing that a glass rod was rotating the plane of polarization in the direction of the positive current generating the magnetic field. So, he *defined* a Faraday rotation in this direction as positive. Therefore, glass has by definition a positive Faraday rotation.

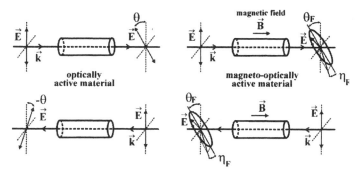

Figure 12-25: Optical activity versus Faraday effect.

We should point out here that there exists another material property which leads to a rotation of the polarization direction of light. It is called *optical activity* and can be observed, e.g., when light is transmitted through a solution of sugar and water. The rotation is then proportional to the sugar concentration. This effect has nothing to do with magnetism. It originates from the shape of the sugar molecules. There are similarities between Faraday effect and optical activity in such a way that they both exhibit the same optical anisotropy of the material. But there is a major difference between them because of the dependence of θ_F on the magnetic field. This is explained in Fig.12-25. In the case of optical activity, the rotation depends on the direction of the light, i.e., it will reverse sign when the light travels in the opposite direction. In the case of Faraday effect, the sign of the rotation depends solely on the direction of the *magnetic field* independent of the light direction. Therefore, by keeping the magnetic-field direction fixed, θ_F will not change sign when the light travels in the opposite direction. Thus, the Faraday effect violates time-reversal invariance. This has some consequences for technical applications. So is it possible to amplify the Faraday effect by letting the light travel multiple times through the material. This can be used to build a device which is able to transmit light in one direction and completely blocks it in the reversed direction. Such a device is called an optical isolator and is widely used in laser applications in order to suppress back-reflected light spots.

However, the blocking effect has already been known in the early days of magneto-optics. Lord Rayleigh [84] invented the so-called Rayleigh trap which was used in those days to entertain the lords and ladies at parties. The principle, which is identical to the one of a modern optical isolator, is plotted in Fig.12-26. The thickness of a Faraday rotator and the magnetic field are chosen such that light traveling through the material experiences a rotation of exactly 45°. A polarizer[4] with polarizing axis at 0° is put on one side of the Faraday rotator and a polarizer with the axis at 45° on the other side. When the light travels from the left, it will be polarized at 0° by the polarizer. After transmission through the Faraday rotator it will be rotated to 45° which is exactly the direction of the axis of the polarizer to the right. Hence the light is transmitted at a polarization of 45°. When the light is incident from the right, it will first be polarized at 45° by the polarizer to the right. Then by traveling through the Faraday rotator - because of the time-reversal violation of the Faraday effect - the polarization of the light will be rotated by another 45° to 90°. But now, having a polarization perpendicular to the axis, the light is blocked by the polarizer to the left.

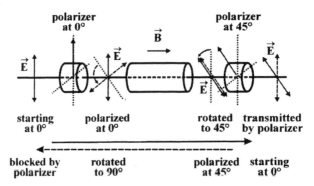

Figure 12-26: The Rayleigh trap. Light traveling from the left is transmitted while light traveling from the right will be completely blocked.

12.4.1.2. Kerr-effect

Besides the Faraday effect, there is another important magneto-optic effect which is widely used in experiments. It is called *magneto-optic Kerr effect* [85] in order to distinguish it from the *electro-optic* Kerr effect which has nothing to do with magnetism. In contrast to the Faraday geometry, the magneto-optic Kerr effect is observed when light is *reflected off* a surface. As an experimental technique, this allows to study thick opaque materials or thin films on opaque substrates. The disadvantage is that the Kerr effect is very small (usually Kerr rotations are less than 1°) and that the effect cannot be magnified by varying the thickness because it does not depend on thickness except for ultrathin films of only a few monolayers.

When linearly polarized light is reflected off the surface of a magnetic material, it will become elliptically polarized as explained in Fig.12-27. The origin of the change in polarization is the same as for the Faraday effect. The magnetic field induces a macroscopic anisotropy in the material which leads to a splitting of the optical properties for *rcp* and *lcp* light such as the index of refraction $n_+ \neq n_- \neq n$. This implies a splitting of the reflection coefficient $\rho_\pm$ due to

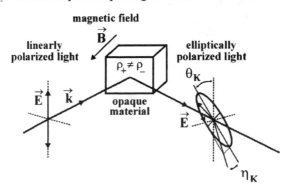

Figure 12-27: Description of the magneto-optic Kerr effect.

Fresnel's formulas [40]. The difference in reflection for *rcp* and *lcp* light leads to the observed elliptical polarization. The rotation of the major axis of the ellipse with respect to the direction of the incoming light is called the *Kerr rotation* θ_K. The ratio between the minor and the major axis is defined as the tangent of the *Kerr ellipticity* η_K. Different from the Faraday effect, η_K is nonzero in the case of nonabsorbing materials and $\theta_K = 0$. There are three different geometries of the Kerr effect depending on the direction of the magnetic field or, more precisely, the direction of the magnetization $\vec{M}$ as shown in Fig.12-28. The effect just described, where the magnetization is perpendicular to the surface, is called the *polar* Kerr effect [Fig.12-28(a)]. If $\vec{M}$ is parallel to the surface and parallel to the plane of incidence[5] the effect is called *longitudinal* Kerr effect [Fig.12-28(b)] and if $\vec{M}$ is parallel to the surface and perpendicular to the plane of incidence the effect is called *transverse* Kerr effect [Fig.12-28(c)]. For small angle of incidence[6], the three effects exhibit a different magnetic-field dependence. While the polar Kerr effect is linear in $\vec{M}$, the longitudinal and transverse Kerr effect are quadratic in $\vec{M}$. This is only true for almost perpendicular incidence. Otherwise, a mixture of linear and quadratic behavior is observed which is sometimes very difficult to interpret. In the other extreme of glancing incidence, the polar Kerr effect is quadratic in $\vec{M}$ while the longitudinal is linear in $\vec{M}$. The transverse Kerr effect is still proportional to the square of $\vec{M}$. The origin of this behavior is due to the anisotropy in the optical properties which is induced by the magnetization. Therefore, the direction of $\vec{M}$ defines the anisotropy axis in the material. Depending on the relative alignment between the wave vector of the light and $\vec{M}$, a different dependence is observed. The component of $\vec{M}$ parallel to produces a linear dependence and the perpendicular components of $\vec{M}$ give rise to a quadratic dependence.

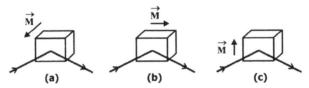

Figure 12-28: Geometries of the magneto-optic Kerr effect. (a) polar (b) longitudinal and (c) transverse Kerr effect.

12.4.1.3. Atomic model of dispersion (Lorentz)

H. A. Lorentz developed a simple atomic theory at the end of the 19th century with the aim of explaining the optical dispersion near absorption lines [86]. His key assumption was that the electrons are bound to an equilibrium position by Hook's law leading to a harmonic oscillation. Taking into account a friction term $\gamma \dot{\vec{r}}$, the periodic electric field of a plane wave $\vec{E} = \vec{E}_0 e^{i(\omega t - gz)}$, and the Lorentz force caused by an external static magnetic field $\vec{B}$, the equation of motion for a bound electron is

$$\ddot{\vec{r}} + \gamma \dot{\vec{r}} + \omega_0^2 \vec{r} = \frac{e}{m} \vec{E}_0 e^{i(\omega t - gz)} + \omega_c \dot{\vec{r}} \times \hat{e}_z \quad . \tag{12.56}$$

Note that the charge e is a negative number for electrons, $\hat{e}_z$ is a unit vector in z direction, and m is the free-electron mass. In Eq. 12.56, the magnetic field is, without loss of generality, assumed to be along the z axis, i. e., $\vec{B}=(0, 0, B)$. In Faraday or polar Kerr geometry. the wave vector $\vec{g} =(0, 0, g)$ is parallel to $\vec{B}$. Thus, the electric field is in the x-y plane, i. e., $\vec{E}_0 = (E_x, E_y, 0)$. In addition, we have defined the resonance frequency ω_0 as

$$\omega_0^2 = \frac{\hat{f}}{m} \quad , \tag{12.57}$$

with $\hat{f}$ being the force constant of Hook's law, and the cyclotron frequency ω_c as

$$\omega_c = \frac{eB}{mc} \quad . \tag{12.58}$$

Looking at Eq. 12.56, we see immediately that the motion in z direction is purely harmonic. The x and y components, however, are coupled through the Lorentz force. As the equation of motion is rotationally invariant along the z axis, it can be solved by transforming it into a circular coordinate system, $\{\hat{e}_+,\hat{e}_-,\hat{e}_z\}$, where

$$\hat{e}_+ = \frac{1}{\sqrt{2}}(\hat{e}_x + i\hat{e}_y), \quad \hat{e}_- = \frac{1}{\sqrt{2}}(\hat{e}_x - i\hat{e}_y),$$

$$\hat{e}_x = \frac{1}{\sqrt{2}}(\hat{e}_+ + \hat{e}_-), \quad \hat{e}_y = \frac{-i}{\sqrt{2}}(\hat{e}_+ - \hat{e}_-). \tag{12.59}$$

Inserting into Eq. 12.56 the ansatz

$$\vec{r}_\pm = \widetilde{A}_\pm \vec{E}_\pm \ , \tag{12.60}$$

where $\vec{E}_\pm = E_{0\pm}\,e^{i(\omega t - gz)}\hat{e}_\pm$ and $\vec{r}_\pm = r_{0\pm}\,\hat{e}_\pm$, yields

$$\widetilde{A}_\pm E_{0\pm}(-\omega^2 + i\gamma\omega + \omega_0^2)\hat{e}_\pm = \frac{e}{m}E_{0\pm}\,\hat{e}_\pm \pm i\omega\omega_c\,\widetilde{A}_\pm E_{0\pm}\hat{e}_\pm \times \hat{e}_z. \tag{12.61}$$

Using the identity $\hat{e}_\pm \times \hat{e}_z = (\hat{e}_x \pm i\hat{e}_y) \times \hat{e}_z = \pm i(\hat{e}_x \pm i\hat{e}_y) = \pm i\hat{e}_\pm$, the amplitude is obtained as

$$\widetilde{A}_\pm = \frac{e}{m(\omega_0^2 - \omega^2 \pm \omega\omega_c + i\omega\gamma)} \ . \tag{12.62}$$

Assuming a concentration, N, of the electrons per unit volume, the total polarization, $\vec{P}_\pm$, amounts to $\vec{P}_\pm = Ne\vec{r}_\pm = Ne\widetilde{A}_\pm\vec{E}_\pm = \widetilde{\alpha}_\pm\vec{E}_\pm$, where $\widetilde{\alpha}_\pm$ is the total polarizability. Finally, the complex dielectric function, $\widetilde{\varepsilon}_\pm(\omega)$, is for rcp ('+') and lcp ('-') light (in cgs units)

$$\widetilde{\varepsilon}_\pm = 1 + 4\pi\widetilde{\alpha}_\pm = 1 + \frac{4\pi Ne^2}{m}\frac{1}{\omega_0^2 - \omega^2 \pm \omega\omega_c + i\omega\gamma} \ . \tag{12.63}$$

By separating $\widetilde{\varepsilon}_\pm = \varepsilon_{1\pm} - i\varepsilon_{2\pm}$ into real and imaginary part, we get

$$\varepsilon_{1\pm} = 1 + \frac{4\pi Ne^2}{m}\frac{\omega_0^2 - \omega^2 \pm \omega\omega_c}{(\omega_0^2 - \omega^2 \pm \omega\omega_c)^2 + \omega^2\gamma^2}, \tag{12.64}$$

$$\varepsilon_{2\pm} = \frac{4\pi Ne^2}{m}\frac{\omega\gamma}{(\omega_0^2 - \omega^2 \pm \omega\omega_c)^2 + \omega^2\gamma^2}.$$

Obviously, the resonance frequency shifts by applying an external magnetic field and the change depends on the polarization state of the light. The new (magnetic-field-dependent) resonance frequency, $\omega_{0\pm}$, satisfies the condition

$$\omega_0^2 - \omega_{0\pm}^2 \pm \omega_{0\pm}\omega_c = 0 \ . \tag{12.65}$$

Using the approximation, $\omega_{0\pm}^2 - \omega_0^2 = (\omega_{0\pm} - \omega_0)(\omega_{0\pm} + \omega_0) \cong 2\omega_{0\pm}(\omega_{0\pm} - \omega_0)$, we obtain

$$\omega_{0\pm} - \omega_0 = \pm\tfrac{1}{2}\omega_c = \pm\omega_L \ , \tag{12.66}$$

where ω_L is the Larmor frequency. Note that $\omega_c < 0$ for electrons.

The Lorentz model is in accord with a positive Faraday rotation of quartz, θ_F^q, in the region of normal dispersion (i. e., $\partial n/\partial\omega > 0$). From Eq. (12.66), it follows that for electrons $\omega_{0+} < \omega_0 < \omega_{0-}$. If $|\omega_{0+}-\omega_{0-}| \ll \omega_0$, then the shape of the energy dispersion of the index of refraction, $n(\omega) \to n_\pm(\omega)$, is to first order unchanged. Therefore, $n_\pm(\omega)$ is simply shifted along the x axis by $\mp\omega_c$. Assuming normal dispersion, we get $n_+(\omega) > n_-(\omega)$ and hence $\theta_F^q \propto (n_+-n_-) > 0$ as defined by Faraday.

12.4.1.4. Origin of the magnetic-field-induced anisotropy

Within the framework of the Lorentz model, it can be demonstrated that a magnetic-field-induced shift of the resonance frequency in a circular basis leads to a macroscopic anisotropy in all optical properties. As an example, let us discuss the magnetic-field-induced anisotropy of the dielectric tensor $\tilde{\boldsymbol{\varepsilon}}$ (neglecting the undisturbed z direction at the moment),

$$\tilde{\boldsymbol{\varepsilon}} = \begin{pmatrix} \tilde{\varepsilon}_{xx} & \tilde{\varepsilon}_{xy} \\ \tilde{\varepsilon}_{yx} & \tilde{\varepsilon}_{yy} \end{pmatrix} . \tag{12.67}$$

In an isotropic material, $\tilde{\boldsymbol{\varepsilon}}$ reduces to a complex number, i.e. $\tilde{\varepsilon}_{xx} = \tilde{\varepsilon}_{yy} = \tilde{\varepsilon}$ and $\tilde{\varepsilon}_{xy} = \tilde{\varepsilon}_{yx} = 0$. In the presence of a magnetic field, the dielectric tensor is determined by Eq. (12.63), i. e., in a circular basis, it has a diagonal form

$$\tilde{\boldsymbol{\varepsilon}}_\pm = \begin{pmatrix} \tilde{\varepsilon}_+ & 0 \\ 0 & \tilde{\varepsilon}_- \end{pmatrix} \tag{12.68}$$

with eigenvalues $\tilde{\varepsilon}_\pm$. Therefore, the electric displacement is given by $\vec{D}_\pm = \tilde{\boldsymbol{\varepsilon}}_\pm \vec{E}_\pm$. In the next step, $\tilde{\boldsymbol{\varepsilon}}_\pm$ is transformed into the x-y basis according to Eq. (12.59). Following the standard procedure of matrix transformation in linear algebra [87], the dielectric tensor, $\tilde{\boldsymbol{\varepsilon}}$, in x-y basis is obtained by

$$\tilde{\boldsymbol{\varepsilon}} = \mathbf{T}\,\tilde{\boldsymbol{\varepsilon}}_\pm\,\mathbf{T}^{-1} , \tag{12.69}$$

where the rows of the transformation matrix, $\mathbf{T}$, are given by the coefficients of expressing the old (circular) basis by the new (x-y) one in Eq. (12.59). So, we find

$$\mathbf{T} = \frac{1}{\sqrt{2}}\begin{pmatrix} 1 & 1 \\ i & -i \end{pmatrix}, \quad \mathbf{T}^{-1} = \frac{1}{\sqrt{2}}\begin{pmatrix} 1 & -i \\ 1 & i \end{pmatrix} . \tag{12.70}$$

Inserting this into Eq. (12.69), the dielectric tensor, $\tilde{\boldsymbol{\varepsilon}}$, in the x-y system is

$$\tilde{\boldsymbol{\varepsilon}} = \frac{1}{2}\begin{pmatrix} \tilde{\varepsilon}_+ + \tilde{\varepsilon}_- & -i\left(\tilde{\varepsilon}_+ - \tilde{\varepsilon}_-\right) \\ i\left(\tilde{\varepsilon}_+ - \tilde{\varepsilon}_-\right) & \tilde{\varepsilon}_+ + \tilde{\varepsilon}_- \end{pmatrix} = \begin{pmatrix} \tilde{\varepsilon}_{xx} & \tilde{\varepsilon}_{xy} \\ \tilde{\varepsilon}_{yx} & \tilde{\varepsilon}_{yy} \end{pmatrix} . \tag{12.71}$$

It follows at once that $\tilde{\boldsymbol{\varepsilon}}$ has finite off-diagonal elements, $\tilde{\varepsilon}_{yx} = -\tilde{\varepsilon}_{xy}$. As a consequence, the dielectric tensor has to be defined as

$$\tilde{\boldsymbol{\varepsilon}} = \begin{pmatrix} \tilde{\varepsilon}_{xx} & \tilde{\varepsilon}_{xy} & 0 \\ -\tilde{\varepsilon}_{xy} & \tilde{\varepsilon}_{xx} & 0 \\ 0 & 0 & \tilde{\varepsilon}_{zz} \end{pmatrix} , \tag{12.72}$$

where

$$\tilde{\varepsilon}_{xx} = \tfrac{1}{2}\left(\tilde{\varepsilon}_+ + \tilde{\varepsilon}_-\right) . \tag{12.73}$$

and $\tilde{\varepsilon}_{xy}$ satisfies the relation

$$\tilde{\varepsilon}_{\pm} = \tilde{\varepsilon}_{xx} \pm i\tilde{\varepsilon}_{xy} \quad . \tag{12.74}$$

Applying Ohm's law, $\vec{J} = \tilde{\sigma}\vec{E} = Ne\dot{\vec{r}}$, a similar anisotropy is obtained for the optical-conductivity tensor, $\tilde{\sigma}$, and we get

$$\tilde{\sigma} = \begin{pmatrix} \tilde{\sigma}_{xx} & \tilde{\sigma}_{xy} & 0 \\ -\tilde{\sigma}_{xy} & \tilde{\sigma}_{xx} & 0 \\ 0 & 0 & \tilde{\sigma}_{zz} \end{pmatrix} , \tag{12.75}$$

where the connection to the circular basis is given by

$$\tilde{\sigma}_{\pm} = \tilde{\sigma}_{xx} \pm i\tilde{\sigma}_{xy} \quad . \tag{12.76}$$

Eqs.(12.74) and (12.76) are not in accordance with recent reviews on magneto-optics [88,89] because they all incorporate a wrong sign due to an incorrect transformation form circular into x-y coordinates [90]. Consistent with the isotropic case, we get

$$\tilde{\varepsilon}_{\pm} = 1 - i\frac{4\pi}{\omega}\tilde{\sigma}_{\pm} , \tag{12.77}$$

and furthermore

$$\tilde{\varepsilon} = 1 - i\frac{4\pi}{\omega}\tilde{\sigma} . \tag{12.78}$$

A description of the optical conductivity tensor, $\tilde{\sigma}$, based on symmetry arguments [91,92], yields that the diagonal element, $\tilde{\sigma}_{xx}$, depends merely on even powers of the magnetic field or magnetization while the off-diagonal element, $\tilde{\sigma}_{xy}$, depends purely on odd powers. As a consequence, to first approximation $\tilde{\sigma}_{xx}$ is a constant and $\tilde{\sigma}_{xy}$ is linear in the applied field or magnetization.

12.4.1.5. Relation between the off-diagonal conductivity and Faraday and Kerr effect

In this section, the relation between the off-diagonal elements $\tilde{\sigma}_{xy}$ of the optical conductivity tensor and the experimental quantities θ_K (polar Kerr rotation), η_K (polar Kerr ellipticity), θ_F (Faraday rotation), and η_F (Faraday ellipticity) will be derived. For this purpose, we define a complex Faraday rotation,

$$\tilde{\theta}_F = \theta_F - i\eta_F , \tag{12.79}$$

and a complex polar Kerr rotation,

$$\tilde{\theta}_K = \theta_K - i\eta_K \quad . \tag{12.80}$$

In the presence of a magnetic field or a spontaneous magnetization, all optical properties become anisotropic. Consequently, the complex index of refraction,

$$\tilde{n}_{\pm} = (n_{\pm} - ik_{\pm}) \quad , \tag{12.81}$$

as well as the complex coefficient of reflection,

$$\tilde{\rho}_\pm = \frac{\tilde{n}_\pm - 1}{\tilde{n}_\pm + 1} \quad , \tag{12.82}$$

are different for *rcp* ('+') and *lcp* ('-') light. The Faraday effect evolves from the different velocity and absorption of *rcp* and *lcp* light when penetrating a magneto-optically active material. The velocity of light in material is then

$$c_\pm = \frac{c}{n_\pm} \quad , \tag{12.83}$$

where c is the speed of light in vacuum, and the absorption constant is

$$K_\pm = \frac{4\pi k_\pm}{\lambda_{vak}}. \tag{12.84}$$

To understand the Faraday effect, we notice that linearly polarized light can be decomposed into equal amounts of *rcp* and *lcp* light. After having traveled through a magneto-optically active material of thickness d, the amplitudes of the *rcp* and *lcp* components have been changed due to Eqs.(12.83) and (12.84). The superposition of the two components yields no longer the original polarization but a polarization which is described by the Faraday effect (see Fig. 12-24):

$$\theta_F = \frac{\omega d}{2c}(n_+ - n_-) \quad , \quad \eta_F = \frac{\omega d}{2c}(k_+ - k_-) \quad . \tag{12.85}$$

This can be combined to a complex relation:

$$\tilde{\theta}_F = \frac{\omega d}{2c}(\tilde{n}_+ - \tilde{n}_-) \quad . \tag{12.86}$$

A phenomenological explanation of the polar Kerr effect is given in terms of Fresnel's formulas at normal incidence. By writing $\tilde{\rho}_\pm$ in a polar coordinate system, $\tilde{\rho}_\pm = \rho_\pm e^{i\phi_\pm}$, the polar Kerr rotation θ_K and the polar Kerr ellipticity η_K are given by

$$\theta_K = -\frac{\phi_+ - \phi_-}{2} \quad , \quad \tan\eta_K = -\frac{\rho_+ - \rho_-}{\rho_+ + \rho_-} \quad . \tag{12.87}$$

The minus sign is due to the sign convention [90]. These two equations can be related to the complex polar Kerr rotation by

$$\tilde{\theta}_K = \theta_K - i\eta_K \cong i\frac{\tilde{\rho}_+ - \tilde{\rho}_-}{\tilde{\rho}_+ + \tilde{\rho}_-} \quad . \tag{12.88}$$

This equation is only valid in the limit of small angles θ_K and η_K. In view of the discovery of polar Kerr rotations in CeSb [93,94] reaching almost 90° at low photon energies, we present the exact relation [95] which is

$$\frac{\sin(2\theta_K)\cos(2\eta_K)}{1 + \cos(2\theta_K)\cos(2\eta_K)} - i\frac{\sin(2\eta_K)}{1 + \cos(2\theta_K)\cos(2\eta_K)} = i\frac{\tilde{\rho}_+ - \tilde{\rho}_-}{\tilde{\rho}_+ + \tilde{\rho}_-} \quad . \tag{12.89}$$

This reduces to Eq.(12.88) for θ_K, $\eta_K \ll 1$. Inserting Eq.(12.82) in Eq.(12.88) and using the approximations $(\tilde{n}_+ + \tilde{n}_-) \cong 2\tilde{n}$ and $\tilde{n}_+\tilde{n}_- \cong \tilde{n}^2$, where $\tilde{n}$ is the complex index of refraction in the absence of magnetic field or magnetization, we get

$$\tilde{\theta}_K = i\frac{\tilde{n}_+^2 - \tilde{n}_-^2}{2\tilde{n}(\tilde{n}^2 - 1)} \quad . \tag{12.90}$$

Using Eq.(12.76) and

$$\tilde{n}_{\pm}^2 = \left(n_{\pm} - i\,k_{\pm}\right)^2 = \tilde{\varepsilon}_{\pm} = 1 - i\frac{4\pi}{\omega}\tilde{\sigma}_{\pm} \quad , \tag{12.91}$$

we finally obtain

$$\tilde{\theta}_K = -\frac{\tilde{\sigma}_{xy}}{\tilde{n}\,\tilde{\sigma}_{xx}} = \frac{4\pi i}{\omega}\frac{\tilde{\sigma}_{xy}}{\tilde{n}\left(\tilde{n}^2 - 1\right)} \quad . \tag{12.92}$$

Separation into real and imaginary part yields

$$\theta_K = -\frac{4\pi}{\omega}\frac{B\sigma_{1xy} + A\sigma_{2xy}}{A^2 + B^2} \quad , \quad \eta_K = -\frac{4\pi}{\omega}\frac{A\sigma_{1xy} - B\sigma_{2xy}}{A^2 + B^2} \tag{12.93}$$

where the prefactors, A and B, are polynomial functions of n and k,

$$A = n^3 - 3nk^2 - n \quad , \quad B = -k^3 + 3n^2k - k \quad . \tag{12.94}$$

Combining Eqs.(12.76), (12.86), and (12.91) we obtain the corresponding relations for the Faraday effect:

$$\tilde{\theta}_F = \frac{2\pi d}{c}\frac{\tilde{\sigma}_{xy}}{\tilde{n}}. \tag{12.95}$$

Separation into real and imaginary part yields

$$\theta_F = \frac{2\pi d}{c}\frac{n\sigma_{1xy} - k\sigma_{2xy}}{n^2 + k^2} \quad , \quad \eta_F = -\frac{2\pi d}{c}\frac{k\sigma_{1xy} + n\sigma_{2xy}}{n^2 + k^2} \quad . \tag{12.96}$$

With these relations it is possible to compute from the experiment the off-diagonal conductivity elements provided the optical functions have been obtained by a supplementary measurement. Note, that Eqs.(12.92) and (12.93) have opposite sign as those given in recent reviews on magneto-optics [88,89]. Using time-dependent perturbation theory, the off-diagonal elements of the conductivity tensor can be related to the quantum-mechanical matrix elements [92]:

$$\sigma_{1xy} = -\frac{e^2}{2\hbar m^2 V}\sum_{\alpha:\,occ}\sum_{\beta:\,unocc}\frac{\left|\langle\beta|\pi_+|\alpha\rangle\right|^2 - \left|\langle\beta|\pi_-|\alpha\rangle\right|^2}{\omega_{\beta\alpha}^2 - \omega^2} \quad . \tag{12.97}$$

$$\sigma_{2xy} = \frac{\pi e^2}{4\hbar\omega m^2 V}\sum_{\alpha:\,occ}\sum_{\beta:\,unocc}\left[\left|\langle\beta|\pi_+|\alpha\rangle\right|^2 - \left|\langle\beta|\pi_-|\alpha\rangle\right|^2\right]\left[\delta\left(\omega_{\beta\alpha} - \omega\right) + \delta\left(\omega_{\beta\alpha} + \omega\right)\right] \tag{12.98}$$

where $\pi_{\pm}$ is the kinetic momentum operator and the summation extends only over occupied initial states $|a\rangle$ and unoccupied final states $|b\rangle$. From Eqs.(12.97) and (12.98) it follows immediately, that the magneto-optic effects are proportional to the difference in the transition probabilities of *rcp* and *lcp* light.

12.4.2. Magneto-optic imaging

The basic objective in magneto-optic microscopy is to image magnetic domains by determining the polarization state of the light after having been transmitted through a material (Faraday effect) or after being reflected off a surface (Kerr effect). As discussed in Chapt. 12.4.1.4, the off-diagonal elements are proportional to the magnetization. It follows from Eqs.(12.92) and (12.95) that the magneto-optic effects are also proportional to the magnetization. Thus, the Faraday and the polar Kerr effect are a direct measure of the magnetization component perpendicular to the sample surface. In the case of magneto-optic magnetometry, Kerr hysteresis loops are determined at a fixed wavelength yielding information on the local magnetic properties.

A variety of detection schemes can be utilized to determine magneto-optic effects. Concerning polar Kerr-effect measurements, rotations as small as a few thousandths of a degree have to be resolved, especially in thin magnetic layers. In order to get the desired resolution, a lock-in technique is applied involving some sort of modulation. There are several ways of doing that. The best one is to modulate the quantity that has to be determined, i. e., the state of polarization. For *SNOM* applications, a photoelastic modulator [96] is favored. Here, the phase between two orthogonal polarization directions is altered by inducing a uniaxial strain in an isotropic glass. Thus, the modulation varies the state of polarization between linear and circular. The advantage of a phase modulation is the high frequency which is in the range of $\Omega_{mod} \cong 40$-80 kHz, depending on the geometry of the modulator, and the large modulation amplitude that can be achieved. Because of space limitation, we will concentrate on one method which is capable to image magnetic domains as well as to determine quantitatively the local magneto-optic rotation. Before we start the discussion, let us say a few words about an alternative method that takes advantage of the fact, that magnet-optic effects violate time-reversal invariance. The method is based on the Sagnac effect.

12.4.2.1. Sagnac SNOM

Commonly, Sagnac interferometers are used in fiber-optic gyroscopes [97]. In 1994, A. Kapitulnik and coworkers [98] presented a Sagnac interferometer for high-resolution magneto-optic measurements. The apparatus was originally built to search for anyon superconductivity in high-temperature superconductors [99]. In their paper, they mentioned the usefulness of the method for *SNOM* as it is based on fiber optics and they presented first results. The angular resolution achieved was $\Delta\theta_F = 3$ µrad and the lateral resolution $\Delta x = 2$ µm. The Sagnac interferometer is not sensitive to linear optical birefringence (which is omnipresent in glass fibers) and optical activity. However, it will detect phase shifts due to broken time-reversal symmetry effects as in magneto-optic effects.

The principle of a Sagnac interferometer in Kerr configuration is plotted in Fig. 12-29. A laser beam is divided in two equal parts by a beam splitter. One component is sent clockwise (*cw*) around the fiber loop while the other one is sent counterclockwise (*ccw*). Disregarding for a moment the optical components that break up the loop, the two light waves will come to the beam splitter after one turn and interfere constructively at the detector because they have traveled the same path. In a gyroscope, a rotation of the Sagnac loop at constant angular velocity will cause a difference in path length of the two counterpropagating beams inducing a small phase shift, the so-called Sagnac effect. This leads to an intensity change when the two beams interfere at the detector. In the configuration shown in Fig. 12-29, the interferometer is at rest. Any optical activity or birefringence that the beams encounter on their way will by canceled due to the opposite direction of the path (as long as any changes are slow as compared to the travel time of the light). However, all effects that violate time-reversal symmetry will cause a finite phase shift $\Delta\varphi_s$ between the two beams yielding an intensity change in the interferometer. Therefore, by breaking the loop as shown in Fig. 12-29 in order to introduce a magneto-optically active sample, the Kerr effect will produce a finite phase shift $\Delta\varphi_{sK}$ that is proportional to the Kerr effect [98].

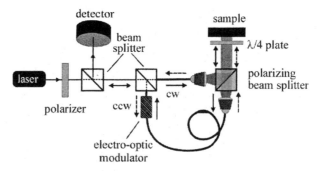

Figure 12-29: Principle of a Sagnac interferometer used in Kerr configuration.

In order to measure in the near-field, a *SNOM* head must be introduced between the λ/4 plate and the sample. This has been proposed by Kapitulnik himself [98] and has been successfully realized a few years later by the group of A. Bauer [100] and since then further optimized [101]. The intriguing feature of the Sagnac *SNOM* is the virtual insensitivity against birefringence in the fiber which is the most severe problem in scanning near-field magneto-optic microscopy as discussed in Section 12.6. A disadvantage is, nevertheless, the shared-aperture mode (see Section 12.2.2.2) which strongly reduces the light intensity. Up to now, only uncoated fiber tips have been used. In their most recent publication, magnetic domains could be imaged in an ultrathin Fe/Cu(100) sample in reflection mode by means of the Kerr effect. An angular resolution of better than 0.02° has been claimed [102] corroborating the advantageous suppression of birefringence by the Sagnac effect.

12.4.2.2. Quantitative scanning near-field magneto-optic microscopy

In this section, a scanning near-field magneto-optic microscope will be introduced that is capable of measuring local magneto-optic effects quantitatively. The *SNOM* head had been discussed in Section 12.3.2.2. Therefore, we will concentrate here on the polarization-sensitive optics. In Fig.12-30 the basic set-up is sketched.

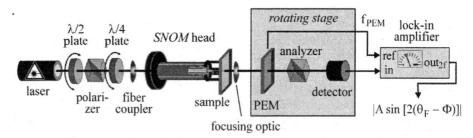

Figure 12-30: Principle of a scanning near-field magneto-optic microscope [60].

The laser light is linearly polarized and coupled into a single-mode fiber with a *SNOM* tip. In front of the polarizer, a λ/2 plate may be introduced to choose a specific polarization direction. Subsequent to the polarizer, a λ/4 plate is used to optimize the polarization within the fiber. After interaction with the near-field, the converted photons are collected by a lens system and analyzed in the detection head which consists of a photoelastic modulator (*PEM*), an analyzer, and a photomultiplier tube (*PMT*) mounted in a fixed relative position onto a rotation stage. The *PEM* modulates the outgoing light with a frequency of 50 kHz.

In order to understand the measurement principle, a short description of polarization states will be given first. There exist two mathematical representations of a polarization state: (i) as a Jones vector and (ii) as Stokes vector [103]. The Jones vector is a 2-dimensional vector and can, therefore, only be used for completely polarized light. Because in fiber optic, one has to deal with a large amount of unpolarized light, the Stokes formalism is better suited to describe the set-up. A Stokes vector has four components:

$$\vec{S} = \begin{pmatrix} I \\ Q \\ U \\ V \end{pmatrix} = \{I, Q, U, V\} = \begin{pmatrix} I_{tot} \\ I_0 - I_{90} \\ I_{+45} - I_{-45} \\ I_+ - I_- \end{pmatrix}, \tag{12.99}$$

where I is the total intensity, $Q = I_0 - I_{90}$ is the difference of the intensities of 0° and 90° linearly polarized light, $U = I_{+45} - I_{-45}$ is the difference of the intensities of 45° and -45° linearly polarized light, and $V = I_+ - I_-$ is the difference of the intensities of *rcp* and *lcp* light. From Eq. (12.99), it follows immediately that a Stokes vector can represent unpolarized light because the total intensity I is independent of the individual polarization components. In the case of a completely polarized beam, we get:

$$I = \sqrt{Q^2 + U^2 + V^2}, \tag{12.100}$$

and for partially unpolarized light, we have

$$I > \sqrt{Q^2 + U^2 + V^2}. \tag{12.101}$$

Partially unpolarized light with elliptical polarization such that the major half axis is located at an angle θ from the x axis and the ellipticity is η, is represented as follows [103]:

$$\vec{S} = E_{pol}^2 \begin{pmatrix} 1 \\ \cos(2\theta)\cos(2\eta) \\ \sin(2\theta)\cos(2\eta) \\ \sin(2\eta) \end{pmatrix} + E_{unpol}^2 \begin{pmatrix} 1 \\ 0 \\ 0 \\ 0 \end{pmatrix} = E_{pol}^2 \begin{pmatrix} 1+\delta \\ \cos(2\theta)\cos(2\eta) \\ \sin(2\theta)\cos(2\eta) \\ \sin(2\eta) \end{pmatrix}, \tag{12.102}$$

where we have defined $\delta = E_{unpol}^2 / E_{pol}^2$. In order to describe the effect of optical elements on the polarization, 4×4 matrices, so-called Müller matrices, are used. For a polarizer (or analyzer) that has the polarizing axis at an angle α with respect to the x axis, the Müller matrix is given by [103]

$$\vec{P}(\alpha) = \frac{1}{2} \begin{bmatrix} 1 & \cos(2\alpha) & \sin(2\alpha) & 0 \\ \cos(2\alpha) & \cos^2(2\alpha) & \cos(2\alpha)\sin(2\alpha) & 0 \\ \sin(2\alpha) & \cos(2\alpha)\sin(2\alpha) & \sin^2(2\alpha) & 0 \\ 0 & 0 & 0 & 0 \end{bmatrix}. \tag{12.103}$$

A photoelastic modulator has a rather complicated Müller matrix. Let $\Delta(t) = \Delta_0 \sin(\Omega_{PEM} t)$ be the phase modulation along the axis that is placed at an angle ϕ with respect to the x axis, then the *PEM* is described as [103]:

$$\ddot{M}(\phi, \Delta(t)) = \hspace{8cm} (12.104)$$

$$\begin{bmatrix} 1 & 0 & 0 & 0 \\ 0 & \cos(4\phi)\sin^2\left(\tfrac{1}{2}\Delta\right)+\cos^2\left(\tfrac{1}{2}\Delta\right) & \sin(4\phi)\sin^2\left(\tfrac{1}{2}\Delta\right) & -\sin(2\phi)\sin\Delta \\ 0 & \sin(4\phi)\sin^2\left(\tfrac{1}{2}\Delta\right) & -\cos(4\phi)\sin^2\left(\tfrac{1}{2}\Delta\right)+\cos^2\left(\tfrac{1}{2}\Delta\right) & \cos(2\phi)\sin\Delta \\ 0 & \sin(2\phi)\sin\Delta & -\cos(2\phi)\sin\Delta & \cos\Delta \end{bmatrix}$$

The Stokes vector of elliptically polarized light as it evolves by transmission through a magneto-optically active sample is given by [103]

$$\vec{S}_F(\theta_F, \eta_F) = E_{pol}^2 \begin{bmatrix} 1+\delta \\ \cos(2\theta_F)\cos(2\eta_F) \\ \sin(2\theta_F)\cos(2\eta_F) \\ \sin(2\eta_F) \end{bmatrix}, \hspace{2cm} (12.105)$$

where we also included an unpolarized component δ that is not changed by the sample. θ_F and η_F are the Faraday rotation and ellipticity, respectively.

The Stokes vector $\vec{S}_{detector}$ at the position of the detector is simply calculated by multiplying the Müller matrices in the order given by the optical set-up of Fig. 12-30 starting with elliptically polarized light coming from the sample:

$$\vec{S}_{det\,ector} = \ddot{P}(\alpha)\ddot{M}(\phi, \Delta(t))\vec{S}_F(\theta_F, \eta_F) \ . \hspace{2cm} (12.106)$$

A detector will only be sensitive to the intensity I, which is the first component of the Stokes vector. Evaluating Eq. (12.106) - which will require a lot of patience and endurance - yields for the intensity at the detector [60]:

$$\begin{aligned} I_{det\,ector}(\alpha, \phi, \theta_F, \eta_F) &= \tfrac{1}{2}E_{unpol}^2 + \tfrac{1}{2}E_{pol}^2\{1 + \cos(2\eta_F)\cos[2(\alpha-\phi)]\cos[2(\theta_F-\phi)]\} \\ &+ \tfrac{1}{2}E_{pol}^2 \cos\Delta(t)\,\cos(2\eta_F)\sin[2(\alpha-\phi)]\sin[2(\theta_F-\phi)] \hspace{1cm} (12.107) \\ &+ \tfrac{1}{2}E_{pol}^2 \sin\Delta(t)\,\sin(2\eta_F)\sin[2(\alpha-\phi)] \end{aligned}$$

Feeding the detector signal into a lock-in amplifier, only the components modulated by the modulation frequency Ω_{PEM} will be measured and all *dc* components suppressed. As a consequence, the measurement is virtually insensitive to the unpolarized component E_{unpol}^2 of the light even if it is dominating. Nevertheless, the *dc* components will limit the dynamic range of the amplifier which is typically -100 dB. Therefore, this method is perfectly suited for scanning near-field magneto-optic measurements with a high amount of unpolarized light. Let us now determine the output signal of the lock-in amplifier. For this purpose, Eq. (12.107) has to be expanded in a series of harmonic functions. Inserting the well-known expansions [104]

$$\sin\Delta(t) = \sin[\Delta_0\sin(\Omega t)] = 2\sum_{n=1}^{\infty} J_{2n-1}(\Delta_0)\sin[(2n-1)\Omega t]\ , \hspace{1cm} (12.108)$$

and

$$\cos\Delta(t) = \cos[\Delta_0\sin(\Omega t)] = J_0(\Delta_0) + 2\sum_{n=1}^{\infty} J_{2n}(\Delta_0)\cos[2n\,\Omega t] \hspace{1cm} (12.109)$$

into Eq. (12.107), we derive the first harmonic component I of the lock-in amplifier

$$I_\Omega(\alpha,\phi,\theta_F,\eta_F) = E_{pol}^2 J_1(\Delta_0) \sin(2\eta_F) \sin[2(\alpha-\phi)]\sin(\Omega t), \qquad (12.110)$$

where J_n are Bessel functions. We notice that I is proportional to $\sin(2\eta_F)$. Thus, the first harmonic is suited to measure the Faraday (or Kerr) ellipticity. In order to maximize I, the term $\sin[2(\alpha-\phi)]$ should be equal to one. This is achieved by setting the analyzer angle

$$\alpha = \phi + 45°. \qquad (12.111)$$

This implies that the analyzer should be kept at a fixed angle of 45° with respect to the *PEM*. We will call this combination the *analyzing unit*. Next, we evaluate the second harmonic component I_2 of the lock-in amplifier

$$I_{2\Omega}(\alpha,\phi,\theta_F,\eta_F) = E_{Pol}^2 J_2(\Delta_0)\cos(2\eta_F)\sin[2(\alpha-\phi)]\sin[2(\theta_F-\phi)]\cos(2\Omega t). \quad (12.112)$$

Again, the signal is maximized by setting $\alpha = \phi + 45°$. The second harmonic $I_{2\Omega}$ is proportional to $\sin[2(\theta_F-\phi)]$. Hence, it is perfectly appropriate to measure the Faraday rotation. Further improvement of the signal is achieved by choosing a proper modulation amplitude Δ_0 in order to maximize the Bessel functions $J_1(\Delta_0)$ and $J_2(\Delta_0)$. Looking at Fig. 12-31, it is easily seen that for the first harmonic I_Ω a value of $\Delta_0 = \pi/2$ (i.e. a phase shift of $\lambda/4$) and for $I_{2\Omega}$ a value of $\Delta_0 = \pi$ (i.e. a phase shift of $\lambda/2$) is most favorable.

For the absolute values of the amplitudes as measured by a two-phase lock-in amplifier we get

$$I_\Omega = |I_\Omega(\eta_F)| \propto |E_{pol}^2 \sin(2\eta_F)|, \qquad (12.113)$$

and

$$I_{2\Omega} = |I_{2\Omega}(\phi,\theta_F,\eta_F)| \propto |E_{pol}^2 \cos(2\eta_F)\sin[2(\theta_F-\phi)]| \qquad (12.114)$$

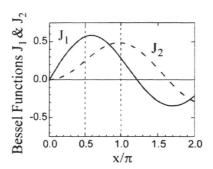

Figure 12-31: The first (J_1) and second (J_2) Bessel function.

12.4.2.3. Contrast formation in magneto-optic imaging

Measuring the second harmonic component $I_{2\Omega}$ of the lock-in amplifier according to Eq. (12.114), two measurement modes can be performed: (i) contrast imaging and (ii) quantitative determination of local Faraday or Kerr rotations. For small differences $|\theta_F - \phi| \ll 1$,

Eq. (12.114) reduces to

$$I_{2\Omega} \propto |\theta_F - \phi| \tag{12.115}$$

Measuring the second harmonic component $I_{2\Omega}$ according to Eq. (12.115), the Faraday rotation θ_F can be measured locally with a high angular resolution by rotating the analyzing-unit (*PEM* plus analyzer at 45°) until the condition $\phi = \theta_F$ is fulfilled [105]. This corresponds to a sharp minimum of the lock-in amplifier output $I_{2\Omega}$. The resulting curve $I_{2\Omega}(\phi)$ has a typical 'V' shape in the vicinity of the minimizing condition. This is explained in Fig. 12-32 where the 'V'-shaped curves of two domains with opposite orientation of the magnetization ('up' and 'down') are plotted as a function of analyzing-unit angle ϕ. The two opposite domains will exhibit a different local Faraday rotation $\theta_{F\uparrow}$ and $\theta_{F\downarrow}$ (the same is true, of course, for the local Kerr rotation as measured in reflection) yielding two 'V'-shaped curves slightly shifted along the ϕ axis. Keeping the analyzing unit at a fixed working position ϕ_1 while scanning the sample will produce a black and white contrast for the 'up' and 'down' domains, respectively (points

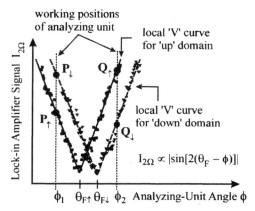

Figure 12-32: Contrast formation and quantitative measurements in magneto-optic *SNOM*. Rotating the analyzing-unit generates typical 'V'-shaped data sets.

$P_\uparrow$ and $P_\downarrow$ in Fig. 12-32). For the corresponding lock-in-amplifier amplitude, we get $I_{2\Omega}(P_\uparrow) < I_{2\Omega}(P_\downarrow)$. Moving the analyzing unit to the working position ϕ_2, which is only a few degrees past the two minimums of the local 'V' curves, will invert the black-and-white contrast because $I_{2\Omega}(Q_\uparrow) > I_{2\Omega}(Q_\downarrow)$ at the points $Q_\uparrow$ and $Q_\downarrow$. For imaging the magnetic contrast of a larger area, the analyzing-unit is maintained at a fixed angle ϕ_0 while I_2 is measured.

In order to get a quantitative value of the local Kerr or Faraday rotation, the *SNOM* tip has to be positioned above the corresponding magnetic domain. Then, the analyzing-unit is rotated through the minimum of the lock-in amplifier amplitude generating a 'V'-shaped curve as shown in Fig. 12-32 by the data points. The minimum position of each curve can be accurately determined by performing a generalized least-square fit [95]. Examples are given in Chapt. 12.5.3.

12.5. CHARACTERIZATION OF MAGNETIC MATERIALS BY *SNOM*

In this chapter a few examples on how to characterize magnetic materials by scanning near-field magneto-optic microscopy will be reviewed. All experiments are performed with a set-up as described in Chapt. 12.3.2.2.

12.5.1. History

The first publication on polarization contrast in near-field optics with applications to magneto-optic imaging was already given by E. Betzig et al. in 1992 [106]. They showed a stripe-domain pattern in Bi-doped yttrium-iron-garnet (*YIG*) films of about 1 µm thickness yielding a peak-to-peak Faraday rotation of roughly 2° which represents an upper limit on the angular resolution[7]. The lateral resolution was not discussed but the domains which could be resolved had a width of 0.7-1 µm. In a subsequent paper, magneto-optic contrast was observed in a 30 nm thick amorphous TbFe film [107] with a peak-to-peak rotation of the order of 1°. Only a couple of months later, Betzig and coworkers demonstrated the possibility of using near-field optics for high density data storage [18]. They accomplished near-field magneto-optic reading of thermally written magnetic bits in a multilayer film consisting of ten bilayers Co(4 Å)/Pt(10 Å) sputtered onto a glass substrate. The domain size was again 1 µm. In a second step, they tried to thermally write bits in the near-field by using a higher laser intensity. The bits had a size of 100 nm and could be resolved by reading in the near-field. However, no contrast reversal was observed when slightly distorting the polarizer. Instead, each bit seemed to consist of a black-white contrast itself. The authors explained this behavior with effects unique to near-field magneto-optics. An other possible explanation could be that the authors had imprinted the bits into the film with the *SNOM* tip. It has been noticed in literature that a *SNOM* tip expands when the laser intensity is increased [108]. An elongation of up to 100 nm per mW of emitted laser power has been observed in collection mode. For a laser power of 1.7 mW coupled into a single-mode fiber, the temperature of the tip region increases by more than 200 K [109]. However, Betzig et al. claim that the shear-force image did not show any evidence of damage within the area of the written domains. In addition, they showed a dependence of bit size on laser power. At 6 mW, domains of only 60 nm could be written and subsequently read. In all the measurements, ordinary single-mode fibers were used, not polarization-preserving ones. Furthermore, they did not even apply a lock-in technique in order to increase resolution. In view of this simple technique, their results are very surprising.

Since this publication in 1992, to our knowledge no other successful attempt has been published on thermally writing magnetic domains by near-field optics except for the so-called solid-immersion-lens (*SIL*) technique as proposed by B. D. Terris et al. in 1996 [23]. The principle of *SIL* is very simple. By focusing light inside a glass lens with large index of refraction n, the spot size can be reduced below the diffraction limit [110,111] because the numerical aperture *N.A.* is increased by a factor n^2. According to Eqs.(12.2) and (12.10) this increases the optical resolution by the same factor. Technically, the *SIL* lens, which consists of a partial sphere, is mounted on a planar slider. At a flying height of 150 nm, a spot size of 360 nm has been obtained using 830 nm light.

The set-up proposed by E. Betzig et al. has been taken over by a lot of other groups imaging magnetic-domains by Faraday effect. Without polarization modulation, domains have been imaged in several µm thick YSmBiGaFe garnet [112] and in 0.8 µm thick Bi-substituted $Y_3Fe_5O_{12}$ garnet (*YIG*) exhibiting a Faraday rotation $\theta_F = 1.8°$ [113]. Adding polarization-modulation with an acousto-optic modulator (*AOM*) and using a bent fiber tip capable of simultaneously performing atomic-force microscopy [114], magnetic domains in a 0.1 µm thick Bi-substituted dysprosium-iron-garnet film with $\theta_F = 1.7°$ were resolved [115]. In a subsequent paper, the same group measured magnetic bits in a 15 nm thick Co/Pt multilayer with a lateral resolution of 150 nm as determined from the width of the transition region between two opposite domains [116]. The bits had previously been recorded using a magnetic-field-modulation method. The angular resolution achieved was approximately 0.5° as estimated from $\theta_F = 0.74°$ and $\eta_F = 0.47°$ measured in far-field at 488 nm. A polarization-modulation technique using a photoelastic modulator (*PEM*) as discussed in Chapt. 12.4.2.2 was implemented for the first time to study sample birefringence [117]. The same scheme can, of-course, be used to measure the Faraday effect as shown on a 3.6 μm thick *YIG* film [118]. With such a set-up, magneto-optic contrast was reported at a wavelength of 632 nm in a 15 nm thick Co/Pt multilayer film which was interpreted as magnetic domains [119]. However, the measurements show significant changes in the domain pattern although the strength of the applied magnetic field pulse was below the coercive field. The authors interpret this behavior with slow relaxation effects in the sample. In our own quantitative studies on a $[Co(3 \text{ Å})/Pt(10 \text{ Å})]_{10}$ multilayer deposited on a glass substrate with a Pt(100 Å) buffer layer, we found a strong birefringence yielding a contrast much larger than the one expected from magnetic domains [71]. These features extend over an area similar to the grain size and resemble magnetic domains but do not show any magnetic-field dependence as shown in Fig. 12-33. The contrast is very sensitive to changes in the sample-tip distance even of a few nanometer. The contrast observed seems not directly be related to topographic features. However, the surface roughness was about 6 nm which is rather large. This might influence the optical image.

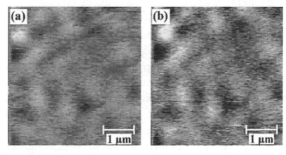

Figure 12-33: Magneto-optic contrast in a Co/Pt multilayer as obtained by reflection-mode *SNOM* [71]: (a) demagnetized state, (b) saturated state.

Only few attempts have been made so far to image magnetic domains by means of the Kerr effect. Using the basic set-up of Betzig without polarization modulation, C. Durkan et al. measured the polar Kerr effect in a $[Co(5 \text{ Å})/Pt(5 \text{ Å})]_{10}$ multilayer film deposited on a glass substrate [120]. Reflection-mode *SNOM* was achieved by placing the tip with the sample in one focus of an elliptical mirror and the photomultiplier detector in the other one. The signal-to-noise ratio reached only 2:1. From the peak-to-peak Kerr rotation, an angular resolution of 0.5° can be estimated. In a succeeding publication, the authors noticed a large influence of the topography on the optical signal [121] which is particularly critical for reflection-mode *SNOM*. In our group, we developed another set-up similar to the one Betzig described using a parabolic and, later, a spherical reflector in order to measure the polar Kerr effect [21,77,79]. Details are given in Chap. 12.3.2.2.

An other approach to image magnetic domains has been chosen by T. S. Silva and S. Schultz [19,122]. As a near-field probe, 20-40 nm silver particles are used which are optically excited near the surface-plasmon resonance frequency. The colloidal silver particles are deposited on a glass hemisphere with a radius of about 6 mm which is mounted on a larger glass hemisphere with 1 inch diameter [122]. Polarized laser light is focused onto the probe particle in a dark-field geometry, i.e. the laser light undergoes total reflection from the bottom of the small hemisphere. Bringing a sample within the evanescent field emanating from the plasmon probes, the incident light will be scattered and can be collected by a microscope objective. Distance control is achieved by a Newton-ring interferometer. They observed magnetic domains in a [Co(3 Å)/ Pt(10 Å)]$_{18}$ multilayer film of total thickness 234 Å deposited on a SiN-coated glass substrate. Besides imaging magnetic domains of 0.5 μm size, a strong reproducible diagonal texturing of unknown origin was found [19]. From the half width of the domain walls, an upper bound for the lateral resolution of 100 nm was derived. In a later publication, the principle was reduced to measuring scattered light from Ag particles deposited on a 200 nm thick $Ni_{0.81}Fe_{0.19}$ layer with a 3-100 nm thick SiO_2 overlayer [123]. A sufficient spacing of the particles of about 2 μm was achieved by adequate dilution. In Kerr geometry, a hysteresis curve could be measured with the scattered light. No near-field microscope was used, just the propagating light of the Ag particles converted in the evanescent field of the illuminated $Ni_{0.81}Fe_{0.19}$ layer was detected. This set-up corresponds exactly to the principle plotted in Fig. 12-10, Chapt. 12.2.2.

The third approach using the Sagnac effect has been discussed in detail in Chapt. 12.4.2.1 and will not be further considered here. With a set-up based on frustrated total reflection as sketched in Fig. 12-13(e), Chapt. 12.2.2.2, magnetic domains of about 3 μm width could be imaged in a 1 μm thick garnet film [124] applying magnetic-field modulation. Yet another approach was chosen by a Dutch group. A photosensitive semiconducting tip served as both, an *STM* tip and a small light detector [125]. Thermomagnetically written bits of 0.8 μm diameter have been imaged with this set-up in a [Co(3.5 Å)/Pt(6 Å)]$_{20}$ multilayer on a glass substrate. A lateral resolution of 250 nm was accomplished.

Besides the techniques using apertures, also apertureless *SNOM* has been applied by one group to measure magneto-optic effects [126]. An electro-chemically etched tungsten wire was used as tip. Magnetic stripe domains were then imaged in $(Y, Gd, Tm, Bi)_3(Fe, Ga)_5O_{12}$ garnet films of 7 μm thickness yielding a large Faraday rotation of 14°.

The last measurements to be mentioned here are those combining near-field optical microscopy with femtosecond-pulsed lasers. A first investigation using such a combination was carried out to image excitonic spin behavior in locally disordered magnetic-semiconductor heterostructures [15, 127]. The set-up is based on the design given by Betzig and the fiber tip was used in collection mode with detection in transmission (see Fig. 12-13(c), Chapt. 12.2.2.2). Besides the expected increase in intensity with decreasing tip-sample distance, a decrease of the polarization has been observed as well. It is explained, first, by the increased effective numerical aperture *N.A.* when the tip is approaching the sample and, second, by a reduced coupling of evanescent waves to circularly polarized far-field light due to the inability of propagating angular momentum. More recently, measurements have been published using a set-up in collection mode to image magnetic domains in a garnet film by means of magnetization-induced second-harmonic generation [128]. The lateral resolution was not very good because uncoated fiber tips had been used.

Due to space limitation we will not review the various theoretical papers on near-field optics. A recent and thorough review is given in the book of M. Ohtsu and H. Hori [38].

12.5.2. Local magnetization measurements

Local magnetization measurements are done by placing the *SNOM* tip at a fixed position and evaluating the local Faraday or Kerr effect as a function of magnetic field. Because the magneto-optic effects are proportional to the magnetization (see Section 12.4.1), the local magnetic properties can be determined in this manner. The procedure has the advantage compared with imaging of magnetic domains that the tip is at rest. Accordingly, the tip-sample distance is kept constant avoiding all topography-related effects. Therefore, even in samples with large birefringence such as Co/Pt multilayers, local hysteresis curves can be determined even though magnetic domains are difficult to image. Obviously, it will not be possible to directly determine the lateral resolution from the measurements because this requires the imaging of magneto-optical features. However, this method is very well suited to investigate local magnetic properties. Limitations arise from the lateral resolution determined by the size of the aperture and from the angular resolution which mainly depends on the size and stability of the polarization at the fiber tip. The limitations will be discussed in Section 12.6.

In order to perform local magnetization measurements, an external field has to be applied. In transmission as well as in reflection mode, special magnets were designed for our equipment which allow to apply a magnetic field of up to 0.45 T [60,71] and 0.2 T [78] in transmission and reflection mode, respectively. In Fig. 12-34 the Faraday hysteresis curve obtained in transmission-mode *SNOM* of a Pt(50 Å)/[Co(3 Å)/Pt(12 Å)]$_{21}$/Pt(8 Å) multilayer deposited on a glass substrate is plotted together with the corresponding far-field curve. The curves look identical. No difference in the coercive field or in the slope of the magnetization-reversal region is observed indicating that the lateral optical resolution is larger than the grain size which is of the order of 10-20 nm in Co/Pt multilayers [129]. In that case, the magnetization reversal measured in the near field is identical to the one in the far field because the *SNOM* aperture averages over many grains. If the lateral resolution were smaller than the grain size, the magnetization reversal of only one grain could be studied. A square hysteresis loop would be expected if the grain size were below the single domain limit. From the scatter in the data, an angular resolution of about 0.1° can be estimated.

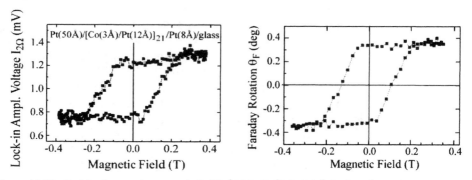

Figure 12-34: Faraday hysteresis curve of a Pt(50 Å)/[Co(3 Å)/Pt(12 Å)]$_{21}$/Pt(8 Å) multilayer as obtained by (a) transmission-mode *SNOM* and (b) in the far field [60]. The contribution to the Faraday rotation due to the glass substrate has been subtracted.

With an optimized fiber tip, a Faraday loop on a very thin Co/Pt sample was measured and compared to a polar Kerr hysteresis loop obtained in the far field as shown in Fig.12-35. The total thickness of the [Pt(11 Å)/Co(3 Å)]$_4$/Pt(100 Å) multilayer was only 170 Å. The peak-to-peak Kerr rotation reached 0.25°. Comparing the near-field loop with the far field one, again, no difference is observed. The angular resolution as estimated from the scatter in the data is better than 0.05°. However, there are long-term drifts of the order of 0.1° as can be seen from structures in the saturated part of the hysteresis which is supposed to be constant.

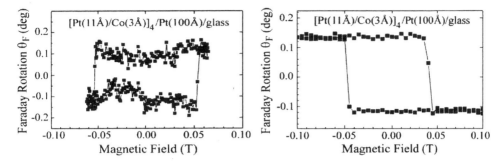

Figure 12-35: (a) Faraday hysteresis curve of a [Pt(11 Å)/Co(3 Å)]$_4$/Pt(100 Å) multilayer as obtained by transmission-mode *SNOM*. (b) Polar Kerr hysteresis loop in the far field of the same sample. The contribution to the Faraday rotation due to the glass substrate has been subtracted.

If the resolution of the *SNOM* aperture is of the order of the domain size of the magnetic material, local effects are expected to be seen in the hysteresis curves. In order to prove this assumption, near-field Faraday hysteresis loops have been made on a EuTm$_2$Fe$_5$O$_{12}$ garnet film. Figure 12-36(a) shows the far-field Faraday hysteresis loop. The peak-to-peak rotation is about 1.2°. The Faraday hysteresis loop recorded in the near-field, as plotted in Fig. 12-36(b), looks completely different. It shows pronounced peak in the transition region. To understand this behavior, more information about the sample has to be given. The garnet film has been grown by liquid-phase epitaxy on a gadolinium-gallium garnet (*GGG*) of several hundred μm thickness. By liquid phase epitaxy, the film grows on both sides of the substrate. Therefore, the sample possesses two magneto-optic layers separated by the transparent *GGG* substrate. Only the layer close to the *SNOM* will interact with the evanescent waves. But, because the converted photons have to propagate through the sample, the adjacent garnet layer will give rise to a far-field Faraday effect which adds to the near-field effect. Therefore, the hysteresis loop has a background which is similar to the far-field curve of Fig.12-36(a).

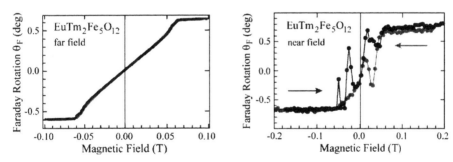

Figure 12-36: (a) Faraday hysteresis curve of a EuTm$_2$Fe$_5$O$_{12}$ garnet film as obtained by transmission-mode *SNOM*. (b) Faraday hysteresis loop in the far field of the same sample [71].

By subtracting the background, we are able to reduce the data in order to evaluate the near-field component. The result is plotted in Fig.12-37. The Faraday rotation in the far-field has been divided by a factor 2 and then subtracted from the near-field curve assuming that the contribution to the Faraday rotation in the far-field is equal to the one in the near-field. What is left is a square hysteresis curve with superimposed sharp peaks. Apparently, by increasing the magnetic field, a domain crosses three times the tip position while this happens only one time on the way back. As can be seen from imaging magneto-optic domains (see next chapter), the width of domains pointing opposite to the applied-field direction decreases while increasing the external magnetic field. Having reached an energetically favorable thickness, the domains start to unwind (see Fig.12-43). Therefore, it may happen that a domain crosses several times the same position and the *SNOM* will record a peak each time a magnetic domain travels through the tip region. In fact, the measurements presented here have been taken after imaging the magnetic domains. This clearly demonstrates that in *SNOM* local magneto-optic properties are recorded.

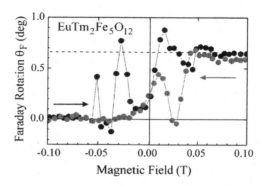

Figure 12-37: Difference between the near field and the far-field Faraday hysteresis curve of a $EuTm_2Fe_5O_{12}$ garnet film. The peaks arise when a stripe domain crosses the tip position while it is unwinding.

12.5.3. Quantitative magneto-optic measurements

The potential of scanning near-field magneto-optic microscopy will be demonstrated by a few examples in this chapter. As already mentioned before, this technique has not yet evolved to an easy-to-use method due to the complexity of the technique. In reviewing the literature (see Section 12.5.1), one will notice that the best results obtained so far have been attained on either transparent garnet films or on amorphous rare-earth transition-metal films. The garnets are epitaxial with a very flat surface and virtually without grain boundaries. Hence, they show very little stress-induced birefringence. In addition, because they are transparent, several μm thick samples can be investigated yielding a large Faraday rotation of usually more than 1°. Amorphous rare-earth transition metal films, on the other hand, have very small grains sizes (a few nanometer or less). Consequently, the stress-induced birefringence is also small. Furthermore, even for a thickness small enough to perform transmission experiments (approximately 10-30 nm) they still have a considerable Faraday rotation of several tenths of a degree. Last but not least, they exhibit a very good surface quality. The third system used are Co/Pt multilayers. However, depending on deposition they might exhibit stress-induced birefringence which will obstruct polarization-sensitive measurements except perhaps for the Sagnac *SNOM* (see Section 12.4.2.1). On the other hand, Co/Pt multilayers or alloys show a large Faraday and Kerr rotation, which makes them well suited for magneto-optic investigations. A disadvantage of this material is that it usually does not show a stripe-domain pattern but rather irregular domains [130] which are connected to the grain size. Because the latter is of the order of a few nanometer, domains are very difficult to observe. In the following, we will present some of our results obtained on a $EuTm_2Fe_5O_{12}$ garnet film of 2.1 μm thickness and on Co/Pt multilayers, were magnetic bits had been created by thermomagnetic writing. The aim of this section is to show how to make *quantitative* measurements with the set-up discussed in Section 3.2.2 using a lock-in method as described in Section 12.4.2.2 .

The first thing to do when a contrast has been measured in polarization-sensitive *SNOM* is to check whether this contrast is due to magneto-optic or birefringence effects. Measuring the magnetic-field dependence would be the preferred procedure. Because this is usually not possible one will verify contrast reversal. This can simply be done by rotation of the analyzing unit, consisting of photoelastic modulator (*PEM*), polarizer, and detector mounted on a common rotating stage, by a few degrees as explained in Fig.12-32, Section 12.4.2.3. However, a contrast reversal by rotating the analyzer unit can be obtained in the case of birefringence as well. So, this is not an irrefutable proof of magnetism. The only definite proof is to verify the magnetic-field dependence. Nevertheless, the test is commonly used and accepted.

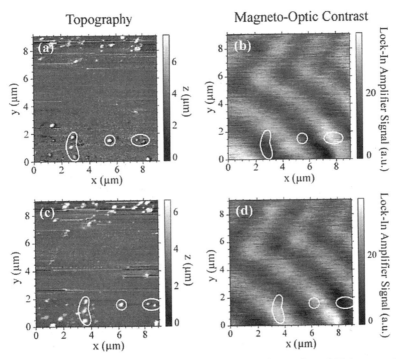

Figure 12-38: Topography (a) and (c) and magneto-optic Faraday image (b) and (d) in transmission mode of a $EuTm_2Fe_5O_{12}$ garnet sample. The images (c) and (d) have been obtained after rotating the analyzing unit a few degrees in order to induce a contrast inversion. The white circles mark topographic features and depict identical locations in the magneto-optic images (b) and (d) [131].

A typical example is shown in Fig.12-38, where the stripe-domain pattern of a 2.1 μm thick $EuTm_2Fe_5O_{12}$ garnet sample is illustrated [131]. The magneto-optic contrast is obtained in transmission mode (Faraday effect). Topography and magneto-optic image have been taken simultaneously. Figures (c) and (d) are obtained after rotation of the analyzing unit by a few degrees. A regular pattern of stripe domains is seen in the magneto-optic images (b) and (d) while the topography (a) and (c) does not show pronounced features. A few inhomogeneities at the surface are marked with white circles to facilitate identifying corresponding locations in the magneto-optic images. By comparing the white circles, a contrast inversion between Fig.12-38 (b) and (d) is clearly visible. The lateral displacement of 0.75 μm between the two measurements is due to the fact that the tip had been withdrawn from the sample between the two measurements.

Figure 12-39 (a) shows another magneto-optic image of the same sample [20]. The white line marks the path of a line scan taken perpendicular to the magnetic domains which is displayed in Fig. 12-39 (b). From the line scan the domain width of 1.2 μm is derived. In addition, the thickness of the domain-wall region is estimated to be of the order of 400 nm. In contrast, a simple calculation of the wall thickness as a result of the isotropic exchange constant and the uniaxial anisotropy constant of garnet samples results in a wall thickness of $d_w = 50$-100 nm for the given domain size of 1.2 μm [132]. The discrepancy is explained by a reduced lateral

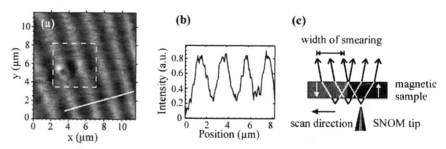

Figure 12-39: (a) Magneto-optic Faraday image of a $EuTm_2Fe_5O_{12}$ garnet film showing a stripe-domain pattern. The dashed square marks the sample area imaged in Fig. 12-40. (b) Line scan along the white solid line plotted in (a). (c) Sketch of domain-wall smearing due to the opening angle of light emission [20].

resolution due to the following reasons: Since the aperture of the tip emits light under a wide angle [59], the light forms a cone within the sample whose maximum width can be predicted from the numerical aperture of the objective and the sample thickness. For the objective used, the numerical aperture is $N.A. = 0.65$ [20] leading to a cone angle of approximately 45°. As a consequence, converted propagating waves emitted at a wide angle will cross two adjacent magnetic domains within a region of the size of the sample thickness. Therefore, the Faraday rotation of two opposite domains will be averaged and the image of the domain wall will be smeared out as explained in Fig.12-39(c). The other parameter limiting the resolution is the diameter of the aperture at the fiber tip which was bigger than 100 nm for the tip used.

Figure 12-40 displays both the topographic (a) and magneto-optic (b) data taken simultaneously from the area marked with a dashed line in Fig.12-39. The analyzing unit was kept at a fixed position [dashed line in (c)]. The topographic irregularity in the lower left part of Fig.12-40(a) is seen as a small bubble domain in the magneto-optic image. It looks as if the surface obstruction pins the bubble domain. Figure 12-40(c) shows 'V' curves taken at several spots of the sample marked with corresponding symbols in (b). The 'V' shape is a consequence of the lock-in technique as explained in Section 12.4.2.3. They are generated by rotating the analyzing unit while keeping the *SNOM* tip with the aperture at a fixed position. According to Eq. (12.114), the minimum of a 'V' curve is a direct measure of the local Faraday or Kerr rotation enabling quantitative determination of local magneto-optic effects. The fits to the data sets are obtained by a generalized least-square procedure explained elsewhere [95,21]. From the minimum position of the 'V' curves, the difference of the Faraday rotation of two domains magnetized in opposite directions can be found to be 3.6°. This implies a specific Faraday rotation of about 9×10^3 deg cm^{-1}, a value common for garnet films [133]. The identical slope of all 'V' curves proves that the contrast is merely an effect of the Faraday rotation, as variations in the local transmission would affect the value of the slope.

By exposing to a strong magnetic field of 2 T parallel to the surface, the stripe domains switched to a bubble-domain pattern. Figure 12-41(a) displays the magneto-optic Faraday data [20]. Small bubble domains have arranged in some areas to a perfect hexagonal pattern. A close-up of Fig. 12-41(a) was chosen to demonstrate the possibility to measure the Faraday rotation quantitatively for an entire area. A sequence of ten pictures was taken from the same area with the analyzing-unit position incremented by 1° after each scan. Image-processing software then performed a ,V'-curve fit for every pixel yielding the local Faraday rotation θ_F displayed as a gray tone. In order to increase the angular resolution, much more scans would have to be done which is a question of time and long-term stability. Nevertheless, this procedure demonstrates the possibility of quantitatively imaging local magneto-optic effects.

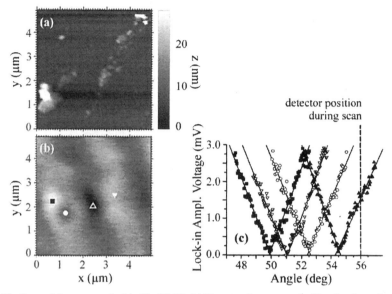

Figure 12-40: Scan of the area marked in Fig.12-39. (a) Topography as acquired by the shear-force feedback loop. (b) Magneto-optic Faraday image simultaneously taken. (c) V-curves obtained at different image locations directly after completion of the scan. The symbols correspond to the positions marked in (b) [20].

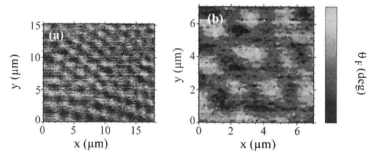

Fig. 12-41: Images taken from a $EuTm_2Fe_5O_{12}$ garnet sample after treatment with an external field of 2 T parallel to the sample surface. (a) Magneto-optic Faraday data. A pattern of hexagonally ordered bubble domains is seen. (b) Quantitative Faraday picture as calculated from ten images taken subsequently from the same sample area with the analyzing unit rotated by 1° after each scan. A ,V'-curve fit for every pixel yields the calculated minimum angle $\varphi_0 = \theta_F$ displayed as a gray tone [20].

Of course, the same treatment can be applied in reflection mode as well which is much more versatile as it enables, in principle, to study any type of sample. For reflection-mode *SNOM*, the set-up is modified by introducing a spherical mirror as discussed in Section 12.3.2.2. The requirements for the apertures are much more severe than for the transmission mode. This will be discussed in detail in Section 12.6. In Fig.12-42(a), thermomagnetically recorded bits on a Co/Pt magneto-optic disk are imaged in reflection mode *SNOM*. The Co/Pt sample was cut from a magneto-optic disk [134]. Magnetic domains had been thermomagnetically written onto the disk with a size of approximately 1×3 μm^2. Polar Kerr hysteresis loops of some pieces were measured in a conventional Kerr spectrometer yielding a coercivity of 0.45 T and an far-field Kerr rotation of $\theta_K = 0.48°$ [21]. The dashed line denotes the position where the analyzing unit had been slightly rotated causing a contrast reversal. In order to determine quantitatively the local Kerr rotation with *SNOM*, the analyzer unit is rotated yielding a 'V'-shaped curve as explained before. The two locations selected for the local measurement are marked with a symbol in Fig.12-42(a). The first one is right on top of a magnetic bit ($\triangle$) and the second one is on the background ($\bullet$), where the magnetization is opposite. According to Eq. (12.114), varying the analyzing-unit angle ϕ yields a 'V'-shaped curve where the local Kerr rotation θ_K corresponds to the angle at the minimum. The experimental result is shown in Fig. 12-42(b). Despite the scattering of individual points, a clear offset is observed between the two 'V'-shaped curves. As a result of the fit [solid and dashed lines in Fig.12-42(b)], a difference of $\Delta\theta_K = 0.43°$ is obtained between the two minimum positions. This is in excellent agreement with $\theta_K = 0.48°$ which had been measured on a conventional high-resolution Kerr spectrometer [21]. The slight difference can be explained by local variations in the Kerr rotation, which are averaged in the conventional set-up. The error of the Kerr angle $\delta\theta_K$ can be estimated from the error of the least-square fit and amounts to $|\delta\theta_K| = 0.11°$. This error could be further reduced by accumulating more data points per 'V' curve provided the thermal drift, the change in laser intensity, and, last but not least, the change in the polarization direction are still negligible.

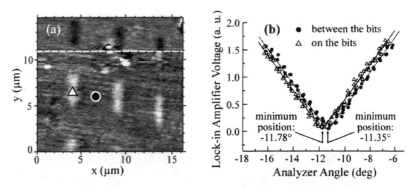

Figure 12-42: (a) Magneto-optic contrast as obtained in reflection mode (Kerr rotation) of a Co/Pt sample. At the dashed line, the analyzing unit had been slightly rotated in order to generate a contrast reversal. The symbols indicate the positions where the local Kerr rotation has been measured. (b) Local 'V'-shaped curves as obtained by rotation of the analyzing unit. The solid and dashed lines are fits as explained in the text. The minimum position defines the local Kerr rotation at the positions marked by corresponding symbols in (a) [21].

As a last example, we present magnetic-field-dependent measurements. The most intriguing prospect of magneto-optic *SNOM* is to image local magnetization-reversal processes. As mentioned before, both microscopes, in transmission as well as in reflection mode, are upgraded with a magnet. In transmission mode, a field of up to 0.45 T can be applied while the maximum field in reflection mode amounts to 0.2 T. A series of Faraday images of a $EuTm_2Fe_5O_{12}$ garnet sample, which have been recorded while sweeping the magnetic field, is plotted in Fig. 12-43 [71]. The magnetization reversal is nicely documented by the pictures. Starting at zero field, in the demagnetized state, the thickness of the magnetic stripe domains with orientation opposite to the applied field first decreases with increasing field. According to the theory of magnetic domains [135], the thickness cannot be greatly reduced. As a consequence, the domains start to unwind in order to reduce the length of the domain walls. During the process of unwinding, especially close to an ending point of a domain (see right section of the Faraday images), a domain may cross the same point several times explaining the local hysteresis behavior in Fig.12-37.

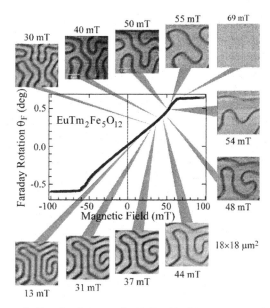

Figure 12-43: Magneto-optic Faraday images of a $EuTm_2Fe_5O_{12}$ garnet sample as a function of applied magnetic field. Note the unwinding of magnetic stripe domains with orientation opposite to the applied field. [71].

In conclusion, it is possible with a set-up as described in this article to determine *quantitatively* the local Kerr and Faraday rotation by scanning near-field optical microscopy. This opens a new way of investigating magnetic domains and domain walls on a nanometer scale in transparent and opaque samples. The limiting factor of the angular resolution is given by the degree of linear polarization of the light exciting the fiber tip and the temporal and thermal stability of the SNOM as discussed in the following chapter.

12.6. FUNDAMENTAL PROBLEMS

In the last chapter we will try to summarize the most important problems encountered with polarization-sensitive *SNOM*. Naturally, some troubles are related to the specific equipment used. Those will not be the issue here. Nevertheless, there exist problems which most set-ups have in common because they are based on the same principle. The major problems discussed here are topographic cross-talk and its influence on optical resolution and polarization stability.

12.6.1. Transmission mode versus reflection mode

Generally, *SNOM* is much easier to carry out in transmission mode than in reflection mode for several reasons. First, in transmission mode the intensity of the signal is enhanced when entering the near-field region because the conversion efficiency of evanescent waves into propagating waves increases. Non-evanescent stray light leaving the tip through imperfections of the tip coating will be partially reflected from the surface (especially on metallic films). So, the ratio between the near field and far-field contribution increases with decreasing tip-sample distance. As a consequence, the tip quality does not have to be extremely good. In reflection mode, the situation is reversed. The signal amplitude decreases upon approaching the sample because the tip is shadowing its own light. When the tip is within the near-field, the distance from the sample surface is of the same order or smaller as the size of the aperture, i. e. 10-50 nm. The thickness of the metal coating is 50-100 nm. As a consequence, the body of the tip is blocking all the reflected light up to a certain angle which is the larger the closer the tip approaches the surface. On the other hand, the non-evanescent stray light coming from imperfections in the coating will not be affected. Hence, the ratio between the near field and far-field contribution decreases with decreasing tip-sample distance. This requires tip apertures of extraordinary good quality which should produce as little stray light as possible. In addition, the well-defined 20 μm large tetrahedral metal-pyramid apertures, introduced in Section 12.3.1.4, cannot be used. Because they are glued onto the flat end of a 125 μm wide fiber, the converted reflected light will be completely blocked when approaching the near field. Finally, the magneto-optic effects are smaller in reflection (Kerr effect) than in transmission (Faraday effect) mainly because the Faraday effect is proportional to the thickness of the magneto-optic layer. Typical values of the Kerr rotation are 0.2–0.5° for the transition metals Co, Fe, and Ni. For ultrathin films of only a few atomic layers, the Kerr rotation will hardly exceed 0.1° and in diluted magnetic semiconductors it amounts to only a few hundredths of a degree. This requires an angular resolution of the Kerr rotation of about 0.001° to 0.005° in order to be able to measure a magnetic contrast in such samples. The lateral resolution of magneto-optic contrast is, however, better in reflection mode than in transmission mode. As explained in Fig.12-39(c) in Section 12.5.3, the image of domain walls is smeared out due to the fact that propagating waves will cross two adjacent magnetic domains within a region of the size of the cone width of the emitted light. The effect is reduced in reflection mode, as the penetration depth is of the order of a few nanometer in metals. For transparent materials, this is of course not the case and the lateral resolution is comparable.

12.6.2. Topographic cross talk

As mentioned before, a major problem in scanning near-field optical microscopy is the influence of topographic features on the optical image. The origin for this topographic cross-talk is the relatively large size of the tip as compared to topographic features of the sample surface and the imperfection of the aperture and the metal coating. An example of a tip-shape effects in the optical image is given in Fig.12-44 [77,105]. It shows the topography (a) and the optical image in reflection-mode *SNOM* (b) of a *CD* master made of nickel which is used to imprint compact disks. The height of the protrusions is 110 nm and the track width is 1.6 μm. The optical image clearly exhibits a substructure on top of the protrusions.

This is better illustrated in a line scan along the white lines in the center of the image as depicted in Fig. 12-44(c). The substructure can be explained in the context of a simple model [77,105]. Provided that the size of the mesa-like protrusions is larger than the aperture of the *SNOM* tip, the reflected light intensity should be the same when the tip is in between the protrusions (region I) and on top of a mesa-like protrusion (region III) as long as the tip-sample distance is kept constant. This is evidenced in the line scan by the two equal minima at region I and III. When the tip is approaching a protrusion, the aperture-sample distance increases temporarily because of the width of the tip leading to an increased intensity in the transition region II. The same happens when the tip is leaving the protrusion (region IV). The asymmetry which is evident in region II and IV is due to imperfections in the *SNOM* tip such as, e.g., an uneven coating.

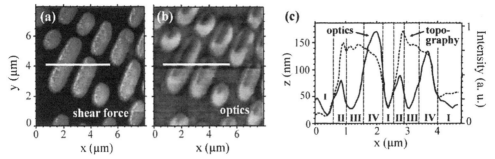

Figure 12-44: (a) Topography and (b) optical contrast of a *CD* master made of nickel. The bits appear as flat mesa-like protrusions in the topography but show a fine structure in the optical image. (c) Line scans along the white lines in (a) and (b). The regions I-IV are explained in the text [77,105].

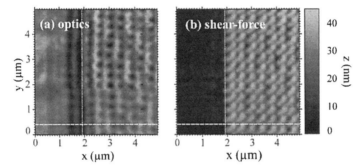

Figure 12-45: Optical (a) and topographic (b) image of a periodic Ni-dot array. The dotted vertical line marks the edge of the patterned area. The line scan shown in Fig. 12-46 has been taken along the dashed horizontal line [136].

To investigate the effects of far-field light on the optical image in reflection-mode *SNOM*, a lithographically patterned Ni-dot array on a silicon substrate was measured [136]. The individual dots have a diameter of 200 nm and a height of 30 nm and are patterned in a 50×50 μm array with a periodicity of 400 nm. Figure 12-45 shows an optical (a) and a topographic (b) image of an area at the edge of the patterned structure. The images have been obtained simultaneously. The topography shows two images of the pattern diagonally shifted by 200 nm which is caused by a double-tip effect. The edge of the patterned area has been marked with a vertical dotted line. In the optical image, a pattern can also be observed on the left-hand side of the dotted line. This means that the optical signal shows contrast due to the pattern while the tip is still located over the unpatterned area of the sample. Line scans taken in Fig. 12-45 along the dashed horizontal line are plotted in Fig. 12-46(a). The optical signal (dashed line) clearly shows an intensity modulation in the unpatterned area while the topography (thin solid line) does not.

This effect can be reproduced by a simulation algorithm using a ray-optics approach (bold line).There are several assumptions made for this simulation as outlined in Fig. 12-46(b) [136]: (i) The focal area d_f of the parabolic mirror is much larger than the tip aperture and may be offset from the aperture by x_0; (ii) the aperture is a point source that emits light homogeneously in all directions and the total emitted intensity is constant and independent of the tip-sample distance; (iii) wherever a ray is incident on the sample surface, a spherical propagating wave will be emitted; (iv) rays which are reflected from the sample surface under the same angle interfere with each other; (v) protrusions at the surface block both part of the incident light and part of the reflected light.

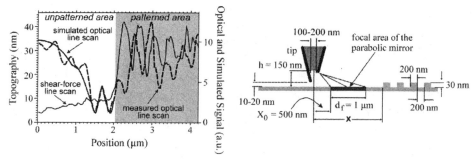

Figure 12-46: (a) Measured topographic (thin solid line), optical (dashed line), and simulated (bold line) line scans taken along the horizontal dashed line in Fig.12-45. The shaded area marks the patterned area. (b) Geometry used for the simulation [136].

To conclude, a major problem in near-field optical imaging in reflection mode arises from the combination of both, a large area on the sample being illuminated by far-field light from the tip and the focal area of the detection optics being much larger than the near-field aperture and in general displaced from it. This allows optical signals to be collected from locations on the sample that are far away from the actual position of the tip. Therefore, the fact that the optical image shows features not present in the topography does not imply that these features are true optical information. Furthermore, any topographic feature will influence the optical data because it forces an adjustment of the tip-sample distance which alters the local evanescent field distribution.

12.6.3. Polarization effects

Another important factor causing problems is the influence of the shape of the aperture or of topographic features of the sample on the polarization of the emitted light. This is especially the case when a metallic interface is involved. The Maxwell equations require the tangential component of the electric-field vector to be continuous at an interface [46]:

$$\vec{n} \times \left(\vec{E}_2 - \vec{E}_1 \right) = 0, \tag{12.116}$$

where $\vec{E}_{1,2}$ is the electric field on both sides of the interface and $\vec{n}$ is the unit vector normal to the interface. Furthermore, no electric field can build up within an ideal conductor. As a consequence, the tangential component of the field has to vanish at the interface of an ideal conductor. By applying that to a light wave passing along a metallic surface, the component of the light with (electric) polarization tangential to the surface will be suppressed. This will substantially influence the state of polarization of the light wave depending on the shape of a conducting surface. As an example, the influence of the aperture shape is demonstrated by analyzing the polarization of the light emitted from the rectangular aperture of a metallic tetrahedral pyramid tip [60]. The aperture is directly illuminated with circularly polarized light, no fiber optic is involved. It is clearly seen from Fig.12-47 that the emitted light is preferentially polarized in the direction perpendicular to the long side of the aperture. As sketched in Fig.12-47(c), the electric field vector parallel to the long side of the aperture is stronger suppressed because for this orientation the edge where the electric field must vanish is much closer. The field component parallel to the narrow side is less suppressed because in this case the edge is much farther away. Consequently, the electric field perpendicular to the long side is preferred. This result can be generalized in that sense, that any non-circular aperture will polarize the emitted light to some extent.

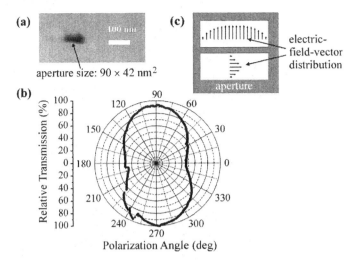

Figure 12-47: (a) Scanning electron micrograph of the rectangular aperture of a metallic tetrahedral pyramid tip. (b) Transmitted light intensity as a function of polarizer angle, the incident light was circularly polarized. (c) Simple sketch of the suppression of the electric field vector through the aperture periphery [60].

Similar interaction can be observed when scanning topographic features on the sample surface such as pinholes in thin metallic films. The experiment has been performed on a pinhole with a diameter below 1 μm that is surrounded by a homogeneous Ni film [137]. A tapered fiber tip coated with aluminum emitting linearly polarized light had been used to scan the sample. The topography image acquired by the shear-force feedback loop exhibited a simple dent within a flat surface. The optical image of the pinhole shown in Fig.12-48(a) has been taken by measuring the I component of the lock-in amplifier which is sensitive to the ellipticity [see Eq.(12.113)]. It shows a complex structure in the area within the pinhole. In an attempt to understand this pattern, a simple model has been employed to simulate the extinction effects induced during the scan. A result of this calculation is shown in Fig.12-48(b) [137]. The comparison with the experimental result in Fig.12-48(a) shows a clear correspondence between both images: within the hole, two bright cones meet with their tips in a dark background. The cones have dark spots in their centers. The structure of the bright cones is surrounded by a dark rim. For the simulation, the tip is regarded as a point source of near-field radiation with a given axis of polarization and ellipticity described by the ratio of $a{:}b$ as sketched in Fig.12-48(c). The emitted near-field radiation is assumed to decay exponentially with distance from the tip. At the rim of the hole, near-field radiation is converted into propagating waves. Any radiation polarized tangential to the hole's rim will be extinguished by the conducting rim. Therefore, the emitted far-field light will have a linear polarization radial to the rim. Details of the calculation can be found in literature [137].

In conclusion, extinction effects can lead to significant polarization contrast in scanning near-field optical microscopy. They can occur within the sample itself as well as within the tip. This contrast mechanism is undesired when imaging magneto-optic contrast. Hence, care has to be taken to avoid the occurrence of metallic structures or particles on the sample. The physical origin of the extinction effects are the boundary conditions as derived from the Maxwell equations. As mentioned in Section 12.5, local birefringence will generate a background contrast as well which can easily exceed the magneto-optic signal. This problem may be solved by using a Sagnac *SNOM* or by making field-dependent measurements.

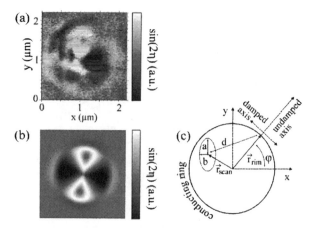

Figure 12-48: (a) *SNOM* image in transmission mode of a pinhole within a nickel film. The optical image is scaled with the ellipticity of the transmitted light. (b) Simulated *SNOM* image with polarization contrast depending on the ellipticity. (c) Scheme of the model used to simulate extinction effects by a circular structure as described in the text [137].

12.6.4. Polarization stability

As mentioned before, it is mandatory to achieve a high angular resolution when measuring magneto-optic effects. Therefore, two criteria have to be fulfilled. First, a high degree of linear polarization and, second, a high stability of the polarization axis. The first condition can be circumvented by using a lock-in method as explained before. This will guarantee that only the polarized fraction of the light is taken into account. However, the unpolarized part will load the lock-in preamplifier limiting the signal sensitivity by the dynamic range which is about -100 dB. The second criterion is much more severe. It can only be neglected to a certain extent when using a Sagnac *SNOM* (see Section 12.4.2.1). Figure 12-49 depicts the drift of the polarization axis of a bare glass fiber [105]. For this measurement, linearly-polarized laser light had been coupled into a bare single-mode fiber with a flat end. While the input polarization was kept constant, the output polarization has been monitored for several hours. Obviously, the shift of the angle of the polarization axis within an hour is much more than the size of the Kerr rotation to be measured. The functional dependence of the curve suggest that the single-mode fiber has to reach thermal equilibrium which takes apparently several hours. The reason for that could be a slight heating by the laser light. However, from time to time plateaus are appearing followed by jumps of several tenths of a degree. This indicates that mechanical strain is relieved within the fiber. But even after several hours, the spread in the minimum positions of consecutive 'V' curves is still more than 0.05° as shown in Fig.12-49(b). The goal of an angular resolution of $\Delta\theta_K = 0.001°$ corresponds to the thickness of the dashed line.

This behavior is not surprising as single-mode fibers are pulled from a massive glass tube which implies a lot of residual mechanical stress leading to birefringence in the fiber which will cause the polarization axis to change. It is known that the polarization of a glass fiber is very sensitive to small temperature changes [138]. The evolution of polarization along a single-mode fiber has been studied before [139]. A review on birefringence in glass fibers can be found in literature [140]. As a solution to the problem, it was proposed to temper the glass fibers in order to relieve the mechanical stress which has build up during manufacturing [141,142]. The disadvantage of such a procedure is that the protecting cover of the fibers has to be completely

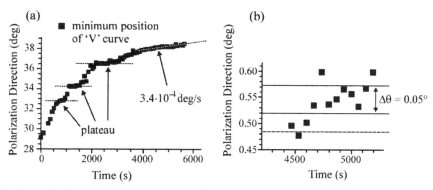

Figure 12-49: (a) Drift of the polarization axis of a bare single-mode fiber. (b) Enlarged section of (a) after having reached thermal equilibrium [105].

removed as it would not survive the high temper temperatures. As a consequence, the fibers become extremely brittle and are very difficult to handle afterwards. Nevertheless, it is possible to temper the single-mode fibers and to form an aperture e.g. by etching. The polarization drift is greatly reduced after tempering as can be seen in Fig.12-50. After one hour, the average drift is about 0.01°. However, this is still a factor 5-10 more than is needed to image magnetic contrast in reflection mode.

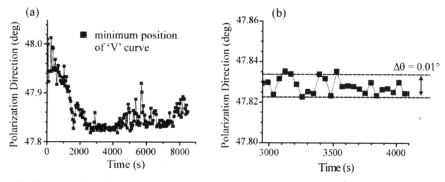

Figure 12-50: (a) Drift of the polarization axis of a tempered single-mode fiber. (b) Enlarged section of (a) after having reached thermal equilibrium [71].

12.7. SUMMARY

In this article, scanning near-field magneto-optic microscopy (*SNOM*) has been reviewed and discussed as a new method for the characterization of magnetic materials. The magnetic information is resolved in the optical near field by means of magneto-optic effects such as the Faraday or the polar Kerr effect. Because the interaction is pure optical, the magnetic domains are not at all influenced while scanning the sample. This is a major advantage compared with magnetic force microscopy. Furthermore, large external magnetic fields can be applied which is not possible for methods using charged particles for imaging such as in electron microscopy.

In order to be able to investigate a broad variety of magnetic materials, Faraday or Kerr rotations of only a few hundredths of a degree must be detected. This implies an angular resolution of a *SNOM* of 0.001-0.005°. For imaging magnetic domains, a lateral resolution of less than 20 nm is desirable. Meeting these requirements is a necessity to turn *SNOM* into an widely used method. State-of the art measurements reach an angular resolution of a few hundredths of a degree which is still a factor of ten above the needs. Lateral pure optical resolution, attained so far, is of the order of 60 nm being three times larger than desired.

The highest angular resolution has been achieved using a Sagnac *SNOM*. The weakness of this method is that only bare glass-fiber tips can be used up to now. Smaller apertures, made by metal coating, will reduce the light transmitted in the shared-aperture mode too much. The chief advantage is certainly the strong suppression of birefringence effects in glass fiber and sample which are ubiquitous.

By using a standard set-up as first designed by Betzig, rotations as small as 0.05° have been resolved. It allows quantitative measurements of local Faraday or Kerr rotations making this method ideal for the study of the local behavior of magnetization reversal. However, the major drawback of this set-up is its sensitivity to birefringence effects in the sample and in the glass fiber. The latter is due to mechanical stress created by the pulling process during manufacturing. Tempering will reduce the birefringence at the cost of increased brittleness of the fibers. In addition, it will diffuse the sharp index-of-refraction change at the interface between core and cladding. Hence, the wave-guide properties are reduced and, ultimately, cause the laser light to leak into the cladding. Another problem encountered with this method is the sensitivity to sample birefringence which is sometimes dominating the magneto-optic contrast. It can, in principle, be subtracted by a reference measurement, which is standard procedure in Kerr microscopy. However, so far we have not succeeded to considerably reduce the birefringence contrast by subtracting two images obtained in different magnetic fields. Apparently, the tip is not at exactly the same distance from the sample when applying an external magnetic field. This will change the local birefringence pattern and the alterations are larger than the magneto-optic effect to be measured.

In conclusion, even though *SNOM* is not a standard technique, it is a valuable new method for characterizing selected magnetic materials. There is still hope that the limitations will be overcome in near future which would turn this technique into one of the most valuable methods for the characterization of magnetic materials.

FOOTNOTES

[1] The de Broglie relation states that any object moving with a momentum p represents a quantum-mechanical material wave with wavelength $\lambda = h/p$, where h is the Planck constant.

[2] Another frequently used acronym is *NSOM* for near-field scanning optical microscope.

[3] The Heaviside function is defined as $\Theta(x) = 0$ if $x < 0$ and $\Theta(x) = 1$ if $x \geq 0$.

[4] A polarizer is an optical device which polarizes light independent of its own polarization along a specific direction defined by the polarizer. This direction is called the axis of the polarizer. As a consequence, it blocks light which is polarized in a direction perpendicular to the axis of the polarizer.

[5] The plane of incidence is the plane which is defined by the surface normal and the wave vector of the incoming light. It describes the plane in which the light travels when it is reflected off the surface.

[6] The angle of incidence is the angle between the wave vector of the light and the surface normal.

[7] Angular resolution means the minimum size of Faraday or Kerr rotation which is still detectable.

REFERENCES

[1]   E. Abbe, in *Archiv für Mikroskopische Anatomie*, vol. **9** (1873), p. 413.

[2]   O. Lummer and F. Reiche, *Die Lehre von der Bildentstehung im Mikroskop von Ernst Abbe*, Vieweg, Braunschweig, 1910.

[3]   Lord Rayleigh, Phil. Mag. (5), **42** (1896) 167.

[4]   E. A. Synge, Phil. Mag. (7), **6** (1928) 356.

[5]   E. A. Ash and G. Nicholls, Nature **237** (1972) 510.

[6]   G. Binnig and H. Rohrer, Helv. Phys. Acta **55** (1982) 726.

[7]   G. Binnig, H. Rohrer, Ch. Gerber, and E. Weibel, Phys. Rev. Lett. **49** (1982) 57.

[8]   A. Lewis, M. Isaacson, A. Harootunian, and A. Muray, Ultramicr. **13** (1984) 227.

[9]   D. W. Pohl, W. Denk, and M. Lanz, Appl. Phys. Lett. **44** (1984) 651.

[10]   E. Betzig and R. J. Chichester: Science **262** (1993) 1422.

[11]   R. X. Bian, R. C. Dunn, X. S. Xie, and P. T. Leung, Phys. Rev. Lett. **75** (1995) 4772.

[12]   A. J. Meixner, D. Zeisel, M. A. Bopp, and G. Tarrach, Optical Engineering **34** (1995) 2324.

[13]   U. Ch. Fischer and D. W. Pohl, Phys. Rev. Lett. **62** (1989) 458.

[14]   B. Hecht, H. Bielefeldt, L. Novotny, Y. Inouye, and D. W. Pohl, Phys. Rev. Lett. **77** (1996) 1889 .

[15]   J. Levy, V. Nikitin, J. M. Kikkawa, D. Awschalom, and N. Samarth, J. Appl. Phys. **79** (1996) 6095.

[16]   C. D. Poweleit, A. Gunther, S. Goodnick, and J. Menéndez, Appl. Phys. Lett. **73** (1998) 2275.

[17]   A. V. Zayats, T. Kalkbrenner, V. Sandoghdar, and J. Mlynek, Phys. Rev. B **61** (2000) 4545.

[18]   E. Betzig, J. K. Trautman, R. Wolfe, E. M. Gyorgy, P. L. Finn, M. H. Kryder, and C.-H. Chang, Appl. Phys. Lett. **61** (1992) 142.

[19]   T. J. Silva, S. Schultz, and D. Weller: Appl. Phys. Lett. **65** (1994) 658.

[20]   G. Eggers, A. Rosenberger, N. Held, and P. Fumagalli, Surf. Interface Anal. **25** (1997) 483.

[21]   P. Fumagalli, A. Rosenberger, G. Eggers, A. Münnemann, N. Held, and G. Güntherodt, Appl. Phys. Lett. **72** (1998) 2803.

[22]   S. Hosaka, T. Shintani, M. Miyamoto, A. Kikukawa, A. Hirotsune, M. Terao, M. Yoshida, K. Fujita, and S. Kämmer, J. Appl. Phys. **79** (1996) 8082.

[23]   B. D. Terris, H. J. Mamin, and D. Rugar, Appl. Phys. Lett. **68** (1996) 141.

[24]   P. J. Moyer and S. B. Kämmer, Appl. Phys. Lett. **68** (1996) 3380.

[25]   E. Betzig and J. K. Trautman, Science **257** (1992) 189.

[26]   S. Wegscheider, A. Kirsch, J. Mlynek, and G. Krausch, Thin Solid Films **264** (1995) 264.

[27]   S. Madsen, M. Müllenborn, K. Birkelund, and F. Grey, Appl. Phys. Lett. **69** (1996) 544.

[28]   R. Riehn, A. Charas, J. Morgado, and F. Cacialli, Appl. Phys. Lett. **82** (2003) 526.

[29]   A. Chimmalgi, T. Y. Choi, C. P. Grigoropoulos, and K. Komvopoulos, Appl. Phys. Lett. **82** (2003) 1146.

[30]   E. Betzig, J. K. Trautman, T. D. Harris, J. S. Weiner, and R. L. Kostelak, Science **251** (1991) 1468.

[31]   J. Koglin, U. C. Fischer, and H. Fuchs, J. Biomed. Opt. **1** (1996) 75.

[32]   J. Koglin, U. C. Fischer, K. D. Brzoska, W. Göhde, and H. Fuchs, in Photons and Local Probes, eds. O. Marti and R. Möller (Kluwer, the Netherlands, 1995) 79.

[33] F. Zenhausern, M. P. O'Boyle, and H. K. Wickramasinghe, Appl. Phys. Lett. **65** (1994) 1623.

[34] F. Zenhausern, Y. Martin, and H. K. Wickramasinghe, Science **269** (1995) 1083.

[35] *Near field optics*, editors: D. W. Pohl and D. Courjon, Kluwer, Dordrecht, 1993.

[36] M. Paesler and P. Moyer, *Near-field Optics: Theory, Instrumentation and Applications*, Wiley, New York, 1996.

[37] *Near-Field Nano/Atom Optics and Technology*, editor: M. Ohtsu, Springer, Tokyo, 1998.

[38] M. Ohtsu and H. Hori, *Near-field nano-optics: from basic principles to nano-fabrication and nano-photonics*, Kluwer, New York, 1999.

[39] D. Courjon and C. Bainier, Rep. Prog. Phys. **57** (1994) 989.

[40] See any text book on optics, e.g., E. Hecht, *Optics*, 2nd ed., Addison-Wesley, Reading, 1987.

[41] For a complete treatment see, e.g., M. Born and E. Wolf, *Principles of Optics*, 4th ed., Pergamon, Oxford, 1970).

[42] W. H. Heisenberg, Z. Phys. **43** (1927) 172.

[43] N. Bohr, Nature **121** (1928) 580.

[44] J. M. Vigoureux and D. Courjon, Appl. Opt. **31** (1992) 3170.

[45] See e.g. K. Krane, *Modern Physics*, 2nd edition (Wiley, New York, 1996).

[46] For a complete treatment of electrodynamics, see J. D. Jackson, *Classical Electrodynamics*, 2nd ed., Wiley, New York, 1975.

[47] M. Spajer, D. Courjon, K. Sarayeddine, A. Jalocha, and J.-M. Vigoureux, J. Phys. III **1** (1991) 1.

[48] H. Bielefeldt I. Horsch, G. Krausch, M. Lux-Steiner, J. Mlynek, and O. Marti, Appl. Phys. A **59** (1994) 103.

[49] H. K. Wickramasinghe and C. C. Williams, U.S. Patent 4,947,034 (April 28,1989).

[50] J. S. Batchelder and M. A. Taubenblatt, Appl. Phys. Lett. **55** (1989) 215.

[51] D. R. Turner, U.S. Patent 4,469,554 (1983).

[52] K. M. Takahashi, J. Colloid Interface Sci. **134** (1990) 181.

[53] T. Hartmann, R. Gatz, W. Wiegräbe, A. Kramer, A. Hillebrand, K. Lieberman, W. Baumeister, and R. Guckenberger, in *Near-fiele optics*, Nato ASI series E, vol. **242** (Kluwer, Dordrecht, 1993) 35.

[54] P. Hoffmann, B, Dutoit, R.-P. Salathé, Ultramicroscopy **61** (1995) 165.

[55] S. Jiang, H. Ohsawa, K. Yamada, T. Pangaribuan, M. Ohtsu, K. Imai, and A. Ikai, Jpn. J. Appl. Phys. **31** (1992) 2282.

[56] T. Pangaribuan, K. Yamada, S. Jiang, H. Ohsawa, and M. Ohtsu, Jpn. J. Appl. Phys. **31** (1992) L1302.

[57] Sutter Instrument Company, Novato, CA, USA.

[58] G. A. Valaskovic, M. Holton, and G. H. Morrison, Appl. Opt. **34** (1995) 1215.

[59] C. Obermüller and K. Karrai, Appl. Phys. Lett. **67** (1995) 3408.

[60] G. Eggers, Ph.D. thesis, Technical University (RWTH) of Aachen (1999). Published by *dissertation.de*, Berlin, 2000, www.dissertation.de.

[61] W. Noell, M. Abraham, K. Mayr, A. Ruf, J. Barenz, O. Hollricher, O. Marti, and P. Güthner, Appl. Phys. Lett. **70** (1997) 1236.

[62] M Abraham, W. Ehrfeld, M. Lacher, K. Mayr, W. Noell, P. Güthner, and J. Barenz, Ultramicroscopy **71** (1998) 93.

[63] S. Münster, S. Werner, C. Mihalcea, W. Scholz, and E. Oesterschulze, J. Microsc. **186** (1997) 17.

[64] E. Oesterschulze, O. Rudow, C. Mihalcea, W. Scholz, and S. Werner, Ultramicroscopy **71** (1998) 85.

[65] E. Oesterschulze, Appl. Phys. A **66** (1998) S3.

[66] S. Werner, O. Rudow, C. Mihalcea, and E. Oesterschulze, Appl. Phys. A **66** (1998) S367.

[67] P. N. Minh, T. Ono, and M. Esashi, Sensors and Actuators **80** (2000) 163.

[68] M. Muranishi, K. Sato, S. Hosaka, A. Kikukawa, T. Shintani, and K. Ito, Jpn. J. Appl. Phys. **36** (1997) L942.

[69] J. A. Veerman, A. M. Otter, L. Kuipers, and N. F. van Hulst, Appl. Phys. Lett. **72** (1998) 3115.

[70] T. Yatsui, M. Kourogi, and M. Ohtsu, Appl. Phys. Lett. **73** (1998) 2090.

[71] U. K. W. Thiele, Ph.D. thesis, Freie Universität Berlin (2003). Published by *dissertation.de*, Berlin, 2003.

[72] E. Betzig, P. L. Finn, and J. S. Weiner, Appl. Phys. Lett. **60** (1992) 2484.

[73] S. Hoppe, J. Paggel, and P. Fumagalli, *to be published*.

[74] M. J. Gregor, P. G. Blome, J. Schöfer, and R. G. Ulbrich, Appl. Phys. Lett. **68** (1996) 307.

[75] K. Karrai and R. D. Grober, Appl. Phys. Lett. **66** (1995) 1842.

[76] R. Toledo-Crow, P. C. Yang, Y. Chen and M. Vaez-Iravani, Appl. Phys. Lett. **60** (1992) 2957.

[77] A. Rosenberger, Ph.D. thesis, Technical University (RWTH) of Aachen, Germany (1999).

[78] S. Hoppe, *unpublished*.

[79] S. Hoppe, Diploma thesis, Technical University of Braunschweig, Germany (1998).

[80] Dr. Volker Klocke Nanotechnik, Aachen, Germany, www.nanomotor.de

[81] M. Faraday, Phil. Trans. **136** (1846) 1.

[82] See for example *Handbook of Optical Constants of Solids*, vol. 1, edited by E. D. Palik (Academic, New York, 1998).

[83] D. Y. Smith, J. Opt. Soc. Am. **66** (1976) 454.

[84] Lord Rayleigh, Phil. Trans. **176** (1885) 343.

[85] J. Kerr, Rep. Brit. Ass. **5** (1876), Phil. Mag. (5) **3** (1877) 321.

[86] H. A. Lorentz, *Versuch einer Theorie der elektrischen und optischen Erscheinungen in bewegten Körpern*, (Leiden, 1895).

[87] See any textbook on linear algebra, as e. g., U. Stammbach, *Lineare Algebra* (Teubner, Stuttgart, 1980).

[88] W. Reim and J. Schoenes, in *Ferromagnetic Materials*, vol. 5, edited by K. H. J. Buschow and E. P. Wohlfarth (North-Holland, Amsterdam, 1990, p. 133.

[89] J. Schoenes, in *Materials Science and Technology*, vol. 3A, edited by R. W. Cahn, P. Haasen, and E. J. Kramer (VCH, Weinheim, 1992), p. 147.

[90] P. Fumagalli, Habilitation thesis, Technical University (RWTH) of Aachen, Germany, 1997 (*unpublished*).

[91] L. M. Roth, Phys. Rev. **133** (1964) A542.

[92] H. S. Bennett and E. A. Stern, Phys. Rev. **137** (1965) A 448.

[93] W. Reim, J. Schoenes, F. Hulliger, and O. Vogt, J. Magn. Magn. Mater. **54-57** (1986) 1401.

[94] R. Pittini, J. Schoenes, O. Vogt, and P. Wachter, Phys. Rev. Lett. **77** (1996) 944.

[95] P. Fumagalli, Ph. D. thesis no. 9082, Swiss Federal Institute of Technology (ETH), Zürich, Switzerland, 1990 (*unpublished*).

[96] For example from Hinds Instruments, Hilsboro, OR, USA.

[97] E. Udd, in *Fiber Optic Sensors*, ed. E. Udd (Wiley, New York, 1991), 233.

[98] A. Kapitulnik, J. S. Dodge, and M. M. Fejer, J. Appl. Phys. **75** (1994) 6872.

[99]   S. Spielman, K. Fesler, C. B. Eom, T. H. Geballe, M. M. Fejer, and A. Kapitulnik, Phys. Rev. Lett. **65** (1990) 123.

[100]   B. L. Petersen, A. Bauer, G. Meyer, T. Crecelius, and G. Kaindl., Appl. Phys. Lett. **73** (1998) 538.

[101]   G. Meyer, T. Crecelius, G. Kaindl, and A. Bauer, J. Magn. Magn. Mater. **240** (2002) 76.

[102]   G. Meyer, T. Crecelius, A. Bauer, I. Mauch , and G. Kaindl, Appl. Phys. Lett. **83** (2003) 1394.

[103]   See for example: D. S. Kliger, J. W. Lewis, and C. R. Randall, *Polarized Light in Optics and Spectroscopy* (Academic Press, San Diego, 1990).

[104]   See for example in I. N. Bronstein and K. A. Scmendjajew, *Taschenbuch der Mathematik* (Harry Deutsch, Frankfurt a.M., 1981).

[105]   P. Fumagalli, in *Advances in Solid State Physics*, **39**, ed. B. Kramer (Vieweg, Braunschweig, 1999) 531.

[106]   E. Betzig, J. K. Trautman, J. S. Weiner, T. D. Harris, and R. Wolfe, Appl. Opt. **31** (1992) 4563.

[107]   J. K. Trautman, E. Betzig, J. S. Weiner, D. J. DiGiovanni, T. D. Harris, F. Hellman, and E. M. Gyorgy, J. Appl. Phys. **71** (1992) 4659.

[108]   Ch. Lienau, A. Richter, and T. Elsaesser, Appl. Phys. Lett. **69** (1996) 325.

[109]   D. I. Kavaldjiev, R. Toledo-Crow, and M. Vaez-Iravani, Appl. Phys. Lett. **67** (1995) 2771.

[110]   S. M. Mansfield and G. S. Kino, Appl. Phys. Lett. **57** (1990) 2615.

[111]   S. M. Mansfield W. R. Studenmund, G. S. Kino, and K. Osato, Opt. Lett. **18** (1993) 305.

[112]   U. Hartmann, J. Magn. Magn. Mater. **157-158** (1996) 545.

[113]   F. Matthes, H. Brückl, G. Reiss, Ultramicr. **71** (1998) 243.

[114]   S. Shalom, K. Lieberman, A. Lewis, and S. R. Cohen, Rev. Sci. Instr. **63** (1992) 4061.

[115]   K. Nakajima, Y. Mitsuoka, N. Chiba, H. Muramatsu, T. Ataka, K. Sato, and M. Fujihira, Ultramicr. **71** (1998) 257.

[116]   T. Ishibashi, T. Yoshida, A. Iijima, K. Sato, Y. Mitsuoka, and K. Nakajima, J. Microscopy **194** (1999) 374.

[117]   M. Vaez-Iravani and R. Toledo-Crow, Appl. Phys. Lett. **63** (1993) 138.

[118]   T. Lacoste, T. Huser, H. Heinzelmann, Z. Phys. B **104** (1997) 183.

[119]   V. Kottler, N. Essaidi, N. Ronarch, C. Chappert, Y. Chen, J. Magn. Magn. Mater. **165** (1997) 398.

[120]   C. Durkan, I. V. Shvets, and J. C. Lodder, Appl. Phys. Lett. **70** (1997) 1323.

[121]   C. Durkan and I. V. Shvets, J. Appl. Phys. **83** (1998) 1171.

[122]   T. J. Silva and S. Schultz, Rev. Sci. Instrum. **67** (1996)715.

[123]   M. R. Pufall, A. Berger, and S. Schultz, J. Appl. Phys. **81** (1997) 5689.

[124]   V. I. Safarov, V. A. Kosobukin, C. Hermann, G. Lampel, C. Marlière, and J. Peretti, Ultramicr. **57** (1995) 270.

[125]   M. W. J. Prins, R. H. M. Groeneveld, D. L. Abraham, H. van Kempen, and H. W. van Kesteren, Appl. Phys. Lett. **66** (1995) 1141.

[126]   H. Wioland, O. Bergossi, S. Hudlet, K. Mackay, and P. Royer, Eur. Phys. J. Appl. Phys. **5** (1999) 289.

[127]   J. Levy, V. Nikitin, J. M. Kikkawa, A. Cohen, N. Samarth, R. Garcia, and D. D. Awschalom, Phys. Rev. Lett. **76** (1996) 1948.

[128]   D. Wegner, U. Conrad, J. Güdde, G. Meyer, T. Crecelius, and A. Bauer, J. Appl. Phys. **88** (2000) 2166.

[129]   D. Weller, H. Notarys, T. Suzuki, G. Gorman, T. Logan, I. McFadyen, and C. J. Chien, IEEE Trans. Magn. **28** (1992) 2500.

[130]   J. Valentin, Th. Kleinefeld, and D. Weller, J. Phys. D: Appl. Phys. **29** (1996) 1111.

[131]   P. Fumagalli, G. Eggers, A. Rosenberger, N. Held, and A. Münnemann, J. Magn. Soc. Jpn. **22** (1998) Suppl. S2, 27.

[132]   A. A. Thiele, J. Appl. Phys. **41** (1970) 1139.

[133]   S. Wittekoek, T J. A. Popma, J. M. Robertson and P. F. Bongers, Phys. Rev. B **12**, (1975) 2777.

[134]   Provided by IBM Research Center at Almaden, CA

[135]   A. Hubert and R. Schäfer, Magnetic Domains: The Analyis of Magnetic Microstructures, Springer, Berlin, Germany (1988).

[136]   A. Rosenberger, A. Münnemann, F. Kiendl, G. Güntherodt P. Rosenbusch, J. A. C. Bland, G. Eggers, and P. Fumagalli, J. Appl. Phys. **89** (2001) 7727.

[137]   G. Eggers, A. Rosenberger, N. Held, G. Güntherodt, and P. Fumagalli, Appl. Phys. Lett. **79** (2001) 3929.

[138]   O. K. Sklyarov, Sov. J. Opt. Technol. **43** (1976) 145.

[139]   A. Simon and R. Ulrich, Appl. Phys. Lett. **31** (1977) 517.

[140]   S. C. Rashleigh, J. Lightwave Technol. **LT-1** (1983) 312.

[141]   D. Tang, A. H. Rose, G. W. Day, S. M. Etzel, J. Lightwave Technol. **9** (1991) 1031.

[142]   A. H. Rose, Z. B. Ren, G. W. Day, in *10th Optical Fiber Sensors Conference*, Proc. SPIE **2360** (1994) 306.

CHAPTER 13

Magnetization dynamics using time-resolved magneto-optic microscopy

13.1. INTRODUCTION

The physics of magnetism in small magnetic elements, where the thickness and lateral dimensions are on the nanometer scale, has become one of the most vigorous research areas in condensed matter physics [1–3]. The subject of nanomagnetism is stimulated both by discoveries of a steadily increasing range of magnetic phenomena and by the increasing interest in advanced magnetic information storage and data process technology. Over the last decades, novel physics has been found in ultrathin magnetic materials. These include the two-dimensional (2D) magnetic phase transition, magnetic interactions, surface magnetic anisotropies and 2D magnetic ordering phenomena, which have all yielded important new fundamental insights. Particularly important findings from the application point of view include the discovery of the giant magnetoresistance (GMR) effect [4], interlayer exchange coupling, and spin-dependent tunneling. Tremendous possibilities of novel magnetic phenomena for technological applications become immediately apparent by considering that most of the information we deal with is processed and stored magnetically, from audio and video products to information storage on computer hard disks. GMR gives one example of such practical applications, where its extreme sensitivity to the magnetic field is used in computer hard disks to enhance information storage density.

Recently, attention has focused on the dynamic behavior of the magnetization in small patterned magnetic elements, fabricated by sophisticated lithography techniques [5–8]. This is mainly because magnetization dynamics on short time scales is different in many aspects from the static case [3,9]. Moreover, the magnetization dynamics in small patterned elements significantly differs from that in continuous films due to the magnetostatics of element edges, thus modifying the equilibrium states of the element in terms of the magnetic moment distribution [10,11]. From a practical point of view, understanding magnetization dynamics on nano- and pico-second time scales in small elements with dimensions in the micrometer size regime and below has become crucial, owing to the increasing demands on conventional storage technologies and for newer approaches such as magnetic random access memories (MRAM) [12,13]. For example, the data rate for current hard disk magnetic recording device is approaching 1 Gbit/s.

Motivated by all of these accumulated interests, dynamic behavior in micro- and nanosized magnets is being studied by a number of groups [14–18]. In order to elucidate magnetization dynamics in small elements, it has long been recognized that direct observations of nonequilibrium magnetization processes with simultaneous spatial and temporal resolutions are most desirable. Imaging of micromagnetic configurations, however, has been carried out mostly by magnetic force microscopy (MFM) [19], Lorentz transmission electron microscopy [20] and ballistic electron magnetic microscopy [21,22] in addition to magneto-optic microscopy [23]. These techniques provide good spatial resolution, but are generally focused on quasi-static magnetic imaging. For the study of dynamic phenomena, it has been demonstrated that very high spatial and temporal resolution can be achieved by employing stroboscopic scanning Kerr microscopy [24–26]. This technique proven to be a powerful tool for dynamic micromagnetic imaging in small structures.

In this chapter, the experimental details of time-resolved magneto-optical microscopy will be described. We also present representative experimental results of ultrafast magnetization reversal, and introduce fundamental dynamic phenomena. These include the precessional motion of the magnetization vector, and complex formation of dynamic magnetic domains, which occur when the magnetic system is abruptly driven into a nonequilibrium state.

13.2. EXPERIMENTAL METHODS

Probing the domain structures at magnetic surfaces is one of the fundamental problems of magnetism. Even though a number of magnetic imaging techniques have been developed and successfully demonstrated over the past decades, the simplest and most widely employed method is the magneto-optical effect, which originates from the interactions of light with magnetic medium [27]. When linearly polarized light is reflected from a magnetic surface, the incident light is transformed into elliptically polarized light. Thus, the final state of polarization can be characterized by both a rotation of the major axis, θ, and an ellipticity, δ, defined as the ratio between minor and major axis. Both θ and δ are proportional to the magnetization of the material. This effect, known as the Kerr effect [3,27], has provided an important tool to visualize the magnetization configuration of a broad range of magnetic materials.

Magneto-optical effect (including the Faraday effect in transmission) can be further used for exploring time-resolved studies of ultrafast magnetic phenomena, by employing short pulses as a light source. This section details the experimental technique of ultrafast scanning Kerr microscopy used for the study of nonequilibrium dynamics in magnetic materials on picosecond time scales. The experimental setup involves the generation of fast magnetic pulses, precise control of the timing between the pump and probe beams, and detection of magnetic signals by means of analysis of the optical polarization.

13.2.1. Pump-and-Probe Method

In time-resolved experiments one measures the time evolution of a physical process, away from or towards equilibrium, in response to sudden perturbations. One of the ways to visualize a fast process is to use a very short flashing light to freeze process and make observations. The stroboscopic technique, was perfected for photographic imaging by H. Edgerton and colleagues in the 1930s. The application of a related approach to magnetic systems was inspired by the first ultrafast pump-probe studies in dilute magnetic systems by Awschalom et al. in 1985 [28].

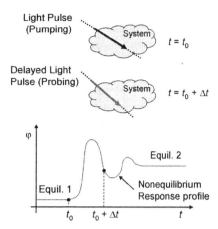

Figure 13-1: Working principle of the "pump-and-probe" technique. At time $t = t_0$, the system is excited out of its equilibrium with the pump pulse. After a short time interval Δt, the system is probed with the probe pulse. The time delay between the probe and pump beam is then changed to give a new Δt temporal position and the signal built-up again. This procedure is repeated until the entire nonequilibrium profile is measured.

Figure 13-1 schematically illustrates the working principle of the "pump-probe" technique. A short pulse beam is split such that one part of the pulses, called the pump pulses, is used to excite a nonequilibrium state in the system, φ, at the time $t = t_0$, while the delayed part, i.e. probe pulses, are used to detect the corresponding change in the system at $t = t_0 + \Delta t$. The time interval Δt can be created, for example, just by making the optical path of the probe beam longer than that of the pump beam. After setting a certain time point t, the perturbation of the system, φ(t), is detected. The probe beam path is then changed to give a new Δt temporal position, and the probe beam detects the corresponding change in the system again. This procedure is repeated until the entire nonequilibrium response profile of the excited system is measured. In the study of magnetization reversal dynamics, for instance, the nonequilibrium state φ(t) corresponds to the magnetization switching process excited by magnetic pulses. In such experiments, a synchronously triggered transient magnetic pulse is propagated past the sample under study, perturbing the magnetization system, and the subsequent evolution of the magnetization configuration is monitored through its interaction with a probe beam.

13.2.2. Experimental Setup

Figure 13-2 illustrates the schematic diagram for the entire microscopy system, including the optical and electronic layouts. Experimental arrangements are based on a scanning Kerr microscopy, including pulsed laser system and optics, a piezo-driven three-axis flexure stage for scanning the sample surface, and electronics controlling the time-delay of probe beam and magnetic pulse generation. Details of the particular parts of the apparatus follow.

13.2.2.1. Optical setup and signal detection

The pulsed light source used here is a mode-locked Ti:Sapphire laser, which provides 70 fs long pulses of near-infrared light ($\lambda = 800$ nm) at a repetition rate of 82 MHz, which is reduced to below

1 MHz by a pulse picker, for use with the pulse electronics shown in Fig. 13-2. The laser beam is split into two (a pump and a probe beam) with a 50/50 beam splitter (BS), which lets through equal amounts of s- and p-polarized light. The probe beam passes through a linear polarizer (POL) and is deflected toward the sample by the second beam splitter (BS) and is focused on to the sample using an infinity-corrected microscope objective (OBJ), while the pump beam is directed for triggering magnetic pulses (as will be described more in detail below). The optical power of the probe beam is reduced before being brought to a sharp focus on a sample in order to avoid permanent damage to the sample surface. Typically, an average power of 30 µW at the 1 MHz range repetition rate is focused onto the sample through a microscope objective. The spatial resolution (d) is determined by the numerical aperture (N.A.) of the objective lens and wavelength (λ) of the laser beam, given by the diffraction limited Rayleigh criterion, $d = (0,82\,\lambda) / \text{N.A.}$ In our experimental setup, a spatial resolution down to 0.9 µm is yielded using the 0.75 N.A. microscope objective and near-infrared light source.

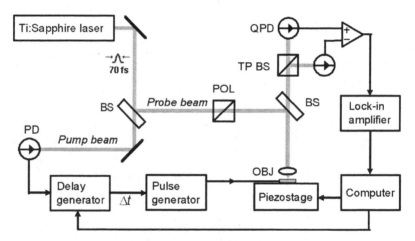

Figure 13-2: A block diagram of the layout of the time-resolved scanning Kerr microscopy. A linear polarized laser beam is split into a pump and a probe beam with a 50/50 beam splitter (BS). The probe beam is focused on to the sample using a microscope objective (OBJ), and then reflected from the sample. The change of the polarization of the reflected light is analyzed using Thomson polarizing beam splitter (TP BS), combined with the signal detection using quadrant photodiodes (QPD). A small polarization rotation of the reflected light induced by the magnetization in the sample is turned into intensity shift by TP BS, in which the intensity is measured. The pump beam is directed for triggering magnetic pulses. After electrical pulses are generated by a photodiode (PD), the output signals are synchronized by a computer controlled variable delay generator (VDG). Magnetic pulses are generated by an electronic pulser, which provides fast electrical pulses of $45 - 50\ V$, $0.25 - 0.5$ ns rise times, and pulse widths of 10 ns. These pulses are delivered to micro stripe lines, which create magnetic field pulses. (*see* inset in Fig. 13-5 showing such stripe lines). Samples are mounted on a computer controlled piezo-driven flexure stage, which enables raster scanning of the sample surface to build a spatial image.

After the probe beam is reflected from the sample surface, magnetization measurements are accomplished through polarization analysis of the reflected light in an optical bridge. A particular detection method using quadrant photodiodes (QPD) has been recently developed to allow for simultaneous detection of all three magnetization components (i.e. vector magnetometry) [8]. This approach is adopted from static Kerr imaging [29,30], and works equally well in time-resolved measurements. The principle of the differential detection method is schematically depicted in Fig. 13-3. The probe beam reflected from the sample surface is split into two orthogonal polarization states by the Thomson polarizing beam splitter (TP BS), which is set at

45° to the incident polarization plane. If there is no polarization rotation in the incident beam, each portion of the split beam will be of equal intensity and differential subtraction of the outputs coming from the two photodiodes will result in a zero signal. A small polarization rotation induced by the magnetization in the sample, however, will be turned into intensity shift by the Thomson analyzer, in which the intensity in one PD increases while decreasing in the other. This differential detection technique, where each PD signal is used as a reference to the other, has the advantage of common mode rejection of laser noise, while doubling the signal [31].

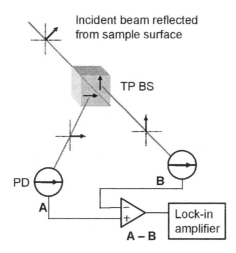

Figure 13-3: Schematic illustration describing differential detection of polarization rotations induced by the magneto-optical effect. Both outputs of a polarizing beam splitter (at 45° from incident polarization) are used in subtraction. The intensity in each arm rises or falls, respectively with rotations of the plane of polarization. The subtracted signal (A–B) is only non-zero when a polarization rotation occurs.

An important issue in the signal detection is the signal-to-noise ratio, which has always been a problem in Kerr imaging due to the weak magneto-optical Kerr effect. In static Kerr measurements, wide-field imaging with digital enhancement is generally used, in which the image with magnetization pattern is digitally subtracted from an image of magnetization saturation [32]. This technique removes edge defect and optical system polarization artifacts. In our experimental setup, signal enhancement is achieved at each pixel of a raster scan using synchronous modulation of the magnetic excitation and lock-in signal detection technique. The modulated magnetic excitation is accomplished by modulating the trigger electronics, which generates magnetic pulses, at a low frequency (1 – 4 kHz). The modulation on the magnetization in the sample itself isolates the signal from artifacts such as depolarizing effects in the system.

13.2.2.2. Synchronization and magnetic field pulse generation

In the stroboscopic method, the temporal excitation ("pumping") of the system must be triggered synchronously with the probe pulses. When a single probe pulse does not yield sufficient signal-to-noise (true in the present case), repetitive events are averaged and represented as a single measurement.

For the experimental setup schematically described in Fig. 13-2, the repetition rate of the pulsed beam is reduced from 82 MHz to 0.8 MHz via pulse picking of the mode-locked laser pulse train. This is required since the maximum trigger rate of the typical delay generator electronics is limited to 1 MHz. On the other hand, the delay electronics creates the propagation delay of the order of 100 ns. Therefore, an additional time delay between the pump and probe beam is required to achieve temporal synchronization. This can be achieved by delaying clock signals by an equivalent amount by propagation, for example, through a length of coaxial cable, until current pulse is actually synchronized with the laser pulse immediately following the one it was triggered by. After setting a time delay Δt, the delay generator sends the clock pulses on to the pulse generator.

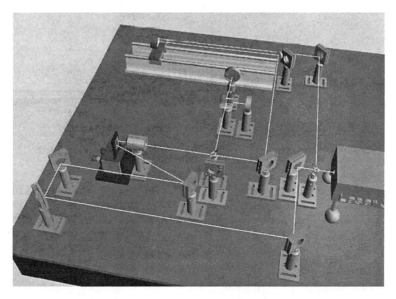

Figure 13-4: Rendering of the experimental setup, in which an optical delay line is used for the synchronization of the probe beam and magnetic pulse. The travel path of probe beam with respect to pump beam is computer-controlled using a retroreflector mounted on a slider (3). Magnetic field pulses are generated by a fast photodiode (6), which is launched by the pump pulse itself. Other components include pulsed laser (1), 50/50 beam splitter (2), linear polarizer (4), infinity-corrected microscope objective (5), Thomson polarizing beam splitter (7), and quadrant photodiodes (8).

The electronic delay method is very convenient, particularly when delay ranges of 10 ns or more are needed, but adds undesirable electronic jitter of about 50 ps rms. This trigger jitter is the main limiter of temporal resolution in this case. Alternatively, an optical delay line can be used for the synchronization of the probe beam and magnetic pulse, in which the travel path of probe beam with respect to pump beam is computer-controlled using a retroreflector mounted on a slider. Figure 13-4 shows an experimental set-up, in which the electrical pulses generating magnetic field pulses are derived directly from a fast photodiode. This technique is inherently jitter-free and is beneficial to measurements for faster (low ps regime) processes. However, in practice the time delay range usually spans only a few nanoseconds due to the limited length of delay line.

For small perturbations as in magnetic resonance experiments, the magnetic pulse generation has been usually made using a photoconductive switch [33], in which the switch is closed by photons above the bandgap energy, freeing up carriers between two biased, metallized regions in a semiconductor substrate. This method generates fast and jitter-free pulses, but it is generally known that the electric pulse shape is hard to control. In most cases in switching experiments, the generation of magnetic pulses relies on the current driver, which is based on the avalanche transistor pulser using the technique of discharging a transmission line. Pulses from this source are delivered to micro stripe lines, which create magnetic field pulses. The inset in Fig. 13-5 shows an image for such stripe lines, on or near which magnetic elements are placed. To excite the sample with an out-of-plane magnetic field pulse, samples are situated between lines, and for an in-plane pulse, on top of a line. Strip lines are fabricated using lithography, and have the width of 20 μm and thickness of 300 nm in this case. This type of strip line creates magnetic pulses as high as 24 kA/m by a current pulse of ~ 1 A. The temporal shape of field pulses can be measured by a commercially available 2 GHz inductive probe or by measuring Faraday rotation in a garnet indicator film [19]. A garnet film allows optical measurement of the current waveforms in a very high bandwidth (over 50 GHz), in addition to providing an absolute time reference for the time-resolved magnetic measurements [26].

13.2.2.3. Sample preparation and magnetic field configuration

The samples investigated in our experiments are polycrystalline Permalloy ($Ni_{80}Fe_{20}$) thin film elements with various dimensions and shapes and typical thicknesses of 15 nm. The Permalloy is a soft magnetic alloy widely used throughout the information storage industry. The $Ni_{80}Fe_{20}$ film was deposited using DC magnetron sputtering onto 300 nm thick Au on sapphire substrate, at a growth rate of 0.1 nm/s in a high-vacuum system with a base pressure of 5×10^{-8} Torr. During deposition an external magnetic field with the strength of 12 kA/m was applied in the plane of the substrate in order to induce the uniaxial magnetic anisotropy. The sapphire substrate has usually been diced to approximately 1×1 cm^2 in size. Magnetic microstructures were fabricated by electron beam lithography and lift-off techniques. The right hand side of Fig. 13-5 shows an optical microscope image of the magnetic structures and transmission lines. The patterned elements are made on (or close to) a transmission line that carries a fast current pulse. In most cases the element shapes are nicely defined, at our current lithographic resolution of about 50 nm. Atomic Force Microscopy (AFM) and Scanning Electron Microscopy (SEM) are also employed to inspect the completed structures. In most cases the lithographic structures have a very smooth surface with minor burrs at the sample edge [34].

The geometric configuration of the biasing and switching magnetic fields is illustrated on the left hand side of figure 13-5, where the strip line carries a current pulse in order to create an in-plane switching field pulse (H_s) along the long axis of the sample. In 180° experiments, the sample is first magnetically saturated in the easy axis direction, parallel to the long sides of the elements, by an in-plane static biasing field ($H_l = 0 - 32$ kA/m). An in-plane switching field pulse, H_s, is then applied in the opposite direction to H_l in order to flip the magnetization direction. The element is optically interrogated while switching is taking place. Additionally, an in-plane static transverse biasing field, H_t, can be applied in order to manipulate the magnetization reversal process.

13.2.2.4. Crossed-wire element

An important complement to the above work utilizes crossed-wire elements consisting of two mutually isolated crossed thin film wires with a ferromagnetic sample deposited at the intersection. Current pulses propagating down each wire allow apply pulsed magnetic field parallel and perpendicular to the easy axis of the sample magnetization. Varying the strength and the shape of both pulses helps one to achieve different switching scenarios.

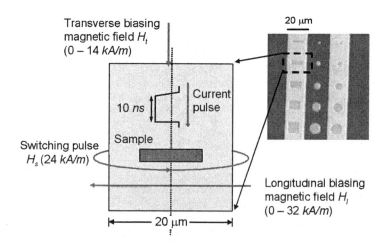

Figure 13-5: Schematic measurement configuration of a 180° dynamic magnetization reversal experiment for microstructure excitation. H_s, H_l and H_t indicate the switching field, longitudinal (easy-axis) biasing field, and transverse (hard-axis) biasing field, respectively. In the inset an image using an optical microscope is given, showing magnetic elements on top or near of the gold transmission lines.

One scenario of particular interest is so called "half-select" switching. In this regime magnetic fields created by each wire are too weak to flip the magnetization of the ferromagnetic element, but their combined action is sufficient. This is a normal operational regime of Magnetic Random Access Memories (MRAMs) [35], built as a grid of isolated orthogonal wires with magnetic memory cells placed at the intersections. Each wire serves multiple cells, but only the element at the crossing can be switched. The wire, creating the easy-axis magnetic field is often called the "word" wire (WW), and the other one the "digital" wire (DW).

In equilibrium, the critical curve separating the switching regime from non-switching is known as the Stoner-Wolfarth astroid [36]. This curve has been observed in numerous quasi-static experiments [35], but switching by short, fast-rising field pulses of finite duration much less investigated. Numerical simulations [37] predict a very complex switching diagram.

The element (Fig.13-6) consists of photo-lithographically patterned wires isolated from each other by two layers of spin-on-glass (SOG) planarizing dielectric. Both gold write wires are 200 nm thick, the width of the DW is approximately 6 μm, and the width of the WW is 3.5 μm. Several 3×1 μm 15-nm thick elliptical permalloy ($Ni_{80}Fe_{20}$) elements are deposited on top of the word wire using e-beam lithography and lift-off with PMMA resist. Ferromagnetic elements are isolated from the wire by three layers of spin-on glass (which ensure uniform coating of the structure by PMMA resist) and a layer of silicon dioxide protecting the SOG from the aggressive PMMA resist remover on the last step of the lithography and improving the surface quality.

13.2.3. Operation Modes in TR-SKM Experiments

Two operation modes are usually employed in TR-SKM experiments. One of these is temporal-resolving mode, where one obtains a majority of the information very efficiently by measuring the dynamic response of the 'local' magnetization of a sample as a function of time. In this mode the probe beam is focused on a particular place (usually at the center) of the sample surface, and then the time delay Δt is changed. This mode is suitable for quick local characterization of

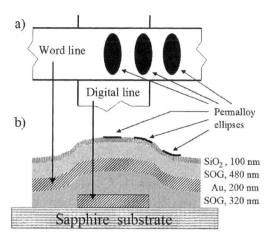

Figure 13-6: Schematic planar view (a) and a cross-section (b) of the crossed wires element. Two 3×1 µm permalloy elliptical samples are placed at the intersection.

the magnetic dynamics. Even though this operation mode ignores spatial information, the time-dependent profiles of the magnetization components collected on various locations on the sample provide valuable data about local magnetic response. The information of local magnetic responses is crucial for properly interpreting the dynamic process, since the dynamic response of the most magnetic systems studied can be spatially non-uniform [8,9]. [*see* Section 3.3] Temporal resolution is ultimately limited by the laser pulse width, but practically limited by the trigger jitter from the delay electronics, as described above. The Kerr signal is detected after each time step, building up the time-dependent profile for selected magnetization components. One example is given in Fig. 13-6, in which time traces for the magnetic responses, *i.e.*, M_x, M_y, and M_z measured for three orthogonal in-plane magnetization components, are shown. The data have been measured at the center of a 15 *nm* thick square (10×10 µm²) $Ni_{80}Fe_{20}$ element, and have been offset for clarity. This example clearly demonstrates that the magnetization is thrown out of its initial saturation state and into magnetization oscillation about the effective magnetic field upon application of the switching pulse. The magnetization oscillations result from magnetic precession about the effective switching field axis, which occurs generally when the magnetization vector feels a torque due to a non-parallel local effective field. In the present case the magnetization vector undergoes a large angle motion, as schematically shown in the inset of Fig. 13-7. Experiments of this kind provide understanding of the local vectorial response in nonequilibrium magnetic systems.

More detailed information on the magnetization dynamics can be obtained from the spatiotemporal-resolving operation mode. After the temporal response of the magnetization is measured, the sample surface can be scanned at a particular fixed time delay in order to obtain two-dimensional images mapping the magnetization configuration of the entire sample. Figure 13-8 displays an example of such spatial-time scanned domain images, captured during magnetization reversal in a thin rectangular (10×2 µm²) Permalloy film element. The images are on a linear gray scale to render the change in magnetization components with black corresponding to no change and white corresponding to the saturation level or maximum possible reversal. The evolution of spatial profiles with increasing time, associated with the magnetic domain nucleation, reveals that the magnetic response occurs locally at different times across the sample. (Details about this experiment will be discussed in section 13.3.1.) During this measurement mode, the sample is placed on computer-controlled piezo-driven three-axis flexure stage providing scanning motion at a typical scan rate of 8 pixels/s (which typically corresponds to about 0.3 µm/s). A time record of few hundred separate sample measurements of the average magnetization was collected for each pixel in the image. Critical in this operation mode is the stability of the piezostage over long time intervals, since the quality of this stage also contributes to the effective spatial resolution of the system. We have implemented software correction to compensate for thermal drift of the stage (typically 0.1 μm/hr in our experiments).

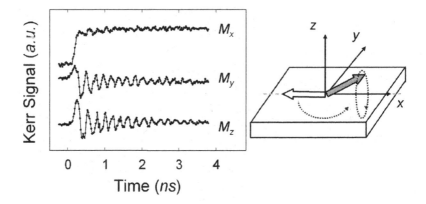

Figure 13-7: Example of time-resolved magnetization components from a Permally ($Ni_{80}Fe_{20}$) rectangular element. Data are from the magnetization response at the center of the sample versus time, and show phase shifted oscillations in x, y, and z magnetization components. These phase shifts indicate precessional motion. Inset shows a schematic cartoon of idealized precessional switching induced by the application of an abrupt magnetic pulse.

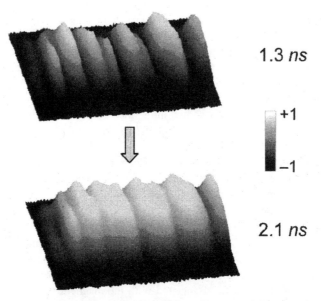

Figure 13-8: Spatial-time scanned data captured during magnetization reversal. Spatial profiles demonstrate that the magnetic response occurs highly local at different times spatially across the sample. Movies showing complete sequences of reversal processes are found on http://www.nanoscale.phys.ualberta.ca. *Also see the color plate.*

13.3. EXPERIMENTAL RESULTS FOR MAGNETIZATION REVERSAL DYNAMICS

In this section, time-resolved scanning Kerr microscopy measurements in patterned Permalloy elements are presented. A major issue here is what happens when magnetic elements undergo large angle magnetization reversal. Such switching experiments are also directly relevant to the magnetic recording, where information is written on a magnetic medium by reversing the magnetization using a recording head. Considering that data rates can now be higher than 1 Gbit/s, understanding the real time magnetization reversal dynamics in magnetic elements has become more crucial for the development of high-density information-storage media.

13.3.1. Domain Nucleation vs. Domain wall motion in nonequilibrium state

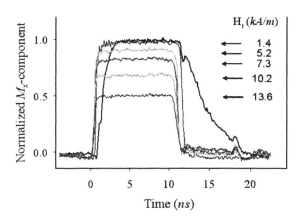

Figure 13-9: Time-resolved magnetization components (M_x) along the easy axis, measured in the center of the element for different transverse biasing field (H_t). The longitudinal biasing field H_l is being held fixed at 4.8 kA/m. The thick line indicates the M_x component measured at H_t = 0 kA/m. The switching pulse begins at 0 ns.

Magnetic domain formation is a complex process and can be drastically changed, for example, by the shape, dimensions, and thickness of the elements [22,26]. As in the cases for the complex magnetic responses in quasi-static cases, the dynamic magnetic properties can also be sensitively modified by sample shapes and experiment parameters. For example, the applied magnetic field conditions can play an essential role for determining for magnetization switching behavior. Figure 13-9 shows experimental data of the local magnetic response of a 15 nm thick $Ni_{80}Fe_{20}$ rectangular element with a dimension of 10×2 μm^2. Measurements were made with the 0.9 μm focus spot positioned at the center of the structure. Note the biasing and switching field configuration as depicted in figure 13-5; i.e. the sample is first magnetically saturated in one direction and then has a short magnetic field pulse applied to the opposite direction in order to reverse the magnetization direction. Finally, the sample magnetization would end up a saturated state again but in the opposite direction. The data represents the time traces of easy axis magnetization components M_x, compared for different transverse biasing fields H_t, while H_l is kept at 4.8 kA/m. Applying no transverse field, i.e. H_t = 0 kA/m as indicated by the thick line, a definite delay in the magnetic response after the beginning of the pulse is observed, and the subsequent dynamics are relatively slow with the magnetization fully reversed after 3.5 ns. When a transverse biasing field H_t is applied perpendicular to the easy axis, a striking change in the magnetic response is observed. The first point to note is that the magnetic response becomes faster with respect to the case without applying H_t. In such cases the magnetization switches within 1 ns after the pulse is given, and a transverse biasing field strength as low as 1.4 kA/m is found to be sufficient to cause such an abrupt switching. The qualitative explanation for this effect is that when a transverse field H_t is applied, the equilibrium position of the magnetization vector **M** is away from the initial easy axis, hence the reversing field exerts torque on the magnetization vector immediately. In addition, the magnetization is away from the minimum anisotropy energy state along the easy axis, so the effective coercive field is lower than for $H_t = 0$ kA/m. Consequently, lower longitudinal Zeeman energy or smaller switching field strength is required to overcome the energy barrier. We note that this is well known from quasistatic studies [26] and modeling, and our study is extending to the fast dynamic regime.

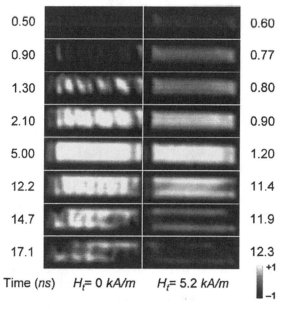

Figure 13-10: Spatial magnetization profiles of the M_x component as a function of time after the magnetic pulse was applied. Each panel corresponds to 12×4 μm field of view, and contains the entire 10×2 μm sample. The numbers by the frames indicate the time at which the measurement was made, relative to the initial application of the switching pulse. The comparison of time domain images clearly reveals a sensitive transition between two distinct magnetization reversal mechanisms, i.e. domain nucleation and domain wall motion, depending on the effective biasing field configuration. *Also see the color plate.*

More detailed insight into the temporal evolution of the magnetization reversal dynamics is obtained by direct time-domain imaging. Figure 13-9 shows a sequence of time-resolved images representing the easy axis magnetization components (M_x) for $H_t = 4.8$ kA/m at selected time points, demarcated in nanoseconds relative to the initial application of the switching pulse. For $H_t = 0$ kA/m (the left column), the reversal is mainly governed by a domain nucleation process. In the very beginning (0.5 ns) a stripelike instability is observed inside the sample, from nucleation occurring in the same regions. The main dynamical reversal, however, is first initiated from the demagnetized edges (0.9 ns), is followed by expansion of the nucleated domains (1.3 and 2.1 ns), and finally leads to a uniform distribution of fully reversed magnetization, excluding the left and right edge regions (5 ns). These edge regions correspond to free magnetic poles related to the demagnetized areas in a ferromagnet of finite size. On the back reversal, the stripe instability is also pronounced (12.2 ns). From the spatiotemporally resolved magnetic images, it becomes clear how the magnetization switching in small elements occurs in this particular case: The finite domain nucleation is the main reversal mechanism, and limits switching time to about 3.5 ns.

This nucleation-dominant reversal process, however, can be drastically manipulated through an application of an additional transverse biasing field. This is demonstrated by time domain images, shown in the right column of Fig. 13-10. Applying an additional transverse biasing field $H_t = 5.2$ kA/m, the 180° domains at the short edges are formed (0.6 ns), but there appears no stripelike distribution inside. The edge domains expand quickly in the easy axis direction to form a long, narrow domain parallel to the easy axis (0.77 ns). In the next stage, this elongated domain expands by parallel shifts in the hard direction towards the long edge (0.8 and 0.9 ns) until saturation is reached (1.20 ns). This type of reversal, which is characteristic of domain wall motion, is considerably faster as revealed in the time dependence of magnetization in Fig. 13-9.

The differences in the time domain sequences demonstrate that the formation of nuclei inside the sample can be easily avoided by the presence of H_t and that the nucleation process is replaced by domain wall motion. This result clearly reveals a sensitive transition between two distinct magnetization reversal mechanisms in the same magnetic element, in which switching occurs over longer times when the stripe domains are involved in the reversal process than if pure domain wall motion occurs.

13.3.2. Complex Domain Pattern Formation

One of the fundamental features associated with fast magnetization reversal process is the spontaneous development of magnetic domain patterns. This concerns the time scale and mechanism for removal of the initial excess Zeeman energy from the nonequilibrium magnetic states. In this section, the observation of the spontaneous domain pattern configuration in a 15 nm thick polycrystalline $Ni_{80}Fe_{20}$ square (10×10 μm^2) element is discussed. The magnetic field configuration used in this experiment is as described in the previous section (*cfr.* figure 13-5), except that the rise time of the switching pulse is varied between 0.24 and 8.6 ns.

Figure 13-10 shows the spatio-temporal evolution of the magnetization component in response to short magnetic pulses with the rise time of 240 ps. The magnetic domain images are measured along the magnetic easy axis at selected time points after applying a switching pulse. The contrast in the images reflects the local degree of magnetization reversal, with white areas corresponding to fully reversed regions. The domain configurations show a complex spatial appearance, which for purposes of compact description we characterize as a labyrinth pattern. A particularly well-developed labyrinth pattern appears for the frame measured at 900 ps. We note that the qualitatively same behavior is observed in other samples. Emergence of the labyrinth formation is seen already at the very beginning of the reversal process. Near the initial state (750 ps), the magnetization switch has started at the element ends and is accompanied by the nucleation of branch-like fine structures visible in the interior regions. Labyrinth domain patterns evolve out of these fine nucleation sites, and once a pattern forms, it is quasi-stable and the completion of reversal is mainly governed by a gradual expansion of the reversed domains. It is remarkable that this level of detail is observed in a stroboscopic (*i.e.*, repetitively averaged) measurement. It remains to be understood how the nonequilibrium dynamics, presumably acting in concert with sample imperfections or pinning sites, select out a pattern that largely recurs from reversal to reversal. Note that the spatial contrast one observes in this experiment represents only the lower limit of what one might find in a "single shot" temporal observation, or with finer spatial resolution in the stroboscopically-averaged case.

A further interesting phenomenon arising in the stroboscopic observation of these patterns is demonstrated in figure 13-12, where time domain images captured during magnetization oscillation are presented. Some of the fine spatial structure has vanished before even the first cycle of precession is complete, but an alternating change of domain pattern contrast is observed clearly during the magnetization precession, and without significant changes of the pattern shape. This becomes immediately apparent by comparing domain images measured at the peak and dip of the magnetization oscillations (compare to the time traces in the inset). During oscillations, the already fully reversed regions, measured at the oscillation peak (a), simply turn into not-fully-reversed regions at the dip (b). In the next oscillation peak (c), one observes a same domain pattern with progressing reversal.

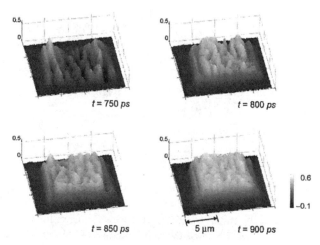

Figure 13-11: Spatio-temporally resolved domain images representing the easy axis magnetization component at selected time points after the onset of a magnetic switching field of 0.24 ns rise time (0 – 90%). The 3D topography and color map both render the magnitude of the magnetic signal from unchanged (black) to fully reversed (white). A complex labyrinth-like domain pattern evolves during the reversal, unlike anything observed at slower switching speeds.

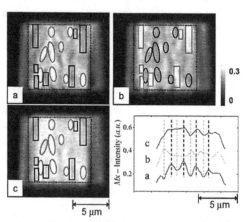

Figure 13-12: (*Top panel*) Time traces of the magnetization at the center of the element, when excited by magnetic field pulses of different rise time (0.24 ns (a), 2.8 ns (b), 5.1 ns (c), and 8.6 ns (d)). Traces show the change of magnetization from one saturated state to another, and magnetic responses show additional delay with increasing rise time, with oscillations following the primary switch. (*Bottom panel*) Dynamic domain patterns revealed as a function of the magnetic pulse rise time. The spontaneous domain configuration is sensitively dependent on the switching speed. A transition from labyrinth- to stripe-like domains occurs with increasing rise time.

The data of Figs. 13-11 and 13-12 prove that the switching mechanism in this case is driven by the multiple local nucleation of reversed domains, and the domain walls seem not to be able to propagate far enough across the sample in order to build bigger reversed domains. Instead, collective precessional motion occurs inside the reversed domains. This is demonstrated in Fig. 13-12 where the alternating magnetic contrast during precession implies that the selected domain pattern is maintained while the remaining excess Zeeman energy introduced by the applied switching field is dissipated by carrying out the precessional magnetization motion.

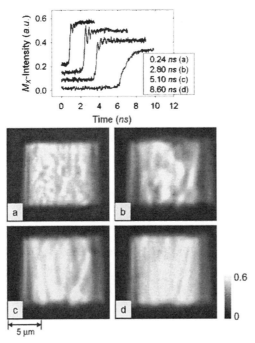

Figure 13-13: Comparing spatio-temporally resolved domain images measured at the peaks (a and c) and dip (b) during magnetization oscillation. A striking feature is the alternation of contrast during the magnetization precession without the spatial pattern changing significantly. Through a cycle of oscillations, the already fully reversed regions at the oscillation peak (a) simply turn into not-fully-reversed regions at the dip (b). In the next oscillation peak (c), one observes the same pattern with progressing reversal. (*inset*) Horizontal cross-sectional views of the domain images.

The direct evidence that the domain wall propagation is effectively involved in the fast switching process is demonstrated in Fig. 13-13, where the transition from quasi-static to dynamic reversal behavior is mapped out, as far as the constraints of our apparatus will allow. In this experiment, evolution of the domain configuration has been measured as a function of the rise time of the magnetic switching pulse. The upper panel of Fig. 13-13 shows the temporal evolution of the magnetization at the center of the same 10×10 μm^2 element, excited by magnetic pulses with different rise time (i.e. t_{rise} = 0.24 ns, 2.7 ns, 5.1 ns, and 8.6 ns). All traces show a change of magnetization from one saturated state to another, and magnetic responses show additional delay with increasing rise time, with oscillations following the primary switch. The bottom panel of Fig. 13-12 shows characteristic domain patterns imaged at time points at which the reversal reaches about 50 % of the full M_x–intensity change. The situations for slower switching are very different from the fast case (t_{rise} = 0.24 ns), for example, for t_{rise} = 5.1 ns the labyrinth domain pattern is no longer observed and instead the domain configuration has a more regular character in which stripe domains predominate. Thus, accelerating the switching process is required to cause the more intricate structure to form, as seen in the domain pattern evolution from $t_{rise} = 8.6$ to 0.24 ns. This observation clearly illustrates that the speed at which the magnetization reversal is driven plays an essential role in determining the domain patterns. We propose that the sensitive dependence of the spatial pattern on switching speed should be understood with the concept of domain wall propagation. Namely, the domain walls of reversed domains will propagate as long as the rise time of the switching pulse allows. Consequently, long travel of locally nucleated domain walls will not be expected when a switching pulse with extremely short rise time is applied. In agreement with this simple concept, the formation of complex configuration with small nucleated domains is most pronounced for the shortest rise time of 0.24 ns.

13.3.3. Large Amplitude Precessional Oscillations

In this section, the direct observation of a novel switching phenomena, *i.e.*, large amplitude precessional oscillations of closure domain magnetization, is discussed. The sample used in this experiment is a patterned 15 nm thick polycrystalline $Ni_{80}Fe_{20}$ rectangular element with the lateral dimension of 10×2 μm^2. First, the temporally resolved magnetic response is measured after the sample magnetization is excited by a short magnetic filed pulse, which has the field strength of 24 kA/m and rise time of 250 ps. As in the previous experiments the sample is first saturated along the long axis by applying an in-plane bias field ($H_l = 5$ kA/m) parallel to its direction. Then a switching pulse H_s is applied antiparallel to H_l in order to reverse the sample

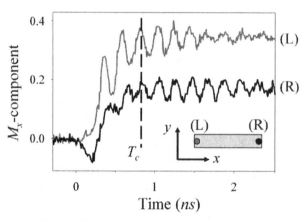

Figure 13-14: Temporal evolution of magnetization components (M_x) along the long-axis direction after a magnetic pulse is applied. The beginning of the pulse corresponds to 0 ns. The schematic illustrates the positions of the focused laser beam on the element during the measurements.

magnetization. Additionally, an in-plane transverse bias field ($H_t = 0.8 - 1.2$ kA/m) is applied along the hard axis of the sample in order to control the initial magnetization state. Figure 13-14 shows the in-plane component (M_x) of the magnetization along the x-axis as a function of time measured with $H_t = 0.8$ kA/m. As schematically illustrated, the laser beam with the spot size of 0.9 μm is focused near the left (L) and right (R) ends of the element, respectively. First, one observes that the magnetization reversal is driven by a precessional motion of the sample magnetization with a typical frequency of about 4 GHz. Inspection of the temporal evolution of the curves measured at both edges reveals, however, that the peaks in the curves (L) and (R) are shifted by about $\pi/2$ with respect to each other at the initial stage, *i.e.*, below a certain critical time $t_c \approx 800$ ps. The time interval between a crest and trough is read about 100 ps for $t < t_c$. After the curves undergo about two cycles of oscillation, the curves reveal no phase shift, indicating that the sample magnetization precesses in-phase at both ends. Interesting phenomenon in this measurement is the out-of-phase magnetization dynamics for $t < t_c$, which we will attribute to the large amplitude precessional oscillations of closure domain magnetization occurring when the magnetization reversal is driven by rapidly time-varying fields (here $\sim$ 1 Oe/ps).

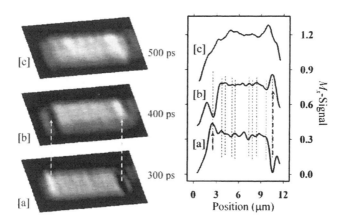

Figure 13-15: (Left column) Time domain images of M_x component taken at different delay times, *i.e.*, 300, 400, and 500 ps after a switching pulse is given. An in-plane transverse bias field ($H_t = 1.1$ kA/m) is applied along the hard axis of the sample. Each panel corresponds to a 12.8 μm × 3.2 μm field of view. The images are on a linear gray scale to render the change in magnetization components with white corresponding to the maximum possible reversal. Note that the magnetic poles seen at 300 ps are flipped at 400 ps. (Right column) Cross sectional views of the time domain images, which show magnetic pole flipping and envelopes of spin waves.

More detailed insight into the magnetic pole flipping is obtained by direct time-domain imaging. The left column of figure 13-15 shows a sequence of time domain images representing the easy axis magnetization components at selected time points, demarcated in nanoseconds relative to the initial application of the switching pulse. As discussed in the previous sections, the magnetization reversal dynamics in small magnetic elements cannot be treated as a process in a single saturated state. Instead, the magnetization reversal occurs in a spatially nonuniform way, and the formation of the stripe domain perpendicular to the applied field is a predominant feature in the very initial spin configuration in response to a sudden change of the external magnetic field [5]. An interesting observation here is that large amplitude precessional oscillations of closure domain magnetization appears to be a 'magnetic pole flipping' with switching occurring at both edges of the element, as though the free magnetic poles are changing sign. However, in contrast to switching scenarios we have investigated in the past, this one involves a significant out-of-plane component in the precession of the closure magnetization. Consequently an admixture of polar Kerr signal appears with the longitudinal signal when the focal spot is near the end of the structure. Here one sees how the participation of the out-of-plane precessing magnetization manifests itself as an apparent "flipping" of the local magnetization at each end of the sample ($t = 300$ ps). The edge magnetizations at 400 ps appear to be fully flipped relative to the image at $t = 300$ ps. The strong gradient in the demagnetizing field underlies this strong spatial variation in the fast reversal process.

The large amplitude precessional oscillations can be more clearly seen when comparing the cross-sections of the time domain images [a] and [b] shown in the right column of figure 13-15.

The contrast of the magnetic signals, however, is reduced very quickly, and the signals from both edges fade away after a single flip. This is due to the rapidly increasing background signal and the nature of scanning process, in which the magnetization signal is averaged across the sample area corresponding to the probe laser beam spot. The large amplitude precessional oscillations are also observed in the transverse in-plane magnetization component (M_y). The left panel of figure 13-16 shows the time-domains of M_y component captured with 100 ps time interval. The images reveal that M_y magnetization is highly localized along and near the long edges, and changes the sign of the magnetization with increasing time. The cross sectional views (right panel) of the time domain images also reveal the precessional oscillations, but the magnetic contrast fades quickly away due to the rapidly decreasing M_y intensity after a single flip of the poles. Large modulation in the sample magnetization is found for the temporal magnetic responses as shown in figure 13-16. In the measurements, the laser beams are focused the near the center at the long edges, as schematically illustrated. The temporal evolution of M_y component is characterized as a strongly damped oscillations with 4 GHz, which terminate about 3 ns after the sample magnetization is excited.

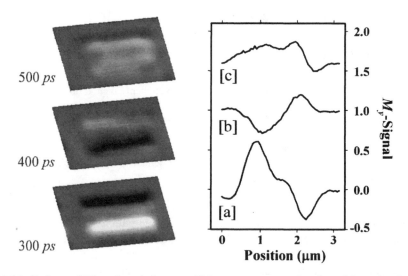

Figure 13-16: (Left panel) Time domain images of M_y components as a function of time after the magnetic pulse is applied. Each panel corresponds to a 12 μm × 3.2 μm field of view. The numbers by the frames indicate the time in picoseconds at which the measurement was made, relative to the initial application of the switching pulse. (Right column) Cross sectional views of the time domain images, captured at 300 ps [a], 400 ps [b], and 500 ps [c].

13.3.4. Switching diagrams for crossed-wire devices

In the next two sections we discuss a time-resolved imaging study of half-select switching between remanent states using the crossed-wires geometry in half-select and precessional regimes respectively.

In order to study pulse duration effects on switching diagrams we send nearly-trapezoidal short pulses down both write wires. The word pulse rise time is approximately 300 ps, digital pulse rise time 250 ps; with fall times 1 ns and 700 ps respectively. The duration of the plateau phase of the word pulse can be varied from approximately 250 ps (in which case the full width half maximum

(FWHM) of the pulse is approximately 900 ps) to 10 ns. In our previous experiments we found that the half-select switching is more reliable when the word pulse turns on during the digital pulse. The pulse timing is designed so that the digital pulse arrives one nanosecond before the word pulse, and both pulses end simultaneously. In order to improve the spatial resolution of our measurements blue light (400 nm of wavelength) was used to probe the sample magnetization.

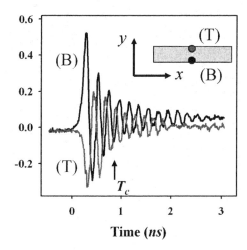

Figure 13-17: Comparison of the temporal evolution of M_y components. Schematic illustrates the laser beams focused near the center at the long edges during the measurements. The temporal evolutions reveal strongly damped oscillations, which are fully decayed about 3 ns after a switching pulse is applied. Note that the M_y components oscillate out-of-phase.

The spatially resolved history of half-select switching by a 5 ns 40 Oe word, 30 Oe digital pulse pair is shown at the figure 13-18. Two ellipses placed at the wire intersection switch simultaneously. The switching starts in the center of each ellipse and then the switched area expand over the sample, while the edges are reluctant to switch. The reluctance of the sample edge to switching has been observed previously [35] and can also be simulated in micromagnetic calculations [36]. This switching scenario is similar to previously observed switching in presence of static magnetic field [5] and it differs significantly from usual easy-axis switching started as a nucleation process at the opposite ends of the sample [37].

Switching diagrams for different pulse durations are shown in the figure 13-19. The final state of the easy-axis magnetization (shown in grayscale) is measured 20 ns after the beginning of the easy-axis pulse at the center of the lower ellipse (Fig.13-18), which is positioned at the very center of the wire intersection. Several features must be noted. First, even the shortest pulse (~900 ps FWHM) is sufficient for complete and reliable half-select switching. Second, switching astroids shrink in the easy-axis direction as the pulse duration increases, as longer pulses are better able to switch the magnetization. Beyond the easy-axis pulse duration of 5 ns, the diagram stays virtually unchanged for longer pulses. Third, for weak easy-axis fields an "incomplete" switching is observed (gray area at the plots). A likely reason is a vortex state formation at the center of the sample. Clearly, this regime should be avoided in technological applications. Stronger easy-axis fields are able to wipe the vortex away and end up in a completely switched state. And finally, in some cases (Fig.13-19(c,d)) "pockets" of switching appear (local spots of switching in the diagram, surrounded by unswitched regions). This might be an indication of the complex structure predicted in simulations [38].

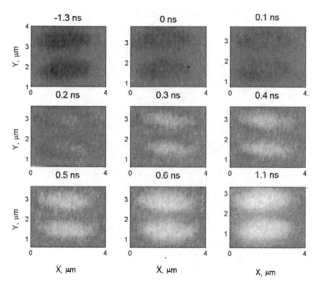

Figure 13-18: Time-resolved history of the switching of the two elliptical samples at the intersection of the crossed wires, by the combined action of word and digital magnetic field pulses in the half-select regime. Nucleation starts in the middle of both samples and propagates towards the edges.

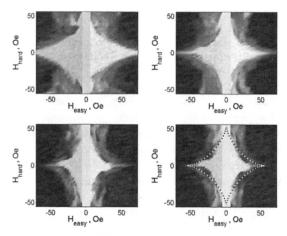

Figure 13-19: Switching diagrams for different easy-axis pulse durations. In all cases the digital pulse precedes the word pulse by one nanosecond and both pulses end at the same time. The gray scale represents the value of the easy-axis magnetization (white – unswitched, black – switched). No data points are taken below 5.8 Oe of the easy-axis field (plain gray stripe in the middle of every diagram). Easy axis pulse (FWHM) durations 900 ps (a); 2 ns (b); 5.5 ns (c), and 7 ns (d).

13.3.5. Precessional switching with the crossed-wire geometry

The ultimate limit of the switching speed is predicted to be achieved in the coherent precessional switching regime [39]. An easy way to induce precessional motion is an application of fast-rising hard-axis field [35, 40–42]. In Fig.13-22 experiment, a 140 Oe 50 ps rise time pulse is applied to the digital wire (Fig.13-22, inset), the magnetization profile is analyzed at the center of the sample. Dual detectors record the out-of-plane and one of the in-plane magnetization change profiles (the easy-axis direction in this case).

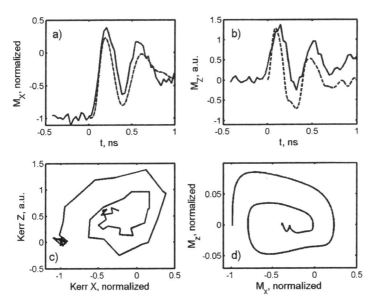

Figure 13-20: Comparison of experimental and simulated temporal traces of in- and out- of-plane magnetization components. (a) Experimental longitudinal Kerr signal normalized to the saturation levels (solid) vs corresponding simulated M_x component (dashed); (b) experimental polar Kerr signal, divided by fifteen to scale sensitivity relative to longitudinal component (solid) vs simulated M_z (dashed); (c) Polar Kerr (b) vs longitudinal Kerr signal (a) to illustrate the precessional nature of the magnetization change; (d) Corresponding M_z vs M_x plot for the simulation.

The measured temporal profiles of both magnetization components during the pulse and comparison to the micromagnetic simulations are shown in the Fig.13-20. The experimental easy-axis magnetization (Fig.13-20(a)) is normalized to the saturation levels, and the magnetic constants used for the simulation are those of the thin film permalloy: M_s=800 emu/cm^3, K_{u2} = 1000 erg/cm^3, A=1.05 μerg/cm. The simulated magnetization profile is in quantitative agreement with the experimental data. The corresponding out-of plane magnetization traces are shown in Fig.13-20(b). The experimental polar Kerr signal was divided by fifteen in order to fit the simulated M_z component. The precessional nature of the switching process is illustrated in Figs.13-20(c) and (d) for the experimental data and simulated magnetization respectively. Contrary to the previously observed precessional switching experiments [35,42,43], the magnetization here does not rotate in the XY plane around a strong induced demagnetization field. Rather, it exhibits an oblate (flattened by demagnetization) XZ rotation around the applied field direction, in agreement with conclusions drawn for non-imaging magnetization measurements of smaller samples [44].

After the end of the pulse the experimental and simulation results diverge. In this instance, the experiment shows stabilization of the magnetization level at approximately 50% of the saturation (zero net magnetization, as high and low saturation states are normalized to plus and minus one), interpreted it a vortex formation at the center of the sample. The possibility of formation of both clockwise and counter-clockwise vortices makes them undetectable by stroboscopic TR-SKM technique. However, simulations predict an almost complete relaxation to the initial state. The vortex formation may depend strongly on edge roughness, possible sample defects and other fine details not reflected properly in the simplified model. Moreover, simulation results differ in details of the vortex formation from one run to another, and the stroboscopic experiment yields only a weighted average of all switch scenarios that actually occur.

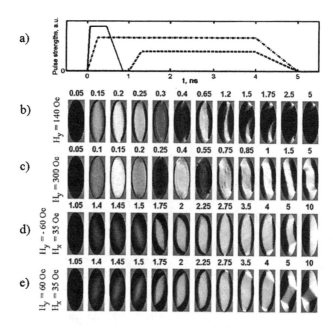

Figure 13-21: Simulated switching histories for 3×1 μm ellipses for different word and digital magnetic pulse profiles. (a) Simple pulse profiles used in the simulations. Linear rise time / plateau duration / linear fall time are as follows: 60/400/400 ps for a fast digital pulse (solid line); 250/3750/1000 ps for a slow digital pulse (dash-dotted line); and 300/2700/1000 ps for word pulse. The word pulse lags one nanosecond with respect to the digital pulse. (b) Coherent rotation caused by fast digital pulse of 140 Oe amplitude and formation of a vortex after the end of a pulse. The grayscale represents the magnetization parallel to the long axis of the ellipse in all plots below. The numbers at the top of the subplots indicate the time passed from the beginning of the digital pulse in nanoseconds. (c) The same as (b) for a stronger 300 Oe digital pulse. A band of switched material and two vortices are formed after the pulse. (d) Incomplete switching by a slow negative digital pulse of -60 Oe combined with a word pulse of 35 Oe in the half-select regime. In this simulation the crystalline anisotropy is tilted fifteen degrees with respect to the long axis of the ellipse. A single vortex at the center of the sample is eventually formed. (e) The same as (d) for a positive 60 Oe slow digital pulse and the same word pulse. Incomplete switching resulting in a double-vortex state.

Several scenarios of the vortex formation are illustrated in Fig.13-21. Simplified magnetic field pulse shapes (Fig.13-21(a)) were used in these simulations to model the magnetic response to fast- and slow- rise time currents produced by pulse generators. The evolution of the easy-axis magnetization component during and after the field pulses is shown in Fig.13-21(b-d). Figure 13-21(b) corresponds to the simulated precessional switching caused by a short, fast-rising digital pulse (Fig.13-21(a), solid line) of 140 Oe of magnitude. This is the same simulation reported in the Fig.13-20. Notably, two oscillations occur during the pulse of magnetization, however, the sample edges demonstrate reluctance to the switching. This reluctance explains the fact that the easy-axis magnetization does not reach the positive saturation level, and also that the center of the precessional trajectory (both in the simulation and in the experiment, Fig.13-20(c,d)) is shifted in negative M_x direction. The central magnetization of the sample is reversed at the end of the pulse (0.65 ns), but the strip of switched material narrows after the pulse (1.2 ns), and breaks (1.5 ns). One of the ends of this broken band eventually disappears and the other forms a vortex near the edge of the sample. The details of this band breaking process vary from one run to the next even when started from the same initial state in these final temperature simulations. In some runs the broken band forms two vortices at the opposite ends. This irreproducible shot-to-shot behavior cannot be observed by the existing stroboscopic TR-SKM technique.

A stronger 300 Oe fast pulse, Fig.13-21(c), causes three precessional rotations of the magnetization of the central part of the sample during the pulse. At the end of the pulse (0.85 ns) a wider band of the switched material is formed. After the pulse this band initially narrows (1.5 ns) and then starts to widen again forming two vortices that propagate slowly towards the ends of the sample. Here the magnetization ends up in a mostly-switched double-vortex state. The key reason for switching is the correct phase of the oscillation at the end of the pulse. Analogous simulations conducted for 250 Oe and 350 Oe pulse amplitude wind up in an unswitched state (similar to Fig.13-21(b)). It is also important to note that although the 300 Oe pulse drives the sample into almost coherent precession during the first half period of rotation (Fig.13-21(c), 0.15 ns), the edges of the sample still remain unperturbed. We observed similar reluctance of the edges to switching in numerous simulations, including simulations on smaller samples down to 600×200 nm ellipses. The edge switching should be taken into account in the analysis of performance limits of MRAM elements.

Figure 13-21(d,e) show the magnetization change under influence of simultaneously acting easy- and hard- axis field (dashed and dash-dotted lines respectively in Fig.13-21(a)). Pulse rise times are 300 and 250 ps respectively, the digital pulse starting one nanosecond prior to the word pulse and both pulses are terminated simultaneously. In this example the anisotropy axis of the permalloy film does not coincide with the long axis of the ellipse, but is tilted 15 degrees with respect to it. Figure 13-21(d) shows the effect of positive 35 Oe word pulse and negative 60 Oe digital pulse (corresponding to the lower part of the astroid). A stripe of the switched material moves to one side of the ellipse effectively reversing the edge magnetization and quickly forming a single vortex that later propagates towards the center of the sample. On the other hand, positive digital pulse combined with the same word pulse (the top part of the astroid, Fig.13-21(e)) causes formation of double-vortex state with a diamond-shaped region magnetized along the short axis of the ellipse. In both cases the net easy-axis magnetization of the sample is nearly zero (corresponding to the gray areas of the astroids in the figure 13-19(c)). Again, in both cases edge reluctance (observed experimentally, Fig.13-18) appears during the pulse. The edge reluctance in the simulation appears to be stronger than observed in the experiment, as these particular field values correspond to complete switching regime (black area of the astroids shown in Fig.13-19(c)).

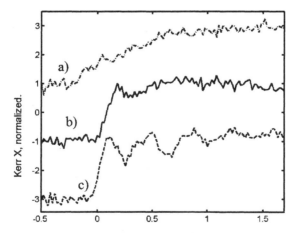

Figure 13-22: Half-select switching. (a) Slow incoherent switching by a strong (70 Oe) easy-axis (Word) magnetic field pulse. (b) Very fast (180 ps 10–90%) half-select switching caused by 90 Oe fast hard-axis (Digital) pulse and 40 Oe Word pulse, properly timed with Digital pulse. (c) 40 Oe Word pulse delayed 170 ps with respect to 90 Oe Digital pulse. Inset: Measured Digital (solid) and Word (dashed) pulse profiles (arbitrary units).

Finally, consider the half-select regime where the fast digital pulse requires the assist of a slow word pulse (Fig.13-22, inset) in order to accomplish switching. The top (dash-dotted) graph of Fig.13-22 shows an incoherent switching of the easy-axis magnetization by a strong easy-axis field to determine the saturation levels. The two bottom plots correspond to the half-select regime. When applied independently, a 90 Oe digital pulse and 40 Oe word pulse leave the magnetization in its initial state (separate measurement, not shown). However, the combined action of the two results in complete very fast switching (180 ps 10–90%). Evidently, the word pulse is able to prevent vortex formation. Moreover, if the pulses are properly timed (Fig.13-22(b)), ringing is suppressed (and virtually non-existent from 0.5 ns beyond the fast pulse leading edge). If the word pulse is deliberately mistimed by 170 ps (roughly half of the oscillation period at Fig.13-22(c)), ringing becomes pronounced, although complete half-select switching still takes place.

13.4. CONCLUSIONS AND PROSPECTS

The combination of temporal and spatial resolution in TR-SKM creates a straighforward approach to the elucidation of ultrafast magnetization dynamics. Examples include complex domain pattern formation and pole flipping.

The main limitation is the spatial resolution, which can be improved for example by incorporation of a solid immersion lens (SIL) [45] into the stroboscopic microscope. A solid immersion lens using a high index material such as $SrTiO_3$ ($n = 2.4$) or GaP ($n = 3.4$), will improve the spatial resolution down to the order of 100 nm. Time-resolved x-ray methods are the avenue of choice for the finer resolution.

An important question related to resolution is whether the entire dynamics measured is perfectly repeatable, since the stroboscopic scanning technique by its nature captures only the repetitive part of the process being imaged. Instabilities and thermal fluctuations will lead to the nonrepetitive response. In addition to varying scan rate, number of averages, etc., the most sensitive test for underlying stochastic behavior so far comes from spectrum analysis of the noise on the magneto-optic signal. An example of noise imaging has been presented [46], in which details of random magnetic switching are exposed in stroboscopically averaged time-resolved experiments. A single-shot measurement of the dynamics would be most desirable.

REFERENCES

[1]   S.A. Wolf et al. Science **294** (2001) 1488.

[2]   P. Grünberg, Physics Today, May 31 (2001).

[3]   B. Heinrich and J.A.C. Bland (Eds.), *"Ultrathin Magnetic Structures"*, Springer Verlag, (1994).

[4]   M.N. Baibich *et al.*, Phys. Rev. Lett. **61** (1988) 2472.

[5]   B.C. Choi, G.E. Ballentine, M. Belov, W.K. Hiebert, and M.R. Freeman, Phys. Rev. Lett. **86** (2001) 728.

[6]   Y. Acremann, C.H. Back, M. Buess, O. Portmann, A. Vaterlaus, D. Pescia, and H. Melchior, Science **290** (2000) 492.

[7]   R.H. Koch, J.G. Deak, D.W. Abraham, P.L. Trouilloud, R.A. Altman, Yu Lu, W.J. Gallagher, R.E. Scheuerlein, K.P. Poche, and S.S.P. Parkin, Phys. Rev. Lett. **81** (1998) 4512 .

[8]   W.K. Hiebert, A. Stankiewicz, and M.R. Freeman, Phys. Rev. Lett. **79** (1997) 1134.
[9]   S.W. Yuan and H.N. Bertram, J. Appl. Phys. **73** (1993) 5992.
[10]   A.F. Popkov, L.L. Savchenko, N.V. Vorotnikova, S. Tehrani and J. Shi, Appl. Phys. Lett. **77** (2000) 277.
[11]   K.J. Kirk, J.N. Chapman, and C.D.W. Wilkinson, Appl. Phys. Lett. **71** (1997) 539.
[12]   J.M. Daughton *et al*, Thin Solid Films **216** (1992) 162.
[13]   W.J. Gallagher *et al*, J. Appl. Phys. **81** (1997) 3741.
[14]   C.H. Back, D. Weller, J. Heidmann, D. Mauri, D. Guarisco, E.L. Garwin, and H.C. Siegmann, Phys. Rev. Lett. **81** (1998) 3251.
[15]   C. Stamm, F. Marty, A. Vaterlaus, V. Weich, S. Egger, U. Maier, U. Ramsperger, H. Fuhrmann, and D. Pescia, Science **282** (1998) 449.
[16]   M. Hehn, K. Ounadjela, J.-P. Bucher, F. Rousseaux, D. Decanini, B. Bartenlian, and C. Chappert, Science **272** (1996) 1782.
[17]   R.P Cowburn, D.K. Koltsov, A.O. Adeyeye, M.E. Welland, and D.M. Tricker, Phys. Rev. Lett. **83** (1999) 1042.
[18]   A.Y. Elezzabi, M.R. Freeman, and M. Johnson, Phys. Rev. Lett. **77** (1996) 3220.
[19]   R.D. Gomez, T.V. Luu, A.O. Pak, K.J. Kirk, and J.N. Chapman, J. Appl. Phys. **85** (1999) 6163.
[20]   J.N. Chapman, P.R. Aitchison, K.J. Kirk, S. McVitie, J.C.S. Kools and M.F. Gillies, J. Appl. Phys. **83** (1998) 5321.
[21]   W.H. Rippard and R.A. Buhrman, Appl. Phys. Lett. **75** (1999) 1001.
[22]   W.H. Rippard, A.C. Perrella, P. Chalsani, F.J. Albert, J.A. Katine, and R.A. Buhrman, Appl. Phys. Lett. **77** (2000) 1357.
[23]   A. Hubert and R. Schäfer, "*Magnetic Domains*", Springer Verlag, (1999).
[24]   T.M. Crawford, T.J. Silva, C.W. Teplin, and C.T. Rogers, Appl. Phys. Lett. **74** (1999) 3386 .
[25]   G. Ju, A.V. Nurmikko, R.F.C. Farrow, R.F. Marks, M.J. Carey, and B.A. Gurney, Phys. Rev. Lett. **82** (1999) 3705.
[26]   G.E. Ballentine, W.K. Hiebert, A. Stankiewicz, and M.R. Freeman, J. Appl. Phys. **87** (2000) 6830.
[27]   S.D. Bader, J. Magn. Magn. Mater. **100** (1991) 440.
[28]   D. D. Awschalom, J.–M. Halbout, S. von Molnar, T. Siegrist, and F. Holtzberg, Phys. Rev. Lett. **55** (1985) 1128.
[29]   W.W. Clegg, N.A.E. Heyes, E.W. Hill, and C.D. Wright, J. Mag. Mag. Mat. **83** (1990) 535, and J. Mag. Mag. Mater. **95** (1991) 49.
[30]   T.J. Silva and A.B. Kos, J. Appl. Phys. **81** (1997) 5015.
[31]   P. Kasiraj, R.M. Shelby, J.S. Best, and D.E. Horne, IEEE Trans. Magn. **22** (1986) 837.
[32]   P. L. Trouilloud, B. Petek, and B. E. Argyle, IEEE Trans. Magn. **30** (1994) 4494.
[33]   D.H. Auston, Appl. Phys. Lett. **26** (1975) 101.
[34]   J. Lohau, S. Friedrichowski, and G. Dumpich, J. Vac. Sci. Technol. B **16** (1998) 1150.
[35]   W. K. Hiebert *et. al.*, J. Appl. Phys. **93** (2003) 6906.
[36]   M. Sheinfein, LLG micromagnetic simulator (TM)
[37]   A.S. Arrott, in "Ultrathin Magnetic Structures IV", B. Heinrich and J.A.C. Bland, editors, Springer Verlag, Berlin (2004).
[38]   M. Bauer, J. Fassbender, B. Hillebrands, R.L. Stamps, Phys. Rev. B **61** (2000) 3410.
[39]   A. Aharoni, "Introduction to the Theory of Ferromagnetism", Clarendon Press, Oxford (1996).
[40]   H. W. Schumacher *e. al.*, IEEE Trans. Mag. **38** Sept. (2002) NO. 5.
[41]   H. W. Schumacher, C. Chappert, R. C. Sousa, P. P. Freitas, and J. Miltat; Phys. Rev. Lett. **90** (2003) 17204.

[42] Th. Gerrits *et. al.*, Nature **418** (2002) 509.

[43] S. Kaka and S. E. Russek, Appl. Phys. Lett. **80** (2002) 2958.

[44] H. W. Schumacher, C. Chappert, P. Grozat, R. C. Sousa. P. P. Freitas, M. Bauer; Appl. Phys. Lett. **80** (2002) 3781.

[45] J.A.H. Stotz, M.R. Freeman, Rev. Sci. Instrum. **68** (1997) 4468.

[46] M. R. Freeman, R. W. Hunt, and G. M. Steeves, Appl. Phys. Lett. **77** (2001) 717.

CHAPTER 14

Brillouin light scattering spectroscopy

14.1. INTRODUCTION

The Brillouin light scattering technique has grown during the last decades into an extremely powerful method for investigating properties of magnetic films and multilayers. This is mainly due to the large amount of magnetic information, which can be obtained using this method.

The primary measured quantity is the frequency of a spin wave supported by the material. Although this might sound like a rather indirect approach to access a material property, spin waves contain a large richness of information. First, in a magnetic system, both the exchange interaction and the dipolar interaction contribute to the spin wave dispersion, and, thus, by measuring the dispersion, information on the coupling constants can be obtained. Second, the primary underlying equation of motion is a torque equation, and therefore Brillouin light scattering, like ferromagnetic resonance or static torque magnetometry, senses all mechanisms, which exert a torque to a magnetic system or subsystem. Prominent examples are volume and bulk anisotropy contributions in magnetic films and interlayer coupling in bilayers and multilayers. Moreover, the experimentalist has free control on a very useful parameter, which is the wavevector. Therefore the measurement of the spin wave dispersion, i.e., the dependence of the frequency of a spin wave mode on the wavevector, yields much additional information. This is of especial advantage in those systems, where the wavelength of spin waves crosses intrinsic length scales, for example the periodicity in a multilayer stack or the lateral extensions of nano- and microstructures.

The fundamental principle of a Brillouin light scattering experiment is rather simple: Monochromatic laser light is focused onto the sample and inelastically scattered from the spin waves. The analysis of the inelastically scattered light yields the spin wave frequencies and, by a fit, the magnetic properties. Sections 14.2 and 14.3 provide a detailed discussion of the experiment and the underlying physics.

Since Brillouin light scattering is an optical method, it is a rather local method. The sampling size is given by the size of the laser spot on the sample. It can be easily implemented into various environments like low and high temperatures, vacuum and ultra-high vacuum. This provides a substantial technical advantage to the experimentalist.

A number of excellent reviews has been published about the field of Brillouin light scattering from spin wave excitations [1-21]. Hillebrands [18] provides an overview on the status of Brillouin light scattering from layered magnetic structures including a survey of existing literature as of winter 1998/99. Carlotti and Gubbiotti provide a more recent overview [21]. Two recent reviews containing studies of confined magnetic structures are found in [20,22]. Overviews about the theory of spin waves in films and multilayers are given by the "classic" review of Wolfram and DeWames [1], as well as by Mills [2], Borovik-Romanov and Kreines [4], Patton [5], Mills [6], Cochran [11] and Cottam et al. [16], the latter two references discussing in particular the light scattering cross section problem. Introductions in the field of magnetism in ultrathin films are found in the two-volume book of Bland and Heinrich [23] as well as in the reviews of Gradmann [24] and of Heinrich and Cochran [10].

This review addresses the measurement technique and the data analysis in particular. After two introductory sections describing the experimental technique (section 14.2) and the underlying basic theory of spin wave properties needed for data analysis (section 14.3), the applicability of Brillouin light scattering is discussed in the following section 14.4 using case studies. These are the determination of magnetic anisotropies (section 14.4.1), of the exchange constant (section 14.4.2), and of the interlayer coupling constants (section 14.4.3). Section 14.4.4 is devoted to the analysis of magnetic inhomogeneities evidenced by line width broadening effects. In section 14.5 we discuss spin waves in patterned magnetic structures, and section 14.6 is devoted to the investigation of wave propagation phenomena, presenting a recent advance in the development of the Brillouin light scattering technique towards spatial and temporal resolution. Section 14.7 contains concluding remarks.

In this review, for the sake of conciseness, we address several aspects of the Brillouin light scattering measurement technique and the underlying theoretical background. We do not, however, address in full depth the physical problems and certain technical aspects (like sample preparation) behind the experiments presented here in the case studies. The reader is referred to the original literature, which is cited in each study.

14.2. THE BRILLOUIN LIGHT SCATTERING TECHNIQUE

14.2.1. Experimental setup

We start by describing the Brillouin light scattering setup as shown in Fig.14-1. The light of a frequency stabilized laser in single mode operation (Δv=20 MHz), which is typically an Argon$^+$-ion laser (λ= 514.5 nm) or a solid state laser, is focused onto the sample with an objective lens. For collecting the scattered light the backscattering geometry is frequently used, in particular for metallic samples. The light backscattered from the sample (elastic and inelastic contributions) is collected by the same objective lens used for focusing the laser beam onto the sample, and it is sent through a spatial filter for suppressing background noise before entering the Fabry-Pérot interferometer used as the monochromator. The frequency selected light transmitted by the interferometer is detected by a photomultiplier or an avalanche photodiode after passing through a second spatial filter for additional background suppression. A prism or an interference filter in front of the second spatial filter and the detector serves for suppression of inelastic light from common transmission orders outside the frequency region of interest (not shown in Fig.14-1). A computer collects the photon counts and displays the data.

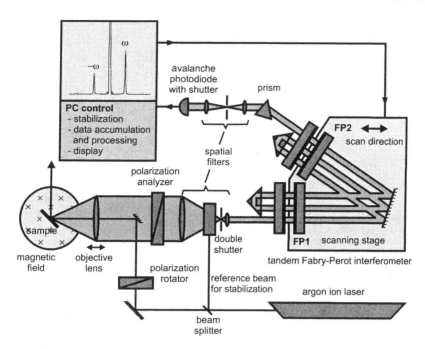

Figure 14-1: Schematic view of a Brillouin light scattering setup.

As the monochromator a tandem Fabry-Pérot interferometer developed by J.R. Sandercock is used. It is a highly sensitive spectrometer with a frequency resolution in the sub-GHz regime and a contrast of better than 10^{10} [3,25]. Therefore it is best suited for studying spin wave excitations. The sensitivity is large enough to detect signals even from magnetic monolayers [26].

The frequency selecting element is an etalon consisting of two parallel optical mirror plates (flatness better than $\lambda/200$) of rather high reflectivity (typically 92-96%). The etalon transmits light of wavelength λ, if the plate distance is an integer multiple of $\lambda/2$. In conventional interferometry using one single etalon, the analysis of inelastic excitations is hampered by the ambiguous assignment to the appropriate transmission order, since the transmission is periodic in $\lambda/2$ in the mirror plate spacing. These ambiguities are avoided in the tandem arrangement.

The central part of the tandem interferometer is displayed in Fig.14-2 (top). The light passes in series through two Fabry-Pérot etalons FP1 and FP2. For both etalons one of the two mirrors is mounted on a common translation stage which is piezo-electrically driven. In order to illustrate the function of the tandem arrangement Fig.14-2 (bottom) displays schematically the transmission curves of the etalon (a) FP2, (b) FP1 and (c) of both etalons in series (tandem operation) as a function of the mirror separation of the first etalon, L_1. Assuming that for a given value L_1 both etalons are in transmission, a change of $\lambda/2$ in L_1 puts FP1 into the next transmission order. Due to the common mounting of the movable mirrors the change in the spacing of the second etalon is smaller by a factor of $\cos\Theta$ with Θ the angle between the optical axes of the two interferometers as displayed in the figure. Thus FP2 is now not in transmission, and the transmission maxima of both etalons lie at different values of L_1. The same arguments applies for inelastic excitations, which are transmitted only if they belong to the common transmission order. The inelastic signal represents closely the scattering cross section of the sample.

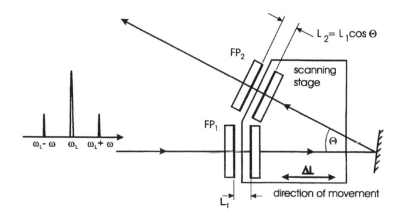

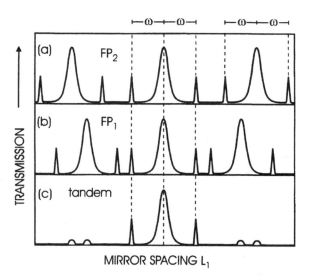

Figure 14-2: Schematic view of the operation of a tandem Fabry-Pérot interferometer. Top: View of the light pass. Bottom: Transmitted intensity of first (FP1), second (FP2) and both etalons in series. The inelastic contributions due to an inelastic light scattering process are indicated by ω (from [18]).

For an experimental realization a planity of the mirror surfaces of better than $\lambda/200$ and a parallelity of $\lambda/100$ of the two mirrors of each etalon are necessary. To maintain the latter, a sophisticated active stabilization of the mirror alignment is needed, as discussed below. In order to obtain the high contrast necessary to detect the weak inelastic signals the light is sent several times through both etalons using a system of retroreflectors and mirrors (see Fig.14-1).

Modern interferometers [3,25] are mostly set up in the (3+3)-pass arrangement. Special measures are taken to protect the detector from overload while scanning through the elastic peak. This is achieved by using an acousto-optic modulator, or, as shown in Fig.14-1, by a shutter system. In the latter case, in each scan, a shutter switches the light, which will enter the interferometer, between the inelastic light from the sample and the light from a reference beam in a window around the central peak (common transmission of both etalons for the laser frequency ω_L). Data collection is performed by a personal computer or by a multichannel analyzer.

Multipass operation allows for a high contrast of better than 10^{10} enabling detection of elastic and magnetic surface excitations even in metallic systems with a sensitivity down to the monolayer regime, as shown for the case of spin waves in Co films [26]. However, the requirements on the alignment of the optical setup are extremely high. Alignment can only be maintained over the time of a measurement by an active stabilization scheme. John R. Sandercock reported first the successful installation of an actively stabilized tandem Fabry-Pérot interferometer [27]. To achieve long-term stability of the etalon mirror alignment, several measures have been proposed and realized: First, a very good isolation from ground vibrations is needed. This is obtained by an optic table and/or by an actively stabilized table as the mount of the etalon mirrors [28]. Second, thermal and acoustic isolation of the etalon mirrors can be obtained by a box surrounding the interferometer setup and coated with sound isolation and thermal isolation material. Third, and most crucial, the active stabilization mechanism for maintaining parallelity and correlation between the etalon mirror pairs is of key importance. For the latter, several solutions using analog circuits [29-31], digital hardware [32,33], or a personal computer [34,35] have been realized.

For achieving a system of high stabilization and flexibility we have developed in our laboratory the computer control TFPDAS3 for the Sandercock-type multipath tandem Fabry-Pérot interferometer [35]. The control box of the Sandercock system is prepared for this upgrade. The TFPDAS3 system offers high versatility, stability and user friendliness, and it extends the applicability of tandem Fabry-Pérot spectroscopy to new areas, which in the past were not easily accessible. A scientist can be trained in its operation within a day. The range of stability is increased due to an adaptive control of the mirror dither amplitudes used in the active stabilization procedures for maintaining the mirror parallelities and synchronization. In preset regions the scan speed is reduced by a preset factor to increase the signal to noise ratio. The alignment is fully automated. The software control enables a programmable series of measurements with control of, e.g., the position and rotation of the sample, the angle of light incidence, the sample temperature or the strength and direction of an applied magnetic field. Build-in fitting routines allow for a precise determination of frequency positions of excitation peaks combined with increased frequency accuracy due to a correction of any residual nonlinearity of the mirror stage drive.

14.2.2. The Brillouin light scattering cross section

The basic light scattering process is illustrated in Fig.14-3. Photons of energy $\hbar\omega_L$ and momentum $\hbar\vec{q}_L$ interact with the elementary quanta of spin waves $(\hbar\omega_L, \hbar\vec{q}_L)$, which are the magnons provided, that time invariance and translational invariance are given, the scattered photon gains an increase in energy, $\hbar(\omega_L + \omega)$, and momentum $\hbar(\vec{q}_L - \vec{q})$, if a magnon is annihilated. A magnon can also be created by an energy and momentum transfer from the photon, which in the scattered state has the energy $\hbar(\omega_L - \omega)$ and momentum $\hbar(\vec{q}_L - \vec{q})$. For finite temperatures $(T \gg \hbar\omega/k_B \approx 5\text{K})$ both processes have about the same probability. In a classic treatment the scattering process can be understood for many materials as follows: Due to the spin-orbit coupling a phase grating is created in the material, which propagates with the velocity of the spin wave. Light is Bragg-reflected from the phase grating with its frequency Doppler-shifted by the spin wave frequency. For a detailed discussion see [18] and references therein.

14.3. SPIN WAVE BASICS

Spin waves as investigated by Brillouin light scattering spectroscopy usually have wavelengths larger than 100 nm, which is much larger than interatomic distances. Therefore we may use a continuum model for the description, see [18] for details.

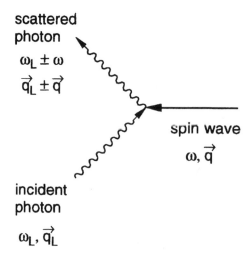

Figure 14-3: Scattering process of photons from spin wave excitations (magnons).

The fundamental equation of motion is the Landau-Lifshitz torque equation of motion:

$$\frac{1}{|\gamma|}\frac{\partial \vec{M}}{\partial t} = -\vec{M} \times \vec{H}_{eff} \quad , \tag{14.1}$$

where $\gamma = \gamma_e g/2$ is the gyromagnetic ratio, $\gamma_e = 1.759 \times 10^7$ Hz/Oe is the value of γ for the free electron and g is the spectroscopic splitting factor. The effective magnetic field acting on the magnetization, $\vec{H}_{eff}$, is given by

$$\vec{H}_{eff} = \vec{H} - \frac{1}{M_s}\vec{\nabla}_{\vec{\alpha}} g_{vol} + \frac{2A}{M_s^2}\nabla^2\vec{M} \quad . \tag{14.2}$$

The first term on the right hand side is a field which includes the external applied field $\vec{H}_0$ and the fluctuating fields $\vec{h}$ generated by the precessing spins. The second term is an effective field due to magnetic anisotropies with g_{vol} the enthalpy density describing the volume anisotropy and $\vec{\nabla}_{\vec{\alpha}}$ the gradient operator for which the differentiation variables are the components of the unit vector $\vec{\alpha}$ pointing into the direction of the saturation magnetization $\vec{M}_s$. We consider the case of small excitations, $|\vec{M}| = |\vec{M}_s|$, i.e., the reduction in $|\vec{M}_s|$ due to the precessing moments is negligibly small. The various anisotropy contributions will be discussed in section 14.4. The last term is the exchange field due to volume exchange interaction with A the exchange stiffness constant.

Inspecting eq.(14.2) it becomes immediately clear, that volume anisotropies and the exchange constant determine the spin wave frequencies and thus can be obtained by a fit of the spin wave dispersion to experimental data with the anisotropy constants and the exchange constant as parameters.

Together with the equation of motion the magnetostatic Maxwell equations need to be fulfilled [36]:

$$\vec{\nabla} \times \vec{H} = 0 \quad , \tag{14.3}$$

$$\vec{\nabla} \cdot (\vec{H} + 4\pi\vec{M}_s) = 0 \quad . \tag{14.4}$$

From the equation of motion, eq.(14.1), six independent partial wave solutions are obtained when exchange interaction is taken into account.

14.3.1. Spin waves in magnetic films

In the presence of a surface or an interface, boundary conditions must be fulfilled by a suitable superposition of the partial waves found in the bulk. These are the Maxwell boundary conditions and a boundary condition, which is partly derived from the equation of motion, eq.(14.1). For the surface of a film, this is the so-called Rado-Weertman boundary condition [37]:

$$\vec{M} \times \left[\vec{\nabla}_{\vec{\alpha}} \sigma_{inter} - \frac{2A}{M_s} \frac{\partial \vec{M}}{\partial n} \right]\Bigg|_{interface} = 0 \quad . \tag{14.5}$$

Equation (14.5) describes the balance of torques acting on the interface. σ_{inter} is the enthalpy density describing the magnetic interface anisotropy. It is evident, that interface anisotropies provide an additional torque to the magnetization at the interface.

If interface anisotropy is nonzero, the corresponding term depends on the symmetry of the interface. For example, for a surface with two-fold symmetry about the surface normal, like a [110]-oriented epitaxial film, σ_{inter} is given by

$$\sigma_{inter} = -k_s \alpha_x^2 + k_p^{(2)} \alpha_y^2 + k_p^{(4)} \alpha_y^2 \alpha_z^2 \quad , \tag{14.6}$$

where α_i are the direction cosines of the direction of magnetization. Here it is assumed that the film normal is along the x-axis and the y- and z-axes are in the film plane along crystallographic high-symmetry axes (usually the [001] and the [$1\bar{1}0$] directions). k_s is the so-called perpendicular interface anisotropy constant, and $k_p^{(2)}$ and $k_p^{(4)}$ are the twofold and fourfold in-plane interface anisotropy constants. By convention we use upper case letters for volume anisotropy constants and lower case letters for interface anisotropy constants.

In general, Brillouin light scattering is capable to find a difference between interface anisotropies of the two surfaces of a film, if sensing surface spin-wave modes (Damon-Eshbach modes, see below). The difference in anisotropy is evidenced by a difference in the absolute values of the spin-wave frequencies of the two Damon-Eshbach modes, each localized at one of the two interfaces, which appear in the spectrum on the energy-loss (Stokes) and the energy-gain (anti-Stokes) side. However this difference is small, in most cases too small for detection. See [38] for a discussion.

Conceptually, volume anisotropy contributions enter the equation of motion, eq.(14.1), and interface anisotropy contributions the boundary condition, eq.(14.5). However, as in static measurements, volume and interface anisotropies can often be joint to an effective volume anisotropy by replacing the interface enthalpy density by a volume enthalpy density,

$$g_{inter} = \frac{2}{d} \sigma_{inter} \quad , \tag{14.7}$$

with d the thickness of the film and the factor of two counting the two interfaces of the film.

Eq.(14.7) is valid for films thinner than the exchange correlation length $(A/2\pi M_s^2)^{1/2}$. Rado [39] and Gradmann et al. [40] give estimates for the range of validity for eq.(14.7). The acting effective anisotropy, g_{eff} is the sum of the volume anisotropy, g_{vol}, and the effective interface anisotropy, g_{inter}.

For data fitting purposes it is most often fully sufficient, to work with effective anisotropies and not to consider anisotropies in the boundary conditions.

The spin wave dispersion depends on the relative orientation of the direction of the wavevector, $\vec{q}$, and the direction of magnetization, $\vec{M}$, with respect to the film orientation. Figure 14-4 illustrates the various modes in the magnetostatic limit, where exchange interaction can be neglected.

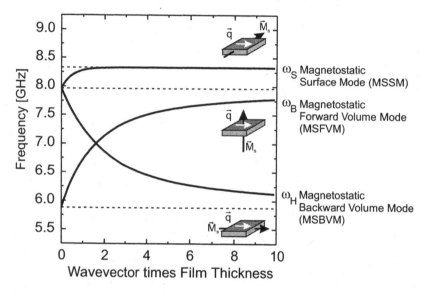

Figure 14-4: Typology of spin wave modes as a function of the directions of the magnetization, $\vec{M}_S$, and the wavevector $\vec{q}$. ω_S: frequency of the surface spin wave (Damon-Eshbach mode). ω_B, ω_H: frequencies of the volume spin waves with the wavevector perpendicular and parallel to the external field, respectively (adapted from [19]).

14.3.1.1. Procedure to determine anisotropy constants

The mode commonly used for the investigation of magnetic anisotropies is the Damon-Eshbach mode (magnetostatic surface mode). Here the direction of the wavevector and the direction of the magnetization are roughly perpendicular to each other and both lie in the film plane. For the case of negligible anisotropy the frequency is given by

$$\left(\frac{\omega}{\gamma}\right)^2 = H_0(\ H_0 + 4\pi M_s \sin\ \varphi_0)\ + (\ 2\pi M_s)^2\left(1 - e^{-2q_\parallel d}\right) \qquad (14.8)$$

with φ_0 the angle between $\vec{M}_S$, and $\vec{q}_\parallel$. In the experiment the external field is applied perpendicular to the light scattering plane (see Fig.14-1). In the presence of in-plane anisotropies the direction of magnetization is defined by the energy minimum of the total magnetic enthalpy density,

$$g_{tot} = g_{eff} - \vec{M}_s \vec{H}_0 \ , \tag{14.9}$$

which is the sum of the enthalpy density of the effective magnetic anisotropy and the Zeeman energy. The direction of $\vec{M}_s$ may deviate from the direction of the external field $\vec{H}_0$. Under these circumstances a fit to the experimental data consists of two steps: First, the static equilibrium direction of $\vec{M}_s$ is determined from eq.(14.9) by minimization. Second, the spin wave frequencies are calculated for this direction.

In general, for arbitrary anisotropy, a numerical procedure must be used (see [18] for details). For in-plane magnetized films and in the limit of small film thicknesses, i.e., the product of film thickness, d, and wavevector, $|\vec{q}_{\|}|$, is small compared to 1, an analytical formula can be used [41]:

$$
\begin{aligned}
\left(\frac{\omega}{\gamma}\right)^2 = &\left(\frac{1}{M_s}\frac{\partial^2 g_{eff}}{\partial\theta^2} + H_0\cos(\varphi - \varphi_H) + \frac{2A}{M_s}q^2 \right. \\
&\left. + 4\pi M_s f \cdot \left(1 - \frac{1}{2}q_{\|}d\right)\right) \times \left(\frac{1}{M_s}\frac{\partial^2 g_{eff}}{\partial\varphi^2} + H_0\cos(\varphi - \varphi_H) \right. \\
&\left. + \frac{2A}{M_s}q^2 + 2\pi M_s f q_{\|} d\sin^2(\varphi - \varphi_q)\right) - \frac{1}{M_s^2}\left(\frac{\partial^2 g_{eff}}{\partial\theta\partial\varphi}\right)^2
\end{aligned}
\tag{14.10}
$$

with θ and φ the polar and azimuthal angle of the magnetization using a conventional spherical coordinate system, the saturation magnetization M_s, the applied magnetic field H_0, the exchange constant A, the demagnetizing factor f (which may deviate from 1 for very thin films, see [42,43]), and the film thickness d. The angles φ, φ_H, φ_q indicate the direction of magnetization, of the external field, and of the wavevector with respect to an in-plane reference direction, which is usually chosen along a high-symmetry crystallographic axis. In eq.(14.10) the directional derivatives of the effective anisotropy enthalpy density, g_{eff}, are written as derivatives with respect to the polar and azimuthal angle.

Eq.(14.10) is often used to fit the spin wave frequencies to deduce anisotropy constants. The corresponding procedure is:

- Choose a suitable enthalpy density expression to describe the acting anisotropies. If this is not formulated in a spherical coordinate system, transform the expression to this system.

- Calculate analytic expressions for the derivatives

$$\frac{\partial^2 g_{eff}}{\partial\theta^2}, \quad \frac{\partial^2 g_{eff}}{\partial\varphi^2}, \quad \text{and} \quad \frac{\partial^2 g_{eff}}{\partial\theta\partial\varphi}$$

and insert them into eq.(14.10).

- Measure the spin wave frequencies as a function of the in-plane angle φ_H by rotating the sample in the magnetic field and taking spectra at various rotation angles. Typical step sizes are between $10°$ and $30°$. Here the angle between the direction of the external field and the transferred wavevector is fixed to 90° by the scattering geometry. Modern Brillouin light scattering systems allow for an automated operation of this step (see e.g. [35]). Analyze the spectra and determine the frequency positions of the spin wave peaks as a function of φ_H.

- Fit the experimental data using the anisotropy constants as fit parameters. In each iteration of the fit procedure first determine the static direction of the magnetization $\vec{M}_S$, i.e., calculate φ by minimizing the total free energy g_{tot} (see eq.(14.9)), then calculate the spin wave frequencies.

Usually, as a function of the rotation angle, φ_H, the spin wave frequencies show a periodic variation. Maxima in the frequencies indicate easy directions of the magnetization. This is easy to understand by inspecting eq.(14.10). Relevant is the projection of the external field onto the direction of the magnetization ($\sim \cos(\varphi - \varphi_H)$ in eq.(14.10)), which is the acting field strength. If the direction of magnetization is not along an easy direction, it is canted away from the direction of the external field, and thus the acting effective field is smaller resulting in a lower spin wave frequency.

When measuring anisotropy constants, the experimenter needs to distribute the total time available on the Brillouin light scattering system to the spectra to be taken. It is clear, that a larger number of spectra, and thus a larger number of data points to be evaluated in the fit procedure, results in less time per spectrum and thus in a lower signal-to-noise ratio of the data in each spectrum. In contrary, the information about the anisotropy constants is contained in the variation of the spin wave frequencies with the angle of sample rotation, and thus a larger number of measured frequencies allows for a better fit. In total, it is best to increase the number of data points up to a value, where the determination of the frequency positions in each spectrum still is possible with sufficient stability. As a rule of thumb, a good compromise is achieved, when the peaks in the spectra indicating a spin wave mode show a signal of at least twice the noise level.

Note that in eq.(14.10) it is assumed, that the magnetization lies in the film plane. For systems with large out of plane anisotropies, where the magnetization tilts out of the plane see, e.g., [18,44].

14.3.1.2. Standing spin waves

In addition to the dipolar dominated modes, exchange dominated modes, so-called standing spin waves exist. They travel forth and back between the front and rear surface of a film. To form a standing wave the film thickness must match an integer multiple of half the spin-wave wavelength, and thus the wavevector is quantized, $q_n = n(\pi/d)$ with d the film thickness and n a positive integer. Outside the crossing regime with the dipolar modes discussed above, the frequency is approximately given by

$$\left(\frac{\omega}{\sigma}\right)^2 = \left(H + D q_n^2\right)\left(H + 4\pi M_S + D q_n^2\right) \tag{14.11}$$

with $D = 2A/M_S$ the so-called exchange stiffness constant. For a calculation of the full problem, including the cross over with the dipolar dominated modes, a numerical model as in [18] must be used.

14.3.2. Spin waves in magnetic double layers and multilayers

In magnetic double and multilayers coupling between magnetic layers plays an important role. Thus information about the coupling constants can be inferred from the spin wave spectra.

First there is the dipolar coupling caused by the dipolar stray fields of the dynamic component of the magnetization. In [18] a full discussion is given. In essence, in a multilayer of N identical films, the dipolar coupling lifts the degeneracy of the N dipolar modes in the N layers, and a band of collective spin waves is formed. From the shape of the band information about the thicknesses of the magnetic and the nonmagnetic layers can be obtained.

14.3.2.1. Enhancement and reduction of layer magnetizations, polarization effects

In multilayers sometimes an enhancement or a reduction of the total magnetic moment is observed. This may have two origins: first the magnetization of the magnetic layer can be enhanced or reduced. Second the spacer layer might gain an induced magnetic moment due to a high magnetic polarizability, as is the case for, e.g., multilayers with Pd and Pt as the spacer materials (see discussion in [45,46]) or dead layers may be formed near the interface due to interdiffusion at the interface. Brillouin light scattering allows for a clear separation of both mechanisms, since an enhancement or reduction of the magnetization is evidenced by increased and decreased spin wave frequencies, whereas a polarization of the spacer layer or a dead layer results in a modification of the shape of the band. Figure 14-5 shows experimental and fit results for a Co/Pd multilayer.

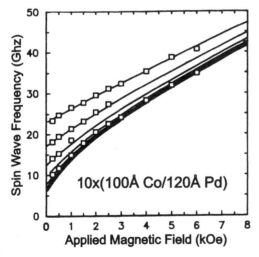

Figure 14-5: Measured (squares) and calculated (solid line) spin wave frequencies vs. applied magnetic field of a multilayer with 10 bilayers of 100 Å Co layer thickness and 120 Å Pd layer thickness each. The fit parameters are $4\pi M_s = 14.52$ kG, $g = 2.03$, $K_V = 3.10 \times 10^6$ erg/cm^3, and $K_S = -0.45$ erg/cm^2 (from [46]).

14.3.2.2. Interlayer exchange coupling

Of particular interest are systems with interlayer exchange coupling. See [47] for a recent review. We consider the case of a double layer. In a phenomenological classification the interlayer exchange coupling can be of bilinear, biquadratic or of non-trigonometric type [48,49]. With 2α the angle between the magnetizations of the two layers (each magnetization is tilted away from the equilibrium direction by α due to interlayer exchange coupling) the free coupling energy surface densities in the static case are

$$\sigma_{bilinear} = -A_{12} \cos 2\alpha \tag{14.12}$$

$$\sigma_{biquadratic} = B_{12} \cos^2 2\alpha \tag{14.13}$$

$$\sigma_{non-trig} = C_+ (2\alpha)^2 + C_- (2\alpha - \pi)^2 \tag{14.14}$$

In the more recent literature a second set of parameters, J_1 and J_2, is more commonly used for the bilinear and biquadratic coupling parameters [47]. Their relation to A_{12} and B_{12} is $J_1 = A_{12}, J_2 = -B_{12}$.

In general the resulting boundary conditions are rather complex. In the case of the bilinear exchange coupling, the boundary condition entering the spin wave calculation is the so-called Hoffmann boundary condition, which includes exchange coupling to the interface of the next magnetic layer at $x = d_n$ [50,51]:

$$
\begin{aligned}
\vec{M}_n \times &\left(\frac{1}{M_{s,n}} \vec{\nabla}_{\vec{\alpha}_n} \sigma_{inter,n} - \frac{2A_n}{M_{s,n}^2} \frac{\partial \vec{M}_n}{\partial n_n} \right)\bigg|_{x=d_n} \\
&- \vec{M}_n \times \frac{2A_{nn'}}{M_{s,n}M_{s,n'}} \left(\vec{M}_{n'} + a_{n'} \frac{\partial \vec{M}_{n'}}{\partial n_{n'}} \right)\bigg|_{x=d_{n'}} = 0
\end{aligned}
\tag{14.15}
$$

where σ_{inter} is the interface anisotropy energy. ∂/∂_n is the partial derivative with respect to the surface normal unit vector, $\vec{n}$. The latter is directed from the interface into the corresponding magnetic layer. The interlayer exchange constant between layers n and n' is $A_{nn'}$. Without loss of generality we call this parameter A_{12} in the following. The lattice constant is denoted by a_n. The first term in Eq.(14.15) is the so-called Rado-Weertman boundary condition, describing the surface torque of a single magnetic film due to interface anisotropies and volume exchange interaction [37], whereas the second term describes the exchange coupling between the two layers.

For biquadratic and non-trigonometric coupling the resulting boundary conditions have large complexity, they can only be reduced to simple expressions in case of parallel or antiparallel alignment of the layer magnetizations. Results are reported for the biquadratic coupling case by Macciò et al. [52] and for the non-trigonometric case by Tschopp et al. [53]. Although the type of coupling determines the remagnetization curves to a large extent, the dynamic properties depend to a much higher degree on the underlying coupling mechanism. In case of non-trigonometric coupling the layer magnetizations show only an asymptotic approach to saturation with increasing field.

14.4. DETERMINATION OF MAGNETIC PARAMETERS

In this section we illustrate the determination of magnetic parameters using Brillouin light scattering in case studies made mostly in our laboratory in Kaiserslautern. We discuss the measurement of magnetic anisotropies, of the exchange constant and of the interlayer coupling constant. Finally we comment on the investigation of magnetic inhomogeneities.

14.4.1 Magnetic anisotropies

The basic properties have been discussed in section 14.3.1. We illustrate the determination of surface and interface anisotropies on two cases, i) epitaxial Co(110) films grown onto Cu(110) substrates i and ii) the modification of surface anisotropies of Fe(001) films by a Pd overlayer of varying thickness.

14.4.1.1 Epitaxial Co(110) films on a Cu(110) substrate

See Refs.[46,54,65] for details of the sample preparation. This system was of particular interest, since the easy axes of the in-plane surface anisotropy and the bulk anisotropy are not collinear, and it was expected to observe a reorientation transition of the easy axis of magnetization as a function of Co film thickness.

For the study several films of various thicknesses are needed. Since in a experiment only a film area of the order of the laser focus diameter is needed (Ø≈40 μm), it is advantageous, to use films with a wedge like or a step like geometry. Here a range of film thicknesses is accessible on the same sample, and Brillouin light scattering measurements are made as a function of film thickness by moving the laser focus across the sample.

For the Co/Cu(110) samples a schematic outline of the sample is shown in Fig.14-6. In this particular study the influence of interface strain was of interest, and thus samples with and without an intervening $Cu_{62}Ni_{38}$ strain relaxation layer were studied.

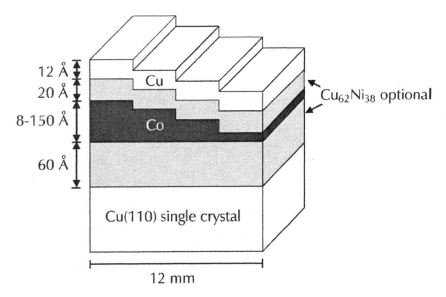

Figure 14-6: Schematic sketch of the samples. To test the strain dependence one series of samples is prepared with $Cu_{62}Ni_{38}$ buffer layers reducing the lattice mismatch from -2% to -1% (from [57]).

For each thickness the Brillouin light scattering spectra have been taken as a function of the in-plane direction of the magnetization by rotating the sample in the external field. The transferred wavevector was kept perpendicular to the field. Figure 14-7 shows the results measured in an external field of 3 kOe in the Damon-Eshbach geometry. The Damon-Eshbach spin-wave frequencies are in the range of 30 to 40 GHz.

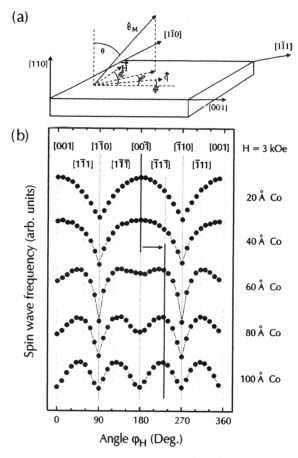

Figure 14-7: (a) Experimental geometry. (b) Measured spin wave frequencies of 20-100 Å thick Co films on a Cu(110) single crystal covered by a 12 Å thick Cu overlayer as a function of the in-plane angle φ_H of the applied field with respect to the [001] direction at an applied field of 3 kOe. The in-plane crystallographic directions are indicated by dashed lines. For clarity the spin wave frequencies of each series are shifted (maximum variation: 30.0-38.6 GHz for d_{Co}=20 Å; 20.0-25.1 GHz for d_{Co}=100 Å) (from [57]).

As a function of the angle between the external field and the in-plane reference direction φ_H, we find for $\varphi_H = 0$ ([001] direction) and $\varphi_H = 90°$ ([1$\bar{1}$0] direction) the highest and lowest spin-wave frequency values, respectively, indicating the easy and hard magnetization direction. By fitting the position of the spin-wave frequencies and plotting them as a function of the in-plane angle θ_H for different Co film thicknesses we obtain data as displayed for the cubic volume anisotropy constant in Fig.14-8. For d_{Co}=20 Å the spin-wave frequencies display a twofold behavior as a function of φ_H. The maxima of the spin wave frequencies, indicating the easy magnetization directions, are found at φ_H =0°, 180° and 360° (along in-plane <001> directions) clearly exhibiting a twofold in-plane symmetry with the easy axis of magnetization along the <001> axes, indicative for a two-fold surface anisotropy as the leading anisotropy contribution. With increasing film thickness the pattern changes drastically. In the thickness regime between 40 and 60 Å the maxima of the spin wave frequencies and therefore the easy magnetization direction switch from the <001> axes toward the <111> axes. For d_{Co}=100 Å the maxima of the spin-wave frequencies found at the in-plane axes are a clear signature for the presence of a dominating magnetocrystalline anisotropy (pseudo-fourfold symmetry). Figure 14-8 shows the obtained results for the four-fold magnetocrystalline anisotropy. As is evident from this figure, the interface strain provides a mechanism to suppress this anisotropy contribution. A test using a strain-releasing intermediate $Cu_{62}Ni_{38}$ buffer layer provides additional evidence. See Ref. [57] for a discussion of the origin.

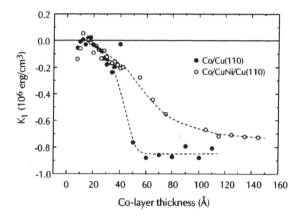

Figure 14-8: Magnetocrystalline anisotropy constant K_1, as obtained from fits to data as displayed in Fig. 14-7, as a function of the Co film thickness for Co films with (full symbols) and without a strain-relaxing $Cu_{62}Ni_{38}$ buffer layer (open symbols) (from [57]).

14.4.1.2. Epitaxial Fe(001) films with Pd cover layer

The central issue here is to determine the modification in the surface anisotropy of epitaxial Fe(001) films by a Pd cover layer. For the sample preparation see [58]. 15 ML thick Fe(001) films have been prepared onto GaAs(001) using a 150 nm thick fcc Ag(001)-buffer layer. The detailed description of the preparation procedure is published in [59]. The films were covered with a Pd film with a wedge-shaped thickness in the range of 0-10 ML and finally capped with a Au cover layer for protection.

The spin wave frequencies for the sample 15 ML Fe/1.5 ML Pd are shown in Fig.14-9(a) measured as a function of the angle φ_H of the external field of $H_0=1$ kOe with the in-plane Fe-[100] axis. It clearly demonstrates, that the four-fold symmetry of Fe films is not broken by the Pd overlayer. Therefore, we write the entropy density of the anisotropy contribution in spherical coordinates as

$$g_{ani} = -K_{out}\cos^2\theta + K_{in}\sin^4\theta\cos^2\varphi\sin^2\varphi \quad , \qquad (14.16)$$

where θ is the polar angle of the direction of magnetization measured against the interface normal and φ is the azimuthal angle measured with respect to the [100] axis of Fe in the film plane. K_{out} and K_{in} are the out of plane and the in-plane effective anisotropy constants containing both bulk and surface contributions, respectively. The spin wave frequencies for H_0 parallel to [100] and for H_0 parallel to [110] as a function of d_{Pd} are shown in Fig.14-9(b). In both cases, the Pd deposition dramatically changes the spin wave frequencies even after the Fe surface is completely covered by the Pd layers for $d_{Pd} > 2$ ML. Using the fit procedure both K_{out} and K_{in} are determined. The results of the fits are shown in Fig.14-10. The derived anisotropy constants show a strong linear change with increasing d_{Pd} up to 5.5 ML with following saturation, with K_{out} showing an additional abrupt change in slope within the first monolayer of Pd coverage. Assuming that the Pd overlayer does not influence the bulk properties of the Fe films, one can calculate the surface anisotropy contributions, their maximum changes being Δk_{out}=0.8 erg/cm² and Δk_{in}=0.04 erg/cm².

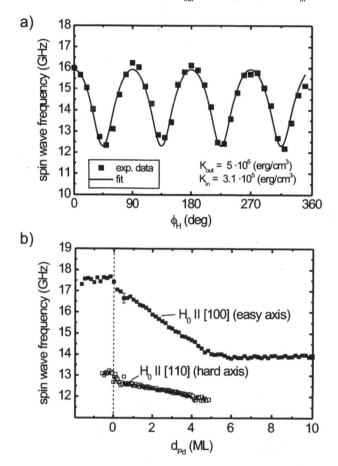

Figure 14-9: (a) Spin wave frequency for the 15 ML Fe/1.5 ML Pd sample, measured at H_0=1 kOe as a function of the in-plane angle φ_H between the field and the Fe[100] direction. The fit is performed on the basis of Eq.(14.16). (b) Spin wave frequency for a Pd wedge on a 15 ML thick Fe film, measured at H_0=1 kOe (full squares: H_0 along the easy [100] direction, open squares: H_0 along the hard [110] direction) as a function of d_{Pd} "$d_{Pd}< 0$" corresponds to the uncovered Fe film and is used as a reference (from [58]).

14.4.1.3. Comparison of Brillouin light scattering with magnetometry for the determination of anisotropy constants

Finally, we conclude this section by a comparison of the Brillouin light scattering technique to conventional magnetometry for the determination of magnetic anisotropy properties. Magnetometry techniques, like vibrating sample magnetometry (VSM), superconducting quantum interference device (SQUID) magnetometry, and magneto-optic Kerr effect magnetometry, usually allow for magnetization reversal measurements, and the investigated quantity is the projection of the magnetization onto the direction of the applied field. Sometimes vector magnetometry is used, where two or all three components of the magnetization are

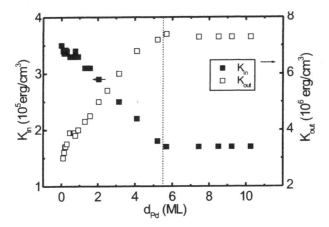

Figure 14-10: Obtained anisotropy constants K_{in} and K_{out} for a Pd wedge on a 15 ML thick Fe film as a function of d_{Pd} (from [58]).

determined. However, in the majority of the magnetometry experiments, anisotropies are inferred from magnetization curves. Frequently, in the analysis of such data a coherent rotation model of the magnetization (Stoner-Wohlfarth model) is assumed to be applicable. Therefore the influence of domain formation and propagation of domain walls during magnetization reversal is neglected. This leads to a systematic underestimation of the involved anisotropy constants using conventional magnetometry.

In a Brillouin light scattering experiment the measurements are made in saturation. Thus domain formation may not affect the determination of anisotropy properties. The frequencies of the spin waves depend on the second derivatives of the free energy with respect to the angles of the magnetization, i.e., on the curvature of the free energy surface. Therefore Brillouin light scattering is sensitive to acting torques and thus particularly sensitive to higher order anisotropies, which are often hard to distinguish using conventional magnetometry, especially if more than one anisotropy contribution is present in the system.

For the same reasons Brillouin light scattering is not very sensitive to the unidirectional anisotropy present in exchange bias systems, which can easily be determined using conventional magnetometry. Thus combining both techniques is extremely helpful and necessary to obtain a more complete picture of the magnetic properties of a system, in particular if several anisotropy contributions are present.

14.4.2. The exchange constant

The exchange constant A is best determined by measuring films, which show simultaneously dipolar and standing spin waves. The dipolar modes are only rather weakly influenced by the exchange interaction, and they serve as a reference, in particular if the saturation magnetization also enters as a fit parameter. The exchange modes, in turn, are largely determined by A, see eq.(14.11).

As an example we discuss the determination of the exchange constant for a $Gd_{13.5} Tb_{6.2} Fe_{80.3}$ film. This material is used in magneto-optic recording. See [60] for a description of the sample preparation.

p-polarized light from a single-mode Ar$^+$-ion laser was focused onto the sample with an angle of incidence of 45°. A magnetic field of 2 kOe was applied parallel to the film plane and perpendicular to the wavevector of the spin waves tested in the experiment. Data accumulation times of up to 20 hours per spectrum were used since the intensity of the scattered light is rather small in this material. The laser power was chosen smaller than 75 mW to avoid any damage of the sample.

Figure 14-11 shows the results of the Brillouin light scattering measurement as a function of film thickness. The measured and calculated spin-wave frequencies are represented by squares and solid lines, respectively. The first standing spin-wave mode can be well identified by its characteristic $1/d^2$ dependence. The fit shows a good agreement between data and simulation.

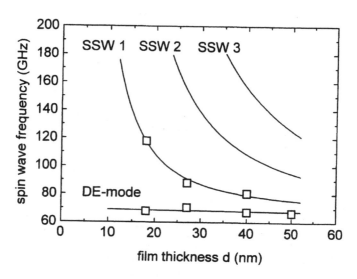

Figure 14-11: Spin-wave frequencies as a function of film thickness for a Gd$_{13.5}$Tb$_{6.2}$Fe$_{80.3}$ sample. Measured and fitted data are represented by symbols and solid lines, respectively. SSW*n* denotes the standing spin wave modes of nth order, DE is the Damon-Eshbach mode (from [60]).

14.4.3. Interlayer coupling

Brillouin light scattering is very well suited for measuring interlayer exchange coupling. In contrary to magnetometry, the coupling constant(s) can be determined not only for an antiparallel or canted arrangement of the layer magnetizations, but also for a parallel arrangement, i.e., positive interlayer coupling.

We start with a double layer system. Here each layer carries a Damon-Eshbach mode. In the coupled system, the two Damon-Eshbach modes hybridize. One becomes the Damon-Eshbach mode of the combined system, which is independent on interlayer exchange coupling, since the precessions of the moments in both layers are in phase (acoustic mode). The other mode, the so-called optic mode, is sensitive to interlayer exchange coupling. The precessions of both modes are out-of-phase, and the mode frequency shifts to larger (smaller) frequency values with increasing (decreasing) value of the interlayer exchange coupling strength, both in the positive and negative regimes of the coupling constant.

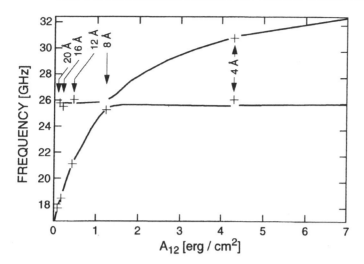

Figure 14-12: Theoretical (full lines) and experimental data for room temperature spin wave frequencies in the Fe/Cr/Fe system. The applied field is 3 kOe applied along the easy in-plane axis. For the calculation the following parameters are used: $q_\parallel = 1.73 \times 10^5$ rad/cm, $4\pi M_s = 20.7$ kG, $A = 1.6 \times 10^{-6}$ erg/cm, $K_1 = 4.0 \times 10^5$ erg/cm^3, $g = 2.1$. The Fe layers are both 106 Å thick (from [61]).

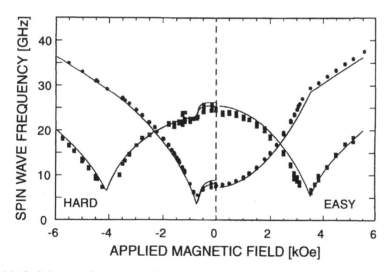

Figure 14-13: Spin wave frequencies of the (100) Fe/Cr/Fe sample for the field applied along the hard and easy directions measured at room temperature. Symbols are experimental points, the line is a fit. For clarity the hard axis results have been plotted along the negative field axis. The results are $A_{12} = 0.57 \pm 0.02$ erg/cm^2 and $B_{12} = 0.003 \pm 0.003$ erg/cm^2. The saturation magnetization of $4\pi M_s = 19.8$ kG and the in-plane anisotropy of 2.2×10^5 erg/cm^3 were determined independently (from [63]).

Figure 14-12 shows one of the first measurements of the two modes by Barnas and Grünberg [61]. By comparing the measured spin wave frequencies with those calculated as a function of the interlayer exchange coupling constant, A_{12}, the value of A_{12} is determined from the experiment. Figure 14-12 demonstrates the independence of the Damon-Eshbach mode on A_{12} and the large dependence of the optic mode, in particular for small A_{12}. A detailed study of the interlayer coupling was performed in the saturated Fe/Cr/Fe(100) layered system using Brillouin light scattering by Demokritov et al. [62].

Current state-of-the-art results [63] on epitaxial Fe/Cr/Fe(100) trilayers are shown in Fig.14-13. Here a fourfold in-plane anisotropy acts on the direction of magnetization in addition to the bilinear and biquadratic interlayer exchange coupling. Measurements are shown for the external field applied along the in-plane easy direction (positive field values) and along the in-plane hard direction (plotted as negative field values). The full lines are a fit to the experimental data (dots and squares) with the bilinear (A_{12}) and biquadratic (B_{12}) exchange coupling constant as fit parameters. The data shows very clearly the properties of the acoustic and the optic spin wave modes in the aligned state ($H_0 > 3.5(4.0)$ kOe in the easy (hard direction) and in the canted state. At a field of 2.4 kOe (easy direction) both modes cross. For measurements along the hard direction (negative field values in Fig.14-13) a characteristic change in frequency at -0.7 kOe is observed, which is due to the compensation of the in-plane anisotropy by the applied field. The excellent agreement between the fitted curves and the experimental data demonstrates the power of the technique for determining the exchange coupling strengths and anisotropies.

14.4.4. Magnetic inhomogeneities

The spin wave spectra do not only contain information on frequency positions. The line width may also allow to draw important conclusions.

Materials which are spatially inhomogeneous show a broadening of the spin wave peaks. Spatial inhomogeneities cause a distribution of the internal field, both in magnitude and in direction. Quantities which may spatially vary include the saturation magnetization, anisotropies and interlayer coupling. Depending on the ratio of the length scale of these variations to the spin wave wavelength, we distinguish between two mechanisms, although the boundary line between them is not sharp:

- If the wavelength of the spin wave is larger than the spatial variation length, so-called two-magnon scattering processes may occur. A magnon with low wavevector is scattered into a magnon of same energy but high wavevector. This process is strongly wavevector dependent, and damping for spin waves with wavevectors, as studied by spectroscopy, seems to be much stronger than damping of $q=0$ modes studied by ferromagnetic resonance. See the review article by Mills and Rezende for reference [64].

- If the length scale of the inhomogeneities is larger than the spin wave wavelength, but smaller than the diameter of the laser focus in the BLS experiment, the experiment averages over a spin wave distribution. We can assume, that locally a well-defined spin wave mode exists with the frequency depending on the local parameters.

For a more detailed discussion of line width broadening see [18].

As one example, Fig.14-14 shows a study of the $Ni_{80}Fe_{20}/Fe_{50}Mn_{50}$ exchange bias system [65]. In such a system the interlayer exchange coupling is locally very strong but frustrated, and can be described by a spatially strongly varying interface anisotropy. Shown is the measured line width as a function of the direction of the applied magnetic field, which is applied in the film plane. The spin wave modes show a large mode broadening of more than a factor of five upon covering the NiFe layers by FeMn. The mode width varies within a factor of two as a function of the azimuthal angle φ_H. The maximum and minimum values correspond to the hard [001] and easy [1$\bar{1}$0] direction of magnetization, respectively. The line width strongly decreases with increasing

NiFe layer thickness converging to the width of the uncovered NiFe films. This is characteristic for an interface effect.

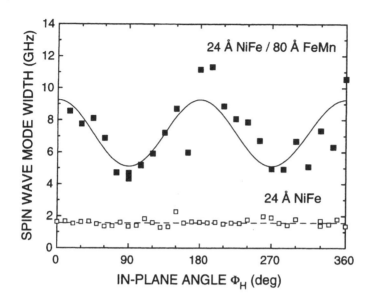

Figure 14-14: Spin wave mode width as a function of the angle of the in-plane applied field, φ_H, with the in-plane [001] direction for the 24 Å thick $Ni_{80}Fe_{20}$ layer in contact to a 80 Å thick $Fe_{50}Mn_{50}$ layer. The full and open squares are the data of the $Ni_{80}Fe_{20}/Fe_{50}Mn_{50}$ bilayer and of a $Ni_{80}Fe_{20}$ reference layer, respectively. The full and dashed lines are guides to the eye (from [65]).

The large spin wave mode broadening and its dependence on the in-plane angle of the external field can be understood as follows: In an exchange bias system the local interlayer exchange coupling varies on the length scale of antiferromagnetic domains in the $Fe_{50}Mn_{50}$ layer [66]. From the local exchange coupling the macroscopic, averaged exchange bias field characteristic for an exchange bias system is generated. The broadening is largest, if not only internal field contributions vary, but also the direction of magnetization, since the spin wave frequency depends largely on the direction of magnetization (see, e.g., Fig.14-4). The latter occurs, if the easy axis of the spatially varying internal field is not collinear with the magnetization. By changing the direction of the external field, minima and maxima in the line width appear at the easy and hard axes of the dominating, exchange coupling induced uniaxial anisotropy. This serves as one example, that Brillouin light scattering studies allow for a detailed characterization of such internally varying fields.

14.5. PATTERNED MAGNETIC STRUCTURES

Patterned magnetic structures are currently an important topic in magnetic research. Of particular interest are their dynamic properties, motivated largely by applications in the field of sensors and data storage, where the speed of operation reaches now into the GHz regime.

Brillouin light scattering has gained a central place in the investigation of the excitation spectrum of patterned structures. The reader is referred to recent review articles to learn about the physical properties [20,22,67]. An essential advantage of the method lies in the fact, that the spin wave wavelength is typically of the same order of magnitude than the lateral dimensions. Therefore effects like discrete spin wave frequencies due to quantization and localization of spin wave modes inside objects are accessible.

14.5.1. Quantized spin-wave modes in magnetic stripes

A well-studied test object are stripes made from Permalloy ($Fe_{20}Ni_{80}$) films of typically 20-40 nm thickness with a typical width in the range of 1-2 μm. Permalloy has the advantage, that magnetic anisotropies are negligibly small. This makes the data analysis easier. We consider surface spin wave modes (Damon-Eshbach modes) which propagate in the film plane perpendicular to the stripes. The geometry and one typical mode are illustrated in Fig.14-15. When the spin wave reaches the lateral boundaries of the stripe it will be reflected and will form a standing wave.

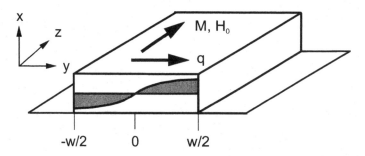

Figure 14-15: Geometry of a stripe of width w. Shown is the (n=1) mode.

Figure 14-16 shows a typical Brillouin light scattering spectrum. Several modes are found, which are the lateral standing waves, as well as the first perpendicular standing spin wave as described by eq.(14.1).

For a full investigation, spectra are taken as a function of the transferred in-plane wavevector, $q_\parallel$. The obtained frequencies as a function of $q_\parallel$ are shown in Fig.14-17. $q_\parallel$ was varied in the range $(0\text{-}2.2)\times 10^5$ rad/cm by changing the angle of light incidence.

For small wavevectors the spin waves split into several modes. Each of these modes shows no dispersion and is observable over a well defined continuous wavevector range. The separation between the modes decreases with increasing frequency. Towards larger wavevectors the modes become indistinguishable. At 14 GHz a dispersionless mode (apart from the crossing regime) is observed, which is identified as a perpendicular standing spin wave.

We discuss the two main properties in some more detail, which are the lack of dispersion and the finite wavevector range of observation. The discrete mode spectrum can be easily understood assuming a quantization of the parallel wavevector, $q_\parallel$, due to the finite width of the stripe:

$$w = n\frac{\lambda_n}{2} \quad ; \quad q_{\parallel,n} = \frac{2\pi}{\lambda_n} = \frac{\pi}{w}n \quad . \tag{14.17}$$

Here n is the mode index and λ_n the corresponding wavelength. We obtain in good approximation the mode frequencies by inserting the discrete wavevector values $q_{\parallel,n}$ into the Damon-Eshbach Eq.(14.8).

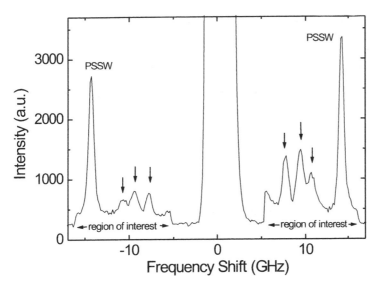

Figure 14-16: Experimental Brillouin light-scattering spectra obtained from the sample with a thickness of 40 nm, a stripe width of 1.8 μm, and separation of the stripes of 0.7 μm. The applied field is 500 Oe oriented along the stripe axis and the incoming wavevector of the light is oriented perpendicular to the stripes. The transferred wavevector is $q_\parallel$=0.33×10⁵ rad/cm. In the region of interest (5-17 GHz) the scanning speed was reduced by a factor of 3 increasing the number of recorded photons by the same factor. Several lateral standing spin waves are observed, indicated by the arrows. The mode indicated by PSSW is a perpendicular standing spin wave (from [68]).

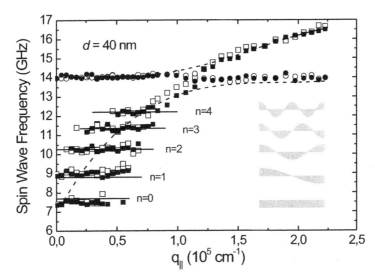

Figure 14-17: Obtained spin-wave dispersion curves for an array of stripes with a stripe thickness of 40 nm, a stripe width of 1.8 μm, and a separation between the stripes of 0.7 μm (open symbols) and 2.2 μm (solid symbols). The external field applied along the stripe axis is 500 Oe. The solid horizontal lines indicate the results of a calculation using eq.(14.5) with the quantized values of $q_\parallel$ and modified boundary conditions taking dynamic stray field effects into account [69]. The dashed lines, showing the hybridized dispersion of the Damon-Eshbach mode and the first perpendicular standing spin-wave mode, were calculated numerically for a continuous film with a thickness of 40 nm. On the right hand side the mode profiles are illustrated (from [68]).

A comparison to the experiment provides already a rather good account. A better agreement, as shown by the full lines in Fig.14-17, is achieved by considering the boundary conditions at the side walls of the stripe. Due to the precession of the spins dynamic stray fields are generated outside the stripe, which act as an additional dynamic surface torque. See [69] for details.

It might appear surprising, that, although the wavevector is discrete, the corresponding modes are observed over a continuous wavevector range. In order to understand this we consider the basics of a scattering experiment. In the scattering process a photon is scattered from a quantum of the spin wave, the magnon, under energy and momentum conservation. Transferred to wave optics, this implies that all three particles are represented by infinitely extended plane waves. In a common Brillouin light scattering experiment this is usually well fulfilled. The smallest size is the focus of the laser beam on the surface with a diameter of ≈ 40 µm, which is large compared to the wavelength. In the case of narrow stripes the smallest size is the stripe width. We must take into account, that the spin wave mode might be described by a plane wave inside the stripe. However, at the side walls of the stripe this wave is truncated since the magnetization is zero outside the stripe by definition. Thus the dynamic magnetization of this mode, $m_n(y)$, is not anymore a periodic function. As the main consequence, the wavevector is not anymore conserved in the scattering cross section. In order to calculate the light scattering cross section we need to perform a Fourier transformation of the mode profile $m_n(y)$. For each of the Fourier components $m_n(q_\parallel)$, wavevector conservation is given. The dynamic part of the magnetization $m_n(q_\parallel)$ is now a continuous function of the wavevector $q_\parallel$ due to the non-periodicity of $m_n(y)$. Thus, in the Brillouin light scattering experiment, we observe the modes over a continuous wavevector regime. The extend of this regime provides directly information about the spatial localization, which here is given by the boundaries of the stripe. Figure 14-18 shows the measured and calculated mode profiles for the data displayed in Fig.14-17.

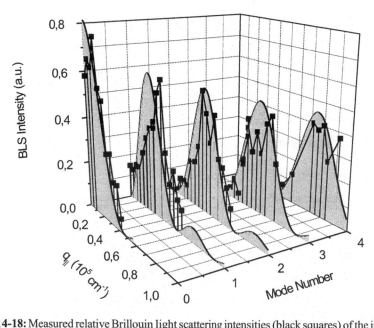

Figure 14-18: Measured relative Brillouin light scattering intensities (black squares) of the in-plane quantized spin-wave modes as a function of the wavevector $q_\parallel$ and the mode number n in comparison to the calculated results (gray colored curves) (from [68]).

14.5.2. Characterization of the spatial distribution of the spin-wave intensity and localization of modes

From the dependence of the cross section of the quantized modes on the wavevector, information on the spatial distribution of spin waves can be obtained. The cross section is proportional to $|m_n(q_\parallel)|^2$. Thus length scales can be deduced from such experiments, but not the spatial position of the modes, since any phase information is lost in squaring $m_n(q_\parallel)$. Nevertheless, there are several applications, where information on the lateral extend of spin wave modes is very useful. As one example we examine spin waves in a rectangular structure as shown in the inset of Fig. 14-19. Due to stray fields the internal field is intrinsically inhomogeneous. At the side walls of the rectangular perpendicular to the external field, the internal field may go to zero. Since the dispersion of spin waves depends on the strength of the internal field, cases, where spin wave propagation is allowed only in certain parts of such structures, can be easily constructed. See, e.g., [70,71] for details.

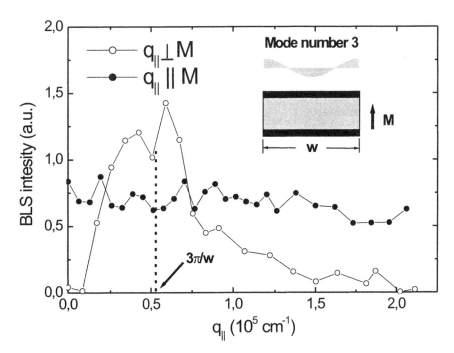

Figure 14-19: Brillouin light scattering study of a rectangular element with dimensions $1 \times 1.75 \ \mu m^2$. Shown is the intensity of the third localized mode as a function of the transferred in-plane wavevector for two scattering geometries. Inset: bottom: sketch of the element under consideration with black marking non-saturated parts; top: a mode profile consistent with the measured Brillouin light scattering intensities (from [72]).

For illustration see Fig.14-19. Here the modes are measured with the wavevector parallel (open symbols) and perpendicular (closed symbols) to the magnetization. In the perpendicular orientation the mode is observed over the entire accessible wavevector range. Since the lateral extend of the mode in real space is inversely proportional to the extend in wavevector space, this indicates a localization of the mode. By performing simulations the mode is identified as a mode localized in the regimes of low internal field close to the boundaries of the rectangular, which are perpendicular to the external field [70,71].

14.6. WAVE PROPAGATION PHENOMENA

This Section is devoted to demonstrate the wide range of applicability of Brillouin light scattering for the investigation of the propagation properties of spin waves and spin wave packets. We consider systems, where spin waves are generated by an antenna mounted on the surface of the magnetic film and connected to a microwave source. A pulsed microwave source allows one to study spin wave packets.

The case studies discussed below have been made on yttrium iron garnet films and related materials, which are known for their low magnetic damping properties. Using these materials the propagation of spin waves can be monitored over several millimeters. The material has good optical transmission allowing for the forward scattering geometry discussed below.

14.6.1. Experimental setups

14.6.1.1. Space resolved Brillouin light scattering

The aim is to realize a system, which allows one to measure the two-dimensional distribution of spin waves radiated from the antenna. In contrary to the experiments in the preceding sections, where wavevectors of the order of 10^5 rad/cm have been considered, here we use spin waves of much smaller wavevectors, which are in the range of 50-300 rad/cm. The corresponding wavelengths are in the range of 0.2-1 mm. A much smaller wavevector is necessary, since the experiments discussed in this section are based on the excitation of spin waves by microwave antenna structures. An antenna has an upper cut-off wavevector which is of the order of the inverse antenna width. In order to access such small wavevectors, the experiments are not made in backscattering geometry. Instead, a forward scattering geometry is used involving a focussing lens to focus the laser beam onto the surface and a collection lens behind the film to collect the transmitted light and to send it to the spectrometer (see Fig.14-20). A small blind behind the collection lens stops the direct light and avoids an overload of the detector. In this geometry no wavevector sensitivity is given, since the collection lens collects all inelastic light within the collection angle. Wavevector sensitivity can be added by a slit device inserted behind the collection lens, allowing only for the transmission with well-defined change in angle (see [73]), although this is much to the expense of the scattering intensity. In the majority of the experiments the wavevector is defined by the microwave frequency and the dispersion curve, so that wavevector selectivity is not an issue.

The small focus size of about 40 µm in diameter allows for an easy spatial mapping of spin wave intensities by scanning the sample under the laser focus using a motorized sample stage and by measuring the intensity of the spin wave at each position. In the experiment we use the digital control TFPDAS3 developed in our laboratory for stabilizing the interferometer (see section 14.2), and a separate PC for data collection and control of the stepper motors of the sample stage.

14.6.1.2. Space-time resolved Brillouin light scattering

In addition to spatial resolution, temporal resolution allows one to monitor the propagation of spin waves. The setup is schematically shown in Fig.14-20. Temporal resolution is achieved by using a time correlated single photon counting method similar to time-of-flight analysis in, e.g., mass spectroscopy. As the frequency resolving device a tandem Fabry-Pérot interferometer as described in section 14.2 is used. A pulse generator, serving as the central clock, generates pulses of typically 10-30 ns duration with a repetition rate of 1 MHz. The pulses are sent to a microwave switching device situated between a microwave generator operating at frequency ω_0 and the

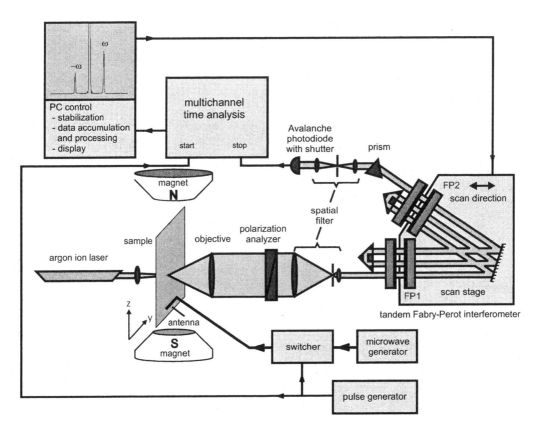

Figure 14-20: Schematic layout of the Brillouin light scattering apparatus with space and time resolution. For a discussion of the components see the main text (from [19]).

antenna to create spin waves pulses at the antenna. The output signal from the pulse generator is also used to start a 24 bit reference counter counting the output pulses of a 1.2 GHz time base. If the spin wave pulse crosses the laser spot, light is inelastically scattered, and the output signal of the photon detector is used to stop the reference counter. The counter content is now a measure of the elapsed time between the launch of the spin wave pulse and the arrival at the position of the laser spot. A cell of a memory array addressed by the content of the counter is incremented by one and the procedure is repeated. After accumulating a large number of events the content of the memory array represents the temporal variation of the spin wave intensity at the current position of the laser spot. By repeating the procedure for different positions of the laser spot on the sample, a two-dimensional map of the spin wave intensity is constructed for each value of delay time. The data is arranged in a digital video animation with each frame representing the spatial distribution of the spin wave intensity for a given delay time. The entire system is realized using a commercially available digital signal processing device chip which interacts with a PC via a RS232 interface. The device can handle up to 2.5×10^6 events per second continuously. A lower limit of about 2 ns on the time resolution is imposed by the intrinsic time resolution of the etalons and the multipass arrangement in the Brillouin light scattering spectrometer. Typical accumulation times are 5 seconds per position of the laser spot. A complete measurement of a two-dimensional spin wave intensity pattern in a YIG film with a sampling area of 2×6 mm^2 and a mesh size of 0.1 mm takes a little more than two hours including dead times caused by sample positioning.

14.6.2. Spin-wave radiation from an antenna and spin wave caustics

We begin by illustrating the spin-wave intensity mapping technique without time resolution. The example is a fundamental one: we study the radiation of spin waves from an antenna, which is driven by a cw current of microwave frequency ω_0.

An in-plane magnetized magnetic film is intrinsically anisotropic, since the direction of magnetization breaks the in-plane rotational symmetry. As one consequence, for finite wavevector, the frequency of the backward volume mode (MSBVM, parallel orientation between magnetization, $\vec{M}_s$, and wavevector, $\vec{q}$) is very different from that of the Damon-Eshbach mode (perpendicular orientation between $\vec{M}_s$, and $\vec{q}$), see Fig.14-4. The radiation of spin waves is anisotropic, and caustic effects may appear.

A microwave antenna is made by a thin and narrow conducting stripe attached to the film. One end is connected to the microwave source, the other is grounded. Such an antenna will predominantly radiate spin waves into the perpendicular direction. In the backward volume mode geometry considered here this is the direction of the external field, $\vec{H}_0$. If the antenna is of finite length, it will radiate spin waves into a cone. The cone angle is defined by the wavevector of the mode radiated perpendicular to the antenna and the antenna length l_a as follows: The driving microwave frequency, ω_0, and the mode dispersion $\omega(\vec{q})$, determine the wavevector component $q_\parallel$ parallel to the mean radiation direction. The length l_a of the antenna determines the perpendicular wavevector component $q_\perp$: $q_{\perp,max} \approx \pi/l_a$. Thus spin waves are radiated into a cone with the boundaries defined by the angle φ_{max}=arctan($q_{\perp,max}/q_\parallel$). However, this is not the range of radiation, which can be experimentally measured. The reason is that the radiation cone defined by φ_{max} applies to the phase velocity, $\vec{v}_{ph} = \omega(\vec{q})\vec{q}/q^2$, which is collinear with the direction of the wavevector $\vec{q}$. Brillouin light scattering detects spin wave intensities, which propagate with the group velocity, $\vec{v}_g = \partial\omega(\vec{q})/\partial\vec{q}$. The phase velocity does not, in general, coincide with the direction of the group velocity of a wave packet. For the magnetostatic backward volume mode discussed here the group velocity is roughly antiparallel to the phase velocity, since the slope of $\omega(q)$ is negative (see Fig.14-4). The angle θ between the direction of $\vec{v}_g$ and the mean radiation direction (direction of the bias field $\vec{H}_{ext}$) is determined by the expression

$$\theta = \arctan\left(\frac{v_{g\perp}}{v_{g\parallel}}\right) = \arctan\left(\frac{\partial\omega/\partial q_\perp}{\partial\omega/\partial q_\parallel}\right) \quad . \tag{14.18}$$

From this equation the relation $\theta(\varphi)$ can be calculated. Fig.14-21 shows the geometry (left) and an example of the relation $\theta(\varphi)$ (right).

Measured two-dimensional distributions of the spin wave intensities for two different carrier wavevector values are displayed in Fig.14-22. The intensities are displayed in a gray scale with white/black indicating high/low intensity. The attenuation is corrected to show clearly the details of the spin wave distribution. The position of the input antenna is marked by the dashed bar on the left. Figure 14-22(a) corresponds to a small ($q_\parallel$=44 rad/cm) and Fig.14-22(b) to a large value ($q_\parallel$=300 rad/cm) of the carrier wavevector. For low values of $q_\parallel$ the spin waves are radiated in a large angular range, and two preferential directions of the wave beam propagation, i.e., spin wave caustics, are clearly visible. The directions agree very well with the calculated values of the angle $\theta_{max} \approx 42°$ (see Fig.14-21) for the conditions of this experiment. These calculated directions are shown by white lines. For large wavevector values the beam is mostly radiated in the direction perpendicular to the antenna (Fig.14-22(b)).

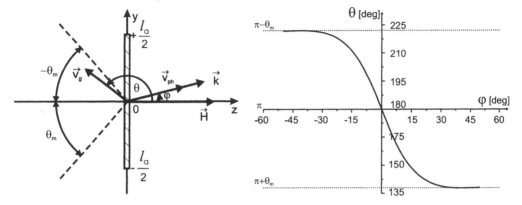

Figure 14-21: Left: Geometry of the excitation of a backward volume magnetostatic wave (MSBVM) by a microwave antenna. The dashed bar in the middle shows the position of the antenna of the length l_a, extended along the y-direction. The bias magnetic field is parallel to the positive z-direction. The angle φ indicates the direction of the wavevector (or phase velocity) of the radiated wave relative to the direction of the bias field $\vec{H}_0$. The angle θ indicates the direction of the group velocity v_g. Dashed lines show the two preferential directions ($\pm\theta_{max}$) of the group velocity in a wide wave beam radiated by the antenna. Right: Dependence of the directional angle of the group velocity θ on the directional angle of the phase velocity φ in a beam of magnetostatic waves excited in a tangentially magnetized $Lu_{0.96}Bi_{2.04}Fe_5O_{12}$ film (film thickness L=1.5 µm, saturation magnetization $4\pi M_s$ =1750 Oe, bias magnetic field H_0 = 2298 Oe). The value of θ_{max} in this case is 42° (from[19]).

14.6.3 Spin wave tunnelling

To demonstrate the power of the space and time resolved Brillouin light scattering technique we discuss the case of the propagation of spin waves across a narrow region with an internal field inhomogeneity, where spin wave propagation is forbidden [74]). Here, due to the different field, the dispersion curve does not allow anymore for a real wavevector to exist for the given carrier frequency of the spin wave packet, ω_0. A sketch of the experimental setup is shown in Fig.14-23. We use an yttrium ion garnet (YIG) film due to its optical transparency and low damping losses. Its geometry is 1.8 mm width, 30 mm length and 5 µm thickness. Spin wave pulses of wavevector $\approx$ 125 rad/cm are generated using a microstrip antenna connected to a pulsed micro-wave source of frequency ω_0 = 7.315 GHz. An external field H_0 = 1898 Oe is oriented parallel to the film and the propagation direction of the spin waves. Here we have the so-called backward volume wave geometry, characterized by a negative group velocity of the spin waves. The pulses travel from left to right in Fig.14-23. The spatial and temporal evolution is measured by space and time resolved Brillouin light scattering.

The field inhomogeneity is created by a wire of 50 µm diameter mounted across the film and carrying a dc current (see Fig.14-23). Depending on the sign of the dc current the local field is either enhanced or reduced by the Oerstedt field generated by the current.

A travelling spin wave packet of carrier frequency ω_0 arriving at a position with increased or decreased internal field may not be allowed anymore to continue propagation, if the local dispersion does not provide for a real wavevector value for the given frequency ω_0. The situation is reminiscent to quantum mechanical tunnelling. Here, in analogy, the width of the field inhomogeneity, and thus the width of the film regime forbidden for propagation, is small compared to the carrier wavelength. Therefore the field inhomogeneity is named a spin wave well.

Figure 14-24 shows the measured propagation of the spin wave packet. For several delay times the spatial distribution is displayed. Indeed, inspecting the time frames in Fig.14-24, left panels, we find that the spin-wave wave-packet is partially reflected and partially transmitted at this well. It should be noted, that the width of the well is smaller than the spin wave wavelength, strengthening the argument of spin wave tunnelling through a potential well. A local enhancement of the field, as displayed in the right panels, does not significantly affect the propagation, since here only a small local change in wavevector takes place. Such a setup, as used here, does not only allow for interesting new experiments, in which the detailed influence of local field variation on the dynamic properties becomes experimentally accessible. New applications of spin-wave propagation in high-frequency devices can be tested, and the manipulation of the propagation properties by a current carrying wire offers a simple tool for fast control in a device.

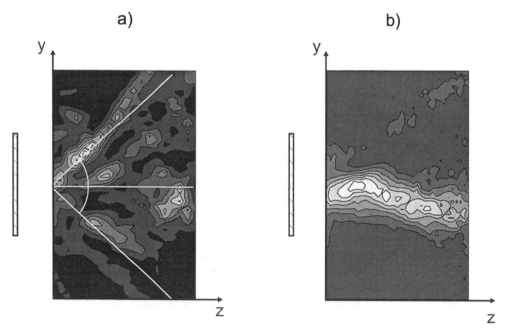

a)   b)

Figure 14-22: Two-dimensional plots of the spin wave intensity distribution in a large 1.5 μm thick $Lu_{0.96}Bi_{2.04}Fe_5O_{12}$ sample for two different values of the carrier wavevector $q_\parallel$ of the excited magnetostatic backward volume mode. The position of the input antenna is marked by the dashed bar on the left of each panel. a) Case of relatively small carrier wavevector ($q_\parallel = 44$ rad/cm$\sim 2\pi/l_a$). Here the waves are exited in a large angular range and two preferential directions of wave group velocity corresponding to $\pm\theta_{max}$ are clearly seen. Theoretical values of $\theta_{max} = \pm42°$ are shown by white lines. b) Case of a relatively large carrier wavevector ($q_\parallel = 300$ rad/cm $\gg 2\pi/l_a$). Waves are mostly radiated perpendicular to the antenna. Film width 8 mm, length 10 mm. The static magnetic field applied parallel to the mean wavevector direction is 2090 Oe and the microwave input power is 10 mW (from [19]).

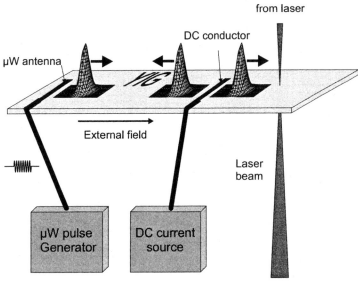

Figure 14-23: Setup for the measurement of the interaction of a spin-wave packet with a local field inhomogeneity. A microwave antenna is connected to a microwave pulse generator and generates spin wave packets. The packets travel from the left to the right across a YIG film. The field inhomogeneity is realized by a dc-current carrying wire mounted across the film. The propagation of the wave packet is observed by scanning a laser beam across the surface and measuring the photons inelastically scattered by the spin waves using a tandem Fabry-Pérot interferometer. Time resolution is obtained by a time-of-flight-type analyzer measuring the elapsed time between the launch of a microwave pulse and the detection of correspondingly inelastically scattered photons (from [74]).

14.6.4. Magnetic solitons and bullets

If large-amplitude microwave pulses are applied to the input antenna in a geometry as used in the two preceding sections, nonlinear spin wave can be generated. As examples, magnetic solitons in one-dimensional magnetic waveguides and magnetic bullets in magnetic films have been observed using space- and time resolved Brillouin light scattering, see [75] for details.

Figure 14-25 shows the experimentally measured two-dimensional distribution of the intensity of a propagating wave packet obtained for an input power of $P_{in}=460$ mW corresponding to five different delay times. The cross sections of the wave packets at the level of half-maximum are shown in the lower part of the figure. The position of the input antenna is also shown at $z=0$. The data clearly demonstrate the existence of spatio-temporal self-focusing of the propagating wave packet with the focal point situated near the position $z=2$-2.5 mm (corresponding to the propagation time $t=50$-60 ns), where the peak amplitude of the wave packet has a maximum, and the packet width along the perpendicular $\hat{y}$ axis shows a minimum. For $t<40$ ns, the wave packet, generated at the antenna, is entering the region of the film accessible by Brillouin light scattering. For $t=40$-45 ns, a rapid collapse-like self-focusing of the packet is observed along both in-plane directions. Subsequently the collapse is stabilized by dissipation, and in the time interval $50<t<100$ ns both in-plane sizes of the propagating packet are almost constant; i.e., a quasi-stable spin wave bullet is propagating in the film. For $t >100$ ns, the transverse size of the packet starts to increase rapidly due to the influence of diffraction.

Such a system provides an excellent testing ground for nonlinear wave propagation. For example, in a follow-up experiment, the collision of solitons and bullets was studied using this technique [76].

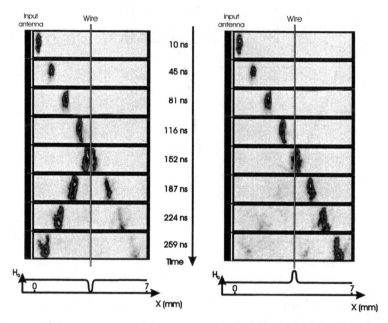

Figure 14-24: Propagation of a spin-wave wave-packet in backward volume mode geometry across a film with local field inhomogeneity, created by a dc-current carrying wire, observed by time and space resolved Brillouin light scattering spectroscopy. The bottom graphs display the external field as a function of the position, the other graphs show the position of the wave packet in time frames with the elapsed time noted in the middle. The left panel shows the case of a local minimum in the internal field, corresponding to a potential well, and the right panel shows the case of a local maximum in the internal field, corresponding to a potential dip (from [74]).

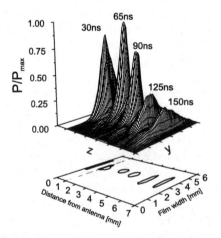

Figure 14-25: Two-dimensional (y,z) distributions of normalized intensity in propagating dipolar spin wave packets, corresponding to five different values of the propagation (delay) time as indicated in the figure. The distributions were experimentally measured by space- and time-resolved Brillouin light scattering technique for a pulse length of $T_{pulse} = 29$ ns and $P_{in} = 460$ mW. The cross sections of the propagating wave packets taken at half-maximum power are shown on the (y,z) plane below (from [75]).

14.7. CONCLUSIONS

Focus of this review was the applicability of the Brillouin light scattering technique to determine magnetic parameters in magnetic films and to study the physics of linear and nonlinear spin wave propagation. To conclude, we shortly comment on the comparison between Brillouin light scattering spectroscopy and related techniques. Being an optical technique, all advantages of an optical approach like non-contact measurements, small sample size, easy implementation in experimental environments apply. In many cases this allows for an over-all easier access to certain material properties then more direct techniques like standard magnetometry methods. In the field of dynamic excitations the wavevector and its related quantity, the wavelength have large importance. In particular in systems with intrinsic length scales, like lateral dimensions, this property is of large advantage.

ACKNOWLEDGMENTS

Brillouin light scattering has been a central technique in the academic life of the author. The author would like to thank all colleagues – inside and outside of his group – who have collaborated with the author and contributed to the success of Brillouin light scattering. In particular the author thanks the involved senior scientists in his group, which are Sergej O. Demokritov, Jürgen Fassbender and Olexandr Serha, for the many discussions and contributions they have made and the joint excitement for the field. Many post-docs, Diploma and Ph.D. students have contributed to the work, they are too many to list them all here. In particular the author would like to thank Martin Bauer, Oliver Büttner and Jörg Jorzick for their contributions. Of particular pleasure was the collaboration with John R. Sandercock. Very enjoyable was the theoretical support by Andrej Slavin. In the field of nonlinear excitations the author would also like to thank in particular Boris Kalinikos and Carl Patton for their input and collaborations on various issues.

REFERENCES

[1]   T. Wolfram and R.E. DeWames, Progr. Surf. Sci. **2** (1972) 233.

[2]   D.L. Mills and K.R. Subbaswamy, Surface and size effects on the light scattering spectra of solids, edited by E.Wolf, page 47, North Holland, 1981.

[3]   J.R. Sandercock, Trends in Brillouin light scattering: studies of opaque materials, supported films and central modes,edited by M. Cardona and G. Güntherodt, volume 51 of *Topics in Appl. Physics*, page 173, Springer Verlag, Berlin, Heidelberg, New York, Tokyo, 1982.

[4]   A.S. Borovik-Romanov and N.M. Kreines, Physics Reports **81** (1982) 351.

[5]   C.E. Patton, Physics Reports **103** (1984) 251.

[6]   D. L. Mills, Surface spin waves on magnetic crystals, edited by V. M. Agranovich and R. Loudon, Elsevier, 1984.

[7]   P. Grünberg, Progr. Surf. Sci. **18** (1985) 1.

[8]   P. Grünberg, Light scattering from spin waves in thin films and layered magnetic structures, edited by M. Cardona and G. Güntherodt, volume 66 of *Topics in Applied Physics*, page 303, Springer Verlag, Berlin, Heidelberg, New York, Tokyo, 1989.

[9]   M. Grimsditch, Brillouin light scattering from metallic superlattices, edited by M. Cardona and G. Güntherodt, volume 66 of *Topics in Applied Physics*, page 285, Springer Verlag, Berlin, Heidelberg, New York, Tokyo, 1989.

[10] B. Heinrich and J. F. Cochran, Advances in Physics **42** (1993) 523.

[11] J. F. Cochran, Light scattering from ultrathin magnetic layers and bilayers, edited by B. Heinrich and J. A. C. Bland, Springer Verlag, Heidelberg, Berlin, London, New York, Tokyo, 1994.

[12] B. Hillebrands and G. Güntherodt, Brillouin light scattering in magnetic superlattices, edited by B. Heinrich and J. A. C. Bland, Springer Verlag, Heidelberg, Berlin, London, New York, Tokyo, 1994.

[13] J. R. Dutcher, Light scattering and microwave resonance studies of spin-waves in metallic films and multilayers, edited by M. G. Cottam, World Scientific, Singapore, 1994.

[14] B. Hillebrands, P. Krams, J. Fassbender, C. Mathieu, G. Güntherodt, R. Jungblut, and M. T. Johnson, Acta Phys. Pol. A **85** (1994) 179.

[15] S. O. Demokritov and E. Tsymbal, J. Phys. C **6** (1994) 7145.

[16] M. G. Cottam, I. V. Rojdestvenski, and A. N. Slavin, Brillouin light scattering from dipole-exchange microwave spin waves in magnetic films, edited by H. G. Srinivasan and A.N. Slavin, World Scientific, Singapore, 1995.

[17] B. Hillebrands, J. Fassbender, P. Krams, C. Mathieu, G. Güntherodt, and R. Jungblut, Scienza e Tecnologia **25** (1996) 24.

[18] B. Hillebrands, Brillouin light scattering from layered magnetic structures, edited by M. Cardona and G. Güntherodt, volume 75 of *Topics in Applied Physics*, page 174, Springer Verlag, Berlin, Heidelberg, 2000.

[19] A. Slavin, S. Demokritov, and B. Hillebrands, Nonlinear spin waves in one- and two-dimensional magnetic waveguides, edited by B. Hillebrands and K. Ounadjela, volume 83 of *Topics in Applied Physics*, page 35, Springer Verlag, Berlin, Heidelberg, New York, Tokyo, 2002.

[20] S. Demokritov and B. Hillebrands, Spin waves in laterally confined structures, edited by B. Hillebrands and K. Ounadjela, volume 83 of *Topics in Applied Physics*, page 65, Springer Verlag, Berlin, Heidelberg, New York, Tokyo, 2002.

[21] G. Carlotti and G. Gubbiotti, J. Phys. C **14** (35), (2002) 8199.

[22] S. Demokritov, B. Hillebrands, and A. Slavin, Physics Reports **348** (2001) 441.

[23] J.A.C. Bland and B. Heinrich, *Ultrathin Magnetic Structures Vol I and II*, Springer Verlag, Heidelberg, Berlin, London, New York, Tokyo, 1994.

[24] U. Gradmann, Magnetism in ultrathin transition metal films, edited by K. H. J. Buschow, volume 7, Elsevier Publishers B. V., Amsterdam, 1993.

[25] R. Mock, B. Hillebrands, and J. R. Sandercock, J. Phys. E **20** (1987) 656.

[26] P. Krams, F. Lauks, R. L. Stamps, B. Hillebrands, and G. Güntherodt, Phys. Rev. Lett. **69** (1992) 3674.

[27] J. Sandercock, Solid State Commun. **26** (1978) 547.

[28] J. Sandercock, SPIE **732** (1987) 157.

[29] J. R. Sandercock, The design and use of a stabilised multipassed interferometer of high contrast ratio, edited by M. Balkanski, page 9, Flammarion, Paris, 1971.

[30] S. Lindsay and I. Shepherd, Review of Scientific Instruments **48** (1977) 1228.

[31] S. Lindsay, M. Anderson, and J. Sandercock, Review of Scientific Instruments **52** (1981) 1478.

[32] A. Yoshihara, Jap. J. Appl. Phys. **33** (1994) 3100.

[33] D. Bechtle, Digital stabilizer for scanning Fabry-Pérot, Review of Scientific Instruments **47** (1976) 1377.

[34] K. Weishaupt, S. Anders, R. Eberle, and M. Pietralla, Review of Scientific Instruments **68** (1997) 3996.

[35] B. Hillebrands, Review of Scientific Instruments **70** (1999) 1589.

[36] M. J. Hurben and C. E. Patton, J. Magn. Magn. Mater. **139** (1995) 263.

[37] G. T. Rado and J. R. Weertman, J. Phys. Chem. Solids **11** (1959) 315.

[38] B. Hillebrands, Phys. Rev. B **41** (1990) 530.

[39] G. T. Rado, J. Appl. Phys. **61** (1987) 4262.

[40] U. Gradmann, J. Korecki, and G. Waller, Appl. Phys. A **39** (1986) 101.

[41] R. L. Stamps and B. Hillebrands, Phys. Rev. B **44** (1991) 12417.

[42] J.F. Cochran, B. Heinrich, and A. S. Arrott, Phys. Rev. B **34** (1986) 7788.

[43] B. Heinrich, S. T. Purcell, J. R. Dutcher, K. B. Urquhart, J. F. Cochran, and A. S. Arrott, Phys. Rev. B **38** (1988) 12879.

[44] R. L. Stamps and B. Hillebrands, Phys. Rev. B **43** (1991) 3532.

[45] B. Hillebrands, A. Boufelfel, C. M. Falco, P. Baumgart, G. Güntherodt, E. Zirngiebl, and J. D. Thompson, J. Appl. Phys. **63** (1988) 3880.

[46] J.V. Harzer, B. Hillebrands, R. L. Stamps, G. Güntherodt, C. D. England, and C. M. Falco, J. Appl. Phys. **69** (1991) 2448.

[47] D. Bürgler, P. Grünberg, S. Demokritov, and M. Johnson, Interlayer exchange coupling in layered magnetic structures, edited by K. Buschow, volume 13, page 1, Elsevier Science B.V., 2001.

[48] M. D. Stiles, Phys. Rev. B **48** (1993) 7238.

[49] J. C. Slonczewski, J. Magn. Magn. Mater. **150** (1995) 13.

[50] F. Hoffmann, A. Stankoff, and H. Pascard, J. Appl. Phys. **41** (1970) 1022.

[51] F. Hoffmann, Phys. Status Solidi **41** (1970) 807.

[52] M. Macciò, M. G. Pini, P. Politi, and A. Rettori, Phys. Rev. B **49** (1994) 3283.

[53] S. Tschopp, G. Robins, R. L. Stamps, R. Sooryakumar, M. E. Filipkowski, C. J. Gutierrez, and G. A. Prinz, J. Appl. Phys. **81** (1997) 3785.

[54] J. Fassbender, C. Mathieu, B. Hillebrands, G. Güntherodt, R. Jungblut, and M. T. Johnson, J. Appl. Phys. **76** (1994) 6100.

[55] J. Fassbender, C. Mathieu, B. Hillebrands, and G. Güntherodt, J. Magn. Magn. Mater. **148** (1995) 156.

[56] B. Hillebrands, J. Fassbender, R. Jungblut, G. Güntherodt, D. J. Roberts, and G. A. Gehring, Phys. Rev. B **53** (1996) 10548.

[57] J. Fassbender, G. Güntherodt, C. Mathieu, B. Hillebrands, R. Jungblut, J. Kohlhepp, M. Johnson, D. Roberts, and G. Gehring, Phys. Rev. B **57** (1998) 5870.

[58] S. O. Demokritov, C. Mathieu, M. Bauer, S. Riedling, O. Büttner, H. de Gronckel, and B. Hillebrands, J. Appl. Phys. **81** (1997) 4466.

[59] P. Grünberg, S. O. Demokritov, A. Fuss, R. Schreiber, J. A. Wolf, and S. T. Purcell, J. Magn. Magn. Mater. **106** (1992) 1734.

[60] D. Raasch, J. Reck, C. Mathieu, and B. Hillebrands, J. Appl. Phys. **76** (1994) 1145.

[61] J. Barnás and P. Grünberg, J. Magn. Magn. Mater. **82** (1989) 186.

[62] S. O. Demokritov, J. A. Wolf, and P. Grünberg, Europhys. Lett. **15** (1991) 881.

[63] M. Grimsditch, S. Kumar, and E. E. Fullerton, Phys. Rev. B **54** (1996) 3385.

[64] D. Mills and S. Rezende, Spin damping in ultrathin magnetic films, edited by B. Hillebrands and K. Ounadjela, volume 87 of *Topics in Applied Physics*, page 27, Springer Verlag, Berlin, Heidelberg, New York, Tokyo, 2003.

[65] C. Mathieu, M. Bauer, B. Hillebrands, J. Fassbender, G. Güntherodt, R. Jungblut, J. Kohlhepp, and A. Reinders, J. Appl. Phys. **83** (1998) 2863.

[66] A. Malozemoff, J. Appl. Phys. **63** (1988) 3874.

[67] S. O. Demokritov and B. Hillebrands, J. Magn. Magn. Mater. **200** (1999) 706.

[68] J. Jorzick, S. O. Demokritov, C. Mathieu, B. Hillebrands, B. Bartenlian, C. Chappert, F. Rousseaux, and A. N. Slavin, Phys. Rev. B **60** (1999) 15194.

[69] K. Guslienko, S. O. Demokritov, B. Hillebrands, and A. N. Slavin, Phys. Rev. B **66** (2002) 132402.

[70] J. Jorzick, S. O. Demokritov, B. Hillebrands, M. Bailleul, C. Fermon, K. Guslienko, A. N. Slavin, D. Berkov, and N. Gorn, Phys. Rev. Lett. **88** (2002) 047204/1.

[71] C. Bayer, S. O. Demokritov, B. Hillebrands, and A. N. Slavin, Appl. Phys. Lett. **82** (2003) 607.

[72] J. Jorzick, C. Kramer, S. O. Demokritov, B. Hillebrands, B. Bartenlian, C. Chappert, D. Decanini, F. Rousseaux, E. Cambril, E. Sondergard, M. Bailleul, C. Fermon, and A. N. Slavin, J. Appl. Phys. **89** (2001) 7091.

[73] W. D. Wilber, W. Wettling, P. Kabos, C. E. Patton, and W. Jantz, J. Appl. Phys. **55** (1984) 2533.

[74] S. O. Demokritov, A. A. Serga, A. Andre, V. E. Demidov, M. P. Kostyler, B. Hillebrands, A. N. Slavin, Phys. Rev. Lett. **93** (2004) 047201.

[75] M. Bauer, O. Büttner, S. Demokritov, B. Hillebrands, V. Grimalsky, Y. Rapoport, and A. N. Slavin, Phys. Rev. Lett. **81** (1998) 3769.

[76] O. Büttner, M. Bauer, S. O. Demokritov, B. Hillebrands, M. Kostylev, B. Kalinikos, and A. N. Slavin, Phys. Rev. Lett. **82** (1999) 4320.

Appendices

Appendix I

Principal Physical Constants

Bohr radius	$5.291772108 \times 10^{-11}$ m	$a_H = 4\pi\varepsilon_0\,\hbar^2\,/\,m_0\,e^2$
speed of light in vacuum	2.99792458×10^{8} m s^{-1}	c
elementary charge	1.602176×10^{-19} C	e
Planck constant	$1.05457168 \times 10^{-34}$ J s	$\hbar = h\,/\,2\pi$
	$6.6260693 \times 10^{-34}$ J s	h
Boltzmann constant	$1.3806505 \times 10^{-23}$ J K^{-1}	k_B
electron mass	$9.1093826 \times 10^{-31}$ kg	m_e
proton mass	$1.67262171 \times 10^{-27}$ kg	m_p
Avogadro number	6.0221415×10^{23} mol^{-1}	N_A
molar gas constant	8.314472 J mol^{-1} K^{-1}	R
Rydberg constant	2.17991×10^{-18} J	$R_\infty = \alpha^2 m_0 c\,/\,2h$
Rydberg constant for hydrogen	13.6056923 eV	R_H
standard molar volume (273.15 K, 101.325 kPa)	22.413996×10^{-3} m^3 mol^{-1}	V_m
electric permittivity of vacuum	$8.8541878 \times 10^{-12}$ F m^{-1}	ε_0
magnetic permeability of vacuum	$4\pi \times 10^{-7}$ N A^{-2}	μ_0
Bohr magneton	$9.27400949 \times 10^{-24}$ A m^2 or J T^{-1}	$\mu_B = e\hbar\,/\,2m_e$
nuclear magneton	$5.05078343 \times 10^{-27}$ A m^2 or J T^{-1}	$\mu_N = e\hbar\,/\,2m_p$
magnetic flux quantum	$2.06783372 \times 10^{-15}$ T m^2	$\varPhi_0$

Appendix 2

Conversion between cgs Units and SI Units for Magnetic Quantities

In cgs units : $B = H + 4\pi M$,   in SI units : $B = \mu_0 (H + M)$, $\mu_0/4\pi = 10^{-7}\ N/A^2$

quantity	symbol	cgs units	conversion factor (C) SI = C (cgs)	SI units
magnetic induction	B	Gauss (G)	10^{-4}	tesla (T) Wb/m^2
magnetic flux	ϕ	G-cm^2 Maxwell (Mx)	10^{-8}	Weber (Wb)
magnetic field intensity	H	Oersted (Oe)	$10^3/4\pi$	A/m
magnetic moment	m	emu	10^{-3}	A-m^2
magnetization	M	emu/cm^3	10^3	A/m
	$4\pi M$	G	$10^3/4\pi$	A/m
mass magnetization	M_g	emu/g	1	A-m^2/kg
			$4\pi\times10^{-7}$	Wb-m/kg
volume susceptibility	χ	dimensionless	4π	dimensionless
mass susceptibility	χ_g	cm^3/g	$4\pi\times10^{-3}$	m^3/kg
permeability	μ	dimensionless	$4\pi\times10^{-7}$	Wb/A-m

Appendix 3

Conversion of Energy Related Units

Energy equivalents for photons $\quad E = h\nu = k_B T = \dfrac{hc}{\lambda}$

E (J)	E (eV)	ν (Hz)	T (K)	λ (m)	λ^{-1} (m^{-1})
1	6.242×10^{18}	1.509×10^{33}	7.243×10^{22}	1.486×10^{-25}	5.034×10^{24}
1.602×10^{-19}	1	2.418×10^{14}	1.160×10^{4}	1.240×10^{-6}	8.066×10^{5}
6.626×10^{-34}	4.136×10^{-15}	1	4.799×10^{-11}	2.998×10^{8}	3.336×10^{-9}
1.381×10^{-23}	8.617×10^{-5}	2.084×10^{10}	1	1.439×10^{-2}	69.50
1.987×10^{-25}	1.240×10^{-6}	2.998×10^{8}	1.439×10^{-2}	1	1.0
1.987×10^{-23}	1.240×10^{-4}	2.998×10^{10}	1.439	0.01	100

Physical Properties of Some Magnetic Substances (measured at room temperature)

substance	structure	lattice parameter nm	T_C (K)	M_s (kA m^{-1})	$\mu_0 M_s$ (T)	σ_s (A m^2 kg^{-1})	K_1 (kJ m^{-3})	density (g cm^{-3})
Fe	bcc	a = 0.287	1043	1714	2.16	221.9	48	7.85
Co	hcp	a = 0.251, c = 0.407	1404	1422	1.72	162	530	8.71
Ni	fcc	a = 0.352	631	485	0.61	57.5	-6.0	8.90
Fe$_3$O$_4$	spinel	a = 0.839	858	477	0.60	91.0	-13	5.19
Fe-80%Ni	fcc	a = 0.354	595	828	1.04	95.7	-2*	8.65
SmCo$_5$	h	a = 0.500, c = 0.397	995	836	0.965	97.4	17000	8.3
Sm$_2$Co$_{17}$	h	a = 0.838, c = 0.815	920	1030	1.29	101	3300	7.60
Nd$_2$Fe$_{14}$B	tetragonal	a = 0.879, c = 1.221	585	1280	1.57		4900	7.60

T_C : Curie temperature
M_s : spontaneous magnetization
μ_0 : permeability of free space
σ_s : specific magnetic moment
K_1 : magnetocrystalline anisotropy coefficient
* K_1 is very sensitive to thermal treatment for Permalloy, and changes from -2kJm^{-3} to a value 10 times smaller after quenching

Subject Index

Subject Index